# Handbook of Experimental Pharmacology

## *Volume 108/II*

# GTPases in Biology II

Contributors

M. Biel, L. Birnbaumer, K.J. Blumer, J. Bockaert, E. Bosse, J.L. Boyer, P. Brabet, M. Camps, M.G. Caron, P.J. Casey, J. Chen, K. Clark, D. Corda, S. Coulter, P.N. Devreotes, D. Donelly, T.D. DuBose, Jr., J.H. Exton, J.B.C. Findlay, R.A. Firtel, V. Flockerzi, M. Forte, C. Gaskins, P. Gierschik, R. Gundersen, R.W. Gurich, J.R. Hadcock, J.A. Hadwiger, T.K. Harden, J.D. Hildebrandt, Y.-K. Ho, F. Hofmann, K.P. Hofmann, V. Homburger, M.D. Houslay, R. Hullin, D. Hyde, R. Iyengar, O. Jacobowitz, S. Jahangeer, K.H. Jakobs, G.L. Johnson, R.L. Johnson, N. Kimura, S.D. Kroll, Y. Kurachi, R.H. Lee, R.J. Lefkowitz, Y. Li, C. Londos, C.C. Malbon, D.H. Maurice, K.R. McLeish, G. Milligan, A.J. Morris, E.J. Neer, A.S. Otero, U. Panten, D. Park, G.S. Pitt, R.T. Premont, F. Quan, M. Rodbell, W. Rosenthal, M. Russell, P. Ruth, C. Schwanstecher, M. Schwanstecher, I.A. Simpson, E. Stefani, G. Szabo, J.A. Thissen, T.D. Ting, L. Toro, G.L. Waldo, A. Welling, T.G. Wensel, M. Whiteway, W. Wolfgang, L. Wu

Editors

Burton F. Dickey and Lutz Birnbaumer

Springer-Verlag
Berlin Heidelberg New York London Paris
Tokyo Hong Kong Barcelona Budapest

Professor Burton F. Dickey, M.D.
Baylor College of Medicine
Department of Medicine Cell Biology and
Experimental Therapeutics, Room 3C-385
Houston VA Medical Center
2002 Holcombe Boulevard
Houston, TX 77030, USA

Professor Lutz Birnbaumer, Ph.D.
Baylor College of Medicine
Department of Cell Biology, Molecular Physiology and Biophysics,
Medicine and Neurosciences
One Baylor Plaza
Houston, TX 77030, USA

With 144 Figures and 22 Tables

Library of Congress Cataloging-in Publication Data. GTPases in biology. (Handbook of experimental pharmacology; v. 108) "Contributors, M. Biel" . . . [et al.], vol. 2—t.p. Includes bibliographical references and index. 1. Guanosine triphosphatase. I. Aktories, K. II. Dickey, Burton F., 1953– . III. Birnbaumer, Lutz. IV. Series. QP905.H3 vol. 108 615′11 s 93-21636 [QP609.G83] [574.19′253]
ISBN-13: 978-3-642-78347-0 e-ISBN-13: 978-3-642-78345-6
DOI: 10.1007/ 978-3-642-78345-6


Softcover reprint of the hardcover 1st edition 1993



Typesetting: Best-set Typesetter Ltd., Hong Kong
27/3130/SPS – 5 4 3 2 1 0 – Printed on acid-free paper

# List of Contributors


Biel, M., Institut für Pharmakologie und Toxikologie der Technischen Universität, Biedersteiner Straße 29, D-80802 München, Germany

Birnbaumer, L., Departments of Cell Biology, Molecular Physiology and Biophysics, and Medicine (Endocrinology) and Division of Neurosciences, Baylor College of Medicine, Texas Medical Center, One Baylor Plaza, Houston, TX 77030, USA

Blumer, K.J., Department of Cell Biology and Physiology, Box 8228, Washington University School of Medicine, 660 South Euclid Avenue, Saint Louis, MO 63110, USA

Bockaert, J., UPR 9023 Centre CNRS-INSERM de Pharmacologie-Endocrinologie, Rue de la Cardonille, F-34094 Montpellier Cedex 5, France

Bosse, E., Institut für Pharmakologie und Toxikologie der Technischen Universität, Biedersteiner Straße 29, D-80802 München Germany

Boyer, J.L., Department of Pharmacology, School of Medicine, CB#7365, The University of North Carolina at Chapel Hill, Chapel Hill, NC 27599–7365, USA

Brabet, P., UPR 9023 Centre CNRS-INSERM de Pharmacologie-Endocrinologie, Rue de la Carodonille, F-34094 Montpellier Cedex 5, France

Camps, M., Molecular Pharmacology Division, German Cancer Research Center, Im Neuenheimer Feld 280/III, D-69120 Heidelberg, Germany

Caron, M.G., Departments of Medicine (Cardiology), Biochemistry and Cell Biology, Howard Hughes Medical Institute, Duke University Medical Center, P.O. Box 3821, Durham, NC 27710

Casey, P.J., Section of Cell Growth, Regulation and Oncogenesis and Department of Biochemistry, Duke University Medical Center, P.O. Box 3686, Durham, NC 27710, USA

Chen, J., Department of Pharmacology, Mount Sinai School of Medicine of the City University of New York, P.O. Box 1215, One Gustave Levy Place, New York, NY 10029, USA

Clark, K., Eukaryotic Genetics Group, Biotechnology Research Institute, National Research Council of Canada, 6100 Royalmount Avenue, Montreal, Quebec, Canada H4P, 2R2

Corda, D., Laboratory of Cellular and Molecular Endocrinology, Istituto di Ricerche Farmacologiche "Mario Negri", Consorzio Mario Negri Sud, Via Nazionale, I-66030 S. Maria Imbaro (Chieti), Italy

Coulter, S., Signal Transduction Section, Laboratory of Cellular and Molecular Pharmacology, Building 7-737, Research Triangle Park, NC 27709, USA

Devreotes, P.N., Department of Biological Chemistry, Johns Hopkins University, School of Medicine, Baltimore, MD 21205, USA

Donnelly, D., Department of Biochemistry and Molecular Biology, The University of Leeds, Leeds LS2 9JT, Great Britain

DuBose, T.D., Jr., University of Texas Health Science Center at Houston, Division of Nephrology, 6431 Fannin, P.O. Box 20708, Houston, TX 77025, USA

Exton, J.H., Howard Hughes Medical Institute and Department of Molecular Physiology and Biophysics, Vanderbilt University, School of Medicine, 802 Light Hall, Nashville, TN 37232, USA

Findlay, J.B.C., Department of Biochemistry and Molecular Biology, The University of Leeds, Leeds LS2 9JT, Great Britain

Firtel, R.A., Department of Biology, Center for Molecular Genetics, University of California, San Diego, La Jolla, CA 92093-0634, USA

Flockerzi, V., Institut für Pharmakologie und Toxikologie der Technischen Universität, Biedersteiner Straße 29, D-80802 München Germany

Forte, M., Vollum Institute for Advanced Biomedical Research, Oregon Health Sciences University, 3181 S.W. Sam Jackson Park Rd., Portland, OR 97201, USA

Gaskins, C., Department of Biology, Center for Molecular Genetics, University of California, San Diego, La Jolla, CA 92093-0634, USA

Gierschik, P., Molecular Pharmacology Division, German Cancer Research Center, Im Neuenheimer Feld 280/III, D-69120 Heidelberg, Germany

Gundersen, R., Department of Biological Chemistry, Johns Hopkins University, School of Medicine, Baltimore, MD 21205, USA

Gurich, R.W., University of Texas Medical Branch, Division of Nephrology, 4.200 Old John Sealy, Galveston, TX 77550, USA

HADCOCK, J.R., American Cyanamid Company, Agricultural Research Division, Molecular and Cellular Biology Group, P.O. Box 400, Princeton, NJ 08543-0400, USA

HADWIGER, J.A., Department of Biology, Center for Molecular Genetics, University of California, San Diego, La Jolla, CA 92093-0634, USA

HARDEN, T.K., Department of Pharmacology, School of Medicine, CB#7365, The University of North Carolina at Chapel Hill, Chapel Hill, NC 27599-7365, USA

HILDEBRANDT, J.D., Department of Cell and Molecular Pharmacology, Medical University of South Carolina, 171 Ashley Avenue, Charleston, SC 29425-2251, USA

HO, Y.-K., Departments of Biochemistry and Ophthamology, University of Illinois at Chicago, Health Sciences Center, m/c 536, 1853 West Polk Street, Chicago, IL 60612, USA

HOFMANN, F., Institut für Pharmakologie und Toxikologie der Technischen Universität, Biedersteiner Straße 29, D-80802 München Germany

HOFMANN, K.P., Institut für Biophysik und Strahlenbiologie, Albert-Ludwigs-Universität, Albertstraße 23, D-79104 Freiburg, Germany

HOMBURGER, V., UPR 9023 Centre CNRS-INSERM de Pharmacologie-Endocrinologie, Rue de la Cardonille, F-34094 Montpellier Cedex 5, France

HOUSLAY, M.D., Molecular Pharmacology Group, Department of Biochemistry, University of Glasgow, Glasgow G12 8QQ, Great Britain

HULLIN, R., Institut für Pharmakologie und Toxikologie der Technischen Universität, Biedersteiner Straße 29, D-80802 München, Germany

HYDE, D., Department of Biological Sciences, University of Notre Dame, Notre Dame, IN 46556, USA

IYENGAR, R., Department of Pharmacology, Mount Sinai School of Medicine of the City of New York, P.O. Box 1215, One Gustave Levy Place, New York, NY 10029, USA

JACOBOWITZ, O., Department of Pharmacology, Mount Sinai School of Medicine of the City of New York, P.O. Box 1215, One Gustave Levy Place, New York, NY 10029, USA

JAHANGEER, S., Signal Transduction Section, Laboratory of Cellular and Molecular Pharmacology, Building 7-733, Research Triangle Park, NC 27709, USA

JAKOBS, K.H., Institut für Pharmakologie, Universitätsklinikum Essen, Hufelandstr. 55, D-45122 Essen, Germany

Johnson, G.L., Division of Basic Sciences, National Jewish Center for Immunology and Respiratory Medicine, 1400 Jackson Street, Denver, CO 80206, USA

Johnson, R.L., Department of Biological Chemistry, Johns Hopkins University, School of Medicine, Baltimore, MD 21205, USA

Kimura, N., Department of Molecular Biology, Tokyo Metropolitan Institute of Gerontology, 35-2, Sakaecho, Itabashi-ku, Tokyo-173, Japan

Kroll, S.D., Department of Pharmacology, Mount Sinai School of Medicine, City University of New York, P.O. Box 1215, One Gustave Levy Place, New York, NY 10029, USA

Kurachi, Y., Division of Cardiovascular Diseases, Department of Internal Medicine, and Department of Pharmacology, Mayo Clinic, Mayo Foundation, Rochester, MN 55905, USA

Lee, R.H., Department of Anatomy and Cell Biology, UCLA School of Medicine and Developmental Neurology Laboratory, Veteran Administration Medical Center, Sepulveda, CA 91343, USA

Lefkowitz, R.J., Department of Medicine (Cardiology), Biochemistry and Cell Biology, Howard Hughes Medical Institute, Duke University Medical Center, P.O. Box 3821, Durham, NC 27710

Li, Y., Department of Molecular Physiology and Biological Physics, Jordan Hall, Box 449, University of Virginia, Charlottesville, VA 22908, USA

Londos, C., Membrane Regulation Section, Laboratory of Cellular and Developmental Biology, National Institute of Diabetes and Digestive and Kidney Diseases, National Institutes of Health, Bldg. 6, Rm. B1-15, Bethesda, MD 20892, USA

Malbon, C.C., Department of Pharmacology, Diabetes and Metabolic Diseases Research Program, School of Medicine, Health Sciences Center, University Stony Brook, Stony Brook, NY 11794-8651, USA

Maurice, D.H., Department of Pharmacology, School of Medicine, CB#7365, The University of north Carolina at Chapel Hill, Chapel Hill, NC 27599-7365, USA

McLeish, K.R., Kidney Disease Program, Departments of Biochemistry and Medicine, University of Louisville and the Veterans Administration Medical Center, 500 South Floyd Street, Louisville, KY 40292, USA

Milligan, G., Molecular Pharmacology Group, Department of Biochemistry, University of Glasgow, Glasgow G12 8QQ, Scotland, Great Britain

MORRIS, A.J., Department of Pharmacology, School of Medicine, CB#7365, The University of North Carolina at Chapel Hill, Chapel Hill, NC 27599-7365, USA

NEER, E.J., Cardiovascular Division, Department of Medicine, Brigham and Women's Hospital and Harvard Medical School, 75 Francis Street, Boston, MA 02115, USA

OTERO, A.S., Department of Molecular Physiology and Biological Physics, Jordan Hall, Box 449, University of Virginia, Charlottesville, VA 22908, USA

PANTEN, U., Institut für Pharmakologie und Toxikologie der Universität Göttingen, Robert-Koch-Straße 40, D-37075 Göttingen, Germany

PARK, D., Laboratory of Biochemistry, National Heart, Lung and Blood Institute, National Institutes of Health, 9000 Rockville Pike, Building 3, Room 125, Bethesda, MD 20892, USA

PITT, G.S., Department of Biological Chemistry, Johns Hopkins University, School of Medicine, Baltimore, MD 21205, USA

PREMONT, R.T., Department of Pharmacology, Mount Sinai School of Medicine of the City University of New York, P.O. Box 1215, One Gustave Levy Place, New York, NY 10029, USA Present address: Department of Medicine, Box 3821, Duke University Medical Center, Durham, NC 27710, USA

QUAN, F., Vollum Institute for Advanced Biomedical Research, Oregon Health Sciences University, 3181 S.W. Sam Jackson Park Rd., Portland, OR 97201, USA

RODBELL, M., Signal Transduction Section, Laboratory of Cellular and molecular Pharmacology, Building 18-01, Research Triangle Park, NC 27709, USA

ROSENTHAL, W., Rudolf-Buchheim-Institut für Pharmakologie der Justus-Liebig-Universität Gießen, Frankfurter Straße 107, D-35392 Gießen, Germany

RUSSELL, M., Division of Basic Sciences, National Jewish Center, for Immunology and Respiratory Medicine, 1400 Jackson Street, Denver, CO 80206, USA

RUTH, P., Institut für Pharmakologie und Toxikologie der Technischen Universität, Biedersteiner Straße 29, D-80802 München, Germany

SCHWANSTECHER, C., Institut für Pharmakologie und Toxikologie der Universität Göttingen, Robert-Koch-Straße 40, D-37075 Göttingen, Germany

SCHWANSTECHER, M., Institut für Pharmakologie und Toxikologie der Universität Göttingen, Robert-Koch-Straße 40, D-37075 Göttingen, Germany

SIMPSON, I.A., Experimental Diabetes, Metabolism and Nutrition Section, Diabetes Branch, National Institute of Diabetes and Digestive Disease and Kidney Diseases, National Institutes of Health, Bethesda, MD 20892, USA

STEFANI, E., Department of Molecular Physiology and Biophysics, Baylor College of Medicine, Texas Medical Center, One Baylor Plaza, Houston, TX 77030, USA

SZABO, G., Department of Molecular Physiology and Biological Physics, Jordan Hall, Box 449, University of Virginia, Charlottesville, VA 22908, USA

THISSEN, J.A., Section of Cell Growth, Regulation and Oncogenesis and Department of Biochemistry, Duke University Medical Center, P.O. Box 3686, Durham, NC 27710, USA

TING, T.D., Department of Biochemistry, University of Illinois at Chicago, Health Sciences Center, 1853 West Polk Street, Chicago, IL 60612, USA

TORO, L., Department of Molecular Physiology and Biophysics, Baylor College of Medicine, Texas Medical Center, One Baylor Plaza, Houston, TX 77030, USA

WALDO, G.L., Department of Pharmacology, School of Medicine, CB#7365, The University of North Carolina at Chapel Hill, Chapel Hill, NC 27599-7365, USA

WELLING, A., Institut für Pharmakologie und Toxikologie der Technischen Universität, Biedersteiner Straße 29, D-80802 München, Germany

WENSEL, T.G., Verna and Marrs McLean Department of Biochemistry, Baylor College of Medicine, Medical Center, One Baylor Plaza, Houston, TX 77030, USA

WHITEWAY, M., Eukaryotic Genetics Group, Biotechnology Research Institute, National Research Council of Canada, 6100 Royalmount Avenue, Montreal, Quebec, Canada H4P 2R2

WOLFGANG, W., Vollum Institute for Advanced Biomedical Research, Oregon Health Sciences University, 3181 S.W. Sam Jackson Park Rd., Portland, OR 97201, USA

WU L., Department of Biological Chemistry, Johns Hopkins University, School of Medicine, 725 N. Wolfe St., Baltimore, MD 21205, USA

# Preface

The question that may be raised is why a book on GTPases, especially as so much has been written lately in so many primary journals and review publications. Moreover, I received Pedro Cuatrecasas' invitation to edit this book after having just completed the editing, with Ravi Iyengar, of *G Proteins* for Academic Press. Understandably, my initial reaction was negative. Prodded by Pedro, and witnessing the breakneck pace at which new information was appearing in journals, I finally agreed. But even so, I decided to delay until I could better define how to organize and focus the book. A review of *G Proteins* by Peter Gierschik, which he titled "Balanced but unpackaged," proved helpful for the selection of topics and details of execution. *G Proteins* had been on trimeric G proteins only. Gierschik asked for "something on the small GTPases" and also more about structure. He also suggested more cross referencing, more serious editing by the editors, and faster publication. The arrival in Houston, from Boston, of my co-editor Burt Dickey, with his interest in small GTPases and GTPase structure, and his willingness to aid me, gave me – us – the impetus of requesting chapters.

It soon proved that Burt was doing all the work, as some of the authors know. Burt would not take a "no" for an answer, and insisted, again and again. The consequence of the broadened scope was that a 30–40 chapter book became what it is now – a two volume compendium. The original deadline for receipt of the manuscripts was July 1992. The last chapters arrived around New Year. The original manuscripts represented the state-of-the-art in mid-late 1992, but thankfully many of the authors have taken advantage of an opportunity to add remarks on important recent developments.

I am not sure that we have been able to cope with all of Peter Gierschik's (valid) criticisms. We hope that errors have been kept to a minimum. But he also wished that authors would not contradict each other, or at least that the editors would edit out the contradictions. Yet, for the most part contradictions are based on the working hypotheses of the individual investigators and represent their legitimate points of view. They are, in fact, an exposition of the evolution of thoughts and arise because the experiments are not yet clear. We have thus left them, as far as they exist. Authors also change their mind. For example, I am sure that if one looked carefully, Henry Bourne no longer supports the idea that $\beta\gamma$ dimers promote GDP release. And I have been swayed to recognize $\beta\gamma$ dimers as regulators of effectors.

We have attempted to standardize nomenclature throughout. However, given the lack of agreement on usage even among workers within a field such as the Ras family, we have not been rigid in imposing terminology. Also, the nomenclature can be expected to evolve. Certain usages, however, were suggested. The term "GTPases" has been used to denote the superfamily because it is informative and short; "G protein" is reserved for members of the heterotrimeric signal-transducing family out of deference to the pleas of workers in this field, "Ras-related" or "small" is applied to the extended family of GTPases of molecular masses 19 to 25 kDa, including ARFs, in accordance with recommendations from a FASEB-sponsored conference, although it is recognized that ARFs possess several structural features which could warrant their designation as a distinct family. Proteins that modulate guanine nucleotide exchange are designated GEPs (guanine nucleotide exchange proteins), with a GDS (guanine nucleotide dissociation stimulator) accelerating exchange, and a GDI (guanine nucleotide dissociation inhibitor) retarding it. A protein that accelerates intrinsic GTPase activity is designated a GAP (GTPase activating protein), and one that retards it a GIP (GTPase inhibitory protein).

To finally answer the question raised in the first line of this preface, we except this book to serve as a compendium which investigators and advanced students alike can go to and find much of what they need if they are interested in regulatory GTPases, be they large or small, be it structural or functional. This must be coupled with the realization that the information covers knowledge and concepts as they stood at mid to late 1992.

Some of the most recent developments could not be included. For example we do not treat the role of molecular diversity of adenylyl cyclases in single cells. Only hints are presented about the complexity of PLC$\beta$'s, which at this time number 4 in the mammalian genome, each responding to a different extent to $\alpha$'s of the $\alpha_q$ family and to $\beta\gamma$ dimers. The specific roles of the many Rab's in the ever-expanding intracellular membrane compartments needs updating. The role of trimeric G proteins in the regulation of vesicle budding is absent, as is an extensive treatment of vesicle coating and uncoating. Heidi Hamm's and Paul Seglar's crystal of transducin $\alpha$ is absent. Maybe it is time to begin editing a follow up?

Houston, TX, USA — LUTZ BIRNBAUMER
November 1993

The obvious answer to the question raised by Lutz Birnbaumer, "Why a book on GTPases?", is that it has been four years since the last book on this still evolving field was published. More significantly, the discovery of entirely new members of the GTPase superfamily and considerable progress in elucidating the functions and molecular structures of individual GTPases rates this an apt time to update the subject. Given the rate at which the superfamily continues to grow, it will become less and less reasonable to consider all its branches in a single book. The size to which this volume grew should be a warning to any future editors!

The unique perspective of the variety and the shared features of GTPases afforded by an inclusive volume such as this makes it almost inevitable to begin to ask fundamental questions about why cells contain so many GTPases, and what the origins are of the basic GTPase mechanism. Answers to such questions can at present only be speculative, but studies of molecular evolution allow us to make some educated quesses. The ribosomal machinery, in which RNA functions both in information storage and in catalysis, may be thought of as an archaic remnant of life's origins in an RNA world. As such, the molecular species which make up this machine appear to be some of the most ancient and highly conserved among all contemporary cellular organisms. Included among the components of the translation apparatus are a number of related GTPase initiation and elongation factors. Several of these GTPases appear to function as proofreaders which neatly reconcile the inherently conflicting demands for speed and for accuracy in protein synthesis. Similarly, the uniquely successful organism which has given rise to all currently known cellular organisms must have been the one that was best able to satisfy the competitive needs for both reproductive speed and accuracy. One can imagine the elegant molecular mechanisms of the cellular synthetic machinery (including the GTPase proofreaders) being forged on the anvil of intense primordial competition.

If the initiation and elongation factors are indeed the ancestral GTPases, then the basic mechanism of GTP binding and hydrolysis has subsequently been appropriated for the regulation of a surprisingly wide array of biochemical processes. Other avenues for speculation include the coincidence that the two great families of intracellular regulatory switches – the protein kinases and the GTPases – both depend on interaction with a purine for their regulatory function. Are their shared, albeit limited, stuctural motifs evidence of a common origin, or of convergent evolution? Study of the superfamily of GTPases, in addition to providing us a panerama of the extraordinary breadth of the regulatory biology of cells, may yet afford us some insight into the very nature and origins of life.

Houston, TX, USA
November 1993

BURTON F. DICKEY

# Contents

CHAPTER 67

**Hormonal Regulation of Phospholipid Metabolism via G-Proteins: Phosphoinositide Phospholipase C and Phosphatidylcholine Phospholipase D**



CHAPTER 68

**Hormonal Regulation of Phospholipid Metabolism via G-proteins II: $PLA_2$ and Inhibitory Regulation of PLC**



CHAPTER 69

**G-Protein Regulation of Phospholipase C in the Turkey Erythrocyte**

CHAPTER 72

**Involvement of Pertussis-Toxin-Sensitive G-Proteins in the Modulation of $Ca^{2+}$ Channels by Hormones and Neurotransmitters**

CHAPTER 73

**Regulation of Cell Growth and Proliferation by $G_o$**



CHAPTER 74

**Role of Nucleoside Diphosphate Kinase in G-Protein Action**

CHAPTER 75

**G-Protein Regulation of Cardiac $K^+$ Channels**

CHAPTER 76

**Modulation of $K^+$ Channels by G-Proteins**



CHAPTER 77

**ATP-Sensitive $K^+$ Channel: Properties, Occurrence, Role in Regulation of Insulin Secretion**

CHAPTER 78

**Modulation of Maxi-Calcium-Activated K Channels: Role of Ligands, Phosphorylation, and G-Proteins**



CHAPTER 79

**Regulation of the Endosomal Proton Translocating ATPase ($H^+$-ATPase) and Endosomal Acidification by G-Proteins**

CHAPTER 80

**cAMP-Independent Regulation of Adipocyte Glucose Transport Activity and Other Metabolic Processes by a Complex of Receptors and Their Associated G-Proteins**

# Contents of Companion Volume 108/I

*D. Regulation of and by Small GTPases*

# Section IV
# Signal Transduction by Trimeric G Proteins

## *A. Cellular Architecture and its Role in Signal Transduction*

CHAPTER 44

# G-Proteins Have Properties of Multimeric Proteins: An Explanation for the Role of GTPases in their Dynamic Behavior

M. RODBELL, S. JAHANGEER, and S. COULTER

## A. Introduction

The possible structural organization of surface membrane receptors, heterotrimeric GTP-binding regulatory proteins (G-proteins), and effectors (adenylyl cyclase, phospholipases, phosphodiesterases, ion channels, etc.) is a major unresolved question. The physical relationship between these elements has been discussed in an insightful review of the role of G-proteins in signal transduction (NEER and CLAPHAM 1988). One of the questions raised is whether the components of signal transduction are free to move ("float") in the plane of the membrane and interact in accordance with their relative affinities for each other. Given nature's propensity to spatially organize and regulate structures in a highly specialized, nonrandom fashion, perhaps the question should be posed: how are receptors, transducers, and effectors spatially organized for rapid, efficient, and reversible interactions that only allow restricted diffusion of the components? There is evidence that receptors are precoupled to G-proteins in the absence of agonists (for example, see WATANABE et al. 1986). As discussed elsewhere in detail (RODBELL 1991; RODBELL 1992), there are many reasons to believe that receptors are coupled with heterotrimeric G-proteins structured in the form of multimers. As in the case of membrane proteins attached to filamental actin, it is likely that such large structures have restricted mobility in the membrane. As for effectors, it has been suggested that adenylyl cyclase is linked, indirectly or directly, to actin (WATSON 1990). Other potential effectors such as sodium and calcium channels also appear to be associated with actin through the actin-binding protein ankyrin (DAVIS and BENNETT 1990; EDELSTEIN et al. 1988), implying organized, fixed locations of effectors associated with the intracellular cytoskeletal matrix. If such organization applies to all signal transduction systems, it must be concluded that some type of signal is emitted from the receptor–G-protein complex that can be rapidly and reversibly conveyed between immobilized receptor–G-protein complexes and immobilized effectors.

## B. Theories

### I. Shuttle Theory

In this theory (GILMAN 1987), agonist-occupied receptors catalyze the exchange of GTP with GDP bound to the receptor-coupled G-protein. The GTP-occupied $\alpha$ subunits then undergo a conformational change which induces release of the $\alpha$ subunits from the $\beta\gamma$ complexes; the latter, because of their inherent lipophilic properties, remain associated with and are presumably mobile in association with the inner aspect of the cell membrane. The released $\alpha$ subunits, generally hydrophilic, are presumed to enter the aqueous cytosolic phase and then "hop" on to the effectors causing changes in activity. This shuttling mechanism has also been applied to the phototransduction system involving the transducing G-protein $G_t$ (CHABRE et al. 1988). There is experimental support for release of $\alpha_s$ subunits from membranes (NEGISHI et al. 1992; RANSNAS and INSEL 1988; RANSNAS et al. 1989, 1991). G-proteins, generally in detergent-containing solutions, also dissociate into $\alpha$ and $\beta\gamma$ subunits when exposed to Mg ions and guanine nucleotide analogs [GTP$\gamma$S or Gpp(NH)p] or aluminum fluoride, agents that cause marked stimulation, for example, of adenylyl cyclase. It should be noted that GTP, the natural activating nucleotide, induces dissociation either weakly or not at all unless very high concentrations of Mg ions are present.

### II. Collision-coupling Theory

This theory, originally applied to the activation of adenylyl cyclase by catecholamines and Gpp(NH)p in turkey erythrocyte membranes (TOLKOVSKY and LEVITZKI 1978), has been recently modified (LEVITZKI and BAR-SINAI 1991). Fundamental to this concept is that $G_s$ is directly coupled to adenylyl cyclase in its GDP-bound form. Following agonist-induced exchange of GTP with bound GDP, $G_s$ bound to the enzyme dissociates into free $\alpha_s$ and $\beta\gamma$, both of which remain associated with the enzyme. Support for the contention that heterotrimeric G-proteins are linked to the enzyme derives from the findings that Gpp(NH)p-preactivated adenylyl cyclase from both turkey erythrocyte and rat brain membranes copurifies with the $\alpha_s$ and $\beta$ subunits of $G_s$ in stoichiometric amounts (LEVITZKI and BAR-SINAI 1991). Not reported, however, is whether the enzymes copurify with heterotrimeric $G_s$ in their nonactivated state, a better test of the assumption that $G_s$ is permanently associated with the enzyme. Nonetheless, this information with the preactivated enzyme implies that heterotrimeric $G_s$ rather than simply the $\alpha_s$ subunit can interact with adenylyl cyclase; activation of the enzyme arises when the $\alpha_s$ subunit dissociates from the heterotrimeric complex. In this respect the collision-coupling and shuttling models are in agreement, although clearly they are incompatible on structural and kinetic grounds.

### III. Disaggregation Theory

This concept (RODBELL 1980) is based on target analysis (SCHLEGEL et al. 1979). Its basic premise is that receptors and G-proteins are precoupled as oligomeric structures; hormone and GTP act in concert to release "monomers" of G-proteins which then interact with adenylyl cyclase to cause stimulation or inhibition (SCHLEGEL et al. 1980). Subtracting the target size mass of adenylyl cyclase($\approx$120 kDa), now confirmed from analysis of its gene construct (KRUPINSKI et al. 1989), from the Gpp(NH)p-activated mass ($\approx$250 kDa), the difference, 130 kDa, is sufficiently close to heterotrimeric $G_s$ (90–95 kDa) to suggest that $G_s$ is attached to the enzyme as a heterotrimer rather than simply as $\alpha_s$, in agreement with the evidence obtained with preactivated adenylyl cyclase in brain and turkey erythrocyte membranes (LEVITZKI and BAR-SINAI 1991). Hence, from this evaluation it is possible that the heterotrimeric form of G-proteins, rather than the free $\alpha$ subunits, are the vectors that communicate between receptors and effectors. Since the heterotrimers are attached to the membrane through the isoprenylated $\gamma$ subunit (SANFORD et al. 1991), G-protein "monomers" released from the multimeric structures can be considered as mobile vehicles that remain membrane-attached during their transfer from receptor–G-protein multimers to effectors.

## C. Evidence for Multimeric Structures of G-Proteins

### I. Properties in Detergents

Sodium cholate and lubrol are the detergents generally used for extraction and purification of heterotrimeric G-proteins, from which arose the premise that heterotrimers are the primary form of G-proteins coupled to receptors. However, one explanation for the lack of multimeric structures in these detergents is that they disrupt multimeric structures of G-proteins. Indeed, when rat brain synaptoneurosome G-proteins ($G_s$, $G_i$, $G_q$, $G_o$) are extracted with octylglucoside, dramatic differences are seen in the range of sedimentation coefficients exhibited on sucrose density gradients compared with those obtained with membranes extracted with cholate, lubrol, or 3-[3-(cholamidopropyl)dimethylammonio]-1-propanesulfonic acid (CHAPS) (JAHANGEER and RODBELL 1993, in press). Whereas cholate, lubrol, or CHAPS extractions yield structures with S values of 4.5–5.0 (peak values) normally found for heterotrimeric G-proteins, octylglucoside, for example, extracts G-proteins displaying a polydisperse range of S values, including material that sediments through 20% sucrose (NAKAMURA and RODBELL 1990). Extraction with digitonin yields structures with even higher S values (at least 50% of the G-proteins sediment through the gradients) than those given by octyglucoside extraction, suggesting even larger structures.

Similarly, it has been shown that digitonin extracts $G_s$, adenylyl cyclase, and $\beta$-adrenergic receptors as structures that sediment through 20%–40% sucrose gradients along with cytoskeletal actin (LEVINE et al. 1982). Digitonin has been successfully used also for extraction of G-proteins coupled to the prostaglandin $E_2$ and vasopressin receptors (FITZGERALD et al. 1986; WATANABE et al. 1986); uncoupling of the prostaglandin receptor–Gi complex occurred when the digitonin-solubilized material was exposed to CHAPS.

In recent experiments (JAHANGEER and RODBELL 1993, in press), incubation of synaptoneurosome membranes with GTP$\gamma$S and Mg ions (2 m$M$) followed by extraction with octylglucoside caused shifts in S values toward the heterotrimeric structure (4.5 S), the sensitivity (minimal necessary concentration) to the nucleotide ranging from 1 $\mu M$ for $G_i$ and $G_o$, 100 $\mu M$ for $G_s$, and 1 m$M$ for $G_q$. Similar sensitivities to GTP$\gamma$S are observed with cholate, lubrol-, or CHAPS-extracted G-proteins. In lubrol, as distinct from all other detergents, the S-values shift from about 4.5 S to about 2.1 S (presumably the free $\alpha$ subunits; IYENGAR et al. 1988) in response to GTP$\gamma$S. By contrast, after incubation of synaptoneurosome membranes with GTP$\gamma$S, G-proteins extracted with either octylglucoside or digitonin exhibit sedimentation values that are three to four times higher than those obtained from GTP$\gamma$S-treated membranes subsequently extracted with lubrol, cholate, or CHAPS. These findings suggest that higher order structures of G-proteins in the membrane are disaggregated in response to the nucleotide to yield smaller species of multimers. Clearly, structures of G-proteins can be extracted from biological membranes in different forms depending on the type of detergent used for extraction. Hence, it is not possible to decipher, based on detergent extracted material alone, the physiologically relevant structures of G-proteins in membranes. Moreover, the different structures of G-proteins induced by the actions of GTP$\gamma$S depending on whether they are heterotrimers in lubrol or multimers in octylglucoside or digitonin present another problem: what are the biologically important products of the actions of activating guanine nucleotides and Mg ions?

## II. Cross-Linking of G-Proteins in Membranes

One means of determining the structures of G-proteins in their membrane environment is to treat the membranes with cross-linking agents followed by extraction and determination of the sizes of the cross-linked material relative to non-cross-linked G-proteins. When synaptoneurosomes are treated with *p*-phenylenedimaleimide and then extracted with sodium dodecyl sulfate, $G_s$, $G_i$, $G_o$ (detected by immunoblotting) are eluted from sizing columns (BioGelA 50m) either as very large structures in the void volume or, after SDS-PAGE and western blotting, as polydisperse structures much larger than expected from cross-linked heterotrimeric G-proteins (COULTER and RODBELL 1992). The distribution and sizes of the cross-linked material resemble those obtained from octylglucoside or digitonin extracts of the same membranes. Moreover, the cross-linked materials eluted in the void volume of BioGelA 50m are sufficiently large to account for a major portion of the 1500-kDa structures determined by target analysis of the cyclase system in hepatic membranes (SCHLEGEL et al. 1979). Hence, target size analysis, cross-linking of G proteins in their membrane environment, and extraction with octylglucoside or digitonin all point to the same conclusion: G-proteins most likely are associated with biological membranes in the form of higher order structures, possibly multimers.

## III. Glucagon Activation of Multimeric $G_s$ in Hepatic Membranes

Extractions of $G\alpha_s$ from hepatic membranes with octylglucoside display polydisperse multimer-like structures comparable to those seen in octylglucoside-extracts of synaptoneurosomes (NAKAMURA and RODBELL 1991). As in the case of synaptoneurosome $G\alpha_s$, hepatic $G\alpha_s$ is relatively insensitive to the GTP$\gamma$S-induced shifts to smaller size structures (4.5 S or $G_s$ heterotrimer), requiring minimally $100\,\mu M$ nucleotide. However, in the presence of $1-10\,\mu M$ GTP$\gamma$S and $2\,\text{m}M$ $Mg^{2+}$, glucagon ($1\,\text{n}M$) induces shifts to smaller forms of $G\alpha_s$ (peak values between 4.0 and 5.0 S), suggesting that the product is predominantly heterotrimeric $G_s$. The conditions for generating this product are strikingly similar to those required for activating hepatic adenylyl cyclase (RODBELL et al. 1974). It was also found that glucagon and the nucleotide only act upon a very large $G\alpha_s$-containing structure as revealed by prelabeling of $G\alpha_s$ using cholera toxin and $^{32}P$-NAD; the preponderance of labeling occurs in structures that sediment through 20% sucrose with S values much larger than 12 S. These studies revealed that: (a) multimeric forms of $G\alpha_s$ are favored substrates of the ADP-ribosylating activity of choleragen, (b) the glucagon receptor preferentially interacts with large, presumably multimeric forms of $G\alpha_s$, and (c) the hormone increases the reactivity of GTP$\gamma$S with multimeric $G\alpha_s$, i.e., relatively high concentrations of the nucleotide ($>100\,\mu M$) are required in the absence of glucagon. These studies, combined with the results obtained with synaptoneurosomes, provide major experimental support for the "disaggregation" theory of hormone and GTP activation of G-proteins. The major contributor to the 1500-kDa structure of the glucagon-sensitive adenylyl cyclase system in hepatic membranes is likely a multimeric structure of $G\alpha_s$. Importantly from the mechanistic standpoint, these studies show that the large multimeric structure is a precursor to a monomeric form of $G_s$, in strong support of the inferences drawn from target size analysis of the fluoride or Gpp(NH)p-preactivated cyclase system (SCHLEGEL et al. 1979). Thus, $G_s$ can be equated with the "monomer" postulated in the "disaggregation" theory. It serves as the activating vehicle between the glucagon receptor–G-protein complex and adenylyl cyclase, not free $\alpha_s$ (or $\beta\gamma$ complexes) as proposed in the "shuttle" theory.

## D. Coupling of Receptors to Multimeric G-Proteins

Both the collision–coupling and shuttle models of hormonal activation of adenylyl cyclase incorporate the notion that the shifts in the affinity states of receptors for agonists are due to the occupation of $G\alpha_s$ by GTP, and that these shifts are primarily the basis for the dissociation of the receptor from $G_s$ and, therefore, for the ability of the receptor to act in catalytic fashion. Mitigating against this thesis, however, is the fact that GDP, not GTP, is the primary inducer of the shifts of the glucagon receptor to its low affinity form, GDP being more potent that GTP by at least an order of magnitude (ROJAS and BIRNBAUMER 1984). Accordingly, it is thermodynamically unlikely that the GDP-bound form of $G_s$ is coupled to the high affinity form of the receptor, which is necessary to allow activation of the receptor by physiological concentrations of glucagon. Numerous studies with membranes or cells have shown that receptors can exist in both low- and high-affinity binding states. A rather special binding state is formed by hormones when binding to isolated membranes is studied in the absence of guanine nucleotides. For example, studies of glucagon binding to the hepatic receptor have shown that steady-state binding of the hormone takes at least 20 min; once bound, the hormone fails to dissociate even after repeated washing (RODBELL et al. 1971). This hormone-induced state is coupled to $G_s$ since it can be extracted in a nucleotide-sensitive, relatively stable state containing bound hormone; treatment with GTP releases the receptor from $G_s$ (WELTON et al. 1977). Whether this hormone-induced state is that which activates the associated G-protein multimer or is a transition state that leads to the activated state are unsolved questions. It is clearly not the low-affinity state induced by the GDP-bound form of the coupled G-protein. A likely possibility is that receptors coupled to G-proteins assume at least three states: a low-affinity hormone-binding state induced by the binding of GDP, an irreversible hormone-binding state when the coupled G-proteins lack bound guanine nucleotides, and an "activated state" of the G-protein containing bound GTP in which the receptor takes a reversible, high-affinity form. From this one can speculate that the hormone-induced exchange reaction is actually a composite of the three states of the coupled receptor–multimeric G-protein complexes. A possible reaction scheme (adapted from RODBELL 1992) between receptor and multimeric $G_s$-protein is depicted in Fig. 1. It is suggested that the $G_s$ multimer is dynamically activated by a combination of the exchange reaction induced by the hormone (H)-activated receptor and the splitting of GTP to GDP and Pi, in a manner similar to the "ratcheting" of F-actin and myosin by the exchange of ATP for bound ADP and the energy derived from cleavage of ATP to ADP+Pi (CARLIER 1990). This scheme provides an important basis for receptor coupling to multimeric G-proteins rather than simply to a single heterotrimer: (a) it explains how a single receptor activates a number of G-proteins, and (b) multimers may be either homogeneous (identical species in the same multimer) or

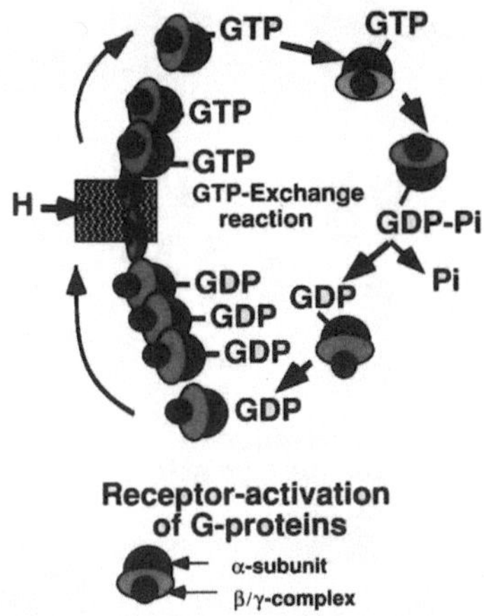


**Fig. 1.** Model for receptor activation of G-proteins. All of the components remain bound to the surface membrane during the entire dynamic process. Multimeric G-proteins are depicted in three states: GDP-bound; unoccupied by nucleotide and interacted with hormone (*H*) receptor; and occupied by GTP. The nucleotide exchange reaction occurs for each "monomer" attached to the high-affinity form of the receptor, resulting in progressive movement, in ratchetlike fashion, of the multimeric structure. Release of monomers occupied by GTP from multimers occurs in "quantal" fashion. On-site hyrolysis of GTP to GDP+Pi, followed by release of Pi controls the dynamics of the process in conjunction with the hormone-activated exchange reaction. See text for other details

heterogeneous (multiple species), the latter providing the means for pleiotropic expression of signals emanating from a single receptor (Abou-Samra et al. 1992; Gudermann et al. 1992).

## E. Hydrolysis of GTP is Fundamental to Signal Transduction Dynamics

The discovery that Gpp(NH)p, a nucleotidase-resistant analog of GTP, stimulates adenylyl cyclase in a slowly reversible fashion (Londos et al. 1974) led to the speculation, subsequently demonstrated (Cassel et al. 1977), that hydrolysis of GTP is associated with $G\alpha_s$. Many investigators have speculated from these findings that binding of GTP causes activation of G-coupled effectors whereas subsequent splitting of GTP is the cause of the decay in activity of effectors. This simple "turn-on, turn-off" mechanism, however, is inconsistent with certain findings. For example, modeling of the kinetics of activation of the adenylyl cyclase system in liver membranes by Gpp(NH)p or GTP and $Mg^{2+}$ fits well with a three-state, transition model (Rendell et al. 1975; Rendell et al. 1977) in which the state containing bound Gpp(NH)p or GTP is a transition state leading to the high-activity

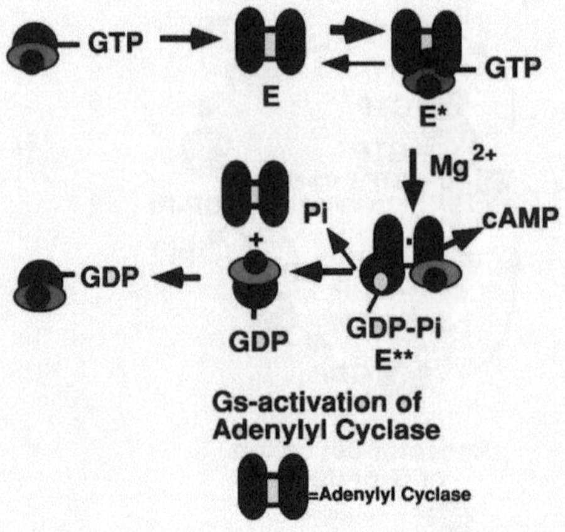


**Fig. 2.** Model for activation of adenylyl cyclase by Gs, the stimulatory GTP-binding protein. All components and their interactions are bound to the surface membrane. The heterotrimeric $G_s$ containing bound GTP that was released from the multimeric structure (see Fig. 1) during hormonal activation interacts with adenylyl cyclase in its ground or basal state (*E*) to form the transition state (*E**). $Mg^{2+}$ induces hydrolysis of GTP to GDP+Pi resulting in transformation of heterotrimer to free $G\alpha_s$ and $\beta\gamma$ complex, each reacting with separate sites on the two distinct "cassettes" of the enzyme. This form of the enzyme (*E***) is a high-activity state resulting in increased production of cyclic AMP (*cAMP*). Release of Pi is a slow, rate-limiting step, resulting in formation of the GDP-bound form of $G\alpha_s$ and its rapid recombination with the $\beta\gamma$ complex. The GDP-bound $G_s$ then recycles through the multimeric structure depicted in Fig. 1. See text for other details

state. When simulations carried out with Gpp(NH)p were substituted with GTP, the transition state model predicted that hydrolysis of GTP to GDP+Pi occurs when the enzyme system is in its highest activity state; only in this manner did the kinetics of GTP-activation display the rapid onset and offset of cyclase activity. By contrast, Gpp(NH)p induced the high-activity state by a slow, pseudo-irreversible process, typical for nonhydrolyzable analogs of GTP. At the time the modeling was carried out, little was known of the structural components of the adenylyl cyclase system. With the knowledge that G-proteins are heterotrimeric proteins, the three-state transition model can be depicted as illustrated in Fig. 2, in which the basal state (E) is reversibly coupled to the GTP-occupied form of heterotrimeric $G_s$ to yield the transition state (E*) which may not have activity greater than that of the basal state. Transit to the high activity state (E**) is driven by hydrolysis of GTP to GDP and Pi, both of which are tightly bound to distinct sites on the $\alpha_s$ subunit. To account for the rapid dynamics of the activation process, it is further postulated that the hydrolytic process leads concomitantly to a rearrangement of the physical relationship between the $\alpha_s$ subunit, the $\beta\gamma$ complex, and adenylyl cyclase to yield the activated state (E**), followed by rapid release of bound Pi and return of the enzyme to its basal state (E). Structural reorganization of bound $G_s$ may involve dissociation of $\alpha_s$ and $\beta\gamma$ (a $Mg^{2+}$-dependent process; CODINA et al. 1984). This raises the possibility that the enzyme has independent binding sites for both $\alpha_s$ and $\beta\gamma$ at separate sites on the two different "cassettes" of adenylate cyclase (KRUPINSKI et al. 1989), as depicted in Fig. 2.

In this regard, certain species of adenylyl cyclase are activated independently but synergistically by $\alpha_s$ and $\beta\gamma$ (TANG and GILMAN 1991). In accordance with the model are the findings that in the rod outer segment photo-transduction system, on-site hydrolysis of GTP to GDP and Pi is largely responsible for initiating the rapid onset of activation of cyclic GMP phosphodiesterase, a major recipient of the photoactivation process; Pi release is the rate limiting step in the overall dynamic process (TING and HO 1991). Rapid release of Pi is promoted by the $\gamma$ subunit of the cyclic GMP phosphodiesterase which plays a major role in the dynamics of phototransduction (ARSHAVSKY and BOWNDS 1992). Thus, although derived from different perspectives, both the activation of adenylyl cyclase by $G_s$ and the activation of phosphodiesterase by $G_t$ appear identical in their overall behavior. In both cases, Pi release is the rate-limiting step in the overall process and seems to be accelerated during activation of the effector (adenylyl cyclase or phosphodiesterase), as predicted (RENDELL et al. 1977). Hence, as normally measured, Pi release by isolated G-proteins is not necessarily a measure of the rate of on-site GTP cleavage and indeed appears to be a slow process relative to on-site cleavage (TING and HO 1991). It is interesting that mutants of $G_s\alpha$ have been identified in various tumor cells (LANDIS et al. 1989; LYONS et al. 1990; WEINSTEIN et al. 1991) that display very low GTPase as measured by Pi release. Possibly the mutations (generally $Arg^{201}$) actually cause $G_s\alpha$ to have very low Pi release rates and high on-site hydrolysis of GTP, leading to constitutively high activity of $G_s\alpha$. $AlF_3$ or $BeF_3$ mimic the terminal phosphate-binding site and act in conjunction with GDP in activating G-proteins, actions that do not occur with small GTP-binding proteins, as exemplified by *ras* $p^{21}$ (KAHN 1991). The long-lasting effects of $AlF_3$, in contrast to GTP, are likely due to its much slower egress from the Pi-binding domain. As for the nonhydrolyzable analogs of GTP [Gpp(NH)p or GTP$\gamma$S], a plausible explanation for their relatively slow onset of activation is that they remain bound long enough to stabilize a favorable conformation for inducing dissociation of the heterotrimers into $\alpha$ and $\beta\gamma$ subunits, a phenomenon not seen with GTP and which may account for their pseudo-irreversible effects.

## F. Conclusions

Evidence has been offered that receptor/G-protein mediated signaling systems contain large, presumably multimeric structures of G-proteins that are controlled in a manner resembling the interactions between myosin and fibrillar actin: (a) both systems contain multimeric components, (b) dynamic relationships between the components are controlled by the binding and cleavage of nucleotide triphosphates, (c) Pi release is a rate-limiting step in the overall dynamics, and (d) both systems contain sites for the binding and actions of $AlF_3$ (Antonny and Chabre 1992 and references therein). In the case of actin/myosin, the result is controlled spatial movement; in the case of receptor/multimeric G-proteins, the result is quantal release of heterotrimers that react with multiple effectors. In the broad sense this concept provides, in an oversimplified manner, a physical basis for how receptors catalytically activate a number of individual G-proteins in a multimeric structure. As stated above, multimers may be homogeneous (identical species of G-proteins) or heterogeneous (multiple species). Finally, as is well established in the coupling of actin filaments to membrane receptor proteins (Bourguignon and Singer 1977), coupling of receptors to multimeric G-proteins provides, in theory, a means of moving the combined structure along the membrane in an energy-dependent (GTPase) and possibly directional fashion, analogous to the ratcheting movements that occur between filamental actin and myosin in muscle.

## References

Abou-Samra A-B, Jüppner H, Force T, Freeman MV, Kong X-F, Schipani E, Urena P, Richards J, Bonventre JV, Potts JT Jr, Kronenberg HM, Segre GV (1992) Expression cloning of a rat bone PTH/PTHrP receptor: a single receptor activates both adenylate cyclase and phospholipase C. Proc Natl Acad Sci USA 89:2732–2736

Antonny B, Chabre M (1992) Characterization of the aluminum and beryllium fluoride species which activate transducin: analysis of the binding and dissociation kinetics. J Biol Chem 267:6710–6718

Arshavsky VY, Bownds MD (1992) Regulation of deactivation of photoreceptor G protein by its target enzyme and cGMP. Nature 357:416–417

Bourguignon L, Singer SJ (1977) Transmembrane interaction and the mechanism of capping surface receptors by their specific ligands. Proc Natl Acad Sci USA 74:5031–5035

Carlier MF (1990) Actin polymerization and ATP hydrolysis. Adv Biophys 26:51–73

Cassel D, Levkovitz H, Selinger Z (1977) The regulatory GTPase cycle of turkey erythrocyte adenylate cyclase. J Cyclic Nucleotide Res 3:393–406

Chabre M, Bigay J, Bruckert F, Bornancin F, Deterre P, Pfister C, Vuong TM (1988) Visual signal transduction: the cycle of transducin shuttling between rhodopsin and cGMP phosphodiesterase. Cold Spring Harbor Symp Quant Biol 53:313–324

Codina J, Hildebrandt JD, Birnbaumer L, Sekura RD (1984) Effects of guanine nucleotides and Mg on human erythrocyte Ni and Ns, the regulatory components of adenylyl cyclase. J Biol Chem 259:11408–11418

Coulter S, Rodbell M (1992) Heterotrimeric G proteins in synaptoneurosome membranes are crosslinked by p-phenylenedimaleimide, yielding structures comparable in size to crosslinked tubulin and F-actin. Proc Natl Acad Sci USA 89:5842–5846

Davis LH, Bennett V (1990) Mapping the binding sites of human erythrocyte ankyrin for the anion exchanger and spectrin. J Biol Chem 265:10589–10596

Edelstein NG, Catterall WA, Moon RT (1988) Identification of a 33-kilodalton cytoskeletal protein with high affinity for the sodium channel. Biochemistry 27:1818–1822

Fitzgerald TJ, Uhing RJ, Exton JH (1986) Solubilization of the vasopressin receptor from rat liver plasma membranes. Evidence for a receptor GTP-binding protein complex. J Biol Chem 261:16871–16877
Gilman AG (1987) G proteins: transducers of receptor-generated signals. Annu Rev Biochem 56:615–649
Gudermann T, Birnbaumer M, Birnbaumer L (1992) Evidence for dual coupling of the murine luteinizing hormone receptor to adenylyl cyclase and phosphoinositide breakdown and $Ca^{2+}$ mobilization. J Biol Chem 267:4479–4489
Iyengar R, Rich KA, Herberg JT, Premont RT, Codina J (1988) Glucagon receptor-mediated activation of Gs is accompanied by subunit dissociation. J Biol Chem 263:15348–15353
Jahangeer S, Rodbell M (1993) The disaggregation theory of signal transduction revisited: further evidence that G-proteins are multimeric and disaggregate to monomers when activated. Proc Natl Acad Sci USA I (in press)
Kahn RA (1991) Fluoride is not an activator of the smaller (20–25 kDa) GTP-binding proteins. J Biol Chem 266:15595–15597
Krupinski J, Coussen F, Bakalyar HA, Tang WJ, Feinstein PG, Orth K, Slaughter C, Reed RR, Gilman AG (1989) Adenylyl cyclase amino acid sequence: possible channel- or transporter-like structure. Science 244:1558–1564
Landis CA, Masters SB, Spada A, Pace AM, Bourne HR, Vallar L (1989) GTPase inhibiting mutations activate the $\alpha$ chain of Gs and stimulate adenylyl cyclase in human pituitary tumours. Nature 340:692–696
Levine HI, Sayhoun NE, Cuatrecasas P (1982) Properties of rat erythrocyte membrane cytoskeletal structures produced by digitonin extraction: digitonin insoluble $\beta$-adrenergic receptor, adenylate cyclase, and cholera toxin substrate. J Membrane Biol 64:225–231
Levitzki A, Bar-Sinai A (1991) The regulation of adenylyl cyclase by receptor-operated G proteins. Pharmac Ther 50:271–283
Londos C, Salomon Y, Lin MC, Harwood JP, Schramm M, Wolff J, Rodbell M (1974) 5′-Guanylylimidodiphosphate, a potent activator of adenylate cyclase systems in eukaryotic cells. Proc Natl Acad Sci USA 71:3087–3090
Lyons J, Landis CA, Harsh G, Vallar L, Grunewald K, Feichtinger H, Duh QY, Clark OH, Kawasaki E, Bourne HR, McCormick F (1990) Two G protein oncogenes in human endocrine tumors. Science 249:655–659
Nakamura S, Rodbell M (1990) Octyl glucoside extracts GTP-binding regulatory proteins from rat brain "synaptoneurosomes" as large, polydisperse structures devoid of beta gamma complexes and sensitive to disaggregation by guanine nucleotides. Proc Natl Acad Sci USA 87:6413–6417
Nakamura S, Rodbell M (1991) Glucagon induces disaggregation of polymer-like structures of the alpha subunit of the stimulatory G protein in liver membranes. Proc Natl Acad Sci USA 88:7150–7154
Neer EJ, Clapham DE (1988) Roles of G protein subunits in transmembrane signalling. Nature 333:129–134
Negishi M, Hashimoto H, Ichikawa A (1992) Translocation of alpha subunits of stimulatory guanine nucleotide-binding proteins through stimulation of the prostacyclin receptor in mouse mastocytoma cells. J Biol Chem 267: 2364–2369
Ransnas LA, Insel PA (1988) Subunit dissociation is the mechanism for hormonal activation of the Gs protein in native membranes. J Biol Chem 263:17239–17242
Ransnas LA, Svoboda P, Jasper JR, Insel PA (1989) Stimulation of beta-adrenergic receptors of S49 lymphoma cells redistributes the alpha subunit of the stimulatory G protein between cytosol and membranes. Proc Natl Acad Sci USA 86:7900–7903
Ransnas LA, Leiber D, Insel PA (1991) Inhibition of subunit dissociation and release of the stimulatory G-protein, Gs, by beta gamma-subunits and somatostatin in S49 lymphoma cell membranes. Biochem J 280:303–307

Rendell M, Salomon Y, Lin MC, Rodbell M, Berman M (1975) The hepatic adenylate cyclase system. III. A mathematical model for the steady state kinetics of catalysis and nucleotide regulation. J Biol Chem 250:4253–4260
Rendell MS, Rodbell M, Berman M (1977) Activation of hepatic adenylate cyclase by guanyl nucleotides. Modeling of the transient kinetics suggests an "excited" state of GTPase is a control component of the system. J Biol Chem 252:7909–7912
Rodbell M (1980) The role of hormone receptors and GTP-regulatory proteins in membrane transduction. Nature 284:17–22
Rodbell M (1991) G proteins may exist as polymeric proteins: a basis for the disaggregation theory of hormone action. In: Bellve AR, Vogel HJ (eds) Molecular mechanisms in cellular growth and differentiation. Academic Press, New York, pp 45–58
Rodbell M (1992) The role of GTP-binding proteins in signal transduction: from the sublimely simple to the conceptually complex. Curr Top Cell Regul 32:1–47
Rodbell M, Krans HM, Pohl SL, Birnbaumer L (1971) The glucagon-sensitive adenyl cyclase system in plasma membranes of rat liver. IV. Effects of guanylnucleotides on binding of $^{125}$I- glucagon. J Biol Chem 246:1872–1876
Rodbell M, Lin MC, Salomon Y (1974) Evidence for interdependent action of glucagon and nucleotides on the hepatic adenylate cyclase system. J Biol Chem 249:59–65
Rojas FJ, Birnbaumer L (1984) Regulation of glucagon receptor binding: lack of effect of Mg and preferential role for GDP. J Biol Chem 260:7829–7835
Sanford J, Codina J, Birnbaumer L (1991) Gamma-subunits of G proteins, but not their alpha- or beta-subunits, are polyisoprenylated. Studies on post-translational modifications using in vitro translation with rabbit reticulocyte lysates. J Biol Chem 266:9570–9579
Schlegel W, Kempner ES, Rodbell M (1979) Activation of adenylate cyclase in hepatic membranes involves interactions of the catalytic unit with multimeric complexes of regulatory proteins. J Biol Chem 254:5168–5176
Schlegel W, Cooper DM, Rodbell M (1980) Inhibition and activation of fat cell adenylate cyclase by GTP is mediated by structures of different size. Arch Biochem Biophys 201:678–682
Tang WJ, Gilman AG (1991) Type-specific regulation of adenylyl cyclase by G protein beta-gamma subunits. Science 254:1500–1503
Ting TD, Ho YK (1991) Molecular mechanism of GTP hydrolysis by bovine transducin: pre-steady-state kinetic analyses. Biochemistry 30:8996–9007
Tolkovsky AM, Levitzki A (1978) Mode of Coupling between the beta-adrenergic receptor and adenylate cyclase in turkey erythrocytes. Biochemistry 17:3795–3810
Watanabe T, Umegaki K, Smith WL (1986) Association of a solubilized prostaglandin $E_2$ receptor from renal medulla with a pertussis toxin-reactive guanine nucleotide regulatory protein. J Biol Chem 261:13430–13439
Watson PA (1990) Direct stimulaton of adenylate cyclase by mechanical forces in S49 mouse lymphoma cells during hyposmotic swelling. J Biol Chem 269:6569–6975
Weinsein LS, Shenker A, Gejman PV, Merino MJ, Friedman E, Spiegel AM (1991) Activating mutations of the stimulatory G protein in the McCune-Albright syndrome. N Engl J Med 325:1688–1695
Welton AF, Lad PM, Newby AC, Yamamura H, Nicosia S, Rodbell M (1977) Solubilization and separation of the glucagon receptor and adenylate cyclase in guanine nucleotide-sensitive states. J Biol Chem 252:5947–5950

# *B. G-Protein Coupled Receptors*

CHAPTER 45

# The Superfamily: Molecular Modelling

J.B.C. Findlay and D. Donnelly

## A. Introduction

G-protein-coupled receptors (GPCRs) are a diverse family of integral membrane proteins that are believed to share many structural and functional properties. Upon stimulation, these receptors can transmit a signal across the bilayer that results in the activation of enzymes or transport systems via a G-protein. The size of the family is increasing rapidly and includes the neurotransmitter receptors (e.g., adrenergic, muscarinic acetylcholine, neurokinins, dopamine, serotonin) as well as the visual and odorant receptors (e.g., rhodopsin).

GPCRs are believed to share a common transmembrane domain made up of seven segments, but to date no three-dimensional (3D) structure has been elucidated for any member of the family. The 3D structure is vital for the design of novel ligands and for a full understanding of the function of these receptors. Protein modelling, the prediction of the 3D structure of a protein from its amino acid sequence, is still a primitive and imprecise science, but it is critical for the future development of molecular biology. This is particularly true for integral membrane proteins, since progress in structure determination has been slow and is likely to remain so for some time. Protein engineering is a valuable tool, especially when structural information is available. However, in the absence of this information, the results from this approach may be harnessed to other data to provide at least crude working models which can serve as the basis for further experimental design.

For integral membrane proteins there are two basic sources of information that can be used in modelling. The first is experimental data which ideally should include the 3D structure of a homologous protein, but more usually comes from the likes of spectroscopy, molecular genetics and protein chemistry. The second includes more theoretical approaches that seek to establish rules, derived from comparisons of protein structures and sequences, that can then be applied to the sequence of the protein of interest. Both approaches are didactic and should be complementary. Several models of GPCRs have been proposed using these and other approaches (e.g., Pappin et al. 1984; Donnelly et al. 1989; Findlay and Eliopoulos 1990; Hibert et al. 1991; Grötzinger et al. 1991;

MALONEYHUSS and LYBRAND 1992; D. DONNELLY, A.M. MACLEOD and T.L. BLUNDELL, unpublished results). This review describes the results of two of these studies that illustrate how the methods outlined above can be employed to arrive at a crude representation of a GPCR (FINDLAY and ELIOPOULOS 1990; D. DONNELLY, A.M. MACLEOD and T.L. BLUNDELL, unpublished results). Such representations are certainly likely to be incorrect, if only in detail, but they do suggest experiments that may provide new data. They can also be refined as these and other data become available.

## B. General Principles – Modelling Integral Membrane Domains

There are four basic pieces of data which are vital for the construction of any convincing representation of an integral membrane protein. The first is the determination of the number of transmembrane segments, i.e., the disposition or topography of the polypeptide chain in the bilayer. There are a number of routes to this information. For rhodopsin, protein sequencing was used to determine residues in the protein subjected to biological (i.e., glycosylation and phosphorylation) or chemical (exogenous hydrophilic probes) modification (reviewed in FINDLAY and PAPPIN 1984). Antipeptide antibodies were employed to deduce a rather similar disposition for the β-adrenergic receptor (β-AR) (WANG et al. 1989). A more genetic approach, not used for the GPCRs, was the splicing of putative transmembrane segments to reporter enzymes, such as lactamase, whose intra- and extracellular location allowed definition of these segments (TADAYYON et al. 1992). Computer-based methods include the use of hydrophilic and/or hydrophobic indices of various kinds (e.g., CORNETTE et al. 1987). Taken together all these strategies should generate a sound appreciation of the basic protein topography, although none, particularly the more theoretical ones, is likely to be infallible. A more contemporary strategy likely to produce more reliable observations is to generate a series of mutants containing a single readily modifiable side chain placed at positions likely to define the bilayer boundary. Protein chemistry and/or electron spin resonance (ESR) spectroscopy could then be used to determine the relative exposure of particular residues (ALTENBACH et al. 1990). A more theoretical approach for determining the point at which a transmembrane segment meets the bilayer border is described later.

The second piece of information required is the conformation of these transmembrane segments and this is best determined experimentally. Circular dichroism (CD) and Fourier transform infrared (FTIR) spectroscopy offer the best prospects for obtaining a rough estimate of the proportion of α helix of β sheet in the protein. In the case of rhodopsin, there is general agreement that about 60% of the protein is α helical with a small amount of $\beta$ sheet (STUBBS et al. 1976). This is enough for a decision to be made as to the likely structure of the transmembrane segments, but their exact positions are not provided by these approaches. Analysis, by these methods or by nuclear magnetic resonance (NMR), of synthetic peptides representing each transmembrane region could generate useful data but is flawed by the loss of influence of the remaining structure on the peptide and by the practical methods (solvents, etc.) used in such studies. In the same way as for the topographic analysis above, protein chemistry or ESR spectroscopy will provide useful general data, but huge amounts of work are required to characterize every position in the transmembrane segment. Finally, secondary structure prediction methods (e.g., GARNIER et al. 1978) have limited usefulness, since high predictions of both $\alpha$ helix and $\beta$ sheet often occur for the same regions (e.g., PAPPIN et al. 1984). A more interesting approach is to carry out Fourier transform analyses in order to detect the periodicity of the particular secondary structural elements from the disposition of hydrophobic or conserved residues in sequence alignments of the transmembrane regions (see below).

The orientation (i.e., sidedness) of each of the transmembrane segments is the third element of information. Again, one can adopt an experimental approach by determining either the availability of particular side chains to modification by hydrophobic probes that partition into the fatty acid milieu of the bilayer (DAVISON and FINDLAY 1986) or the environment of spin labels attached at particular residue positions (ALTENBACH et al. 1990). Both should give consistent results and more interestingly have the potential to indicate the positions of regions that intriguingly depart from secondary structural regularity. The amount of effort involved in comprehensive surveys of this kind is very large. It is much easier to make assumptions nevertheless based on the more hydrophobic or variable face (exposed to lipid) and on the location of conserved or charged residues (internally orientated), such as outlined below. Although probably correct, these assumptions require experimental verification.

The fourth and most difficult element of information to obtain is some idea of the 3D packing arrangement of the transmembrane segments. This can only be obtained by X-ray or electron diffraction studies. Fortunately in the case of this work, the medium resolution structure of bacteriorhodopsin (HENDERSON and UNWIN 1975; HENDERSON et al. 1990) provides an insight as to how seven transmembrane helices can be arranged. Other less definitive biophysical methods (reviewed in FINDLAY and PAPPIN 1984) suggested that the overall shape and intramembranous cross-sectional area of rhodopsin and bacteriorhodopsin were similar. Therefore, although the two proteins share no sequence homology and despite the fact that bacteriorhodopsin is not a member of the GPCR family, the two proteins may share a similar structural fold. This has been used as a justification for basing 3D models of GPCRs on that of bacteriorhodopsin, though it must be remembered that this may not be entirely correct. Some experimental data for example, are also consistent with numerous other arrangements (e.g., where helix 1 of rhodopsin is equivalent to helix 2 of bacteriorhodopsin, helix 2 of rhodopsin is equivalent to helix 3 of bacteriorhodopsin, etc.).

## I. Summary of Information Available for G-Protein-Coupled Receptor Modelling Studies

From the approaches outlined above, therefore, we now have sound data indicating that GPCRs consist of seven transmembrane helices, whose orientation can be deduced from an amalgamation of protein chemistry data and from analyses of the positions of hydrophobic, charged and conserved residues. The packing arrangement and connectivity of these helices is based on the structure of bacteriorhodopsin.

Likely ligand-binding regions in rhodopsin are indicated by the identification of the attachment sites on opsin for the chromophore 11-cis (WANG et al. 1980). Site-directed mutagenesis on the $\beta$-AR has provided additional ligand-binding data (STRADER et al. 1989, 1991). Protein chemistry has revealed the presence of a single critical disulphide bond and has located the positions of two palmitic acid residues (KARNIK and KHORANA 1990; OVCHINNIKOV et al. 1988). Oligomeric associations need not be considered in the modelling studies, since cross-linking studies determined its monomeric status (BRETT and FINDLAY 1979). Finally, a wealth of natural and site-directed mutants provide data that can be incorporated into the structural representation to obtain a better appreciation of the structure of the protein.

**Table 1.** The most conserved residues in the putative transmembrane helices (residues in backets are occasional substitutions). Residue numbers correspond to ovine rhodopsin (FINDLAY et al. 1981; BRETT and FINDLAY 1983)

| Helix | Residue |
|---|---|
| 1 | N-55 |
| 2 | D-83 |
| 3 | C-110 |
| | D(E)-133 |
| | R-134 |
| | Y(F,W)-135 |
| 4 | W-161 |
| 5 | P-215 |
| | Y-223 |
| 6 | W-265 |
| | P-267 |
| 7 | N(D)-302 |
| | P-303 |
| | Y-306 |

# C. Modelling G-Protein-Coupled Receptors From Sequence Alignments

## I. Sequence Comparisons

There are several residues that are highly conserved across the GPCR family. Most of these are located in the seven putative transmembrane helices and can be used to align the sequences of these regions (Table 1). Sequence alignments of the transmembrane regions are attempts to emulate structural alignments and therefore should not contain insertions or deletions if they are indeed regular $\alpha$ helices. There are members of the family that do not possess all of the conserved residues (e.g., the odorant receptors do not have the conserved proline or tryptophan in helix 6) and this makes sequence alignments more difficult. The transmembrane helices form the protein's structural core and are probably also responsible for ligand binding and signal transduction in most, if not all, members of the family. The terminal regions and those that connect the transmembrane domains are more variable, both in length and sequence, and as such it is difficult to produce reliable sequence alignments for these regions.

## II. Fourier Transform Analysis of G-Protein-Coupled Receptor Sequence Alignments

### 1. Prediction of Structural Environments from Sequence Alignments

The residues that form the core of a protein have very different properties from those that are located on the surface. Due to the increased packing constraints in the protein interior, these residues mutate at a slower rate and undergo a more limited range of substitutions. Therefore, the amino acid substitution pattern at each position in a sequence alignment can provide information on the probability of that position being buried in the protein core. This can be achieved by either quantifying the extent of variability at each position in the sequence alignment (KOMIYA et al. 1988) or by using environment-dependent substitution tables (calculated from protein structures) in order to predict the environment from the sequence patterns in the alignment (DONNELLY et al. 1993). In this case "environment" refers to either "internal" or "surface" residues.

The hydrophobicity of residues in the interior of a protein differs from that of the surface residues. In water soluble proteins (and in the water soluble regions of membrane proteins) the interior is more hydrophobic than the surface. In integral membrane proteins, however, the surface that contacts the lipid fraction of the bilayer is more hydrophobic than the interior (REES et al. 1989). In general, therefore, hydrophobicity can be used to discriminate between internal and surface residues. In addition, it can also be used to distinguish between those regions buried in the bilayer and those exposed to an aqueous environment.

A transmembrane helix that has one face buried in the protein core and the opposite face exposed to the lipid environment of the bilayer will have a repeat of buried/exposed residues compatible with the 3.6 residues per turn of the helix. Thus, if the environment of the residues at each position in a sequence alignment can be deduced, the periodic change in the environment can be used to predict helical structures.

The environment of each position in the alignment is described by a profile $U$ that consists of elements $U_j$. $U_j$ is a measure of the extent to which position $j$ in the alignment is buried or exposed. This can either be variability ($V$; KOMIYA et al. 1988), hydrophobicity ($H$; EISENBERG et al. 1984) or a profile calculated from residue substitution patterns (S; DONNELLY et al. 1993).

## 2. Detection of Periodicity and the Discrimination of the Different Sides of the Helix

The periodicity in the profile $U$ can be calculated using a Fourier transform procedure (e.g., EISENBERG et al. 1984). Prior to this, $U$ can be modified in several ways to improve the reliability of detecting periodicity (DONNELLY et al. 1993). The power spectrum ($P(\omega)$) is then calculated by

$$P(\omega) = [x]^2 + [y]^2 \quad (1)$$

where

$$x = \sum_{j=1}^{N} U_j^n \sin(j\omega) \quad (2)$$

$$y = \sum_{j=1}^{N} U_j^n \cos(j\omega) \quad (3)$$

and where

$$U_j^n = U_j - \bar{U}\,(j = 1,2,\ldots N) \quad (4)$$

$\bar{U}$ is the average value of $U_j$ over the window. $U^n$ is therefore a normalized version of the profile $U$ adjusted so that $\Sigma_{j=1}^{N} U_j^n = 0$ and hence $\bar{U}^n = 0$. The new profile $U^n$ consists of elements $U_j^n$ in which the periodicity in internal/external residues is predicted by the periodicity in positive/negative values (or negative/positive values) of $U_j^n$. This results in cleaner power spectra (since $P(\omega) = 0$ when $\omega = 0$) so that the alpha periodicity index AP can be calculated as

$$\mathrm{AP} = \left[(1/30)\int_{90^\circ}^{120^\circ} P(\omega)d\omega\right] \Big/ \left[(1/180)\int_{0^\circ}^{180^\circ} P(\omega)d\omega\right] \quad (5)$$

AP is analogous to $\psi$ used in KOMIYA et al. (1988) and to amphipathic index AI used in CORNETTE et al. (1987) (although the precise boundaries of the helical regions of the power spectrum differ in the latter). AP is a ratio of the extent of the periodicity in the helical region of the spectrum compared with that over the whole spectrum. KOMIYA et al. (1988) suggest that a value of AP greater than 2 indicates that the helical periodicity is significant. This points to the presence of a helix within the window of sequence used, but it is difficult to predict its precise beginning and end.

The direction of the internal face of a predicted helix can be estimated from the direction of the moment $\sqrt{P(\omega)}$ where $\omega = 100°$. This is the moment produced by the profile $U^n$ when the sequences form an ideal alpha helix. $\theta$ is the angle describing the direction of the moment relative to the first residue ($j = 1$). $\theta$ can be calculated by

$$\gamma = \arccos[y/\sqrt{P(\omega)}] \tag{6}$$

where $\gamma$ is greater than 0° and less than 180°.

$$\theta = \begin{cases} \gamma & \text{if } x > 0 \\ 360 - \gamma & \text{if } x < 0 \end{cases} \tag{7}$$

## 3. Detection of the Ends of the Transmembrane Regions of the Helices

Comparison of the direction of the vector representing the predicted internal face of the helix with that representing the hydrophobic face indicates whether the helix is in a lipid-buried or water-soluble region of the protein. In a lipid-buried region these faces are opposite, whereas in a water-soluble region they are equivalent. Therefore, the point at which a transmembrane helix passes from the lipid region of the bilayer into the more polar region of the phosphate–head groups can be detected by a shift in the direction of the hydrophobic face by 180° relative to the direction of the internal face. This will only be detected if a significant portion (say, 7–10 residues) of the helix protrudes into the polar environment outside the membrane.

Another approach that can help to locate the end of the transmembrane region of the helix is the detection of the point at which charged residues appear on the external face of the helix. This should not occur in the lipid-spanning region and hence there should be a charge-free zone over about 20 residues on this side of the helix in each member of the family. Of course, internal charged residues can exist in membrane proteins just as they do in water soluble proteins.

## 4. Summary of Methodology

Therefore, the predicted environment of each position in the sequence alignment is summarized in a profile $U$. The Fourier transform analysis of this profile allows us firstly to confirm the presence of a helix and secondly to predict which face of the helix faces the protein core. This can be carried out independently using different profiles ($H$, $V$ or $S$; see above). The combination of the results can provide a third piece of information, since it can be used to predict whether it is typical of a helix in a lipid-buried or water-soluble region of the protein. This may be useful in identifying where and when a transmembrane helix protrudes from the membrane into the more polar aqueous environments. The position of charged residues on the outside face of a transmembrance helix can also help to identify the point at which the helix reaches the polar/lipid border of the bilayer.

## 5. Application to G-Protein-Coupled Receptors

An alignment of 53 GPCR sequences was used with the Fourier transform approach described above (DONNELLY et al. 1989; D. DONNELLY, A.M. MACLEOD and T.L. BLUNDELL, unpublished results). Each of the seven putative helical regions was scanned with a window range of 10–18 and AP

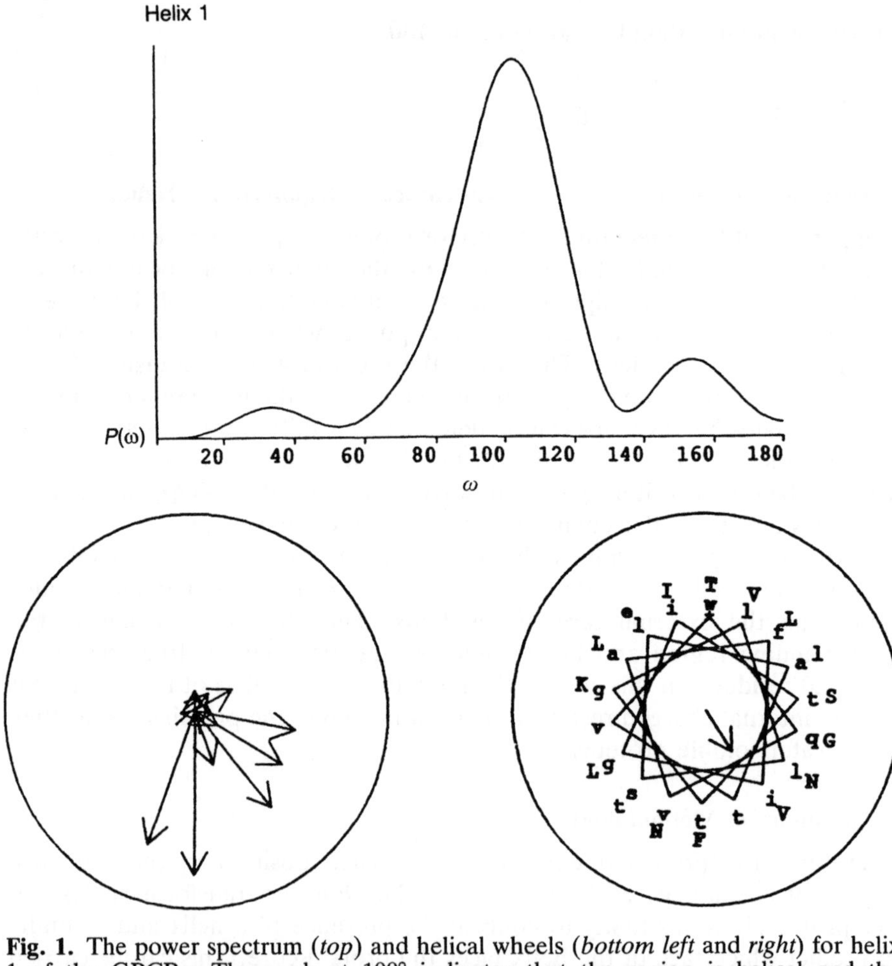


**Fig. 1.** The power spectrum (*top*) and helical wheels (*bottom left* and *right*) for helix 1 of the GPCRs. The peak at 100° indicates that the region is helical and the distribution of the *arrows* in the helical wheel indicates the internal face of the helix (sequence corresponds to the M1 subtype of the human muscarinic receptor)

was calculated for each window using the profile S. The optimal values of AP for each helix ranged between 2.72 and 3.61, indicating that these regions are indeed likely to be helical. Figure 1 shows the power spectrum for helix 1. The peak at about 100° indicates the presence of a helix, and AP for this spectrum is 3.39. The inside face of each helix was also determined and is indicated by the arrow at the centre of the helical wheel. Figure 2 shows a vertical representation of this helical wheel, in which internal residues are indicated by horizontal lines to the right and surface residues are indicated by lines to the life. Asterisks (*) represent positions in the sequence alignment where there is at least one charged residue (D, E, H, K or R). There is a region of 24 residues in which there are no charges on the outside of the helix and this may represent the membrane-spanning region. This was repeated for each of the seven putative transmembrane helices.

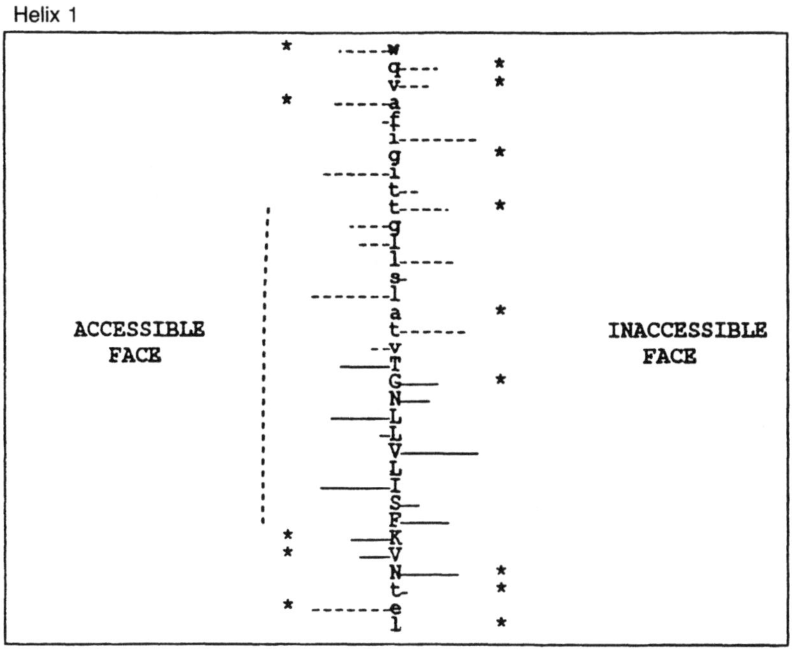


**Fig. 2.** A vertical representation of the helical wheel shown in Fig. 1 with the cytoplasmic side of the membrane at the bottom of Fig. 2. The *horizontal lines* indicate the extent to which each residue position is buried or exposed. *Lines to the right* indicate buried residues whereas *lines to the left* indicate exposed residues. Residues in *uppercase* and with *solid horizontal lines* indicate the region that shows the greatest helical periodicity. The *asterisks* represent the positions where there is at least one charged residue in the sequence alignment, and the positions where these appear on the exposed face of the helix indicate the polar regions on each side of the bilayer

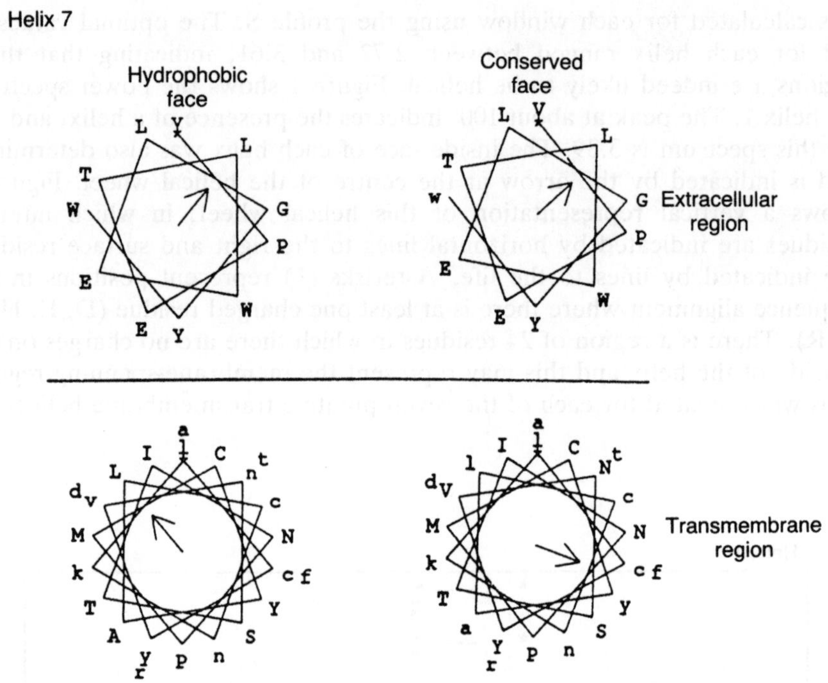


**Fig. 3.** The point at which helix 7 enters the bilayer can be seen by the change in the direction of the hydrophobic face relative to the internal face. In the extracellular region of the helix, the hydrophobic and internal faces are euivalent, but they are opposite in the transmembrane region. Additionally, the internal face of the helix changes by approximately 80° as it enters the bilayer, indicating a distortion in this region of the helix

The hydrophobic face of each helix was also calculated and compared to the predicted internal face. This comparison indicated the positions at which helix 5 protrudes into the cytoplasm and helix 7 protrudes from the extracellular side of the membrane. Figure 3 shows the case for helix 7. The internal and hydrophobic faces in the lipid-buried region are opposite, whereas they are equivalent in the extracellular region. In addition, the internal face shifts by about 80° as the helix protrudes from the membrane, indicating an irregularity in this region. A similar distortion is observed in the case for the cytoplasmic protrusion of helix 5. The distortion in helix 7 is on the N-terminal side of the position equivalent to the retinal-binding lysine (Lys-296) of the rhodopsins. An alternative distortion in this helix was predicted by ELIOPOULOS et al. (1982), but in this case the irregularity is on the C-terminal side of Lys-296.

# D. Three-Dimensional Models of G-Protein-Coupled Receptors

## I. Construction of G-Protein-Coupled Receptor Models Based on the Fourier Transform Predictions

The positions in the helices predicted from the Fourier transform analysis described above were aligned with those of bacteriorhodopsin. This was carried out by aligning the predicted internal residues in the GPCRs that reside within the bilayer, with the known internal residues of the bacteriorhodopsin transmembrane helices. Therefore, the alignment was based only on structural comparisons.

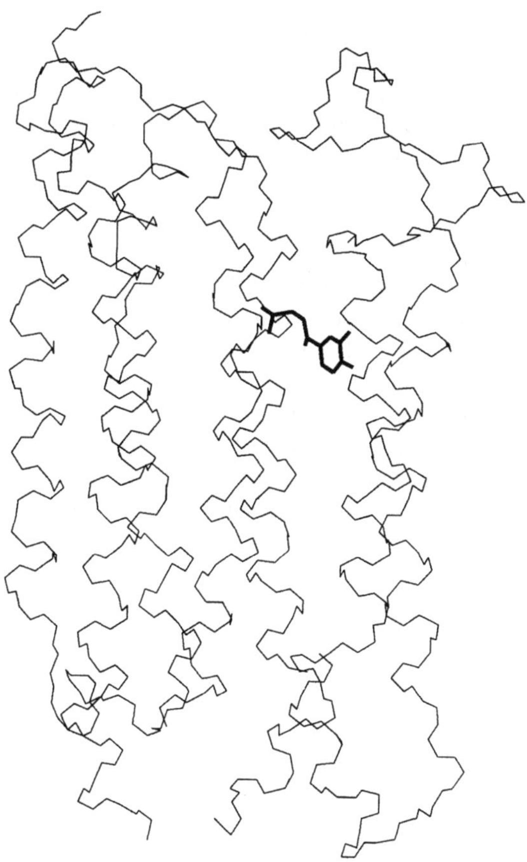

**Fig. 4.** The model of the $\beta$-adrenergic receptor based on the Fourier transform predictions and structure of bacteriorhodopsin. The agonist isoproterenol (*bold*) is shown in the putative ligand-binding site

The model-building program COMPOSER (SUTCLIFFE et al. 1987a,b; BLUNDELL et al. 1988) was then used to build a model of the human β-2 adrenergic receptor based on the above alignment. The helices in the GPCRs were extended, since the intervening loop regions are longer than those in bacteriorhodopsin. Manual interventions were carried out using interactive graphics in order to alter side chain conformations and to build in the predicted kinks at the termini of helices 5 and 7. Interactive graphics were also used to dock adrenergic ligands into the putative ligand-binding site. Steric clashes and unfavourable local conformations were removed using energy minimization programs. The model is shown in Fig. 4 with the agonist isoproterenol docked into the binding site proposed by STRADER et al. (1989). The model of the human β-adrenergic receptor may then be used as a template from which other members of the superfamily can be constructed using COMPOSER.

An alternative modelling approach based on experimental data was used to construct 3D models of rhodopsin (PAPPIN et al. 1984; FINDLAY and ELIOPOULOS 1990). Since the attachment site of the ligand 11-cis-retinal was known to be Lys-296, it was possible to show that the binding site was located in the transmembrane region between helices 3 and 7. Attention has since been focused on this region in other members of the family.

## II. Analysis of the Models

In analysing the models of the GPCRs, the most relevant residues to consider are those that are conserved across the whole family (see Table 1). These residues mostly face the interior of the bundle where they can participate in structural and functional interactions. Likewise, the residues known to be involved in ligand binding should also be internally orientated.

The conserved asparagine in helix 1 is capable of forming a hydrogen bond to the almost invariant aspartate in helix 2. This may be either structurally important or involved in signal transduction. There are other well-conserved residues at the interface of these two helices that further suggest it as a key region.

The invariant cysteine in helix 3 is involved in a crucial disulphide bond to Cys-187. The "conserved" DRY sequence at the cytoplasmic end of this helix may be critical to G-protein coupling. The ligand-binding aspartic acid residue in the β-AR is also on the internal face, which suggests that the Fourier transform method correctly orientates this helix. No residues in this helix were labelled by the hydrophobic probe (DAVISON and FINDLAY 1986).

The two conserved tryptophan residues in helices 4 and 6 have no known function, but they can interact with each other and so form a stabilizing interaction right in the core of the protein. They may also form the bottom of a generalized ligand-binding site across the family. Likewise, the three conserved proline residues in helices 5, 6 and 7 have no known function, but they may produce distortions in the helices that are critical to the formation of the ligand-binding pocket.

The invariant tyrosine in helix 5 cannot be placed on the same face of the helix as the two ligand-binding serine residues of the $\beta$-AR unless the distortion predicted by the Fourier transform analysis is applied. Only then can it participate in the critical structural or functional role that has led to it being so well conserved.

Sequence alignments between bacteriorhodopsin and the GPCRs have been used to approximate the positions of the GPCR residues in 3D space. This is not the case for the two modelling studies described in this review, since there is no clear evidence that the sequence of bacteriorhodopsin and the GPCRs derive from a common ancestral gene. Therefore, sequence alignments are ambiguous, perhaps to the point of being misleading. This also applies to helix 7, which has often been modelled by aligning the position equivalent to Lys-296 with the retinal attachment point in bacteriorhodopsin Lys-216.

The Fourier transform analysis suggests that the retinal-binding lysine is positioned between helix 7 and helix 1 very close to the extracellular side of the membrane, since it is located on the C-terminal side of the distortion at the membrane interface described in Fig. 3. However, the study by Findlay and Eliopoulos (1990) suggests that Lys-296 resides between helix 7 and helix 6 and is more buried in the bilayer (though not as deeply as Lys-216 in bacteriorhodopsin). This study, like the Fourier transform approach, suggests a distortion in the helix, but on the C-terminal, rather than the N-terminal, side of the lysine. In this orientation, the conserved Asn in helix 7 can interact with the conserved Asn in helix 1 and the conserved Asp in helix 2. Alternatively, the conserved asparagine, proline and tyrosine in helix 7, which are all predicted to be internally orientated by the Fourier transform analysis, may interact with each other. The Asn can hydrogen-bond both to the phenolic group of the Tyr and also to the main-chain carbonyl oxygen four residues before the Pro. The latter hydrogen-bond replaces the classical $i$ to $i + 4$ hydrogen-bond that is missing due to the absence of a free main-chain – NH function in Pro. The Tyr can also undergo a favourable ring-stacking interaction with the ring of the proline.

We have not dealt with any of the "loop" regions that connect the transmembrane segments since their modelling is less secure due to the greater sequence variability. Clearly, however, as shown by human retinitis pigmentosa mutations at Pro-23 and Gly-106 (which are conserved in the rhodopsins), there are critical elements in these regions.

The principal usefulness of the GPCR models lies in predicting the potential binding-site residues and suggesting possible roles for the conserved residues that can be investigated using mutagenesis and cross-linking experiments. The results of such experiments can then in turn be used to test and refine the model.

## References

Altenbach C, Marti T, Khorana HG, Hubbell WL (1990) Structural studies on transmembrane proteins 2. Spin labeling of bacteriorhodopsin mutants at unique cysteines. Science 248:1088–1092

Brett M, Findlay JBC (1979) Investigation of the organisation of rhodopsin in sheep photoreceptor membranes using cross-linking agents. Biochem J 117:215–223

Brett M, Findlay JBC (1983) isolation and characterization of the CNBr peptides from the proteolytically derived N terminal fragment of ovine rhodopsin. Biochem J 211:661–670

Cornette JL, Cease KB, Margalit H, Spouge JL, Berzofsky JA, DeLisi C (1987) Hydrophobicity scales and computational techniques for detecting amphipathic structures in proteins. J Mol Biol 195:659–685

Davison MD, Findlay JBC (1986a) Modification of ovine opsin with the photosensitive hydrophobic probe 1-azido-4[$^{125}$I]iodobenzene. Biochem J 234:413–420

Davison MD, Findlay JBC (1986b) Identification of the sites in opsin modified photoactivated 1-azido-4[$^{125}$I]iodobenzene. Biochem J 236:389–395

Donnelly D, Johnson MJ, Blundell TL, Saunders J (1989) An analysis of the periodicity of conserved residues in sequence alignments of G-protein coupled receptors. FEBS Lett 251:109–116

Donnelly D, Overington JP, Ruffle SV, Nugent JHA, Blundell TL (1993) Modelling $\alpha$-helical transmembrane domains – the calculation and use of substitution tables for lipid facing residues. Protein Sci 2:55–70

Eisenberg D, Weiss RM, Terwilliger TC (1984) The hydrophobic moment detects periodicity in protein hydrophobicity. Proc Natl Acad Sci USA 81:140–144

Findlay JBC, Brett M, Pappin DJC (1981) Primary structure of C-terminal functional sites in ovine rhodopsin. Nature 293:314–316

Findlay JBC, Pappin DJC (1986) The opsin family of proteins. Biochem J 238: 625–642

Findlay J, Eliopoulos E (1990) Three-dimensional modelling of G protein-linked receptors. TIPS 11:492–499

Garnier J, Osguthorpe DJ, Robson B (1978) Analysis of the accuracy and implications of simple methods for predicting the secondary structure of globular proteins. J Mol Biol 120:97–120

Grötzinger J, Engelss M, Jacoby E, Wollmer A, Straßburger W (1991) A model for the C5a receptor and for its interaction with the ligand. Protein Sci 4:767–771

Henderson R, Unwin PTN (1975) Three-dimensional model of purple membrane obtained by electron microscopy. Nature 257:28–32

Henderson R, Baldwin JM, Ceska TA, Zemlin F, Beckmann E, Downing KH (1990) Model for the structure of bacteriorhodopsin based on high resolution electron cyro-microscopy. J Mol Biol 213:899–929

Hibert MF, Trumpp-Kallmeyer S, Bruinvels A, Hoflack J (1991) Three-dimensional models of neurotransmitter G-binding protein-coupled receptors. Molecular Pharmacology 40:8–15

Karnik SS, Khorana HG (1990) Cysteine residues 110 and 187 are essential for the formation of correct structure in bovine rhodopsin. J Biol Chem 265:17520–17524

Komiya H, Yeates TO, Rees DC, Allen JP, Feher G (1988) Structure of the reaction centre from Rhodobacter sphaeroides R-26 and 2.4.1: symmetry relations and sequence comparisons between different species. Proc Natl Acad Sci USA 85:9012–9016

MaloneyHuss K, Lybrand TP (1992) Three-dimensional structure for the $\beta 2$ adrenergic receptor protein based on computer modeling studies. J Mol Biol 225:859–871

Ovchinnikov YuA, Abdulaev NG, Bogachuk AS (1988) Two adjacent cysteine residues in the C-terminal cytoplasmic fragment of bovine rhodopsin are palmitylated. FEBS Lett 230:1–5

Pappin DJC, Eliopoulos E, Brett M ,Findlay JBC (1984) A structural model for ovine rhodopsin. Int J Biol Macromol 6:73–76
Rees DC, DeAntonio L, Eisenberg D (1989) Hydrophobic organization of membrane proteins. Science 245:510–513
Strader CD, Candelore MR, Hill WS, Sigal IS, Dixon RAF (1989) Identification of two serine residues involved in agonist activation of the β-adrrenergic receptor. J Biol Chem 264:13572–13578
Strader CD, Gaffney T, Sugg EE, Candlemore MR, Keys R, Patchett AA, Dixon RAF (1991) Allele specific activation of genetically engineered receptors. J Biol Chem 266:5–8
Stubbs GW, Smith HG, Litman BJ (1976) Alkyl glucosides as effective solubilizing agents for bovine rhodopsin – a comparison of several commonly used detergents. Biochim Biophys Acta 246:46–56
Sutcliffe MJ, Haneef I, Carney D, Blundell TL (1987a) Knowledge-based modelling of homologous proteins 1. Three-dimensional frameworks derived from the simultaneous superposition of multiple structures. Prot Engng 1:377–384
Sutcliffe MJ, Hayes FRF, Blundell TL (1987b) Knowledge-based modelling of homologous proteins 2. Rules for the conformation of substituted sidechains. Prot Engng 1:385–392
Tadayyon M, Zhang Y, Gnaneshan S, Hunt L, Mehraein-Ghomi F, Broome-Smith JK (1992) β-Lactamase fusion analysis of membrane protein assembly. Biochem Soc Trans 20:598–601
Wang JK, McDowell JHM, Hargrave PA (1980) Site of attachment of 11-*cis*-retinal in bovine rhodopsin. Biochem 19:5111–5117
Wang H-y, Lipfert L, Malbon CC, Bahouth S (1989) Sire-directed anti-peptide antibodies define the topography of the β-adrenergic receptor. J Biol Chem 264:14424–14431

CHAPTER 46

# The Role of Receptor Kinases and Arrestin-Like Proteins in G-Protein-Linked Receptor Desensitization

R.J. LEFKOWITZ and M.G. CARON

G-protein-linked receptors constitute a diverse array of transmembrane signal transduction molecules. They represent the first in a series of protein components that mediate cellular responses to stimuli as varied as neurotransmitters, hormones, chemoattractants, cytokines, odorants, and even photons of light (DOHLMAN et al. 1991). Over the past several years it has become clear that the functional and regulatory properties of these receptors are, like their structural features, remarkably similar. Agonist occupation of G-protein-linked receptors leads to intramolecular conformational changes that render the receptors "active," and there are at least two major pathways along which the activated receptors may then proceed. One involves coupling with G-protein, which subsequently activates a biochemical effector, ultimately generating a physiological response. The other pathway involves interaction with regulatory proteins that mediate receptor desensitization. In this review, we will briefly describe findings pertaining to a biochemical mechanism only recently appreciated by which activated G-protein-coupled receptors become desensitized. This mechanism involves two families of proteins: receptor-specific kinases and the arrestin-like proteins. Other aspects of receptor regulation, including regulation of receptor function by other protein kinases, such as cAMP-dependent protein kinase, and regulation by sequenstration/internalization of the receptors, have recently been reviewed (PERKINS et al. 1991) and will not be developed here.

Figure 1 illustrates schematically the known components of two model G-protein-coupled receptor signal transducing systems, the adenylyl-cyclase-coupled $\beta_2$-adrenergic receptor ($\beta_2$AR) and the retinal phototransduction system, as well as proteins involved in their respective desensitization pathways. The similarities betwen these two systems are remarkable. In each case, a receptor protein initially interacts with a stimulating agent. A number of the genes encoding members of this receptor family have been cloned and sequenced and the salient features of the proteins described (DOHLMAN et al. 1991). They all appear to comprise seven hydrophobic membrane-spanning regions interconnected by hydrophilic extra- and intracellular loops. In general, it appears to be the membrane-spanning domains which form the ligand-binding pocket of the various receptors. The regions of the receptors that interact with G-proteins appears to constitute

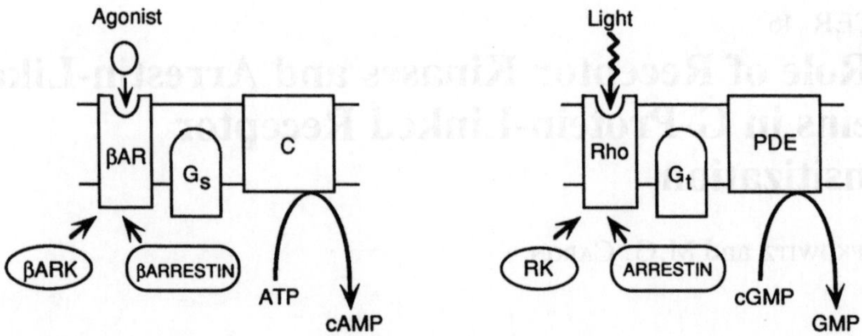


**Fig. 1.** Schema depicting known signal transducing and desensitizing components of the adenylyl cyclase/$\beta_2$AR system and the retinal phototransduction system

those segments of the intracellular loops which are in proximity to the plasma membrane. Regions in the third intracellular loop appear to be particularly important. Heterotrimeric G-proteins, $G_s$ in the case of the $\beta_2$AR system and $G_t$ (transducin) in the case of the phototransduction system, interact with the stimulus-modified receptors. The characteristics of these ubiquitous signal-transducing proteins have been recently reviewed (BOURNE et al. 1991). They consist of an $\alpha$ subunit that binds and hydrolyzes GTP and a $\beta\gamma$ complex that helps to anchor the $\alpha$ subunit in the membrane and performs a number of other functions during signal transduction and desensitization.

The third component of G-protein-linked receptor transmembrane signaling systems is an effector molecule, such as adenylyl cyclase and cGMP phosphodiesterase. Such effectors help maintain prescribed intracellular levels of second messengers. In the case of adenylyl cyclase, generated cAMP goes on to activate the cAMP-dependent protein kinase. In the case of retinal phototransduction, cGMP is responsible for gating the sodium channel that controls electrical signaling from the photon-activated retina.

Figure 1 also depicts the major components of a pathway that functions to desensitize the agonist- or photon-stimulated receptors. In each case, a cytoplasmic protein kinase (rhodopsin kinase, RK; LORENZ et al. 1991; PALCEWSKI et al. 1992) in the case of phototransduction and the $\beta$-adrenergic receptor kinase ($\beta$ARK; BENOVIC et al. 1986) in the case of the $\beta_2$AR recognizes the activated conformation of the receptor. These kinases phosphorylate their respective receptor substrates on multiple serine and threonine residues located in the distal portion of their C-terminal cytoplasmic tails. These phosphorylation reactions lead to the binding of a second cytosolic factor, termed arrestin (WILDEN et al. 1986) in the retinal system and $\beta$-arrestin (LOHSE et al. 1990) in the $\beta_2$AR system. These proteins bind to the $\beta$ARK- or RK-phosphorylated forms of the receptors

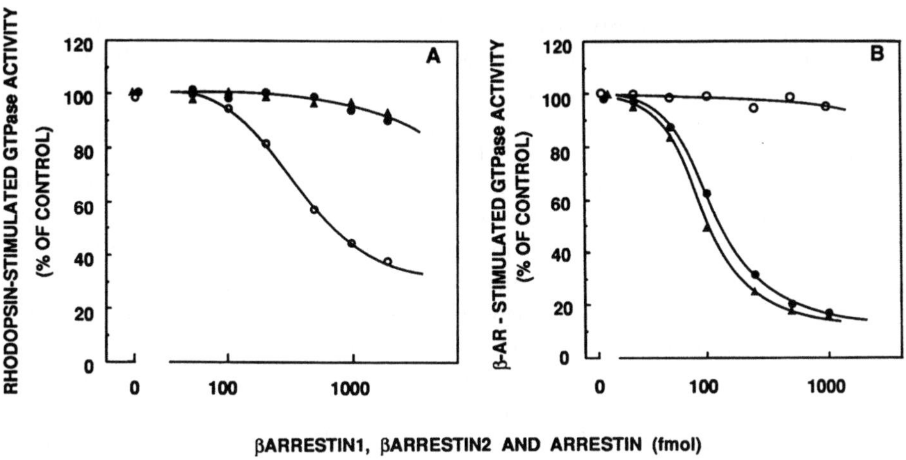


**Fig. 2A,B.** Inhibition of reconsituted signal transduction system by $\beta$-arrestin-1, $\beta$-arrestin-2, and arrestin. **A** Inhibition of RHO-stimulated $G_t$ GTPase activity by purified $\beta$-arrestin-1, $\beta$-arrestin-2, and arrestin. The assay was performed with 200 fmol of phosphorylated RHO reconstituted into phospholipid vesicles with 100 fmol $G_t$ and the addition of increasing amounts of arrestin (○), $\beta$-arrestin-1 (●), or $\beta$-arresting-2 (▲), as indicated. The activities are calculated as percent control activity (i.e., light-dependent GTPase activity in the absence of arrestin proteins). **B** Inhibition of $\beta_2$AR-stimulated $G_s$ GTPase activity by purified arrestins. Receptor-stimulated $G_s$ GTPase turnover rates were measured as the difference between isoproterenol (50 $\mu M$)- and propranolol (50 $\mu M$)-stimulated $G_s$ GTPase activities. The assay was performed using 100 fmol phosphorylated $\beta_2$AR, 50 fmol $G_s$, and increasing amounts of $\beta$-arrestin-1 (●), $\beta$-arrestin-2 (▲), or arrestin (○)

and appear to sterically hinder receptor–G-protein interactions, thus leading to functional desensitization of the system.

All of the components described above have been purified, cloned, and sequenced. To date, two forms of $\beta$ARK ($\beta$ARK1, Benovic et al. 1989); $\beta$ARK2, Benovic et al. 1991) and one form of RK (Lorenz et al. 1991) are known. Although each is a "typical" protein kinase, with centrally located and highly conserved catalytic domain, the group appears to form a distinct subfamily of kinases, sharing approximately 45% sequence identity within their catalytic regions. The noncatalytic portions of the proteins, however, show much less sequence similarity. Similarly, two forms of $\beta$-arrestin ($\beta$-arrestin-1, Lohse et al. 1990; and $\beta$-arrestin-2, Attramadal et al. 1992), as well as retinal arrestin (Shinohara et al. 1987), have been cloned and sequenced, sharing about 60%–70% overall sequence identity.

It has been documented that the components described above are necessary and sufficient to reconstitute both the transmembrane signaling and desensitization pathways in vitro. For example, as shown in Fig. 2, upon purification and reconstitution of rhodopsin (RHO) and $G_t$ into phospholipid vesicles, followed by phosphorylation with RK, arrestin potently desensitizes the phosphorylated form of RHO, whereas the $\beta$-

arrestins are considerably weaker. Conversely, when the analogous reconstitution experiment is done with $G_s$ and $\beta$ARK1-phosphorylated $\beta_2$AR, the addition of $\beta$-arrestin-1 or $\beta$-arrestin-2 leads to a dose-related inhibition of signalling through the phosphorylated receptor, but arrestin is then much weaker in comparison. Half-maximal inhibition in either system (i.e., RHO/$G_t$/RK/arrestin or $\beta_2$AR/$G_s$/$\beta$ARK/$\beta$-arrestin) appears to require about a 1:1 to a 2:1 molar ratio of arrestin protein to receptor protein. These data illustrate the marked system specificity in the ability of $\beta$-arrestins versus arrestin to desensitize their respective receptors.

Following development of polyclonal antibodies raised against $\beta$ARK and $\beta$-arrestin that distinguish between the various distinct forms of the molecules, it has been shown that all four proteins are distributed throughout the central nervous system (CNS; Arriza et al. 1992; Attramadal et al. 1992). These data suggest that the function(s) of the $\beta$ARK–$\beta$-arrestin system is not limited to desensitization of $\beta_2$AR, or even adrenergic receptors in general. Rather, they suggest that many different neurotransmitter receptor systems may be regulated by $\beta$ARK–$\beta$-arrestin.

A related finding, revealed by immunoelectron microscopy, indicates that both forms of $\beta$ARK and $\beta$-arrestin are mainly synaptic in their distributions (Arriza et al. 1992; Attramadal et al. 1992), occurring either postsynaptically or presynaptically. These results confirm speculation that the $\beta$ARK–$\beta$-arrestin system might be of particular importance in regulation of synaptic neurotransmitter receptors. This notion is based on the known high expression of $\beta$ARK and $\beta$-arrestin proteins in neuronal tissues, as well as the observation that $\beta$ARK functions most efficaciously at high agonist occupancy levels of receptor. Such levels of occupancy are typically observed at neurotransmitter synapses, as opposed to peripheral receptor sites.

A characteristic property of $\beta$ARK and RK is that, although they are cytosolic enzymes, they rapidly migrate to the plasma membrane upon stimulation of the relevant receptor system. However, the mechanism(s) orchestrating this compartmentalization of the kinases was, until recently, completely unknown.

Following the molecular cloning of RK, it became clear that the kinase contained a "CAAX box" at its C terminus (i.e., CVLS) (Lorenz et al. 1991), where C is cysteine, A is aliphatic amino acid, and X is any amino acid. This motif is known to direct a series of three sequential, posttranslational modifications of proteins. Initially, isoprenylation of the cysteine residue occurs via thioether bond formation with either a farnesyl (C15) or geranylgeranyl (C20) isoprenoid. The exact nature of the C-terminal sequence directs which isoprenyl group is added. In the case of RK, CVLS directs farnesylation (Inglese et al. 1992a). Isoprenylation is followed by proteolysis of the three C-terminal residues which, in turn, is followed by carboxyl methylation of the now C-terminal cysteine residue. Although the full biological import of such modifications is incompletely understood even in the most studied case, that of the *ras* proto-oncogene product, it has been shown that farnesylation is required both for membrane association of RAS, as well as for RAS-mediated cellular transformation. Accordingly, we formulated the hypothesis that isoprenylation is required for stimulus-mediated translocation of RK.

To test this notion, we constructed two forms of RK-bearing point mutations in the CAAX box (INGLESE et al. 1992b). One comprised a cysteine to serine mutation (i.e., SLVS), which should block isoprenylation, whereas the other comprised a serine to leucine mutation (i.e., CVLL), which should direct geranylgeranylation, rather than farnesylation found with the wild-type kinase. Following transfection into COS-7 cells, each kinase was found to undergo the correct isoprenyl modification. When tested for their biological activities, it was observed that the C15- and C20-modified forms of the enzyme were approximately equiactive, whereas the nonprenylated form had markedly reduced biological activity (Fig. 3).

Further examination of the involvement of isoprenylation in the translocation of RK indicated that the wild-type, C15 form of the enzyme was mainly cytosolic, but translocated almost quantitatively to the plasma membrane following stimulation (Fig. 4). In contrast, the geranylgeranylated form of RK was almost exclusively membrane associated even in the absence of light, and showed little further translocation when illuminated. Finally, the nonprenylated form was found to be incapable of

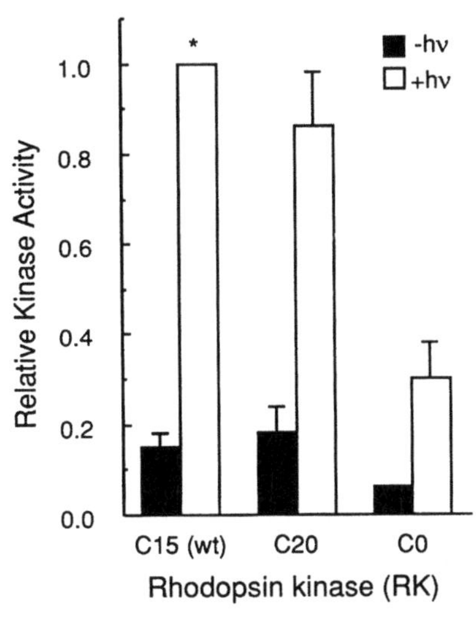


**Fig. 3.** Relative activities of expressed recombinant RKs. Bovine retina rod outer segment (ROS) membranes were incubated and phosphorylated with COS-7 cell extracts expressing wild-type farnesylated RK (C15), mutant geranylgeranylated RK (C20), and mutant nonprenylated RK (C0). The activity values are normalized to the wild-type kinase (*). The values shown are the means ± SEM from five separate experiments

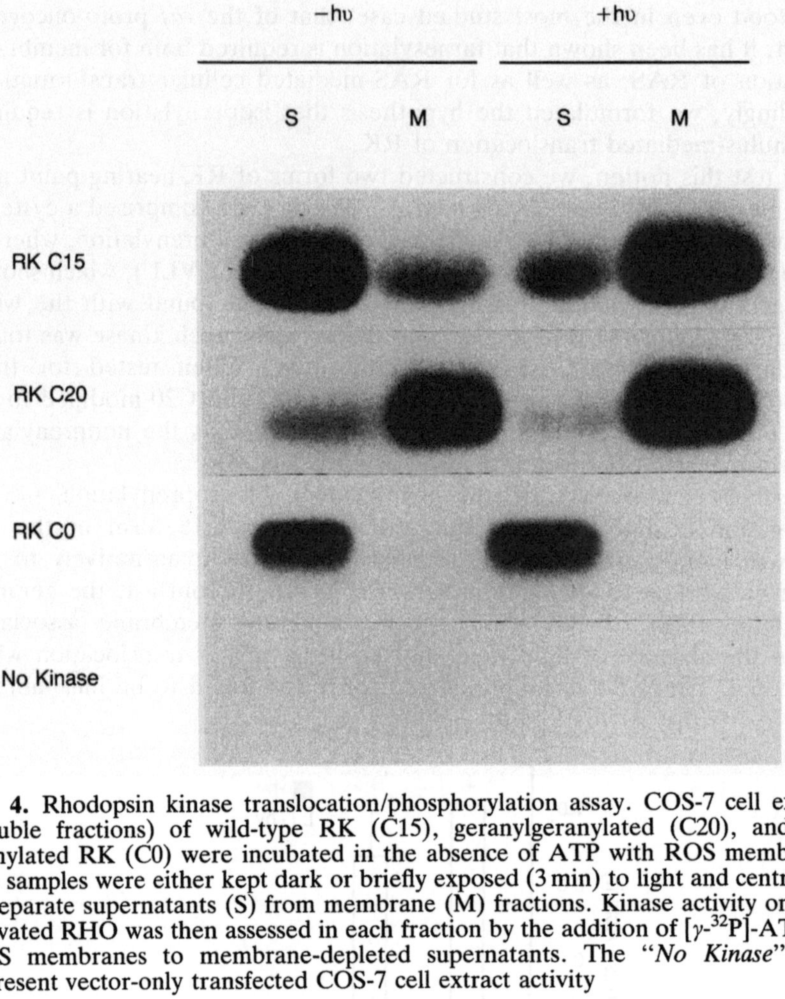


**Fig. 4.** Rhodopsin kinase translocation/phosphorylation assay. COS-7 cell extracts (soluble fractions) of wild-type RK (C15), geranylgeranylated (C20), and non-prenylated RK (C0) were incubated in the absence of ATP with ROS membranes. The samples were either kept dark or briefly exposed (3 min) to light and centrifuged to separate supernatants (S) from membrane (M) fractions. Kinase activity on light-activated RHO was then assessed in each fraction by the addition of [$\gamma$-$^{32}$P]-ATP and ROS membranes to membrane-depleted supernatants. The "*No Kinase*" lanes represent vector-only transfected COS-7 cell extract activity

translocation under any circumstances (INGLESE et al. 1992b). These data indicate the essential requirement of a specific isoprenyl moiety for stimulus-mediated translocation of RK to its receptor substrate.

Unlike RK, neither form of $\beta$ARK possesses a CAAX box, yet, like RK, both $\beta$ARK1 and $\beta$ARK2 translocate from cytosol to membrane-bound receptor upon agonist exposure. We speculated that isoprenylation might also be involved in this process, but perhaps in a more circuitous fashion. To test the feasibility of this notion, we constructed two mutant forms of $\beta$ARK in which its C-terminal sequence was altered to direct farnesylation (i.e., CVLS) and geranylgeranylation (i.e., CVLL) in vitro. Data from experiments with these mutant forms of $\beta$ARK were similar to those obtained with our RK mutants (Fig. 4) in that they were both more active

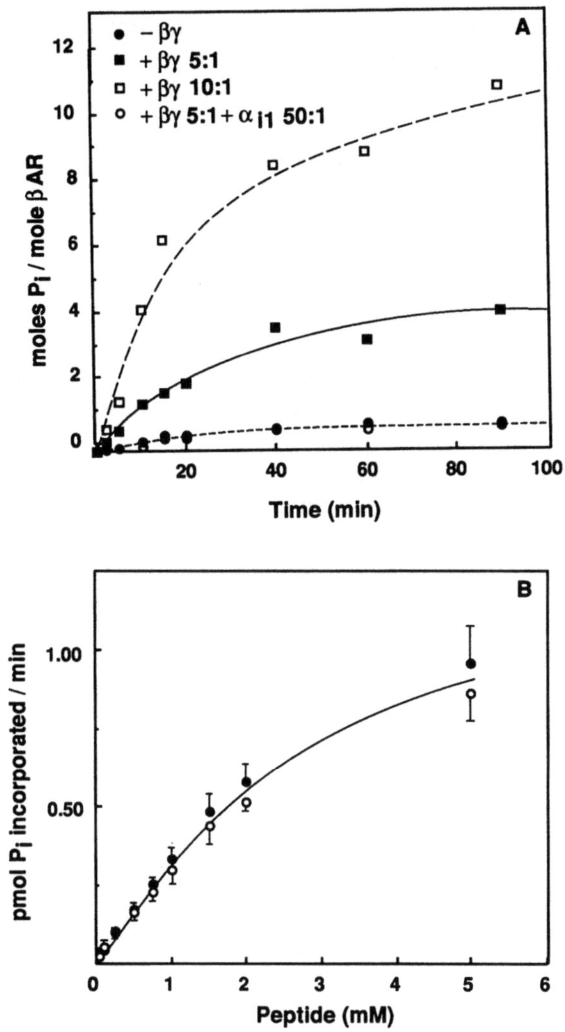


**Fig. 5A,B.** Effects of $\beta\gamma$ on $\beta$ARK activity. **A** Enhancement of $\beta$ARK-mediated $\beta_2$AR phosphorylation by bovine brain $\beta\gamma$. Reconstituted $\beta_2$AR (20 n$M$) was incubated with $\beta$ARK (30 n$M$) in buffer containing 100 $\mu M$ [$\gamma$-$^{32}$P]-ATP and isoproterenol (100 $\mu M$) (○). Alternatively, 150 n$M$ $\beta\gamma$ (■), 300 n$M$ $\beta\gamma$ (□), or 150 n$M$ $\beta\gamma$ + 1.5 $\mu M_{\mathrm{ai1}}$ (●) was included in the incubation mixture. Phosphorylation stoichiometries were determined assuming that all reconstituted receptor was accessible to kinase. **B** $\beta$ARK phosphorylation of peptide substrates in the presence of $\beta\gamma$ subunits. The synthetic peptide RRREEEEESAAA, at the concentrations indicated, was incubated with purified $\beta$ARK (25 n$M$) in 100 $\mu M$ [$\gamma$-$^{32}$P]-ATP, 10 $\mu M$ cAMP-dependent protein kinase inhibitor, and BSA (0.5 mg/ml) at 30°C for 10 min. Phosphorylation reactions were performed in the absence (○) or presence of 255 n$M$ $\beta\gamma$ subunits (●). The results are the means ± SEM from three separate experiments

and more highly membrane associated than the nonprenylated enzyme. This indicates that isoprenyl groups have the *potential* to function as membrane anchors for βARKs as well as RK (INGLESE et al. 1992b).

Although βARK is not prenylated in vivo, a recent finding suggests a novel mechanism by which isoprenylation can orchestrate βARK translocation (HAGA and HAGA 1992). These workers found that the βγ subunits of G-proteins activate a muscarinic receptor kinase that has properties similar, if not identical, to βARK. As shown in Fig. 5, we confirmed this observation and documented a greater than ten fold stimulation of βARK1 activity by bovine brain βγ that was only observed when the substrate was receptor (PITCHER et al. 1992). When a model peptide substrate was used, βγ had minimal, if any, effect on kinase activity, suggesting that βγ does not directly activate βARK, but appears to operate through some other mechanism.

Since the γ subunit of βγ is geranylgeranylated in vivo, we speculated that βγ subunits might "donate" isoprenyl groups to βARK, thus allowing it to anchor in the membrane during receptor phosphorylation. Supporting evidence for this hypothesis is that βγ subunits are only free (i.e., dissociated

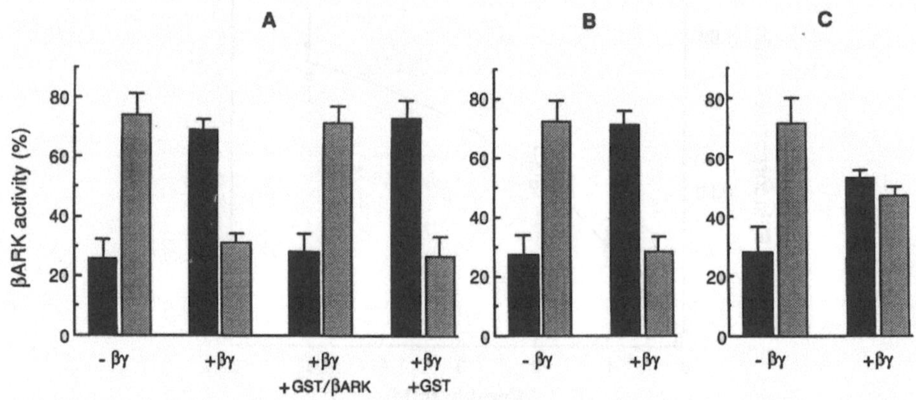


**Fig. 6A–C.** Membrane targeting of the βARK–βγ complex. Purified βARK (15 n*M*) wasincubated with native (**A** and **B**) or heat-inactivated (**C**) ROS membranes (final concentration of RHO ~5 μ*M*). Incubations were performed at 22°C for 3 min under bright illumination (**A** and **C**) or in the dark (**B**). βγ subunits purified from bovine brain (0.6 m*M*), GST/βARK fusion protein (3 μ*M*), or GST (3 μ*M*) were included in the incubations as indicated. Samples were subsequently centrifuged, the supernatants rapidly removed, supplemented with ROS membranes and [$\gamma$-$^{32}$P]-ATP (final concentration 100 μ*M*), and incubated under illumination at 30°C for 5 min. The sedimented ROS membranes were resuspended in buffer containing 100 μ*M* [$\gamma$-$^{32}$P]-ATP and incubated under identical conditions. The βγ subunits (0.3 μ*M*) were added to both the supernatant and sedimented fractions of samples that lacked βγ during the first incubation. The total radioactivity incorporated into RHO in both the supernatant and pellet fractions for a given sample was taken as 100%. The data shown are the means ± SEM from three separate experiments. *Solid bars* is activity in membranes, *stippled bars* is activity in supernatants

from the $\alpha$ subunit and available to perform such a putative function) upon receptor stimulation, thus providing a mechanism for stimulus-dependent, $\beta\gamma$-mediated translocation of $\beta$ARK to the plasma-membrane-bound receptors. To test this notion, we examined the ability of $\beta\gamma$ to promote the association of $\beta$ARK with membrane-bound receptors (Fig. 6) (Pitcher et al. 1992). The data suggest that $\beta\gamma$ subunits may associate not only with $\beta$ARK, but with the receptor component of the system as well. This finding is consistent with the notion tht this pathway is entirely "receptor dependent" in vivo, since receptor stimulation is required to dissociate $\beta\gamma$ into a form in which it is capable of binding and translocating $\beta$ARK, as well as interacting with the receptor.

In other experiments, we have shown that bovine brain $\beta\gamma$ subunits bind tightly to an immobilized fusion protein in which the last 220 amino acids of $\beta$ARK have been fused to glutathione-*S*-transferase. Interestingly, $\beta\gamma$ subunits derived from $G_t$, which are much weaker than brain $\beta\gamma$ in stimulating $\beta$ARK activity, are not able to bind to the fusion protein under the assay conditions used (Pitcher et al. 1992), suggesting the possibility that there may be specificity in the actions of $\beta\gamma$ from different G-proteins to interact with and translocate different isoforms of $\beta$ARK.

These novel findings concerning the role of isoprenylation in the stimulus-directed targeting of regulatory G-protein-coupled receptor kinases to their receptor substrates provide new insights into the mechanisms by which this form of posttranslational protein modification may function to direct trafficking of enzymes to their crucial cellular substrates. Moreover, they suggest additional avenues of experimental approach to the problem of developing inhibitors for $\beta$ARK-like enzymes, the ultimate goal of which is to attenuate desensitization reactions and prolong the therapeutic efficacy of drugs that function by stimulation of G-protein-linked receptors.

*Acknowledgements.* This work was supported in part by a grant from the NIH #HL16037.

## References

Arriza JL, Dawson TM, Simerly RB, Martin LJ, Caron MG, Snyder SH, Lefkowitz RJ (1992) The G protein-coupled receptor kinases $\beta$ARK1 and $\beta$ARK2 are widely distributed at synapses in rat brain. J Neurosci 12(10):4045–4055

Attramadal H, Arriza JL, Aoki C, Dawson TM, Codina J, Kwatra MM, Snyder SH, Caron MG, Lefkowitz RJ (1992) $\beta$Arrestin2 – a novel member of the arrestin/$\beta$arrestin gene family. J Biol Chem 267:17882–17890

Benovic JL, Strasser RH, Caron MG, Lefkowitz RJ (1986) $\beta$-Adrenergic receptor kinase: identification of a novel protein kinase which phosphorylates the agonist-occupied from of the receptor. Proc Natl Acad Sci USA 83:2797

Benovic JL, DeBlasi A, Stone WC, Caron MG, Lefkowitz RJ (1989) $\beta$-Adrenergic receptor kinase: primary structure delineates a multigene family. Science 246:235

Benovic JL, Onorato JJ, Arriza JL, Stone WC, Lohse M, Jenkins NA, Gilbert DJ, Copeland NG, Caron MG, Lefkowitz RJ (1991) Cloning, expression, and chromosomal localization of $\beta$-adrenergic receptor kinase 2. A new member of the receptor kinase family. J Biol Chem 266:14939

Bourne HR, Sanders DA, McCormick F (1991) The GTPase superfamily: conserved structure and molecular mechanism. Nature 349:117

Dohlman HG, Thorner J, Caron MG, Lefkowitz RJ (1991) Model systems for the study of seven-transmembrane-segment receptors. Annu Rev Biochem 60:653

Haga K, Haga T (1992) Activation by G protein $\beta\gamma$ subunits of agonist- or light-dependent phosphorylation of muscarinic acetylcholine receptors and rhodopsin. J Biol Chem 267:2222

Inglese J, Koch WJ, Caron MG, Lefkowitz RJ (1992a) The role of isoprenylation in the regulation of signal transduction via the G protein-coupled receptor kinases. Nature 359:147–150

Inglese J, Glickman JF, Lorenz W, Caron MG, Lefkowitz RJ (1992b) Isoprenylation of a protein kinase. Requirement of farnesylation/$\alpha$-carboxyl methylation for full enzymatic activity of rhodopsin kinase. J Biol Chem 267:1422

Lohse MJ, Benovic JL, Codina J, Caron MG, Lefkowitz RJ (1990) $\beta$-Arrestin: a protein that regulates $\beta$-adrenergic receptor function. Science 248:1547

Lorenz W, Inglese J, Palczewski K, Onorato JJ, Caron MG, Lefkowitz RJ (1991) The receptor kinase family: primary structure of rhodopsin kinase reveals similarities to the $\beta$-adrenergic receptor kinase. Proc Natl Acad Sci USA 88:8715

Palczewski K, Rispoli G, Detwiler PB (1992) The influence of arrestin (48K protein) and rhodopsin kinase on visual transduction. Neuron 8:117

Perkins JP, Hausdorff WP, Lefkowitz RJ (1991) Mechanisms of ligand-induced desensitization of the $\beta$-adrenergic receptor. In: Perkins JP (ed) The beta-adrenergic receptors. Humana, Clifton, pp 125–180

Pitcher JA, Inglese J, Higgins JB, Arriza JL, Casey PJ, Kim C, Benovic JL, Kwatra MM, Caron MG, Lefkowitz RJ (1992) $\beta\gamma$ subunits of G proteins target the $\beta$-adrenergic receptor kinase to membrane bound receptors. Science 257:1264

Shinohara T, Dietzschold B, Craft CM, Wistow G, Early JJ, Donoso LA, Horwitz J, Tao R (1987) Primary and secondary structure of bovine retinal S antigen (48 kDa protein). Proc Natl Acad Sci USA 84:6975

Wilden U, Hall SW, Kuhn H (1986) Phosphodiesterase activation by photoexcited rhodopsin is quenched when rhodopsin is phosphorylated and binds the intrinsic 48-kDa protein of rod outer segments. Proc Natl Acad Sci USA 83:1174

# *C. Trimeric G-Proteins*

CHAPTER 47

# Qualitative and Quantitative Characterization of the Distribution of G-Protein $\alpha$ Subunits in Mammals

G. MILLIGAN

## A. Introduction

A large family of heterotrimeric guanine nucleotide binding proteins (G-proteins) have now been identified by a combination of biochemical and molecular biological approaches (SIMON et al. 1991; KAZIRO et al. 1991). In a number of cases the proteins have been purified essentially to homogeneity, and much is known about the receptor and effector systems which interact with them. However, in a number of other examples the protein has yet to be identified, and little more than information on the predicted protein sequence and tissue distribution of corresponding mRNA is currently known (STRATHMANN and SIMON 1991). Details of the tissue and cellular distribution of these G-protein polypeptides (MILLIGAN 1989) would clearly assist in efforts to delineate their likely functions. For example, knowledge of the tissue distribution of relevant mRNA and of the CNS sites of action of pharmaceutical agents assisted in the identification of an orphan G-protein linked receptor cDNA as one encoding a cannabinoid receptor (MATSUDA et al. 1990). While it is generally accepted that the central role for heterotrimeric G-proteins is in the transmission of information across the limiting external membrane of a cell, it is now clear that the subcellular location of G-proteins is not restricted to the plasma membrane (MILLIGAN 1989; ERCOLANI et al. 1990; BURGOYNE 1992; KTISKATIS et al. 1992). For example, it has been reported that in LLC-PK1 cells $G_{i3}\alpha$ displays a distinct perinuclear pattern of immunostaining suggesting location on the Golgi membranes (ERCOLANI et al. 1990), and a similar pattern of $G_{i3}\alpha$ immunoreactivity has been noted in BALB/c3T3 fibroblasts (LAMORTE et al. 1992). A $G_i$-like polypeptide has also been reported in rat liver nuclei (TAKEI et al. 1992), and agonist activation of receptors can, at least in certain circumstances, alter the cellular distribution and/or amounts of G-proteins (MCKENZIE and MILLIGAN 1990b; GREEN et al. 1992; NEGISHI et al. 1992; see MILLIGAN and GREEN 1991 for review). Knowledge of the cellular localization of G-proteins may thus assist in the identification of further functions for these polypeptides (ERCOLANI et al. 1990; KTISKATIS et al. 1992; see BURGOYNE 1992 for review). Furthermore, at least in certain polarized cells, G-proteins are not evenly distributed in the plasma membrane (ALI et al. 1989; GABRION et al. 1989; ERCOLANI et al. 1990; van den BERGHE et al.

1991), and the possibility of a regulatory role for cytoskeletal control of G-protein distribution in a cell is gaining credence (WANG et al. 1990).

While useful, qualitative physical identification of the distribution of the G-protein polypeptides is insufficient in isolation, and quantitative analyses of cellular G-protein polypeptide amounts are beginning to provide information on the stoichiometry of interactions of the proteinaceous components of signal transduction cascades. This chapter focuses on strategies which have been used to address both qualitiative and quantitative aspects of G-protein $\alpha$ subunit distribution and provides examples to demonstrate how such data can provide insights into the detailed mechanisms of receptor–G-protein–effector interactions. The relative infancy of the field precludes useful analysis, at this time, of the qualitative and quantitative distribution of genetically distinct G-protein $\beta$ and $\gamma$ subunits, although it is anticipated that this area will become of major importance in the near future.

## B. Identification of G-Protein $\alpha$ Subunits

### I. [$^{32}$P]ADP Ribosylation

With the knowledge that genes encoding at least 17 distinct G-protein α subunits are expressed in mammalian systems (SIMON et al. 1991) and the realization that G-proteins are expressed widely in both animals and plants it has become imperative to be able to identify these proteins unambiguously. Until relatively recently the most widely used approach was to take advantage of the ability of exotoxins produced by either *Vibrio cholerae* (cholera toxin) or *Bordetella pertussis* (pertussis toxin) to catalyse the mono-[$^{32}$P]ADP-ribosylation of the $\alpha$ subunits of a limited repetoire of these G-proteins (MILLIGAN 1988). Of course, no information as to the existence, amounts or function of G-protein $\alpha$ subunits which are not substrates for the ADP-ribosyltransferase activity of these toxins can be gained from such studies, and while arguements have been advanced to indicate that the state of association of G-proteins can be detected due to differences in the ability of associated and dissociated G-protein $\alpha$ subunits to act as substrates for modification by the toxins, only very limited qualitative information on the presence and/or amounts of $\beta\gamma$ subunits can be inferred.

Indications of the likely role of these toxins in regulation of G-protein function had initially been deduced from observations on the disruptive effect these agents had on receptor control of the adenylyl cyclase cascade. While cholera toxin is able to catalyse ADP-ribosylation of a variety of proteins in in vitro assays, its key substrates are the various splice forms of the $\alpha$ subunit of the stimulatory G-protein ($G_s$) of adenylyl cyclase. Pertussis toxin was initially believed to modify only the inhibitory G-protein ("$G_i$") of the adenylyl cyclase cascade, but it is now clear that any of the natural G-protein $\alpha$ subunits which have a cysteine residue located four amino acids from the C-terminus can act as a substrate for the ADP-ribosyltransferase activity of this toxin. To date, at least seven G-protein $\alpha$ subunits fall into this category (TD1, TD2, $G_{i1}$, $G_{i2}$, $G_{i3}$, $G_{o1}$ and $G_{o2}$). It is clear that the presence of the target cysteine residue, while essential for pertussis toxin-catalysed ADP-ribosylation, is not sufficient in isolation. A mutant form of a $G_s\alpha$ cDNA which should encode a cysteine residue in this position was unable to produce a protein which was an effective target for pertussis toxin-catalysed ADP-ribosylation (FREISSMUTH and GILMAN 1989).

Initial indications that pertussis toxin was able to modify only a single G-protein $\alpha$ subunit turn out with hindsight to reflect both that of the various pertussis toxin-sensitive G-proteins, only two ($G_{i2}$ and $G_{i3}$) are universally expressed in detectable levels (see below), and that all of these polypeptides migrate in SDS-PAGE with an apparent $M_r$ close to 40 KDa.

Despite these caveats it would appear that, when using [$^{32}$P]NAD as substrate, cholera and pertussis toxin-catalysed [$^{32}$P]ADP-ribosylation should be useful for both qualitative detection and quantitative estimation of G-protein $\alpha$ subunits which are modified by these agents. In qualitative applications two distinct limitations on the use of pertussis toxin-catalysed ADP-ribosylation are evident. The first is the need to resolve the individual pertussis toxin-sensitive G-protein $\alpha$ subunit polypeptides in SDS-PAGE. A number of approaches have been used. The two most useful are (a) to use gels in which the concentration of bis-acrylamide is reduced (usually from 0.27% w/v to about 0.06% w/v; see MITCHELL et al. 1989 for example). Under such conditions what appears to be a single radiolabelled band in SDS-PAGE (10% acrylamide) in the presence of the higher concentration

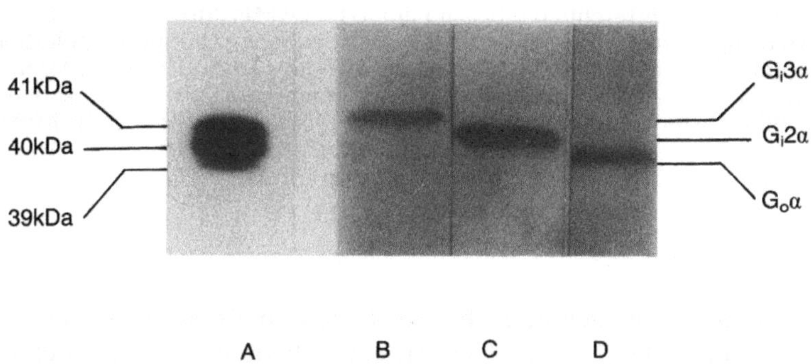


**Fig. 1.** Resolution and identification of pertussis toxin-sensitive G-proteins. The use of pertussis toxin-catalysed [$^{32}$P]ADP-ribosylation and immunodetection. Membranes of NG108-15 neurobastoma x glioma hybrid cells (100 $\mu$g) were treated with thiol-activated pertussis toxin and [$^{32}$P]NAD for 90 min at 37°C and separated by SDS-PAGE (10% acrylamide, 0.06% bisacrylamide). Sample were transferred to nitrocellulose and immunoblotted to detect the presence of the $\alpha$ subunits of $G_{i3}$ (*B*), $G_{i2}$ (*C*) or $G_o$ (*D*) and subsequently autoradiographed (*A*). Under these conditions, $G_{o1}$ and $G_{o2}$ cannot be resolved

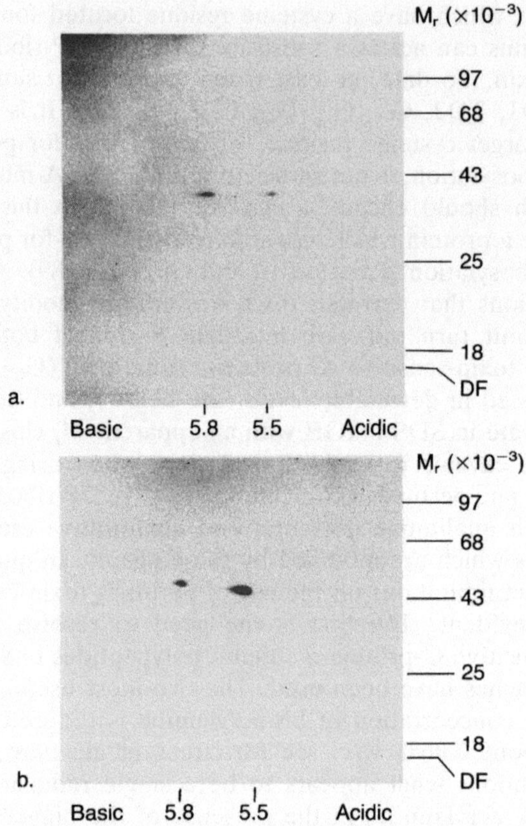


**Fig. 2a,b.** Two-dimensional resolution and separation of isoforms of $G_o\alpha$. Membranes (150 μg) of undifferentiated (**a**) and dibutyryl cAMP-differentiated (**b**) NG108-15 cells were resolved by two-dimensional electrophoresis and immunoblotted with an antiserum which identifies the $\alpha$ subunits of both $G_{o1}$ and $G_{o2}$. Differentiation resulted in the presence of markedly increased immunoreactivity corresponding to the more acidic form (pI = 5.5) but not the more basic form (pI = 5.8); MULLANEY and MILLIGAN (1990). The acidic form has subsequently been defined as $G_{o1}\alpha$, using an antiserum able to discriminate between the two forms

of bis-acrylamide can generally be resolved into three bands which correspond to $G_{i1}$ + $G_{i3}$, $G_{i2}$ and $G_o$ (if the cell or membrane preparation contains these individual G-proteins; Fig. 1) or (b) to include deionized urea in the resolving gel (SCHERER et al. 1987). Individual protocols call for urea concentrations between 4 and 8 *M* to be present. Such conditions have been particularly useful for the resolution of individual forms of $G_o$ (MULLANEY and MILLIGAN 1990), as has two-dimensional electrophoresis (Fig. 2). At least three distinct species of $G_o$ can be separated in the presence of urea, and these conditions also provide good resolution between $G_{i2}$ and $G_{i1}$ +

$G_{i3}$. We have not been reproducibly successful in the resolution of $G_{i1}$ and $G_{i3}$ by either of these approaches.

The second potential limitation on the use of pertussis toxin-catalysed [$^{32}$P]ADP-ribosylation in qualitative assessments of the distribution of individual pertussis toxin-sensitive G-proteins is that the addition of ADP-ribose to the G-protein $\alpha$ subunit reduces the mobility of the polypeptide through SDS-PAGE (GOLDSMITH et al. 1987). It has been argued that this mobility shift is not dependent upon the addition of ADP-ribose (RIBIERO-NETO and RODBELL 1989), but this is clearly incorrect. The alteration in mobility is presumably a reflection of the additonal molecular mass provided rather than alteration in charge. The reduction in mobility of pertussis toxin-sensitive G-proteins following modification is very useful for confirmation that in vivo treatment of cells or tissue with pertussis toxin has successfully modified the G-protein, as immunological detection (see below) of the polypeptides demonstrates this shift in position in the gel. Thus, for qualititative assessment of the presence of a particular pertussis toxin-sensitive G-protein, the standards used (either membranes of cells in which the G-protein population has been defined rigourously or purified or exogenously expressed G-protein) must be modified to completion with pertussis toxin and [$^{32}$P]NAD to provide comparability of mobility in gels.

It should be possible to obtain good quantitiative assessments of levels of the $\alpha$ subunits of cholera and pertussis toxin-sensitive G-protein species in membrane preparations. This is particularly so as in vivo treatment of cells with either cholera or pertussis toxin can cause relatively rapid ADP-ribosylation of the total cellular population of relevant G-proteins. However, quantitative estimates of G-protein level based on scintillation counting of gel sections subsequent to [$^{32}$P]ADP-ribosylation and SDS-PAGE are often considerably lower than estimates produced by ELISA or quantitative immunoblotting methods. As such, toxin-catalysed ADP-ribosylation is likely to provide a minimum estimate of G-protein levels. Furthemore, the relative intensity of [$^{32}$P]ADP-ribosylation of a mixture of pertussis toxin-sensitive G-proteins present in a membrane preparation is ofter markedly different from the relative amount of those G-proteins measured by quantitative immunoblotting with purified or recombinant standards (MCCLUE and MILLIGAN 1991; MCCLUE et al. 1992).

While cholera toxin-catalysed [$^{32}$P]ADP-ribosylation is often stated to modify only $G_s$ of the widely expressed G-protein $\alpha$ subunits (and also $G_{olf}$, TD1 and TD2 which are expressed with highly restricted distributions; See Table 1), this reaction can also modify other G-proteins (GIERSCHIK and JAKOBS 1987; MILLIGAN and MCKENZIE 1988; GIERSCHIK et al. 1989; IIRI et al. 1989; MILLIGAN et al. 1991). To date, this has been recorded only for the pertussis toxin-sensitive G-proteins. In our own hands cholera toxin-catalysed [$^{32}$P]ADP-ribosylation of these "inappropriate" G-protein $\alpha$ subunits is generally an extremely inefficient process and hence is not useful for either qualitative or quantitative analysis. Cholera toxin-catalysed [$^{32}$P]ADP-

**Table 1.** Distribution and function of the $\alpha$ subunits of the heterotrimeric G-proteins

| G-protein | Sensitivity to ADP-ribosylation by bacterial toxins | Function | Distribution |
|---|---|---|---|
| $G_s$ | Yes, cholera toxin | Stimulation of adenylyl cyclase, activation of dihydropyridine-sensitive $Ca^{2+}$ channels | Universal |
| $G_{olf}$ | Yes, cholera toxin | Stimulation of adenylyl cyclase | Olfactory sensory neurones |
| $G_{i1}$ | Yes, pertussis toxin | Undefined | Limited; high levels in brain |
| $G_{i2}$ | Yes, pertussis toxin | Inhibition of adenylyl cyclase | Universal |
| $G_{i3}$ | Yes, pertussis toxin | Regulation of $K^+$ channels | Universal |
| $G_{o1}$ | Yes, pertussis toxin | Regulation of $Ca^{2+}$ channels | Limited to nervous and endocrine tissue |
| $G_{o2}$ | Yes, pertussis toxin | Regulation of $Ca^{2+}$ channels | As for Go1 |
| TD1 | Yes, cholera and pertussis toxins | Activation of cGMP phosphodiesterase | Rod outer segments |
| TD2 | Yes, cholera and pertussis toxins | Activation of cGMP phosphodiesterase | Cone outer segments |
| $G_z$ | No | Undefined | Undefined, but highly restricted |
| $G_q$ | No | Activation of phosphoinositidase C | Universal |
| $G_{11}$ | No | Activation of phosphoinositidase C | Universal |
| $G_{12}$ | No | Undefined | Universal? |
| $G_{13}$ | No | Undefined | Universal? |
| $G_{14}$ | No | as for $G_q$/$G_{11}$ | Undefined but restricted, e.g. spleen, lung, kidney |
| $G_{15}$ | No | Undefined | Certain haemopoietic cells |
| $G_{16}$ | No | As for $G_q$/$G_{11}$ | Certain haemopoietic cells |

ribosylation of pertussis toxin-sensitive G-protein $\alpha$ subunits can, however, provide novel qualititative information on the identity of the G-protein(s) which become activated by agonist/receptor interaction (at least at a semi-quantitative level, agonist enhancement of binding of GTP[$^{35}$S]$\gamma$S may represent a more appropriate route to measure the amount of G-proteins activated by receptors although this strategy has not been rigorously evaluated). In membranes of HL60 cells GIERSCHIK et al. (1989) have noted that the presence of the chemotactic peptide fMet-Leu-Phe allowed cholera toxin-catalysed incorporation of [$^{32}$P]ADP-ribose into two polypeptides of 41 and 40 kDa. These polypeptides were assumed to be the $\alpha$ subunits of $G_{i2}$ and $G_{i3}$ as previous studies had indicated expression of these two G-proteins

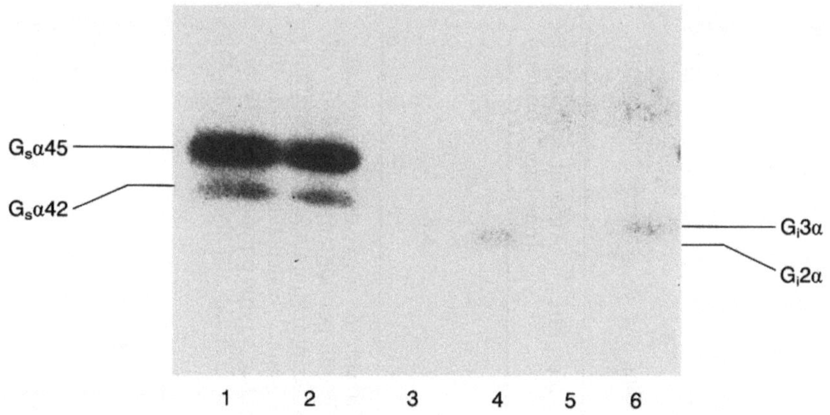


**Fig. 3.** Activation of multiple pertussis toxin-sensitive G-proteins in Rat 1 fibroblasts transfected with the $\alpha_{2C10}$-adrenergic receptor. Cholera toxin-catalysed [$^{32}$P]ADP-ribosylation was performed on membranes of a clone of Rat 1 fibroblasts which express permanently the $\alpha_{2C10}$-adrenergic receptor. Samples in lanes 2, 4 and 6 contained the $\alpha_2$-adrenergic agonist UK14304 while those in lanes 1, 3 and 5 did not. Samples in lanes 1 and 2 were subsequently immunoprecipitated with an antiserum against $G_s\alpha$, those in lanes 3 and 4 with an antiserum which identifies $G_{i2}\alpha$, and those in lanes 5 and 6 with an antiserum which identifies $G_{i3}\alpha$. Immunoprecipitation of radiolabelled polypeptides in lanes 4 and 6 but not lanes 3 and 5 defines that the receptor caused the activation of both $G_{i2}$ and $G_{i3}$ (see MILLIGAN et al. 1991 for further details)

in HL60 cells but no detectable expression of other pertussis toxin-sensitive G-proteins (MURPHY et al. 1987). Using a similar strategy, subsequent to transfection of $\alpha_{2c10}$-adrenergic receptor DNA into Rat 1 fibroblasts, MILLIGAN et al. (1991) were able to demonstrate, following SDS-PAGE and autoradiography, $\alpha_2$-adrenergic agonist-dependent cholera toxin-catalysed [$^{32}$P]ADP-ribosylation of a broad band of some 40 kDa. Immunoprecipitation of such samples with mutually exclusive anti-$G_{i2}$ and $G_{i3}$ antisera demonstrated incorporation of [$^{32}$P]ADP-ribose into both of these polypeptides and thus indicated direct interaction of this single receptor gene product with two distinct pertussis toxin-sensitive G-proteins (MILLIGAN et al. 1991; Fig. 3). However, because it is not clear whether cholera toxin is able to modify $G_{i2}$ and $G_{i3}$ equally well under these conditions, it was not possible to use the incorporation of radioactivity into the immunoprecipitated G-protein $\alpha$ polypeptides as an adequate measure of the relative degree of interaction between the receptor and these two G-proteins. This question has been addressed by measuring the amounts of each of $G_{i2}$ and $G_{i3}$ present in the membranes of these transfected cells by quantitative immunoblotting (see below) and by assessing the contribution of each of these G-proteins to $\alpha_{2c10}$-adrenergic receptor stimulation of high-affinity GTPase activity by the ability of the antisera noted above to block this activity. MCCLUE et al. (1992; Table 2) thus determined that essentially all of the

**Table 2.** Agonist regulation of G-protein activation in membranes of Rat 1 fibroblasts expressing the $\alpha_{2c10}$-adrenergic receptor (adapted from McCLUE et al. 1992)

| G-protein | Amount ($pmol\ mg^{-1}$ protein) | GTPase blocked by antiserum ($pmol\ min^{-1}\ mg^{-1}$ protein) | G-protein activated by receptor (%) | Amount activated ($pmol\ mg^{-1}$ protein) |
|---|---|---|---|---|
| $G_{i1}\alpha$ | 5.1 | | Unknown | |
| $G_{i2}\alpha$ | 53.8 | 17.3 | 16 | 8.1 |
| $G_{i3}\alpha$ | 4.8 | 10.4 | 100 | 4.8 |
| $G_o$ | Not detected | | | |

Level of the individual pertussis toxin-sensitive G-proteins expressed in Rat 1 fibroblasts transfected with the $\alpha_{2c10}$-adrenergic receptor (clone 1C; see MILLIGAN et al. 1991) were assessed by quantitative immunoblotting using specific anti-peptide antisera and *E. coli* expressed recombinant G-protein $\alpha$ subunits as standards. The same antisera were used to blockade receptor stimulation of high-affinity GTPase activity in these membranes. As the measured kcat for both $G_{i2}\alpha$ and $G_{i3}\alpha$ is close to $2\ min^{-1}$, calculations indicate that essentially all the membrane $G_{i3}\alpha$ would have to be activated to produce the GTPase activity ascribed to this polypeptide while only some 15% of the theoretically available $G_{i2}$ would have to be activated to provide the level of GTPase activity ascribed to this G-protein.

theoretically available $G_{i3}\alpha$ but only some 15% of the membrane amount of $G_{i2}\alpha$ was activated by stimulating the entire membrane population of $\alpha_{2c10}$ receptors with a saturating concentration of a full agonist. Such an approach provides novel and valuable quantitative information on the relative interactions of receptors and G-proteins in native membranes.

## II. Immunological Determination of G-Protein Distribution

The initial generation of polyclonal antisera against purified G-proteins and individual G-protein subunits offered the opportunity for analysis of the distribution of G-proteins and also predicted that some early concepts in this area must represent an oversimplification. Production of a panel of polyclonal antisera in rabbits against purified rod transducin (TD1) by Spiegel and co-workers provided a series of reagents able to identify both TD$\alpha$ and G-protein $\beta$-subunit. Cross-reactivity of antibodies within these antisera with a 35- to 36-kDa polypeptide in a wide range of cells and tissues (Gierschik et al. 1985) and in a wide range of species indicated that $\beta$-subunit must be highly conserved and essentially universally expressed (subsequent appreciation of the genetic diversity of $\beta$ subunits has led to the generation of immunological probes able to discriminate between these forms (see Law et al. 1991 for example) and a better understanding of the relative distribution of these polypeptides). Only one of these antisera showed cross-reactivity in tissues other than outer segments, with polypeptides migrating in the position anticipated for other G-protein $\alpha$ subunits, confirming the tissue specificity of TD$\alpha$ expression. However, a single antiserum (CW6; with hindsight, CW6 was probably a Gil$\alpha$ specific antiserum) from this panel identified a 40-kDa pertussis toxin-sensitive G-protein purified from bovine brain ("$G_i$") but did not identify a 39-kDa pertussis toxin-sensitive G-protein ($G_o$) which had essentially co-purified with "$G_i$" (Pines et al. 1985). Such information provided strong evidence that $G_o$ was unlikely to be a proteolytic cleavage product of "$G_i$". This conclusion was confirmed with the generation of an antiserum (RV3) against fractions of the purified bovine brain G-protein preparation which identified $G_o$ but not $G_i$ (Gierschik et al. 1986a). Evidence for further multiplicity of pertussis toxin-sensitive G-proteins was provided when it was shown that neither antisera RV3 or CW6 could identify the major pertussis toxin-sensitive G-protein(s) in neutrophils (Gierschik et al. 1986b) or in either glioma C6 or neuroblastoma x glioma hybrid NG108-15 cells (Milligan et al. 1986).

With the isolation of cDNA species corresponding to two distinct forms of the $\alpha$ subunit of TD, generation of selective anti-peptide antisera based on the predicted amino acid sequence of the polypeptides demonstrated the selective localization of TD2 in cones rather than rods (Lerea et al. 1986) and provided an elegant demonstration of the usefulness of anti-peptide antisera. While polyclonal antisera against purified G-protein preparations are still used and are often more effective in immunoprecipitation protocols and in immunocytochemical applications (TERISHIMA et al. 1987; BRABET et al. 1988), the majority of studies on G-protein distribution and quantitation now make use of polyclonal anti-peptide antisera. To some extent this is a reflection that a number of these are now available commercially.

Armed with the knowledge that the anti-TD1 antiserum CW6 identified "$G_i$" but not $G_o$, and that epitope mapping indicated the major site of interaction to be close to the C-terminal of TD1$\alpha$, antiserum AS7 was generated against a synthetic peptide which represented the last ten amino acids of TD1 (FALLON et al. 1986; GOLDMITH et al. 1987). This was a reflection that the transducins were the only G-proteins for which cDNA clones, and hence predicted sequences, were available at that time. It was hoped that this antiserum would, like antiserum CW6, identify $G_i$ and not $G_o$. This was indeed the case, and, further, antiserum AS7 also identified the major pertussis toxin-sensitive G-protein in neutrophils (FALLOON et al. 1986) and C6 glioma cells (GOLDSMITH et al. 1987), indicating that while these species were not identical to brain "$G_i$", they must contain an epitope which was either highly similar or identical. Subsequent cDNA cloning of $G_{i1}\alpha$ and $G_{i2}\alpha$ confirmed that these species are identical in the region targeted by antiserum AS7, and that there is only a single amino acid difference between these two G-proteins and TD1 in this region. A number of laboratories now automatically generate new anti-peptide antisera with the publication of novel G-protein subunit cDNA information. Many of these antisera are based (as with AS7) on sequences which should represent the extreme C-terminal of the G-protein $\alpha$ subunit both because the C-terminal regions are generally hydrophillic and would be predicted to be highly antigenic, and because AS7 and a number of other antisera of this class have proved to be able to interfere with receptor–G-protein interactions and hence have been invaluable in defining the specificity of receptor–G-protein interactions (SIMONDS et al. 1989a,b; MCKENZIE and MILLIGAN 1990; MCFADZEAN et al. 1989; MCCLUE et al. 1991, 1992; GUTOWSKI et al. 1991; SHENKER et al. 1991). Antipeptide antisera have recently been of particular use for the identification of the protein products of two novel families ($G_q$/$G_{11}$ and $G_{12}$/$G_{13}$) of G-protein $\alpha$ subunits which had been identified in the initial instance only by the isolation of partial and subsequently full-length cDNA sequences (GUTOWSKI et al. 1991; MITCHELL et al. 1991; MILLIGAN et al. 1992). While, with the power of polymerase chain reaction technology it is probably unwise to state that any cell or tissue does not express a particular gene product, at least at the level of general immunological detection the $\alpha$ subunits of $G_{14}$, $G_{15}$ and $G_{16}$ of the $G_q$/$G_{11}$ family are restricted in tissue distribution (WILKIE et al. 1991) while $G_q$ and $G_{11}$ are believed to be widely, if not universally, expressed (see Table 1). The available evidence to date indicates that $G_{12}$ and $G_{13}$ are also probably ubiquitously expressed (STRATHMANN and SIMON 1991), but levels of $G_{13}$ in brain appear to be low (STRATHMANN and SIMON 1991; MILLIGAN et al. 1992), and the range of cells and tissues which have been examined immunologically for expression of $G_{12}$ and $G_{13}$ is currently small. The availability of anti-peptide antisera against the C-terminal region of $G_q/G_{11}$ (and which are also likely to cross-react with $G_{14}$ in cells in which it is expressed, but not with $G_{12}$, $G_{13}$, $G_{15}$ or $G_{16}$; MILLIGAN 1992) has allowed demonstration that one or more G-proteins of this family are involved in receptor-regulation of phosphoinositidase C activity (GUTOWSKI et al. 1991), and given the essentially universal occurrence of this signalling mechanism it would be anticipated that the relevant G-protein(s) should also be expressed universally.

Of the pertussis toxin-sensitive G-proteins, we have noted immunological expression of both $G_{i2}$ and $G_{i3}$ in every cell and tissue that we have examined (although levels of authentic $G_{i3}$ are very low in some regions of brain). By contrast cellular distribution of the other pertussis toxin-sensitive G-proteins is considerably more limited. Both $G_{i1}$ and forms of $G_o$ are highly expressed and concentrated in nerve cells and thus also in tissues which are highly innervated. However, distribution of these G-proteins is not limited to these sites. $G_{i1}\alpha$, for example, is highly expressed in white adipocytes although the functional significance of this is not understood (MITCHELL et al. 1989), and while not expressed in detectable levels in many commonly studied fibroblastic cell lines such as NIH 3T3 cells (MILLIGAN et al. 1989a), it is clearly expressed in others such as Rat 1 fibroblasts (MILLIGAN et al. 1991). $G_o$ is also expressed outwith nervous tissue, particularly in cells of endocrine glands (BOCKAERT et al. 1990). It should be noted at this point that a commercially available antiserum (GC/2, Dupont/NEN) which identifies $G_o\alpha$ also displays weak cross-reactivity with $G_{i3}\alpha$ (both the suppliers and the workers who generated the antiserum have commented on the cross-reactivity of this antiserum; SPIEGEL 1990), and it is possible that some reports of the detection of $G_o$ immunoreactivity in unanticipated locations such as cells of the haemopoietic system may be due to this cross-reactivity.

## C. Immunological Determination of G-Protein α Subunit Levels

### I. Quantitative and Relative Intensity Immunoblotting

Although the term "quantitative immunoblotting" might appear to be an oxymoron, variants of this strategy represent the most used and potentially most useful approach currently widely available for detection and quantitation of G-protein subunits. In most cases what is actually performed is relative-intensity immunoblotting, in which potential alterations in amounts of a G-protein in a cell or membrane preparation are assessed before and after some experimental manipulation. As long as adequate preliminary controls are performed, such as assessment of the range of linearity of immunological signal with amount of G-protein subunit, such an approach can provide valuable information on the regulation of G-protein expression, location and degradation in health and disease. Such relative intensity immunoblotting is restricted in being unable to provide quantitative information on G-protein subunit amounts, and as interest moves towards an understanding of the stoichiometry and selectivity of interaction of proteins involved in G-protein-mediated signal transduction, it is vital that quantitative estimates of the components can be produced. Estimates of receptor level and of the proportion of those receptors coupled to G-proteins can be produced via ligand-binding analyses of the binding of [$^3$H]antagonists and [$^3$H]agonists, but, as noted above, radioisotopic analysis of G-protein amount are restricted by the lack of specificity of toxin-dependent [$^{32}$P]ADP-ribosylation or the binding of [$^{35}$S]GTP$\gamma$S. As such, quantitative immunoblotting, which requires the co-incidental immunological detection of authentic G-protein standards along with the unknown samples to allow an internal standard curve to be constructed is the most appropriate method. The G-protein standards may be produced either by protein purification from (usually) a mammalian tissue source expressing high levels of the G-protein of interest or more recently by purification following production of the G-protein in either *Escherichia coli* (LINDER et al. 1990, 1991) or *baculovirus* (GRABER et al. 1992) expression systems.

In our own laboratory we have utilized both relative-intensity and quantitative immunoblotting to examine the selectivity and quantitative elements of interactions between receptors and G-proteins in a variety of systems. One system which we have examined in detail is the interaction between an IP prostanoid receptor and $G_s$ in neuroblastoma x glioma hybrid, NG108-15, cells (MCKENZIE and MILLIGAN 1990b; ADIE et al. 1992). Specific binding of [$^3$H]prostaglandin $E_1$ ([$^3$H]PGE$_1$) indicates that membranes of these cells express some 1.5 pmol/mg membrane protein of a single high-affinity population of this receptor. Following sustained activation of this receptor in whole cells with either PGE$_1$ or with iloprost, which is a stable analogue of prostacyclin, cellular levels of $G_s\alpha_{45}$ (which corresponds in mobility in SDS-PAGE to the long form of $G_s\alpha$) are reduced by some 40%–70% when assessed by relative-intensity immunoblotting. Quantitative immunoblotting using purified $G_s\alpha_{45}$ derived from the expression of the long form of Gs in *E. coli* was used to demonstrate that membranes of NG108-15 cells express some 10 pmol/mg protein of this polypeptide (i.e., some 0.05% of the total membrane protein), and that activation of the IP prostanoid receptor results in the co-temporal downregulation of some 8 mol $G_s\alpha_{45}$ for each mole of the prostanoid receptor. As recombinant G-protein subunits become more widely available, it is likely that true quantitative immunoblotting will become a method of choice for G-protein quantitation.

One potential limitation to the use of antisera for quantitative measurement of G-protein subunits relates to the possibility that covalent modification may alter the affinity of interaction between antibody and the G-protein antigen. A number of concerns exist. The first of these is of particular importance for quantitation of the $\alpha$ subunit of pertussis toxin-sensitive G-proteins, as the majority of easily available antisera are raised to the C-terminal region of these polypeptides. The peptides used to generate the antisera invariably include the cysteine residue which is the site for pertussis toxin-catalysed ADP-ribosylation. It might thus be anticipated that addition of ADP-ribose to this site would be likely to hinder antibody recognition of G-proteins of this class. However, we normally observe a somewhat enhanced immunological signal following this modification (GOLDSMITH et al. 1987). Such results highlight the fact that noted alterations in relative-intensity immunoblotting experiments may not be a reflection of changing molar amounts of the antigen but of other, more subtle modifications. In a number of cases it is clear that well-established covalent modifications do

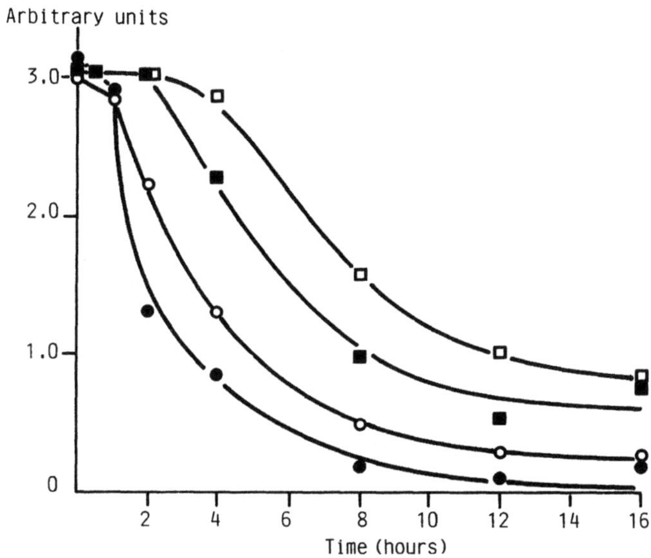


**Fig. 4.** Cholera toxin-catalysed downregulation of cellular levels of $G_s\alpha$ requires ADP-ribosylation of the $G_s\alpha$ polypeptide. L6 skeletal myoblasts were treated with either 1000 ng/ml (*filled symbols*) or 100 ng/ml (*open symbols*) cholera toxin for varying times. Membranes were then prepared from the cells, and detection by both back titration of the remaining $G_s\alpha$ using fresh cholera toxin and [$^{32}$P]NAD to detect non-ADP-ribosylated $G_s\alpha$ (*circles*) and immunoblotting with a specific anti-$G_s\alpha$ antiserum (*squares*) was performed. The higher concentration of cholera toxin produced a more rapid loss of $G_s\alpha$ as assessed by both immunodetection and ADP-ribosylation, but loss of immunodetectable $G_s\alpha$ occurred more slowly than the reduction in $G_s\alpha$ measured by ADP-ribosylation. These data indicate that ADP-ribosylation is required to produce cholera toxin-induced cellular downregulation of $G_s\alpha$ and also demonstrates that immunodetection of the $G_s\alpha$ polypeptide is not compromised by the covalent modification (after 4-h treatment with 100 ng/ml cholera toxin 60% of the $G_s\alpha$ has been ADP-ribosylated but only 5% down-regulated). (Data from MILLIGAN et al. 1989b)

not alter the effectiveness of immunological detection of a G-protein polypeptide. For many years it was assumed that cholera toxin-catalysed ADP-ribosylation maintains the modified and, as such, activated population of $G_s\alpha$ at the plasma membrane. However, relative-intensity immunoblotting of membranes of untreated and cholera toxin-treated cells demonstrated markedly lower detection of $G_s$ in membranes derived from the toxin-treated cells (MILLIGAN et al. 1989b; CHANG and BOURNE 1989). It was, of course, possible that this observation was simply a reflection that the modified form of $G_s$ was detected much more poorly by the antiserum; however, careful time courses which demonstrated that the G-protein must be ADP-ribosylated prior to the loss of immunological signal also indicated that greater than 60% of the G-protein could be ADP-ribosylated without any loss of immunological signal and hence demonstrated that the reduction in immunological signal was a true indication of physical loss of the polypeptide (MILLIGAN et al. 1989b; Fig. 4).

## II. ELISA

Given the availability of a wide range of high-titre polyclonal and anti-peptide antisera against specific G-protein subunits, it may be a surprise that G-protein quantitation based on ELISA type procedures has not yet been widely adopted. Indeed apart from a measurement of plasma membrane levels of $Gs\alpha$ in S49 lymphoma cells (estimated to be $18.9 \pm 2.3$ pmol/mg membrane protein; RANSNAS and INSEL 1988), the use of this assay to assess the effectiveness of $\beta$-adrenergic induced cellular redistribution of $G_s$ (RANSNAS et al. 1989) and quantitation of cellular levels of $G_o$ (ASANO et al. 1987), relatively little has been achieved. Again, the availability of purified proteins and concerns as to the relative interaction of the antibody with G-protein polypeptides at different degrees of denaturation make these experiments more difficult than might be anticipated. Another minor limitation with the use of such approaches is the likely need to purify the antibodies prior to use. Many immunoblots using crude G-protein antisera as the primary reagent show reactivity of the antiserum with polypeptides other than the G-protein. In many cases figures of such data are conveniently cropped for publication purposes, but such immunoreactivity cannot be as easily obscured in an ELISA assay. Furthermore, it may well be that other cellular proteins do contain regions of sequence similar to the peptides used for generation of the G-protein antisera, and it is useful to search the available protein sequence data bases prior to production of new antisera with the aim of limiting potential non-specific or at least unwanted immunoreactive interference from such assays.

## III. Other Approaches

The availability of both antisera and nucleotide sequence information has allowed, at least at a relatively gross anatomical level, both immunocytochemical (TERASHIMA et al. 1987; LAD et al. 1987) and in situ hybridization patterns (WORLEY et al. 1986; BRANN et al. 1987) of the distribution of both G-protein polypeptides and mRNA to be obtained. While such studies do not provide quantitative evaluation of G-proteins, they can be used to provide hints as to the possible function of G-proteins of undefined function (ERCOLANI et al. 1990), but it must be stressed that such studies based on co-localization of elements of signalling cascades can suggest only possibilites (WORLEY et al. 1986) which must be addressed by other approaches.

Quantitation of levels of mRNA encoding various G-protein polypeptides has also been achieved by the use of DNA excess solution hybridization strategies (see HADCOCK et al. 1990, 1991 for examples). While providing a quantitative estimate of mRNA levels may be helpful by right, the most useful element of such studies is to provide information on either cellular regulation of G-protein gene expression and/or regulation of G-protein polypeptide mRNA stability as it appears that hormonal control at these levels is likely to contribute to the regulation of cellular cross-talk between different transmembrane signalling cascades (MALBON et al. 1990). Furthermore, mRNA detection by DNA excess solution hybridization is likely to be more informative quantitatively than northern blotting although it is clearly more difficult to use if it is not in routine use in a laboratory.

## D. Asymmetric Distribution of G-Proteins in the Plasma Membrane

The distibution of G-proteins in the limiting membrane of the cell is often non-uniform. This has been examined by relative-intensity immunoblotting and bacterial toxin catalysed [$^{32}$P]ADP-ribosylation in a variety of polarized cells in which methods have been estalished for the isolation of different sections of the plasma membrane. Hepatocytes maintain substantially higher levels of $G_{i2}\alpha$ and -$\beta$ subunits in the basolateral section of the plasma membrane than at the sinusoidal face (ALI et al. 1989), and while this is also true of other elements associated with signal transduction cascades, such as enzymes involved in inositol phospholipid metabolism (SHEARS et al. 1988), the reason for this is unclear as it might be anticipated that signal transduction would be restricted to the blood-bathed sinusoidal face. Such an arrangement is not restricted to hepatocytes. Enterocytes also have a markedly asymmetric distibution of G-protein subunits in their plasma membrane, with the $\alpha$ subunits of $G_{i2}$ and $G_{i3}$ being located predominantly in the basolateral membrane while the $\alpha$ subunit of $G_s$ and $\beta$ subunit are more evenly distributed between apical and basolateral faces (VAN DEN BERGHE et al. 1991). Strict control of the delivery of newly synthesized G-protein polypeptides would thus appear to be a prerequisite for such distribution, and in polarized renal epithelial LLC-PK1 cells in which the distibution of the $\alpha$ subunits of $G_{i2}$ and $G_{i3}$ are extremely asymmetric overexpression of $G_{i2}$ resulted in correct targetting of the polypeptide to the basolateral membrane (Ercolani et al. 1990).

## E. Conclusions

Qualitative analysis of the distribution of G-protein $\alpha$ subunits indicates that while many of these polypeptides are widely expressed, others show highly specific tissue and cellular distributions. Information such as that Golf is expressed only in nasal neuroepithelium (Jones and Reed 1989) clearly assists in the allocation of function to the individual members of the family. However, such qualititative information is of restricted use in understanding the relative importance of co-expressed closely related G-proteins. In these cases quantitative assessment of the levels of the polypeptides is of central importance. The more widespread availability of purified and particularly recombinant G-protein $\alpha$ subunits will lead to a rapid increase in information in this area.

*Acknowledgements*. Work in the authors' laboratory is supported by the Medical Research Council, the Agriculture and Food Research Council, The Wellcome Trust and the British Heart Foundation. I thank Fergus R. McKenzie, Ian Mullaney, Fiona M. Mitchell, Steven J. McClue, Catherine Clark, Elaine J. Adie, Craig Carr, Matthew Dodd and Fraser McCallum, who as past and present members of my laboratory have contributed significantly to the work. I also thank Allen M. Spiegel. Many of our studies have developed from initial collaborative interactions with his laboratory.

## References

Adie EJ, Mullaney I, McKenzie FR, Milligan G (1992) Concurrent downregulation of IP prostanoid receptors and the $\alpha$ subunit of the stimulatory guanine nucleotide binding protein (Gs) during prolonged exposure of neuroblastoma x glioma cells to prostanoid agonists. Quantitation and functional implications. Biochem J 285:529–536

Ali N, Milligan G, Evans WH (1989) Distribution of G-proteins in rat liver plasma membrane domains and endocytic pathways. Biochem J 261:950–912

Asano T, Semba R, Ogasawara N, Kato K (1987) Highly sensitive immunoassay for the alpha subunit of the GTP-binding protein Go and its regional distribution in bovine brain. J Neurochem 48:1617–1623

Bockaert J, Brabet P, Gabrion J, Homberger V, Rouot B, Toutant M (1990) Structural, immunobiological, and functional characterization of guanine nucleotide-binding protein Go. In: Iyengar R, Birnbaumer L (eds) G-proteins. Academic Press, San Diego, pp 81–113

Brabet P, Dumuis A, Sebben M, Pantaloni C, Bockaert J, Homberger V (1988) Immunocytochemical localization of the guanine nucleotide-binding protein Go in primary culture of neuronal and glial cells. J Neurosci 8:701–708

Brann MR, Collins RM, Spiegel AM (1987) Localization of mRNAs encoding the $\alpha$ subunit of signal transducing G-proteins within rat brain and among peripheral tissues. FEBS Lett 222:191–198

Burgoyne RD (1992) Trimeric G proteins in golgi transport. Trends Biochem Sci 17:87–88

Chang F-H, Bourne HR (1989) Cholera toxin induces cAMP-independent degradation of Gs. J Biol Chem 264:5352–5357

Ercolani L, Stow JL, Boyle JF, Holtzman EJ, Lin H, Grove JR, and Ausiello DA (1990) Membrane localization of the pertussis toxin-sensitive G-protein subunits $\alpha$i2 and $\alpha$i3 and expression of a metallothionein-$\alpha$i2 fusion gene in LLC-PK1 cells. Proc Natl Acad Sci USA 87:4635–4639

Falloon J, Malech H, Milligan G, Unson C, Kahn R, Goldsmith P, Spiegel A (1986) Detection of the major pertussis toxin substrate of human leukocytes with antisera raised against synthetic peptides. FEBS Lett 209:352–356

Freissmuth M, Gilman AG (1989) Mutations of Gs$\alpha$ designed to alter the reactivity of the protein with bacterial toxins. Substitutions at Arg 187 results in the loss of GTPase activity. J Biol Chem 264:21907–21914

Gabrion J, Brabet P, Nguyen-Than Dao B, Homburger V, Dumuis A, Sebben M, Rouot B, Bockaert J (1989) Ultrastructural localization of the GTP binding protein Go in neurons. Cellular Signalling 1:107–123

Gierschik P, Codina J, Simons C, Birnbaumer L, Spiegel A (1985) Antisera against a guanine nucleotide binding protein from retina cross-react with the beta subunit of the adenylyl cyclase associated guanine nucleotide binding proteins, Ns and Ni. Proc Natl Acad Sci USA 82:727–731

Gierschik P, Milligan G, Pines M, Goldsmith P, Codina J, Klee W, Spiegel A (1986a) Use of specific antibodies to quantitate the guanine nucleotide binding protein Go in brain. Proc Natl Acad Sci USA 83:2258–2262

Gierschik P, Falloon J, Milligan G, Pines M, Gallin JI, Spiegel A (1986b) Immunochemical evidence for a novel pertussis toxin substrate in human neutrophils. J Biol Chem 261:8058–8062

Gierschik P, Jakobs K-H (1987) Receptor mediated ADP-ribosylation of a phospholipase C-stimulating G-protein. FEBS Lett 224:219–223

Gierschik P, Sidiropoulos D, Jakobs K-H (1989) Two distinct Gi-proteins mediate formyl peptide receptor signal transduction in human leukemia (HL-60) cells. J Biol Chem 264:21470–21473

Goldsmith P, Gierschik P, Milligan G, Unson C, Vinitsky R, Malech HL, Spiegel AM (1987) Antibodies directed against synthetic peptides distinguish between GTP-binding proteins in neutrophils and brain. J Biol Chem 262:14683–14688

Graber SG, Figler RA, Garrison JC (1992) Expression and purification of functional G-protein $\alpha$ subunits using a baculovirus expression system. J Biol Chem 267: 1271–1278

Gutowski S, Smrcka A, Nowak L, Wu D, Simon M, Sternweis PC (1991) Antibodies to the $\alpha$q subfamily of guanine nucleotide binding regulatory protein $\alpha$ subunits attenuate activation of phosphatidylinositol 4,5-bisphosphate hydrolysis by hormones. J Biol Chem 266:20519–20524

Hadcock JR, Ros M, Watkins DC, Malbon CC (1990) Cross-regulation between G-protein-mediated pathways. Stimulation of adenylyl cyclase increases expression of the inhibitory G-protein, Gi$\alpha$2. J Biol Chem 265:14784–14790

Hadcock JR, Port JD, Malbon CC (1991) Cross-regulation between G-protein mediated pathways. Activation of the inhibitory pathway of adenylylcyclase increases the expression of $\beta$2-adrenergic receptors. J Biol Chem 266:11915–11922

Iiri T, Tohkin M, Morishima N, Ohoka Y, Ui M, Katada T (1989) Chemotactic peptide receptor-supported ADP-ribosylation of a pertussis toxin substrate GTP-binding protein by cholera toxin in neutrophil-type HL 60 cells. J Biol Chem 264:21394–21400

Jones DT, Reed RR (1989) Golf: an olfactory neuron specific G-protein involved in odorant signal transduction. Science 244:790–795

Kaziro Y, Itoh H, Kozasa T, Nakafuka M, Satoh M (1991) Structure and function of signal-transducing GTP-binding proteins. Annu Rev Biochem 60:349–400

Ktiskatis NT, Linder M, Roth MG (1992) Action of brefeldin A blocked by activation of a pertussis-toxin sensitive G-protein. Nature 356:344–346

Lad RP, Simons C, Gierschik P, Milligan G, Woodward C, Griffo M, Goldsmith P, Ornberg R, Gerfen CR, Spiegel A (1987) Differential distribution of signal transducing G-proteins in retina. Brain Res 423:237–246
LaMorte VJ, Goldsmith PK, Spiegel AM, Meinkoth JL, Feramisco JR (1992) Inhibition of DNA synthesis in living cells by microinjection of Gi2 antibodies. J Biol Chem 267:691–694
Law SF, Manning D, Reisine T (1991) Identification of the subunits of GTP-binding proteins coupled to somatostatin receptors. J Biol Chem 266:17885–17897
Lerea CL, Somers DE, Hurley JB, Klock IB, Bunt-Milan AH (1986) Identification of specific transducin alpha subunits in retinal rod and cone photoreceptors. Science 234:77–80
Linder ME, Ewald DA, Miller RJ, Gilman AG (1990) Purification and characterization of Go$\alpha$ and three types of Gi$\alpha$ after expression in Escherichia coli. J. Biol Chem 265:8243–8251
Linder ME, Pang I-H, Duronio RJ, Gordon JI, Sternweis PC, Gilman AG (1991) Lipid modifications of G-protein subunits. Myristoylation of Go$\alpha$ increases its affinity for $\beta\gamma$. J Biol Chem 266:4654–4659
Malbon CC, Hadcock JR, Rapiejko PJ, Ros M, Wang HY, Watkins DC (1990) Regulation of transmembrane signalling elements: transcriptional, post-transcriptional and post-translational controls. Biochem Soc Symp 56:155–164
Matsuda LA, Lolait SJ, Brownstein MJ, Young AC, Bonner TI (1990) Structure of a cannabinoid receptor and functional expression of the cloned cDNA. Nature 346:561–564
McClue SJ, Milligan G (1991) Molecular interaction of the human $\alpha$2-C10 adrenergic receptor, when expressed in Rat-1 fibroblasts, with multiple pertussis toxin-sensitive guanine nucleotide binding proteins: studies with site directed antisera. Mol Pharmacol 40:627–632
McClue SJ, Selzer E, Freissmuth M, Milligan G (1992) Gi3 does not contribute to the inhibition of adenylate cyclase when stimulation of an $\alpha$2 adrenergic receptor causes activation of both Gi2 and Gi3. Biochem J 284;565–568
McFadzean I, Mullaney I, Brown DA, Milligan G (1989) Antibodies to the GTP binding protein, Go, antagonize noradrenaline-induced calcium current inhibition in NG108-15 hybrid cells. Neuron 3:177–182
McKenzie FR, Milligan G (1990a) $\delta$ Opioid receptor-mediated inhibition of adenylate cyclase is transduced specifically by the guanine nucleotide binding protein Gi2. Biochem J 267:391–398
McKenzie FR, Milligan G (1990b) Prostaglandin E1-mediated, cyclic AMP-independent, downregulation of Gs$\alpha$ in neuroblastoma X glioma hybrid cells. J Biol Chem 265:17084–17093
Milligan G (1988) Techniques used in the identification and analysis of function of pertussis toxin-sensitive guanine nucleotide binding proteins. Biochem J 255:1–13
Milligan G (1989) Tissue distribution and subcellular location of guanine nucleotide binding proteins: implications for cellular signalling. Cell Signalling 1:411–419
Milligan G (1992) Multiple heterotrimeric guanine nucleotide binding proteins: roles in the determination of cellular signalling specificity. Biochem Soc Trans 20:135–140
Milligan G, Gierschik P, Spiegel AM, Klee WA (1986) The GTP-binding regulatory proteins of neuroblastoma x glioma, NG108-15, and glioma, C6, cells. Immunochemical evidence of a pertussis toxin substrate that is neither Ni nor No. FEBS Lett 195:225–230
Milligan G, McKenzie FR (1988) Opioid peptides promote cholera toxin catalysed ADP-ribosylation of the inhibitory guanine nucleotide binding protein (Gi) in membranes of neuroblastoma x glioma hybrid cells. Biochem J 22:369–373
Milligan G, Davies S-A, Houslay MD, Wakelam MJO (1989a) Identification of the pertussis and cholera toxin substrates in normal and N-*ras* transformed NIH3T3 fibroblasts and an assessment of their involvement in bombesin-stimulation of inositol phospholipid metabolism. Oncogene 4:659–663

Milligan G, Unson CG, Wakelam MJO (1989b) Cholera toxin treatment produces down-regulation of the $\alpha$-subunit of the stimulatory guanine-nucleotide binding protein (Gs). Biochem J 262:643–649

Milligan G, Green A (1991) Agonist control of G-protein levels. Trends Pharmacol Sci 12:207–209

Milligan G, Carr C, Gould GW, Mullaney I, Lavan BE (1991) Agonist-dependent, cholera toxin-catalysed ADP-ribosylation of pertussis toxin-sensitive G-proteins following transfection of the human $\alpha$2-C10 adrenergic receptor into Rat 1 fibroblasts. Evidence for the direct interaction of a single receptor with two pertussis toxin-sensitive G-proteins, Gi2 and Gi3. J Biol Chem 266:6447–6455

Milligan G, Mullaney I, Mitchell FM (1992) Immunological identification of the $\alpha$ subunit of $G_{13}$, a novel guanine nucleotide binding protein. FEBS Lett 297: 186–188

Mitchell FM, Griffiths SL, Saggerson ED, Houslay MD, Knowler JT, Milligan G (1989) Guanine nucleotide binding proteins expressed in rat white adipose tissue. Identification of both mRNAs and proteins corresponding to $G_{i1}$, $G_{i2}$ and $G_{i3}$. Biochem J 262:403–408

Mitchell FM, Mullaney I, Godfrey PP, Arkinstall SJ, Wakelam MJO, Milligan G (1991) Widespread distribution of Gq$\alpha$/G11$\alpha$ detected immunologically by an antipeptide antiserum directed against the predicted C-terminal decapeptide. FEBS Lett 287:171–174

Mullaney I, Milligan G (1990) Identification of two distinct isoforms of the guanine nucleotide binding protein $G_o$ in neuroblastoma x glioma hybrid cells: independent regulation during cyclic AMP-induced differentiation. J Neurochem 55:1890–1898

Murphy PM, Edie B, Goldsmith P, Brann M, Gierschik P, Spiegel A, Malech HL (1987) Detection of multiple froms of $G_i$ alpha in HL60 cells. FEBS Lett 221:81–86

Pines M, Gierschik P, Milligan G, Klee W, Spiegel A (1985) Antibodies against the C-terminal 5 kDa peptide of the alpha subunit of transducin crossreact with the 40 but not the 39 kDa guanine nucleotide binding protein from brain. Proc Natl Acad Sci 82:4095–4099

Ransnas LA, Insel PA (1988) Quantitation of the guanine nucleotide binding regulatory protein $G_s$ in S49 cell membranes using antipeptide antibodies to $\alpha_s$. J Biol Chem 263:9482–9485

Ransnas LA, Svoboda P, Jasper JR, Insel PA (1989) Stimulation of $\beta$-adrenergic receptors of S49 lymphoma cells redistributes the $\alpha$ subunit of the stimulatory G-protein between cytosol and membranes. Proc Natl Acad Sci USA 86: 7900–7903

Ribiero-Neto FAP, Rodbell M (1989) Pertussis toxin induces structural changes in G$\alpha$ proteins independently of ADP-ribosylation. Proc Natl Acad Sci USA 86:2577–2581

Scherer NM, Toro M-J, Entman ML, Birnbaumer L (1987) G-protein distribution in canine cardiac sarcoplasmic reticulum and sacrolemma: comparison to rabbit skeletal muscle membranes and to brain and erythrocyte G-proteins. Arch Biochem Biophys 259:431–440

Shears SB, Evans WH, Kirk CJ, Michell RH (1988) Preferential localization of rat liver D-myo-inositol 1,4,5-triksphosphate/1,3,4,5-tetraksphosphate 5-phosphatase in bile-canalicular plasma membrane and "late" endosomal vesicles. Biochem J 256:363–369

Shenker A, Goldsmith P, Unson CG, Spiegel AM (1991) The G-protein coupled to the thromboxane A2 receptor in human platelets is a member of the novel Gq family. J Biol Chem 266:9309–9313

Simon MI, Strathmann MP, Gautam N (1991) Diversity of G proteins in signal transduction. Science 252:802–808

Simonds WF, Goldsmith PK, Woodward CJ, Unson CG, Spiegel AM (1989a) Receptor and effector interactions of $G_s$: functional studies with antibodies to the $\alpha_s$ carboxyl-terminal decapeptide. FEBS Lett 249:189–194

Simonds WF, Goldsmith PK, Codina J, Unson CG, Spiegel AM (1989b) Gi2 mediates $\alpha$2 adrenergic inhibition of adenylate cyclase in platelet membranes. In situ identification with G$\alpha$ C-terminal antibodies. Proc Natl Acad Sci USA 86:7809–7813
Spiegel AM (1990) Immunologic probes for heterotrimeric GTP-binding proteins. In: Iyengar.R, Birnbaumer L (eds) G-proteins. Academic, San Diego, pp 116–143
Strathmann MP, Simon MI (1991) G$\alpha$12 and G$\alpha$13 subunits define a fourth class of G protein $\alpha$ subunits. Proc Natl Acad Sci USA 88:5582–5586
Takei Y, Kurosu H, Takahashi K, Katada T (1992) A GTP-binding protein in rat liver nuclei serving as the specific substrate of pertussis toxin-catalysed ADP-ribosylation. J Biol Chem 267:5085–5089
Terashima T, Katada T, Oinuma M, Inoue Y, Ui M (1987) Immuno-histochemical localization of guanine nucleotide binding proteins in rat retina. Brain Res 410:97–100
van den Berghe N, Nieuwkoop NJ, Vaandrager AB, de Jonge HR (1991) Asymmetrical distribution of G-proteins among the apical and basolateral membranes of rat enterocytes. Biochem J 278:565–571
Wang N, Yan K, Rasenick MM (1990) Tubulin binds specifically to the signal-transducing proteins, Gs$\alpha$ and Gi$\alpha$1. J Biol Chem 265:1239–1242
Wilkie TM, Scherle PA, Strathmann MP, Slepak VZ, Simon MI (1991) Characterization of G-protein $\alpha$ subunits in the Gq class: expression in murine tissues and in stromal and hematopoietic cell lines. Proc Natl Acad Sci USA 88:10049–10053
Worley PF, Baraban JM, Van Dop C, Neer E, Snyder SH (1986) Go, a guanine nucleotide binding protein: immunochemical localization in rat brain resembles distribution of second messenger systems. Proc Natl Acad Sci USA 83:4561–4565

CHAPTER 48

# Subunit Interactions of Heterotrimeric G-Proteins

E.J. NEER

## A. Signalling by $\alpha$ and $\beta\gamma$ Subunits

A defining feature of the GTP-binding $\alpha$ subunit of the heterotrimeric G-proteins is its interaction with the $\beta\gamma$ subunit. Although the G proteins are heterotrimers, they are functional dimers because the $\beta$ and $\gamma$ subunits appear not to dissociate under native conditions. The equilibrium between the associated and dissociated states of $\alpha$ and $\beta\gamma$ subunits is controlled by the conformational changes in the $\alpha$ subunit that accompany guanine nucleotide binding and hydrolysis. As diagrammed in Fig. 1, when a guanine nucleoside diphosphate is bound to $\alpha$, the associated form is favored. When the diphosphate is exchanged for a triphosphate, the equilibrium between $\alpha$ and $\beta\gamma$ shifts toward dissociation.

The GDP-liganded heterotrimer is the inactive form of the G-protein. Dissociation produces two active signaling molecules: $\alpha$-GTP and $\beta\gamma$-. The activated $\alpha$ subunit controls a number of enzymes, such as adenylyl cyclase (NORTHUP et al. 1983a), phospholipases (SMRCKA et al. 1991), and ion channels (CODINA et al. 1987). Activation of an effector by $\beta\gamma$ was first shown by LOGOTHETIS et al. (1987) who found that $\beta\gamma$ activated the cardiac $K^+$ channel, probably by a combination of direct and indirect actions (KIM et al. 1989; ITO et al. 1992; KURACHI, Chap. 75, this volume). The proposal that $\beta\gamma$ could activate effectors was initially controversial, but the controversy has abated with the finding that this is a common regulatory mechanism. Like some other effectors, the cardiac $K^+$ channel is regulated by $\alpha$ subunits (CODINA et al. 1987), as well as by $\beta\gamma$ (LOGOTHETIS et al. 1988). Effectors regulated by $\beta\gamma$ are phospholipase $A_2$ (JELSEMA and AXELROD, 1987), adenylyl cyclases (TANG and GILMAN, 1991), receptor kinases (HAGA and HAGA, 1992; PITCHER et al. 1992), phospholipase C$\beta$ (GIERSCHIK and CAMPS, Chap. 59, this volume; KATZ et al. 1992). The $\beta\gamma$ subunit can activate or inhibit different subtypes of the same effector (TANG and GILMAN, 1991).

The $\alpha$ and $\beta\gamma$ subunits modulate each other's function. An excess of $\beta\gamma$ drives the equilibrium in Fig. 1 to the left, favoring the deactivation of $\alpha$ (NORTHUP et al. 1983b; NEER et al. 1987). The activity of a can thus be titrated by the amount of $\beta\gamma$. Conversely, the GDP-liganded $\alpha$ subunit turns off the action of $\beta\gamma$ (see, for example, LOGOTHETIS et al. 1987). Therefore, even though $\beta\gamma$ does not bind GTP, its activity is ultimately controlled by GTP binding and hydrolysis.

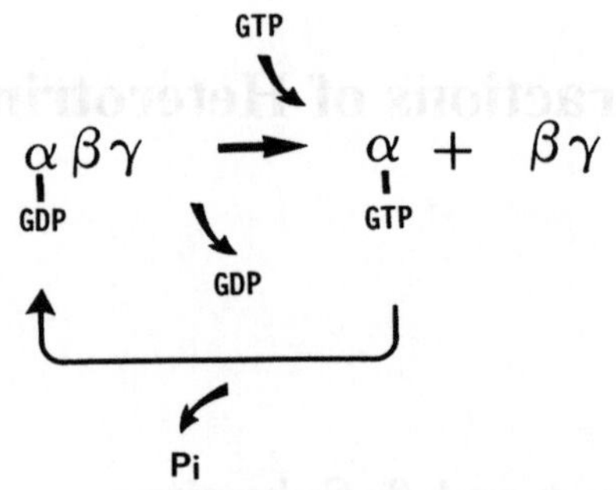


**Fig. 1.** Dissociation of the $\alpha\beta\gamma$ heterotrimer. Binding of GDP to the $\alpha$ subunit stabilizes the heterotrimeric form of the G-protein. Exchange of GDP for GTP, stimulated by an agonist-liganded receptor, causes activation and dissociation of the protein into $\alpha$ and $\beta\gamma$ subunits. Hydrolysis of GTP on the active site restores the inactive heterotrimeric state

The formation of the inactive heterotrimer is probably an obligate part of the regulatory cycle in the cell. Once GTP has been hydrolyzed to GDP at the active site of an $\alpha$ subunit, the GDP binds very tightly, and the $\alpha$ subunit must reassociate with a hormone receptor to release GDP and bind a new GTP. Although $\alpha$ subunits are able to interact with receptors as monomers, the association is greatly enhanced by $\beta\gamma$ subunits (FUNG 1983; HEKMAN et al. 1984; YAMAZAKI et al. 1987; FLORIO and STERNWEIS 1989). Both subunits appear to bind to the receptor, although as yet nothing is known about the order of binding or the sites at which the $\beta\gamma$ subunit might bind to the receptor (IM et al. 1988).

## I. Effect of Subunit Association on the Guanine Nucleotide Binding and GTPase Activity

Just as guanine nucleotide binding affects the equilibrium between $\alpha$ and $\beta\gamma$, subunit interactions modify the affinity of $\alpha$ for nucleotides. The $\beta\gamma$ subunit decreases the affinity of $\alpha$ subunits for GTP analogues but increases their affinity for GDP (HIGASHIJIMA et al. 1987; BRANDT and ROSS 1985). The increased affinity for GDP results both from an increase in the GDP association rate and from a decrease in the GDP dissociation rate. $Mg^{2+}$ ion modulates the effect of $\beta\gamma$ on the affinity of $\alpha$ for GDP, probably because the divalent cation influences the ability of $\beta\gamma$ to associate with $\alpha_o$. Without $Mg^{2+}$ the $K_d$ for GDP is approximately 40 n$M$ for $\alpha_o$ and approximately 0.1 n$M$ for $G_o$ ($\alpha_o$ $\beta\gamma$). In the presence of 10 m$M$ $Mg^{2+}$ $\beta\gamma$ increases the affinity of $\alpha_o$ for GDP. The $K_d$ drops from about 100 n$M$ to about 10 n$M$, still a tenfold change but less than the 400-fold change caused by $\beta\gamma$ without the divalent cation. $Mg^{2+}$ affects the dissociation only of GDP from $G_o$ ($\alpha_o\beta\gamma$), not that from $\alpha_o$, consistent with the idea that a major part of the effect of $Mg^{2+}$ is on subunit interactions.

Since $\beta\gamma$ modulates the dissociation of GDP from the $\alpha$ subunit, it also affects the GTPase activity. During steady-state hydrolysis of GTP most of the $\alpha$ subunit is in the GDP-bound form, and $k_{cat}$ is typically an order of magnitude greater than the dissociation rate for GDP ($0.3\,min^{-1}$). Therefore, the reaction is limited by the release of the product, GDP. At low $Mg^{2+}$ concentration $\beta\gamma$ inhibits GTPase activity by further slowing GDP release (HIGASHIJIMA et al. 1987). At high $Mg^{2+}$ concentrations the situation is more complex, and the GTPase activity of $\alpha\beta\gamma$ may be greater than that of $\alpha$ (HIGASHIJIMA et al. 1987). The concentration of $Mg^{2+}$ required for this effect is different for different heterotrimers (BRANDT and ROSS 1985; HIGASHIJIMA et al. 1987).

## II. Physical Properties of Associated and Dissociated G-Protein Subunits

Even before the heterotrimeric nature of the G-proteins was known, physical analysis of the factor that stimulates adenylyl cyclase activity suggested that activation involves a dissociation of protein components (PFEUFFER 1979; HOWLETT and GILMAN 1980). Once the proteins were purified, dissociation of the $\alpha$ and $\beta\gamma$ subunits was readily shown by a change in molecular weight upon activation by poorly hydrolyzable GTP analogues (for example, STERNWEIS et al. 1981; CODINA et al. 1983; NORTHUP et al. 1983b; CODINA et al. 1984; HUFF et al. 1985; HUFF and NEER 1986).

In detergent solution, $\alpha$ can be dissociated from $\beta\gamma$ only by nonhydrolyzable analogues of GTP, not by GTP itself (HUFF and NEER 1986). The reason for this is probably that GTP is rapidly hydrolyzed to GDP, which then remains bound at the active site. Modification of $\alpha_s$ (the $\alpha$ subunit of $G_s$, the G protein that stimulates adenylyl cyclase) by cholera toxin inhibits its GTPase activity and decreases the affinity of $\alpha_s$ for $\beta\gamma$ (KAHN and GILMAN 1984). The effect of GTP on dissociation was not compared to GTP$\gamma$S in that study.

Although the equilibrium is pushed either toward the monomer or the heterotrimer according to the guanine nucleotide bound, the rate of association and dissociation of the subunits is rapid whether a poorly hydrolyzable analogue of GTP such as GTP$\gamma$S, GTP, or GDP is on the active site. This conclusion was first reached by FUNG et al. (1983) from analysis of the activation of $\alpha$ subunits of transducin. Further evidence for this conclusion came from analysis of pertussis toxin-catalyzed ADP-ribosylation of $\alpha$ subunits. Although pertussis toxin modifies the $\alpha$ subunit, the true substrate for the toxin appears to be the $\alpha\beta\gamma$ heterotrimer (NEER et al. 1984; TSAI et al. 1984). Therefore, the reaction depends on the presence of $\beta\gamma$. CASEY et al. (1989) showed that the action of $\beta\gamma$ in supporting ADP-ribosylation of $\alpha$ is catalytic, so that a given concentration of $\beta\gamma$ can support the ADP-ribosylation of a much larger concentration of $\alpha$ subunits. Rapid association and dissociation of $\alpha\beta\gamma$ heterotrimers is also the likely explanation for the observation that poorly hydrolyzable guanine nucleotides have little effect on the rate of chemical cross-linking of $\alpha$ to $\beta\gamma$ subunits (YI et al. 1991). The irreversible cross-linking reaction traps the heterotrimer as it forms, even in the presence of GTP$\gamma$S.

## III. The $\alpha$ and $\beta\gamma$ Interface

### 1. Analysis by Site-Directed Mutagenesis of Requirements for $\alpha$ and $\beta\gamma$ Interactions

The sites of interaction that are essential for the rapid and reversible association of $\alpha$ and $\beta\gamma$ subunits are not yet known in detail. The amino terminal 2 kDa of the $\alpha$ subunit appears to be important for interaction with $\beta\gamma$, either because this region interacts directly with $\beta\gamma$, or because it is required to maintain a conformation that can bind to $\beta\gamma$. Removal of this portion of the molecule from transducin $\alpha$, $\alpha_i$, $\alpha_o$, and $\alpha_s$ by limited digestion with trypsin produces a truncated molecule that is unable to bind to $\beta\gamma$, as documented both by an analysis of its hydrodynamic properties and by its failure to be a substrate for pertussis toxin-catalyzed ADP-ribosylation (NAVON and FUNG 1987; NEER et al. 1988; JOURNOT et al. 1991). The truncated molecule can bind and hydrolyze guanine nucleotides normally, so there does not appear to be a gross change in overall conformation (NEER et al. 1988).

The amino terminus of some $\alpha$ subunits is the site of modification by myristic acid. The $\alpha_o/\alpha_i$ family but not the $\alpha_s$ family of $\alpha$ subunits contains this modification (BUSS et al. 1987). Myristoylation is essential for attachment of this group of $\alpha$ subunits to the plasma membrane but is also important for ability of the $\alpha$ subunits to interact with $\beta\gamma$ (MUMBY et al. 1990; JONES et al. 1990). Unmyristoylated $\alpha$ subunits, in which the glycine that is myristoylated has been mutated to an alanine, have a lower affinity for $\beta\gamma$ (JONES et al. 1990; LINDER et al. 1991; DENKER et al. 1992a). Since the prevention of myristoylation diminishes but does not completely abolish the ability of $\alpha$ to interact with $\beta\gamma$, other structural features must also play a role.

Removal of 2 kDa from the amino terminus of $\alpha$ subunits abolishes interaction with $\beta\gamma$ altogether. In $\alpha_o$ the site of tryptic cleavage is at lysine 21 (HURLEY et al. 1984b). Since removal of the amino terminal peptide causes two changes (loss of protein and loss of myristate), DENKER et al. (1992a) created a series of mutations that segregated the variables. The cDNAs encoding the mutated proteins were translated in a rabbit reticulocyte in vitro translation system (DENKER et al. 1992a). The rabbit reticulocyte lysate is capable of carrying out the myristoylation of proteins (DEICHAITE et al. 1988). DENKER et al. (1992a) used two methods to evaluate the ability of mutated $\alpha_o$ proteins to interact with $\beta\gamma$; sucrose density-gradient centrifugation to measure changes in the sedimentation coefficient and ADP-ribosylation by pertussis toxin. Maintenance of a heterotrimer through

**Table 1.** Summary of effects of mutation in the amino terminus of $\alpha_o$[a]

| | | Interacts with $\beta\gamma$ | |
|---|---|---|---|
| | | Gradient | Ribosylation |
| Myristoylated wild type | | | |
| $\alpha_o$ | GCTLSAEERA ALERSKAIEK NLKEDG.... | + | + |
| Myristoylated | | | |
| D[11–14] | GCTLSAEERA - - - - -SKAIEK NLKEDG.... | + | + |
| D[7–10] | GCTLSA- - - - - ALERSKAIEK NLKEDG.... | − | + |
| D[6–15] | GCTLS- - - - - - - - - - - -KAIEK NLKEDG.... | − | + |
| D[2–20]E25S | G- - - - - - - - - - - - - - - - - - - - -NLKSDG.... | − | − |
| R9P | GCTLSAEEPA ALERSKAIEK NLKEDG.... | + | + |
| R9E | E | + | ND |
| R9A | A | + | ND |
| EER[7–9]QQQ | GCTLSAQQQA ALERSKAIEK NLKEDG... | + | + |
| Nonmyristoylated | | | |
| GLA | ACTLSAEERA ALERSKAIEK NLKEDG.... | − | + |
| D[2–4] | G- - -SAEERA ALERSKAIEK NLKEDG.... | − | + |
| D[2–9] | G- - - - - - - - -A ALERSKAIEK NLKEDG.... | − | − |
| D[2–20] | G- - - - - - - - - - - - - - - - - - - - - -NLKEDG.... | − | − |

A "+" indicates interaction with $\beta\gamma$ on sucrose density gradients or by pertussis toxin-catalyzed ADP-ribosylation. Dashes in sequence indicate deletions; dots represent continuation of sequence. See DENKER et al. (1992a) for a description of the experiments and results.
[a] The $\alpha$ subunit of the predominantly neural G protein, $G_o$.

sucrose density-gradient centrifugation requires that the equilibrium be strongly shifted to the associated form so that the heterotrimer is stable over the period of centrifugation. The capacity to be a substrate for a pertussis toxin-catalyzed ADP-ribosylation is a more sensitive assay for heterotrimer formation because it uses an irreversible reaction as a marker of what may be weak or transient interactions.

The mutations that were made and summary of the analysis is shown in Table 1. The wild type myristoylated $\alpha_o$ protein made by in vitro translation was indistinguishable in its sedimentation behavior from $\alpha_o$ purified from bovine brain. The observation that all of the control in vitro translated $\alpha_o$ behaves normally suggests that the rabbit reticulocyte lysate is able quantitatively to myristoylate the $\alpha$ subunits. The mutations created by DENKER et al. (1992a) can be divided into two groups, those in which the protein can be myristoylated and those in which it cannot. Nonmyristoylated $\alpha_o$ can interact weakly with $\beta\gamma$. Removal of three amino acids at positions 2–4 does not diminish this weak interaction, but removal of a further five amino acids eliminates it. These results point to the particular importance of amino acids 5–9 in allowing $\alpha$ and $\beta\gamma$ to interact.

This suggestion is supported by comparisons within the set of myristoylated proteins. The transducin $\alpha$ subunit, which interacts normally with $\beta\gamma$, is four amino acids shorter at the amino terminus than is $\alpha_o$ or $\alpha_i$. Removing amino acids 11–14 (D[11–14]) generates a transducinlike amino terminus in an $\alpha_o$ background. The $\alpha_o$ containing this deletion binds strongly to $\beta\gamma$ and has a sedimentation coefficient characteristic of the heterotrimer. In contrast, removing four amino acids closer to the amino terminus (D[7–10]) prevents strong interactions with $\beta\gamma$ (detected by sucrose density-gradient centrifugation) and leaves only weak interactions with $\beta\gamma$ (detected by pertussis toxin-catalyzed ADP-ribosylation). The amino acid sequence at positions 7–10 includes a highly conserved acidic, basic pattern that is seen with little or no change in the amino terminus of all known $\alpha$ subunits, except that of the $\alpha$ subunit from *Saccharomyces cereviciae*. However, it is unlikely that this site makes contact with a reciprocal site on the $\beta\gamma$ subunit since these amino acids can be changed to neutral ones or to the opposite charge with no effect on the ability of $\alpha_o$ to bind $\beta\gamma$ (EER [7–9] QQQ; R9E). It is possible that these amino acids are necessary to maintain a secondary structure in the amino terminal region that is essential for stable heterotrimer formation. For example, such a precise orientation might be necessary to direct the myristate group to an appropriate site on the $\beta$ subunit. Finally, deletion of amino acids 7–10 may have a general effect on the protein conformation and affect a contact site at some distance from the modification.

The amino terminus of $\alpha$ is clearly not the only site of interaction with $\beta\gamma$. Analysis of chimeric proteins that combine $\alpha_i$ and $\alpha_s$ sequences suggest that disruption of sequences between residues 54 and 122 of $\alpha_i$ diminishes interaction with $\beta\gamma$ (OSAWA et al. 1990). Further evidence for additional sites comes from cross-linking experiments described below.

## 2. Analysis of $\alpha\beta$ Contact Regions by Cross-Linking

One way to characterize contact regions between polypeptides is to cross-link them chemically and then identify the sites of linkage. Analysis of $\alpha_o$ $\beta\gamma$ cross-linked with the homobifunctional cross-linking reagent bismaleimide hexane suggests that the $\alpha$ subunit interacts with the carboxyl terminal two-thirds of the $\beta$ subunit. The site of the cross-link on the $\alpha$ subunit has not yet been identified but seems not to involve the amino terminal 2-kDa fragment analyzed above. Therefore, there must be additional sites on the molecule that contact $\beta\gamma$ (YI et al. 1991).

Covalent cross-linking of $\alpha$ to $\beta\gamma$ does not prevent a GTP$\gamma$S-induced conformational change in $\alpha$. Normally, GTP$\gamma$S inhibits pertussis toxin-catalyzed ADP-ribosylation of $\alpha$ by dissociating the heterotrimer. In addition, the activated conformation of $\alpha$ may also be intrinsically less susceptible to ADP-ribosylation by pertussis toxin. The $\alpha\beta\gamma$ heterotrimer can be ADP-ribosylated by pertussis toxin after it is cross-linked (Fig. 2A). GTP$\gamma$S inhibits ADP-ribosylation of cross-linked $\beta\gamma$ even though the subunits cannot be separated (Fig. 3). YI et al. (1991) concluded that proximity of the $\beta\gamma$ subunit does not prevent $\alpha_o$ from changing conformation upon binding

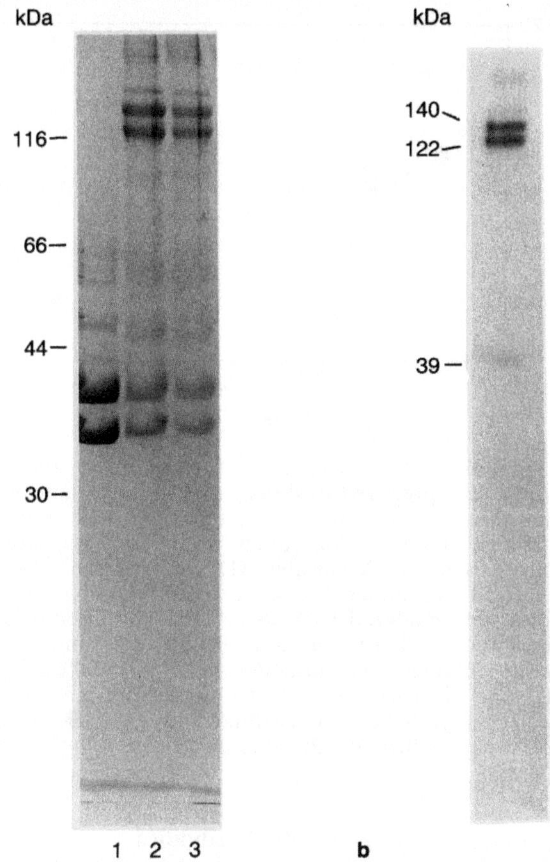


**Fig. 2. a** The $\alpha$ and $\beta$ subunits of $G_o$ can be cross-linked by bismaleimide hexane. Bovine brain $G_o$ (150 $\mu$g/ml) protein was cross-linked with 2 m*M* bismaleimide hexane for 10 min (*lane 2*) and 60 min (*lane 3*). Non-cross-linked protein is shown in *Lane 1*. Protein samples were analyzed on 11% SDS-PAGE gels stained with Coomassie blue. The evidence that the 140-kDa and 122-kDa bands contain both $\alpha$ and $\beta$ subunits is found in Yi et al. 1991. **b** The cross-linked products of $G_o$ can be ADP-ribosylated by pertussis toxin. Bovine brain $G_o$ protein (100 $\mu$g/ml) was cross-linked by 2 m*M* bismaleimide hexane for 30 min at 4°C. The cross-linked $G_o$ samples were then ADP-ribosylated by pertussis toxin as described by Yi et al. (1991). The apparent molecular weight of residual $\alpha_o$ and the cross-linked products are indicated

of GTP$\gamma$S. It is possible that the interface between $\alpha$ and $\beta\gamma$ separates upon activation even though the molecule is pinned at the site of the cross-link. The existence of an activated but incompletely dissociated form of G-protein was also suggested by Mattera et al. (1987) on the basis of hydrodynamic studies.

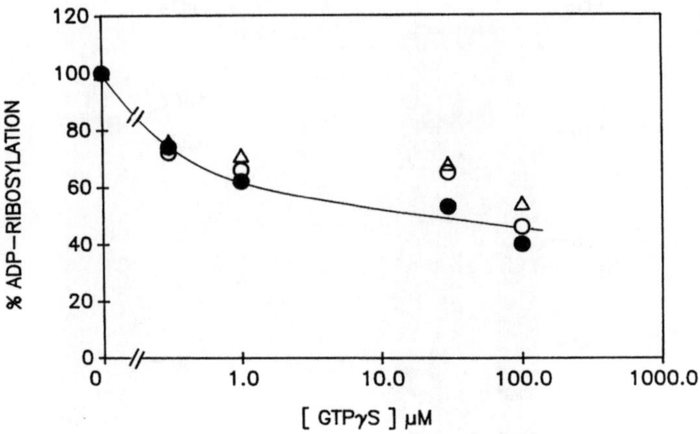


**Fig. 3.** GTPγS inhibits ADP-ribosylation of cross-linked $\alpha\beta\gamma$. Cross-linked $G_o$ proteins (100 μg/ml) were prepared as described in Fig. 2a and by Yi et al. (1991). After cross-linking the protein samples were incubated with various concentrations of GTPγS for 15 min at 23°C. Subsequently, the protein samples were ADP-ribosylated by pertussis toxin. A parallel ADP-ribosylation assay of untreated $G_o$ (20 μg/ml) was carried out under the same conditions as described above. The ribosylated samples were analyzed by SDS-PAGE. The dried gels were exposed to Kodak X-AR films, and the density of the peaks was determined by densitometry of the films. The value obtained with no added GTPγS is 100%. *Points*, the mean of four experiments. The SEM for each point ranged from ±3% to ±10%. △, Non-cross-linked $G_o$; ○, upper band of cross-linked product; ●, lower band of cross-linked product. The data shown are the average of four experiments (from Yi et al. 1991)

### 3. Probing the $\alpha$ and $\beta\gamma$ Interface with Antibodies

Analysis of the epitopes recognized by monoclonal antibodies to the $\alpha$ subunit of transducin suggests that one such antibody recognizes both the amino terminus and the carboxyl terminus (Mazzoni et al. 1991; Navon and Fung 1988). The monoclonal antibody disrupts the transducin heterotrimer and dissociates $\alpha$ from $\beta\gamma$ (Mazzoni et al. 1991). Such a result would be consistent with a model that places the two termini close to each other. Indeed, in models of $\alpha$ subunit structure based on its homology to the *ras* protein, the two termini are on the same surface of the molecule (Holbrook and Kim 1989). If the amino terminus of $\alpha$ makes contact with $\beta\gamma$, the carboxyl terminus might do so as well. However, deletion of ten amino acids from the carboxyl terminus of $\alpha_o$ does not prevent heterotrimer formation (Denker et al. 1992b). It is not clear whether the antibody needs to bind to both the N and the C terminus to disrupt the heterotrimer.

Recently, Murakami et al. (1992) raised antibodies against various regions of the $\beta$ protein to determine which one would interfere with $\alpha\beta\gamma$ trimer formation. All four antibodies interfered to some extent; therefore, it was difficult to reach precise conclusions about the parts of $\beta$ that interact with $\alpha$.

### IV. Does Dissociation of $\alpha$ and $\beta\gamma$ Occur in the Plasma Membrane?

It is very clear that in solution the $\alpha\beta\gamma$ heterotrimer dissociates upon activation by GTR$\gamma$S. For this observation to be helpful in understanding agonist activation of intracellular enzymes and channels it must also be shown to occur in the plasma membrane where the local concentration of subunits may be high enough to dampen dissociation. There the evidence is less clear, although two lines of evidence suggest that subunit dissociation can occur even in the membrane. First, addition of purified $\beta\gamma$ subunits to membranes inhibits GTP$\gamma$S stimulation of adenylyl cyclase (KATADA et al. 1984). Presumably, the subunits must dissociate for the exogenous $\beta\gamma$ to deactivate the endogenous $\alpha_s$. Second, a mutant $\alpha$ subunit that cannot dissociate from $\beta\gamma$ does not activate adenylyl cyclase when expressed in S49 cyc$^-$ cells that lack $\alpha_s$ but do have $\beta\gamma$. The mutant $\alpha$ subunit is quite competent to activate adenylyl cyclase when tested in vitro in the absence of $\beta\gamma$. Therefore, the simplest explanation for the failure to act in the cell is the inability of the subunit to dissociate from $\beta\gamma$ (LEE et al. 1992).

While the evidence suggests that dissociation can occur in the membrane, the subunits need not go very far. Indeed, the cross-linking experiments described above suggest that a limited motion at the interface may be sufficient for activation. Thus, the requirement for dissociation is not inconsistent with the mounting evidence (reviewed in Chap. 71, this volume, and by NEER 1993) that the components of the signaling system are localized to subdomains of the plasma membrane and to intracellular membranes.

## B. Interaction of $\beta$ and $\gamma$ Subunits

At least four types of $\beta$ subunit (FONG et al. 1987; GAO et al. 1987; LEVINE et al. 1990; VON WEIZSACKER et al. 1992) and four $\gamma$ subunits (HURLEY et al. 1984a; VAN DOP et al. 1984; YATSUNAMI et al. 1985; GAUTAM et al. 1989; ROBISHAW et al. 1989) have been identified. The $\beta$ and $\gamma$ subunits apparently do not dissociate once $\beta\gamma$ is formed. The subunits of purified $\beta\gamma$ can be separated only after denaturation. Multiple attempts to renature the subunits into active $\beta\gamma$ dimers have been unsuccessful (NEER, unpublished). Recent studies using recombinant $\beta$ and $\gamma$ subunits have shown that although structure of the $\beta_1$, $\beta_2$ and $\beta_3$ subunits is very highly conserved, they are not capable of forming $\beta\gamma$ dimers with any $\gamma$ subunit. SCHMIDT et al. (1991, 1992) expressed different $\beta$ and $\gamma$ subunits in vitro using the rabbit reticulocyte lysate translation system. Four assays were used to evaluate the ability of different pairs of $\beta$ and $\gamma$ subunits to form $\beta\gamma$ dimers: gel filtration to determine the Stokes radius, production of a characteristic native tryptic cleavage pattern, ability to be cross-linked by bismaleimide hexane, and ability of the dimers to interact with $\alpha$. By these criteria, the $\beta_1$ subunit can interact with both $\gamma_1$ and $\gamma_2$, the $\beta_2$ subunit can interact with $\gamma_2$ but not $\gamma_1$, the $\beta_3$ subunit is incapable of forming a $\beta\gamma$ dimer with either $\gamma_1$ or $\gamma_2$ (see

**Table 2.** Specificity of formation of $\beta\gamma$ dimers

| | $\gamma_1$ | $\gamma_2$ |
|---|---|---|
| $\beta_1$ | + | + |
| $\beta_2$ | − | + |
| $\beta_3$ | − | − |

A "+" indicates that the pair of proteins can form a dimer based on the studies of Schmidt et al. (1992) and Pronin and Gautam (1992).

Table 2). Pronin and Gautam (1992) independently reached the same conclusions by analysis of association of $\beta$ and $\gamma$ subunits cotransfected into fibroblasts.

### I. Site of Interaction of $\beta_1$ with $\gamma_1$ and $\gamma_2$

The $\beta_1$ subunit interacts with $\gamma_1$ and $\gamma_2$ through its amino terminal region. Bubis and Khorana (1990) cross-linked $\gamma_1$ to $\beta_1$ from bovine retina with copper ortho phenanthroline and found that Cys 25 of $\beta_1$ was cross-linked to Cys 36 or 37 of $\gamma_1$. Thomas et al. (1992) cross-linked bovine brain $\beta_1$ to $\gamma_2$ with bismaleimidehexane. The $\gamma$ subunit was cross-linked to an amino terminal 14-kDa tryptic fragment of $\beta$. The only cysteine available for cross-linking in $\gamma_2$ is Cys 41.

## C. Specificity of Interaction Between Particular $\alpha$ and $\beta\gamma$ Combinations

The observation that any $\beta$ does not interact with any $\gamma$ suggests that there may be more subtype specificity to the action of $\beta\gamma$ than had been previously thought. The $\beta\gamma$ subunits purified from all nonretinal sources have been considered functionally interchangeable. Dimers with different pairs of $\beta$ and $\gamma$ subunits have been isolated (Gautam et al. 1990; Fawzi et al. 1991). On closer analysis these do not have equivalent abilities to couple $\alpha$ subunits to receptors or to modulate effectors. For example, rhodopsin couples better to retinal transducin $\beta\gamma$ ($\beta_1$, $\gamma_1$) than it does to placental $\beta\gamma$ ($\beta_2$, $\gamma$?; Fawzi et al. 1991). The brain $\beta\gamma$ subunit (predominantly $\beta_1$, $\gamma_2$) can inhibit $\alpha_s$ activation of adenylyl cyclase better than retinal transducin $\beta\gamma$ (Cerione et al. 1987). The $\beta\gamma$ subunit from brain can stimulate the cardiac $K^+$ channel, but the retinal $\beta\gamma$ is inactive (Logothetis et al. 1987; Kim et al. 1989). Finally, the type II adenylyl cyclase can be stimulated by brain $\beta\gamma$, while transducin $\beta\gamma$ has no effect (Tang and Gilman 1991). Taken together, these results suggest that cellular responses to receptor activation may depend on the $\beta\gamma$ subunit combinations that are found in particular target cells.

An important unanswered question is whether or not particular $\alpha$ subunits associate preferentially with particular $\beta\gamma$ combinations. Analysis of G-proteins after solubilization suggested that different $\alpha$ subunits do not pair with identifiably different $\beta\gamma$ subunits (HILDEBRANDT et al. 1985). However, it is likely that the rapid association and dissociation of the subunits in solution would scramble and obscure any specificity that might exist in vivo. Determining whether or not specific $\alpha$ and $\beta\gamma$ combinations exist stably in the cell, and whether there are functional differences among such stable complexes are challenges for the future.

*Acknowledgments.* I would like to thank Ms. Paula McColgan for expertly typing the manuscript. I am grateful to Drs. C. Schmidt and T. Thomas for a critical reading of the paper. This work was supported by NIH grants GM36259 and GM64370.

## References

Brandt DR, Ross EM (1985) GTPase activity of the stimulatory GTP-binding regulatory protein of adenylate cyclase, $G_s$. Accumulation and turnover of enzyme-nucleotide intermediates. J Biol Chem 260:266–272

Bubis J, Khorana HG (1990) Sites of interaction in the complex between $\beta$- and $\gamma$-subunits of transducin. J Biol Chem 265:12995–12999

Buss JE, Mumby SM, Casey PJ, Gilman AG, Sefton BM (1987) Myristoylated $\alpha$ subunits of guanine nucleotide-binding regulatory proteins. Proc Natl Acad Sci USA 84:7493–7497

Casey PJ, Graziano MP, Gilman AG (1989) G protein $\beta\gamma$ subunits from bovine brain and retina: equivalent catalytic support of ADP-ribosylation of $\alpha$ subunits by pertussis toxin but differential interactions with $G_{s\alpha}$. Biochemistry 28:611–616

Cerione RA, Gierschik P, Staniszewski C, Benovic JL, Codina J, Somers R, Birnbaumer L, Spiegel AM, Lefkowitz RJ, Caron MG (1987) Functional differences in the $\beta\gamma$ complexes of transducin and the inhibitory guanine nucleotide regulatory protein. Biochemistry 26:1485–1491

Codina J, Hildebrandt J, Iyengar R, Birnbaumer L, Sekura RD, Manclark CR (1983) Pertussis toxin substrate, the putative $N_i$ component of adenylyl cyclase, is often an $\alpha\beta$ heterodimer regulated by guanine nucleotide and magnesium. Proc Natl Acad Sci USA 80:4276–4280

Codina J, Hildebrandt JD, Sekura RD, Birnbaumer M, Bryan J, Manclark CR, Iyengar R, Birnbaumer L (1984) $N_s$ and $N_i$, the stimulatory and inhibitory regulatory components of adenylyl cyclase. J Biol Chem 259:5871–5886

Codina J, Yatani A, Grenet D, Brown AM, Birnbaumer L (1987) The $\alpha$ subunit of $G_k$ opens atrial potassium channels. Science 236:442–445

Deichaite I, Casson LP, Long HP, Resh MD (1988) In vitro synthesis of pp60v$^{-src}$: myristoylation in a cell-free system. Mol Cell Biol 8:4295–4301

Denker BM, Neer EJ, Schmidt CJ (1992a) Mutagenesis of the amino terminus of the $\alpha$ subunit of the G protein $G_o$. In vitro characterization of $\alpha_o\beta\gamma$ interactions. J Biol Chem 267:6272–6277

Denker BM, Schmidt CJ, Neer EJ (1992b) Promotion of the GTP-liganded state of the $G_{o\alpha}$ protein by deletion of the C terminus. J Biol Chem 267:9998–10002

Fawzi AB, Fay DS, Murphy EA, Tamir H, Erdos JJ, Northup JK (1991) Rhodopsin and the retinal G-protein distinguish among G-protein $\beta\gamma$ subunit forms. J Biol Chem 266:12194–12200

Florio VA, Sternweis PC (1989) Mechanisms of muscarinic receptor action on $G_o$ in reconstituted phospholipid vesicles. J Biol Chem 264:3909–3915

Fong HKW, Amatruda TT III, Birren BW, Simon MI (1987) Distinct forms of the $\beta$ subunit of GTP-binding regulatory proteins, identified by molecular cloning. Proc Natl Acad Sci USA 84:3792–3796

Fung BKK (1983) Characterization of transducin from bovine retinal rod outer segments. I. Separation and reconstitution of the subunits. J Biol Chem 258: 10495–10502

Fung BKK, Nash CR (1983) Characterization of transducin from bovine retinal rod outer segments. II. Evidence for distinct binding sites and conformational changes revealed by limited proteolysis with trypsin. J Biol Chem 258:10503–10510

Gao B, Gilman AG, Robishaw JD (1987) A second form of the $\beta$ subunit of signal-transducing G proteins. Proc Natl Acad Sci USA 84:6122–6125

Gautam N, Baetscher M, Aebersold R, Simon MI (1989) A G protein gamma subunit shares homology with ras proteins. Science 244:971–974

Gautam N, Northup J, Tamir H, Simon MI (1990) G protein diversity is increased by associations with a variety of $\gamma$ subunits. Proc Natl Acad Sci USA 87:7973–7977

Haga K, Haga T (1992) Activation by G protein $\beta\gamma$ subunits of agonist- or light-dependent phosphorylation of muscarinic acetylcholine receptors and rhodopsin. J Biol Chem 267:2222–2227

Hekman M, Feder D, Keenan AK, Gal A, Klein HW, Pfeuffer T, Levitzki A, Helmreich EJ (1984) Reconstitution of $\beta$-adrenergic receptor with components of adenylate cyclase. EMBO J 3:3339–3345

Higashijima T, Ferguson KM, Sternweis PC, Smigel MD, Gilman AG (1987) Effects of $Mg^{2+}$ and the $\beta\gamma$-subunit complex on the interactions of guanine nucleotides with G proteins. J Biol Chem 262:762–766

Hildebrandt JD, Codina J, Rosenthal W, Birnbaumer L, Neer EJ, Yamazaki A, Bitensky MW (1985) Characterization by two-dimensional peptide mapping of the $\gamma$ subunits of $N_s$ and $N_i$, the regulatory proteins of adenylyl cyclase, and of transducin, the guanine nucleotide-binding protein of rod outer segments of the eye. J Biol Chem 260:14867–14872

Holbrook SR, Kim S-H (1989) Molecular model of the G protein $\alpha$ subunit based on the crystal structure of the HRAS protein. Proc Natl Acad Sci USA 86: 1751–1755

Howlett AC, Gilman AG (1980) Hydrodynamic properties of the regulatory component of adenylate cyclase. J Biol Chem 255:2861–2866

Huff RM, Axton JM, Neer EJ (1985) Physical and immunological characterization of a guanine nucleotide binding protein purified from bovine cerebral cortex. J Biol Chem 260:10864

Huff RM, Neer EJ (1986) Subunit interactions of native and ADP-ribosylated $\alpha_{41}$ and $\alpha_{39}$, two guanine nucleotide-binding proteins from bovine cerebral cortex. J Biol Chem 261:1105–1110

Hurley JB, Fong HK, Teplow DB, Dreyer WJ, Simon MI (1984a) Isolation and characterization of a cDNA clone for the $\gamma$ subunit of bovine retinal transducin. Proc Natl Acad Sci USA 81:6948–6952

Hurley JB, Simon MI, Teplow DB, Robishaw JD, Gilman AG (1984b) Homologies between signal transducing G proteins and *ras* gene products. Science 226: 860–862

Im MJ, Holzhofer A, Bottinger H, Pfeuffer T, Helmreich EJM (1988) Interactions of pure $\beta\gamma$-subunits of G-proteins with purified $\beta_1$-adrenoceptor. FEBS Lett 227:225–229

Ito H, Tung RT, Sugimoto T, Kobayashi I, Takahashi K, Katada T, Ui M, Kurachi Y (1992) On the mechanism of G protein $\beta\gamma$ subunit activation of the muscarinic $K^+$ channel in guinea pig atrial cell membrane. Comparison with the ATP-sensitive $K^+$ channel. J Gen Physiol 99:961–983

Jelsema CL, Axelrod J (1987) Stimulation of phospholipase $A_2$ activity in bovine rod outer segments by the $\beta\gamma$ subunits of transducin and its inhibition by the $\alpha$ subunit. Proc Natl Acad Sci USA 84:3623

Jones TLZ, Simonds WF, Meredino JJ Jr, Brann MR, Spiegel AM (1990) Myristoylation of an inhibitory GTP-binding protein $\alpha$ subunit is essential for its membrane attachment. Proc Natl Acad Sci USA 87:568–572

Journot L, Pantaloni C, Bockaert J, Audigier Y (1991) Deletion within the amino terminal region of $G_{s\alpha}$ impairs its ability to interact with $\beta\gamma$ subunits and to activate adenylate cyclase. J Biol Chem 266:9009–9015

Kahn RA, Gilman AG (1984) ADP-ribosylation of $G_s$ promotes the dissociation of its $\alpha$ and $\beta$ subunits. J Biol Chem 259:6235–6240

Katada T, Bokoch GM, Smigel MD, Ui M, Gilman AG (1984) The inhibitory guanine nucleotide-binding regulatory component of adenylate cyclase. Subunit dissociation and the inhibition of adenylate cyclase in S49 lymphoma $cyc^-$ and wild type membranes. J Biol Chem 259:3586–3595

Katz A, Wu D, Simon MI (1992) Subunits $\beta\gamma$ of heterotrimeric G protein activate $\beta 2$ isoform of phospholipase C. Nature 360:686–689

Kim D, Lewis DL, Graziadei L, Neer EJ, Bar-Sagi D, Clapham DE (1989) G-protein $\beta\gamma$ subunits activate the cardiac muscarinic $K^+$-channel via phospholipase $A_2$. Nature 337:557–560

Lee E, Taussig R, Gilman AG (1992) The G226A mutant $G_{s\alpha}$ highlights the requirement for dissociation of G protein subunits. J Biol Chem 267:1212–1218

Levine MA, Smallwood PM, Moen PT Jr, Helman LJ, Ahn TG (1990) Molecular cloning of beta 3 subunit, a third form of the G protein beta-subunit polypeptide. Proc Natl Acad Sci USA 87:2329–2333

Linder ME, Pang I-H, Duronio RJ, Gordon JI, Sternweis PC, Gilman AG (1991) Lipid modifications of G protein subunits. Myristoylation of $G_{o\alpha}$ increases its affinity for $\beta\gamma$. J Biol Chem 266:4654–4659

Logothetis DE, Kurachi Y, Galper J, Neer EJ, Clapham DE (1987) The $\beta\gamma$ subunits of GTP-binding proteins activate the muscarinic $K^+$ channel in heart. Nature 325:321–326

Logothetis DE, Kim D, Northup JK, Neer EJ, Clapham DE (1988) Specificity of action of guanine nucleotide-binding regulatory protein subunits on the cardiac muscarinic $K^+$ channel. Proc Natl Acad Sci USA 85:5815–5818

Mattera R, Codina J, Sekura RD, Birnbaumer L (1987) Guanosine 5′-0′ (3-thiotriphosphate) reduces ADP-ribosylation of the inhibitory guanine nucleotide-binding regulatory protein of adenylyl cyclase ($N_i$) by pertussis toxin without causing dissociation of the subunits of $N_i$. J Biol Chem 262:11247–11251

Mazzoni MR, Hamm HE (1989) Effect of monoclonal antibody binding on $\alpha\beta\gamma$ subunit interactions in the rod outer segment G protein, $G_t$. Biochemistry 28:9873–9880

Mazzoni MR, Malinski JA, Hamm HE (1991) Structural analysis of rod GTP-binding protein, $G_t$. Limited proteolytic digestion pattern of $G_t$ with four proteases defines monoclonal antibody epitope. J Biol Chem 266:14072–14081

Mumby SM, Heukeroth RO, Gordon I, Gilman AG (1990) G-protein $\alpha$-subunit expression, myristoylation, and membrane association in COS cells. Proc Natl Acad Sci USA 87:728–732

Murakami T, Simonds WF, Spiegel AM (1992) Site-specific antibodies directed against G protein $\beta$ and $\gamma$ subunits: effects on $\alpha$ and $\beta\gamma$ subunit interaction. Biochemistry 31:2905–2911

Navon SE, Fung BD (1987) Characterization of transducin from bovine retinal rod outer segments. Participation of the amino-terminal region of $T\alpha$ in subunit interaction. J Biol Chem 262:15746–15751

Navon SE, Fung BK-K (1988) Characterization of transducin from bovine retinal rod outer segments. Use of monoclonal antibodies to probe the structure and function of the $\alpha$ subunit. J Biol Chem 263:489–496

Neer EJ, Lok JM, Wolf LG (1984) Purification and properties of the inhibitory guanine nucleotide regulation unit of brain adenylate cyclase. J Biol Chem 259:14222–14229

Neer EJ, Pulsifer L, Wolf LG (1988) The amino terminus of G protein $\alpha$ subunits is required for interaction with $\beta\gamma$. J Biol Chem 263:8996–9000
Neer EJ, Wolf LG, Gill DM (1987) The stimulatory guanine-nucleotide regulatory unit of adenylate cyclase from bovine cerebral cortex. Biochem J 241:325–336
Neer EJ (1993) G proteins: Critical control points for transmembrane signals. Protein Sci (In Press)
Northup JK, Sternweis PC, Gilman AG (1983a) The subunits of the stimulatory regulatory component of adenylate cyclase. J Biol Chem 258:11361–11368
Northup JK, Smigel H, Sternweis PC, Gilman AG (1983b) The subunits of the stimulatory regulatory component of adenylate cyclase. J Biol Chem 258: 11369–11376
Osawa S, Dhanasekaran N, Woon CW, Johnson GL (1990) $G\alpha_i$-$G\alpha_s$ chimeras define the function of $\alpha$ chain domains in control of G protein activation and $\beta\gamma$ subunit complex interactions. Cell 63:697–706
Pfeuffer T (1979) Guanine nucleotide-controlled interactions between components of adenylate cyclase. FEBS Lett 101:85–89
Pitcher JA, Inglese J, Higgins JB, Arriza JL, Casey PJ, Kim C, Benovic JL, Kwatra MM, Caron MG, Lefkowitz RJ (1992) Role of $\beta\gamma$ subunits of G proteins in targeting the $\beta$-adrenergic receptor kinase to membrane-bound receptors. Science 257:1264–1267
Robishaw JD, Kalman VK, Moomaw CR, Slaughter CA (1989) Existence of two $\gamma$ subunits of the G proteins in brain. J Biol Chem 264:15758–15761
Schmidt CJ, Neer EJ (1991) In vitro synthesis of G protein $\beta\gamma$ dimers. J Biol Chem 266:4538–4544
Schmidt CJ, Thomas TC, Levine MA, Neer EJ (1992) Specificity of G protein $\beta\gamma$ subunit interactions. J Biol Chem 267:13807–13810
Smrcka AV, Hepler JR, Brown KO, Sternweis PC (1991) Regulation of polyphosphoinositide-specific phospholipase C activity by purified $G_q$. Science 251: 804–807
Sternweis PC, Northup JK, Smigel MD, Gilman AG (1981) The regulatory component of adenylate cyclase. Purification and properties. J Biol Chem 256: 11517–11526
Tang WJ, Gilman AG (1991) Type-specific regulation of adenylyl cyclase by G protein $\beta\gamma$ subunits. Science 254:1500–1503
Thomas TC, Sladek T, Yi F, Smith T, Neer EJ (1993) Structural analysis of the G protein $\beta\gamma$ subunit, Biochemistry (In Press)
Tsai SC, Adamik R, Kanaho Y, Hewlett EL, Moss J (1984) Effects of guanyl nucleotides and rhodopsin on ADP-ribosylation of the inhibitory GTP-binding component of adenylate cyclase by pertussis toxin. J Biol Chem 259:15320–15323
Van Dop C, Medynski D, Sullivan K, Wu AM, Fung BKK, Bourne HR (1984) Partial cDNA sequence of the $\gamma$ subunit of transducin. Biochem Biophys Res Commun 124:250–255
von Weizsäcker E, Strathmann MP, Simon MI (1992) Diversity among the beta subunits of heterotrimeric GTP-binding proteins: characterization of a novel beta-subunit cDNA. Biochem Biophys Res Commun 183:350–356
Yamazaki A, Tatsumi M, Torney DC, Bitensky MW (1987) The GTP-binding protein of rod outer segments. I. Role of each subunit in the GTP hydrolytic cycle. J Biol Chem 262:9316–9323
Yatsunami K, Pandya BV, Oprian DD, Khorana HG (1985) cDNA-derived amino acid sequence of the $\gamma$ subunit of GTPase from bovine rod outer segments. Proc Natl Acad Sci USA 82:1936–1940
Yi F, Denker BM, Neer EJ (1991) Structural and functional studies of cross-linked $G_o$ protein subunits. J Biol Chem 266:3900–3906

CHAPTER 49

# G-Protein $\alpha$ Subunit Chimeras Reveal Specific Regulatory Domains Encoded in the Primary Sequence

M. RUSSELL and G.L. JOHNSON

## A. Background

Within the GTPase family of proteins the members referred to as G-proteins provide a signal transduction coupling mechanism for many cell surface receptors as described in other chapters in this volume. G-proteins are responsible for regulating an intracellular effector, such as an ion channel or an enzyme, in response to an activated receptor (JOHNSON and DHANASEKARAN 1989). G-proteins exist as heterotrimers composed of $\alpha$, $\beta$, and $\gamma$ subunits. The G-protein $\alpha$ subunit is the component that binds GDP and GTP. Receptors coupled to specific G proteins catalyze GDP dissociation, allowing GTP to bind. The GTP$\alpha$ complex in turn regulates the activity of specific effectors. The lifetime of the activated GTP$\alpha$ complex is controlled by an intrinsic GTPase encoded in the $\alpha$ subunit which hydrolyzes the bound GTP to GDP ($\alpha$GDP) returning the $\alpha$ subunit to an inactive state.

The $\beta\gamma$ subunit complex exists as a tightly associated dimer and has three defined functions in regulating the activation of $\alpha$ subunits: (a) in the absence of activated receptor $\beta\gamma$ subunits attenuate G-protein activity by inhibiting GDP dissociation, thereby stabilizing the inactive GDP bound $\alpha$ subunit, (b) the complex is required for efficient coupling of receptors to the $\alpha$ subunit for catalysis of GDP dissociation, and (c) pertussis toxin recognition of $\alpha_i$-like polypeptides is greatly enhanced in the presence of the $\beta\gamma$ complex. A growing awareness of signaling functions for $\beta\gamma$ has recently developed, suggesting that the roles for $\beta\gamma$ complexes in signal transduction pathways is probably much greater than previously appreciated (BLUMER and THORNER 1991; KIM et al. 1989; LOGOTHETIS et al. 1987; TANG and GILMAN 1991).

Much more is known about the role of G-protein $\alpha$ subunits in regulating effector enzyme activities. Among the 20 or so G-protein $\alpha$ subunits that have been cloned there are both common and unique functions encoded in the structure of the $\alpha$ chain polypeptide. Common features include the functions involved in regulation of the $\alpha$ subunit itself: (a) GDP/GTP binding, (b) intrinsic GTPase activity, and (c) binding sites for association with $\beta\gamma$ subunits. Unique functions for each $\alpha$ chain include: (a) selectivity for regulating specific effectors and (b) selectivity for coupling to specific receptors. The unique functions for each $\alpha$ subunit obviously allow selectivity in cellular responses to the environment (hormones and neurotransmitters) and the regulation of specific physiological signal functions (ion channels and enzymes).

Predictably, $\alpha$ subunit sequences involved in two of the common functions for all G-proteins (GDP/GTP binding, GTPase activity) are highly conserved at the amino acid level. The sequences involved in contact and regulation by $\beta\gamma$ complexes are less apparent in the primary sequence of G-protein $\alpha$ subunits. The sequences involved in the unique functions of receptor and effector interaction and regulation are also predictably diverse and nonconserved.

Two G-protein members, $G_s$ and $G_i$, regulate the common effector adenylyl cyclase. $G_s$ activates and $G_i$ inhibits adenylyl cyclase activity. $G_s$ and $G_i$ also preferentially couple to different receptors. For this reason the $\alpha_s$ and $\alpha_i$ polypeptides have proven extremely valuable for characterizing the common and unique function of G protein $\alpha$ subunits by chimera and amino acid mutation analysis. In this chapter we summarize the understanding of $\alpha$ subunit domains and regulation that has been achieved by genetic manipulation of the $\alpha_s$ and $\alpha_i$ subunit sequences.

## B. Mutational Analysis of the GDP/GTP Binding Domain

The *ras* proto-oncogenes, encoding 21-kDa proteins designated as $p21^{ras}$, share sequence homology with a family of GTP-binding proteins which includes the G-proteins. Typically the members of this family are active with GTP bound and inactive with GDP bound (GILMAN 1987). Since there is currently no crystal structure for G-protein $\alpha$ polypeptides, understanding the $\alpha$ chain GTP/GDP binding site three-dimensional structure has relied upon the conservation between G-proteins $p21^{ras}$ and EF-Tu where crystal structure does exist for $p21^{ras}$ and EF-Tu (JURNACK 1985; DE VOS et al. 1988; PAI et al. 1988; SCHLICHTING et al. 1989). Consensus sequences induced in GTP/GDP binding, termed G1 through G5, are found dispersed throughout the GTPase polypeptide (BOURNE et al. 1991a,b). These sequences are highly conserved among all G-protein $\alpha$ subunits identified to date.

The activated GTP-bound $\alpha$ subunit hydrolyzes the bound GTP to GDP, returning the $\alpha$ subunit to an inactive state. Mutations that block this GTPase activity stabilize the $\alpha$ subunit polypeptide in a GTP-bound state. Frequently, GTPase inhibiting mutations result in constitutive activation of the $\alpha$ subunit. GTPase-inhibiting mutations in $p21^{ras}$, for example, are associated with many forms of cancer (BOS 1989). These mutations stabilize $p21^{ras}$ in an activated, GTP-bound state that promotes proliferation (BARBACID 1987). In mammals activation of $p21^{ras}$ produced by oncogenic mutations appears to be an early step in malignant transformation of cells. Activated $p21^{ras}$ proteins promote cellular proliferation that in turn allows

**Table 1.** Common mutations in GDP-/GTP-binding domains of p21*ras* and $\alpha_s$

| Domain | | GTPase activity | Transformation | Adenylyl cyclase stimulation |
|---|---|---|---|---|
| G1 | | | | |
| p21$^{ras}$ | Gly-12 → Val | Inhibited | ++ | NA |
| $\alpha_s$ | Gly-49 → Val | Inhibited | NA | ++ |
| G3 | | | | |
| p21$^{ras}$ | Ala-59 → Thr | Inhibited | ++ | NA |
| $\alpha_s$ | Gly-225 → Thr | Unable to assume proper GTP conformation | NA | Inhibited (behaves as a competitive inhibitory $\alpha_s$) |
| p21$^{ras}$ | Gly-60 → any a.a. | Unable to assume GTP conformation | – (inactive) | NA |
| $\alpha_s$ | Gly-225 → Ala | Unable to assume GTP conformation | NA | Inhibited (behaves as a competitive inhibitory $\alpha_s$ or inactive) |
| p21$^{ras}$ | Gln-61 → Leu | Inhibited | ++ | NA |
| $\alpha_s$ | Gln-227 → Leu | Inhibited | NA | ++ |
| G4 | | | | |
| p21$^{ras}$ | Asp-119 → Ala | Decreases affinity for GDP and GTP, GTPase normal | ++ | NA |
| $\alpha_s$ | Asp-295 → Ala | Decreases affinity for GDP and GTP, GTPase normal | NA | Null |
| G5 | | none | – | – |

NA, Not applicable. Mutations in the G2 domain have not been compared between p21$^{ras}$ and $\alpha_s$; however, mutation of $\alpha_s$ Arg-201 → Cys or Pro inhibits GTPase activity.

further genetic defects to accumulate, resulting in eventual transformation (KUMAR et al. 1990). Mutations in G-protein coupled systems, including but not limited to adenylyl cyclase and cAMP, can similarly induce hyperplasia and contribute to neoplastic transformation of certain cell types. The mutations in p21*ras* and G-protein $\alpha$ subunits leading to constitutive activation and contributing to neoplasia occur at conserved amino acids in the polypeptides.

Specific mutations in the GDP-/GTP-binding domains of p21*ras* and analogous mutations in $\alpha_s$ are outlined in Table 1. In the G1 domain the consensus sequence is GXXXXGK(S/T), and it contains the binding loop for $\alpha$ and $\beta$ phosphates of GDP and GTP. Mutation of p21*ras* Gly-12 to Val (G12V) in G1 inhibits GTPase activity and causes transformation (GIBBS et al. 1984; MCGRATH et al. 1984). The analogous mutation in $\alpha_s$ Gly-49 to Val (G49V) also inhibits GTPase activity and stimulates adenylyl cyclase activity (GRAZIANO and GILMAN 1989; JOHNSON et al. 1991; MASTERS et al. 1989; WOON et al. 1989). The G2 domain contains a highly conserved threonine residue which may interact with the magnesium ion bound to the $\beta$ and $\gamma$ phosphates of the guanine nucleotide. The G3 domain contains the sequence DVGGQ encoded in all $\alpha$ subunits and contains an invariant aspartate which binds the catalytic $Mg^{+2}$ ion through an $H_2O$ molecule and a glycine which forms a hydrogen bond with the GTP$\gamma$ phosphate. Also the second glycine functions as a "pivot" for conformational change induced by GTP.

Of the several mutations made in the G3 domain of p21*ras* and $\alpha_s$ the p21*ras* Gln-61 to Leu (Q61L) and $\alpha_s$ Gln-227 to Leu (Q227L) mutations both strongly inhibit GTPase activity. This inhibition is much stronger than that seen with p21*ras* G12V and $\alpha_s$ G49V mutations within the G1 domain of the respective polypeptides. Another mutation in the G3 domain of p21*ras* at residue Ala-59, A59T, inhibits GTPase activity and causes transformation while the analogous $\alpha_s$ mutation at residue Gly-225, G225T, is not an activating mutation (OSAWA and JOHNSON 1991). This $\alpha_s$ mutation behaves as a competitive inhibiting mutant polypeptide which decreases adenylyl cyclase activity; thus not all activating mutations in p21*ras* are activating in $\alpha_s$. A third mutation in the G3 domain of p21*ras* Gly-60 to any other amino acid and an analogous $\alpha_s$ mutation at residue Gly-226, G226A, are unable to assume proper GTP conformations and are therefore unable to activate their respective effectors. The mutant $\alpha_s$ Q226A polypeptide behaves as a selective inhibitory mutant polypeptide (OSAWA and JOHNSON 1991).

The G4 domain which contains the sequence NKXD has an aspartate which forms a hydrogen bond with the guanine ring and amides of asparagine and lysine to help stabilize GDP/GTP binding. Both the mutant p21*ras* D119A and the analogous mutant $\alpha_s$ D295A have decreased affinity for GDP and GTP, but their GTPase activities are normal (OSAWA and JOHNSON 1991; SIGAL et al. 1986). The fact that this mutation in p21*ras* causes transformation even though there is reduced affinity for GDP and GTP can be attributed to an increased dissociation rate for GDP, allowing GTP to bind and activate $p21^{ras}$ (SIGAL et al. 1986). The corresponding mutation in $\alpha_s$ D295A has no significant influence on the ability of the mutant $\alpha_s$ to stimulate adenylyl cyclase activity relative to wild type $\alpha_s$ (OSAWA and JOHNSON 1991). The failure of the $\alpha_s$ D295A mutation to activate adenylyl cyclase, unlike the activating nature of the $p21^{ras}$ D119A mutation, is probably related to the intrinsically high GTPase activity of $\alpha_s$ relative to $p21^{ras}$ (GRAZIANO and GILMAN 1989), maintaining the mutant $\alpha_s$ in a predominantly inactive, GDP-bound state. The G5 domain is less conserved but has a general consensus of TCAXDT in $\alpha$ subunits and is predicted to interact indirectly with the guanine nucleotide to stabilize its binding. Mutations in the G5 region have not been characterized.

As mentioned above, several of the differences seen in analogous mutations between $p21^{ras}$ and $\alpha_s$ may result from intrinsic differences in the GTPase regulatory properties of $p21^{ras}$ and G-protein $\alpha$ chains. The $p21^{ras}$ polypeptide has a low intrinsic GTPase activity but in the presence of the GTPase activating protein (GAP) the GTPase activity is increased (TRAHEY and MCCORMICK 1987). In contrast, the G-protein $\alpha$ chain has a high intrinsic GTPase activity that does not require a GAP-like protein. The region in G-protein $\alpha$ chains responsible for the high GTPase activity has been putatively mapped to a region that surrounds Arg-201, which is the site ADP-ribosylated in $\alpha_s$ by cholera toxin (CASEY and GILMAN 1988). ADP-ribosylation of Arg-201 results in inhibition of the intrinsic $\alpha_s$ GTPase activity. Additionally, mutation of Arg-201 to almost any other amino acid also results in inhibition of the intrinsic GTPase activity despite the fact that this region is not directly involved in GTP/GDP binding (CASEY and GILMAN 1988). It has been postulated that this region of the polypeptide is controlling a GAP function which is intrinsic to G-protein subunit polypeptides (LANDIS et al. 1989). Supporting this notion is the finding that residues which surround Arg-201 (189–203) share homolgy with a putative GAP-binding site in $p21^{ras}$ (MCCORMICK 1989).

The similarities between Ras and G-protein $\alpha$ subunits led to the idea that $p21^{ras}$ normally functions in regulating signal transducers (BARBACID 1987). In this model, signals originate at some form of a detector element, such as a tyrosine kinase encoded growth factor receptor, which organizes a proliferative signal, via $p21^{ras}$, to a downstream effector, which in turn generates a signal or second messenger. Mutations in $p21^{ras}$ would leave this system, and its signals would be constitutively activated. Similarly in the G-protein system signals transduced from an activated receptor require a G-protein coupled to an intracellular effector. Mutations which activate the G-protein also leave the receptor-regulated signal pathway constitutively activated. Several mutations which inhibit the intrinsic GTPase activity of the $\alpha_s$ polypeptide predictably produce a constitutively activated G-protein coupled system. Just as with activating mutations of $p21^{ras}$, GTPase-inhibiting mutations in G-protein $\alpha$ subunits can have major phenotypic consequences. In fact, experimental evidence indicates that GTPase-inhibited mutant $\alpha_s$ and $\alpha_{i2}$ polypeptides can function as oncoproteins in specific cell types similar to activated p21$^{ras}$. Experimental evidence to support the hypothesis that GTPase deficient $\alpha$ subunits of G-proteins can behave as oncogenes includes: (a) GTPase-inhibiting mutations in the $\alpha$ subunit of $G_s$ have been found in clonally derived tumors of pituitary somatotroph (LANDIS et al. 1989; CLIMENTI et al. 1990) and thyrotroph (LYONS et al. 1990; SUAREZ et al. 1991) origin. Similar mutations in the $\alpha$ subunit of $G_{i2}$ have been identified in tumors derived from adrenal cortex and ovarian endocrine cells (LYONS et al. 1990). (b) GTPase-deficient $\alpha_{i2}$ subunits induce a transformed phenotype in Rat 1 cells (PACE et al. 1991; GUPTA et al. 1991a). (c) NIH3T3 cells may be transformed by the ectopic expression of hormone or neurotransmitter receptors that couple to the hydrolysis of phosphatidylinositol via G-proteins (presumably the $G_q$ family), but require the continued presence of receptor activation to maintain the transformed phenotype (JULIUS et al. 1989; GUTKIND et al. 1991).

Activating mutations in specific G-protein $\alpha$ subunits are probably involved in several other human diseases. For example, McCune-Albright syndrome results from somatic mutation of $\alpha_s$ in affected tissues (WEINSTEIN et al. 1991). The mutation characterized in tissues of McCune-Albright patients is $\alpha_s$ R201C which inhibits the intrinsic GTPase activity of the $\alpha_s$ polypeptide.

## C. Competitive Inhibitory Mutations

The sequence Asp-223 to Gln-227 (GDP-/GTP-binding domain G3) in the $\alpha_s$ polypeptide is predicted to interact with the $\gamma$ phosphate of GTP and be involved in mediating a conformational change required for activation of $\alpha_s$. Mutation of Gly-225 to Thr produces a mutant $\alpha_s$ polypeptide ($\alpha_s$G225T) which when expressed in COS-1 or CHO cells markedly diminishes the ability of the $\beta$-adrenergic receptor to activate adenylyl cyclase. The mutant $\alpha_s$ inhibits, in a concentration-dependent manner, $\beta$-adrenergic receptor stimulation of cAMP synthesis (OSAWA and JOHNSON 1991). The G225T mutation in $\alpha_s$ apparently alters the ability of GTP to activate the polypeptide and when overexpressed this inactive mutant $\alpha_s$ exerts an inhibitory phenotype by competing with the endogenous wild type $\alpha_s$ subunit for interaction and activation of adenylyl cyclase. Another mutation in this region, Gly-226 to Ala, also produces a mutant $\alpha_s$ which when expressed in COS-1 cells is incapable of activating adenylyl cyclase (OSAWA and JOHNSON 1991). The mutation G226A was originally identified as a spontaneous $\alpha_s$ mutant in S49 mouse lymphoma cells which was incapable of dissociating from $\beta\gamma$ and activating adenylyl cyclase in the presence of GTP (MILLER et al. 1988). The glycine at residue 226 has a side chain proton which is necessary for allowing the predicted conformational change induced by GTP binding immediately adjacent to this glycine, which is essential for assuming an activated $\alpha$ subunit state capable of regulating effector activity. Replacement of Gly-226 with an amino acid having a bulky side chain apparently causes a loss in the ability to assume the proper GTP conformation because the polypeptide is prevented from "pivoting" in response to the binding of GTP. The steric hindrance inhibits the conformational change necessary for dissociation from the $\beta\gamma$ subunit complex and activation of adenylyl cyclase. In contrast to the $\alpha_s$G226A mutation is the adjacent Q227L mutation, which strongly activates adenylyl cyclase. The Q227L mutation helps to stabilize the conformation induced by GTP and inhibits GTPase activity, conferring a constitutively activated state. The difference between the inhibitory phenotype of the G226A and G225T mutations and the strong activating property of the Q227L mutation is the influence of the different amino acid changes within the G3 domain to prevent or stabilize the GTP-activated conformation.

The properties of the $\alpha_s$G225T mutation were distinguished from the G226A mutation by secondary mutations in the $\alpha_s$ N terminus. Mutation of the $\alpha_s$ N terminus was achieved using a chimeric $\alpha$ subunit ($\alpha_{i(54)/s}$) which has the first 61 amino acid of $\alpha_s$ substituted with the N terminal 54 residues of $\alpha_{i2}$ (OSAWA et al. 1990a). The $\alpha_{i(54)/s}$ chimera behaves as a constitutively active $\alpha_s$ polypeptide which has an enhanced rate of GDP dissociation resulting from the mutations introduced at the N terminus of the polypeptide (OSAWA et al. 1990a; GUPTA et al. 1991b). The mutations leading to enhanced GDP dissociation from the $\alpha_{i(54)/s}$ sequence in combination with the $\alpha_s$G225T mutant polypeptide ($\alpha_{i(54)/sG225T}$) results in an $\alpha$ chain which functionally activates adenylyl cyclase to levels similar to that for $\alpha_{i(54)/s}$ alone. In contrast, the $\alpha_{i(54)/s}$ sequence in combination with $\alpha_s$G226A ($\alpha_{i(54)/sG226A}$) is unable to rescue the inactive phenotype of the mutant $\alpha_s$G226A polypeptide. The failure of the $\alpha_{i(54)/s}$ sequence to rescue the activity of the Gly-226 to Ala mutation indicates that Gly-226 is essential for $\alpha_s$ to assume a GTP-induced active conformation. In contrast, Gly-225 is not critical as a pivot point in the GTP-induced conformational changes of the polypeptide because a second mutation compensates for the amino acid substitution. Rather, Gly-225 in the $\alpha_s$ polypeptide must be involved in the tertiary structural interactions that regulate GTP binding and activation of adenylyl cyclase but is not an essential pivot for GTP-induced conformational changes.

## D. Regulatory Properties of the $\alpha_s$ N Terminus

To identify regions of $\alpha$ subunit primary sequence important for regulation and that impart the unique properties of the different G-proteins, a series of $\alpha_{i2}/\alpha_s$ chimeras were generated (OSAWA et al. 1990a). The $\alpha_{i2}$ and $\alpha_s$ polypeptides, respectively, inhibit and stimulate adenylyl cyclase.

Experimentally the chimeric proteins are transiently expressed in COS-1 cells as a simple, reproducible measure of the ability of mutant $\alpha_s$ polypeptides to regulate adenylyl cyclase activity. Regulatory functions controlling GTP activation of adenylyl cyclase can then be readily compared among the different chimeras and the wild-type $\alpha_{i2}$ and $\alpha_s$ polypeptides.

The $G_{i2}$- and $G_s$-proteins also couple to different receptors, and the receptor selectivity is a function of the $\alpha_s$ and $\alpha_{i2}$ polypeptides. In addition, cholera toxin ADP-ribosylates $\alpha_s$ but not $\alpha_{i2}$, whereas pertussis toxin ADP-ribosylates $\alpha_{i2}$ but not $\alpha_s$. For these reasons, appropriate $\alpha_{i2}/\alpha_s$ chimeras have the potential to switch functional domains as well as to introduce multiple nonconserved amino acid substitutions within unique domains of the two $\alpha$ subunit polypeptides.

## E. $\alpha_{i2}/\alpha_s$ Chimeras Reveal the Regulatory Function of the $\alpha$ Subunit N Terminus

A series of $\alpha_{i2}/\alpha_s$ chimeras were constructed having specific $\alpha_s$ N terminal sequences substituted with the corresponding region of the $\alpha_{i2}$ polypeptide. For example, $\alpha_{i(7)/s}$ has the first seven amino acids of $\alpha_{i2}$ replacing the corresponding sequence in $\alpha_s$, where the remaining sequence encodes the wild-type $\alpha_s$ polypeptide; $\alpha_{i(17)/s}$ has the first 17 amino acids of $\alpha_{i2}$ replacing the corresponding N terminal sequence of $\alpha_s$, etc. Figure 1 shows the ability of the different $\alpha_{i2}/\alpha_s$ chimeras to activate adenylyl cyclase relative to the wild $\alpha_s$ and $\alpha_{i2}$ polypeptides. A rather striking result is apparent in comparison of the N terminal $\alpha_{i2}/\alpha_s$ chimeras. Substitution of the first 34 residues of $\alpha_s$ with the corresponding 27 residues of $\alpha_{i2}$ (there are seven unique $\alpha_s$ amino acids at the N terminus which are absent in $\alpha_{i2}$) has no discernible influence on the ability of the $\alpha_{i2}/\alpha_s$ chimeras ($\alpha_{i(7)/s}$, $\alpha_{i(17)/s}$, and $\alpha_{i(27)/s}$) to activate adenylyl cyclase relative to that observed with the wild-type $\alpha_s$ polypeptides. Further mutation of the $\alpha_s$ N terminus by $\alpha_{i2}$ sequence substitution with the $\alpha_{i(34)/s}$, $\alpha_{i(54)/s}$, and $\alpha_{i(64)/s}$ chimeras leads to a strong activation of adenylyl cyclase activity in the absence of hormonal stimulation. Additional $\alpha_{i2}$ sequence beyond the first 64 amino acids encoded in $\alpha_{i2}/\alpha_s$ chimeras $\alpha_{i(94)/s}$, $\alpha_{i(109)/s}$, $\alpha_{i(122)/s}$, and $\alpha_{i/s(Bam)}$ return the chimeric polypeptides to basal wild-type $\alpha_s$ activity in their ability to stimulate adenylyl cyclase and increase cAMP synthesis. The activated phenotype of the $\alpha_{i(34)/s}$, $\alpha_{i(54)/s}$, and $\alpha_{i(64)/s}$ chimeras contrasts with the chimeras of shorter or longer $\alpha_{i2}$ substitutions in the $\alpha_s$ polypeptide. The expression of each of the chimeras relative to the wild-type $\alpha_s$ polypeptide is similar, indicating that the increased adenylyl cyclase activity is due to the activated state of the $\alpha_{i(34)/s}$, $\alpha_{i(54)/s}$, and $\alpha_{i(64)/s}$ polypeptides and not due to differences in expression levels.

We have shown that the activated character of the $\alpha_{i(54)/s}$ and $\alpha_{i(64)/s}$ chimeras is the result of enhanced GDP dissociation, allowing GTP

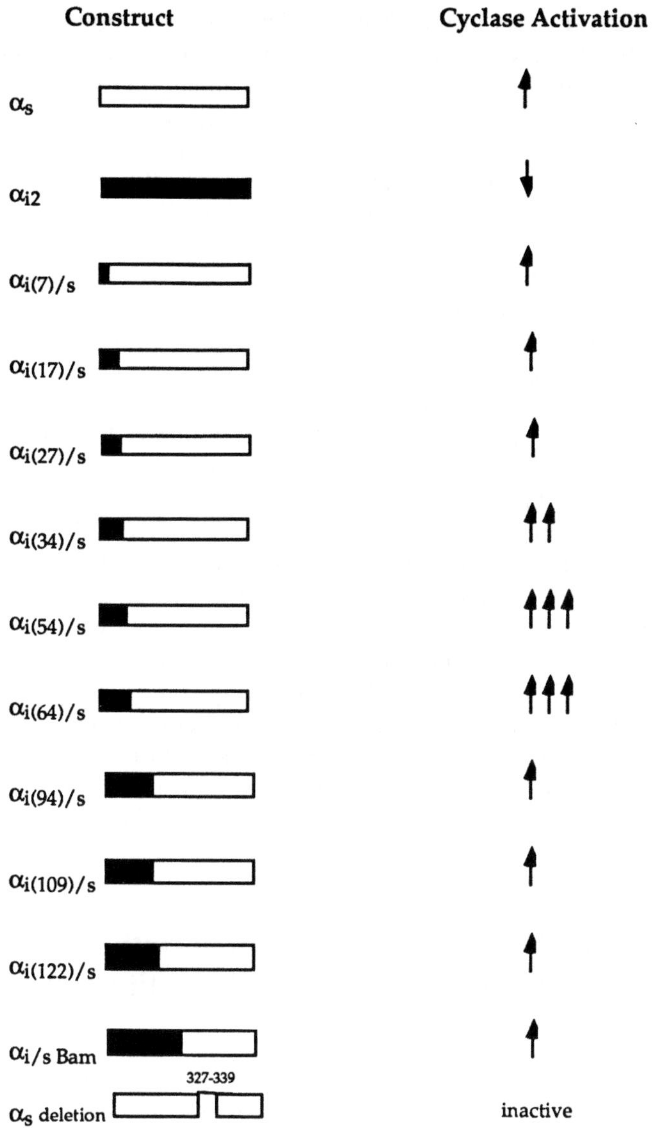


**Fig. 1.** Effect of $\alpha_{i2}/\alpha_s$ chimeras on adenylyl cyclase activation. The relative contribution of $\alpha_s$ (*open bars*) and $\alpha_{i2}$ (*black bars*) is shown for each $\alpha_{i2}/\alpha_s$ chimera. Chimeras were transiently expressed in COS cells, and cAMP levels were measured as an indication of adenylyl cyclase activation in the transfected cells. The ability of each construct to regulate adenylyl cyclase activity is designated inhibition (*down arrow*) or stimulation (*up arrow*). *Multiple arrows*, relative ability of stimulatory mutants to activate adenylyl cyclase in comparison to $\alpha_s$ (*single arrow*). The $\alpha_{i(34)/s}$, $\alpha_{i(54)/s}$, and $\alpha_{i(64)/s}$ chimeras have greater intrinsic adenylyl cyclase activation properties relative to the $\alpha_s$ polypeptide. All of the other chimeras are functionally similar to $\alpha_s$. Levels of expression for each construct were similar as determined by immunoblotting (not shown)

activation of the $\alpha$ chain in the absence of hormonal stimulation (OSAWA et al. 1990a; DHANASEKARAN et al. 1991). Cumulatively, the activity of the various chimeras indicate that the N terminal moiety of $\alpha_{i2}$ and $\alpha_s$ may be interchanged with normal maintenance of intrinsic $\alpha_s$ regulation (i.e., $\alpha_{i/s(Bam)}$ chimera). Appropriate mutation within the $\alpha$ subunit N terminal moiety results in loss of an attenuator function controlling GDP dissociation, a limiting step necessary for subsequent GTP binding and stimulation of adenylyl cyclase.

The findings with the $\alpha_{i2}/\alpha_s$ chimeras demonstrate that deletion of the unique $\alpha_s$ sequences Leu-4 to Gln-10 and Gly-72 to Gly-86, which are absent in the corresponding region of $\alpha_{i2}$, are not responsible for the $\alpha_{i(34)/s}$, $\alpha_{i(54)/s}$, and $\alpha_{i(64)/s}$ phenotypes. In addition, the $\alpha_{i(94)/s}$ chimera behaves similar to $\alpha_s$ and $\alpha_{i/s(Bam)}$ in its ability to activate adenylyl cyclase, indicating that the $\alpha_s$ sequences distal to Pro116 are not directly involved in the activated phenotype observed with the $\alpha_{i(34)/s}$, $\alpha_{i(54)/s}$, and $\alpha_{i(64)/s}$ chimeras. Thus, the domain controlling GDP dissociation maps to within the boundaries of the $\alpha_{i(27)/s}$ and $\alpha_{i(94)/s}$ chimeras that correspond to $\alpha_s$ residues Gly-35 to Pro-116. Within this region, residues corresponding to $\alpha_s$Arg-42 to Arg-62 and $\alpha_{i2}$Lys-35 to Lys-55 are identical except for the two Arg to Lys conserved substitutions. The G1 phosphate-binding sequence lies within this region. Our unpublished studies indicate that residues directly N terminal and including $\alpha$Arg-42 are critical in generating the activated phenotype observed with the $\alpha_{i(34)/s}$, $\alpha_{i(54)/s}$, and $\alpha_{i(64)/s}$ chimeras, suggesting this region is dominant in the regulation of GDP dissociation. It remains to be determined how the $\beta\gamma$ subunit complex and activated hormone receptors influence this region to regulate GDP dissociation from the wild-type $\alpha_s$ polypeptide.

## F. Mutations that Influence GDP Dissociation and GTPase Activity Create Strong Constitutively Active $\alpha_s$ Polypeptides

The two rate-limiting steps controlling the activity of G protein $\alpha$ subunits are GDP dissociation allowing GTP binding and activation and the intrinsic GTPase activity encoded in the polypeptide that returns the activated $\alpha_{.GTP}$ complex to an inactive $\alpha_{.GDP}$ state. If the regulation of both processes can be disrupted in a single polypeptide chain, a strong constitutively active $\alpha_s$ would be predicted. The point mutations and chimeras described above provide a strategy to disrupt both the regulation of GDP dissociation and inhibit GTPase activity in a single $\alpha_s$ polypeptide. To accomplish this, the single amino acid mutation Q227L in the $\alpha_s$ polypeptide was used to inhibit the GTPase activity (see above). The GTPase-inhibiting $\alpha_s$Q227L mutation was placed in combination with a number of $\alpha_{i2}/\alpha_s$ N terminal chimeras (Fig. 2), one of which we have characterized ($\alpha_{i(54)/s}$) to have an accelerated rate

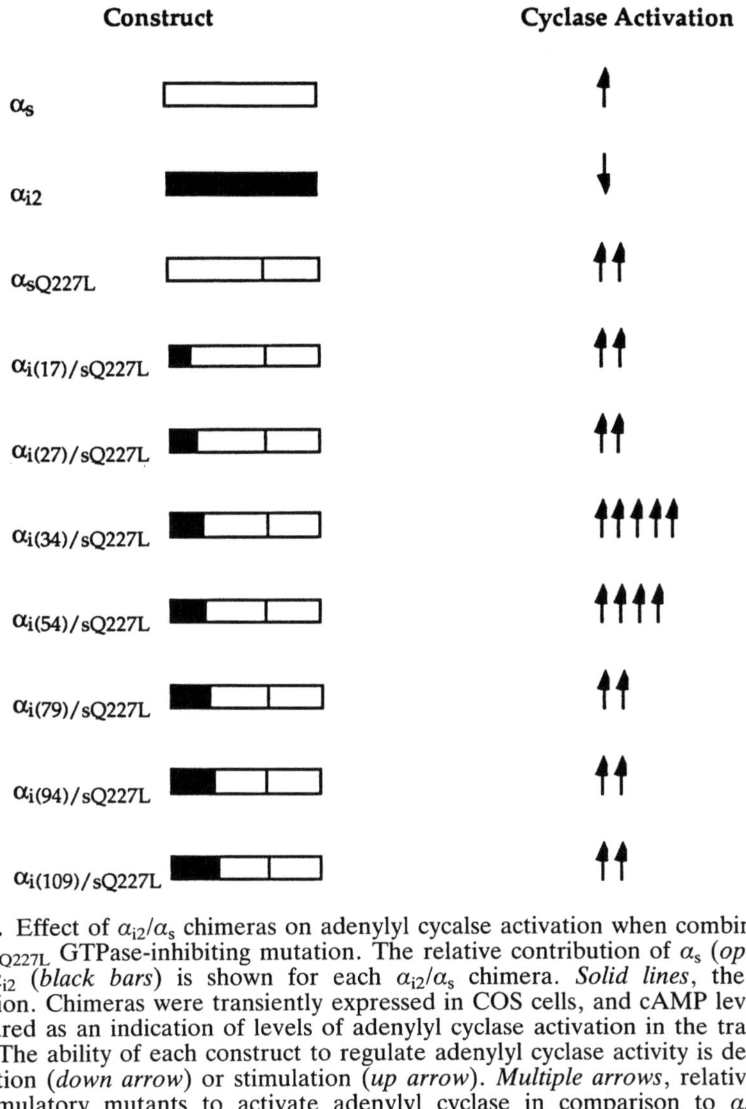


**Fig. 2.** Effect of $\alpha_{i2}/\alpha_s$ chimeras on adenylyl cycalse activation when combined with the $\alpha_{sQ227L}$ GTPase-inhibiting mutation. The relative contribution of $\alpha_s$ (*open bars*) and $\alpha_{i2}$ (*black bars*) is shown for each $\alpha_{i2}/\alpha_s$ chimera. *Solid lines*, the $\alpha_{sQ227L}$ mutation. Chimeras were transiently expressed in COS cells, and cAMP levels were measured as an indication of levels of adenylyl cyclase activation in the transfected cells. The ability of each construct to regulate adenylyl cyclase activity is designated inhibition (*down arrow*) or stimulation (*up arrow*). *Multiple arrows*, relative ability of stimulatory mutants to activate adenylyl cyclase in comparison to $\alpha_s$ (*single arrow*). The $\alpha_{i(54)/s}$ and $\alpha_{sQ227L}$ mutations are basically additive in their ability to activate cAMP synthesis when placed in the same polypeptide. The $\alpha_{i(34)/s}$ chimera is weakly activating by itself, but when placed with the Q227L mutation in the same $\alpha_s$ polypeptide strongly potentiates activation of cAMP synthesis

of GDP dissociation and a wild-type GTPase activity (Osawa et al. 1990b, and unpublished results). The $\alpha_{i(34)/s/Q227L}$ and $\alpha_{i(54)/sQ227L}$ chimeric polypeptides activate adenylyl cyclase to a level that is equal to or greater than the sum of the stimulation observed with similar levels of expression of the $\alpha_{i(54)/s}$ or $\alpha_s$Q227L polypeptides. Other N terminal $\alpha_{i2}/\alpha_s$ chimeras that do not disrupt the control of GDP dissociation when encoded in the same $\alpha_s$ polypeptide as the Q227L mutation fail to enhance adenylyl cyclase activity beyond that which is measured with $\alpha_s$Q227L alone. The findings confirm that the activated nature of the $\alpha_{i(34)/s}$, $\alpha_{i(54)/s}$, and $\alpha_{i(64)/s}$ chimeras influences limiting regulatory events that are independent of the GTPase functions inhibited by the $\alpha_s$Q227L mutation. Direct measurement of the adenylyl cyclase activation kinetics prove that the $\alpha_{i(54)/s}$ chimera influences GDP dissociation without influencing intrinsic GTPase activity (Gupta et al. 1991b).

## G. Sites of $\beta\gamma$ Subunit Interactions

Pertussis toxin catalyzes the ADP-ribosylation of $\alpha_{i2}$ but not $\alpha_s$ (Gilman 1987). Efficient pertussis toxin-catalyzed $\alpha_{i2}$ ADP-ribosylation requires the presence of the $\beta\gamma$ subunit complex. Within the $\alpha_{i2}$ polypeptide Cys-351, four amino acids from the C terminus is the residue ADP-ribosylated by pertussis toxin. The C terminal $\alpha_s/\alpha_{i2}$ chimera, referred to as $\alpha_{s/i(38)}$, has the last 38 amino acids of $\alpha_s$ replaced with the last 36 residues of $\alpha_{i2}$ (Woon et al. 1988). This chimera was shown to be a functional $\alpha_s$ polypeptide (Woon et al. 1988) that encodes the pertussis toxin ADP-ribosylation site normally found in the $\alpha_{i2}$ polypeptide. The $\alpha_{s/i(38)}$ chimera is not ADP-ribosylated by pertussis toxin (Osawa et al. 1990), indicating that the 36 amino acid $\alpha_{i2}$ C terminal sequence is not sufficient for efficient pertussis toxin recognition of a G-protein $\alpha$ subunit. The $\alpha_{i(Bam)/s/i(38)}$ and $\alpha_{i(64)/s/i(38)}$ chimeric polypeptides were excellent substrates for pertussis toxin-catalyzed ADP-ribosylation. This finding suggests that sequences within the N-terminal moiety of $\alpha_{i2}$ are necessary for efficient pertussis toxin-catalyzed ADP-ribosylation of G protein $\alpha$ subunits. Somewhat surprisingly, the $\alpha_{i(54)/s/i(38)}$ and $\alpha_{i(122)/s/i(38)}$ polypeptides are not substrates for pertussis toxin, even though they are functional G-protein $\alpha$ subunits. The phenotypes of the different chimeras suggest that the N terminus plays a major structural role in pertussis toxin recognition of G-protein $\alpha$ subunits. Presumably, part of this recognition function for pertussis toxin is $\beta\gamma$ interactions with specific $\alpha$ subunit N terminal sequences which have not been rigorously mapped to date.

In contrast to pertussis toxin, cholera toxin recognition of $\alpha_s$ does not require $\beta\gamma$ subunit complexes. Arg-201 is the residue ADP-ribosylated in $\alpha_s$ by cholera toxin. This site is in the middle of the $\alpha_s$ polypeptide. Interestingly, the $\alpha_{i(54)/s}$ chimera is a poor substrate for cholera toxin-catalyzed ADP-ribosylation, and $\alpha_{i/s(Bam)}$ is not recognized by cholera toxin (Osawa et al. 1990). The $\alpha_{s/i(38)}$ chimera also is not recognized by cholera toxin. Thus, mutation of both the N and C termini disrupt cholera toxin-catalyzed ADP-ribosylation even though the mutant $\alpha_s$ polypeptides are functionally capable of activating adenylyl cyclase (Osawa et al. 1990). The

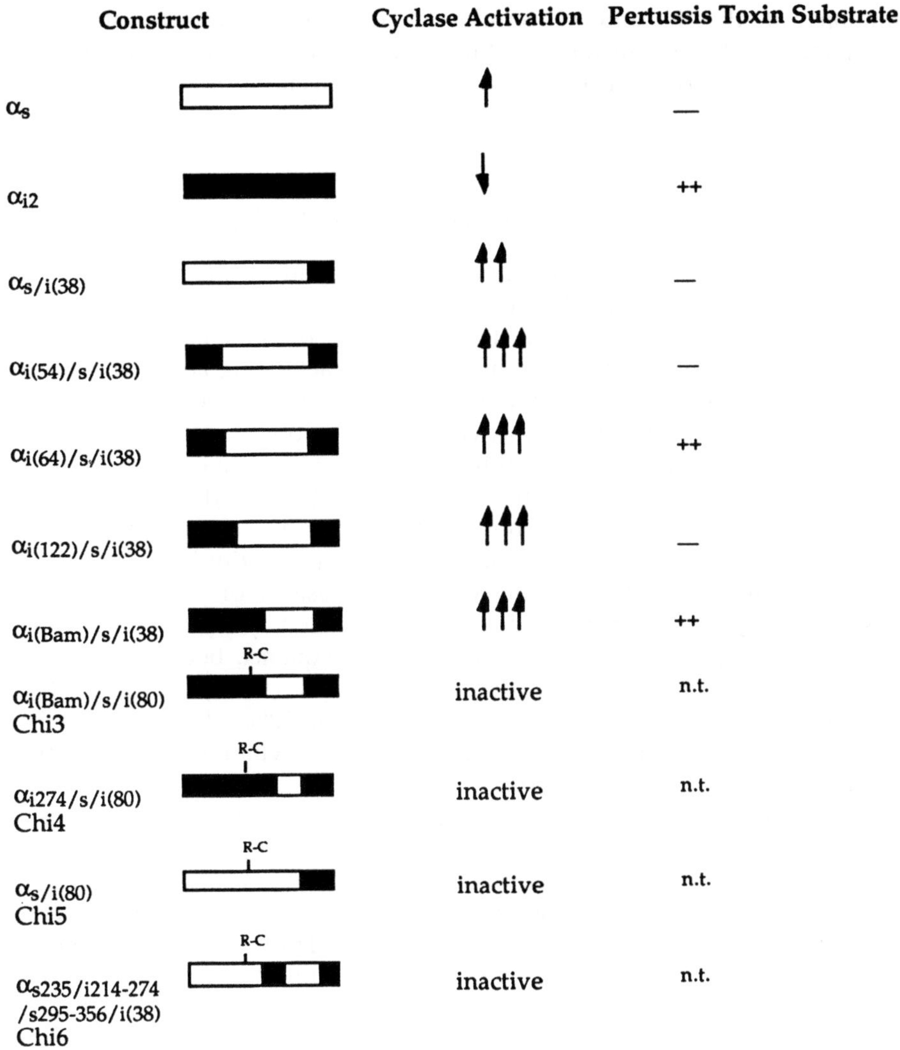


**Fig. 3.** Properties of $\alpha_{i2}/\alpha_s$ chimeras having an $\alpha_{i2}$ C terminus. A diagram showing the relative contribution of $\alpha_s$ (*open bars*) and $\alpha_{i2}$ (*black bars*) is shown for each $\alpha_{i2}/\alpha_s$ chimera. Chimeras were transiently expressed in COS cells, and cAMP levels were measured as a indication of levels of adenylyl cyclase activation in the transfected cells. The ability of each construct to regulated adenylyl cyclase activity is designated inhibition (*down arrow*) or stimulation (*up arrow*). *Multiple arrows*, relative ability of stimulatory mutants to activate adenylyl cyclase in comparison to $\alpha_s$ (*single arrow*). For pertussis toxin substrate data (++) and (−−) indicates whether the mutant construct is or is not a substrate for ADP-ribosylation. Chi3–Chi6 are mutants made by BERLOT and BOURNE (1992). *R-C*, mutation R179C in $\alpha_s$; *n.t.*, not tested

mutations at the N and C termini of $\alpha_s$ must therefore introduce intramolecular changes in the structure of the $\alpha_s$ polypeptide, diminishing the ability of cholera toxin to catalyze ADP-ribosylation of Arg-201 in the middle of the polypeptide.

## H. Mapping of the $\alpha_s$ Adenylyl Cyclase Activation Domain

The chimera $\alpha_{i/s(Bam)}$ replaces the first 234 amino acids of the 394 residue $\alpha_s$ polypeptide with the first 212 residues of $\alpha_{i2}$, yielding a polypeptide encoding the first 60% of $\alpha_{i2}$ and the last 40% of $\alpha_s$ (Fig. 1). The $\alpha_{i/s(Bam)}$ polypeptide is a functional $\alpha_s$ subunit capable of activating adenylyl cyclase and coupling to the $\beta$-adrenergic receptor (MASTERS et al. 1988; OSAWA et al. 1990a). Additionally, the $\alpha_{s/i(38)}$ chimera (see above and Fig. 3) which has the last 38 residues of $\alpha_s$ substituted with the C terminal 36 residues of $\alpha_{i2}$ behaves as a functional $\alpha_s$ capable of activating adenylyl cyclase (WOON et al. 1988). The combination of both mutations in the same polypeptide to yield the $\alpha_{i(Bam)/s/i(38)}$ chimera (Fig. 3) yields a constitutively active $\alpha_s$ polypeptide that strongly stimulates cAMP synthesis (GUPTA et al. 1991b). These three chimeras originally defined a core adenylyl cyclase activation domain that mapped within a 122 amino acid sequence between residues Ile-235 to Arg-356 of the 394 residue $\alpha_s$ polypeptide (OSAWA et al. 1990a). Amazingly, the N terminal 60% and C terminal 10% of the $\alpha_s$ can be substituted with the analogous $\alpha_i$ sequences, and the chimera maintains the ability to activate adenylyl cyclase.

Within the 122 residue core adenylyl cyclase activation domain there is a 12 amino acid sequence (Thr-325 to Arg-336) that is unique to $\alpha_s$ and absent in all other G-protein $\alpha$ subunits characterized to date (DHANASEKARAN and JOHNSON 1989). Deletion of this sequence in the $\alpha_s$ deletion construct (see Fig. 1) results in an $\alpha_s$ polypeptide that is incapable of stimulating adenylyl cyclase. Loss of adenylyl cyclase stimulation resulting from deletion of this sequence and its absence in G-protein $\alpha$ subunits that do not stimulate cAMP synthesis strongly indicate that this unique $\alpha_s$ sequence is critical for stimulation of adenylyl cyclase activity by the core $\alpha_s$ activation domain.

Recent work has even more precisely mapped the regions of $\alpha_s$ necessary for stimulation of adenylyl cyclase. BERLOT and BOURNE (1992) mutated small clusters of amino acids in the $\alpha_s$ adenylyl cyclase activation domain. In their mutagenesis protocol they either replaced $\alpha_s$ amino acids with the corresponding $\alpha_{i2}$ residues or with alanine. Their mutagenesis strategy revealed clusters of charged amino acids critical for activation of adenylyl cyclase. Six groups of clustered residues (Fig. 4) were found in the $\alpha_s$ activation domain of the $\alpha_{i/s(Bam)}$ chimera, referred to as Chil by BERLOT and BOURNE (1992), that inhibit the ability of the mutant $\alpha_s$ polypeptide to activated adenylyl cyclase. Interestingly, the three-dimensional model proposed by BERLOT and BOURNE (1992) predicts that three of these clusters

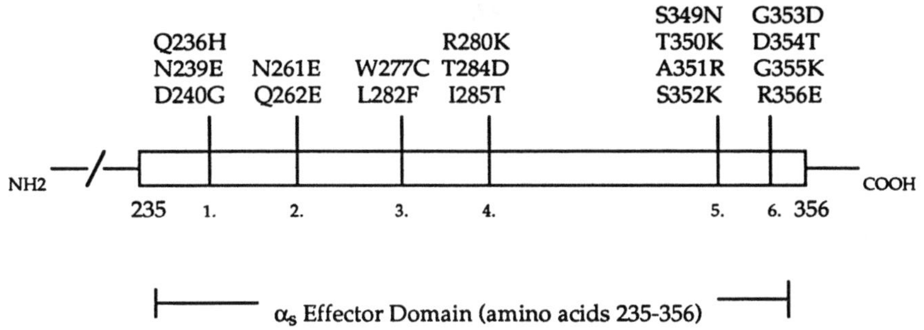


**Fig. 4.** Mutations in $\alpha_{i/s(Bam)}$ (Chi1) that inhibit activation of adenylyl cyclase. Location of the six groups of point mutations is shown in the effector domain of $\alpha_s$ in $\alpha_{i/sBam}$ (= Chi1; BERLOT and BOURNE 1992) that inhibits the ability of the mutant construct to activate adenylyl cyclase

(1, 3, and 4 in Fig. 4) are located on the $\alpha_s$ surface facing the plasma membrane. This has led to the hypothesis that orientation of these $\alpha_s$ sequences toward the membrane would allow interaction with the membrane bound effector adenylyl cyclase (BERLOT and BOURNE 1992), but little is known at present regarding the $\alpha_s$ contact sites on the adenylyl cyclase molecule.

# I. Conclusions

Functional domains of G-protein $\alpha$ subunit polypeptides have been defined using $\alpha_{i2}/\alpha_s$ chimeras, specific point mutations, and analysis of cholera and pertussis toxin-catalyzed ADP-ribosylation reactions. Figure 5 shows the functional G-protein $\alpha$ subunit domains defined using mutational analysis of $\alpha_s$ polypeptide sequence. The $\alpha_s$ polypeptide may be roughly divided in halves based on the properties of the $\alpha_{i/s(Bam)}$ chimera (Chi1 using the Berlot and Bourne nomenclature). The C terminal moiety encodes the adenylyl cyclase activation domain and receptor selectivity domain. The G4 and G5 sequences within the GTP-/GDP-binding domain are also encoded within the C terminal $\alpha_s$ moiety. Consistent with the receptor selectivity site mapping to the extreme C terminus, the cysteine four residues from the C terminal end of $\alpha_i$-like polypeptides is the amino acid ADP-ribosylated by pertussis toxin. The ADP-ribosylation of this residue functionally uncouples receptor activation of the covalently modified G-protein $\alpha$ subunit.

The N terminal moiety of $\alpha_s$ controls GDP dissociation and interaction with $\beta\gamma$ subunit complexes. The attenuator function of $\alpha_{i2}$ and $\alpha_s$ are similar, allowing the N terminal half of $\alpha_s$ to be substituted with the corresponding region of $\alpha_{i2}$. Disruption of the attenuator domain, as observed with the $\alpha_{i(34)/s}$, $\alpha_{i(54)/s}$, and $\alpha_{i(64)/s}$ chimeras, results in an activated $\alpha_s$ polypeptide,

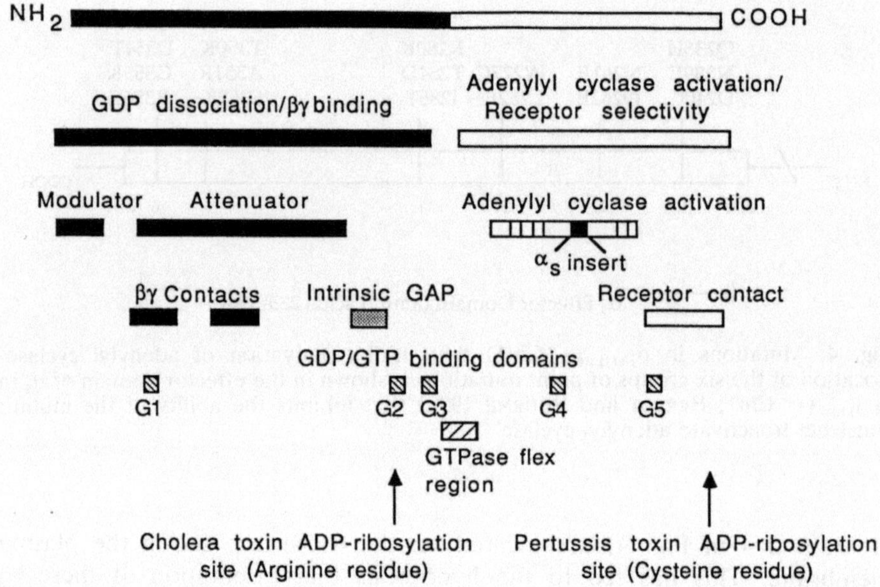


**Fig. 5.** The $\alpha_s$ subunit polypeptide domains. The $\alpha_s$ polypeptide can be divided into halves. The N terminal moiety encodes regulatory regions dominant in controlling GDP dissociation and $\beta\gamma$ interactions. The C terminal moiety encodes the adenylyl cyclase activation and receptor selectivity domains. Within the adenylyl cyclase activation domain the six lines denote the clusters of amino acids whose mutation inhibits $\alpha_s$ stimulation of adenylyl cyclase. *Black bar* (in this domain), the unique $\alpha_s$ amino acid sequence. The G1–G5 sequences comprising the GTP-/GDP-binding domain are dispersed throughout the polypeptide. See text for discussion of the $\alpha_s$ sequence and determination of functional domains

presumably due to an enhanced rate of GDP dissociation and activation by GTP (DHANASEKARAN et al. 1991; GUPTA et al. 1991b).

Finally, mutations at both the N and C termini influence the ability of cholera toxin to ADP-ribosylate Arg-201 in the middle of the $\alpha_s$ polypeptide. N terminal sequences are required for efficient pertussis toxin-catalyzed ADP-ribosylation of a Cys-351 four residues from the $\alpha_{i2}$ C terminus. These findings indicate that the N and C terminal $\alpha_s$ sequences function as modulators of the core domains including attenuator, GTP-/GDP-binding, and effector activation. The modulatory functions are, in turn, influenced by interactions with $\beta\gamma$ subunits and activated receptors. Additional mechanisms of regulation beyond what we have presented in terms of mutational analysis of $\alpha$ subunit polypeptide sequence exists for some G-proteins. For example, N terminal myristoylation of $\alpha_{i2}$ is required for its signaling and transformation functions when expressed in cells (GALLEGO et al. 1992). This finding has been interpreted to suggest that the myristic acid modified $\alpha_{i2}$ polypeptide interacts with specific "binding proteins" in the plasma membrane required for $G_{i2}$ signaling. In contrast, $\alpha_s$

and $\alpha_o$ are not N terminally myristoylated. How the myristic acid modification of $\alpha_i$-like proteins changes their interaction with other proteins is currently a major focus of attention. Myristoylation does not seem to dramatically influence the major functional domains that we identify in Fig. 5. In the subset of G-protein $\alpha$ subunits that are myristoylated, it is possible that myristic acid in combination with $\alpha$ subunit primary amino acid sequence creates a new domain required for proper organization of the $\alpha$ subunit in a signal complex at the plasma membrane. We are currently using the panel of chimeras that we have characterized to determine $\alpha$ subunit sequences involved in functions complemented by myristoylation.

## References


Barbacid M (1987) *ras* genes. Annu Rev Biochem 56:779–827

Berlot CH, Bourne HR (1992) Identification of effector-activating residues of $G_{s\alpha}$ Cell 68:911–922

Birnbaumer L, Abramowitz J, Brown AM (1990) Receptor-effector coupling by G proteins. Biocim Biophys Acta 1031:163–224

Blumer KL, Thorner J (1991) Receptor-G protein signaling in yeast. Annu Rev Physiol 53:37

Bos JL (1989) ras oncognes in human cancer: a review. Cancer Res 49:4682–4689

Bourne HR, Sanders DA, McCormick F (1991a) The GTPase superfamily: a conserved switch for diverse cell functions. Nature 348:125–132

Bourne HR, Sanders DA, McCormick F (1991b) The GTPase superfamily: conserved structure and molecular mechanism. Nature 349:117–127

Casey PJ, Gilman AG (1988) G protein involvement in receptor-effector coupling. J Biol Chem 263:2577–2580

Climenti E, Malgaretti N, Meldolesi J, Toramelli R (1990) A new constitutively activating mutation of $G_s$ protein a subunit-gsp oncogene is found in human pituitary tumors. Oncogene 5:1059–1061

de Vos AM, Tong L, Milburn MV, Matias PM, Jancarik J, Noguchi S, Nishimura S, Miura K, Dhtsaka E, Kim SH (1988) Three-dimensional structure of an oncogene protein: catalytic domain of human c-H-ras p21. Science 239:888–893

Dhanasekaran N, Osawa S, Johnson GL (1991) The $NH_2$-terminal $\alpha$ subunit attenuator domain confers regulation of G protein activation by $\beta\gamma$ complexes. J Cell Biochem 47:352–358

Gallego C, Gupta SK, Winitz S, Eisfelder BJ, Johnson GL (1992) Myristoylation of the $G\alpha_{i2}$ polypeptide is required for its signalling and transformaiton functions. Proc Natl Acad Sci USA (in press)

Gibbs JB, Sigal IS, Poe M, Scolnick EM (1984) Intrinsic GTPase activity distinguishes normal and oncogenic ras p21 molecules. Proc Natl Acad Sci USA 81:5704–5708

Gilman AG (1987) G proteins: transducers of receptor generated signals. Annu Rev Biochem 56:615–649

Graziano MP, Gilman AG (1989) Synthesis in Escherichia coli of GTPase-deficient mutants of $G_{s\alpha}$. J Biol Chem 264:15475–15482

Gupta SK, Gallego C, Lowndes JM, Pleiman CM, Sable C, Eisfelder BJ, Johnson GL (1991A) Analysis of the fibroblast transformation potential of GTPase-deficient gip2 oncogenes. Mol Cell Biol 12:190–197

Gupta SK, Dhanasekaran N, Heasley LE, Johnson GL (1991b) Activating mutations in the $NH_2$ and COOH-terminal moieties of the $G_{s\alpha}$ subunit have dominant phenotypes and distinguishable kinetics of adenylyl cyclase stimulation. J Cell Biochem 47:357–368

Gutkind JS, Novotony EA, Brann MR, Robbins KC (1991) Muscarinic acetylcholine receptor subtypes as agonist-dependent oncogenes. Proc Natl Acad Sci USA 88:4703–4707

Johnson GL, Dhanasekaran N (1989) The G-protein family and their interaction with receptors. Endo Rev 10:317–333

Johnson GL, Dhanasekaran N, Gupta SK, Lowndes JM, Vaillancourt RR, Ruoho AE (1991) Genetic and structural analysis of G protein $\alpha$ subunit regulatory domains. J Cell Biochem 47:136–146

Julius D, Tivelli TJ, Jessell TM, Axel R (1989) Ectopic expression of the serotonin 1c receptor and the triggering of malignant transformation. Science 244:1057–1062

Jurnak F (1985) Structure of the GDP domain of EF-Tu and location of the amino acids homologous to the ras oncogene proteins. Science 230:32–36

Kim D, Lewis DL, Graziadei L, Neer EJ, Bar-Sagi D, Clapham DE (1989) G-protein $\beta\gamma$-subunits activate the cardiac muscarinic $K^+$-channel via phospholipase A2. Nature 337:557–560

Kumar R, Sukumar S, Barbacid M (1990) Activation of ras oncogenes preceeding the onset of neoplasia. Science 248:1101–1104

Landis CA, Masters SB, Spada A, Pace AM, Bourne HR, Vallar L (1989) GTPase inhibiting mutations activate the $\alpha$ chain of $G_s$ and stimulate adenylyl cyclase in human pituitary tumors. Nature 340:692–696

Logothetis DE, Kurachi Y, Galper J, Neer EJ, Clapham DE (1987) The beta gamma subunits of GTP-binding proteins activate the muscarinic $K^+$ channel in heart. Nature 325:321–326

Lyons J, Landis CA, Harsh G, Vallar L, Grunewald K, Feichtinger H, Such QY, Clark OH, Kawasaki E, Bourne HR, McCormick F (1990) Two G protein oncogenes in human endocrine tumors. Science 249:655–659

Masters SB, Miller RT, Chi MH, Chang FH, Beiderman B, Lopez NG, Bourne HR (1989) Mutations in the GTP binding site of $G_s$ alpha alter stimulation of adenylyl cyclase. J Biol Chem 264:15467–15474

Masters SB, Sullivan KA, Miller RT, Beiderman B, Copey NG, Ramachandran J, Bourne HR (1988) Carboxyl terminal domain of $G_{\alpha s}$ specifies coupling of receptors to stimulation of adenylyl cyclase. Science 241:448–451

McCormick F (1989) Gasp: not just another oncogene. Nature 340:678–679

McGrath JP, Capon DJ, Goeddel DV, Levinsen AD (1984) Comparative biochemical properties of normal and activated human ras p21 protein. Nature 310:644–649

Miller RT, Masters SB, Sullivan KA, Beiderman B, Bourne HR (1988) A mutation that prevents GTP-dependent activation of $\alpha$ chain of $G_s$. Nature 334:712–715

Osawa S, Johnson GL (1991) A dominant negative $G\alpha_s$ mutant is rescued by secondary mutation of the $\alpha$ chain amino terminus. J Biol Chem 266:4673–4676

Osawa S Dhanasekaran N, Woon CW, Johnson GL (1990A) $G_{\alpha i}$-$G_{\alpha s}$ chimeras define the function of $\alpha$ chain domains in control of G protein activation and $\beta\gamma$ subunit complex interactions. Cell 63:697–706

Osawa S, Heasley LE, Dhanasekaran N, Gupta SK, Woon CW, Berlot C, Johnson GL (1990b) Mutation of the $G_s$ protein $\alpha$ subunit $NH_2$ terminus relieves an attenuator function, resulting in constitutive adenylyl cyclase stimulation. Mol and Cell Biol 10:2931–2940

Pace AM, Wong YH, Bourne HR (1991) A mutant $\alpha$ subunit of $G_{i2}$ induces neoplastic transformation of Rat-1 cells. Proc Natl Acad Sci USA 88:7031–7035

Pai, EF, Kabash W, Krengel U, Holmes KG, John J, Wittinghofer A (1989) Structure of the guanine-nucleotide-binding domain of the Ha-ras oncogene product p21 in the triphosphate conformation. Nature 341:209–214

Schlichting I, Almo SC, Parr G, Wilson K, Petratos K, Lentfer A, Wittinghofer A, Kabash W, Pai EF, Petsho GA, Goody RS (1990) Time-resolved X-ray crystallographic study of the conformational change in Ha-ras p21 protein on GTP hydrolysis. Nature 345:309–315

Sigal IS, Gibbs JB, D'Alonzo JS, Temeles GL, Wolanski BS, Socker SH, Scolnick EM (1986) Mutant ras-encoded proteins with altered nucleotide binding exert dominant biological effects. Proc Natl Acad Sci USA 83:952–956

Stryer L, Bourne H (1986) G proteins: a family of signal transducers. Annu Rev Cell Biol 2:391–419

Suarez HG, du Vullard JA, Caillou B, Schumberger M, Parmentier C, Monier R (1991) gsp mutations in human thyroid tumors. Oncogene 6:677–679

Tang W-J, Gilman AG (1991) Type-specific regulation of adenylyl cyclase by G protein $\beta\gamma$ subunits. Science 254:1500–1503

Trahey M, McCormick F (1987) A cytoplasmic protein stimulates normal N-ras p21 GTPase, but does not affect oncogenic mutants. Science 238:542–545

Weinstein LS, Shenker A, Gejman PV, Merino MJ, Friedman E, Spiegel AM (1991) Activating mutations of the stimulatory G protein in the McCune-Albright syndrome. N Engl J Med 325:1688–1695

Woon CW, Heasley L, Osawa S, Johnson GL (1989) Mutation of glycine 49 to valine in the $\alpha$ subunit of $G_s$ results in constitutive elevation of cyclic AMP synthesis. Biochemistry 28:4547–4551

Woon CW, Soparkar S, Heasley L, Johnson GL (1988) Expression of a $G_{\alpha s}/G_{\alpha i}$ chimera that constitutively activates cyclic AMP synthesis. J Biol Chem 264:5687–5693

CHAPTER 50
# The GTPase Cycle: Transducin

T.D. TING, R.H. LEE, and Y.-K. HO

## A. The Retinal cGMP Cascade and Visual Excitation

Visual excitation in vertebrate rod photoreceptor cells involves the light activation of a cGMP enzyme cascade. The rod outer segment (ROS) is comprised of approximately 2000 flattened membrane disks which contain the protein components of the cGMP cascade, such as rhodopsin (R), transducin (T), and cGMP phosphodiesterase (PDE). Surrounding the ROS disk membranes is the plasma membrane which contains cation channels whose permeability are controlled by intracellular cGMP concentration. In the dark-adapted state, the cGMP cascade is inactive, and the cytosolic concentration of cGMP is relatively high ($\sim 60 \mu M$; WOODRUFF and FAIN 1982). This maintains a large fraction of the cGMP-sensitive channels in the open state (FESENKO et al. 1985) and keeps the cell relatively depolarized. Excitation of rhodopsin by light ultimately leads to the activation of PDE, which rapidly hydrolyzes cGMP. As the level of cGMP in the ROS is reduced, the cGMP-sensitive channels start to close and the cell hyperpolarizes. Signal coupling between photolyzed rhodopsin (R*) and PDE is mediated by a G-protein called transducin (FUNG 1983; for review see HO et al. 1989; STRYER 1991). A single R* is capable of catalyzing the GDP/GTP exchange reaction in hundreds of transducin molecules thereby converting them to the active form of T-GTP. Subsequently, T-GTP activates the latent PDE by removing the inhibitory $P_\gamma$ subunit from the PDE complex. Activated PDE hydrolyzes thousands of cGMP molecules per second. A signal amplification factor of $10^6$ is built into the cGMP cascade (FUNG and STRYER 1980; FUNG et al. 1981; HURLEY and STRYER 1982). In order for the rod cell to be repeatedly responsive to continuous cycles of stimulation by light, the cascade must be turned off rapidly and the cGMP level restored to the dark-adapted level. The inhibitory regulation at the rhodopsin level is mediated by the phosphorylation of R* by rhodopsin kinase. The phosphorylated form of R* is capable of binding to a 48-kDa protein called arrestin (LIEBMAN and PUGH 1980; WILDEN et al. 1986; SITARAMAYYA and LIEBMAN 1983b). The binding of arrestin blocks further activation of transducin by R*. Deactivation at the transducin level is coupled to the hydrolysis of transducin-bound GTP by the intrinsic GTPase activity (FUNG et al. 1981). Once T-GTP is removed, the $P_\gamma$ subunit recombines with the catalytic $P_{\alpha\beta}$ subunits of PDE and turns off the hydrolysis of cGMP (Hurley and Stryer 1982). The depleted cytosolic cGMP is replenished by an increase in the activity of guanylate cyclase, and the photoreceptor cell is ready for another cycle of light activation.

This chapter reviews the reaction dynamics of the transducin coupling cycle and its relationship to its signal transducing function. Discussion is focused on the subunit interaction of transducin, the kinetics of the GTP hydrolysis reaction, and the regulation of the transducin cycle. Specific interactions of transducin with rhodopsin and PDE are discussed in other chapters and are not emphasized here.

## B. The Coupling Cycle of Transducin

Transducin is a heterotrimer composed of three polypeptides: $T_\alpha$ (39 kDa), $T_\beta$ (37 kDa), and $T_\gamma$ (8.5 kDa) and performs its coupling action as two functional subunits, $T_\alpha$, and $T_{\beta\gamma}$. In the dark-adapted state, $T_\alpha$ contains a bound GDP ($T_\alpha$-GDP) and interacts with $T_{\beta\gamma}$. The $T_\alpha$-GDP·$T_{\beta\gamma}$ complex tightly associates with rhodopsin. Upon photoexcitation, R* catalyzes a GTP/GDP exchange reaction converting $T_\alpha$-GDP to the active form of $T_\alpha$-GTP. $T_\alpha$-GTP dissociates from R* and $T_{\beta\gamma}$. $T_\alpha$-GTP exhibits high affinity for the PDE complex and alone can activate PDE by relieving the restraint exerted by the $P_\gamma$ inhibitory subunit. The coupling cycle is completed when the $T_\alpha$-bound GTP is hydrolyzed to GDP. $T_\alpha$-GDP recombines with $T_{\beta\gamma}$ and is ready for another cycle of activation. The transducin coupling cycle is depicted in Fig. 1. The basic mechanism of signal transduction utilized by transducin is a subunit dissociation and association cycle regulated via guanine nucleotide binding and hydrolysis. $T_\alpha$ functions as the signal carrier. By switching between GDP-bound and GTP-bound states, $T_\alpha$ can exist in two conformations which differ in their interaction with the receptor (R*) and effector (PDE) molecules. $T_{\beta\gamma}$ acts as a modulator which facilitates the binding of $T_\alpha$-GDP to rhodopsin.

Rod photoreceptor cells possess the ability to detect a single photon. Specific mechanisms are built into each of the biochemical steps of the signal transduction pathway to ensure a high signal to noise ratio. At the rhodopsin level, the thermal isomerization of the 11-*cis* retinal to all-*trans* retinal is approximately one in $10^7$ molecules per minute at room temperature, which generates extremely low noise in the absence of light. At the transducin level, purified transducin in solution does not exchange its bound nucleotide in the absence of R*. The spontaneous nucleotide exchange rate of transducin has been estimated to be $5 \times 10^{-6}\,s^{-1}$ (Matesic and Liebman 1992), a rate which does not generate significant noise in the system. This implies that the guanine nucleotide binding site of transducin is in a closed conformation which prevents spontaneous nucleotide exchange. GDP cannot be incorporated into $T_\alpha$ in the absence of $T_{\beta\gamma}$ and R*. During photoexcitation,

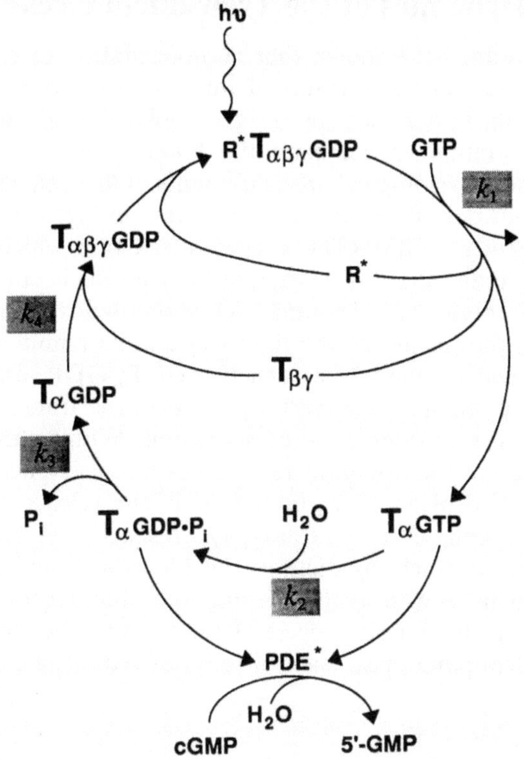


**Fig. 1.** The signal coupling cycle of transducin in the retinal cGMP cascade. $k_1$, $k_2$, $k_3$, and $k_4$ represent the apparent rate constants for the activation of transducin, on-site hydrolysis of $T_\alpha$-bound GTP release of Pi from $T_\alpha$-GDP·Pi, and recycling of $T_\alpha$-GDP and $T_{\beta\gamma}$, respectively

$T_\alpha$-GDP has high affinity for $T_{\beta\gamma}$ and R*. Interaction with R* opens the GTP-binding site on $T_\alpha$ for nucleotide exchange. Upon exchanging bound GDP for GTP, $T_\alpha$-GTP dissociates from $T_{\beta\gamma}$ and R*. The dissociation of the three components of the nucleotide exchange complex, R*, $T_\alpha$, and $T_{\beta\gamma}$, commits the transducin coupling cycle to operate unidirectionally. The GTP-binding site of dissociated $T_\alpha$-GTP is again in a closed conformation, which prevents any further nucleotide exchange from occurring to deactivate $T_\alpha$-GTP. $T_\alpha$ must hydrolyze its bound GTP in order to return to its latent state and reassociate with $T_{\beta\gamma}$. These unique biochemical properties of the transducin coupling cycle allow it to behave as a low-noise but high-gain switch for visual excitation.

## C. The Reaction Dynamics of the Transducin Cycle

Electrophysiological studies have shown that photoexcitation of rod cells is completed within 200 ms, and the turn-off of the response occurs in a few seconds. To further our understanding on the molecular mechanism of photoexcitation it is essential to correlate the kinetics of the transducin coupling cycle with the physiological measurements. One can dissect the transducin cycle into four partial reactions, as indicated in Fig. 1: (a) the R*-catalyzed GTP binding via a GDP/GTP exchange reaction which leads to the formation of $T_\alpha$-GTP and the dissociation of the transducin subunit from R*, (b) the on-site hydrolysis of $T_\alpha$-bound GTP which leads to the formation of $T_\alpha$-GDP·Pi, (c) the release of the tightly bound inorganic phosphate (Pi) from $T_\alpha$-GDP·Pi which regenerates deactivated $T_\alpha$-GDP, and (d) the recycling of $T_\alpha$-GDP by recombining with $T_{\beta\gamma}$ to form the transducin complex which rebinds R* for another cycle of activation. Within this scheme, $k_1$, $k_2$, $k_3$, and $k_4$ represent the apparent rate constants for the activation of transducin, the on-site hydrolysis of $T_\alpha$-bound GTP, the release of Pi from $T_\alpha$-GDP·Pi, and the recycling of $T_\alpha$-GDP and $T_{\beta\gamma}$, respectively. The relative rates of these four partial reactions dictates the kinetics of activation and deactivation of transducin as well as the lifetime and abundance of each of the transducin species in the coupling cycle. To determine the rate-limiting step of the transducin coupling cycle each of the four partial reactions need to be examined.

### I. Transducin Subunit Interaction

Based on our current understanding of the cGMP cascade, the first step of transducin activation occurs quickly, on a time scale of milliseconds, and is definitely not rate limiting. The rates of the R*-catalyzed GTP/GDP exchange reaction and the dissociation of $T_\alpha$-GTP and $T_{\beta\gamma}$ from ROS membrane have been measured by light-scattering changes due to the release of protein complexes from the disk membrane (KUHN et al. 1981; BENNETT and DUPONT 1985). These studies suggest that the exchange of guanine nucleotides and the dissociation of $T_\alpha$-GTP from ROS disk membranes occurs in less than 100 ms. Biochemical studies demonstrate that cGMP hydrolysis due to the activation of PDE occurs in less than 1 s, which implies that the formation of $T_\alpha$-GTP must occur within 100 ms. This rapid activation of transducin fulfills its functional role as a signal carrier for visual excitation, which is completed in 200 ms. However, the overall turnover of the transducin molecule as monitored by the GTPase activity is very slow and occurs two or three times per minute. The low turnover rate of transducin may be due to a slow deactivation of $T_\alpha$-GTP via hydrolysis of bound GTP. Alternatively, the recombination of $T_\alpha$-GDP and $T_{\beta\gamma}$ for the initiation of another activation cycle may be rate limiting. Several lines of evidence indicate that the recombination of $T_\alpha$-GDP with $T_{\beta\gamma}$ is not rate limiting. It has been demonstrated that in the R*-catalyzed GTP/GDP exchange reaction, all of the transducin-bound GDP can be replaced by GTP, and that GTP incorporation reaches 1 mol GTP per mole of transducin at steady state (FUNG et al. 1981). This result suggests that the reassociation of $T_\alpha$-GDP with $T_{\beta\gamma}$ occurs rapidly and the recombined transducin complex can be immediately reactivated by R* without significant delay. Otherwise, a sizable fraction of transducin exists as $T_\alpha$-GDP at steady state. Further evidence that recombination of $T_\alpha$-GDP with $T_{\beta\gamma}$ is not rate limiting is demonstrated by the reconstitution study by FUNG (1983), who measured the GTPase activity of reconstituted transducin complexes at various ratio of $T_\alpha$ and $T_{\beta\gamma}$. The results are shown in Fig. 2.

When $T_{\beta\gamma}$ concentration is kept constant, the GTPase activity remains linearly proportional to the $T_\alpha$ subunit concentration even at a $T_\alpha:T_{\beta\gamma}$ ratio of 12:1. However, when $T_\alpha$ concentration is kept constant, the GTPase

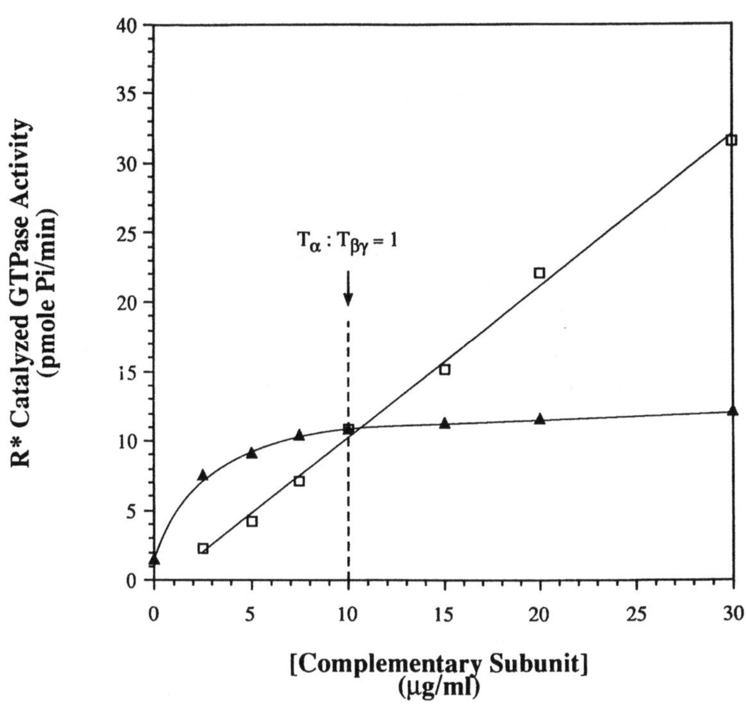


**Fig. 2.** Reconstitution of the GTPase activity with purified subunits of transducin. The GTPase activity is measured in the presence of R*, [$\gamma$-$^{32}$P]GTP, and reconstituted transducin. ▲, A constant amount of 10 $\mu$g/ml $T_\alpha$ subunit is titrated with increasing amounts of $T_{\beta\gamma}$; □, a constant amount of 10 $\mu$g/ml $T_{\beta\gamma}$ subunit is titrated with increasing amounts of $T_\alpha$. *Dashed line*, the point of a 1:1 stoichiometric ratio between $T_\alpha$ and $T_{\beta\gamma}$

activity becomes saturated at a less than a 1:1 ratio of $T_\alpha:T_{\beta\gamma}$. Further increases in $T_{\beta\gamma}$ has no effect on the GTPase activity. In contrast, the binding of the transducin heterotrimer to R* requires a stoichiometric ratio of 1:1 in $T_\alpha$-GDP and $T_{\beta\gamma}$. The results shown in Fig. 2 suggest that $T_{\beta\gamma}$ has a catalytic role in turning over the $T_\alpha$ subunit for GTP binding and hydrolysis. Less than equimolar amount of $T_{\beta\gamma}$ to $T_\alpha$ is sufficient to maintain the recycling of all the $T_\alpha$ molecules at steady state. This low molar requirement of $T_{\beta\gamma}$ during steady-state turnover also suggests that the majority of $T_\alpha$ exists in the $T_\alpha$-GTP form. When some of $T_\alpha$-bound GTP is hydrolyzed to GDP, the small population of $T_\alpha$-GDP can be immediately recycled by a small amount of $T_{\beta\gamma}$ and R*. This implies that the activated form of $T_\alpha$ which remains dissociated from $T_{\beta\gamma}$ and R* has a relatively long half-live before it is deactivated and recombines with $T_{\beta\gamma}$. This reconstitution study clearly demonstrates that the rate-limiting step of the transducin coupling cycle is the deactivation of $T_\alpha$-GTP via GTP hydrolysis and is not the nucleotide exchange reaction or the recombination of deactivated subunits.

## II. Pre-Steady-State Kinetic Analysis of the GTP Hydrolysis Reaction

The deactivation of $T_\alpha$-GTP is governed by the two partial reactions shown in Fig. 1, the on-site hydrolysis of the bound GTP and the release of the hydrolytic product, inorganic phosphate (Pi). The relative rates of these two reactions have been measured by studying the pre-steady-state kinetics of GTP hydrolysis using a rapid acid quenching method (TING and HO 1991). In a reconstituted system, sample containing purified transducin and excess R* is mixed rapidly with excess [$\gamma$-$^{32}$P]GTP at time zero. The presence of excess R* and [$\gamma$-$^{32}$]GTP ensures that the incorporation of [$\gamma$-$^{32}$P]GTP into all of the nucleotide binding sites of transducin is synchronized. The reaction mixture is then allowed to react for a predetermined time between 2 and 60 s. At the end of the incubation period a third solution containing perchloric acid is rapidly added to the mixture to quench the reaction, and the amount of [$^{32}$P]Pi formed at each sample time point is quantitated. A typical result of the rapid quenching experiment is shown in Fig. 3A. The rate of Pi formation in the pre-steady-state is biphasic; it begins with a rapid phase of "burst" of Pi formation from 1–4 s followed by a slow steady-state rate. This biphasic characteristic of Pi formation can be explained if the release of the hydrolytic product Pi from the $T_\alpha$-GDP·Pi complex is the rate limiting step of the cycle. If the rate of Pi release is slower than the rate of on-site GTP hydrolysis, the rapid phase of Pi formation represents the accumulation of the $T_\alpha$-GDP·Pi complex resulting from the fast on-site hydrolysis of bound GTP. The perchloric acid in the quenching solution denatures the protein and releases the tightly bound Pi, which can be detected as a burst of Pi formation. Since the breakdown of the $T_\alpha$-GDP·Pi complex is slow, all

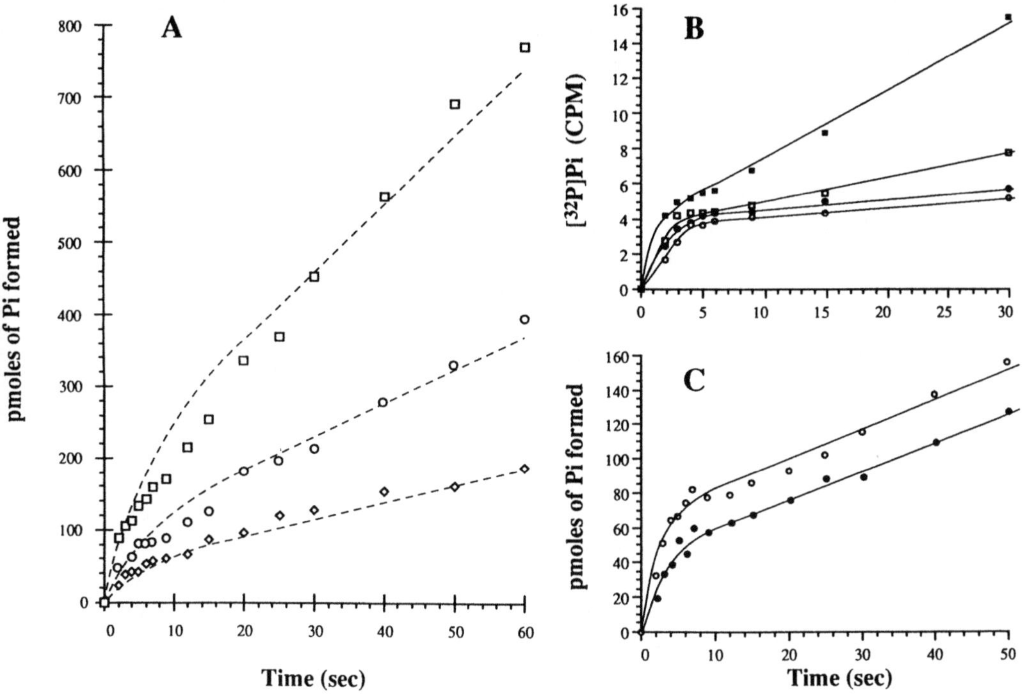


**Fig. 3A-C.** The effects of transducin concentration, temperature, and $D_2O$ solvent isotope on the pre-steady-state kinetics of GTP hydrolysis. The pre-steady-state kinetics of GTP hydrolysis is measured according to the rapid acid quenching method as described in the text. **A** The effect of transducin concentration. The GTPase assays were carried out at 22°C, and the reaction volume was 60 $\mu$l. Each sample contained 120 $\mu M$ [$\gamma$-$^{32}$P]GTP, excess R* (20 $\mu M$), and various amount of T. *Dashed lines*, computer-generated fitting of the kinetic data according Eq. 1. The best fits were found using $k_2 = 0.16\,s^{-1}$, $k_3 = 0.04\,s^{-1}$, and with different amounts of transducin (□, 72 pmol; ●, 144 pmol; ○, 288 pmol). Each data point represents the average of quadruplet samples. **B** The effect of temperature. Each sample contained 6 $\mu M$ R*, 0.6 $\mu M$ T, and 30 $\mu M$ [$\gamma$-$^{32}$P]GTP in a volume of 60 $\mu$l. The GTPase assays were carried out at various temperatures: 4°C (○), 10°C (●), 22°C (□), and 38°C (■). **C** The $D_2O$ solvent isotope effect. The GTPase assays were carried out at 22°C, and each sample contained 9 $\mu M$ R*, 0.9 $\mu M$ T, and 30 $\mu M$ [$\gamma$-$^{32}$P]GTP in a volume of 60 $\mu$l in either $H_2O$ (○) or $D_2O$ (●)

the transducin molecules in the sample are rapidly transformed to the $T_\alpha$-GDP·Pi form, and the burst phase is over in approximately 4 s. As the tightly bound Pi is slowly released from $T_\alpha$-GDP·Pi, deactivated $T_\alpha$-GDP recombines with $T_{\beta\gamma}$, and enters another hydrolysis cycle. Thus, the slow phase of Pi formation is dominated by the Pi release rate of the $T_\alpha$-GDP·Pi complex and represents the steady-state rate of GTP hydrolysis of the transducin cycle.

To further confirm the proposed interpretation, the effects of transducin concentration, temperature, and replacement of $H_2O$ with $D_2O$ in the reaction buffer on the biphasic characteristic of GTP hydrolysis were examined. If the initial burst of Pi formation is due to the rapid on-site hydrolysis of $T_\alpha$-bound GTP and the accumulation of T-GDP·Pi complexes, an increase in transducin concentration should cause a corresponding increase in the size of the initial burst and the steady-state rate of GTP hydrolysis. Since the on-site GTP hydrolysis and the release of Pi are intrinsic properties of $T_\alpha$ and are essentially unimolecular in nature, the time required for $T_\alpha$-bound GTP to be hydrolyzed and the time that $T_\alpha$ retains the tightly bound Pi should be independent of transducin concentration. As shown in Fig. 3A, at constant temperature an increase in transducin concentration generates a corresponding increase in the burst and the steady state rate but has no affect on the time required to reach the transition point between the rapid and slow phase, which remains constant at approximately 4–6 s.

When the temperature dependence of the pre-steady-state kinetics of GTP hydrolysis was investigated, it was found that both the initial rate of the on-site GTP hydrolysis and the steady-state rate of Pi release increase with respect to increasing temperature. The size of the initial burst, which is due mainly to the total transducin concentration, is not affected significantly at different temperatures. However, the time required to reach the transition point between the initial burst and the slow steady-state rate increases as temperature decreases as shown in Fig. 3B. This can be explained if both the rates of on-site GTP hydrolysis and the release of Pi from $T_\alpha$ are decreased at lower temperatures, and the system requires more time to reach the steady state rate. Moreover, the temperature dependence of the rates of the two partial reactions are different. At temperatures below 10°C the initial burst is easily detected, indicating that the on-site GTP hydrolysis still proceeds at a relatively fast rate. However, the steady-state rate of Pi release has diminished significantly at low temperatures. To quantitate this observation, the temperature dependence of the two partial reactions is analyzed by the Arrhenius plot of reaction rate versus inverse temperature. Both partial reactions exhibited linear Arrhenius plot. The results indicate that the rate of the on-site GTP hydrolysis reaction is less temperature dependent with an activation energy ($E_a$) of 28.1 kJ per mole of GTP hydrolyzed; the release of Pi from $T_\alpha$-GDP·Pi is more temperature dependent, with an estimated $E_a$ of 44.7 kJ per mole of Pi released. These results provide further support to the interpretation of the kinetic data. Furthermore, it clearly demonstrates that the two partial reactions of transducin GTPase activity can be studied independently by the pre-steady-state method.

Finally, the $D_2O$ solvent isotope effect on the pre-steady-state kinetics further supports the view that the rate-limiting step in the reaction pathway is the release of Pi from $T_\alpha$-GDP·Pi and not the on-site hydrolysis of GTP. Since water molecule directly participates in the on-site GTP hydrolysis step as a reactant, substituting $D_2O$ for $H_2O$ may exhibit a primary isotope effect on the hydrolysis reaction, which may be observed as a decrease in the initial burst of the pre-steady-state kinetics. However, water molecules may not participate in the slow step of Pi release, thus the steady-state rate of the GTPase activity may remain unchanged in $H_2O$ and $D_2O$ buffers. The effect of $D_2O$ on the pre-steady-state kinetics is shown in Fig. 3C. The rate of Pi formation in the rapid phase is slower in $D_2O$ buffer with a solvent isotope effect of approximately 1.7, while no solvent isotope effect can be detected in steady-state rate. These results are in complete agreement with the above interpretation.

## III. Quantitative Analysis of the Pre-Steady-State Kinetics

The burst phenomenon in the pre-steady-state kinetics is similar to that observed in enzyme catalysis in which a covalent intermediate is involved, and the breakdown rather than the formation of the intermediate is rate limiting. This "burst" phenomenon can be analyzed quantitatively with a simple kinetic model, as described by CORNISH-BOWDEN (1976). Under the experimental conditions, the following assumptions can be made: (a) the concentration of GTP is large enough to be treated as a constant, (b) the rates of the reverse reactions are insignificant, and (c) $k_1$ and $k_4$ are large as compared to $(k_2 + k_3)$. Under these assumptions, the concentration of T-GDP can be considered negligible because it is converted to T-GTP as soon as it is made. The reaction then simplifies to a simple reversible first-order reaction, as depicted in Fig. 4. Since the rate of GTP exchange ($k_1$) and the recycling rate of $T_\alpha$-GDP ($k_4$) are fast, they can be ignored in the kinetic scheme.

In this case, if $T_0$ is the amount of total transducin concentration, $q$ is the concentration of T-GDP·Pi, and $T_0 - q$ is the concentration of T-GTP, a rate equation for the formation of T-GDP·Pi can be written:

$$v = dq/dt = k_2(T_0 - q) - k_3q = k_2T_0 - (k_2 + k_3)q$$

Integration of the rate equation gives:

$$-\frac{\ln[k_2T_0 - (k_2 + k_3)q]}{k_2 + k_3} = t + \alpha$$

Where $\alpha$ is the integration constant. At $t = 0$, $q = 0$, thus $\alpha = -\dfrac{\ln(k_2T_0)}{k_2 + k_3}$

Substituting for $\alpha$ and rearranging yields:

$$q = [\text{T-GDP}\cdot\text{Pi}] = \frac{k_2T_0\,(1 - e^{-(k_2+k_3)t})}{k_2 + k_3}$$

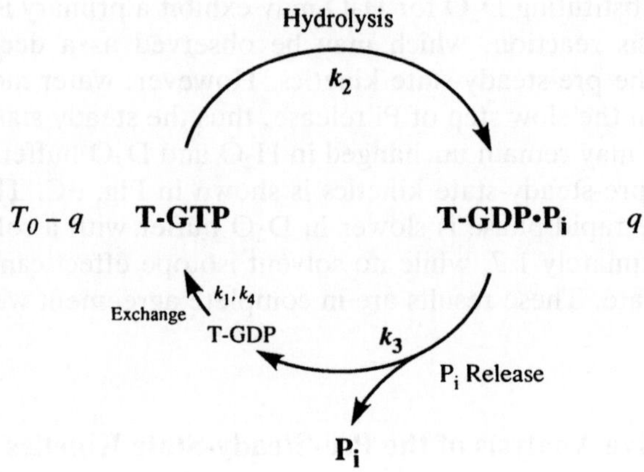


**Fig. 4.** Reversible first-order reaction

Similarly, the rate of formation of free Pi $= \dfrac{dPi^{FREE}}{dt} = k_3[\text{T-GDP·Pi}]$, and

$$\frac{dPi^{FREE}}{dt} = k_3 \frac{k_2 T_0(1 - e^{-(k_2+k_3)t})}{k_2 + k_3} = \frac{k_2 k_3 T_0}{k_2 + k_3} - \frac{k_2 k_3 T_0 e^{-(k_2+k_3)t}}{k_2 + k_3}$$

After integration,

$$Pi^{FREE} = \frac{k_2 k_3 T_O}{k_2 + k_3} t + \frac{k_2 k_3 T_0 e^{-(k_2+k_3)t}}{(k_2 + k_3)^2} + \alpha$$

Where $\alpha$ is the integration constant. At $t = 0$, $Pi^{FREE} = 0$, thus

$$\alpha = -\frac{k_2 k_3 T_0}{(k_2 + k_3)^2}$$

Therefore,

$$Pi^{FREE} = \frac{k_2 k_3 T_0}{k_2 + k_3} t - \frac{k_2 k_3 T_0 (1 - e^{-(k_2+k_3)t})}{(k_2 + k_3)^2}$$

Since the rapid quenching technique measures total $P_i$ which includes both the Pi bound in T-GDP·Pi as well as free Pi, thus

$$Pi^{TOTAL} = Pi^{FREE} + [\text{T-GDP·Pi}]$$

or

$$\mathrm{Pi}^{\mathrm{TOTAL}} = \frac{k_2 k_3 T_0}{k_2 + k_3} t - \frac{k_2 k_3 T_0 (1 - e^{-(k_2+k_3)t})}{(k_2 + k_3)^2} + \frac{k_2 T_0 (1 - e^{-(k_2+k_3)t})}{k_2 + k_3}$$

and

$$\mathrm{Pi}^{\mathrm{TOTAL}} = \frac{k_2 k_3 T_0}{k_2 + k_3} t - \frac{k_2^2 T_0 (1 - e^{-(k_2+k_3)t})}{(k_2 + k_3)^2} \tag{1}$$

Some useful experimental parameters can be obtained from Eq. 1. The rate of the on-site GTP hydrolysis can be obtained from the initial slope of the rapid phase. This rate can be mathematically treated by taking the derivative of the above equation with respect to time, which yields the following equation:

$$\frac{d\mathrm{Pi}^{\mathrm{TOTAL}}}{dt} = \frac{k_2 T_0 (k_2 e^{-(k_2+k_3)t} + k_3)}{k_2 + k_3}$$

The initial rate can be obtained by setting $t = 0$, and the equation simply becomes:

$$\frac{d\mathrm{Pi}^{\mathrm{TOTAL}}}{dt} = k_2 T_0$$

Thus, the initial rate of the rapid phase is equivalent to the rate of GTP hydrolysis. The slow phase of the data of Fig. 3A represents the steady-state rate of Pi formation. The equation describing this linear phase can be obtained from Eq. 1 at large values of $t$ by omitting the exponential term from Eq. 1:

$$\mathrm{Pi}^{\mathrm{TOTAL}} = \frac{k_2 k_3 T_0}{k_2 + k_3} t + \frac{k_2^2 T_0}{(k_2 + k_3)^2}$$

A linear plot of this equation gives the slope $= \dfrac{k_2 k_3 T_0}{k_2 + k_3}$

and the $y$-intercept $= \dfrac{k_2^2 T_0}{(k_2 + k_3)^2}$.

The values for the slope and $y$-intercept can be obtained by measuring the rate of Pi formation in the steady state and by quantitating the size of the initial burst, respectively. Both parameters are proportional to the amount of transducin present in the reaction mixture as shown in Fig. 3A. A transition point between the rapid burst and the steady-state rate is defined by the intersection of two straight lines drawn from the rapid and slow phases of the data. Based on the proposed reaction scheme, the time required to reach this transition point represents the time required to activate all of the transducin molecules to the $T_\alpha$-GDP·Pi form and is also the onset of the slow release of $T_\alpha$-bound Pi. This simple kinetic model provides a quantitative means to analyze the biphasic characteristic of the pre-steady-state GTP hydrolysis by transducin. The results from Fig. 3A indicate that the on-site hydrolysis rate is approximately four times faster than the rate of release of the tightly bound Pi. The size of the initial burst is estimated to be 67% of the amount of T added. According to the above scheme, a ratio of 0.67 between the burst and the amount of enzyme indicates that the rate of GTP hydrolysis is approximately 4.5 times faster than the rate of Pi release. The concentration of T in the experiment is defined, and the kinetic data can be fitted according to Eq. 1 using a nonlinear curve fitting computer program. The best fits of the data are found by using $k_2 = 0.16\,s^{-1}$ and $k_3 = 0.04\,s^{-1}$ and are shown by the dashed lines in Fig. 3A. As can be seen at high transducin concentration, the fitting of Eq. 1 deviates from the data points with a slower steady state rate of Pi formation. This could be due to the allosteric characteristics of transducin GTPase activity exhibited at higher concentration of transducin (WESSLING-RESNICK and JOHNSON 1987). Nevertheless, this quantitative treatment provides a description of the pre-steady-state kinetics of the transducin GTPase cycle.

## D. Relationship of GTP Hydrolysis and PDE Deactivation

The pre-steady-state kinetic analyses described above indicate that Pi release from $T_\alpha$-GDP·Pi is the rate-limiting step of the transducin coupling cycle. This results in the accumulation of $T_\alpha$-GDP·Pi which becomes the dominant species in the system at steady state. Several questions concerning the activation of PDE by transducin can be raised in light of this implication. Does $T_\alpha$-GDP·Pi have the same conformation as $T_\alpha$-GTP? Is $T_\alpha$-GDP·Pi capable of activating PDE? Which of the partial reactions, the on-site GTP hydrolysis or the release of Pi from $T_\alpha$-GDP·Pi, actually controls the deactivation of transducin? Answers to these questions will have significant consequence in understanding the dynamics of the cGMP cascade. It has been shown that the light-activated PDE activity can be turned off in less than 2 s under physiological condition (SITARAMAYYA and LIEBMAN 1983a). If $T_\alpha$-GDP·Pi is capable of activating PDE, its long half-life in the order of 20 s will make the deactivation of transducin so slow as to preclude it from being the major turn-off mechanism for the cGMP cascade. This may mean that the cGMP cascade is shut off via other mechanisms such as the phosphorylation of R* or an increase in guanylate cyclase activity to compensate for the cGMP hydrolyzed by PDE. However, if $T_\alpha$-GDP·Pi is incapable of activating PDE, it would imply that the rapid on-site hydrolysis of the $T_\alpha$-bound GTP may function as the deactivation step for transducin. Since on-site GTP hydrolysis occurs between 100 ms and 3 s, the on-site hydrolysis of GTP by transducin may still play a significant role in shutting off the cascade. A comparison between the rate of Pi release from $T_\alpha$-GDP·Pi and the rate of PDE deactivation associated with the turnover of transducin has been investigated. If these two rates are similar, $T_\alpha$-GDP·Pi is capable of activating PDE. However, if the rate of PDE deactivation is much faster than the rate of Pi release, the deactivation of transducin is associated with the on-site hydrolysis of $T_\alpha$-bound GTP.

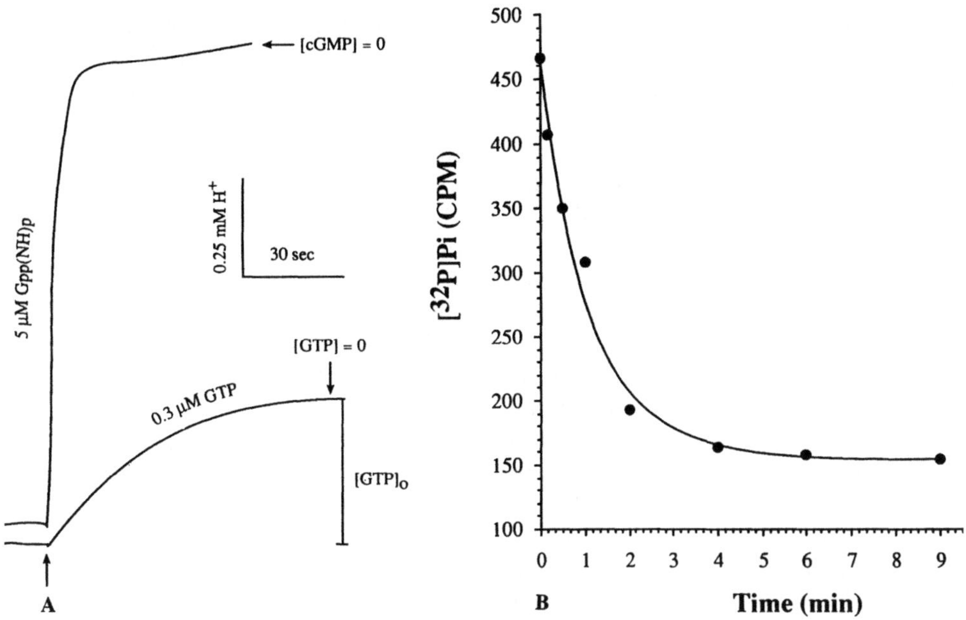


**Fig. 5A,B.** The rates of PDE deactivation associated with the turnover of transducin and Pi release from $T_\alpha$-GDP·Pi. **A** Results of the PDE activation assay as monitored by the rate of $H^+$ release due to cGMP hydrolysis using a pH electrode. Each sample contained 25 $\mu M$ R*, 10 $\mu$g T, 20 $\mu$g PDE, and 5 m$M$ cGMP. The reactions were initiated by the addition of 0.3 $\mu M$ GTP or 5 $\mu M$ Gpp(NH)p. *Arrow*, addition of GTP or Gpp(NH)p. *Upper curve*, total hydrolysis of the cGMP in the reaction mixture, where the cascade was activated by the addition of nonhydrolyzeable Gpp(NH)p; *lower curve*, deactivation of PDE activity due to the hydrolysis of bound GTP by transducin. **B** Result of the experiments measuring the rate of Pi release from $T_\alpha$-GDP·Pi. A 30-$\mu$l sample containing 30 $\mu M$ R* and 10 $\mu$g T was mixed with 30 $\mu$l of excess [$\gamma$-$^{32}$P]GTP (60 $\mu M$) at time =0. The reaction was allowed to reach steady state and was subsequently filtered through a nitrocellulose filter. The filters were incubated for various time intervals and washed again with buffer. The radioactivity retained by the filter was then measured

The rate of PDE deactivation associated with the turnover of transducin can be measured in a system containing R*, purified transducin, PDE, and excess cGMP (Frey et al. 1988). The results are shown in Fig. 5A. A pH electrode is used to monitor the PDE activity, which can be detected as a decrease in medium pH due to cGMP hydrolysis. When a limited amount of GTP is added at time zero to initiate the reaction, hydrolysis of cGMP due to PDE activation increases but slowly diminishes back to the basal level. The diminishing PDE activity is due to the turnover of transducin's GTPase activity consuming all the added GTP. From the decay of PDE activity a first-order rate constant of 0.018 $s^{-1}$ for the deactivation of PDE is obtained. The rate of Pi release from $T_\alpha$-GDP·Pi can be measured by a filtering-washing method (Navon and Fung 1984); the results are shown in Fig. 5B.

A reconstituted sample containing R* and transducin is mixed with excess [$\gamma$-$^{32}$P]GTP (30 $\mu M$) at time =0. The reaction mixture is incubated for 15 s to allow all the transducin to become $T_\alpha$-GDP·[$^{32}$P]Pi. The reaction mixture is then diluted with buffer and immediately filtered through a nitrocellulose filter. $T_\alpha$-GDP·[$^{32}$P]Pi is retained by the filter, whereas excess [$\gamma$-$^{32}$P]GTP and free [$^{32}$P]Pi are removed in the filtrate. The removal of [$\gamma$-$^{32}$P]GTP from the filter prevents any additional GTP/GDP exchange reactions from occurring. However, the $T_\alpha$-GDP.[$^{32}$P]Pi complex associated on the moistened nitrocellulose remains active and slowly releases its tightly bound Pi. The nitrocellulose filters containing $T_\alpha$-GDP·[$^{32}$P]Pi are allowed to stand for various time intervals and are subsequently washed with cold buffer to remove any free [$^{32}$P]Pi released from $T_\alpha$-GDP·[$^{32}$P]Pi. Thus, the decrease of radioactivity associated with the nitrocellulose filters as a function of time between the loading $T_\alpha$-GDP·[$^{32}$P]Pi onto the filter, and the removal of free [$^{32}$P]Pi by washing the filter should represent the rate of Pi release from the $T_\alpha$-GDP·[$^{32}$P]Pi. The [$^{32}$P]Pi release rate also follows first-order kinetics with a first-order rate constant of 0.0173 $s^{-1}$ which is similar to the rate of PDE deactivation ($k = 0.018\,s^{-1}$) when measured at the same temperature. This comparison conclusively demonstrates that the release of Pi from $T_\alpha$-GDP·Pi is the deactivation step of transducin and $T_\alpha$-GTP and $T_\alpha$-GDP·Pi should have similar ability to activate PDE. Judging from the slow deactivation rate of PDE as regulated by transducin's GTPase activity, one must conclude that the retinal cGMP cascade is probably turned off by other processes such as R* phosphorylation and guanylate cyclase activity enhancement, which occur in a faster time scale. After releasing the tightly bound Pi, $T_\alpha$-GDP resumes an inactive conformation and recombines with $T_{\beta\gamma}$ to be ready for another R*-catalyzed activation cycle. Using [$^3$H]GTP as the radioactive tracer, the stability of the $T_\alpha$-[$^3$H]GDP is examined. The [$^3$H]GDP remains tightly bound to $T_\alpha$ for more than 20 min in the filtration assay, as described above. This observation implies that $T_\alpha$-GDP is a very stable complex, and GDP is never released from $T_\alpha$-GDP to leave an empty nucleotide binding site. The bound GDP can be released only via another cycle of the GTP/GDP exchange reaction in the presence of GTP, $T_{\beta\gamma}$, and R*.

The suggestion that $T_\alpha$-GDP·Pi is capable of activating PDE is consistent with the observed effect of $AlF_4^-$ on G-protein activation. It has been proposed that $AlF_4^-$ functions as a phosphate analogue which occupies the $\gamma$-phosphate position of the GTP binding site on G-proteins. Indeed, studies have shown that in the presence of GDP, $AlF_4^-$ is capable of activating transducin (BIGAY et al. 1987). It is likely that the $T_\alpha$-GDP·$AlF_4^-$ complex is mimicking the structure of $T_\alpha$-GDP·Pi, and as a results, $T_\alpha$-GDP·$AlF_4^-$ activates PDE. The turnover time for transducin estimated from the reconstituted system is approximately 30 s. This is consistent with results from previous studies which used more intact sample preparations, such as electromicroscopic studies of intact rod outer segment membranes (ROOF et al. 1982), who have shown that transducin particles dissociate from the surface of disk membranes after light activation and reappear on the disk membrane 1 min after incubating in the dark.

## E. Regulation of the Transducin Coupling Cycle by Phosducin

In the transducin coupling cycle, $T_{\beta\gamma}$ apparently plays a role in assisting the binding of $T_\alpha$ to R*. Whether dissociated $T_{\beta\gamma}$ carries out additional regulatory functions is still unclear. Recently, a retinal phosphoprotein called phosducin has been isolated and shown to form a specific complex with $T_{\beta\gamma}$ (LEE et al. 1987, 1990b). It is possible that the interaction of $T_{\beta\gamma}$ with phosducin can modulate the availability of $T_{\beta\gamma}$ needed for the recycling of activated $T_\alpha$ and directly affect the continuous activation of the cGMP cascade. Thus, the formation of the phosducin/$T_{\beta\gamma}$ complex may represent a potential site of regulation for the visual excitation process in rod outer segments. Addition of phosducin to photolyzed ROS membrane reduces the GTP hydrolysis activity of transducin as well as the subsequent activation of PDE. The inhibitory effects of phosducin can be reversed by the addition of exogenous $T_{\beta\gamma}$ (LEE et al. 1992). These results imply that phosducin is capable of regulating the amount of $T_{\beta\gamma}$ available to interact with $T_\alpha$ to form the active transducin complex and thereby functions as a negative regulator. Phosducin also inhibits the pertussis toxin-catalyzed ADP-ribosylation of transducin, indicating that the interaction between the $T_\alpha$ and $T_{\beta\gamma}$ subunits of transducin is interrupted upon binding of phosducin. A conformational change is induced on $T_{\beta\gamma}$ upon the binding with phosducin that leads to the dissociation of $T_\alpha$. Since phosducin is a soluble protein, the interaction with transducin may occur mainly when transducin is dissociated from ROS disk membrane. Indeed, phosducin failed to dissociate ROS membrane-bound transducin.

This observation has important implications for the physiological role of phosducin. It has been well documented that the initial amplification of the cGMP cascade involves the activation of hundreds of transducin by a single R* via lateral diffusion on the disk membrane to activate membrane-associated transducin. If phosducin has no effect on membrane-bound transducin, it should not be involved in regulating the initial activation of transducin. However, phosducin may inhibit the recycling of activated transducin which are dissociated from ROS membrane. To illustrate this point, the effect of phosducin on the pre-steady-state kinetics of GTP hydrolysis by transducin has been examined. The results are shown in Fig. 6 and indicate that phosducin has little effect on the rate of the initial burst of Pi formation. However, the steady-state rate of GTP hydrolysis of the phosducin-containing sample is decreased. This demonstrates that phosducin can regulate the transducin coupling cycle by modulating the turnover of soluble transducin without reducing the initial activation of the transducin

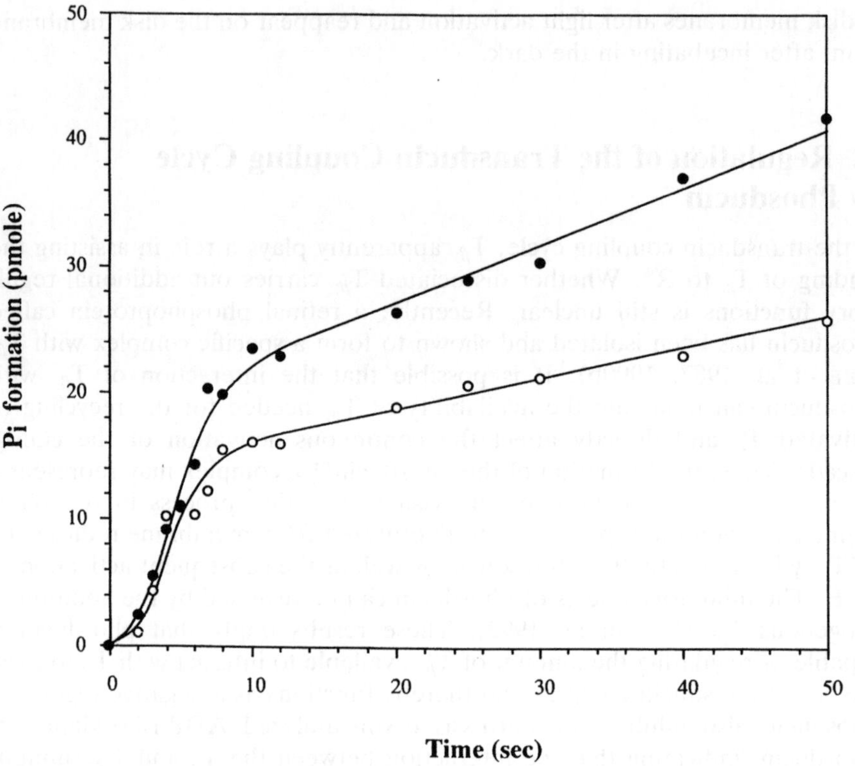


**Fig. 6.** The effect of phosducin on the pre-steady-state kinetics of transducin GTPase activity. The GTPase assays were conducted at 22°C. The reaction mixture contained 30$\mu$l solution of 6$\mu M$ R* and 20$\mu$g T. The reaction was initiated by the addition of 30$\mu$l 60m$M$ [$\gamma$-$^{32}$P]GTP. Time courses of Pi formation due to GTP hydrolysis by transducin in the absence of phosducin (●) and in the presence of 10$\mu$g phosducin (○) are shown. The initial burst indicates the rapid hydrolysis of transducin-bound GTP which leads to the formation of a tightly bound T-GDP·Pi complex. The steady rate of Pi formation represents the steady rate of GTP hydrolysis which is related to the Pi release rate from the T-GDP·Pi complexes

cycle. In the dark, phosducin is phosphorylated by protein kinase A at Ser-73 and its levels of phosphorylation is modulated by light (LEE et al. 1990a). Upon illumination, phosducin is rapidly dephosphorylated by phosphatase. Although both the phosphorylated and dephosphorylated forms of phosducin are capable of binding to $T_{\beta\gamma}$, the dephosphorylated form demonstrates a higher affinity. Thus, the light-dependent phosphorylation/dephosphorylation cycle of phosducin may represent another mechanism of regulating the transducin coupling cycle.

# F. Concluding Remarks

GTP-binding proteins have been shown to participate in various biological coupling functions. Examples include elongation factors in protein synthesis, tubulin in microtubule formation, signal-transducing G-proteins in hormonal and sensory coupling, and small G-proteins in controlling cell growth and secretion. The basic mechanism utilized by these proteins is a subunit association and dissociation cycle regulated by binding and hydrolysis of guanine nucleotide (Ho et al. 1989). The overall GTP hydrolysis mechanism among these GTP-binding proteins is similar. The bound GTP is directly hydrolyzed without forming a phosphorylated intermediate as observed in $Na^+/K^+$ and $Ca^{2+}$ ATPases. However, the time scales of various coupling functions carried out by these GTP-binding proteins differ greatly, from the milliseconds for transducin, to seconds in microtubule formation, to minutes in protein synthesis and hormonal regulation, and to hours in cell transformation. Although these GTP-binding proteins share a similar structure at the GTP binding site and the ability to exchange and hydrolyze nucleotide, their reaction kinetics must be quite different in order to accommodate the wide range of time scales of their coupling functions. Similar kinetic analyzes described in this chapter for transducin has been applied to other GTP-binding proteins. In the case of $G_o$, the on-site hydrolysis of bound GTP is slow and is followed by rapid Pi release without the accumulation of $G_o$-GDP·Pi complex (Higashijima et al. 1987). In the case of *ras*-p21 protein, the rate limiting step of the GTP hydrolysis reaction is a GTP conformational change rather on the hydrolysis reaction itself (Neal et al. 1990). In elongation factor Tu, a transient Tu-GDP·Pi complex can be detected in the presence of $Mn^{2+}$ (Kalbitzer et al. 1990). In tubulin the release of tightly bound Pi again becomes rate limiting and leads to the formation of tubulin-GDP·Pi complex (Melki et al. 1990). A comparison of the reaction dynamics and energetics of the coupling cycles among these GTP-binding proteins will elucidate the common features of these molecular switches in cellular regulation.

## References

Bennett N, Dupont Y (1985) The G-protein of retinal rod outer segments (transducin): mechanism of interaction with rhodopsin and nucleotides. J Biol Chem 260:4156–4168

Bigay J, Deterre P, Pfister C, Chabre M (1987) Fluoride complexes of aluminum or beryllium act on G-proteins as reversibly bound analogues of the gamma phosphate of GTP. EMBO J 6:2907–2913

Cornish-Bowden A (1976) Principles of enzyme kinetics. Butterworth & Co., England, p 156

Fesenko EE, Kolesnikov SS, Lynbarsky AL (1985) Induction of cGMP of cationic conductance on plasma membrane of retinal rod outer segment. Nature (London) 313:310–313

Frey SE, Hingorani VN, Su-Tsai S-M, Ho Y-K (1988) Chromium (III) $\beta, \gamma$-Bidentate guanine nucleotide complexes as probes of the cGMP cascade of retinal rod outer segments. Biochemistry 27:8209–8218
Fung BK-K (1983) Characterization of transducin from bovine retinal rod outer segments. I. Separation and reconstitution of the subunits. J Biol Chem 258: 10495–10502
Fung BK-K, Hurley JB, Stryer L (1981) Flow of information in the light-triggered cyclic nucleotide cascade of vision. Proc Natl Acad Sci USA 78:152–156
Fung BK-K, Stryer L (1980) Photolyzed rhodopsin catalyzes the exchange of GTP for GDP in retinal rod outer segment membranes. Proc Natl Acad Sci USA 78:2500–2504
Higashijima T, Ferguson KM, Sternweis PC, Smigel MD, Gilman AG (1987) Effects of $Mg^{2+}$ and the beta gamma-subunit complex on the interactions of guanine nucleotides with G proteins. J Biol Chem 262:762–766
Ho Y-K, Hingorani VN, Navon SE, Fung BK-K (1989) Transducin: a signaling switch regulated by guanine nucleotides. In: Chock TB (ed) Current topics in cellular regulation. Academic Press, New York, p 171
Hurley JB, Stryer L (1982) Purification and characterization of the gamma regulatory subunit of the cyclic GMP phosphodiesterase from retinal rod outer segments. J Biol Chem 257:11094–11099
Kalbitzer HR, Feuerstein J, Goody RS, Wittinghofer A (1990) Stereochemistry and lifetime of the GTP hydrolysis intermediate at the active site of elongation factor Tu from Bacillus stearothermophilus as inferred from the 170–55 Mn super-hyperfine interaction. Eur J Biochem 188:355–359
Kuhn H, Bennett N, Michel-Villaz M, Chabre M (1981) Interactions between photoexcited rhodopsin and GTP-binding protein: kinetics and stoichiometric analysis from light-scattering changes. Proc Natl Acad Sci USA 78:6873–6877
Lee R-H, Brown BM, Lolley RN (1990a) Protein kinase A phosphorylates retinal phosducin on serine 73 in situ. J Biol Chem 265:15860–15866
Lee R-H, Fowler A, McGinnis JF, Lolley RN, Craft CM (1990b) Amino acid and cDNA sequence of bovine phosducin, a soluble phosphoprotein from photoreceptor cells. J Biol Chem 265:15867–15873
Lee R-H, Lieberman BS, Lolley RN (1987) A novel complex from bovine visual cells of a 33 000-dalton phosphoprotein with beta- and gamma-transducin: purification and subunit structure. Biochemistry 26:3983–3990
Lee R-H, Ting TD, Lieberman BS, Tobias DE, Lolley RN, Ho Y-K (1992) Regulation of retinal cGMP cascade by phosducin in bovine rod photoreceptor cells: interaction of phosducin and transducin. J Biol Chem 267:25104–25112
Liebman PA, Pugh EN Jr (1980) ATP mediates rapid reversal of cyclic GMP phosphodiesterase activation in visual receptor membranes. Nature (London) 287:734–736
Matesic D, Liebman PA (1992) Spontaneous G-protein nucleotide exchange rate: implications for rod dark noise. Invest Ophthal Visual Sci 33(4):1102
Melki R, Carlier M-F, Pantaloni D (1990) Direct evidence for GTP and GDP·Pi intermediates in microtubule assembly. Biochemistry 29:8921–8932
Navon SE, Fung BK-K (1984) Characterization of transducin from bovine retinal rod outer segments. J Biol Chem 259:6686–6693
Neal SE, Eccleston JF, Webb MR (1990) Hydrolysis of GTP by p21NRAS, the NRAS protooncogene product, is accompanied by a conformational change in the wild-type protein: use of a single fluorescent probe at the catalytic site. Proc Natl Acad Sci USA 87:3562–3565
Roof DJ, Korenbrot JI, Heuser JE (1982) Surfaces of rod photoreceptor disk membranes: light-activated enzymes. J Cell-Biol 95:501–509
Sitaramayya A, Liebman P (1983a) Mechanism of ATP quench of phosphodiesterase activation in rod disc membranes. J Biol Chem 258:1205–1209

Sitaramayya A, Liebman PA (1983b) Phosphorylation of rhodopsin and quenching of cyclic GMP phosphodiesterase activation by ATP at weak bleaches. J Biol Chem 258:12106–12109
Stryer L (1991) Visual excitation and recovery. J Biol Chem 266:10711–10714
Ting TD, Ho Y-K (1991) Molecular mechanism of GTP hydrolysis by bovine transducin: pre-steady-state kinetic analyses. Biochemistry 30:8996–9007
Wessling-Resnick M, Johnson GL (1987) Allosteric behavior in transducin activation mediated by rhodopsin: inital rate analysis of guanine nucleotide exchange. J Biol Chem 262:3697–3705
Wilden U, Hall SW, Kuhn H (1986) Phosphodiesterase activation by photoexcited rhodopsin is quenched when rhodopsin is phosphorylated and binds the intrinsic 48-kDa protein of rod outer segments. Proc Natl Acad Sci USA 83:1174–1178
Woodruff ML, Fain GL (1982) $Ca^{2+}$-dependent changes in cyclic GMP levels are not correlated with opening and closing for the light-dependent permeability of toad photoreceptors. J Gen Physiol 80:537–555

CHAPTER 51

# Transcriptional, Posttranscriptional, and Posttranslational Regulation of G-Proteins and Adrenergic Receptors

J.R. HADCOCK and C.C. MALBON

## A. Introduction

Transmembrane signaling via G-protein-linked pathways is regulated dynamically. Catecholamines regulate fundamental physiological pathways, such as respiration, cardiac rate, and lipolysis. Consequently, adrenergic receptors and their G-protein partners through which adenylyl cyclase, phospholipase C, and other effectors are controlled have been adopted as models for intensive study of the regulation of G-protein-mediated pathways. The epertoire of mechanisms for regulation includes transcriptional, posttransriptional, and posttranslational components. Acute, short-term (seconds to minutes) regulation, typified by agonist-induced desensitization, virtually precludes significant transcriptional and posttranscriptional components which are capable of changing steady-state mRNA levels and thereby protein expression. Posttranslational mechanisms, such as protein phosphorylatiion, are most prominent in the acute phases, whereas longer term (hours) regulation may include transcriptional, posttranscriptional, as well as posttranslational components (HAUSDORFF et al. 1990). Both receptors (GEORGE et al. 1988) and G-proteeins (RAPIEJKO et al. 1989) have been shown to be the locus of physiological regulation at which an entire pathway can be influenced. Furthermore, G-protein-linked pathways do not operate in isolation and are subject not only to agonist-induced homologous regulation but also to heterologous regulation by ligands operating via other G-protein-linked pathways, gene regulaation (steroids, retinoids), and/or tyrosine kinases (insulin and growth factors). This chapter highlights mechanisms by which G-protein-mediated pathways are regulated , drawing heavily upon our knowledge of adrenergic receptor-mediated responses.

# B. Agonist-Induced Regulation of Transmembrane Signaling

## I. Transcriptional and Posttranscriptional Regulation

Stimulating cells with $\beta$-adrenergic agonist alters in a biphasic manner the steady-state mRNA levels of the $\beta_2$-adrenergic receptor. Acute challenge of cells (<4 h) with agonist (or with agents such as for kolin which increase intracellular cyclic AMP) transiently increases receptor mRNA and the relative rate of transcription of the receptor gene (COLLINS et al. 1989). The $\beta_2$-adrenergic receptor gene harbors a cyclic AMP-responsive element (CRE), and nuclear run-on assays reveal a 70% increase in the relative rate of transcription in nuclei isolated from agonist-treated cells. This transcriptional activation is transient; however, relative rates of transcription decline to control levels within 4 h. Reporter genes driven by this receptor promoter display a cyclic AMP-sensitive increase in transcriptional activity. DNase I footprinting analysis has identified a specific palindromic sequence (TGACGTCA) in the $\beta_2$-adrenergic receptor gene to which the CRE-binding (CREB) protein binds. Interestingly, no increase in expressed receptor has been observed following short-term challenge with agonist. The physiological significance of the transient up-regulation of receptor mRNA by transcriptional activation via a CRE remains to be established.

Following a transient, modest increase in receptor mRNA, agonists induce a sharp decline in receptor mRNA (HADCOCK and MALBON 1988a; HADCOCK et al. 1989b; COLLINS et al. 1989; BOUVIER et al. 1989). Whereas transcriptional activation of the $\beta_2$-adrenergic receptor gene appears to be dependent solely on elevating intracellular cyclic AMP and activation of A-Kanase, agonist-promoted down-regulation of receptor mRNA appears to be more multifaceted. A-Kanase activity is a prominent, but not the sole, determinant for agonist-induced down regulation of $\beta_2$-adrenergic receptor mRNA. HADCOOCK and MALBON (1988a) and COLLINS et al. (1989) observed a greater down-regulation of receptor mRNA with agonist than forskolin or cyclic AMP analogs, even though intracellular cyclic AMP was comparable. Further evidence of "cyclic AMP-independent" effect came from analysis agonist regulation of $B_2$-adrenergic receptor mRNA from a series of S49 mouse lymphoma variants with mutations in the G-protein $G_s$ (HADCOCK et al. 1989a). One such mutant, *H21A*, expresses a $G_{s\alpha}$ which has normal coupling between receptor and G-protein but an impaired coupling between G-protein and adenylyl cyclase. Although no cyclic AMP accumulated in response to agonist in H21A mutants, down-regulation (approximately 50% of that observed in S49 wild-type cells) of $\beta_2$-adrenergic receptor mRNA was observed. In S49 *kin*$^-$ cells deficient of A-Kanase activity, no agonist-promoted down-regulation of receptor mRNA was observed, demonstrating an obligate role of A-Kanase activity in agonist-induced down-regulation.

How do agonists down-regulate receptor mRNA levels? Transcription of the receptor gene was not repressed by agonist treatment of the cells, yet analysis of receptor mRNA revealed an approximate 50% decline (HADCOCK et al. 1989a). This decline in receptor mRNA was shown to reflect an agonist-induced destabilization of preexisting receptor mRNA (HADCOCK et al. 1989b). BOUVIER et al. (1989) also concluded from additional studies that the agonist-induced decline in receptor mRNA was probably due to destabilization of its mRNA rather than a decrease in transcription.

A destabilization consensus sequence (SHAW et al. 1986) in the 3′ untranslated region (UTR) of the $\beta_2$-adrenergic receptor mRNA has been identified (PORT et al. 1992). Several other G-protein-linked receptors ($D_1$ dopamine, FSH, LH-CG) also contain this consensus sequence (COLLINS et

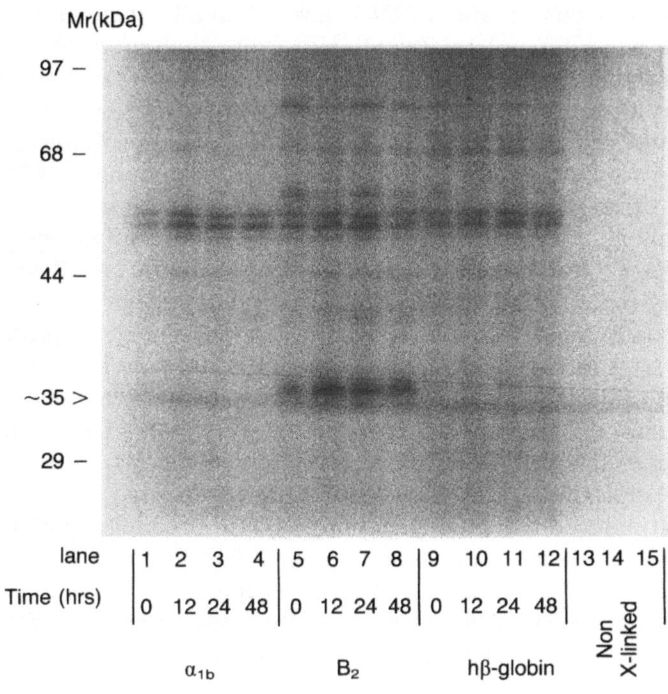


**Fig. 1.** UV cross-linking of a $M_r$ 35 000 protein(s) that selectively binds $\beta$-adrenergic receptor mRNAs. Representative autoradiogram of UV cross-linking between S100 cytosolic fractions from $DDT_1$-MF2 cells and full-length, capped, and uniformly labeled in vitro transcribed mRNAs corresponding to the $\alpha_{1b}$- (*lanes 1–4, 13*) and $\beta_2$-adrenergic receptors (*lanes 5–8, 14*) and to human $\beta$-globin (*lanes 9–12, 15*); *lanes 13–15*, non-UV-cross-linked controls. Cells were untreated or treated for up to 48 h with the cAMP analog CPT-cAMP (10 $\mu M$) as indicated. Fresh media and CPT were added to cells daily. Equal amounts of S100 cytosolic protein and CPMs for each radiolabeled mRNA were added into the appropriate lane. A distinct band appears at approximately $M_r$ 35 000 in lanes 5–8 corresponding to the $\beta_2$-adrenergic receptor mRNA. (From PORT et al. 1992)

al. 1991). More recently a $M_r$ 35 000 RNA-binding protein that specifically binds mRNA of receptors (such as $\beta_1$- and $\beta_2$- but not $\alpha_1$-adrenergic) that display agonist-induced down-regulation of mRNA was discovered (Fig. 1; PORT et al. 1992). This protein, termed the $\beta$-adrenergic receptor mRNA-binding ($\beta$ARB) protein, is induced in cells treated with agonist and displays a AUUUA and U-rich binding motif (HUANG et al. 1992). The precise role which the AUUUA pentamer-containing sequence plays, if any, in the destabilization of receptor mRNA remains to be elucidated.

Depending on the tissue or cell type, the expression of the stimulatory G-protein ($G_{s\alpha}$) is also regulated by stimulatory agonists. Regulation can be observed both at the level of mRNA and that of protein turnover. For example, persistent activation of adenylyl cyclase in S49 cells by $\beta$-adrenergic agonists or forskolin decreases the steady-state levels of $G_{s\alpha}$ 25% (HADCOCK et al. 1991). The mechanisms by which $G_{s\alpha}$ levels decline was not immediately obvious as the mRNA levels actually increased slightly in response to isoproterenol treatment. However, analysis of the half-life of $G_{s\alpha}$ by metabolic labeling and immunoprecipitation revealed a decrease in the half-life of $G_{s\alpha}$ in the treated cells. The half-life of $G_{s\alpha}$ declined from 52 h in control cells to 32 h in cells treated with forskolin. Cholera toxin has been shown to alter steady-state levels of $G_{s\alpha}$, promoting an 80% decline within 12 h of intoxication (CHANG and BOURNE 1989). Based upon the data described above, elevated cyclic AMP is unlikely to explain the decline in $G_{s\alpha}$. Perhaps the posttranslational modification (ADP-ribosylation) catalyzed by the toxin enhances turnover (degradation) of $G_{s\alpha}$.

Little data exist on the regulation of inhibitory G-protein-linked receptors, such as the $\alpha_2$-adrenergic receptor. MONTMAYEUR and BORRELLI (1991) reported that activation of the inhibitory pathway of adenylyl cyclase via dopamine $D_2$ receptors can reduce the activity of a CRE coupled to a reporter gene. Though speculation, inhibitory receptors with a functional CRE in their promoter may also display auto regulation.

Agonist-induced regulation of phospholipase C coupled pathways is less well understood. Izzo et al. (1990) reported that norepinephrine promoted a transient reduction in $\alpha_{1b}$-adrenergic receptor mRNA levels in rabbit aortic smooth muscle cells. The $\alpha_{1b}$-adrenergic receptor mRNA levels declined to 20% of control levels by 4 h of treatment with norepinephrine. By 24 h receptor mRNA levels were at that observed in control cells. Inhibition of transcription with the antibiotic actinomycin D revealed that the down-regulation of $\alpha_{1b}$-adrenergic receptor mRNA was probably due to a change in stability of the message. Changes in the transcription rate of the gene could not, however, be excluded. MORRIS et al. (1991) reported that activation of $\alpha_1$-adrenergic receptors of smooth muscle cells in culture promoted a down-regulation of receptor expression and mRNA that was preceded by a cross-regulatory response that actually increased $\alpha_1$-receptor mRNA and binding (see below). Thus, agonist-induced down-regulation of receptor expression employs several mechanisms including transcriptional, posttranscriptional, and posttranslational.

### II. Posttranslational Regulation

Homologous regulation of G-protein-mediated responses includes a prominent role for posttranslational mechanisms, especially protein phosphorylation, sequestration of signal transducing elements, and changes in protein turnover of the elements that compose the system (HAUSDORFF et al. 1990). Each of these topics is discussed elsewhere in this volume. Specific examples of posttranslational control of G-proteins and their receptors are integrated in the sections that follow.

## C. Cross-Regulation in Transmembrane Signaling

### I. Stimulatory to Inhibitory Adenylyl Cyclase

The existence of multiple G-protein-linked pathways within a single cell forces the consideration of how pathways communicate or "cross-regulate" each other and integrate incoming signals. Some cells express not only $\beta_2$- (linked to stimulatory adenylyl cyclase) and $\alpha_2$- (linked to inhibitory adenylyl cyclase), but also $\alpha_1$- (linked to phospholipase C) adrenergic receptors, which are all activated by epinephrine. Recent studies have established the existence of cross-regulation among G-protein-linked pathways. The strategy employed has been to activate one pathway and explore transmembrane signaling propagated via other G-protein-mediated pathways.

Adenylyl cyclase is under stimulatory and inhibitory control, competing pathways with distinct receptors and G-proteins. Persistent activation of stimulatory adenylyl cyclase has been shown to increase $G_{i\alpha 2}$ expression (RICH et al. 1984; FELDMAN et al. 1988; REITHMAN et al. 1990). Persistent activation (24 h) of the stimulatory pathway with a $\beta$-adrenergic agonist or the diterpene forskolin enhanced the inhibitory response to somatostatin (HADCOCK et al. 1990). G-proteins were the targets for the cross-regulation of inhibitory by stimulatory adenylyl cyclase. $G_{i\alpha 2}$, which mediates the inhibitory adenylyl cyclase response, increased threefold, while $G_{s\alpha}$ levels declined 25% following persistent activation of the stimulatory pathway. $G_{i\alpha 2}$ mRNA levels increased fourfold within 12 h, while mRNA levels for $G_{s\alpha}$ rose initially and then declined to 75% of control. Stimulatory to inhibitory cross-regulation was not observed in S49 *kin*⁻ cells lacking A-Kanase, demonstrating again an obligate role of A-Kanase. Metabolic labeling with [$^{35}$S]methionine revealed increased synthesis rates for $G_{i\alpha 2}$ and no change in half-life ($t_{1/2}$ 80 h). Steady-state levels of $G_{s\alpha}$ declined in spite of a modest increase in the synthetic rate (reflecting increased levels of mRNA) due to a decrease in the $t_{1/2}$ from 55 to 34 h in cells persistently activated. The existence of a CRE in the $G_{i\alpha 2}$ gene suggests that stimulatory to inhibitory cross-regulation of adenylyl cyclase involves transcriptional activation and perhaps posttranscriptional control. The gene encoding $G_{s\alpha}$, in contrast, is devoid of CREs (KOZASA et al. 1988). In addition, alterations in protein turnover appear to dominate regulation of the steady-state levels of $G_{s\alpha}$.

## II. Inhibitory to Stimulatory Adenylyl Cyclase

The discovery of cross-regulation from stimulatory to inhibitory adenylyl cyclase prompted exploration for other examples of cross-regulation. Phenomenology suggesting cross-regulation from the inhibitory to stimulatory adenylyl cyclase was reported early (HOFFMAN et al. 1986; GREEN 1987; LONGABAUGH et al. 1989). LONGABAUGH et al. reported that persistent activation of the inhibitory pathway of rat fat cells with $A_1$ adenosine agonist appeared to increase $G_{s\alpha}$ while decreasing $G_{i\alpha2}$. Interestingly, no changes in mRNA levels of the G-protein subunits were observed. Recent work in smooth muscle cells in culture revealed that persistent activation of the inhibitory pathway with $A_1$ adenosine agonists enhanced by two-fold the stimulatory response of the adenylyl cyclase to $\beta$-adrenergic agonist (HADCOCK et al. 1991). The $ED_{50}$ for stimulation by isoproterenol declined from 50 to 1 n*M*. $\beta$-Adrenergic receptor mRNA and expression were increased; the $t_{1/2}$ for the receptor protein was not altered. $A_1$ adenosine receptor number and $G_{i\alpha2}$ declined by half in cells that were persistently activated by $A_1$ agonists. Metabolic labeling with [$^{35}$S]methionine showed a small increase in synthetic rate of $G_{i\alpha2}$, coupled to a dramatic decline in the $t_{1/2}$ of $G_{i\alpha2}$. The steady-state levels of $G_{s\alpha}$ and its mRNA, in contrast, were not altered. Thus, transcriptional and posttranslational mechanisms appear to predominate in inhibitory to stimulatory cross-regulation.

In the short term, acute activation of the inhibitory pathway of adenylyl cyclase was shown to enhance the stimulatory response, but through an entirely different mechanism (PORT et al. 1992a). Activation (15–60 min) of the inhibitory adenylyl cyclase pathway of smooth muscle cells in culture by $A_1$ adenosine agonists enhanced the stimulatory response at the level of the $\beta$-adrenergic receptor; $G_{s\alpha}$ and the adenylyl cyclase activities were unaltered. Metabolic labeling of cells with [$^{32}$P]orthophosphate and immuneprecipitation of the $\beta$ receptors revealed basal phosphorylation in unstimulated cells and a marked phosphorylation in epinephrine-stimulated cells. When the inhibitory response was activated in the short term, however, the basal phosphorylation state of the receptor declined by 75%. Thus, short-term cross-regulation from the inhibitory to stimulatory adenylyl cyclase pathway displays decreased intracellular cyclic AMP, reduced A-Kanase activity, decreased basal phosphorylation of the $\beta$-adrenergic receptor, and enhanced receptor function (PORT et al. 1992a).

## III. Stimulatory Adenylyl Cyclase to Phospholipase C

Studies of homologous regulation of $\alpha_1$-adrenergic receptors by epinephrine not only identified a down-regulation of receptor mRNA and binding much like that observed for $\beta_2$-adrenergic receptors, but a fascinating cross-regulation between two G-protein-linked pathways – the stimulatory adenylyl cyclase and phospholipase C (Morris et al. 1991; Fig. 2). Short-term (2–8 h) challenge of smooth muscle cells in culture with epinephrine led to an enhanced $\alpha_1$-adrenergic response, increased receptor mRNA, and binding. A latter phase of epinephrine challenge resulted in a down-regulation of receptor mRNA and binding. Interestingly, the initial up-regulation of $\alpha_1$ receptors by epinephrine was blocked by $\beta$- but not by $\alpha$-adrenergic antagonists. The down-regulatory phase was the opposite, being sensitive to $\alpha_1$- but not to $\beta$-adrenergic antagonists. This was the first report of a heterologous up-regulation of receptor mRNA levels for adrenergic receptors. The existence of cross-regulation from stimulatory adenylyl cyclase to phospholipase C pathways was established by these observations, and cyclic AMP appeared to be the mediator. The existence of a CRE in the $\alpha_1$-receptor gene provides the basis for the increase in receptor mRNA, presumably via transcriptional activation following stimulation by epinephrine. The down-regulation of receptor mRNA, which is $\alpha_1$ agonist

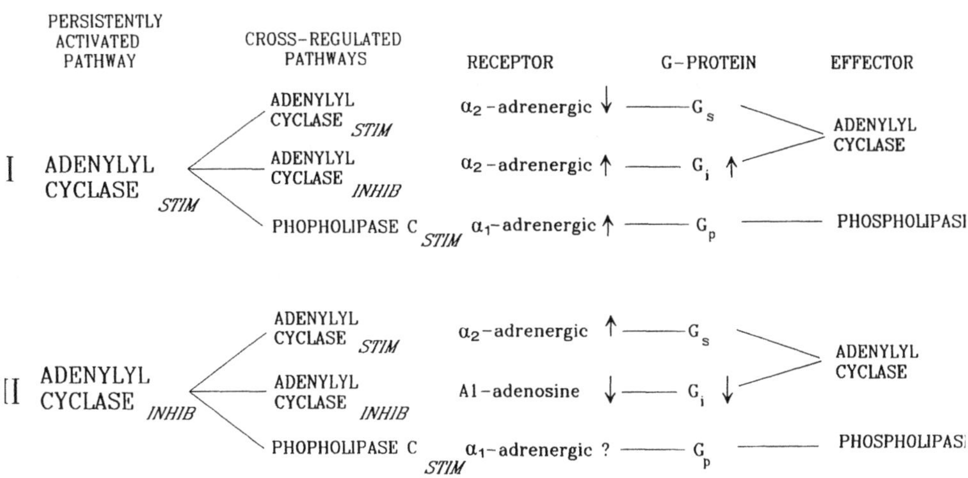


**Fig. 2.** Cross-regulation between G-protein-mediated pathways: regulation of expression. For the stimulatory pathway of adenylyl cyclase (*pathway I*), cyclic AMP is the output, activating the cAMP-dependent protein kinase (protein kinase A). For the inhibitory adenylyl cyclase pathway (*pathway II*), attenuation of protein kinase A activity represents the major effector response. For the stimulatory pathway of phospholipase C, activation of protein kinase C and increases in intracellular $Ca^{2+}$ mobilization are the two outputs. For details of the regulation of system elements in response to activations of the $G_{s\alpha}$-mediated and $G_{i\alpha}$-mediated pathways, see the text. *STIM*, stimulatory; *INHIB*, inhibitory. (From Hadcock et al. 1991)

specific, also appears to be transcriptional. Unlike the agonist-induced down-regulation of $\beta_2$-adrenergic receptor via receptor mRNA destabilization, agonist-induced down-regulation of $\alpha_1$-adrenergic receptors and their mRNA did not involve a change in the $t_{1/2}$ of the receptor mRNA in agonist-treated cells (MORRIS et al. 1991).

### IV. Tyrosine Kinase to Stimulatory Adenylyl Cyclase

Integration of transmembrane signaling would require cross-talk between tyrosine kinase and G-protein-linked receptors (HADCOCK and MALBON 1991). Elements of transmembrane signaling systems not only propagate signals but also may be substrates for phosphorylation. Only recently has cross-talk between G-protein-linked pathways and tyrosine kinase receptors been explored. Insulin and catecholamines are counter-regulatory with respect to carbohydrate, protein, and lipid metabolism. Treating smooth muscle cells in culture with insulin has been shown to blunt catecholamine-stimulated adenylyl cyclase activity at the level of the receptor but not that of G-proteins (HADCOCK et al. 1992). The decrease in $\beta_2$-adrenergic receptor-stimulated activity correlated with an enhanced phosphorylation of $\beta_2$-receptor in the insulin-treated cells. Phosphorylation of $G_{i\alpha 2}$, in contrast, was unaffected. Phosphoamino acid analysis of receptor from metabolically labeled ([$^{32}$P]orthophosphate) cells revealed basal phosphorylation of tyrosine, threonine, and serine. Phosphotyrosine content increased and phosphothreonine content declined in receptors isolated from insulin-treated cells. The extent to which cross-talk exists between elements of G-protein-linked and tyrosine kinase receptor systems remains to be established, but these early studies suggest the existence of a fascinating and novel mode of regulating transmembrane signaling.

## D. Permissive Hormone Regulation of Transmembrane Signaling

Transmembrane signaling via G-proteins is regulated not only via agonist stimulation and cross-regulation, but also indirectly via permissive hormones such as the steroids that can increase or decrease the through-put of signals propagated via G-protein-mediated pathways (MALBON et al. 1989). Thyroid hormones, glucocorticoids, and retinoids provide excellent examples of permissive hormones that regulate complex, hormonally sensitive pathways indirectly through effects exerted upon G-proteins and/or the receptors to which they couple. SAGGERSON (1992) has reviewed in depth the permissive effects of thyroid hormones on G-protein-linked pathways. Readers should consult this work for an exhaustive treatment of the topic; only highlights are provided here. Permissive hormone effects exerted at the level of G-proteins are prominent in adipose, liver, and heart. In adipose tissue, hypothyroidism blunts catecholamine-stimulated lipolysis and elevated (two- to fivefold) expression of $G_{i\alpha 2}$ (MALBON et al. 1985) with no change in subunit mRNA (RAPIEJKO et al. 1989). $G_{\beta}$-subunit expression is also enhanced in hypothyroidism (MILLIGAN et al. 1987). The blunted stimulatory response of the adenylyl cyclase and lipolysis in hypothyroidism appears to reflect an enhanced inhibitory component. Similar observations have been made in heart (LEVINE et al. 1990). In the liver, hypothyroidism potentiates rather than blunts $\beta$-adrenergic responses, exerting its influence most prominently at the level of the receptor, with little effect upon G-proteins (MALBON 1980).

The permissive effects of glucocorticoids on G-protein-linked pathways are well known and have been reviewed recently (HADCOCK and MALBON 1992). For $\beta_2$-adrenergic receptors, glucocorticoids increase receptor expression at the protein and mRNA levels (COLLINS et al. 1988; HADCOCK and MALBON 1988a), providing a basis for enhanced $\beta$-adrenergic responses following glucocorticoid exposure. A glucocorticoid response element (GRE) has been identified in the receptor gene, and this GRE located in the 5′-untranslated region appears to be obligate for transcriptional enhancement of the receptor gene (MALBON and HADCOCK 1989). At the G-protein level, in adipose tissue glucocorticoids increase and adrenalectomy decreases the expression of $G_{s\alpha}$, without altering subunit mRNA levels (ROS et al. 1989). These observations suggest a posttranslational mode of regulation of $G_{s\alpha}$ by glucocorticoids. In the brain, adrenalectomy decreases $G_{s\alpha}$ and increases $G_i$ expression, changes reflecting the steady-state levels of subunit mRNA (SAITO et al. 1989). From these few examples it is clear that glucocorticoids express permissive effects on G-proteins at the transcriptional and posttranscriptional levels.

Retinoids, such as the morphogen retinoic acid, have been shown to control differentiation of F9 teratocarcinoma cells in culture (STRICKLAND and MAHDAVI 1978). Recently it has been shown that retinoic acid induced differentiation of F9 stem cells to primitive endoderm results in enhanced stimulatory adenylyl cyclase responses (GALVIN-PARTON et al. 1990). $G_{s\alpha}$ levels remained constant while $G_{i\alpha 2}$ expression declined precipitously as the stem cells differentiated to primitive endoderm. RNA antisense to $G_{i\alpha 2}$ blocked $G_{i\alpha 2}$ expression in stem cells and induced differentiation in the absence of retinoids (WATKINS et al. 1992). The mechanism(s) by which $G_{i\alpha 2}$ expression is controlled by retinoic acid remains to be elucidated but may well include transcriptional and posttranslational components. How G-proteins such as $G_{i\alpha 2}$ and $G_{s\alpha}$ can control the rate of differentiation (WANG et al. 1992) also remains to be defined.

## E. Perspectives

Although fragments of the way in which G-protein-linked transmembrane signaling is regulated have been revealed by recent studies, the puzzle remains largely unsolved. Research often forces one to examine single pathways in isolation in order to simplify the process. The existence of several major motifs for transmembrane signaling necessitates integration of information not only among tyrosine kinase or steroid hormone and G-protein-linked pathways but also among the many G-protein-linked pathways themselves. Overlaid upon these considerations are other regulatory aspects of transmembrane signaling, for example, agonist-induced desensitization and down-regulation. During differentiation and development the complexion of transmembrane signaling may also change. These regulatory phenomena make use of several molecular mechanisms, including transcriptional activation, posttranscriptional regulation, posttranslational modification, and perhaps translational controls. Our task is to define in greater detail the intracellular network of which G-protein-linked pathways are but one, yet a major, component.

*Acknowledgements*. This work was supported in part by United States Health Services grants DK25410 and DK30111 from the National Institutes of Health and American Heart Association grant 900663. C.C.M. is a recipient of career development award K04 AM00786 from the NIH.

## References


Bouvier M, Collins S, O'Dowd BF, Campbell PT, De Blasi, Kobilka BK, MacGregor GP, Caron MG, Lefkowitz RL (1989) Two distinct pathways of Cyclic AMP mediated down-regulation of the $\beta_2$-adrenergic receptor: phosphorylation of the receptor and regulation of its mRNA. J Biol Chem 264:16786–16792

Chang F-H, Bourne, HR (1989) Cholera toxin induces cAMP-independent degradation of $G_s$. J Biol Chem 264:5452–5357

Collins S, Bouvier M, Bolanowski MA, Caron MG, Lefkowitz RJ (1989) cyclic AMP stimulates transcription of the $\beta_2$-adrenergic receptor gene in response to short term agonist exposure. Proc Natl Acad Sci USA 86:4853–4857

Collins S, Altschmied J, Herbsman O, Caron MG, Mellon PL, Lefkowitz RJ (1990) A cyclic AMP response element in the $\beta_2$-adrenergic receptor gene confers transcriptional autoregulation by cyclic AMP. J Biol Chem 265:19330–19335

Collins S, Caron MG, Lefkowitz RJ (1991) Regulation of adrenergic responsiveness through modulation of receptor expression. Annu Rev Physiol 53:497–508

Galvin-Parton PA, Watkins DC, Malbon CC (1990) Retinoic acid modulation of transmembrane signaling. J Biol Chem 265:17771–17779

George ST, Berrios M, Hadcock JR, Wang HY, Malbon CC (1988) Receptor density and cAMP accumulation: analysis in CHO cells exhibiting stable expression of a cDNA that encodes the $\beta_2$-adrenergic receptor. Biochem Biophys Res Commun 150:665–672

Hadcock JR, Malbon CC (1988a) Down-regulation of $\beta$-adrenergic receptors: agonist-induced reduction in receptor mRNA levels. Proc Nat Acad Sci USA 85:5021–5025

Hadcock JR, Malbon CC (1988b) Regulation of $\beta$-adrenergic receptors by "permissive" hormones: glucocorticoids increase steady-state levels of receptor mRNA. Proc Nat Acad Sci USA 85:8415–8419

Hadcock JR, Ros M, Malbon CC (1989a) Agonist regulation of $\beta$-adrenergic receptor mRNA: analysis in S49 mouse lymphoma mutants. J Biol Chem 264:13956–13961

Hadcock JR, Wang HY, Malbon CC (1989b) Agonist-induced destabilization of $\beta$-adrenergic receptor mRNA. Attenuation of glucocorticoid-induced up-regulation of $\beta$-adrenergic receptors. J Biol Chem 264:19928–19933
Hadcock JR, Williams DW, Malbon CC (1989c) Physiological regulation at the level of mRNA. Am J Physiol 257:C457–C465
Hadcock JR, Ros M, Watkins DC, Malbon CC (1990) Cross-regulation between G-protein-mediated pathways. Stimulation of adenylyl cyclase increases expression of the inhibitory G-protein, $G_{i\alpha}$-2. J Biol Chem 265:14784–14790
Hadcock JR, Malbon CC (1991) Regulation of receptor expression by agonists: transcriptional and post-transcriptional controls. Trends in Neurological Sciences 14:242–247
Hadcock JR, Malbon CC (1992) Adrenal dysfunction and G-protein-mediated pathways. In: Milligan G, Wakelam M (eds) G-proteins: signal transduction and disease. Academic, London, p 109
Hadcock JR, Port JD, Malbon CC (1991) Cross-regulation between G-protein-mediated pathways: activation of the inhibitory pathway of adenylyl cyclase increases expression of $\beta_2$-adrenergic receptors. J Biol Chem 266:11915–11922
Hadcock JR, Port JD, Malbon CC (to be published) Cross-talk between tyrosine kinase and G-protein-linked receptors: phosphorylation of $\beta_2$-adrenergic receptors in response to insulin. J Biol Chem
Hausdorff WP, Caron MG, Lefkowitz RJ (1990) Turning off the signal: desensitization of $\beta$-adrenergic receptor function. FASEB J 4:2881–2889
Izzo NJ, Seidman CE, Collins S, Colucci WS (1990) Alpha-1 adrenergic receptor mRNA level is regulated by norepinephrine in rabbit aortic smooth muscle cells. Proc Natl Acad Sci USA 87:6268–6271
Kozasa T, Itoh H, Tsukamoto T, Kaziro Y (1988) Isolation and characterization of the human $G_{s\alpha}$ gene. Proc Natl Acad Sci USA 85:2081–2085
Levine MA, Feldman AM, Robishaw JD, Ladenson PW, Ahn TG, Moroney JF, Smallwood PM (1990) Influence of thyroid hormone status on expression of genes encoding G-protein subunits in the rat heart. J Biol Chem 265:3553–3560
Longabaugh JP, Didsbury J, Spiegel AM, Stiles GL (1989) Modification of the rat adipocyte A1 adenosine receptor-adenylate cyclase system during chronic exposure to an A1 adenosine receptor agonist: alterations in the quantity of $G_{s\alpha}$ and $G_{i\alpha2}$ are not associated with changes in their mRNA. Mol Pharmacol 36:681–688
Malbon CC (1980) Liver cell adenylate cyclase and $\beta$-adrenergic receptors: increased $\beta$-adrenergic receptor number and responsiveness in the hypothyroid rat. J Biol Chem 255:8692–8699
Malbon CC, Rapiejko PJ, Mangano TJ (1985) Fat cell adenylate cyclase system: enhanced inhibition by adenosine and gtp in the hypothyroid rat. J Biol Chem 260:2558–2564
Milligan G, Spiegel AM, Unson CG, Saggerson ED (1987) Chemically induced hypothyroidism produces elevated amounts of the $\alpha$ subunit of the inhibitory guanine nucleotide binding protein ($G_i$) and the $\beta$ subunit common to all G-proteins. Biochem J 247:223–227
Montmayeur JP, Borrelli E (1991) Transcription mediated by a cAMP-responsive element is reduced upon activation of dopamine $D_2$ receptors. Proc Natl Acad Sci USA 88:3135–3139
Morris GM, Hadcock JR, Malbon CC (1991) Cross-regulation between G-protein coupled receptors: activation of $\beta_2$-adrenergic receptors increases $\alpha_1$-adrenergic receptor mRNA levels. J Biol Chem 266:2233–2238
Port JD, Hadcock JR, Malbon CC (1992) Cross-regulation between G-protein-mediated pathways: acute activation of the inhibitory pathway of adenylyl cyclase reduces $\beta_2$-adrenergic receptor phosphorylation and increases $\beta$-adrenergic receptor responsiveness. J Biol Chem 267:8468–8472
Rapiejko PJ, Watkins DC, Ros M, Malbon CC (1989) Thyroid hormones regulate G-protein $\beta$-subunit mRNA expression in vivo. J Biol Chem 264:16183–16189

Rapiejko PJ, Watkins DC, Ros M, Malbon CC (1990) G-protein subunit mRNA levels in rat heart, liver, and adipose tissues: analysis by DNA-excess solution hybridization. Biochim Biophys Acta 1052:348–350

Ros M, Northup JK, Malbon CC (1988) Steady-state levels of G-proteins and $\beta$-adrenergic receptors in rat fat cells: permissive effects of thyroid hormones. J Biol Chem 263:4362–4368

Ros M, Watkins DC, Rapiejko PJ, Malbon CC (1989) Glucocorticoids modulate mRNA levels of G-protein $\beta$-subunits Biochem J 260:271–275

Saito N, Guitart X, Hayward M, Tallman JF, Duman RS, Nestler EJ (1989) Corticosterone differentially regulates the expression of $G_{s\alpha}$ and $G_{i\alpha2}$ messenger RNA and protein in rat cerebral cortex. Proc Natl Acad Sci USA 86:3906–3910

Saggerson D (1992) Thyroid disorders. In: Milligan G, Wakelam M (eds) G-proteins: signal transduction and disease. Academic Press, London

Shaw G, Kamen R (1986) A conserved AU sequence from the 3′ untranslated region of GM-CSF mRNA mediates selective mRNA degradation. Cell 46:659–667

Wang HY, Lipfert L, Malbon CC, Bahouth SS (1989) Site-directed antibodies define the topography of the $\beta$-adrenergic receptor. J Biol Chem 264:14424–14434

Wang HY, Berrios M, Malbon CC (1989) Indirect immunofluorescence localization of $\beta$-adrenergic receptors and G-proteins in human A431 cells. Biochem J 263:519–532

Wang HY, Berrios M, Malbon CC (1989) Localization of $\beta$-adrenergic receptors in A431 cells in situ: effect of chronic exposure to agonist. Biochem J 263:533–538

Wang HY, Watkins DC, Malbon CC (1992) Antisense oligodeoxynucleotides to $G_s$ protein $\alpha$-subunit sequence accelerate differentiation of fibroblasts to adipocytes. Nature 358:334–337

Watkins DC, Johnson GL, Malbon CC (1992) $G_{i\alpha2}$ regulates differentiation of stem cells to primitive endoderm in F9 teratocarcinoma cells. Science (to be published)

CHAPTER 52

# G-Protein Subunit Lipidation in Membrane Association and Signaling

J.A. THISSEN and P.J. CASEY

## A. Introduction

The heterotrimeric GTP-binding proteins (G-proteins), comprised of $\alpha$, $\beta$, and $\gamma$ subunits, are localized to the inner surface of the plasma membrane where they serve as receptor-mediated signal transducers. It is the $\alpha$ subunit, which exists as many different subtypes, that confers identity to the oligomer and in most systems governs the specificity of the interaction with receptor and effector. The $\beta\gamma$ subunits, of which multiple forms also exist, appear to function as a complex, and individual forms may be shared among the multiple $\alpha$ subunits (ROOF et al. 1985; FUNG 1983; ROBISHAW et al. 1989; SCHMIDT and NEER 1991; TAMIR et al. 1991). A model for G-protein signal transduction is described in Fig. 1. Receptor-mediated activation of the G-protein results in the exchange of GTP for GDP on the $\alpha$ subunit, promoting its separation from $\beta\gamma$. Subsequent interaction with a membrane-associated effector protein in turn generates an intracellular response. The model depicted in Fig. 1 shows the $\alpha$ subunit interacting with the effector protein, as most available evidence points to the GTP-bound form of the $\alpha$ subunit as activating effectors such as adenylyl cyclase, retinal phosphodiesterase, and a phosphoinositide-specific phospholipase C (NORTHUP et al. 1983; KATADA et al. 1984; CERIONE et al. 1988; ROBISHAW et al. 1986; SMRCKA et al. 1991). However, recent evidence has indicated that the $\beta\gamma$ complex may play a more direct role in G-protein-linked signal transduction than previously thought (GILMAN 1987; KIM et al. 1989; TANG and GILMAN 1991; WHITEWAY et al. 1989). This has been most clearly demonstrated in the mating factor signaling pathway in yeast (WHITEWAY et al. 1989, and references therein).

The association of G protein subunits with the plasma membrane is critical to coordinate the receptor, G-protein components, and effector proteins to ensure efficient signal transduction (GILMAN 1987; BOURNE et al. 1991). However, until recently the molecular basis of the association of G-proteins with cellular membranes was unclear. Although the G-protein subunits have been shown to behave similarly to integral membrane proteins in their requirement (in most cases) for detergents to be solubilized, none of the subunit polypeptide sequences contain hydrophobic sequences which would account for this membrane interaction (BUSS et al. 1987; JONES and

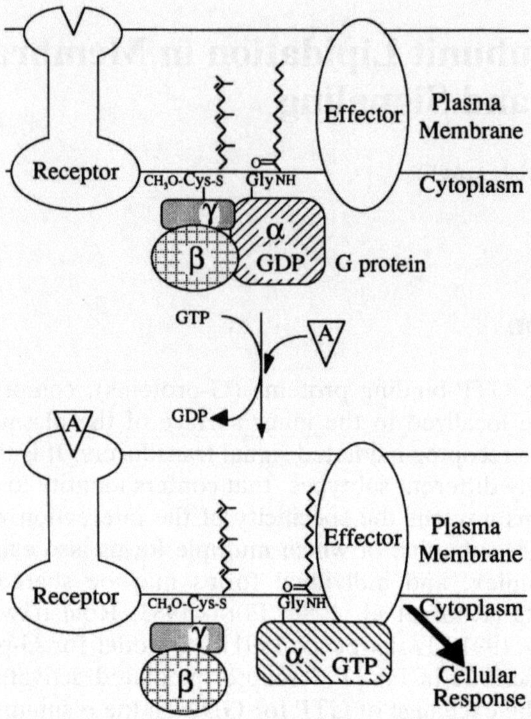


**Fig. 1.** Model of G-protein-dependent transmembrane signaling. Agonist binding to its appropriate receptor stimulates activation of the G-protein, which in turn promotes the exchange of GTP for GDP on the $\alpha$ subunit, its subsequent separation from the $\beta\gamma$ complex, and stimulation of a membrane-bound effector protein. The lipid groups associated with the G-protein subunits, shown as $\alpha$ subunit myristoylation and $\gamma$ subunit prenylation, mediate the association of these proteins with the membrane. The interaction of the protein-bound lipids with the plasma membrane is illustrated as an interaction with the lipid bilayer; however, membrane proteins may be involved. The nature of the interaction between the $\beta$ and $\gamma$ subunits has not yet been determined. (Graphics by Joyce B. HIGGINS)

REED 1987; GAUTAM et al. 1990; HURLEY et al. 1984; YATSUNAMI at al. 1985; ROBISHAW et al. 1989).

The attachment of a lipid to an otherwise soluble protein is now recognized as an important mechanism in stabilizing protein binding to membranes (JAMES and OLSON 1990; GORDON et al. 1991; GLOMSET et al. 1990; SPIEGEL et al. 1991). Lipid modifications of many of the heterotrimeric G protein subunits have now been identified which mediate the membrane association of these polypeptides (LINDER et al. 1991; BUSS et al. 1987; YAMANE et al. 1990; MUMBY et al. 1990). Two distinct types of lipid modifications are now known to play a critical role in G-protein–membrane interactions: myristoylation (attachment of a 14-carbon saturated fatty acid) and isoprenylation (attachment of 15- and 20-carbon unsaturated lipids).

Myristoylation of many of the G-protein $\alpha$ subunits has been shown to mediate the interaction of these proteins with the plasma membrane (MUMBY et al. 1990; JONES et al. 1990) as well as play a role in their interactions with the $\beta\gamma$ complex (LINDER et al. 1991). The more recently identified modification of G-protein $\gamma$ subunits by isoprenoids has provided insight into the means by which this subunit is anchored to the plasma membrane (SIMONDS et al. 1991; YAMANE et al. 1990; MUNTZ et al. 1992; MALTESE and ROBISHAW 1990). The first part of this chapter concentrates on the role of myristoylation of G-protein $\alpha$ subunits, while the isoprenylation of G-protein $\gamma$ subunits and its effect on membrane association and function are discussed in the second part of this chapter.

## B. Myristoylation and Membrane Association of G-Protein $\alpha$ Subunits

Chemical analysis of the G-protein $\alpha$ subunits termed $\alpha_i$ and $\alpha_o$ revealed that these proteins contain myristic acid (BUSS et al. 1987). The protein: myristoyl linkage was determined to be base resistant but acid labile, suggesting that this fatty acid was linked via an amide bond (BUSS et al. 1987), similar to the other myristoylated proteins studied to date (GORDON et al. 1991). Later studies confirmed that several G-protein $\alpha$ subunits (including $\alpha_{i1}$, $\alpha_{i2}$, $\alpha_{i3}$, and $\alpha_o$) are modified by myristic acid via an amide bond to the amino-terminal glycine residue of the polypeptide (Table 1). These studies explained early results that showed that the amino-terminus of many of the G-protein $\alpha$ subunits was important in their interaction with membranes (NEER et al. 1988; NAVON and FUNG 1987). Under certain conditions trypsin digestion of $\alpha$ subunits removes a 1- to 2-kDa amino-terminal peptide and results in a truncated $\alpha$ subunit. In studies conducted with $\alpha_i$ and $\alpha_o$ the truncated $\alpha$ subunits no longer associated with the $\beta\gamma$ complex (NAVON and FUNG 1987; NEER et al. 1988), nor did they serve as substrates for pertussis toxin-catalyzed ADP-ribosylation (which requires $\alpha$-$\beta\gamma$ interaction; NAVON and FUNG 1987; NEER et al. 1988), but they did retain normal GTPase activity (NEER et al. 1988).

### I. Cotranslational Processing of G-protein $\alpha$ Subunits

The process by which a myristic acid is linked to a $NH_2$-terminal glycine residue via an amide bond protein (*N*-myristoylation) is a cotranslational event (WILCOX et al. 1987). The reaction is catalyzed by myristoyl-CoA: protein *N*-myristoyltransferase (NMT; RUDNICK et al. 1990). This enzyme has a pronounced substrate specificity for myristoyl-CoA as well as an absolute requirement for an $NH_2$-terminal glycine on the protein acceptor (TOWLER et al. 1988). Although no strict consensus sequence for myristoylation exists, the glycine residue (initially in position 2 of the polypeptide,

**Table 1.** Sites of lipid modifications on G-protein subunits

| G protein subunit | Modified terminus | Sequence[a] | Lipid modification (on terminus) | References |
|---|---|---|---|---|
| $\alpha_s$ | $NH_2$– | MGCLGNS | None detected | ROBISHAW et al. (1986), MUMBY et al. (1990) |
| $\alpha_{i1}$ | $NH_2$– | MGCTLSA | Myristoyl | MUMBY et al. (1990), JONES et al. (1990) |
| $\alpha_{i1}$ (mutant) (G → A) | $NH_2$– | MACTLSA | None detected | JONES et al. (1990) |
| $\alpha_{i2}$ | $NH_2$– | MGCTVSA | Myristoyl | MUMBY et al. (1990), JONES and REED (1987) |
| $\alpha_{i3}$ | $NH_2$– | MGCTLSA | Myristoyl | MUMBY et al. (1990), JONES and REED (1987) |
| $\alpha_o$ | $NH_2$– | MGCTLSA | Myristoyl | MUMBY et al. (1990), JONES and REED (1987) |
| $\alpha_z$ | $NH_2$– | MGCRQSS | Myristoyl[b] | MUMBY et al. (1990), FONG et al. (1988) |
| $\alpha_t$ | $NH_2$– | MGAGASA | Myristoyl[b] | MUMBY et al. (1990), LOCHRIE et al. (1987) |
| $\alpha_{olf}$ | ($NH_2$–) | MGCLGNS | Unknown | JONES and REED (1989) |
| $\gamma_2$ | COOH– | CAIL | Geranylgeranyl | GAUTAM et al. (1989), ROBISHAW et al. (1989) |
| $\gamma_2$ (mutant) (C → S) | COOH– | SAIL | None detected | MUNTZ et al. (1992), SIMONDS et al. (1991) |
| $\gamma_{t\text{-}1}$ | COOH– | LKGG | None detected | OHGURO et al. (1991) |
| $\gamma_{t\text{-}2}$ | COOH– | CVIS | Farnesyl | OHGURO et al. (1991), LAI et al. (1990) |

[a] Modified amino acid indicated by underlined residue.
[b] Myristoylation detected only in COS cell transfection experiments (MUMBY et al. (1990)).

thus necessitating the removal of the $NH_2$-terminal Met residue) is essential, and a hydroxy-amino acid in position 6 is preferred (Table 1; GORDON 1990).

The discovery that many G-protein $\alpha$ subunits contain this fatty acyl modification provided evidence for a molecular basis by which these relatively hydrophilic proteins associate with membranes, and prompted numerous studies designed to elucidate its role in membrane association and cellular function (STERNWEIS 1986; LINDER et al. 1991; Ross et al. 1978; MUMBY et al. 1990; JONES et al. 1990). The major questions that have been addressed in the studies are described below and include: (a) Does the $\alpha$ subunit interact directly with the membrane lipid bilayer, or does it require $\beta\gamma$ to mediate this interaction? (b) What role does myristoylation of the $\alpha$ subunit play in this interaction?

## II. The Role of Myristoylation in $\alpha$ Subunit–Membrane Association

Transient expression in COS cells has been used to examine the role played by the covalent attachment of myristic acid to G-protein $\alpha$ subunits, specifically $\alpha_{i1}$, $\alpha_{i2}$, $\alpha_{i3}$, $\alpha_o$, $\alpha_s$, $\alpha_t$, and $\alpha_z$ (MUMBY et al. 1990; JONES et al. 1990). Of these $\alpha$ subunits all but $\alpha_s$ were shown to be myristoylated. To assess the role of myristoylation in the membrane association of these proteins, site-directed mutagenesis was used to change the second residue (the site of myristoylation) in $\alpha_{i1}$ and $\alpha_o$ from a glycine to an alanine residue. Transfection of the mutant $\alpha$ subunits into COS cells produced proteins which failed to incorporate [$^3$H]myristate and were localized primarily in the soluble fraction, while the control transfection with wild-type $\alpha$ subunits resulted in myristoylated proteins which were found associated with the membrane fraction (Fig. 2; MUMBY et al. 1990; JONES et al. 1990). Both wild-type and mutant $\alpha$ subunits were capable of interacting with $\beta\gamma$ subunits, as shown by their ability to be ADP-ribosylated with pertussis toxin, although this process with the wild-type $\alpha$ subunit required much lower amounts of $\beta\gamma$ than the mutant (JONES et al. 1990). In contrast, transfection of COS cells with the $\alpha_s$ subunit resulted in its accumulation in the membrane fraction, even though, as noted above, this subunit did not incorporate myristic acid (MUMBY et al. 1990; JONES et al. 1990).

A different approach to studying the effect of $\alpha$ subunit myristoylation has employed the use of coexpression of myristoyltransferase, NMT, and the $\alpha$ subunits in *Escherichia coli*, which results in the production of myristoylated proteins in an organism which normally lacks this capability (LINDER et al. 1991). Both myristoylated recombinant $\alpha_o$ (myr-r$\alpha_o$) and nonmyristoylated recombinant $\alpha_o$ (r$\alpha_o$) proteins were produced and examined for their ability to interact with the $\beta\gamma$ complex by a variety of techniques. The myr-r$\alpha_o$ behaved similar to purified native $\alpha_o$ in its ability to interact with $\beta\gamma$ in these assays; however, nonmyristoylated r$\alpha_o$ interacted poorly with $\beta\gamma$ (Fig. 3; LINDER et al. 1991). These studies indicate that myristoylation of $\alpha$ subunits influences not only their membrane interactions

**Fig. 2.** Myristoylation of an $\alpha$ subunit influences its membrane association. COS cells were transfected with cDNA encoding wild-type $\alpha_o$, mutant $\alpha_o$ containing a $NH_2$-terminal Gly to Ala substitution which prevents myristoylation, or no DNA (*mock*). Membrane and cytoplasmic fractions of the cells were prepared 2 days after transfection, and proteins from each fraction were resolved by SDS-PAGE and transferred to nitrocellulose, and $\alpha_o$ was detected using $\alpha_o$-specific antiserum (MUMBY et al. 1990). See text for further analysis

but is also important in crucial protein–protein (i.e., subunit) interactions of G-proteins.

The myristoyl group attached to the $\alpha$ subunit may serve to anchor this polypeptide to the membrane after G-protein activation, i.e., when the $\alpha$-GTP has dissociated from $\beta\gamma$. It is not known, however, whether this interaction is with the lipid bilayer, as illustrated in Fig. 1, or is mediated by another protein in the membrane. Interestingly, purified $\alpha$ subunits (which presumably contain myristic acid) associate poorly with artificial phospholipid vesicles unless $\beta\gamma$ complexes are also present (STERNWEIS 1986). This suggests that membrane association of the $\alpha$ subunit is not solely a hydrophobic interaction mediated by myristic acid, but that protein–protein interactions are also important. Furthermore, lipid modification of $\alpha$ subunits in COS cells by the less hydrophobic myristate analog 11-oxa myristate resulted in

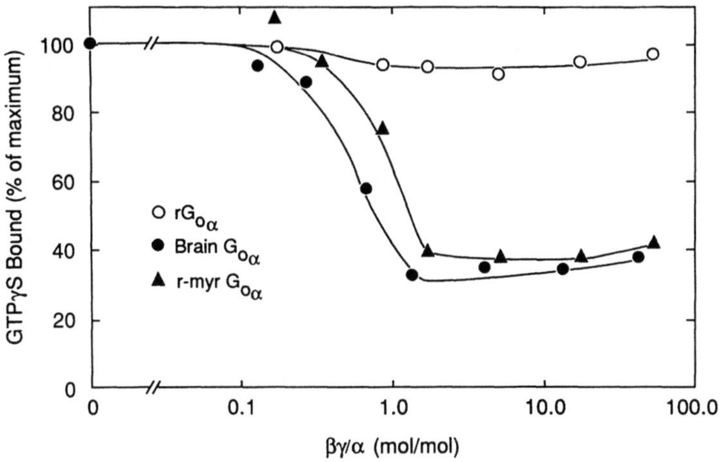


**Fig. 3.** Myristoylation of an $\alpha$ subunit influences its interaction with $\beta\gamma$. Recombinant $\alpha_o$ ($r\alpha_o$, ○) which is not myristoylated, its myristoylated counterpart (r-myr $\alpha_o$, (▲) or brain $\alpha_o$ (●), were incubated with bovine brain $\beta\gamma$ at the indicated molar ratios. At time zero, [$^{35}$S]GTP$\gamma$S was added, the reaction was quenched after 2 min, and bound GTP$\gamma$S was measured. The amount of GTP$\gamma$S bound as a function of the $\beta\gamma$ concentration is indicated (Linder et al. 1991)

some redistribution of the $\alpha_i$ subtypes and of $\alpha_o$ to the cytosolic fraction, while $\alpha_z$ and $\alpha_t$ associated equally well with the membrane whether myristate or the 11-oxa analog was attached (MUMBY et al. 1990). The accumulated data indicate that the myristoyl group does play a role in targeting the acylated $\alpha$ subunits to the membrane but suggests that myristate is not the only signal required to direct these proteins to the plasma membrane (GORDON et al. 1991). It seems likely that additional determinants on the $\alpha$ subunit are required to direct these proteins to, and mediate their interaction with, cellular membranes.

Two $\alpha$ subunits do not appear to be dependent on myristoylation for membrane attachment. The first, $\alpha_t$, can be myristoylated when expressed in COS cells (MUMBY et al. 1990); however, no myristic acid has been detected by chemical analysis of this purified protein. Interestingly, the association of the $\alpha_t$ subunit with membranes is much weaker than for the other $\alpha$ subunits, and it does not require detergent for solubilization (KUHN 1980; Fung et al. 1981). In fact, the $\alpha_t$ can be released from the membranes by treatment with GTP$\gamma$S (NAVON and FUNG 1987). One possibility for the discrepancy between the transfection experiments and chemical analysis is that the environment of $\alpha_t$ in the retinal cell does not facilitate myristoylation. Alternatively, $\alpha_t$ may be modified by a fatty acid other than myristic acid. The lack of in vitro data comparing the properties of myristoylated and nonmyristoylated $\alpha_t$ prevents a more detailed interpretation of the $\alpha_t$–membrane interaction.

Chemical analysis of $\alpha_s$ revealed that it too is not myristoylated; however, unlike $\alpha_t$, $\alpha_s$ is also not myristoylated when transfected into COS cells (MUMBY et al. 1990). It seems likely that a lipid moiety other than myristic acid may be attached to the $\alpha_s$ subunit which could account for this interaction. However, evidence for this is not presently available, and a rigorous chemical analysis is necessary to confirm this conjecture. Interestingly, protease digestion indicates that it is the COOH-terminal portion of the $\alpha_s$ subunit which is responsible for anchoring this protein to the membrane (AUDIGIER et al. 1990).

## C. Prenylation and Membrane Association of G-Protein $\gamma$ Subunits

The G-protein $\beta$ and $\gamma$ subunits are tightly associated and are bound to the plasma membrane as a $\beta\gamma$ complex (TAMIR et al. 1991; FUKADA et al. 1989; SCHMIDT and NEER 1991). The $\beta\gamma$ complex appears to function as a unit in G-protein signal transduction (Fig. 1), and, similar to the $\alpha$ subunits, their polypeptide amino acid sequences do not reveal domains which could account for their tight association with the plasma membrane (ROBISHAW et al. 1989; GAUTAM et al. 1990; HURLEY et al. 1984; YATSUNAMI et al. 1985). The discovery that the $\gamma$ subunits of the G-protein are modified by isoprenoids (long-chain unsaturated fatty acids) has provided crucial evidence as to the hydrophobic properties of this subunit complex and the mechanism by which these proteins interact with cellular membranes (GLOMSET et al. 1991; SIMONDS et al. 1991; YAMANE et al. 1990; CASEY et al. 1991; MUNTZ et al. 1992; MALTESE and ROBISHAW, 1990; GLOMSET et al. 1991).

### I. Posttranslational Processing of G-Protein $\gamma$ Subunits

The $\gamma$ subunits of the heterotrimeric G-protein are members of a family of proteins distinguished by a COOH-terminal Cys-AAX motif (where A and X signify any of several amino acids) which is the site of a series of three posttranslational processing events to produce the mature forms of these polypeptides (SIMONDS et al. 1991; CASEY et al. 1991; OHGURO et al. 1991; GUTIERREZ et al. 1989; HANCOCK et al. 1991). The first processing event, prenylation, involves the enzymatic addition of either a 15-carbon, farnesyl, or 20-carbon, geranylgeranyl, isoprenoid group to the cysteine residue of the Cys-AAX sequence. Prenylation is a cytosolic event, and two separate enzymes have been identified which are specific for either geranylgeranyl or farnesyl addition (YOKOYAMA et al. 1991; CASEY et al. 1991; REISS et al. 1990). The specificity for which isoprenoid is attached to the protein lies in the terminal residue ("X") of the Cys-AAX motif (REISS et al. 1990; CASEY et al. 1991; YOKOYAMA et al. 1991). When X is a leucine residue, a geranylgeranyl group is added by a protein geranylgeranyltransferase, whereas a methionine, serine, or glutamine residue at this position signals attachment of a farnesyl group by a protein farnesyltransferase. The prenyl substrates for these enzymes have been identified as the corresponding prenyl diphosphates (i.e., farnesyl diphosphate and geranylgeranyl diphosphate) which are derived from mevalonate via the cholesterol biosynthetic pathway (GOLDSTEIN and BROWN 1990).

The two processing events which follow prenylation include the proteolytic removal of the three terminal amino acids (the "AAX"; ASHBY et al. 1992; GUTIERREZ et al. 1989; HANCOCK et al. 1991; HRYCYNA and CLARKE 1992) and the methylation of the now carboxyl-terminal, S-prenylated cysteine residue (CLARKE et al. 1988; GUTIERREZ et al. 1989; HANCOCK et al. 1991; HRYCYNA et al. 1991; STEPHENSON and CLARKE 1990). Both of these terminal processing activities have been identified as being associated with the microsomal membrane fraction, although neither of these enzymes has yet been purified (VOLKER et al. 1991; HRYCYNA et al. 1991; VOLKER et al. 1991). Studies have shown, however, that the protease involved in this reaction appears to be an endoprotease, cleaving specifically on the COOH-terminal side of the prenylated Cys residue (ASHBY et al. 1992), while the methyltransferase is specific for the free COOH of the now terminal prenylated Cys residue (GUTIERREZ et al. 1989; STEPHENSON and CLARKE, 1990; VOLKER et al. 1991; VOLKER et al. 1991). Thus, the prenylation of Cys-AAX containing proteins ultimately results in a carboxyl-terminal cysteine residue which is S-prenylated and $\alpha$-methyl esterified (GLOMSET et al. 1991; see Figure 1). The result of these processing events is believed to be part of a specific address which in most cases serves to direct and/or anchor these proteins to their appropriate membrane compartment(s) in the cell (GLOMSET et al. 1991).

Both the farnesyl and geranylgeranyl isoprenoids have been found attached to G-protein $\gamma$ subunits (LAI et al. 1990; YAMANE et al. 1990; MALTESE and ROBISHAW 1990; MUMBY et al. 1990; Table 1). The nonretinal $\gamma$ subunits terminate in Leu and are geranylgeranyl modified, while the $\gamma$ subunit of the retinal G-protein, transducin, terminates in a Ser residue and is modified by a farnesyl group (Table 1). Interestingly, although many of the G-protein $\alpha$ subunits also end in a Cys-AAX motif, these proteins are not prenylated and do not serve as substrates for protein prenyltransferases (JONES and SPIEGEL 1990). Apparently, the Gly residue which immediately follows the Cys residue in the $\alpha$ subunit Cys-AAX motif prevents interaction with protein prenyltransferases (REISS et al. 1991; JONES and SPIEGEL 1990).

## II. The Role of Prenylation in $\gamma$ Subunit–Membrane Association

Evidence accumulated to date suggests that all of the nonretinal $\gamma$ subunit subtypes (which now number at least four) are geranylgeranyl modified (YAMANE et al. 1990; MALTESE and ROBISHAW 1990) while the farnesyl isoprenoid is found associated with retinal $\gamma$ subunits (FUKADA et al. 1990;

LAI et al. 1990). Studies designed to compare the cellular distribution of prenylated versus unprenylated versions of $\gamma$ subunits indicate that these two isoprenoids may confer different functional properties to their respective $\gamma$ subunits. Therefore, the geranylgeranylated subset of $\gamma$ subunits is discussed separately from farnesylated $\gamma$ subunits.

## 1. Geranylgeranyl-Modified $\gamma$ Subunits

That nonretinal G-protein $\gamma$ subunits are prenylated was demonstrated by analyzing proteins after biosynthetic radiolabeling with [$^3$H]mevalonolactone ([$^3$H]MVA; a precursor of cellular isoprenoids; MUMBY et al. 1990). Following the purification of the G-protein $\beta\gamma$ complex from [$^3$H]MAV-labeled cells by affinity chromatography, the $\gamma$ subunits were found to contain the $^3$H label (Fig. 4). HPLC analysis of these proteins revealed that a tritiated lipid which comigrated with the 20-carbon isoprenoid geranylgeraniol was released following a thioether-specific chemical cleavage (MUMBY et al. 1990). A parallel study using a mass analysis approach for

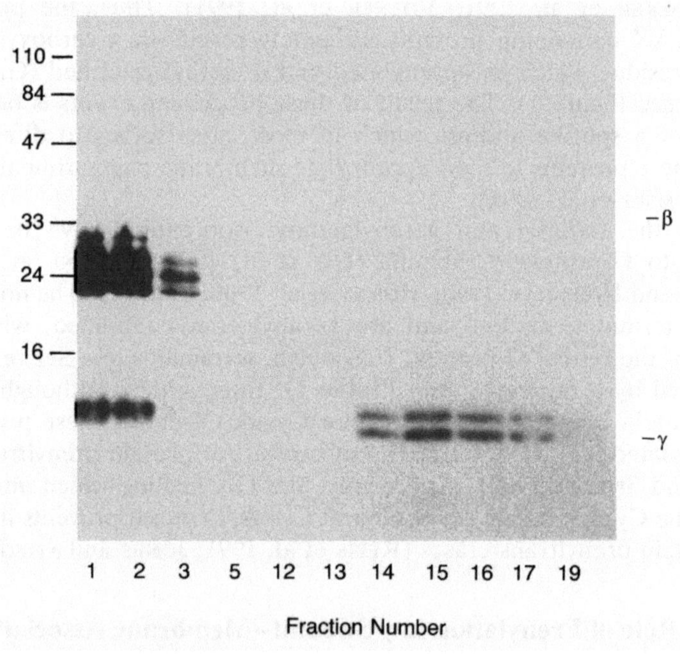


**Fig. 4.** G-protein $\gamma$ subunits modified by a geranylgeranyl isoprenoid. Affinity chromatography of detergent extracts from [$^3$H]MVA-labeled cells on $\alpha_o$-agarose (an affinity column for $\beta\gamma$) were analyzed by SDS-PAGE and fluorography. *Lane numbers*, column fraction number. *Right*, migration of the $\beta$ and $\gamma$ subunits; *left*, molecular weight standards (Mumby et al. 1990)

identification also demonstrated geranylgeranyl modification of $\gamma$ subunits from brain G-proteins (YAMANE et al. 1990).

To address whether the geranylgeranyl group is required for the membrane association of the $\gamma$ subunit, cDNA encoding the major nonretinal form of the $\gamma$ subunit was transfected into COS cells, and the expressed protein was examined for its cellular distribution (MUNTZ et al. 1992; SIMONDS et al. 1991). When the wild-type $\gamma$ subunit was transfected into COS cells, it became prenylated and was found associated with the membranes. However, in COS cells transfected with the same $\gamma$ subunit with a Cys to Ser substitution in the Cys-AAX domain, prenylation of the $\gamma$ subunit was prevented, and the $\gamma$ subunit no longer associated with the plasma membrane. Additional evidence that geranylgeranylation is required for $\gamma$ subunit membrane attachment was shown by inhibiting the synthesis of mevalonate in Neuro 2A cells coexpressing $\beta$ and $\gamma$ subunits with compactin (MUNTZ et al. 1992). A fraction of the $\gamma$ (and $\beta$) subunits was found in the cytoplasmic faction after a 2-day treatment with compactin, whereas all of the $\beta\gamma$ complex was found in the membrane fraction in untreated cells, again supporting the requirement of prenylation for $\gamma$ subunit–membrane binding.

The influence of prenylation of the $\gamma$ subunit on its interactions with the $\beta$ subunit has also been examined. Transient expression of the $\beta$ subunit alone or coexpression of $\beta$ and $\gamma$ subunits both resulted in the localization of the $\beta$ subunit to the membrane fraction (MUNTZ et al. 1992). However, coexpression of the $\beta$ subunit with the unprenylated (Cys to Ser mutant) $\gamma$ subunit resulted in the redistribution of some of the $\beta$ subunit to the cytoplasmic fraction, where this unprenylated $\gamma$ subunit was found. These results suggest that the $\beta$ subunit interacts with the $\gamma$ subunit in the absence of the isoprenoid. To further assess the requirement of the prenyl group in the interaction of $\beta$ subunits with $\gamma$ subunits, extracts from COS cells which were transiently expressing $\beta$ alone or coexpressing $\beta$ subunits with either wild-type (prenylated) or Cys-to-Ser mutant (unprenylated) $\gamma$ subunits were subjected to limited typtic digestions (MUNTZ et al. 1992). The presence of a characteristic 23-kDa fragment which is generated upon limited tryptic digestion only when the $\beta$ subunit is associated with the $\gamma$ subunit was not detected in the cytoplasmic fraction and was barely detectable in the membrane fractions, of cells expressing $\beta$ alone. In contrast, this fragment was detected primarily in the membrane fraction in cells overexpressing wild-type $\gamma$, while it was primarily in the cytosolic fraction in cells coexpressing both $\beta$ and the $\gamma$ subunit (Cys to Ser) mutant. These data indicate that the $\gamma$ subunit can influence the cellular distribution of the $\beta$ subunit and provide additional evidence that prenylation is not an absolute requirement for interaction of $\beta$ and $\gamma$. The mode by which the $\beta$ subunit localizes to the membrane fraction in the absence of $\gamma$ is unclear. However, it should be noted that the $\beta$ subunit found in the membrane fraction may represent an aggregation of $\beta$ subunits which independently sediments with the membranes (MUNTZ et al. 1992).

### 2. Farnesyl-Modified $\gamma$ Subunits

The $\gamma$ subunit of the retinal G-protein, transducin, is modified by the 15-carbon, farnesyl isoprenoid (LAI et al. 1990). However, unlike the non-retinal $\gamma$ subunits, $\gamma_t$ has been isolated from retinal cell membranes in both unfarnesylated ($\gamma_{t-1}$) and farnesylated ($\gamma_{t-2}$) forms (FUKADA et al. 1990; OHGURO et al. 1991; FUKADA et al. 1989). The COOH-terminal amino acid of the purified $\gamma_{t-1}$ is a Gly residue instead of a farnesylated Cys residue (Table 1) which represents a truncation of the normal protein. The fact that both farnesylated and unfarnesylated forms of the $\gamma_t$ subunit are found associated with the membrane suggest that determinants other than the farnesyl group mediate the interaction of transducin $\beta\gamma$ with the membrane (FUKADA et al. 1990; FUKADA et al. 1989). These and related findings (see below) suggest that the role of the farnesyl group on the $\gamma_t$ subunit may include functions other than, or in addition to, mediating its membrane attachment. Interestingly, the transducin $\beta\gamma$ complex can be released from the membrane in the absence of detergent, indicating that this complex associates much less avidly with the membrane than its nonretinal counterparts.

The only data available which address the requirement for prenylation of the $\gamma$ subunit in the function of G-protein signal transduction are from studies using transducin $\gamma$. To assess the functional differences between the unfarnesylated and farnesylated forms of $\gamma_t$ ($\gamma_{t-1}$ versus $\gamma_{t-2}$, respectively), the ability of light-activated rhodopsin to stimulate GTP binding to $\alpha_t$ was tested in the presence of each of these proteins (FUKADA et al. 1990). The $\gamma_{t-2}$ was shown to be essential for GTP binding to the $\alpha_t$ subunit in this reaction, and this farnesylated subunit complex was 30-fold more efficient than that containing the unfarnesylated $\gamma_t$ subunit (FUKADA et al. 1990). Further studies showed that carboxyl-methylation of the farnesylated $\gamma_{t-2}$ enhances its coupling to activated rhodopsin (OHGURO et al. 1991), indicating a role for this additional modification.

## D. Future Directions

Substantial progress has recently been made to improve our understanding of how the G-protein subunits, particularly the $\alpha$ and $\gamma$ subunits, interact with cellular membranes. However, while myristoylation of the $\alpha$ subunits and prenylation of the $\gamma$ subunits are involved in the interaction of these proteins with the plasma membrane, the specific nature of these interactions remains unclear. For example, it has not yet been established whether the myristoyl group on the $\alpha$ subunits interacts directly with the phospholipid bilayer of the membrane, whether the $\beta\gamma$ complex at least partially mediates this interaction, or whether a yet unidentified membrane protein is involved in this interaction. In addition, the extent of a functional role of the myristoyl group in G-protein signal transduction has not yet been established. The tools are now available, however, to further assess the functional aspects of $\alpha$ subunit myristoylation from both molecular and biochemical approaches.

Many of the same questions remain unanswered for the role of $\gamma$ subunit isoprenylation, for example: What role (if any) does prenylation play in its interaction with either the $\beta$ subunit or the interaction of $\beta\gamma$ with the $\alpha$ subunit? Is the interaction of the $\gamma$ subunit with the plasma membrane directly with the lipid bilayer, or is it mediated by a specific protein receptor? The ability to synthesize both unprenylated recombinant $\gamma$ subunits will provide critical tools for answering these questions.

Finally, the mechanism by which the subunits of the heterotrimeric G-protein assemble at the plasma membrane has not yet been established. Another possible role for these lipid modifications could be to serve, probably in conjunction with other determinants on the protein, to specify the trafficking pathways through which these proteins are directed to their appropriate membrane compartments in the cell. This conjecture is supported by the finding that the unlipidated forms of both the $\alpha$ and $\gamma$ subunits are not properly targeted to the plasma membrane in COS cell transformation experiments. Additionally, since the same lipid groups that are attached to the G-protein subunits are also found linked to proteins which localize to many different cellular membranes, it appears that the lipid moieties are only part of the complete address necessary for accurate protein targeting. Further studies should elucidate the involvement of these lipid groups, in concert with other protein domains, in both cellular targeting of these proteins and in G-protein signal transduction.

*Acknowledgements.* Work from the authors' laboratory was supported by research grants from the National Science Foundation and the American Cancer Society. P.J.C. is an Established Investigator of the American Heart Association.

## References

Ashby MN, King DS, Rine J (1992) Endoproteolytic processing of a farnesylated peptide in vitro. Proc Natl Acad Sci USA 89:4613–4617

Audigier Y, Journot L, Pantaloni C, Bockaert J (1990) The carboxy-terminal domain of $G_{s\alpha}$ necessary for anchorage of the activated form in the plasma membrane. J Cell Biol III:1427–1435

Bourne HR, Sanders DA, McCormick F (1991) The GTPase superfamily: conserved structure and molecular mechanism. Nature 349:117–127

Buss JE, Mumby SM, Casey PJ, Gilman AG, Sefton BM (1987) Myristoylated $\alpha$ subunits of guanine nucleotide-binding regulatory proteins. Proc Natl Acad Sci USA 84:7493–7497

Casey PJ, Thissen JA, Moomaw JF (1991) Enzymatic modification of proteins with a geranylgeranyl isoprenoid. Proc Natl Acad Sci USA 88:8631–8635

Cerione RA, Kroll S, Rajaram R, Uson C, Goldsmith P, Spiegel AM (1988) An antibody directed against the carboxy-terminal decapeptide of the $\alpha$ subunit of the retinal GTP-binding protein, transducin. Effects on transducin function. J Biol Chem 263:9345–9352

Clarke S, Vogel JP, Deschenes RJ, Stock J (1988) Posttranslational modification of the Ha-ras oncogene protein:evidence for a third class of protein carboxyl methyltransferases. Proc Natl Acad Sci USA 85:4643–4647
Fong HK, Yoshimoto KK, Eversole-Cire P, Simon MI (1988) Identification of a GTP-binding protein $\alpha$ subunit that lacks an apparent ADP-ribosylation site for pertussis toxin. Proc Natl Acad Sci USA 85:3066–3070
Fukada Y, Ohguro H, Saito T, Yoshizawa T, Akino T (1989) $\beta\gamma$ subunit of bovine transducin composed of two components with distinctive $\gamma$ subunits. J Biol Chem 264:5937–5943
Fukada Y, Takao T, Ohguro H, Yoshizawa T, Akino T, Shimonishi Y (1990) Farnesylated $\gamma$ subunit of photoreceptor G protein indispensable for GTP-binding. Nature 346:658–660
Fung BK-K (1983) Characterization of transduction from bovine retinal rod outer segments. I. Separation and reconsititution of subunits. J Biol Chem 258:10495–10502
Fung BK-K, Hurley JB, Stryer L (1981) Flow of information in the light-triggered cyclic nucleotide cascade of vision. Proc Natl Acad Sci USA 78:152–156
Gautam N, Baetsher M, Aebersold R, Simon MI (1989) A G protein $\gamma$ subunit shares homology with ras proteins. Science 245:971–974
Gilman AG (1987) G proteins: transducers of receptor-generated signals. Ann Rev Biochem 56:615–649
Glomset J, Gelb M, Farnsworth C (1991) The prenylation of proteins. Curr Opinion Lipidology 2:118–124
Glomset JA, Gelb MH, Farnsworth CC (1990) Prenyl proteins in eukaryotic cells: a new type of membrane anchor. Trends Biochem Sci 15:139–142
Goldstein JL, Brown MS (1990) Regulation of the mevalonate pathway. Nature 343:425–434
Gordon JI (1990) Protein N-myristoylation: simple questions, unexpected answers. Clin Res 38:517–528
Gordon JI, Duronio RJ, Rudnick DA, Adams SP, Goke GW (1991) Protein N-myristoylation. J Biol Chem 266:8647–8650
Gutierrez L, Magee AI, Marshall CJ, Hancock JF (1989) Post-translational processing of p21ras is two-step and involves carboxyl-methylation and carboxy-terminal proteolysis. EMBO J 8:1093–1098
Hancock JF, Cadwallader K, Marshall CJ (1991) Methylation and proteolysis are essential for efficient membrane binding of prenylated p21K-ras(B). EMBO J 10:641–646
Hrycyna CA, Sapperstein SK, Clarke S, Michaelis S (1991) The Saccharomyces cerevisiae STE14 gene encodes a methyltransferase that mediates C-terminal methylation of a-factor and ras proteins. EMBO J 10:1699–1709
Hrycyna CH, Clarke S (1992) Maturation of isoprenylated proteins in Saccharomyces cerevisiae:multiple activities catalyze the cleavage of the three carboxyl-terminal amino acids from farnesylated substrates in vitro. J Biol Chem 267:10457–10464
Hurley JB, Fong HKW, Teplow DB, Dreyer WI, Simon MI (1984) Isolation and characterization of a cDNA clone for the $\gamma$ subunit of bovine retinal transducin. Proc Natl Acad Sci USA 81:6948–6952
James G, Olson EN (1990) Fatty acylated proteins as components of intracellular signaling pathways. Biochemistry 29:2623–2634
Jones DT, Reed RR (1987) Molecular cloning of five GTP-binding protein cDNA species from rat olfactory neuroepithelium. J Biol Chem 262:14241–14249
Jones DT, Reed RR (1989) $G_{olf}$, an olfactor neuron specific G protein involved in odorant signal transduction. Science 244:790–795
Jones TLZ, Simonds WF, Merendino JJ, Brann MR, Spiegel AM (1990) Myristoylation of an inhibitory GTP-binding protein $\alpha$ subunit is essential for its membrane attachment. Proc Natl Acad Sci 87:568–572

Jones TLZ, Spiegel AM (1990) Isoprenylation of an inhibitory G protein $\alpha$ subunit occurs only upon mutagenesis of the carboxyl terminus. J Biol Chem 265:19389–19392
Katada T, Northup JK, Bokoch GM, Ui M, Gilman AG (1984) The inhibitory guanine nucleotide-binding regulatory component of adenylate cyclase. Subunit dissociation and guanine nucleotide-dependent hormonal inhibition. J Biol Chem 259:3578–3585
Kim D, Lewis DL, Graziadei L, Neer EJ, Bar-Sagi D, Clapham DE (1989) G-protein $\beta\gamma$ subunits activate the cardiac muscarinic $K^+$-channel via phospholipase $A_2$. Nature 337:557–560
Kuhn H (1980) Light- and GTP-regulated interaction of GTPase and other proteins with bovine photoreceptor membranes. Nature 283:587–589
Lai RK, Perez-Sala D, Canada FJ, Rando RR (1990) The $\gamma$ subunit of transducin is farnesylated. Proc Natl Acad Sci USA 87:7673–7677
Linder ME, Pang I-H, Duronio RJ, Gordon JI, Sternweis PC, Gilman AG (1991) Lipid modifications of G protein subunits: myristoylation of $G_{o\alpha}$ increases its affinity for $\beta\gamma$. J Biol Chem 266:4654–4659
Lochrie MA, Hurley JB, Simon MI (1985) Sequence of the $\alpha$ subunit of photoreceptor G protein: homologies between transducin, ras and elongation factor. Science 228:96–99
Maltese WA, Robishaw JD (1990) Isoprenylation of C-terminal cysteine in a G-protein $\gamma$ subunit. J Biol Chem 265:18071–18074
Mumby SM, Casey PJ, Gilman AG, Gutowski S, Sternweis PC (1990) G protein $\gamma$ subunits contain a 20-carbon isoprenoid. Proc Natl Acad Sci USA 87:5873–5877
Mumby SM, Heukeroth RO, Gordon JI, Gilman AG (1990) G protein $\alpha$ subunit expression, myristoylation, and membrane association in COS cells. Proc Natl Acad Sci USA 87:728–732
Muntz KH, Sternweis PC, Gilman AG, Mumby SM (1992) Influence of $\gamma$ subunit prenylation on association of guanine nucleotide-binding regulatory proteins with membranes. Mol Biol of the Cell 3:49–61
Navon SE, Fung BK-K (1987) Characterization of transducin from bovine retinal rod outer segment. Participation of the amino terminal region of $T_\alpha$ in subunit interactions. J Biol Chem 262:15746–15751
Neer EJ, Pulsifer L, Wolf LG (1988) The amino terminus of G protein $\alpha$ subunits is required for interaction with $\beta\gamma$. J Biol Chem 263:8996–9000
Northup JK, Smigel MD, Sternweis PC, Gilman AG (1983) The subunits of the stimulatory regulatory component of adenylate cyclase. Resolution of the activated 45,000-dalton ($\alpha$) subunit. J Biol Chem 258:11369–11376
Ohguro H, Fukada Y, Takao T, Shimonishi Y, Yoshizawa T, Akino T (1991) Carboxyl methylation and farnesylation of transducin gamma subunit synergistically enhance its coupling with meterhodopsin II. EMBO 10:3669–3674
Reiss Y, Goldstein JL, Seabra MC, Casey PJ, Brown MS (1990) Inhibition of purified p21ras farnesyl:protein transferase by Cys-AAX tetrapeptides. Cell 62:81–88
Reiss Y, Stradley SJ, Gierasch LM, Brown MS, Goldstein JL (1991) Sequence requirements for peptide recognition by rat brain p21ras farnesyl:protein transferase. Proc Natl Acad Sci USA 88:732–736
Robishaw JD, Kalman VK, Moomaw CR, Slaughter CA (1989) Existence of two $\gamma$ subunits of the G proteins in brain. J Biol Chem 264:15758–15761
Robishaw JD, Smigel MD, Gilman AG (1986) Molecular basis for two forms of the G protein that stimulates adenylate cyclase. J Biol Chem 261:9587–9590
Roof DJ, Applebury ML, Sternweis PC (1985) Relationships within the family of GTP-binding proteins isolated from bovine central nervous system. J Biol Chem 260:16242–16249

Ross EM, Howlett AC, Ferguson KM, Gilman AG (1978) Reconstitution of hormone-sensitive adenylate cyclase acitvity with resolved components of the enzyme. J Biol Chem 253:6401–6412

Rudnick DA, McWherter CA, Adams SP, Ropson IJ, Duronio RJ, Gordon JI (1990) Structural and functional studies of Saccharomyces cervisiae myristoyl-CoA:protein N-myristoyltransferase produced in Escherichia coli. J Biol Chem 265:13370–13378

Schmidt CJ, Neer EJ (1991) In vitro synthesis of G protein $\beta\gamma$ dimers. J Biol Chem 266:4538–4544

Simonds WF, Butrynski JE, Gautam N, Unson CG, Spiegel AM (1991) G protein $\beta\gamma$ dimers: Membrane targeting requires subunit coexpression and an intact $\gamma$ CAAX domain. J Biol Chem 266:5363–5366

Smrcka AV, Hepler JR, Brown KO, Sternweis PC (1991) Regulation of poly-phosphoinositide-specific phospholipase C activity by purified $G_q$. Science 251: 804–807

Spiegel AM, Backlund PS, Butyrinski JE, Jones TLZ, Simonds WF (1991) The G protein connection: molecular basis of membrane association. Trends Biochem Sci 16:338–341

Stephenson RC, Clarke S (1990) Identification of a C-terminal protein carboxyl methyltransferase in rat liver membranes utilizing a synthetic farnesyl cysteine-containing peptide substrate. J Biol Chem 265:16248–16254

Sternweis PC (1986) The purified $\alpha$ subunits of $G_o$ and $G_i$ from bovine brain require $\beta\gamma$ for association with phospholipid vesicles. J Biol Chem 261:631–637

Tamir H, Fawzi AB, Tamir A, Evans T, Northup JK (1991) G-protein $\beta\gamma$ forms: identity of $\beta$ and diversity of $\gamma$ subunits. Biochemistry 30:3929–3936

Tang W-J, Gilman AG (1991) Type-specific regulation of adenylyl cyclase by G protein $\beta\gamma$ subunits. Science 254:1500–1503

Towler DA, Gordon JI, Adams SP, Glaser L (1988) The biology and enzymology of eukaryotic protein acylation. Ann Rev Biochem 57:69–99

VanMeurs KP, Angus CW, Lavu S, Kung H, Czarnecki SK, Moss J, Vaughan M (1987) Deduced amino acid sequence of bovine brain retinal $G_{o\alpha}$: similarities to other guanine nucleotide-binding proteins. Proc Natl Acad Sci USA 84:3107–3111

Volker C, Lane P, Kwee C, Johnson M, Stock J (1991) A single activity carboxyl methylates both farnesyl and geranylgeranyl cysteine residues. FEBS Lett 295: 189–194

Volker C, Miller RA, McCleary WR, Rao A, Poenie M, Backer JM, Stock JB (1991) Effects of farnesylcysteine analogs on protein carboxyl methylation and signal transduction. J Biol Chem 266:21515–21522

Whiteway M, Hougan L, Dignard D, Thomas DY, Bell L, Saari GC, Grant FJ, O'Hara P, MacKay VL (1989) The STE4 and STE18 genes of yeast encode potential $\beta$ and $\gamma$ subunits of the mating factor receptor-coupled G protein. Cell 56:467–477

Wilcox C, Hu J-S, Olson EN (1987) Acylation of proteins with myristic acid occurs cotranslationally. Science 238:1275–1278

Yamane HK, Farnsworth CC, Xie H, Howald W, Fung BK-K, Clarke S, Gelb MH, Glomset JA (1990) Brain G protein $\gamma$ subunits contain an all-trans-geranylgeranyl-cysteine methyl ester at their carboxyl termini. Proc Natl Acad Sci USA 87:5868–5872

Yatsunami K, Pandya BV, Oprian DD, Khorana HG (1985) cDNA-derived amino acid sequence of the $\gamma$ subunit of GTPase from bovine rod outer segments. Proc Natl Acad Sci USA 82:1936–1940

Yokoyama K, Goodwin GW, Ghomashchi F, Glomset JA, Gelb MH (1991) A protein geranylgeranyltransferase from bovine brain: implications for protein prenylation specificity. Proc Natl Acad Sci USA 88:5302–5306

CHAPTER 53

# Phosphorylation of Heterotrimeric G-Protein

M.D. HOUSLAY

## A. Introduction

### I. Nature of G-Proteins

Many cell-surface receptors control the activity of their effector, signal-generating, systems through specific members of a family of guanine nucleotide-binding regulatory proteins (G-proteins) (BIRNBAUMER et al. 1988; SPIEGEL 1990). These are heterotrimeric species consisting of $\alpha$, $\beta$ and $\gamma$ subunits. It is the $\alpha$ subunits which are used to define the particular G-protein, and these are responsible for interaction with both the receptor and signal generator as well as binding and hydrolysing GTP. Upon the interaction of an occupied receptor with an appropriate G-protein, it binds GTP, dissociates, and the GTP-bound $\alpha$-subunit subsequently interacts with the effector system. The "turn-off" reaction is effected by the hydrolysis of GTP to GDP and the reassociation of the G-protein complex.

G-proteins play a pivotal role in mediating the action of a wide variety of hormone and neurotransmitter ligands. These effects occur as a consequence of the binding of the ligand to an appropriate receptor. It is now clear that G-protein-linked receptors are integral transmembrane proteins with a unique signature which is characterized by the occurrence of seven transmembrane segments and can be classed as R7G receptors. The linkage region between the fifth and sixth transmembrane domains (third cytosolic loop), together with the C-terminal tail, occur at the cytosol surface of the plasma membrane and are believed to play the primary role in interaction with G-protein $\alpha$ subunits (see HOUSLAY 1992 for review). However, more recently it has been mooted that receptors outside the R7G family may also be capable of interacting with G-proteins. For this to occur the presence of a motif, predominantly of positively charged residues, akin to that found in the third cytosolic loop of various G-protein-linked receptors is apparently required. Such interactions have been mooted for the insulin-like growth factor II, platetet-derived growth factor, and insulin receptors.

The receptor-dependent production of cyclic AMP is effected by adenylyl cyclases whose activity is regulated by two distinct heterotrimeric G-proteins termed $G_s$ (stimulatory) and $G_i$ (inhibitory). These G-proteins are characterized by unique $\alpha$-subunits which are the products of unique genes (BIRNBAUMER et al. 1990). A family of "$G_i$-like" proteins has been identified which is comprised of three highly related pertussis toxin-sensitive G-proteins termed $G_i$-1, $G_i$-2 and $G_i$-3. The precise functional role of these three heterotrimeric G-proteins remains to be elucidated. It is believed that at least $G_i$-3 and possibly all three of these species can couple to $K^+$-channels, and that one of these species may provide the pertussis toxin sensitive G-protein involved in regulating the phosphatidyl inositol 4,5 bisphosphate-specific phospholipase C. Whether all or just certain forms of $G_i$ provide the "true $G_i$", mediating inhibition of adenylyl cyclase, is unknown. However, there is an accumulating body of evidence which suggests that at least $G_i$-2 can achieve this end (see HOUSLAY 1991 for review).

Inhibition of adenylyl cyclase by $G_i$ can apparently ensue through two routes. Firstly, through the direct action of the $\alpha$-$G_i$, presumably on adenylyl cyclase itself, as exemplified by the ability of $G_i$ to inhibit adenylyl cyclase in cyc- cells which lack $G_s$ (BIRNBAUMER et al. 1990) and through the ability of the $\beta$ subunits released from $G_i$ to inhibit the dissociation of $G_s$ through mass action (BIRNBAUMER et al. 1990). The relative roles of these two processes may well vary depending on the circumstances and cell type. Indeed, it has proved a long-standing and rather thorny problem to quantify this. However, it has been suggested that the use of the local anaesthetic benzyl alcohol may provide a useful tool in this regard as it may uncouple $\beta$ but not $\alpha$ subunit mediated $G_i$ inhibition (SPENCE and HOUSLAY 1990).

Other examples of major G-proteins with defined functions are transducin, which serves to couple rhodopsin to the stimulation of cyclic GMP phosphodiesterase; $G_o$, which may regulate $Ca^{2+}$ channels and, in some instances, phospholipase C; and $G_q$, which provides the pertussis toxin-insensitive linkage to phospholipase C (see BIRNBAUMER et al. 1990).

## II. Modulation of G-Protein Action

G-proteins clearly play a pivotal role in regulating cell functioning, and one might reasonably expect that aberrant activities and expression could have deleterious effects. This is indeed the case, and the underlying molecular pathology of a number of disease processes can be attributed to defects at the level of G-proteins (HOUSLAY 1990a).

Activation of G-proteins involves the binding of GTP and then, for heterotrimeric G-proteins, their activation involving the dissociation of the complex to yield a GTP-bound $\alpha$ subunit and a $\beta\gamma$ complex. This dissociation occurs within seconds for GTP when the G-protein couples to an agonist-occupied receptor. However, the hysteretic nature of this event can clearly be seen using GTP analogues which, unlike GTP itself, lead to the time-dependent, constitutive activation of the G-protein over a period of minutes in the absence of agonist (see BIRNBAUMER et al. 1990). This rate of activation occurs within seconds if agonist is supplied. The ability then to modulate the GTPase activity and dissociation rate of G-proteins can be expected to have important functional consequences. This is quite clearly evident for the mini-G-protein *ras* and is also apparent for heterotrimeric G-proteins. For example, point mutations in or close to the GTP-binding site of $G_s$ have been shown to cause the constitutive activation and subsequent transformation of pituicytes, and ADP-ribosylation of the $\alpha$ subunits of $G_i$ and $G_o$ forms by pertussis toxin and $G_s$ by cholera toxin causes profound functional changes (see HOUSLAY 1991 for review).

Analysis of heterotrimeric G-proteins shows that they are plasma membrane associated, where they provide the linkage between transmembrane receptors and effector systems. However, it is quite clear that none of the G-protein subunits is a transmembrane entity, and none has extensive hydrophobic surfaces allowing for membrane association. Thus their membrane association is something of an enigma. A partial explanation for the association of $G_i$ species has been proffered as a result of the observation that the $\alpha$ subunits of these species are normally acylated. While this can reasonably be expected to cause membrane association, it does not explain why they are not found randomly associated with all intracellular membranes. Certainly, the major fraction of these heterotrimeric G-proteins is found within the plasma membrane, suggesting that such specificity is driven by association with a unique integral plasma membrane protein located there. This offers the possibility that modifications of either the G-protein or the anchor could cause release of the G-protein from the membrane. In this regard it is of interest that the chronic exposure of cells to cholera toxin, which causes the ADP-ribosylation of the $\alpha$-subunit of $G_s$ leads to its slow release (see BIRNBAUMER et al. 1990 for review) and, similarly for the mini-G-protein rap1A upon its phosphorylation (see WHITE et al. 1990).

One of the key mechanisms through which cells alter the functioning of specific proteins is by phosphorylation. This offers a rapid and reversible mechanism of control. Indeed, over the past few years evidence has been accumulating which suggests that certain G-proteins may be subject to this post-translational modification.

## 1. Phosphorylation

Jakobs and co-workers were the first to suggest that in cases where phorbol esters lead to the sensitization of adenylyl cyclase activity, this might result from the inactivation of $G_i$ (KATADA et al. 1985). This concept is certainly consistent with a variety of studies which have shown that pertussis toxin treatment of cells, performed to inactivate $G_i$, leads to a potentiation in the ability of stimulatory hormones to activate adenylyl cyclase, presumably caused by the removal of a "tonic" inhibitory input. Indeed, the degree of augmentation of the corticotropic-releasing hormone stimulation of adenylyl cyclase in anterior pituitary cell by phorbol esters was similar in magnitude to the potentiating effect of pertussis toxin. Similar results have been obtained in other systems as well (see HOUSLAY 1991, for review). In each of these instances the effects of pertussis toxin and phorbol esters were not additive. Such a finding, taken together with observations under conditions whereby phorbol esters potentiate hormonal activation of adenylyl cyclase, indicates that the action of this agent is similar in magnitude to the effect achieved by pertussis toxin. From this one could implicate a similar target for action in each case; presumably the inactivation of a tonic inhibitory effect exerted upon adenylyl cyclase by $G_i$.

The concept that phorbol esters might negate a "tonic" inhibitory effect mediated by $G_i$ upon adenylyl cyclase was also elaborated upon by Bell and Brunton (1986), who suggested that phorbol esters act to decrease the rate of activation of $G_i$ by guanine nucleotides. Thus they were able to show that while GTP-mediated inhibitory effects are abolished by phorbol ester treatment of S49 lymphoma cells, inhibitory effects mediated by non-hydrolysable GTP analogues are retained, albeit in a slightly reduced state. Indeed, the phorbol ester-mediated activation of protein kinase C appeared to abolish the "tonic" inhibition of adenylyl cyclase, mediated by either GTP or by non-hydrolysable GTP analogues, in hepatocytes, adenoma pituicytes, Leydig's cells and anterior pituitary cells and partially ablates this activity in S49 lymphoma cells (see HOUSLAY 1991, for review). As BELL and BRUNTON (1986) have suggested that phorbol esters might attenuate the rate of activation of $G_i$ by guanine nucleotides, the varying degrees of loss of "tonic $G_i$ function" observed in these investigations may reflect absolute differences in "activation rate" kinetics of $G_i$ in the various experimental systems.

## B. Phosphorylation of Heterotrimeric G-Proteins in Intact Cells

### I. Hepatocytes

Both Houslay's and Kahn's groups have shown that a fraction of the $\alpha$-$G_i$-2 occurring in hepatocytes is phosphorylated under basal conditions (ROTHENBERG and KAHN 1988; PYNE et al. 1989; BUSHFIELD et al. 1990a,b,c). This was markedly increased upon challenge of the cells with the phorbol ester (TPA) or hormones which stimulate inositol phospholipid metabolism, such as angiotensin and vasopressin (PYNE et al. 1989; BUSHFIELD et al. 1990a,b,c). These effects occurred in a non-additive fashion, indicating that the action is mediated by protein kinase C. Immunoprecipitation experiments with various antisera showed that while $\alpha$-$G_i$-2 is modified, there was no labelling of either $\alpha G_i$-3 or $\alpha$-$G_s$. Further to this, as there was

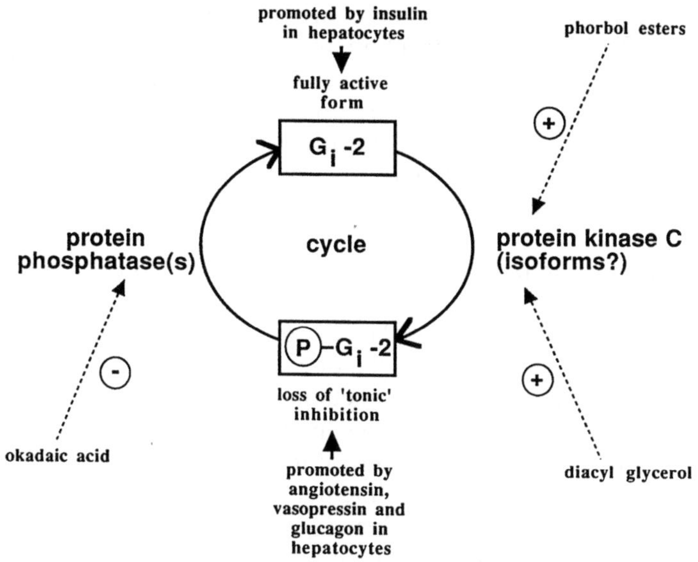


**Fig. 1.** A phosphorylation/dephosphorylation cycle for the control of $G_i$-2 in certain cells

no expression or significant levels of mRNA for $\alpha$-$G_i$-1 in hepatocytes, this species could not contribute to the labelled protein which was immunoprecipitated by the C-terminal antisera. Indeed, $^{32}$P-labelled $G_i$-2 could be specifically immunoprecipitated using antisera directed towards the C-terminal decapeptide of either transducin or $G_i$-2 with the addition of excess C-terminal peptide from $G_i$-2, but not that of $G_z$, preventing the immunoprecipitation of $G_i$-2. As expected, the phosphorylation of $\alpha$-$G_i$-2 led to a more acidic pI being noted for this species. That a fraction of $\alpha$-$G_i$-2 is found to be phosphorylated under basal conditions (i.e. the absence of hormone) indicates that an active phosphorylation/dephosphorylation cycle may be ensuing (HOUSLAY 1991a; Fig. 1). Consistent with this we have noted that treatment of cells with the protein phosphatase inhibitor okadaic acid leads to a rapid increase in the labelling of this G-protein (BUSHFIELD 1991). The stoichiometry of phosphorylation is difficult to gauge precisely as the amount of $\alpha$-$G_i$-2 was determined by quantitative western blotting. However, it would seem that in hepatocytes incubated under basal conditions there was some 0.3 mol $^{32}$P incorporated per mole of $\alpha$-$G_i$-2. This level was increased to over 1 mol $^{32}$p/mol $\alpha$-$G_i$-2 after challenge of intact cells with either TPA or hormones, such as vasopressin and angiotensin, able to stimulate phospholipid signalling pathways. In all of these cases only phosphoserine was observed as the labelled amino acid in the immunoprecipitated $G_i$-2.

To try to ascertain which residue(s) are modified, two-dimensional analyses were carried out on phosphopeptides obtained using trypsin, chymotrypsin and V8 protease either alone or in concert. Knowing the sequence of $G_i$-2 and the specificity of the proteases allowed us (N. MORRIS and M.D.H., unpublished) to calculate, for the expected peptides produced, the mass charge ratio, and thus estimate electrophoretic mobility, as well as estimating their mobility upon chromatography in a defined solvent system. Tryptic phosphopeptide analysis identified one major and two minor spots for $\alpha$-$G_i$-2 immunoprecipitated from hepatocytes incubated in the absence of any added ligands (basal condition) and after challenge with TPA. Such studies suggest that the protein kinase C mediated phosphorylation of $\alpha$-$G_i$-2 occurs at one major site, probably Ser-306 or Ser-144. These residues all lie within what have been postulated as functionally important domains (see BIRNBAUMER et al. 1990). Thus Ser-144 lies in the secondary effector region which is thought to influence nucleotide affinity and Ser-306 lies at the junction of the proposed effector binding/nucleotide binding site. In this regard it is interesting to note that when $G_i$ has bound non-hydrolysable GTP analogues, phosphorylation by protein kinase C is ablated (KATADA et al. 1985); BELL and BRUNTON (1986) have suggested that phorbol esters de rease the rate of activation of $G_i$ by guanine nucleotides.

Considering the ambiguity of the "recognition" sequence for protein kinase C, it is apparent that there are a large number of potential sites for phosphorylation of $\alpha$-$G_i$-2. That they become phosphorylated presumably depends upon the precise specificity of the isoforms of protein kinase C present and the steric availability of residues. This latter feature explains why partially denatured proteins often show aberrant susceptibility to phosphorylation in vitro. However, we should note that, except for Ser-306, the sites tentatively identified on $\alpha$-$G_i$-2 are also present in $\alpha$-$G_i$-3, which does not seem to be phosphorylated in hepatocytes (BUSHFIELD et al. 1990a,b). This might imply conformational differences in these two species for serines which are unique to $\alpha$-$G_i$-2 and might well provide substrates, for protein kinase C do not appear to be modified.

The phosphorylation of $\alpha$-$G_i$-2 in intact hepatocytes correlates well with the loss of ability of GTP itself to mediate an inhibitory effect upon cyclase (BUSHFIELD et al. 1990a,b). This would be consistent with the notion that the phosphorylation of $G_i$ stabilizes the inactive state of the G-protein, and that treatment of a variety of cells with phorbol esters ablate $G_i$ function (see HOUSLAY 1991a,b).

It has also been noted (PYNE et al. 1989; BUSHFIELD et al. 1990a,b,c) that increases in hepatocyte cAMP levels, achieved by either glucagon or 8-bromo-cAMP increase the phosphorylation of $G_i$-2. Additivity experiments suggested the presence of a phosphorylation site on $\alpha$-$G_i$-2 in addition to that (those) for protein kinase A. This was confirmed by phosphopeptide analysis (N. MORRIS and M.D.H., unpublished). The functional significance of this phosphorylation is not known; however, we have tentatively localized the target residue to Ser-207, which is within the putative GTP-binding domain. This cAMP-mediated phosphorylation did not appear to be mediated directly by protein kinase A as treatment of hepatocyte membranes with the purified catalytic unit of this enzyme failed to effect phosphorylation of $\alpha$-$G_i$-2, suggesting an indirect action with another kinase. It should be noted, however, that both of these ligands also elicit an increase in the labelling at the protein kinase C site(s), thus precluding any functional analyses on the site phosphorylated upon elevation of intracellular cAMP. The reason for the labelling of the protein kinase C site(s) is that glucagon can increase intracellular diacylglycerol levels, and that $Ca^{2+}$ and permeant cAMP analogues can elevate intracellular $Ca^{2+}$, events which lead to protein kinase C activation (see HOUSLAY 1991b for review).

We have suggested that the modification of $G_i$-2 in this fashion may be useful in tuning the cellular responsiveness to stimulatory hormones as inactivation of $G_i$ input, for example by pertussis toxin, leads to an enhanced action of stimulatory hormones upon adenylyl cyclase (HOUSLAY 1991a). This process, however, may be subverted in pathological states such as are seen in tissues exhibiting insulin resistance. Thus we have noted that GTP-elicited $G_i$ inhibition of adenylyl adenylyl cyclase is apparently lost in hepatocytes (GAWLER et al. 1987; BUSHFIELD et al. 1990b,c; HOUSLAY 1989) and adipocytes (STRASSHEIM et al. 1990) of streptozotocin-diabetic rats and in also fatty Zucker rats, compared to their lean litermates, concomitant with an increase in the phosphorylation status of $\alpha$-$G_i$-2 at the putative protein kinase C site(s). The defect which underlies these insulin-resistant states may thus be related to either enhanced protein kinase C activity or decreased protein phosphatase activity; our present data (E.Y. TANG and M.D.H., unpublished) suggests that at least the former occurs. Indeed, enhanced protein kinase C activity can be expected to lead to alter insulin receptor tyrosyl kinase functioning as well as that of $G_i$-2 (HOUSLAY 1989, 1991b).

Interestingly, in these instances loss of GTP-elicited inhibition was not paralleled by loss of receptor-mediated inhibition of adenylyl cyclase. This suggests that while GTP alone can elicit a weak, partial inhibition of adenylyl cyclase, the physical coupling of an inhibitory receptor to $G_i$ engenders such a powerful conformational change in this G-protein that it overcomes the action of phosphorylation, presumably by enhancing the binding of GTP and dissociation of $G_i$.

Our observations of altered $G_i$ funtion in various states requires that care be taken in membrane preparation from tissues as any dephosphorylation of $G_i$-2 would lead to the reappearance of GTP-elicited inhibition and thus the failure to observe the changes in $G_i$ function. It may, we think, also offer an explanation for the confusion that has existed in the literature over many years concerning the ability or not of insulin to inhibit adenylyl cyclase activity (see HOUSLAY 1986). This is based upon our observations (N. MORRIS, and M.D.H.) that, in hepatocytes, insulin acts to lower the degree of phosphorylation of $G_i$-2 effected through protein kinase C. This might be expected to lead to a small increase in the tonic GTP-mediated inhibition of adenylyl cylase as it would increase the level of active, dephosphorylated $G_i$-2. Such a model might explain (a) the pertussis toxin dependency of the process; (b) that insulin-mediated inhibition might only be seen in certain cells and even then under highly defined conditions, for example, there would be a requirement for pre-existing phosphorylated $G_i$-2, any insulin- mediated decrease in phosphorylated $G_i$-2 would have to be at a level able to achieve a measurable activity change; and (iii) that some but not others could report insulin inhibition of adenylyl cyclase in membrane preprations if such an action was mediated by a phosphatase loosely associated with the insulin receptor and thus capable of being displaced during certain membrane prepartive procedures.

## II. Promonocytic Cell Line U937

This promonocytic cell line can be induced to differentiate into a monocyte/ macrophage by the action of tumour-promoting phorbol esters. Such an action leads to the increased expression of $G_i$-2 (DANIEL-ISSAKANI et al. 1989). However, these investigators noticed that the isoelectric point of the intracellular $G_i$-2 was shifted slightly to a more acidic region. Immunoprecipitation of $G_i$-2 from these cells showed that the $\alpha$ subunit of $G_i$-2 rapidly became phosphorylated, within 5 min, upon challenge of these cells with TPA. Interestingly, these investigators suggested that such a modification might even enhance the ability of pertussis toxin to cause the ADP-ribosylation of $\alpha$-$G_i$-2. The fact that $G_i$-2 could become both ADP-ribosylated and also phosphorylated was consistent with a further shift in the acidity of the doubly modified G-protein. Consistent with the notion that $G_i$-2 becomes phosphorylated in these cells were experiments showing that treatment of membranes from U937 cells with purified protein kinase C (type III) also led to the phosphorylation of $G_i$-2.

The antipeptide antiserum used in these studies was directed against the C-terminal decapeptide of transducin. This can be used to immunoprecipitate or blot $G_i$-1 and $G_i$-2 but neither $G_i$-3 nor $G_z$ (SPIEGEL 1990; BUSHFIELD et al. 1990). As these cells, however, had high levels of $G_i$-2, did not exhibit $G_i$-1, and showed low levels of $G_i$-3, it would appear that the immunoprecipitation of $G_i$-2 was occurring specifically.

In further experiments these investigators showed that the ability of pertussis toxin to cause the ADP-ribosylation of this G-protein was increased for phosphorylated $G_i$-2 (DANIEL-ISSAKANI et al. 1989). As pertussis toxin ADP-ribosylation requires $G_i$ species to be in an associated state, this suggests that the holomeric state may be stabilized by phosphorylation.

## III. Platelets

### 1. $G_i$-2

KATADA et al. (1985) have suggested that $G_i$ may become phosphorylated in intact platelets, as they noted that treatment of cells with either thrombin or TPA led to the rapidly labelling of a 40-kDa protein. This was later taken up by CROUCH and LAPETINA (1986) who confirmed these results and, furthermore, using immunoprecipitation approaches showed that a phosphorylated form of $G_i$-2 can be obtained from intact platelets challenged with either of these two ligands. The immunoprecipitated material, under analysis by western blotting, confirmed the presence of $G_i$-2. Such results are in contrast to those of CARLSON et al. (1989; see below) who failed to identify phosphorylated $G_i$-2 in phorbol ester treated platelets but showed exhaustively that the related G-protein $G_z$ can be so modified. The saga continues in that recent work by Ui and colleagues (YATOMI et al. 1992) who have shown that the phosphorylation of $G_i$-2 occurs in platelets (and neutrophils) as a result of challenge of the cells not only with the phorbol ester TPA but also with thrombin and with a thromboxane mimetic. Such observations are consistent with the idea that this G-protein can be phosphorylated in intact platelets. One possible explanation for the failure of CARLSON et al. (1989) to note this event could be related to the fact that they used an anti-peptide directed at an internal sequence rather than the C-terminal decapeptide used by the others. While there are no potential phosphorylation sites at or close to the C-terminus of $G_i$-2, this is not true for the internal sequences. Indeed, the serum (1398) used by CARLSON et al. (1989), which could discriminate between $G_i$-2 and $G_z$, was based upon a peptide which contains two serine residues, both of which are potential substrates for protein kinase C. Thus it is possible that in the CARLSON et al. (1989) study any phosphorylation of $G_i$-2 may have further reduced the already rather weak ability of this antiserum to immunoprecipitate $G_i$-2.

### 2. $G_z$

Sequence inspection shows that $G_z$ is closely related to the "$G_i$ family" of G-proteins except that it lacks the C-terminal cysteine residue which provides the target for ADP-ribosylation by pertussis toxin (BIRNBAUMER et al. 1990). This G-protein is found in a limited number of tissues, such as platelets and brain, but is not expressed ubiquitously, as exemplified by its absence from hepatocytes. While the functional role of $G_z$ has yet to be established, it not thought to be involved in mediating the activation of phospholipase C.

Manning and co-workers (CARLSON et al. 1989) discovered that this heterotrimeric G-protein became highly phosphorylated in human platelets which had been treated with agents which activate or lead to the activation of protein kinase C, such as the phorbol ester phorbol myristate acetate

(PMA), and $\alpha$-thrombin. Modification of $G_z$ triggered by both of these agents occured in permeabilized and in intact cells. However, phosphorylation of $G_z$ triggered by the thromboxane $A_2$ analogue U46619, strangely, only appeared to ensue in permeabilized platelets. The most satisfying data which they showed used $G_z$ specific antisera to immunoprecipitate a labelled phosphoprotein, an effect which was competeted out when done in the presence of the "$G_z$ peptide" used to prepare the antiserum. The PMA-mediated phosphorylation of $G_z$ appeared to occur very rapidly, with maximal labelling being achieved within 1 min.

A number of outstanding questions still need to be addressed, such as the site of phosphorylation upon $\alpha$-$G_z$ and the effect that such a modification has upon the functioning of this protein. Certainly, a more detailed analysis of the phosphorylation site of $G_z$, by phosphopeptide sequencing, would also provide the underpining necessary to confirm that it is $G_z$ which is modified in such a fashion (now shown to be $Ser^{27}$ and $Ser^{16}$; Manning et al., unpublished).

## IV. Yeast

Haploid *Saccharomyces cerevisiae* cells of opposite mating type conjugate to form diploid cells. In preparing for such a conjugation event a pheromone signalling system which is transduced by a heterotrimeric G-protein is involved. The mating pheromone response is, remarkably, mediated by the $\beta\gamma$ complex with the $\alpha$ subunit generating an adaptive (desensitizing) signal (see Cross et al. 1988; Stone and Reed 1990; Cole and Reed 1991).

Western blot analysis of haploid cells using an antibody to the $\beta$-subunit (Ste4 gene product) identified a 47-kDa protein which, upon challenge of the cells with pheromone, showed decreased mobility to give two species of 49- and 56-kDa (Cole and Reed 1991). This reduced mobility was shown to be due to the phosphorylation of the $\beta$ subunit as $^{32}P$-labelled species could be immunoprecipitated from cells incubated with [$^{32}P$]Pi. Removal of label, using phosphatases, restored the aberrant mobility of the $\beta$ subunit on SDS-PAGE. Modification occurred exclusively upon serine residues, although there was an indication that multiple sites were being labelled. Preliminary phosphopeptide mapping was done with *N*-chlorosuccinimide, which cleaves at tryptophan residues. The Ste protein, which contains eight tryptophans, yielded four labelled bands. This might indicate that more than one serine residue was modified. As such, it will take detailed analyses to determine which residues are functionally important.

The phosphorylation of this G-protein $\beta$ subunit appears to play a crucial role in the regulation of the pheromone response. This was elegantly demonstrated using various mutant forms of the Ste4 gene aimed at deleting the phosphorylation site (Cole and Reed 1991). The result of which led to enhanced sensitivity to pheromone, did not affect the actual ability to mount a signal, but did affect the ability to recover from pheromone challenge.

Thus the G-protein β subunit appears to be the primary transducer of the mating signal with phosphorylation of this subunit presumably playing a key control role in ensuring the prompt inactivation of the signal by promoting desensitization, that is, the reassociation with the α subunit.

## V. Dictyostelium

Extracellular cAMP plays a variety of key roles in the development of the slime mold dictyostelium. It acts as a chemo-attractant and as a differentiation inducer. These actions are transduced by a transmembrane G-protein-linked cell surface receptor (JANSSENS and VAN HAASTERT 1987). Involved in mediating the action of the cAMP receptor is a heterotrimeric G-protein called Gα2 (this should not be confused with the $\alpha$-$G_i$-2 found in mammalian cells). This G-protein was discovered through molecular genetical analyses of mutant dictyostelium, in particular the Frigid mutants. These fail to respond to extracellular cAMP and were found to contain point mutations or deletions in the Gα2 gene (GUNDERSON and DEVREOTES 1990). It had been noted (GUNDERSON and DEVREOTES 1990), however, that Gα2 undergoes a rapid time-dependent transition in mobility upon SDS-PAGE when immunoprecipitated from cells which have been challenged with cAMP. This takes the form of a transient increase to an apparent molecular size of 43 kDa from the normal, and predicted, size of 40 kDa. Such a change also follows the desensitization of the system and its reversal, occurring upon the removal of extracellular cAMP, follows the pattern of resensitization (GUNDERSON and DEVREOTES 1990). The molecular basis of this reversible transition was found to be due to the phosphorylation of Gα2 (GUNDERSON and DEVREOTES 1990). Thus a phosphorylated form of this G-protein could be immunoprecipitated from dictyostelium challenged with cAMP. In addition, the time course of such a covalent modification paralleled the mobility change observed for Gα2. Consistent with this covalent modification being responsible for the mobility change was the observation that if immunoprecipitated, phosphorylated Gα2 is treated with alkaline phosphatase, the removal of $^{32}P$ from this protein leads to the appearance of a species which migrates normally at 40 kD.

The precise site of modification of Gα2 remains to be determined. However, $^{32}P$ accumulates specifically on serine to a ratio of 1 mol/mol Gα2. Furthermore, on the basis that *N*-chlorosuccinimide cleaves at a unique tryptophan, this residue(s) is within the first 119 amino acids of Gα2. However, as this region contains some 12 serine residues further work is needed to locate precisely the site of covalent modification.

While the details of the modification and its regulation remain to be established, it would seem from the available data that it is responsible for the desensitization of this signal transduction system. The molecular basis of this blockade, however, requires definition. Phosphorylation could block signal transduction at a number of stages. Examples of this might be the interaction of the $G\alpha 2$ with the effector system, it might prevent dissociation of this G-protein, or it might stimulate its GTPase activity to such an extent that the lifetime of the activated state is vanishingly small.

## C. In Vitro Phosphorylation of Isolated Heterotrimeric G-Proteins

### I. Transducin

The G-protein transducin, which is found in retinal rods and cones, is highly homologous to the $G_i$ family of proteins although its physiological function is to mediate the stimulation of a specific cyclic GMP phosphodiesterase upon the photo-activation of rhodopsin. It appears both in vitro and in intact cells to provide a substrate for phosphorylation by both protein kinase C, on serine residues, and by the IGF-I receptor on tyrosine residues (SAGI-EISENBERG 1989). Transducin serves as a substrate for such phosphorylation only when it is in its GDP-bound inactive form, and it is believed that this modification serves to stabilize the inactive holomeric G-protein. While this may serve to attenuate functioning through this signalling system, no analyses have addressed this issue.

### II. $G_i$ and $G_o$

Jakobs and co-workers (KATADA et al. 1985), using in vitro approaches, were the first investigators to show that G-proteins can be phosphorylated. They used a protein kinase C preparation, which we can presume to be a mixture of isoforms, to phosphorylate a purified preparation of rabbit liver "$G_i$", which presumably was a mixture of the various forms of $G_i$. The labelling of this $G_i$ preparation occurred in a $Ca^{2+}$- and phospholipid-dependent form, as would be expected for the action of protein kinase C. Interestingly, while they found that the free $\alpha$ subunit of $G_i$ provided the best substrate for phosphorylation, the binding of a non-hydrolysable GTP analogue to this dissociated subunit severely attenuated its ability to be modified. This might imply that the "GTP-bound" state of $G_i$ is in some way resistant to modification, perhaps because the site of modification by protein kinase C is at or close to the GTP binding site.

Studies on the phosphorylation of $G_i$ by both protein kinase C and the human insulin receptor tyrosyl kinase have suggested that this G-protein can be phosphorylated only in its holomeric, GDP-bound state (KRUPINSKI et al. 1988; O'BRIEN et al. 1987). This suggests that phosphorylation might stabilize the inactive, GDP-bound form of $G_i$ and would be consistent with the hypothesis of BELL and BRUNTON (1986) that phorbol esters serve to attenuate the rate of activation of $G_i$. Certainly, that TPA action on $G_i$ was due to its stabilizing the holomeric state of $G_i$ is consistent with the observation of CHOI and TOSCANO (1988) who noted that the ability of pertussis toxin to mediate the ADP-ribosylation of $\alpha$-$G_i$, which can occur only on the intact GDP-bound holomeric form of this G-protein, was considerably greater in cells treated with phorbol ester for 15 min. Nevertheless, the studies of KATADA et al. (1985) indicate that we should be cautious in the interpretation of such data where it has been suggested that $G_i$ dissociation attenuates phosphorylation, as deduced merely from observations that the ability of $G_i$ to provide a substrate for phosphorylation was reduced in the presence of non-hydrolysable GTP analogues.

Mammalian heterotrimeric G-proteins do not appear to have consensus sequences for phosphorylation by protein kinase A. This would be consistent with our failure and that of many others to observe the phosphorylation of purified and membrane-bound $G_i$ and $G_s$ forms by protein kinase A. Nevertheless, there has been one report of the ability of protein kinase A to phosphorylate both the $\alpha$ and $\beta$ subunits of a partially purified brain $G_i$ preparation (WATANABE et al. 1988). The stoichiometry of this reaction was estimated as 0.34 and 0.18 mol $^{32}P$/mol G-protein $\alpha$ and $\beta$ subunits, respectively. These observations have yet to be explored further. In view of the fact that $G_i$-2 has been shown to become phosphorylated in hepatocytes upon challenge with agents which increase intracellular cAMP, these observations deserve further attention. However, in that system it was suggested that the actual phosphorylation of $G_i$-2 was likely mediated by a kinase other than protein kinase A, although such a species might presumably be activated by protein kinase A. As only a partially purified G-protein preparation was used in the studies of WATANABE et al. (1988), we cannot exclude the possibility of such a kinase being found in their preparation.

Purified $G_i$ and $G_o$ have also been found to be substrates for tyrosyl phosphorylation by the purified human insulin receptor (O'BRIEN et al. 1987a; KRUPINSKI et al. 1988). In this instance the insulin-stimulated phosphorylation of both the $\alpha$ and $\beta$ subunits of these G-proteins was noted. Interestingly, as with protein kinase C, phosphorylation by the insulin receptor tyrosyl kinase was enhanced if these G-proteins were incubated with GDP analogues and inhibited in the presence of GTP analogues. The rates of phosphorylation observed were slow and yielded poor levels of incorporation. Nevertheless, that enhanced labelling was seen if the reactants were incorportaed into lipid bilayers might indicate the importance of environment and conformation to these actions. To date, however, no such phosphorylation has been identified as occurring in intact cells (PYNE et al. 1989b; ROTHENBERG and KAHN 1988). However, it is possible that more rigorous analyses need to be carried out to explore various times of incubation with insulin and also the possibility that the rapid dephosphorylation at tyrosyl residues is occurring during the isolation and subsequent analytical procedure.

With regard to the observation that the purified insulin receptor tyrosyl kinase can phosphorylate the $\beta$ subunit of $G_i$ it is of interest that the epidermal growth factor receptor, another tyrosyl kinase, has been shown to cause the phosphorylation of a 35-kDa protein in placental membranes. This protein reacted with an antiserum able to recognize a G-protein $\beta$ subunit of the same size (Valentine-Braun et al. 1986). The identity of this species, however, remains an unresolved issue although the authors indicated that they were convinced that it was an immunologically related species rather than authentic G-$\beta$.

## III. $G_s$

Recently, Pyne and co-workers (1992) have claimed to be able to show the phosphorylation of purified recombinant $\alpha$-$G_s$ using a purified preparation of brain protein kinase C. They analysed two ($\alpha$-1$G_s$ and $\alpha$-4$G_s$) of the four splice variants of $G_s$ which have been found to occur in mammalian cells. $\alpha$-1$G_s$ and $\alpha$-3$G_s$ differ only in that $\alpha$-3$G_s$ lacks a stretch of 15 amino acids; $\alpha$-4$G_s$ is like $\alpha$-3$G_s$ except that it has three additional nucleotides at the 5′ end of exon 4, and $\alpha$-2$G_s$ is like $\alpha$-1$G_s$ except for three additional nucleotides at the 5′ end of exon 4. Recombinant $\alpha$-1$G_s$ and $\alpha$-4$G_s$ were purified form *Escherichia coli* as 52-kDa and 45-kDa peptides and both incubated with protein kinase C under phosphorylating conditions. Intriguingly, however, only the $\alpha$-4$G_s$ variant became significantly phosphorylated. Stringent analysis of the maximal level of phosphorylation and any functional effect was not possible to determine as the $G_s$ species were extremely thermolabile. Nevertheless it appeared that in excess of 0.5 mol $^{32}$P/mol $\alpha$-4$G_s$ was incorporated. Interestingly, however, the inclusion of a 20-fold excess of $\beta\gamma$ subunits together with GDP$\beta$-S appeared to completely inhibit the phosphorylation of recombinant $G_s$. This contrasts with evidence which suggests that $G_i$ is phosphorylated in its holomeric form by protein kinase C, and that in vitro reactions with the insulin receptor tyrosyl kinase clearly show that the phosphorylation of both $G_i$ and $G_s$ occur only when the G-proteins are in their holomeric forms, preferably with GDP bound. Such studies thus suggest that $G_s$ needs to be in a dissociated state for phosphorylation to be achieved. Strangely, however, the addition of a non-hydrolysable GTP analogue, which would stabilize the activated state of the G-protein, also appeared to inhibit phosphorylation. This may imply that $\alpha$-4$G_s$ may serve as a substrate for protein kinase C in vivo only under very specific conditions. These may take the form of maximal susceptibility at a point when the activated $\alpha$-subunit has begun its deactivation cycle by hydrolysing its bound GTP to GDP and immediately prior to its reassociation with $\beta\gamma$ subunits. Indeed, it may even be the nucleotide-free state of this G-protein which is the favoured species for phosphorylation.

The putative phosphorylation site on $\alpha$-4$G_s$ has yet to be defined. However, Pyne et al. (1992) have suggested that possibilities are Ser-286 or in the splice junction Ser-72.

To date, however, there have been no reports of $G_s$ phosphorylation occurring in intact cells. Indeed, two groups failed to see the phosphorylation of $\alpha$-$G_s$ occurring under conditions where $\alpha$-$G_i$-2 phosphorylation was reported (ROTHENBERG and KAHN 1988; PYNE et al. 1989; BUSHFIELD et al. 1990). However, it should be noted that in the cells studied (hepatocytes), the $\alpha$-4$G_s$ variant is not found (BUSHFIELD et al. 1990).

Nevertheless, their still remains the possibility that the labelling of $G_s$ seen in these studies is purely an in vitro observation due either to the high concentrations of reactants or that the proteins, being bacterially expressed species, are partially denatured and thus subject to aberrant phosphorylation by kinases.

MARTIN and co-workers (FARNDALE et al. 1987), however, are enamoured of the possibility that $G_s$ may be phosphorylated in human platelets as a result of the action of elevated cAMP concentrations. As has been shown by many investigators (see BIRNBAUMER et al. 1990), the GTP analogue p(NH)ppG activates adenylyl cyclase constitutively after a well-defined lag period. They demonstrated that this process is slowed down in the presence of ATP and cAMP, and that such an action is attenuated by an inhibitor of protein kinase A. Their contention is that activation of protein kinase A leads to the phosphorylation of $G_s$, and that this attenuates its activity. However, although they showed that various platelet membrane proteins do become phosphorylated over time, no data have been forthcoming to show that it is $G_s$ which is the phosphorylation target. Thus the molecular basis of these interesting experiments remains to be elucidated. Indeed, it is even possible that adenylyl cyclase itself provided the target for phosphorylation.

Purified $G_s$ has also been found to be substrates for tyrosyl phosphorylation by the purified human insulin receptor (KRUPINSKI et al. 1988). While the rates of phosphorylation observed are slow and yield poor levels of incorporation, as with $G_i$, enhanced labelling is seen if the reactants are incorportaed into lipid bilayers. To date, however, no such phosphorylation has been identified as occurring in intact cells.

## IV. Unidentified "G-Proteins"

SAUVAGE et al. (1991) have purified both a 36-kDa and a 50-kDa G-protein $\alpha$ subunit from rat brain which they called $G_{36}$ and $G_{50}$, respectively. These have not been characterized as regards cross-reactivity with established G-proteins. However, these investigators did show that the constitutively activated form of protein kinase C, protein kinase M, and not a mixed isoform preparation (brain) of protein kinase C itself, causes the phosphorylation of $G_{36}$. This resulted in a marked change in both the GTPase activity and the guanine nucleotide binding properties of this species. Analysis of $G_{50}$ showed that protein kinase A and not protein kinase C causes its phosphorylation. This did not result in any changes in either the GTPase activity of the guanine nucleotide binding properties of this species.

O'BRIEN et al. (1989) also reported the presence of a 60-kDa protein able to bind to a guanine nucleotide affinity columnn whose phosphorylation was increased by insulin challenge of intact mouse fibroblasts which overexpressed the human insulin receptor. The identity of this species remains to be ascertained.

## D. Conclusion

It is clear that G-proteins do not excape from the clutches of phosphorylation, that ubiquitous regulation system so well exploited by the cell. This mechanism appears to affect many, but by no means all of the G-proteins, be they heterotrimeric species or mini-G-proteins. In many instances much has been done to indicate the importance of such modifications. Nevertheless, we have as yet no definitive molecular details concerning the actual residues modified and the mechanism through which the resultant molecular switch alters activities and functioning. A further complexity relates to those G-proteins which appear to be phosphorylated by protein kinase C. We still await information on which of the isoforms of this enzyme family can modify G-proteins and effect functional changes. If specific isoforms are required, this factor together with a requirement for the G-proteins to be found in an appropriate state of association and complexed with GDP may lead us to expect that a G-protein such as $G_i$ is phosphorylated only in certain cells, an observation for which there is some supporting data.

Hopefully it will not be too long before more detailed information on this complex regulatory system is available. This should give considerable insight into the mechanisms of action of G-proteins and the cellular actions which they transduce.

*Acknowledgements*. Work in the author's laboratory was supported by grants from the MRC, BDA and BHF.

### References

Bell JD, Brunton LL (1986) Phorbol ester action on $G_i$ function. J Biol Chem 261:12036–12041

Birnbaumer L, Abramowitz J, Brown AM (1990) Insulin and its interactions with G proteins. Biochim Biophys Acta 1031:163–224

Bushfield M, Murphy GJ, Lavan BE, Parker PJ, Hruby VJ, Milligan G, Houslay MD (1990a) Hormonal regulation of $G_i$-2 $\alpha$-subunit phosphorylation in intact hepatocytes. Biochem J 268:449–457

Bushfield M, Griffiths SL, Murphy GJ, Pyne NJ, Knowler JT, Milligan G, Parker PJ, Mollner S, Houslay MD (1990b) Diabetes-induced alterations in the

expression, functioning and phosphorylation state of the inhibitory guanine nucleotide regulatory protein $G_i$-2 in hepatocytes. Biochem J 271:365–372
Bushfield M, Pyne NJ, Houslay MD (1990c) Changes in the phosphorylation state for the inhibitory guanine nucleotide binding protein $G_i$-2 in hepatocytes from lean (Fa/Fa) and obese (fa/fa) Zucker rats. Eur J Biochem 192:537–542
Bushfield M, Lavan BE, Houslay MD (1991) Okadaic acid identifies a phosphorylation/dephosphorylation cycle controlling $G_i$-2. Biochem J 274:317–321
Carlson KE, Brass LF, Manning DR (1989) Thrombin and phorbol esters cause the selective phosphorylation of a G-protein other than $G_i$ in human platelest. J Biol Chem 264:13298–13305
Choi EJ, Toscano WA (1988) Modulation of adenylate cyclase activity in human keratinocytes py protein kinase. CJ Biol Chem 263:17167–17172
Cole GM, Reed SI (1991) Pheromone-induced phosphorylation of a G-protein $\beta$ subunit in S. cerevisiae is associated with an adaptive response to mating. Cell 64:703–716
Cross F, Hartwell LH, Jackson C, Konopka JB (1988) Conjugation in S. cerevisiae. Annu Rev Cell Biol 4:429–457
Crouch MF, Lapetina EG (1988) A role for $G_i$ in control of thrombin receptor phospholipase C coupling in human platelets. J Biol Chem 263:3363–3371
Daniel-Issakani S, Spiegel AM, Strulovic B (1989) Lipopolysaccharide response is linked to the GTP binding protein $G_i$-2 in the promonocytic cell line U937
Farndale RW, Wong SKF, Martin BR (1987) Activation of adenylyl cyclase in human platelets membranes by 5'-[$\beta\gamma$-imido]triphosphate is inhibited by cyclic AMP-dependent phosphoryltion. Biochem J 242:637–643
Gawler D, Milligan G, Spiegel AM, Unson CG, Houslay MD (1987) Abolition of the expression of inhibitory guanine nucleotide regulatory protein $G_i$ activity in diabetes. Nature 327:229–232
Griffiths SL, Knowler JT, Houslay MD (1990) Diabetes-induced changess in guanine nucleotide regulatory protein (G-protein) mRNA as detected using synthetic oligonucleotide probes. Eur J Biochem 193:367–374
Gunderson RE, Devreotes PN (1990) In vivo receptor-mediated phosphorylation of a G-protein in Dictyostelium. Science 248:591–593
Houslay MD (1986) Insulin, glucagon and the receptor-mediated control of cyclic AMP concentrations in liver (Colworth Medal Lecture). Biochem Soc Trans 14:183–193
Houslay MD (1989) Distict functional domains on the insulin receptor $\beta$-subunit: do they provide a molecular basis for "selective" insulin resistance. Trends Endocrinol Metabolism 1:83–99
Houslay MD (1990a) Altered expression and functioning of guanine nucleotide regulatory proteins during growth, transformation, differentiation and in pathological states. In: Houslay MD, Milligan G (eds) G-proteins as mediators of cellular signalling processes. Molecular pharmacology of cell regulation, vol 1. Wiley, Chichester, pp 197–230
Houslay MD (1990b) Insulin and its interaction with the G-protein system. In: Birnbaumer L, Iyengar R, (eds) G-proteins. Academic, New York, pp 521–553
Houslay MD (1991a) $G_i$-2 is at the centre of an active phosphorylation/dephosphorylation cycle in hepatocytes: the fine-tuning of stimulatory and inhibitory inputs into adenylate cyclase in normal and diabetic states. Cellular Signalling 3:1–10
Houslay MD (1991b) 'Cross-talk': a pivotal role for protein kinase C in modulating relationships between signal transduction pathways. Eur J Biochem 195:9–27
Houslay MD (1992) G-protein linked receptors: a family probed by molecular cloning and mutagenesis procedures. Clin Endocrinol 36:525–534
Houslay MD, Gawler DJ, Milligan G, Wilson A (1989) Multiple defects occur in the guanine nucleotide regulatory protein system in liver plasma membranes of

obese (fa/fa) but not lean (Fa/Fa) Zucker rats: loss of functional $G_i$ and abnormal $G_s$ function. Cellular Signalling 1:9–22
Janssens PMW, Van Haastert PJM (1987) Microbiol Rev 51:396
Katada T, Gilman AG, Watanabe Y, Bauer S, Jakobs KH (1985) Protein kinase C phosphorylates the inhibitory guanine nucleotide regulatory component and apparently suppresses its function in the hormonal inhibition of adenylyl cyclase. Eur J Biochem 151:431–437
Krupinski J, Rajaram R, Lakonuhok M, Benovic JL, Cerione RA (1988) Phosphorylation of G-proteins by the insulin receptor tyrosyl kinase. J Biol Chem 263:12333–12341
Murphy GJ, Gawler DJ, Milligan G, Wakelam MJO, Pyne NJ, Houslay MD (1989) Glucagon desensitization of adenylate cyclase does not involve the inhibitory guanine nucleotide regulatory protein $G_i$ which is inactivated upon challenge of hepatocytes with glucagon. Biochem J 259:191–197
O'Brien RM, Houslay MD, Milligan G, Siddle K (1987) The insulin receptor tyrosyl kinase phosphorylates holomeric forms of the guanine nucleotide regulatory proteins $G_i$ and $G_o$. FEBS Lett 212:281–288
O'Brien RM, Siddle K, Houslay MD, Hall A (1987b) Interaction of the human insulin receptor with the ras oncogene product p21. FEBS Lett 217:253–259
O'Brien R, Houslay MD, Brindle NPG, Milligan G, Whittaker J, Siddle K (1989) Binding to GDP-agarose identifies a novel 60 kDa substrate for the insulin receptor tyrosyl kinase in mouse NIH-3T3 cells expressing high concentrations of the human insulin receptor. Biochem Biophys Res Commun 158:743–748
Pyne NJ, Murphy GJ, Milligan G, Houslay MD (1989) Treatment of intact hepatocytes with either the phorbol ester TPA or glucagon elicits the phosphorylation and functional inactivation of the inhibitory guanine nucleotide regulatory protein $G_i$. FEBS Lett 243:77–82
Pyne NJ, Murphy GJ, Milligan G, Houslay MD (1989a) Treatment of intact hepatocytes with either the phorbol ester TPA or glucagon elicits the phosphorylation and functional inactivation of the inhibitory G-protein $G_i$. FEBS Lett 243:77–82
Pyne NJ, Heyworth CM, Balfour NW, Houslay MD (1989b) Insulin affects the ability of $G_i$ to be ADP-ribosylated but does not elicit its phosphorylation in intact hepatocytes. Biochem Biophys Res Commun 165:251–256
Pyne NJ, Freissmuth M, Palmer S (1992) Phosphorylation of the spliced variant forms of the recombinant guanine nucleotide-binding regulatory protein $G_s$ by protein kinase C. Biochem J (in press)
Rothenberg PL, Kahn R (1988) Insulin alters the ADP-ribosylation of $G_i$ by pertussis toxin. J Biol Chem 263:15546–15552
Sagi-Eisenberg R (1989) Phosphorylation of $G_i$. Trends Biochem Sci 14:355–357
Sauvage C, Rumigny JF, Maitre M (1991) Purification and characterisation of G-proteins from human braine: modification of GTPase activity upon phosphorylation. Mol Cell Biochem 107:65–77
Spence S, Houslay MD (1989) The local anaesthetic benzyl alcohol attenuates the alpha$_2$-adrenoceptor-mediated inhibition of human platelet adenylate cyclase stimulated by $PGE_1$ but not that stimulated by forskolin. Biochem J 264:483–488
Spiegel AS (1990) Structure and identification of G-proteins. In: Houslay MD, Milligan G (eds) G-proteins as mediators of cellular signalling processes. Molecular pharmacology of cell regulation, vol 1. Wiley, Chichester, pp 15–30
Stone DE, Reed SI (1990) G protein mutations that alter the pheromone response in S. cerevisiae. Mol Cell Biol 10:4439–4446
Strassheim D, Milligan G, Houslay MD (1990) Diabetes abolishes the GTP- but not receptor-dependent inhibitory function of $G_i$ on adipocyte adenylate cyclase activity. Biochem J 266:521–526

Valentine-Braun KA, Northup JK, Hollenberg MD (1986) Epidermal growth factor-mediated phosphorylation of a 35 kDa substate in human placental membranes: relationship to the $\beta$ subunit of G-proteins. Proc Natl Acad Sci USA 83:236–240

Watanabe Y, Imaizumi T, Misaka N, Iwakura K, Yoshida H (1988) Effects of phosphorylation of inhibitory GTP-binding protein by cAMP-dependent protein kinase on its ADP-ribosylation by pertussis toxin, islet activating protein. FEBS Lett 236:372–374

White TE, Lacal JC, Reep B, Fischer TH, Lapetina EG, White III GC (1990) Thrombolamban, the 22 kDa platelet substrate of cAMP-dependent protein kinase is immunologically homologous with the ras family of GTP-binding proteins. Proc Natl Acad Sci USA 87:758–762

Yatomi Y, Arata Y, Tada S, Kume S, Ui M (1992) Phosphorylation of the inhibitory guanine nucleotide regulatory protein as a possible mechanism of inhibition by protein kinase C of agonsist-induced $Ca^{2+}$ mobilization in human platelets. Eur J Biochem 205:1003–1009

CHAPTER 54

# Receptor to Effector Signaling Through G-Proteins: $\beta\gamma$ Dimers Join $\alpha$ Subunits in the World of Higher Eukaryotes

L. Birnbaumer

## A. Introduction

Heterotrimeric signal transducing G-proteins are a complex and diverse set of highly homologous proteins involved in coupling of a large number of receptors to a variety of effectors (for details of molecular diversity in G-protein subunits and functional correlates see Chap. 11, this volume). It is now clear that in the animal kingdom $\alpha$ subunits of G-proteins are carriers of the signal generated by G-protein-coupled receptors. Affected systems are at least six adenylyl cyclases (ACs), at least three of the type C, phosphatidylinositol (PIP) specific phospholipases (PLCs), two and possibly three types of $K^+$ channels, and voltage-gated $Ca^{2+}$ channels of the L and N types. Regulation by G-protein $\alpha$ subunits may be of the positive (stimulatory) or negative (inhibitory) type. In in vitro reconstitution assays only subnanomolar concentrations of the various $\alpha$ subunits are needed to measure the desired regulatory activity. But what about $\beta\gamma$ dimers, the second product of the activation reaction catalyzed by receptors?

The $\beta\gamma$ dimers are the second product of the activation reaction catalyzed by receptors. Previous indications were that $\beta\gamma$ dimers play a role in the dynamics of the signal transuction process, being essential for the forward motion of the receptor-induced activation cycle and the reactivation of the deactivated $\alpha$ subunit (Birnbaumer et al. 1985, 1990; Birnbaumer 1990) (Table 1). $\beta\gamma$ dimers are products of gene families. Four genes are known to encode highly homologous $\beta$ subunits, and at least five and possibly as many as nine genes encode homologous but more diverse forms of $\gamma$ subunits. This gives rise to a considerable degree of molecular diversity in $\beta\gamma$ dimer combinations. Indications are that, as it is the case for $\alpha$ subunits, $\beta\gamma$ dimer's also are able to signal. When they may be doing so and when not is the subject of this chapter.

**Table 1.** Recognized roles of $\beta\gamma$ dimers in signal transduction

| |
|---|
| Signaling |
| Primary: stimulation of PLC ($\beta_2 \gg$ others) |
| Conditional |
| Stimulation of type II and IV AC |
| Inhibition of Ca/CaM-stimulated type I AC |
| Signal amplification: recycling of receptor after dissociation of activated $\alpha$ |
| Presentation of $\alpha$ subunit to specific receptor(s) |

## B. $\beta\gamma$ Dimers and Adenylyl Cyclase

### I. Hormonal Inhibition of Adenylyl Cyclase and Stimulation of $K^+$ Channels: Controversies that Settled Mostly in Favor of $\alpha$ Subunits

In the early to middle 1980s $\beta\gamma$ dimers were postulated by GILMAN and collaborators to mediate hormonal inhibition of AC. They, as well as other laboratories including ours, subsequently failed to obtain inhibition with activated $\alpha$ subunits of putative inhibitory G-proteins ($G_i$) but observed that $\beta\gamma$ dimers have an inhibitory effect on AC activity. Relatively high (nanomolar) concentrations of $\beta\gamma$ dimers are needed for this action. A much disputed mechanism of hormonal inhibition based on this phenomenon was proposed whereby $\beta\gamma$ dimers generated by the inhibitory receptor quenches the stimulatory activity of $\alpha_s$. Inhibition thus comes about by removal of a stimulatory signal (reviewed and summarized by GILMAN 1984, 1987). Counterarguments were raised on the basis of kinetic reasonings (reviewed in BIRNBAUMER et al. 1985) and on the finding that *cyc*$^-$ S49 cells, which lack a functional $\alpha_s$, exhibit normal inhibitory regulation of AC by GTP (HILDEBRANDT et al. 1983) as well as unaltered hormonal inhibition (JAKOBS et al. 1983). Although the reconstitution of inhibitory regulation of AC in a cell-free system has not yet been accomplished, subsequent studies have shown further differences in the kinetic properties of inhibition by $\beta\gamma$ dimers and hormones (TORO et al. 1987; HILDEBRANDT and KOHNKEN 1990). The most recent study on this subject showed a receptor-independent inhibitory activity of $\alpha$ subunits that carry an activating Q $\rightarrow$ L mutation for $\alpha_{i1}$, $\alpha_{i2}$, $\alpha_{i3}$, and $\alpha_z$ but not for $\alpha_o$ (WONG et al. 1991). This may have cast the die in favor of the $\alpha$ subunits and against the $\beta\gamma$ dimers in the question as to which mediates the hormonal inhibition of AC (but see below).

In the later 1980s, when it was demonstrated in cell-free systems that the cardiac "muscarinic" inwardly rectifying $K^+$ channel can be activated by a G-protein (YATANI et al. 1987a; LOGOTHETIS et al. 1987), it was again postulated that $\beta\gamma$ dimers, acting at nanomolar concentrations, and not $\alpha$ subunits are the mediators of the receptor signal (LOGOTHETIS et al. 1987;

BOURNE 1987). The case for $\beta\gamma$ was weakened because results showed that $G_i$ type $\alpha$ subunits activate the channel at concentrations as low as those at which $\alpha_s$ stimulates ACs, i.e., 10–100 p$M$ (CODINA et al. 1987; BIRNBAUMER and BROWN 1987). Although $\alpha$ subunits were subsequently confirmed by several laboratories to activate the muscarinic $K^+$ channel, the signal-mediating role (or not) of $\beta\gamma$ dimers is still an unsettled issue. In fact, one and the same preparation of $\beta\gamma$ dimers stemming from a third-party laboratory (EVANS et al. 1987) stimulated channel activity in one laboratory (KURACHI et al. 1989; ITOH et al. 1992) while being inhibitory in another (OKABE et al. 1990). Channel stimulation (LOGOTHETIS et al. 1987) or inhibition (OKABE et al. 1990) both required nanomolar concentrations of $\beta\gamma$ and were best seen above the 10 n$M$ level. Further studies are needed to clarify this impasse, especially in light of the new developments on roles of $\beta\gamma$ dimers in stimulating subtypes of both AC and PLC.

## II. Conditional and Subtype-Specific Regulation of Adenylyl Cyclase Activity by $\beta\gamma$ Dimers

The first AC cloned, AC-I, was of neuronal expression and is stimulated by $\alpha_s$ and calcium-calmodulin (Ca/CaM; KRUPINSKI et al. 1989). In the presence of $\alpha_s$ but not in its absence, AC-I is inhibited by $\beta\gamma$. Seven additional mammalian AC genes have thus far been discovered, and the sequences of five of these have been or are being published (REED and BAKALYAR 1990; FEINSTEIN et al. 1991; GAO and GILMAN 1991; ISHIKAWA et al. 1992; YOSHIMURA and COOPER 1992; PREMONT et al. 1992). A comparison of their properties, the key features of which are summarized in Fig. 1, shows that all are stimulated by $\alpha_s$, only two are stimulated by Ca/CaM, and two are inhibited by low concentrations of $Ca^{2+}$ by an as yet undefined mechanism.

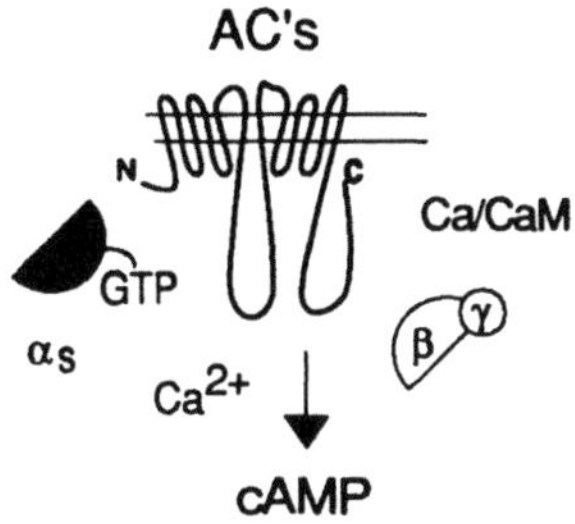


| | $\alpha_s$ | $\beta\gamma$ | Ca/CaM | Ca (μM) | |
|---|---|---|---|---|---|
| Type I | ↑ | ↓ | ↑ | ∅ | brain |
| Type II | ↑ | ↑* | ∅ | ∅ | brain, lung |
| Type III | ↑ | ∅ | ↑ | ∅ | olfactory epithelium, other |
| Type IV | ↑ | ↑* | ∅ | ∅ | ubiquitous |
| Type V | ↑ | ∅ | ∅ | ↓ | ubiquitous, high in brain, heart |
| Type VI | ↑ | ∅ | ∅ | ↓ | ubiquitous, high in lung, spleen, brain |
| Type VII... | | | | | |

*, $\alpha_s$-dependent; ∅, no effect

**Fig. 1.** Predicted topographic structure of adenylyl cyclases and summary of regulations of different adenylyl cyclases. (Adapted from KURPINSKY et al. 1989; REED and BAKALYAR 1990; FEINSTEIN et al. 1991; GAO and GILMAN 1991; ISHIKAWA et al. 1992; YOSHIMURA and COOPER 1992; PREMONT et al. 1992; TANG and GILMAN 1991)

It is striking that while $\beta\gamma$ dimers inhibit AC-I activity, they stimulate the activity of AC-II and AC-IV and have no effect on the others (AC-III, -V, and -VI). Stimulation by $\beta\gamma$, but not inhibition is conditional on $\alpha_s$, i.e., it has as its underlying prerequisite the stimulation of the enzyme by $\alpha_s$ (TANG and GILMAN 1991).

Bourne's group determined that $\beta\gamma$ stimulation of type II AC can occur in the intact cell (FEDERMAN et al. 1992). This was shown in transient expression studies in which human embryonal kidney (HEK) 293 cells were cotransfected with the luteinizing hormone (LH) receptor, which mediates stimulation of AC, and combinations of the $D_2$ dopamine (DA) receptor, which mediates pertussis toxin (PTX) sensitive inhibition of AC, and the type II AC, which is stimulated by $\beta\gamma$. LH-stimulated activity is inhibited by DA in transfections without AC-II but was enhanced by DA in transfections with AC-II, as predicted if activation of the $D_2$ receptor generated an inhibitory $\alpha_i$ and a $\beta\gamma$ that is either neutral or inhibitory to the endogenous AC, but strongly stimulatory to the exogenous AC-II. Consistent with this interpretation, the DA-induced activation of AC-II was PTX sensitive, as was the inhibition of cAMP accumulation in the absence of AC-II. Further, coexpression of the $\alpha$ subunit of transducin ($\alpha_t$), which is able to scavenge $\beta\gamma$ dimers, suppressed the DA-induced stimulation of cAMP formation in AC-II expressing cells (FEDERMAN et al. 1992).

The results with $\beta\gamma$ dimers on different ACs could lead to reconsideration of the interpretation of results in which mutationally activated $\alpha_i$ subunits show an inhibitory activity (WONG et al. 1991) and by which $G_i$ affects activity in $\alpha_s$-negative cells. Results with mutationally activated $\alpha_1$ subunits (Q $\rightarrow$ L mutations) have also been obtained in transient transfections with HEK 293 cells and consist of a reduction in the ability of LH receptor to stimulate cAMP accumulation that – at equal amounts of cDNA transfected – is more pronounced for the Q $\rightarrow$ L $\alpha_i$ subunits than for wild-type counterparts. With the knowledge at hand at that time, it was difficult to see how overexpression of an $\alpha$ subunit could lead to inhibition of AC through generation of $\beta\gamma$, which in turn was the model that was implicitly being tested. However, with new knowledge about the existence of $\beta\gamma$-stimulated ACs it is conceivable that the decrease in AC activity may have been due to changes in levels of $\beta\gamma$ if it is assumed that HEK 293 cells have a $\beta\gamma$-stimulated AC (e.g., type II), if these cells have resting $\beta\gamma$ levels that fully activate its AC-II complement, if overexpression of $\alpha_i$ (even if mutated) scavenges $\beta\gamma$ dimers, and if the higher efficiency of the mutated $\alpha_i$ subunits is due to differences in the levels of protein attained rather than the type of protein made. Likewise, hormonal inhibition of AC in $\alpha_s$-negative S49 *cyc*$^-$ cells could be due to presence of a $\beta\gamma$-inhibitable AC (e.g., type I-like). While these are many conditional statements, and the kineticists among us would still argue in favor of hormonal inhibition being mediated by an $\alpha_i$ – for example, in systems in which a receptor triggers both inhibition of AC and activation of PLC, with the former occurring at much lower concentrations of agonist than the latter (see below) – it is clear that the missing in vitro reconstitution of an (inhibitory) effect on AC is a sore that needs to be cured before the issue of participation or not of $\beta\gamma$ subunits can be settled.

The overall picture emerging in the field of AC regulation, including the relative importance of regulation by $\alpha$ subunit versus $\beta\gamma$ dimer, is one of extreme complexity that requires further investigation. Probably, regulation of cAMP formation in peripheral tissues such as liver, fat, and heart, where $\beta\gamma$-insensitive ACs predominate, will for the most part be as thus far suspected, i.e., $\alpha$ subunits mediating hormonal effects and $\beta\gamma$ dimers ensuring the recycling of the $\alpha$ subunits for renewed activation by receptros. However, in the central nervous system, as well as in some endocrine, exocirne, and epithelial cells, variations in AC subtype expression may lead to responses that vary in kind with receptor occupancy – $\alpha$ subunit-dominant at low receptor occupancy and $\beta\gamma$-dominant at high receptor occupancy.

## C. $\beta\gamma$ Dimers and Phospholipase C: Subtype-Specific Stimulation of Type $\beta$ Phospholipase C by $\beta\gamma$ Dimers

Like AC, PIP-specific PLCs constitute a complex and diverse family of proteins (recently reviewed by Rhee and Choi 1992). They are structurally subclassified into $\beta$, $\gamma$, and $\delta$, and for each of them there are at least three closely related subtypes. Only the $\beta$-class appears to be G-protein sensitive (Wu et al. 1992), and full-length sequence is available for three of these, the PLC-$\beta$1, -$\beta$2 and -$\beta$3. As schematized in Fig. 2, PLC-$\beta$1 is activated by all of the $\alpha$ subunits of the $G_q$ class of G-proteins, $\alpha_q$, $\alpha_{11}$, $\alpha_{14}$, and $\alpha_{16}$ (Lee et al.

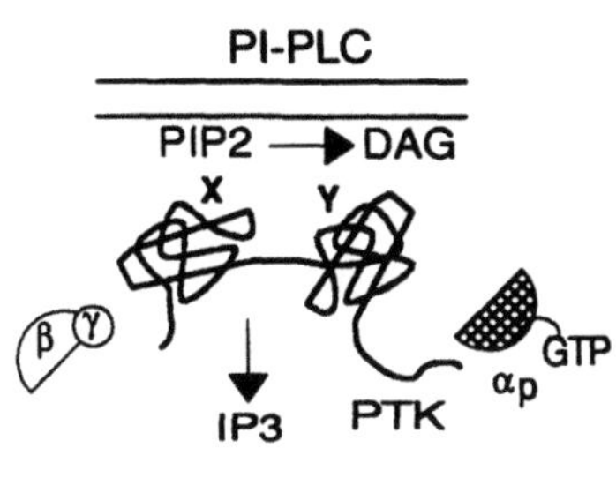


| Type | *src* Homology Domains | Stimulated by |
|---|---|---|
| PI-PLC β1 | None | αq, α11, α14 >> α16 |
| PI-PLC β2 | None | βγ >> α16 >> α11 |
| PI-PLC β3 | None | αq, α11, βγ |
| PI-PLC γ | SH2 & SH3 | PTK |
| PI-PLC δ | None | ? |

**Fig. 2.** Regulation of phospholipase C (*PLC*) subtypes. Designation of PLC subtypes is as in Rhee and Choi (1992). The figure depicts the general structure of phosphatidylinositol-specific phospholipases C (*PI-PLCs*) with homology domains X and Y (present in all) separated by a variable-length linker and the G-protein subunits that regulate different PLC-$\beta$ subtypes. *PTK*, stimulation of activity by protein tyrosine kniase; *SH2/SH3*$^{+}$, presence of *src* homology 2 and 3 domains in the structure X-Y linker region of the PLC-$\gamma$ subtype. No specific form of regulation of PLC-$\delta$ has been discovered as yet

1992). Of these, $\alpha_{16}$ (as well as its murine homologue $\alpha_{15}$) is restricted to bone marrow derived cells (AMATRUDA et al. 1991; WILKIE et al. 1991). PLC-$\beta$2 (PARK et al. 1992) is activated by $\alpha_{16}$ but not by any of the other members of the $G_q$ class $\alpha$ subunits, except perhaps $\alpha_{11}$ (LEE et al. 1992). Although the tissue distribution of PLC-$\beta$2 has not been studied extensively, it is highly expressed in HL-60 cells which are of the neutrophil/macrophage type. Little is known about PLC-$\beta$3 at the time of this writing.

CAMPS et al. (1992a) resolved HL-60 cell PIP-specific PLC into two fractions, PLC-QI and PLC-QII, and noted that while PLC-QI activity is unaffected by $\beta\gamma$, PLC-QII is strongly stimulated by $\beta\gamma$. Using recombinant PLCs, it has now been shown that PLC-$\beta$2 is specifically activated by $\beta\gamma$, and that PLC-$\beta$1 is either not (KATZ et al. 1992) or much less (CAMPS et al. 1992a) affected by $\beta\gamma$. Although this still needs to be documented, it is likely that the PLC-QII enzyme of HL-60 cells is PLC-$\beta$2. While stimulation of PLC by $\alpha_q/\alpha_{11}$ mixtures appears to occur with $\alpha$ concentrations in the low nanomolar and possibly the picomolar range (TAYLOR et al. 1991; SMRCKA et al. 1991), high nanomolar and even micromolar concentrations of $\beta\gamma$ are needed to evoke stimulation of PLC-QII (CAMPS et al. 1992a) and PLC-$\beta$2 (CAMPS et al. 1992b).

The finding that $\beta\gamma$ dimers stimulate a PLC-$\beta$ subtype in HL60 cells needs to be viewed in the context of stimulation of neutrophil and HL-60 cell PLC by the formyl-Met-Leu-Phe (fMLP) receptor. This effect is blocked by PTX (OKAJIMA and UI 1984; OHTA et al. 1985; KIKUCHI et al. 1986) and hence not likely to be mediated by any of the PTX-insensitive G-proteins of the $G_q$ class recently shown to stimulate $\beta$-type PLCs (TAYLOR et al. 1991; SMRCKA et al. 1991; WU et al. 1982; LEE et al. 1992). Previous work by GIERSCHIK and JAKOBS (GIERSCHIK and JAKOBS 1987; GIERSCHIK et al. 1989) had shown that fMLP stimulates cholera toxin mediated ADP-ribosylation of the two main PTX substrates of HL-60, $G_{i2}$ and $G_{i3}$, and had raised the possibility that in this cell $G_i$ types rather than the "classical" $G_p$ may signal PLC activation. Inhibition of AC in intact neutrophils and HL-60 cells may either not occur or, in analogy to regulation of the platelet AC by thrombin, be suppressed by a PKC-mediated inhibitory feedback mechanism secondary to PLC stimulation (JAKOBS et al. 1985). Both, the levels of $G_i$ proteins (>100 pmol/mg protein) and the fMLP receptor density (>100 000/cell) are very high in HL-60 cells, making it plausible that $\beta\gamma$ dimers rather than $\alpha$ subunits might mediate PLC activation in this cell. This is now supported by the work of CAMPS et al. (1992a).

## D. $\beta\gamma$ Dimers and Receptors: Exquisite Specificity of Receptors for $\beta\gamma$ Subtypes

Can $\beta\gamma$ signaling be hormone specific? The most recent studies by KLEUSS et al. (1992, 1993) with rat pituitary GH3 cells strongly suggest that the answer is yes. Earlier studies in GH3 cells on the specificity of the interaction of receptors with G-protein $\alpha$ subunits had shown that selective inhibition of biosynthesis of $\alpha_{o1}$ and $\alpha_{o2}$ by intranuclear injection of subunit-specific antisense oligonucleotides results in selective loss of, respectively, the transduction of the $M_4$ acetylcholine and somatostatin (SST) receptor signals, respectively (receptor signaling interference assay; KLEUSS et al. 1991). Since there are at least four $\beta$ subunits, $G_o$ proteins could have subunits composition $\alpha_{o1}\beta_1$, -$\beta_2$, -$\beta_3$, and -$\beta_4$, and $\alpha_{o2}\beta_1$, -$\beta_2$, -$\beta_3$, and -$\beta_4$ if they are all expressed in the cell. In addition, each of these complexes could come with any one of over five $\gamma$ subunits, less pairing restrictions. The subsequent studies by KLEUSS et al. (1992a,b) showed that the $M_2$ receptor signals through a $G_o$ of subunit composition $\alpha_{o1}\beta_3\gamma_4$ to the exclusion of other $\alpha_{o1}\beta_1$- or $\alpha_{o1}\gamma_3$-containing forms of $G_o$, and that in turn the SST receptor signals through a $G_o$ protein formed of $\alpha_{o2}\beta_1\gamma_3$. Studies on the subunit composition of human erythrocyte G-proteins (GRAF et al. 1992) indicate that single $\alpha$ subunits are about equivalent in their interactions with several $\beta\gamma$ dimers. With four $\beta$ subunit genes and five $\gamma$ subunit genes, there exists the possibility that a cell may express 20 distinct combinations of $\beta\gamma$ dimers. Although this number should be reduced, because not all of the $\beta$ and $\gamma$ genes are expressed in any given cell, and because there are restrictions as to which $\beta$ can pair with which $\gamma$ (PRONIN and GAUTHAM 1992; SCHMIDT et al. 1992), it needs to be increased because there is evidence for the existence of seven to nine $\gamma$ subunits and possibly a fifth $\beta$, making 20 a reasonable number to argue with. Since a receptor may be able to select among these combinations, it follows that unique $\beta\gamma$ dimers generated upon receptor activation may be a small subset of the total and even unique, and thus

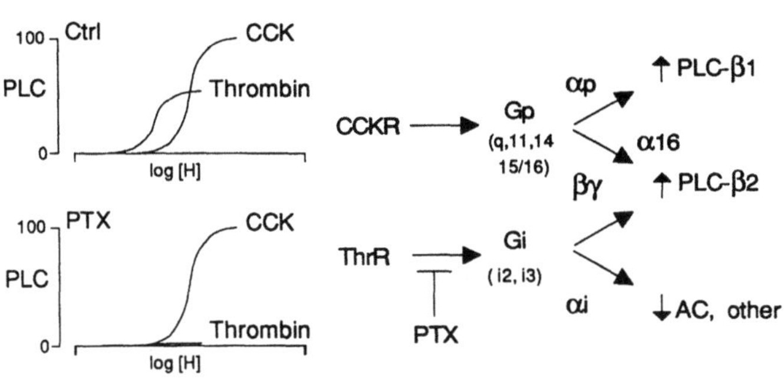


**Fig. 3.** Model of PTX-sensitive and PTX-insensitive stimulation of PLCs as described in CHO cells by ASHKENAZI et al. (1989). *Left*, diagrammatic presentation of experimental data; *Right*, the model. PLC-$\beta$1 is stimulated by a PTX-insensitive $\alpha$ subunit of one of the $G_p$-proteins ($G_q$, $G_{11}$, $G_{14}$, or $G_{15/16}$), which is activated by the CCK receptor but not the thrombin receptor. PLC-$\beta$2 is stimulated by the $\beta\gamma$ dimer of one of the PTX-sensitive $G_i$-proteins ($G_{i2}$ and/or $G_{i3}$) activated by the thrombin but not the CCK receptor. Additivity or lack of additivity of effects depends on whether the cell expresses and the CCK receptor is coupled by $G_{16}$, and whether separate $\beta\gamma$ complements are recognized by the two receptor systems and the PLC-$\beta$2

contribute to the type of cellular response that one or another receptor may elicit if effector functions show specificity for $\beta\gamma$ subtype.

An explanation can now be provided for the finding that both a PTX-sensitive stimulation of PLC by thrombin and a PTX-insensitive stimulation of PLC by cholecystokinin (CCK) can coexist in a single cell line (ASHKENAZI et al. 1989; Fig. 3). Indeed, it PLC-$\beta 2$ is activated by the subset of $\beta\gamma$ dimers with which thrombin receptor interacts, and CCK interacts with a $G_p$ with a different $\beta\gamma$ dimer that does not activate PLC-$\beta 2$, the effects of the two receptor systems can easily be seen to have the potential to act in an additive manner.

## E. Dual Signaling of Single Receptors: Mediation by One or by Two G-Proteins?

Dual signals elicited by single receptor ligands in normal or established cell lines have been known for some time. These include the ability of thrombin to stimulate PLC activity and promote inhibition of AC in platelet membranes (JAKOBS et al. 1985; BRASS et al. 1991), the effect of parathyroid hormone to elevate cyclic AMP and mobilize $Ca^{2+}$ in a variety of cells (reviewed in ABOU-SAMRA et al. 1992), the effects of thyroid-stimulating hormone (TSH) in thyroid cells and LH in luteal and granulosa cells to stimulate cAMP formation and phosphoinositide turnover (TAGUCHI and FIELD 1988; Davis et al. 1984, 1987), and the dual actions of acetylcholine in gastric D cells (CHIBA et al. 1987). However, it could not be established whether these effects were mediated by single or separate receptors. Quite naturally, most of the evidence for dual coupling stems from studying coupling potentials of cloned receptors expressed in model "tester" cell lines. Cases can be made for both mediation by a single G-protein involving its $\alpha$ and $\beta\gamma$ components and for mediation by separate G-proteins, involving separate $\alpha$ subunits.

### I. Inhibition of Adenylyl Cyclase and Stimulation of Phospholipase C

The first report of dual coupling of a cloned receptor was that of ASHKENAZI et al. (1987) who noted that the $M_2$ acetylcholine receptor expressed in CHO cells inhibited cAMP accumulation and stimulated $PIP_2$ breakdown. Curiously, while AC inhibition occurred at low concentrations of agonist ($EC_{50}$ ca. $1\,\mu M$) and was unaffected by receptor density over a range of 150000–2500000 sites per cell, PLC stimulation occurred at much higher concentrations and increased in degree and required progressively lower concentrations of agonist to reach 50% of maximum as receptor abundance increased. Several other cloned receptors that mediate inhibition of AC were subsequently also shown to stimulate cellular PIP-PLC activity, including the serotonin 1A receptor (FARGIN et al. 1989), the histamine $H_1$

receptor (RAYMOND et al. 1991), and the $D_2$ dopamine receptor (VALLAR et al. 1990).

One hallmark of PLC stimulation by $\beta\gamma$ dimers is the already mentioned fact that it requires high nanomolar levels of $\beta\gamma$, which does not appear to be the case for stimulation by the $\alpha_q$ subunits. The reasons for low PLC senstivitity to $\beta\gamma$ dimers could be twofold. One is that the $\beta\gamma$ signal is teleologically meant to be secondary to an $\alpha$ signal, and this is accomplished by a low affinity of the effector for the $\beta\gamma$ dimer. This would explain why inhibitory regulation of AC by the $M_2$ receptor, here assumed to be an $\alpha_i$ signal, was fully developed at low receptor density, while PLC stimulation, here assumed to be a $\beta\gamma$ signal, became more and more intense as receptor density increased (ASHKENAZI et al. 1987). The second reason is that the low sensitivity to $\beta\gamma$ dimers in reconstitution assays is artifactual and but a reflection of a low relative abundance of an active $\beta\gamma$ subtype among $\beta\gamma$ dimers that are either inactive or active only at very high concentrations. The results obtained with the transfected $M_2$ receptor could then be explained by assuming that the receptor selects $G_i$ proteins with $\beta\gamma$ complement(s) of low intrinsic PLC-stimulating activity, or that perhaps the high selectivity of the receptor for the $\beta\gamma$ complement associated with $\alpha$ is not absolute and lost at high occupancy and massive overexpression.

## II. Signaling Quality Through Receptor Quantity?

In intact platelets, thrombin activates PLC without affecting cAMP levels, even though in isolated membranes it elicits both the activation of PLC and the inhibition of AC (JAKOBS et al. 1985; BRASS et al. 1992). In contrast, platelet $\alpha_2$-adrenergic receptors lower cAMP levels without affecting phosphoinositide turnover. The mechanism of suppression of the inhibitory effect of thrombin on AC in the intact platelets has not been clarified but is thought to involve protein kinase C mediated phosphorylation of one or more components of the $G_i$-AC complex (JAKOBS et al. 1985; KATADA et al. 1985). Since platelets lack $G_o$ (J. CODINA and L. BIRNBAUMER, unpublished) and appear to have as their only PTX-sensitive G-proteins $G_{i2}$ and $G_{i3}$, it has been concluded that the PTX-sensitive signaling of the thrombin receptor is mediated by the $G_{i2}$- and/or $G_{i3}$-protein (BRASS et al. 1986; CROUCH and LAPETINA; 1988; ASHKENAZI et al. 1989). The platelet $\alpha_2$-adrenergic receptor is also thought to mediate its action through a PTX-sensitive mechanism, i.e., $G_{i2}$ and/or $G_{i3}$ (SIMONDS et al. 1989). Figure 4 presents a model that could explain the differing quality in the signals generated by the two receptors, based on the idea the PLC stimulation by receptors that are coupled by PTX-substrates is mediated by $\beta\gamma$ dimers rather than their $\alpha$ subunits, and on the experimental finding that thrombin receptor density in platelet membranes is 1500–2000 sites per platelet (BRASS et al. 1992), while that of $\alpha_2$-adrenergic receptors is around only 200 per platelet (STEER et al. 1979; KERRY et al. 1984). In addition, the

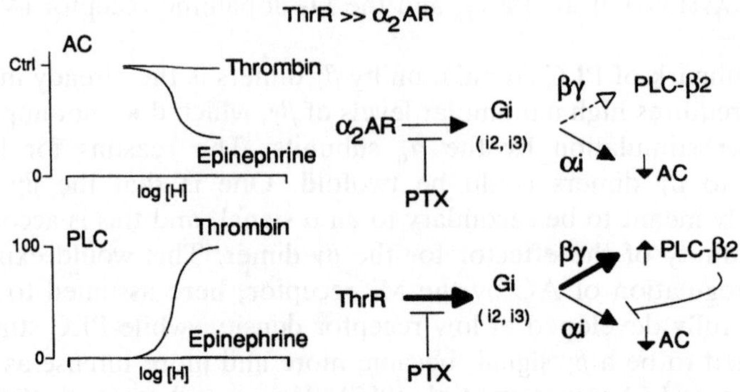


**Fig. 4.** Receptor density as determinant of signaling path. Low levels of receptors such as human platelet $\alpha_2$-adrenergic receptors. (0.1–0.2 fmol/mg membrane protein) activate $G_i$ and either not enough or the wrong type of $\beta\gamma$ dimer to elicit PLC activation. Thrombin receptors (ca. 1500 per platelet) acting on the same $G_i$-proteins are assumed to generate higher levels of $\beta\gamma$ dimers, leading to stimulation of PLC-$\beta$2 and feedback inhibition of $\alpha_i$ function (mechanisms as yet unknown) so that signaling is PLC based rather than AC based

interaction of epinephrine with the $\alpha_2$-adrenergic receptor is rapidly reversible and short lived, while that of thrombin with its receptor leads to a persistent activation (VU et al. 1991). This further strengthens on a temporal basis the signal of the thrombin receptor over than of the catecholamine receptor (see below). The model predicts that less $G_i$ is activated by the $\alpha_2$-adrenergic receptor than by the thrombin receptor, and that as a consequence the levels of $\beta\gamma$ dimer achieved upon stimulation of the $\alpha_2$-adrenergic receptor are not sufficient to trigger PLC stimulation. Thus, in this case the dualistic effect of the thrombin receptor is likely to be the expression of the different activities of the two executive arms of the G-protein that it activates.

## III. Dual Stimulation of Adenylyl Cyclase and Phospholipase C

Several cloned receptors that activate the $G_s$-AC system have also been found to have the potential to stimulate PLC. Included in this group are the TSH and LH receptors (VANSANDE et al. 1990; GUDERMANN et al. 1992), and the calcitonin (CT) and (PTH) receptors (CHABRE et al. 1992; ABOU-SAMBRA et al. 1992). As was the case for the $M_2$ receptors, the dose-response relations for PLC stimulation of AC-stimulating receptors are also up-shifted (by factors of 20–50) with respect to AC stimulation (VANSANDE et al. 1990; GUDERMANN et al. 1992; CHABRE et al. 1992). None of the effects is PTX sensitive. As was the case for $M_2$ receptors, PLC stimulation by AC-stimulatory receptors can be interpreted either in terms of coupling to $G_s$ plus weak coupling to a $G_p$, i.e., a $G_q$ class of G-proteins, or in terms of

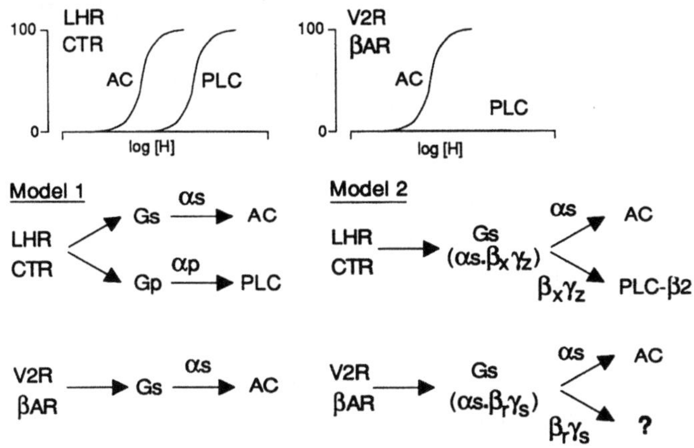


**Fig. 5.** Models for dual coupling of $G_s$-activating receptors based on results of VANSANDE et al. (1990) with the TSH receptors, GUDERMANN et al. (1992) with the LH, $V_2$ vasopressin, and $\beta$-adrenergic receptors, CHABRE et al. (1992) with the calcitonin receptor, and ABOU-SAMRA et al. (1992) with the PTH/PTHrp receptor. *Left*, experimental results in schematic form. *Top*, dose-response curves for stimulation of AC and PLC through LH or CT receptors, also observed for TSH and PTH receptors. *Bottom*, lack of PLC response to stimulation of $\beta$-adrenergic or $V_2$ vasopressin receptors (see GUDERMANN et al. 1992). *Right*, two models to account for the data. *Model 1* simply assumes differential intrinsic activity of receptors to interact with $G_s$ and one of the $G_p$ types; *model 2*, ascribes PLC stimulation to a specific $\beta\gamma$ ($\beta_x\gamma_z$) recognized in the context of $G_s$ by the PLC-activating receptor but not by the receptor that activates only AC. Such $\beta\gamma$ subtype dependence in the stimulation of PLC-$\beta$2 has not yet been demonstrated

generation of $G_s$-derived $\beta\gamma$. $G_s$ is a low-abundance G-protein, so that the high hormone requirement for PLC stimulation through $\beta\gamma$ would have to come about either because of a naturally lower sensitivity of PLC to $\beta\gamma$ concentrations (e.g., high picomolar to low nanomolar) as compared to the exquisitely high sensitivity of AC to $\alpha_s$ ($EC_{50}$ ca. 10 p$M$; NORTHUP et al. 1983; CODINA et al. 1984) or by any one of the arguments presented above in relation to the need of high receptor level or high occupancy for the $M_2$ receptor signal to activate PLC.

If PLC stimulation were mediated indistinctly by all of the $\beta\gamma$ dimers derived from $G_s$, one would predict that all receptors, if expressed at sufficient density, would have to exhibit the same type of dual effects of stimulating AC at low receptor occupancy and PLC at high receptor occupancy. As is schematized in Fig. 5, a comparison of the effect of LH to that of vasopressin or an adrenergic ligand places a further restriction on the types of model to consider. Thus, in the same test system in which an effect of LH on both AC and phosphoinositide hydrolysis was observed, it was found that stimulation of the $\beta$-adrenergic receptor or the $V_2$ vasopressin receptor by their respective ligands was without effect on PLC activity, even though in the case of the $V_2$ receptor, receptor density was ten times that of the LH receptor (GUDERMANN et al. 1992). Again, explanations for this dichotomy in the actions of receptors invoke either involvement of two G-proteins, a $G_s$ and a $G_p$ in the actions of LH (or TSH, PTH, and CT) receptors but not in those of the $V_2$ vasopressin or the $\beta$-adrenergic receptors, or a differential specificity of the receptors of the LH and CT type versus those of the $V_2$ and $\beta$-adrenergic type, where the former are able to interact with $G_s$ molecules having a PLC-stimulating $\beta\gamma$ and the latter are unable to interact with $G_s$ molecules having the PLC-stimulating $\beta\gamma$. The receptor signaling interference assays of KLEUSS et al. (1992a,b) indicate that receptor-specific $\beta\gamma$ dimers may be formed.

## IV. Evidence for Physical Interaction of a Single Receptor with Two Distinct Types of G-Proteins

In spite of all the evidence accumulating for dual signaling being due to a separate actions of $\alpha$ subunits acting on one effector system and $\beta\gamma$ dimers acting on another, Ligget and coworkers (EASON et al. 1992) have provided evidence for G-protein-mediated coupling according to model 1 of Fig. 5, i.e., in which a single receptor affects two distinct effector systems by interacting with G-proteins with differing $\alpha$ subunits. The case in point is the mechanism by which the $\alpha_2$-C10 (platelet) adrenergic receptor expressed in CHO cells mediates inhibition of AC at low receptor occupancy and the stimulation of AC at high receptor occupancy, leading to a V-shaped dose-response curve. Upon immunoprecipitating agonist treated $\alpha_2$-adrenergic receptor, EASON et al. (1992) found the presence in the precipitates of both a PTX-positive $G_i$ and a cholera toxin-positive $G_s$. Immunoprecipitates of the receptor were devoid of both $G_i$ and $G_s$ if obtained from detergent extracts of control membranes (no agonist added) or in extracts of membranes pretreated with $\alpha_2$-adrenergic receptor antagonist.

Figure 6 depicts a different case in which a single receptor signals by interacting with two G-proteins. This is the case of $G_i$-mediated stimulation of $K^+$ channels and $G_o$-mediated inhibition of L and N type $Ca^{2+}$ channels in endocrine and neuronal cells. The first indications for this came from several independent and seemingly unrelated studies. These studies showed that SST receptors in pituitary cells (GH and AtT-20) and opioid receptors in neuroblastoma-glioma (NG) cells act through one or more PTX-sensitive G-proteins to inhibit high-threshold $Ca^{2+}$ currents (e.g., LEWIS et al. 1986; HESCHELER et al. 1987; KLEUSS 1991) and to stimulate an inwardly rectifying $K^+$ current (YATANI et al. 1987b). In the case of the NG cells the PTX-interrupted inhibition of $Ca^{2+}$ channels is preferentially reconstituted with $G_o$, while in the case of GH cells the PTX-inhibited stimulation of $K^+$ channels is reconstituted by $G_i$ but not $G_o$ (YATANI et al. 1987b). Stimulation of azido-GTP binding to NG cell membranes by opioid agonist leads to photolabeling of both $G_i$ and $G_o$ (OFFERMANS et al. 1990). As in all dual signaling cases, the involvement of a single receptor in the regulation of

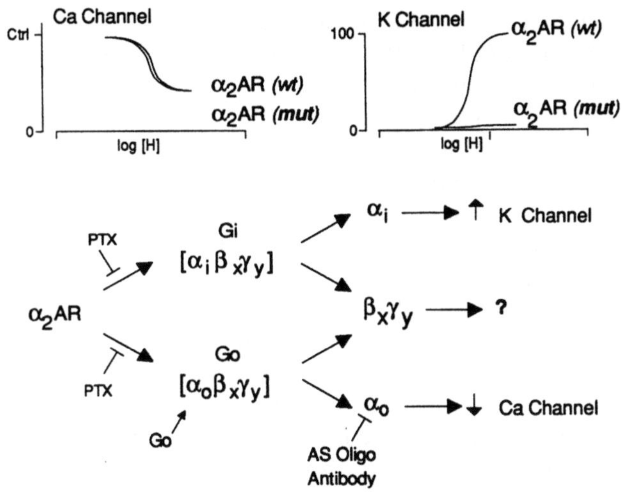


**Fig. 6.** Regulation of two ion channel systems by a single receptor acting by activation of two G-proteins. *Top*, schematic representation of results obtained by SUPRENANT et al. (1992) and others described in the text. *Bottom*, proposed coupling mechanism involving two G-proteins. The $\beta\gamma$ complement of the $G_i$- and $G_o$-proteins recognized by the $\alpha_2$-adrenergic receptor are assumed to be the same on the face of the results obtained by KLEUSS et al. (1992, 1993) in the GH3 cell using the antisense oligonucleotide-based receptor signaling interference assay to probe the differential specificity for $\beta\gamma$ dimers of the SST and $M_2$ acetylcholine (*M2-ACh*) receptors

the two effector systems was established with a cloned receptor, in this case the $\alpha_2$-adrenergic receptor. Stable expression of this receptor in AtT-20 cells led to receptor-dependent stimulation of $K^+$ currents and inhibition of $Ca^{2+}$ currents. Involvement of separate G proteins was shown with the mutant (Asp79Asn) $\alpha_2$-AR receptor. This receptro fails to stimulate the $K^+$ current but retains its ability to regulates the $Ca^{2+}$ influx pathway (SUPRENANT et al. 1992).

This then indicates that while formation of an $\alpha$ plus a $\beta\gamma$ dimer having different activities is a very attractive model of dual receptor signaling, there is biochemical and electrophysiological evidence to the effect that dual signaling may also be the result of bonafide dual coupling, i.e., the interaction of a receptor with more than one G-protein.

# F. The Puzzle of the Up-Shifted Dose-Response Curves for Phospholipase C Elicited by Adenylyl Cyclase Stimulating Agonists

It is necesary to find an explanation not only for the fact that the concentrations of AC-stimulating hormones needed to stimulate PLC are higher but also for the fact that the dose-response curves are parallel to each other and do not show evidence for a cooperative activation of the effector system (e.g., VANSANDE et al. 1990; GUDERMANN et al. 1992; CHABRE et al. 1992). Explanations based on mere quantitative arguments, in which PLC stimulation by a single receptor requires 20–50 times higher concentrations of hormone because AC is more sensitive to $\alpha_s$ than PLC is to an $\alpha_p$, or because of involvement of a nondiscriminating $\beta\gamma$, are not satisfactory. For one, and especially in the case of the LH receptor, the dependence of AC stimulation on hormone concentration goes hand in hand with receptor occupation ($EC_{50} \approx K_d$). As a consequence, at 20 times $EC_{50}$ for AC stimulation (the concentration that is threshold for PLC stimulation), receptor occupation is ca. 95%, and it seems unreasonable to assume that a further 5% increase in receptor occupancy would lead to the experimentally observed and kinetically normal noncooperative activation of PLC by hormone. Depending on actual rate constants, the only parameter that may be proportional, albeit inversely, to hormone concentration at concentrations above tenfold $K_d$ is the average time during which receptor is unoccupied, since reoccupation is directly proportional to the product of the association rate constant ($k_{on}$) and ligand concentration. This leads to the speculation that PLC stimulation depends on a time-dependent conformational change of the G-protein that is separate and slower than the one responsible for mediation of AC activation. A molecular view of this scenario could be that AC can be activated not only by free $\alpha_s$ but also by activated holo-$G_s$ (possibly even associated with the hormone-receptor complex), while PLC is activated only by the free $\beta\gamma$ dimer. By necessity, $\beta\gamma$ would take longer to arise. In support of this possibility, trimeric $G_s$ has been recovered from purifid GMP-P(NH)P-stimulated AC by MARBACH et al. (1990). This topic is discussed further in Chap. 3, this volume.

## G. Concluding Remarks

At this time it is not possible to decide unequivocally for or against an involvement of a single versus two G-proteins in any of the dual signaling situations. It is clear that research will be stimulated by the new discoveries that $\beta\gamma$ dimers can regulate ACs in quite distinct ways, that they can stimulate phospholipases, and that receptors have the intrinsic capability of generating not only specific $\alpha$ subunits but also specific $\beta\gamma$ dimers. Especially interesting should be studies on the capabilities of specific $\beta\gamma$ dimer combinations to regulate effector functions, for example, as mediators of PTX sensitive stimulation of $Ca^{2+}$ channels by angiotensin II in $Y_1$ adrenal cells, and by LH-releasing hormone and thyrotropin-releasing hormone in GH cells (HESCHELER et al. 1988; ROSENTHAL et al. 1988; GOLLASCH et al. 1991). This is not to say that $\alpha$ subunits have lost importance. On the contrary, all indications are that inhibition of AC is due to $\alpha$ rather $\beta\gamma$, that $\beta\gamma$ dimer regulation of AC is dependent on an underlying stimulation by $\alpha_s$, and that all type $\beta$ phospholipases are true effectors of $\alpha_p$-type subunits.

Bourne (1987) has labeled as coincidence detectors those types of effectors that respond to $\beta\gamma$ dimers dependent on $\alpha$ subunit regulation as ACs do. Conditional effectors might be another way of referring to them. Whether this also applies to PLCs, i.e., whether an $\alpha_p$ modulation is required for stimulation of PLC-$\beta$2 by a $\beta\gamma$, is not known but is of importance in the hierarchical ordering of the regulations elicited by receptor occupation. Studies with specific $\beta\gamma$ combinations may also settle the somewhat disturbing requirement for concentrations of $\beta\gamma$ dimers that are about 100 times those of $\alpha$ subunits. If regulatory effects by $\beta\gamma$ are found to depend on underlying regulations by $\alpha$ subunits, they are clearly secondary. On the other hand, in those cases in which $\beta\gamma$ regulation does not require participation of an $\alpha$ subunit, it will be equivalent to that of an $\alpha$ subunit. Secondary or not, if nature has kept $\beta\gamma$ regulation alive throughout evolution, it must be important, and it behooves us to continue with the exciting task of wildcat drilling in nature's landscape to extract from its grounds the riches that she is hiding from us.

## References

Abou-Sambra AB, Jüpmer H, Force T, Freeman MW, Kong XF, Schipani E, Urena P, Richards J, Bonventre JV, Potts JT, Jr, Kronenberg HM, Segre GV (1992) Expression cloning of a common receptor for parathryoid hormone and parathyroid hormone-related peptide from rat osteoblast-like cells: a single receptor stimulates intracellular accumulation of both cAMP and inositol trisphosphates and increases intracellular free calcium. Proc Natl Acad Sci USA 89:2732–2736

Amatruda TT, III, Steele DA, Slepak VZ, Simon MI (1991) G$\alpha$16, a G protein $\alpha$ subunit specifically expressed in hematopoietic cells. Proc. Natl Acad Sci USA 88:5587–5591

Ashkenazi A, Winslow JW, Peralta EG, Peterson GL, Schimerlik MI, Capon DJ, Ramachandran J (1987) An M2 muscarinic receptor subtype coupled to both adenylyl cyclase and phosphoinositide turnover. Science 238:672–675

Ashkenazi A, Peralta EG, Winslow JW, Ramachandran J, Capon DJ (1989) Functionally distinct G proteins selectively couple different receptors to PI hydrolysis in the same cell. Cell 56:487–493

Birnbaumer L, Hildebrandt JD, Codina J, Mattera R, Cerione RA, Hildebrandt JD, Sunyer T, Rojas FJ, Caron MG, Lefkowitz RJ, Iyengar R (1985) Structural basis of adenylyl cyclase stimulation and inhibition by distinct guanine nucleotide regulatory proteins. In: Cohen P Houslay MD (eds) Molecular mechanisms of signal transduction. Elsevier/North Holland, Amsterdam, pp 131–182

Birnbaumer L, Brown AM (1987) G protein opening of $K^+$ channels. Nature 327:21–22

Birnbaumer L, Abramowitz J, Brown AM (1990) Signal transduction by G proteins. Biochim Biophys Acta (Reviews in Biomembranes) 1031:163–224.

Birnbaumer L (1990) G proteins in signal transduction. Annu Rev Pharmacol Toxicol 30:675–705

Bourne HR (1987) Wrong subunit regulates cardiac potassium channels. Nature 325:296–297

Brass LF, Laposata M, Banga HS, Rittenhouse SE (1986) Regulation of the phosphoinositide hydrolysis in thrombin-stimulated platelets by a pertussis toxin-sensitive guanine nucleotide binding protein. Evaluation of its contribution to platelet activation and comparisons with the adenylate cyclase inhibitory protein, $G_i$ J Biol Chem 261:16838–16847

Brass LF, Manning DR, Williams AG, Woolkalis MJ, Poncz M (1991) Receptor and G protein-mediated responses to thrombin in HEL cells. J Biol Chem 266:958–965

Brass LF, Vassallo Jr, RR, Belmonte E, Ahuja M, Cichowski K, Hoxie JA (1992) Structure and function of the human thrombin receptor. Studies using monoclonal antibodies directed against a defined domain within the receptor N terminus. J Biol Chem 267:13795–13798

Burch RM, Luini A, Axelrod J (1986) Phospholipase $A_2$ and phospholipase C are activated by distinct GTP-binding proteins in response to alpha$_1$-adrenergic stimulation in FRTL-5 cells. Proc Natl Acad Sci USA 83:7201–7205.

Camps M, Hou C, Sidiropoulos D, Stock JB, Jakobs KH, Gierschik P (1992a) Stimulation of phospholipase C by guanine-ncleotide-binding protein $\beta\gamma$ subunits. Eur J Biochem 206:821–831

Camps M, Carozzi A, Schnabel P, Scheer A, Parker PJ, Gierschik P (1992b) Isozyme-selective stimulation of phospholipase C-$\beta_2$ by G protein $\beta\gamma$-subunits. Nature (in press)

Chabre O, Conklin BR, Lin HY, Lodish HF, Wilson E, Ives HE, Catanzariti L, Hemmings BA, Bourne HR (1992) A recombinant calcitonin receptor independently stimulates 3′,5′-cyclic adenosine monophosphate and $Ca^{2+}$/inositol phosphate signaling pathways. Mol Endocrinol 6:551–556

Chiba T, Raffouol K, Yamada T (1987) Divergent stimulatory and inhibitory actions of carbamylcholine on gastric D-cells. J Biol Chem 262:8467–8469

Codina J, Hildebrandt JD, Sekura RD, Birnbaumer M, Bryan J, Manclark CR, Iyengar R, Birnbaumer L (1984) $N_s$ and $N_i$, the stimulatory and inhibitory regulatory components of adenylyl cyclases. Purification of the human erythrocyte proteins without the use of activating regulatory ligands. J Biol Chem 259:5871–5886

Codina J, Yatani A, Grenet D, Brown AM, Birnbaumer L (1987) The alpha subunit of the GTP binding protein $G_k$ opens atrial potassium channels. Science 236:442–445

Crouch MF, Lapetina EG (1988) A role for $G_i$ in control of thrombin receptor-phospholipase C coupling in human platelets. J Biol Chem 263:3362–3371

Davis JS, West LA, Farese RV (1984) Effects of luteinizing hormone on phosphoinositide metabolism in rat granulosa cells. J Biol Chem 259:15028–15034

Davis JS, Weakland LL, Farese RV, West LA (1987) Lutenizing hormone increases inositol trisphosphate and cytosolic free $Ca^{2+}$ in isolated bovine luteal cells. J Biol Chem 262:8515–8521

Eason MG, Kurose H, Holt BD, Raymond JR, Liggett SB (1992) Simultaneous coupling of $\alpha_2$-adrenergic receptors to two G-proteins with opposing effects. Subtype selective coupling of $\alpha_2$C10, $\alpha_2$C4 and $\alpha_2$C2 adrenergic receptors to $G_i$ 1n3 $G_s$. J Biol Chem 267:15795–15801

Evans T, Fawzi A, Fraser ED, Brown ML, Northup JK (1987) Purification of a beta$_{35}$ form of the beta-gamma complex common to G-proteins from human placental membranes. J Biol Chem 262:176–181

Fargin A, Reymond JR, Regan JW, Cotecchia S, Lefkowitz RJ, Caron MG (1989) Effector coupling mechanisms of the cloned 5-HT1A receptor. J Biol Chem 264:14949–14852

Federman AD, Conklin BR, Schrader KA, Reed RR, Bourne HR (1992) Hormonal stimulation of adenylyl cyclase through $G_i$-protein $\beta\gamma$ subunits. Nature 356:159–161

Feinstein PG, Schrader KA, Bakalyar HA, Tang W-J, Krupinski J, Gilman AG, Reed RR (1991) Molecular cloning and characterization of a $Ca^{2+}$/calmodulin-insensitive adenylyl cyclase from rat brain. Proc Natl Acad Sci USA 88:10173–10177

Fisher JK, Aronson NN (1992) Characterization of the cDNA and genomic sequence of a G protein $\gamma$ subunit ($\gamma_5$). Mol Cell Biol 12:1585–1591

Gao B, Gilman AG (1991) Cloning and expression of a widely distributed (type IV) adenylyl cyclase. Proc Natl Acad Sci USA 88:10178–10182

Gierschik P, Jakobs KH (1987) Receptor mediated ADP-ribosylation of a phopholipase C-stimulating G protein. FEBS Lett 224:219–223

Gierschik P, Sidiropoulos D, Jalobs KH (1989) Two distinct $G_i$-proteins mediate formyl peptide receptor signal transduction in human leukemia (HL-60) cells. J Biol Chem 264:21470–21473

Gilman AG (1984) G proteins and dual control of adenylate cyclase. Cell 36:577–579

Gilman AG (1987) G proteins: transducers of receptor-generated signals. Annu Rev Biochem 56:615–649

Gollasch M, Haller H, Schultz G, Hescheler J (1991) Thyrotropin releasing hormone induces opposite effects on $Ca^{2+}$ channel currents in pituitary cells by two pathways. Proc Natl Acad Sci USA 88:10262–10266

Graf R, Mattera R, Codina J, Evans T, Ho Y-K, Estes MK, Birnbaumer L (1992) Studies on the interaction of $\alpha$ subunits of G proteins with $\beta\gamma$ dimers. Eur J Biochem (in press)

Gudermann T, Birnbaumer M, Birnbaumer L (1992) Evidence for dual coupling of the murine LH receptor to adenylyl cyclase and phosphoinositide breakdown/$Ca^{2+}$ mobilization. Studies with the cloned murine LH receptor expressed in L cells. J Biol Chem 267:4479–4488

Hescheler J, Rosenthal W, Hinsch K-D, Wulfern M, Trautwein W, Schultz G (1988) Angiotensin II-induced stimulation of voltage-dependent $Ca^{2+}$ currents in an adrenal cortical cell line. EMBO J 7:619–624

Hildebrandt JD, Sekura RD, Codina J, Iyengar R, Manclark CR, Birnbaumer L (1983) Stimulation and inhibition of adenylyl cyclases is mediated by distinct proteins. Nature 302:706–709

Hildebrandt JD, Kohnken RE (1990) Hormone inhibition of adenylyl cyclase. J Biol Chem 265:9825–9830

Ishikawa Y, Katsushita S, Chen L, Hanlon NJ, Kawabe J, Honcy C (1992) Isolation and characterization of a novel cardiac adenylylcyclase. J Biol Chem 267:13553–13557

Itoh H, Sugimoto T, Kobayashi I, Takahashi K, Katada T, Ui, M, Kurachi Y (1991) On the mechanism of basal and agonist-induced activation of the G-protein gated musarinic $K^+$ channel in atrial myocytes of guinea pig heart. J Gen Physiol 98:517–533

Jakobs KH, Aktories K, Schultz G (1983) A nucleotide regulatory site for somatostatin inhibition of adenylate cyclase in S49 lymphoma cells. Nature 303:177–178

Jakobs KH, Bauer S, Watanabe Y (1985) Modulation of adenylate cyclase of human platelets by phorbol ester. Impairment of the hormone-sensitive inhibitory pathway. Eur J Biochem 151:425–430

Katada T, Gilman AG, Watanabe Y, Bauer S, Jakobs KH (1985) Protein kinase C phosphorylates the inhibitory guanine-nucleotide-binding regulatory component and apparently suppresses its function in hormonal inhibition of adenylate cyclase. Eur J Biochem 151:431–437

Katz A, Wu D, Simon MI (1992) $\beta\gamma$ Subunits of the heterotrimeric G protein activate the $\beta 2$ isoform of phospholipase C. Nature (in press)

Kaziro Y, Itoh H, Kozasa T, Nakafuku M, Satoh T (1991) Structure and function of signal-transducing GTP-binding proteins. Ann Rev Biochem 60:349–400

Kerry R, Scrutton MC, Wallis RB (1984) Mammalian platelet adrenoceptors. Br J Pharmacol 81:91–102

Kikuchi A, Kozawa O, Kaibuchi K, Katada T, Ui, M, Takai Y (1986) Direct evidence for involvement of a guanine nucleotide-binding protein in chemotactic peptide-stimulated formation of inositol bisphosphate and trisphosphate in differentiated human leukemic (HL-60) cells. Reconstitution with $G_i$ or $G_o$ of the plasma membranes ADP-ribosylated by pertussis toxin. J Biol Chem 261:11558–11562

Kim D, Lewis DL, Graziadei L, Neer EJ, Bar-Sagi D, Clapham DE (1989) G protein $\beta\tau$-subunits activate the cardiac muscarinic $K^+$ channel via phospholipase $A_2$. Nature 337:557–560

Kleuss C, Hescheler J, Ewel C, Rosenthal W, Schultz G, Wittig B (1991) Assignment of G-protein subtypes to specific receptors inducing inhibition of calcium currents. Nature 353:43–48

Kleuss C, Scherübl H, Hescheler J, Scultz GB, Wittig (1992) Different $\beta$-subunits determine G protein interaction with transmembrane receptors. Nature 358:424–426

Kleuss C, Scherübl H, Hescheler J, Schultz G, Wittig B (1993) Selectivity in signal transduction determined by $\gamma$ subunits of heterotrimeric G proteins. Science 259:832–834

Krupinski J, Coussen F, Bakalyar HA, Tang W-J, Feinstein PG, Orth K, Slaughter C, Reed RR, Gilman AG (1989) Adenylyl cyclase amino acid sequence: possible channel- or transporter-like structure. Science 244:1558–1564

Kurachi Y, Ito H, Sugimoto T, Shimizu T, Miki I, Ui, M (1989) Arachidonic acid metabolites as intracellular modulators of the G protein-gated cardiac $K^+$ channel. Nature 337:555–557

Lee CH, Park D, Wu D, Rhee SG, Simon MI (1992) Members of the $G_q$ alpha subunit gene family activate phospholipase C-beta isozymes. J Biol Chem 267:16044–16047

Levine MA, Smallwood PM, Moen PT, Helman LJ, Ahn TG (1990) Molecular cloning of $\beta_3$ subunit, a third form of the G protein $\beta$-subunit polypeptide. Proc Natl Acad Sci USA 87:2329–2333

Lewis DL, Weight FF, Luini A (1986) A guanine nucleotide-binding protein mediates the inhibition of voltage-dependent calcium current by somatostatin in a pituitary cell line. Proc Natl Acad Sci USA 83:9035–9039

Gierschik P, Stisslinger M, Sidiropolous D, Herrmann E, Jakobs KH (1989) Dual $Mg^{2+}$ control of formyl peptide receptor – G protein interaction in HL-60 cells. Evidence that the low agonist affinity receptor interacts with and activates the G protein. Eur J Biochem 183:97–105

Florio VA, Sternweis PC (1989) Mechanism of muscarinic receptor action on $G_o$ in reconstituted phospholipid vesicles. J Biol Chem 264:3909–3915

Logothetis DE, Kurachi Y, Galper J, Neer EJ, Clapham DE (1987) The $\beta\tau$ subunits of GTP-binding proteins activate the muscarinic $K^+$ channel in heart. Nature 325:321–326

Marbach I, Bar-Sinai A, Minich M, Levitzki A (1990) $\beta$ Subunit copurifies with GppNHp-activated adenylyl cyclase. J Biol Chem 266:999–1004

Northup JK, Smigel MD, Sternweis PC, Gilman AG (1983) The subunits of the stimulatory regulatory component of adenylate cyclase. Resolution of the activated 45,000-dalton (alpha) subunit. J Biol Chem 258:11369–11376

Offermanns S, Schultz G, Rosenthal W (1991) Evidence for opioid receptor-mediated activation of the G proteins $G_o$ and $G_{i2}$ in membranes of neuroblastoma X glioma (NG108–15) hybrid cells. J Biol Chem 266:3365–3368

Ohta H, Okajima F, Ui, M (1985) Inhibition by islet-activating protein of a chemotactic peptide-induced early breakdown of inositol phospholipids and $Ca^{2+}$ mobilization in guinea pig neutrophils. J Biol Chem 260:15771–15780

Okabe K, Yatani A, Evans T, Ho Y-K, Codina J, Birnbaumer L, Brown AM (1990) $\beta\gamma$ Dimers of G proteins inhibit muscarinic $K^+$ channels in heart. J Biol Chem 265:12854–12858

Okajima, F, Ui, M (1984) ADP-ribosylation of the specific membrane protein by islet-activating protein, pertussis toxin, associated with inhibition of a chemotactic peptide-induced arachidonate release in neutrophils. A possible role of the toxin substrate in $Ca^{2+}$-mobilizing biosignaling. J Biol Chem 259:13863–13871

Park D, Jhon DY, Kritz R, Knopf J, SG Rhee (1992) Cloning, expression and $G_q$-independent activation of phospholipase C-$\beta$2. J Biol Chem 267:16048–16055

Premont RT, Chen J, Ma, HW, Ponnapalli M, Iyengar R (1992) Two members of a new subfamily of hormone-stimulated adenylyl cyclases. Proc Natl Acad Sci USA 89 (in press)

Pronin AN, Gautham N (1992) Interaction between G protein $\beta$ and $\gamma$ subunit types is selective. Proc Natl Acad Sci USA 89:6220–6224

Raymond JR, Albers FJ, Middleton JP, Lefkowitz RJ, Caron MG, Obeid LM, Dennis VW (1991) 5-$HT_{1A}$ and histamine $H_1$ receptors in HeLa cells stimulate phosphoinositide hydrolysis and phosphate uptake via distinct G protein pools. J Biol Chem 266:372–379

Reed RR, Bakalyar HA (1990) Identification of a specialized adenylyl cyclase that may mediate odorant detection. Science 250:1403–1406

Rhee SG, Choi KD (1992) Regulation of inositol phospholipid-specific phospholipase C isozymes. J Biol Chem 267:12393–12396

Rosenthal W, Hescheler J. HInsch K-D, Spicher K, Trautwein W, Schultz G (1988) Cyclic AMP independent, dual regulation of voltage-dependent $Ca^{2+}$ currents by LHRH and somastostatin in a pituitatry cell line. EMBO J 7:1627–1633

Schmidt CJ, Thomas TC, Levine MA, Neer EJ (1992) Specificity of G protein $\beta$ and $\gamma$ subunit interactions. J Biol Chem 267:13807–13810

Simon MI, Strathmann MP, Gautam N (1991) Diversity of G proteins in signal transduction. Science 252:802–808

Simonds WF, Goldsmith PK, Codina J, Unson CG, Spiegel AM (1989) $G_{i2}$ mediates $\alpha_2$-adrenergic inhibition of adenylyl cyclase in platelet membranes: in situ identification with $G_\alpha$ C-terminal antibodies. Proc Natl Acad Sci USA 86:7809–7813

Smrcka AV, Helper JR, Brown KO, Sternweis PC (1991) Regulation of polyphosphoinositide-specific phospholipase C activity by purified $G_q$. Science 251:804–807

Steer ML, Khorana J, Galgoci B (1979) Quantitation and characterization of human platelet alpha-adrenergic receptors using [$^3$H]phentolamine. MOl Pharmacol 16:719–728

Strathmann MP, Simon MI (1991) G$\alpha$12 and G$\alpha$13 subunits define a fourth class of G protein $\alpha$ subunits. Proc Natl Acad Sci USA 88:5582–5586

Suprenant A, Horstman DA, Akbarali H, Limbird LE (1992) A point mutation of the $\alpha_s$-adrenoceptor that blocks coupling to potassium but not calcium currents. Science 257:977–980

Taguchi M, Field JB (1988) Effects of thyroid-stimulating hormone, carbachol, norepinephrine, and adenosine 3′,5′-monophosphate on polyphosphatidylinositol phosphate hydrolysis in dog thyroid slices. Endocrinology 123:2019–2026

Tang WJ, Gilman AG (1991) Type specific regulation of adenylyl cyclase by G protein $\beta\gamma$ subunits. Science 254:1500–1503

Taylor SJ, Chae HZ, Rhee SG, Exton JH (1991) Activation of the $\beta$1 isozyme of phospholipase C by $\alpha$ subunits of the $G_q$ class of G proteins. Nature 350:516–518

Toro M-J, Montoya E, Birnbaumer L (1987) Inhibitory regulation of adenylyl cyclases. Evidence against a role for beta/gamma complexes of G proteins as mediators of $G_i$-dependent hormonal effects. Mol Endocrinol 1:669–676

Vallar L, Muca C, Magni M, Albert P, Bunzow J, Meldolesi J, Civelli O (1990) Differential coupling of dopaminergic $D_2$ receprotrs expressed in different cell types. Stimulation of phosphatidylinositol 4,5-bisphosphate hydrolysis in Ltk- fibroblasts, hyperpolarization, and cytosolic free $Ca^{2+}$ concentration decrease in $GH_4C_1$ Cells. J Biol Chem 265:10320–10326
VanSande J, Raspe E, Perret J, Lejeune C, Manhaut C, Vassart G, Dumont JE (1990) Thyrotropin activates both the cAMP and the $PIP_2$ cascade in CHO cells expressing the human cDNA of the TSH receptor. Mol Cell Endo 74:R1–R6
von Weizsäcker E, Strathmann MP, Simon MI (1992) Diversity among the beta subunits of heterotrimeric GTP-binding proteins: characterization of a novel beta-subunit cDNA. Biochem Biophys Res Commun 183:350–356
Vu T-KH, Hung DT, Wheaton VI, Coughlin SR (1991) Molecular cloning of a functional thrombin receptor reveals a novel proteolytic mechanism of receptor activation. Cell 64:1057–1068
Wilkie TM, Scherle PA, Strathmann MP, Slepak VZ, Simon MI (1991) Characterization of G-protein $\alpha$ subunits in the $G_q$ class: expression in murine tissues and in stromal and hematopoietic cell lines. Proc Natl Acad Sci USA 88:10049–10053
Wong UH, Federman A, Pace AM, Zachary I, Evans T, Pouysségur, Bourne HR (1991) Mutant $\alpha$ subunits of $G_{i2}$ inhibit cyclic AMP accumulation. Nature 351:63–65
Wu D, Lee CH, Rhee SG, Simon MI (1992) Activation of phospholipase C by the $\alpha$ subunit of the $G_q$ and $G_{11}$ protein in transfected COS-7 cells. J Biol Chem 267:1811–1817
Yatani A, Codina J, Brown AM, Birnbaumer L (1987b) Direct activation of mammalian atrial muscarinic potassium channels by GTP regulatory protein $G_k$. Science 235:207–211
Yatani A, Codina J, Sekura RD, Birnbaumer L, Brown AM (1987b) Reconstitution of somatostatin and muscarinic receptor mediated stimulation of $K^+$ channels by isolated $G_k$ protein in clonal rat anterior pituitary cell membranes. Mol Endocrinol 1:283–289
Yoshimura M, Cooper DMF (1992) Cloning and expression of a $Ca^{2+}$-inhibitable adenylyl cyclase from NCB-20 cells. Proc Natl Acad Sci USA 89:6716–6720

# *D. Effectors of G-Proteins*

CHAPTER 55

# Molecular Diversity of Mammalian Adenylyl Cyclases: Functional Consequences

R.T. PREMONT, J. CHEN, O. JACOBOWITZ, and R. IYENGAR

## A. Introduction

Adenylyl cyclase (EC 4.6.1.1) converts MgATP into 3′,5′-cyclic AMP (cAMP) and pyrophosphate. Since the initial characterization of the enzyme by SUTHERLAND and coworkers in 1962 adenylyl cyclase activities have been identified in organisms ranging from bacteria to man. The structure and regulation of these adenylyl cyclase enzymes as well as the regulatory role of the cAMP produced varies considerably across this evolutionary distance.

The original interest in adenylyl cyclase stemed from studies searching for the mechanisms underlying the stimulatory effect of hormones such as glucagon and epinephrine on the liver (SUTHERLAND and RALL 1958). These studies identified cAMP as an intracellular second messenger produced by the particulate preparation in response to epinephrine stimulation. Subsequently, adenylyl cyclase activity has been localized to the plasma membranes of most mammalian tissues. Here it produces intracellular cAMP in response to agonist occupancy of receptor proteins for numerous hormones, neurotransmitters, and autocrine, and paracrine factors. The cAMP then acts as an *intracellular* messenger to activate the cAMP-dependent protein kinase (also known as protein kinase A), which phosphorylates cellular substrates such as enzymes or channels which then directly produce or set in motion the cascade that eventually produces the desired biochemical or physiological response to the hormonal signal.

The prototypical mammalian adenylyl cyclase system is composed of three distinct protein components: the adenylyl cyclase enzyme, receptor proteins for extracellular agonists, and the coupling protein (heterotrimeric G-protein) which links the binding of hormone ligand to its receptor to the activation of the effector enzyme. Studies on transmembrane signal transduction systems over the past 30 years have resulted in the identification of numerous additional messenger systems similarly regulated by specific receptors through the action of members of the family of heterotrimeric G-proteins (IYENGAR and BIRNBAUMER 1990; SIMON et al. 1991).

Recent studies have demonstrated that there are multiple distinct families of mammalian G protein-regulated adenylyl cyclases. The unique behavioral characteristics as well as the specific tissue distributions of these various forms of adenylyl cyclase now allows us to develop an understanding of many unique features of hormone-stimulated adenylyl cyclases in different tissues. This chapter focuses on our current understanding of these different adenylyl cyclases and their functional properties and regulation.

## B. Stimulation and Inhibition of Adenylyl Cyclases

Adenylyl cyclase activity is found in virtually all mammalian tissues. In most tissues the activity of the enzyme is controlled mainly by hormones. Such control can either be stimulatory or inhibitory and can be membrane delimited (requiring only the cell surface signaling components) or may involve diffusible intracellular messengers as well.

Hormonal stimulation of adenylyl cyclase requires the heterotrimeric G-protein called $G_s$ (ROSS and GILMAN 1977; MAY et al. 1985). Hormone-occupied receptors activate the G-protein by facilitating both the release of bound GDP from the G-protein and the binding of GTP to the vacant guanine nucleotide binding site (CASSEL and SELINGER 1978; ASANO et al. 1984). $G_s$ is a heterotrimeric G-protein, and its activation involves dissociation of the GTP-bound $\alpha$ subunit from the $\beta\gamma$ subunits. The GTP-bound $G_s$-$\alpha$-protein is the active species which is capable of binding to and stimulating the adenylyl cyclase enzyme (NORTHUP et al. 1983). Activation is terminated by the hydrolysis of the bound GTP to GDP by the intrinsic GTPase activity of the $\alpha$ subunit. Binding of nonhydrolyzable GTP analogs, such as Gpp(NH)p or GTP$\gamma$S, results in a persistently activated state of the $G_s$-protein by locking the $\alpha$ subunit in the GTP-bound conformation (LONDOS et al. 1974). This results in extensive and persistent stimulation of adenylyl cyclase even in the absence of hormone (SCHRAMM and RODBELL 1975). All five members of the $G_s$-$\alpha$-protein family are capable of stimulating adenylyl cyclase. Four of these arise from alternative splicing of the $G_s$-$\alpha$ gene (BRAY et al. 1986). The other, closely related protein is $G_{olf}$-$\alpha$, which is presumed to be an olfactory neuroepithelium-specific version of $G_s$ (JONES and REED 1987). Kinetic differences in the rates of activation of these various $G_s$-proteins by guanine nucleotides appear to be the only observable difference in their ability to stimulate adenylyl cyclase (GRAZIANO et al. 1989; MATTERA et al. 1989; JONES et al. 1990).

In brain, micromolar concentrations of $Ca^{2+}$ have been known to stimulate adenylyl cyclase (BROSTROM et al. 1975; CHEUNG et al. 1975). This stimulation is mediated through calmodulin (CaM), and is observed only for an adenylyl cyclase enzyme that appears to be specifically found in the brain. The effects of $Ca^{2+}$/CaM and G-proteins are synergistic, and function on the same form of the enzyme (HARRISON et al. 1989; TANG et al. 1991). Stimulation of adenylyl cyclase by $Ca^{2+}$/CaM is not observable in most other mammalian tissues, although some $Ca^{2+}$-stimulated activity has been sporadically reported in pancreatic islets, heart, smooth muscle, and kidney.

Most mammalian adenylyl cyclases are stimulated by forskolin, a diterpine isolated from *Coleus* roots (SEAMON et al. 1981). This stimulation results from the direct interaction of forskolin with adenylyl cyclase. The construction of forskolin-agarose represented a breakthrough that led to the isolation and purification of mammalian adenylyl cyclases (PFEUFFER and METZGER 1982). Although forskolin stimulation of adenylyl cyclase does not require $G_s$ (SEAMON and DALY 1981), it appears that only $G_s$-stimulated adenylyl cyclases are regulated by forskolin. Forskolin greatly synergizes $G_s$ stimulation of the enzyme (CLARK et al. 1982), and $G_s$ increases the binding affinity of forskolin on adenylyl cyclase (NELSON and SEAMON 1986). Thus in native systems forskolin stimulation should be viewed as a measure of $G_s$–adenylyl cyclase interactions.

An adenylyl cyclase activity that is not regulated by G-proteins and CaM has been observed in testes (BRAUN and DODS 1975). This enzyme is not stimulated by forskolin but is stimulated by bicarbonate ions. The testis enzyme has been recently purified and shown to be a 43-kDa protein (OKAMURA et al. 1991).

In most instances hormonal inhibition of adenylyl cyclase activity also appears to be mediated by a heterotrimeric G-protein. As with hormonal stimulation of adenylyl cyclase through the $G_s$-protein, hormone binding to specific cell surface receptors leads to activation of a G-protein through GDP dissociation and GTP binding. Unlike stimulation by $G_s$, however, the mechanism by which this G-protein inhibits adenylyl cyclase is not yet clear. G-protein involvement in hormonal inhibition of adenylyl cyclase activity was initially established through studies using pertussis toxin, an ADP-ribosyltransferase which specifically recognizes the $\alpha$ subunits of the $G_i$- and $G_o$-proteins as substrates. Pertussis toxin treatment of cells leads to modification of these G-protein subunits and to loss of hormone receptor coupling to inhibition of adenylyl cyclase activity (JAKOBS et al. 1983; HILDEBRANDT et al. 1983). By the techniques utilized at that time, only one observable G-protein $\alpha$ subunit was modified by ADP-ribosylation in the murine S49 lymphoma cells, and this was called $G_i$. Three distinct forms of $G_i$-$\alpha$ have been identified by molecular cloning (JONES and REED 1987). None of these three $G_i$-$\alpha$-proteins, when persistently activated with a nonhydrolyzable guanine nucleotide analog, inhibits purified bovine brain adenylyl cyclase (LINDER et al. 1990). $G_o$, the other known pertussis toxin substrate G-protein which could thus be a candidate for the inhibitory G-protein, is also unable to inhibit purified adenylyl cyclase. A model for inhibition of adenylyl cyclase mediated through the $\beta\gamma$ subunits released upon hormonal activation of the $G_i$-protein has been proposed to account for this discrepancy (KATADA et al. 1984; GILMAN 1984).

A molecular approach, however, has recently established that the $G_i$-$\alpha$-proteins are directly involved in hormonal inhibition of adenylyl cyclase. A Gln residue conserved among GTPase enzymes, when converted to Leu, greatly lowers the catalytic rate of the GTPase activity and leads to a persistently activated G-$\alpha$-protein. Expression of GTPase-deficient mutants of the three $G_i$-$\alpha$-proteins (but not of $G_o$-$\alpha$) leads to observable inhibition of adenylyl cyclase in human kidney 293 cells (WONG et al. 1992). Further studies are clearly required to clarify the mechanisms responsible for hormonal inhibition of adenylyl cyclase.

In addition to guanine nucleotide and $G_i$-mediated inhibition of adenylyl cyclases, there are two other intracellular agents that inhibit adenylyl cyclases. In most tissues, adenosine and some of its analogs inhibit adenylyl cyclase activity through direct interaction with the enzyme at a site called the P-site (WOOLF et al. 1981). This site is called the P-site because an intact purine ring is essential for inhibition, and it is distinct from cell surface purinergic receptors. Stimulated forms of adenylyl cyclase are more suceptible to inhibition than the basal form. JOHNSON and colleagues (1989) have characterized several specific adenosine analogs that specifically interact with this site with a rank order of potency of 2′-deoxy-3′-AMP > 3′-AMP > 2′-deoxyadenosine > adenosine. The inhibition by adenosine and its analogs is quite extensive, resulting in a 90% decrease in adenylyl cyclase activity at saturating concentrations of agonists.

Some cell lines and tissues exhibit a 30%–40% inhibition of adenylyl cyclase by low ($0.1-1\,\mu M$) concentrations of free $Ca^{2+}$. Hormones that elevate intracellular $Ca^{2+}$ appear to be able to utilize this mechanism to inhibit intracellular cAMP production (BOYAJIAN et al. 1991). This inhibition is additive with inhibition by $G_i$ and does not require CaM (CALDWELL et al. 1992). Since the concentration range in which this inhibition occurs is well within the intracellular concentrations of $Ca^{2+}$ achieved by $Ca^{2+}$ mobilizing hormones, it is likely that such inhibition can be biologically relevant. Like all MgATP-dependent enzymes, adenylyl cyclase is inhibited by high (micromolar) concentrations of $Ca^{2+}$ through substrate depletion.

The various stimulatory and inhibitory agents that regulate adenylyl cyclase and the cellular entities through which they have their biological effects are summarized in Table 1.

**Table 1.** Post receptor regulation of adenylyl cyclase by various ligands

| Ligand | Mode of regulation | Binding site |
|---|---|---|
| GTP analogs | Stimulatory | $G_s$ |
| | Inhibitory | $G_i$ |
| $AlF_4^-$ | Stimulatory | $G_s$ |
| | Inhibitory | $G_i$ |
| Forskolin | Stimulatory | Adenylyl cyclase |
| Adenosine | Inhibitory | Adenylyl cyclase |
| $Ca^{2+}$ | Stimulatory | Calmodulin |
| | Inhibitory | Probably adenylyl cyclase |

# C. Molecular Diversity of Adenylyl Cyclases

## I. Multiple Families of Adenylyl Cyclases

From various studies of adenylyl cyclase regulation, it was known for over 15 years ago that there are at least three distinct forms of adenylyl cyclase activity in mammalian tissues (NEER 1978). Recent cloning studies have unveiled a surprisingly greater degree of heterogeneity.

The development of forskolin-affinity chromatography allowed for the purification of a 120-kDa $G_s$- and $Ca^{2+}$/calmodulin-stimulated enzyme from bovine brain membranes (PFEUFFER et al. 1985; SMIGEL 1986). Partial amino acid sequences of proteolytic peptides were used to generate probes for isolation of the cDNA coding for this enzyme (KRUPINSKI et al. 1989). Expression of the complete cDNA, called type 1 adenylyl cyclase, generates enzymatic activity which is stimulated by hormone receptors through the $G_s$-protein as well as by forskolin and by $Ca^{2+}$/calmodulin (KRUPINSKI et al. 1989; TANG et al. 1991).

Using this brain adenylyl cyclase as a probe, full or partial cDNAs encoding seven additional mammalian adenylyl cyclase forms (types 2–8) have since been identified. Type 2 adenylyl cyclase cDNA was cloned from rat brain (FEINSTEIN et al. 1991) and type 4 cDNA from rat testes (GAO and GILMAN 1991). A partial type 2 cDNA sequence has also been obtained from human brain, and the human gene has been localized to chromosomal band 5p15 (STENGEL et al. 1992). In addition, Kuprinski and coworkers have obtained a cDNA that encodes a portion of an adenylyl cyclase related to the type 2 enzyme from mouse S49 lymphoma cells, which has been named the type 7 enzyme (J. KRUPINSKI, personal communication). Type 3 adenylyl cyclase was cloned from rat olfactory tissue and is thought to be involved in the mediation of olfactory sensory transduction (BAKALYAR and REED 1990). Type 5 adenylyl cyclase cDNAs have been cloned from dog heart (ISHIKAWA et al. 1992) and rat liver and kidney (PREMONT et al. 1992a). Type 6 cDNAs have been cloned from dog heart (KATSUSHIKA et al. 1992), rat heart (KRUPINSKI et al. 1992), rat liver and kidney (PREMONT et al. 1992a), mouse S49 lymphoma cells (PREMONT et al. 1992b), and mouse/hamster hybrid NCB-20 cells (YOSHIMURA and COOPER 1992). Hanoune and coworkers have isolated partial cDNA from a human brain library that appears to encode a distinct species of adenylyl cyclase (PARMA et al. 1991). Corresponding rat cDNA sequences have also been obtained for this form (J. KRUPINSKI, personal communication), which has been termed type 8.

The deduced amino acid sequence of types 1–6 and the partial sequence of type 8 are shown in Fig. 1. Type 7 partial sequence, which is still unpublished in the primary literature, is not shown. Comparision of the amino acid sequences indicates that the overall similarity between the various mammalian adenylyl cyclases is about 50%. However, several

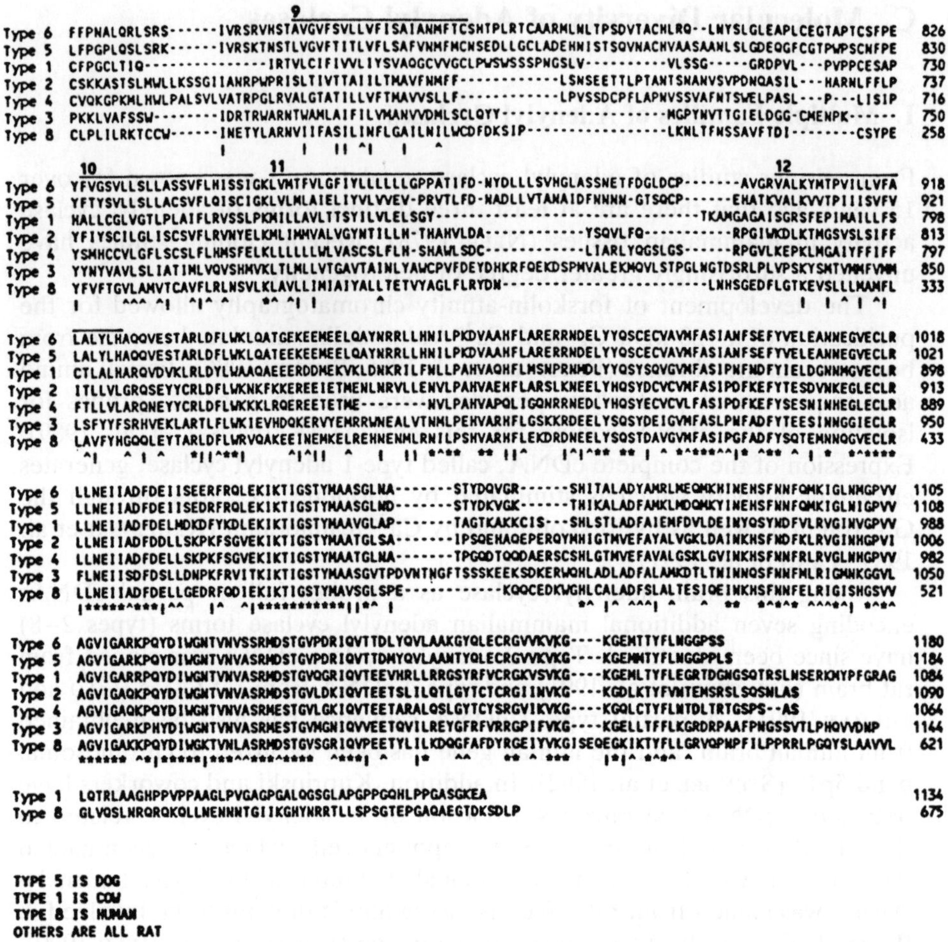


**Fig. 1.** Alignment of mammalian adenylyl cyclase sequences. Bovine type 1 (KUPRINSKI et al. 1989), rat type 2 (FEINSTEIN et al. 1991), rat type 3 (BAKALYAR and REED 1990), rat type 4 (GAO and GILMAN 1991), dog type 5 (ISHIKAWA et al. 1992), rat type 6 (PREMONT et al. 1992a), and the partial human type 8 (PARMA et al. 1991) sequences were aligned using the PC/GENE CLUSTAL program. Some manual adjustments were made in the less conserved regions. *, Conserved residues (indicated under the position); ^, conservative substitution between all of the sequences compared; !, conservative substitutions in five of six or six of the seven sequences compared. Conserved amino acids are: A, G, P, S, T; D, E, N, Q; F, Y, W; H, R, K; I, L, M, V

individual sequences have considerably higher homology with each other. This information has allowed the construction of the homology trees shown in Fig. 2. Such analysis indicates that there are at least five distinct subfamilies of mammalian adenylyl cyclases. Two of these subfamilies have

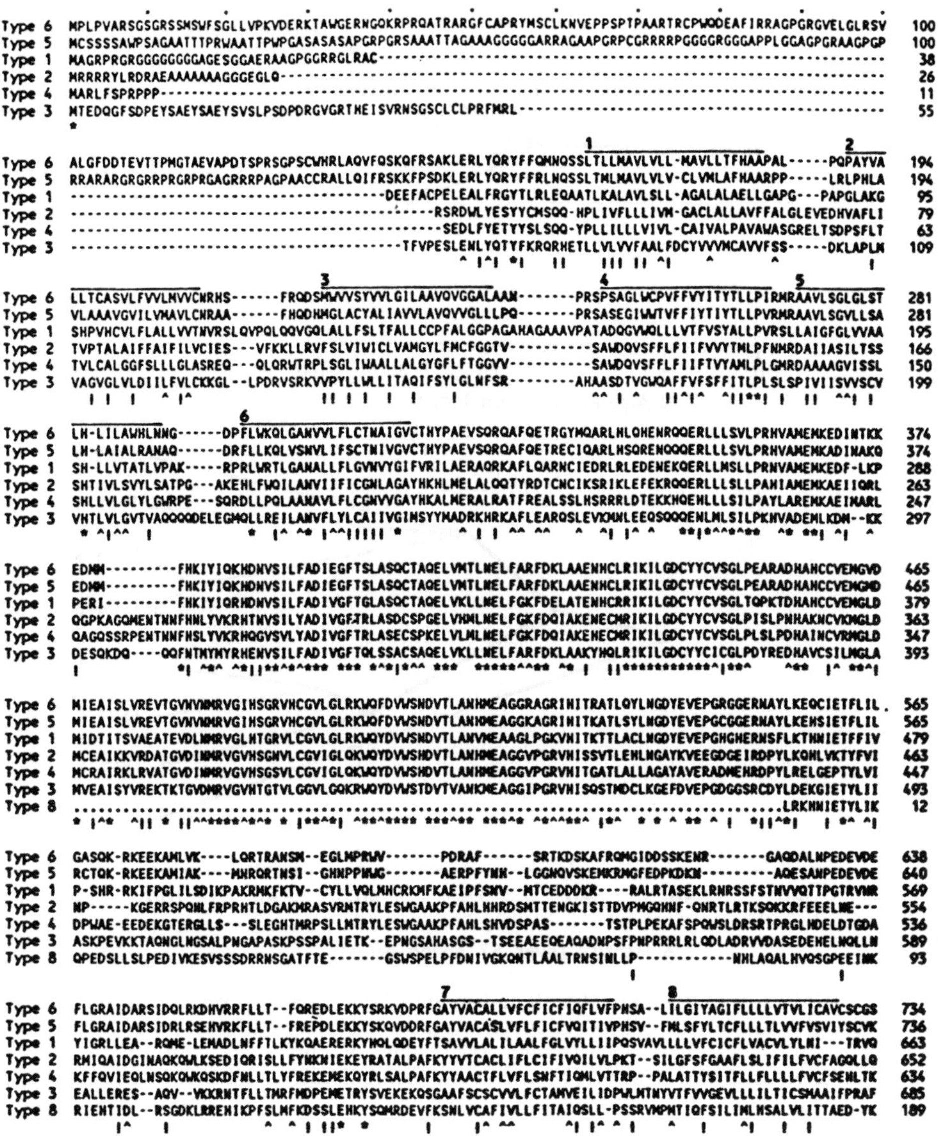

**Fig. 1.** (*Continued*)

multiple members: the type 2 family has three members while the type 6 family has two members. Additional members of these five subfamilies may be identified, and further subfamilies defined, as additional tissues are examined.

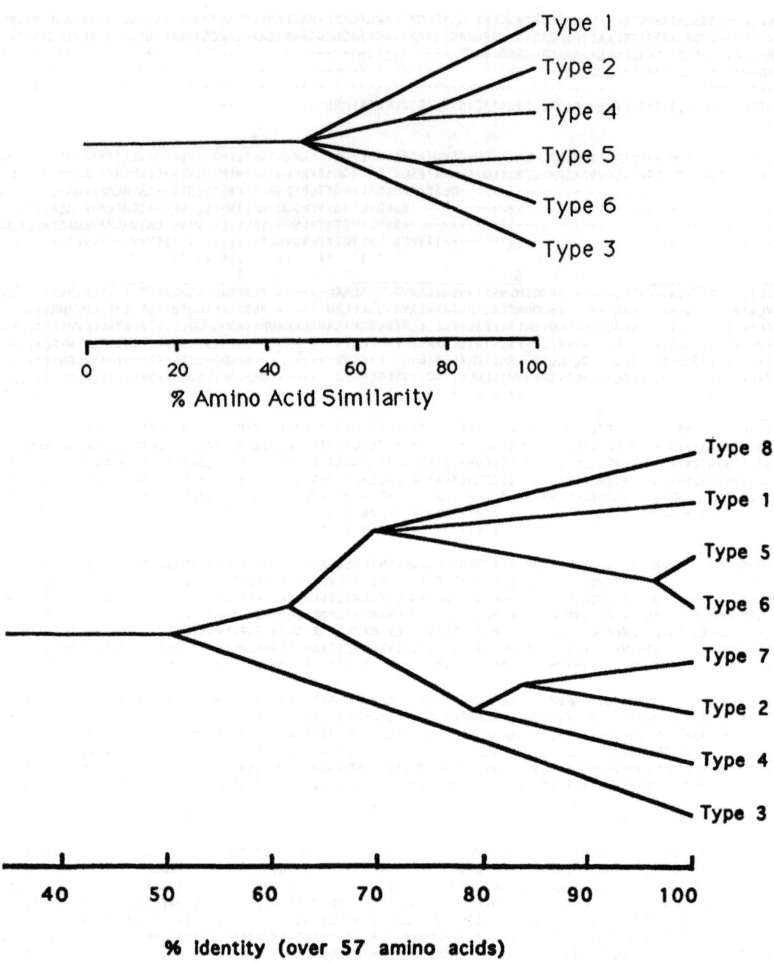


**Fig. 2.** Relationships among the cloned mammalian adenylyl cyclases. a Similarity tree for the full-length colned adenylyl cyclases. Similarity scores were computed from pairwise alignments. Although the range of similarities among the four subfamilies vary from 48% to 53%, an average value of 50% was used to indicate the divergence between the four sub-families. b Comparison of a 57 amino acid segment within the central cytoplasmic loop of eight rat adenylyl cyclase sequences. (Values kindly supplied by Dr. J. KRUPRINSKI)

## II. Secondary Structure and Topography

All the mammalian adenylyl cyclases are predicted to be transmembrane proteins with a complex topological structure similar to that of transporters and ion channels, with 12 transmembrane spans in two domains of six spans each. Both the N- and C-termini are predicted to be cytoplasmic. Two large cytoplasmic domains, a 350 amino acid loop between the first and second

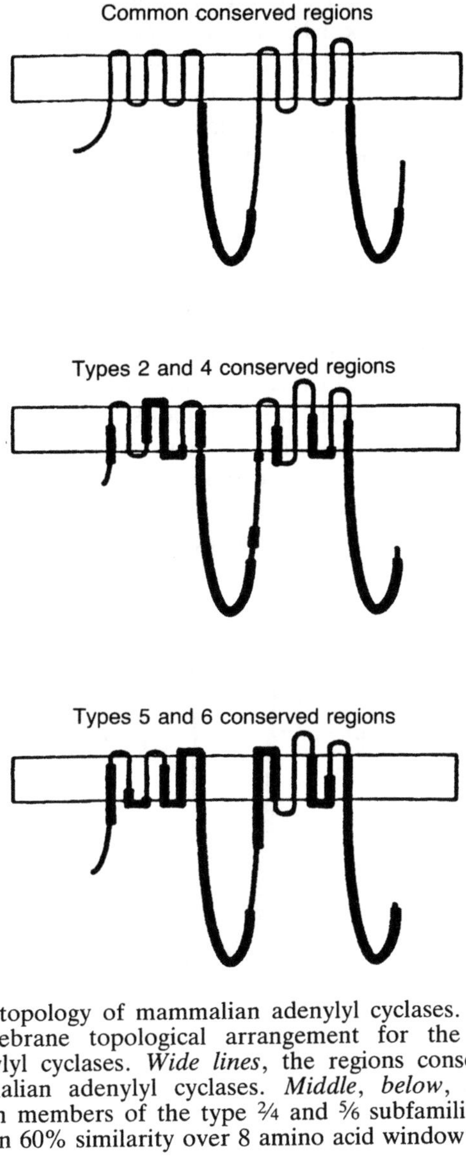


**Fig. 3.** Predicted topology of mammalian adenylyl cyclases. *Above*, the predicted consensus transmebrane topological arrangement for the hormone responsive mammalian adenylyl cyclases. *Wide lines*, the regions conserved between all six full-length mammalian adenylyl cyclases. *Middle*, *below*, additional regions of homology between members of the type 2/4 and 5/6 subfamilies. Conserved regions have a greater than 60% similarity over 8 amino acid window

transmembrane domains and a 250–300 amino acid tail following the second membrane domain, were predicted to encode the catalytic core of the enzyme based on sequence similarity with cloned guanylyl cyclases and (to a lesser degree) yeast adenylyl cyclases (KRUPINSKI et al. 1989). These two large intracellular domains share similarity with each other, and both are highly conserved among the known mammalian adenylyl cyclase sequences. The predicted secondary structure of the mammalian adenylyl cyclases is shown in Fig. 3. The conserved domains in large cytoplasmic loop and tail domains are displayed by thicker lines. Each of these domains is also similar to the single domain of cloned guanylyl cyclases (KOESLING et al. 1991). Many residues in these domains appear to be absolutely conserved among the members of both the adenylyl and guanylyl cyclase families. In contrast, the membrane spanning regions of these adenylyl cyclases appear to be distinct among the various forms.

## III. Putative Catalytic Sites

The cloned mammalian adenylyl cyclases do not have any obvious signature sequences for ATP binding such as the P-loop. However the regions of homology with cloned guanylyl cyclases in the central cytoplasmic loop and the C-terminal tail suggest that these regions are involved in catalysis. Further evidence supporting this line of reasoning comes from studies on *Drosophila*. A homolog of the bovine type 1 enzyme has been cloned from *D. melanogaster* (LEVIN et al. 1991). This *Drosophila* type 1 adenylyl cyclase maps to the rutabaga genetic locus (LEVIN et al. 1991), which is known to

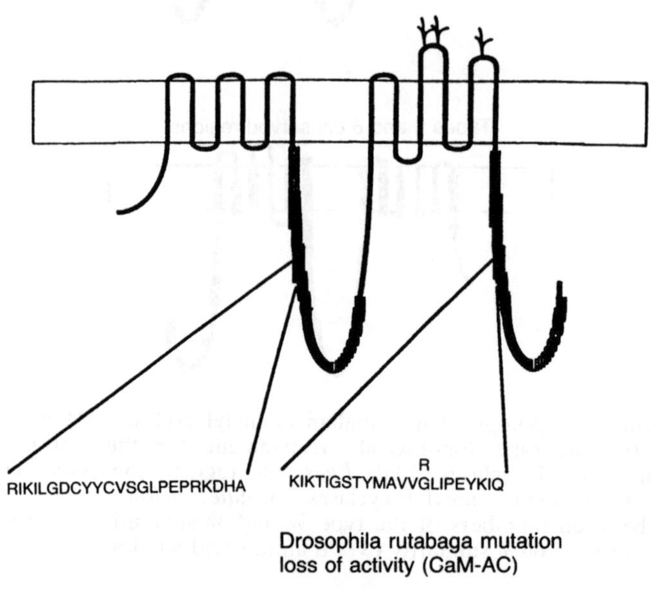


**Fig. 4.** Highly homologous regions of $G_s$-responsive adenylyl cyclase. *Solid boxes*, regions that have greater than 80% similarity (over 8 amino acids); *thick hatched regions*, regions with 60%–80%. The sequence of quasireplicated regions of the *Drosophila* adenylyl cyclase encoded by the rutabaga locus is shown. *, all mammalian adenylyl cyclase have identitical amino acids in this position; ^, conservative substitutions as defined in legend to Fig. 1. Conversion of G to R in the rutabaga mutant results in loss of enzymatic activity. These observations are interpreted as a indications that these regions may constitute parts of the catalytic domain

cause a defect in associative learning in flies and is correlated with a loss of CaM-stimulated adenylyl cyclase activity in the heads of affected flies (LIVINGSTONE et al. 1984; LIVINGSTONE 1986). The loss of adenylyl cyclase activity is due to point mutation in the type 1 adenylyl cyclase, which changes a Gly residure absolutely conserved among known G-protein-regulated adenylyl cyclases to an Arg and leads to a loss of activity (LEVIN et al. 1991). The location of this mutation in the *Drosophila* adenylyl cyclase and the conserved residues are shown in Fig. 4.

Studies from Gilman and coworkers have shown that both the central cytoplasmic loop and the C-terminal tail are essential for catalytic activity. Expression of either half of the molecule does not yield a catalytically active enzyme. However, when the two halves are coexpressed, substantial catalytic activity is restored, and this activity can be regulated by a number of stimulators (TANG et al. 1991). That the two halves do not need to be covalently linked to recover catalytic function suggests that interaction between the cytoplasmic domains of the two halves is the primary requirement for catalysis. Since coexpression of the two halves does not result in full restoration of activity, it is tempting to speculate that the primary function of the transmembrane domains may be to provide the optimal molecular distance between the two cytoplasmic domains for efficient catalysis. The role of the $G_s$-protein can then be envisaged as promoting the association of the two domains to form an active catalytic center.

## IV. Tissue Distribution of the Various Forms

The type 1 adenylyl cyclase appears to be present only in the brain (J. KUPRINSKI, personal communication). Messenger RNA encoding the type 1 adenylyl cyclase has been localized by in situ hybridization to areas in rat brain thought to be involved in associative learning, such as the neocortex, hippocampus, and dentate gyrus (XIA et al. 1991).

mRNA for the type 2 enzyme has been detected only in brain and lung (FEINSTEIN et al. 1991; PREMONT et al. 1992a). Two other members of this subfamily have varying distributions. The mRNA encoding the type 4 enzyme appears to be widely distributed (GAO and GILMAN 1991; R.T.P. and R.I., unpublished). A form of the type 4 enzyme with an extended N-terminal sequence has been recently isolated by D. Cooper and coworkers (Personal communication). The distributions of these long and short forms of the type 4 adenylyl cyclase remains to be determined. The mRNA for the type 7 enzyme is found in S49 lymphoma cells, and in rat brain and heart (J. KRUPINSKI, personal communication).

The type 3 enzyme was cloned from an olfactory epithelial cDNA library, and significant levels of the type 3 mRNA were found only in olfactory tissue and olfactory bulb (BAKALYAR and REED 1990). The mRNA encoding type 3 as well as the type 3 protein were found to be specific to the

**Table 2.** Tissue distribution of the various adenylyl cyclase mRNAs

| Adenylyl cyclase type | Tissue/cells | | | | | | |
|---|---|---|---|---|---|---|---|
| | Liver | Brain | Heart | Kidney | Lung | Testis | S49 cells |
| 1 | − | + | − | − | − | − | − |
| 2 | − | + | − | − | + | − | − |
| 3 | − | +[a] | − | − | − | − | − |
| 4 | (+) | + | (+) | + | (+) | (+) | − |
| 5 | + | + | + | + | + | (+) | − |
| 6 | + | + | + | + | + | (+) | + |
| 7 | − | (+) | (+) | − | − | − | + |
| 8 | − | + | − | − | − | − | − |

Detection is by northern blotting or solution hybridization unless otherwise specified. A − indicates absence and + presence of mRNA of the specified type of adenylyl cyclase; (+) Indicates detection of mRNA by methods based on polymerase chain reaction.
[a] Type 3 adenylyl cyclase appears to be specifically present in the olfactory bulb.

olfactory neuroepithelial cells using a neuronal ablation technique (BAKALYAR and REED 1990).

Both the type 5 and 6 mRNAs are found at high levels in brain and heart, and at lower levels in several other tissues, including kidney, liver, lung, and testes (ISHIKAWA et al. 1992; KATSUCHIKA et al. 1992; PREMONT et al. 1992a; YOSHIMURA and COOPER 1992). Solution hybridization studies with probes specific for types 5 and 6 messages confirm a widespread distribution (D. MILLER, R.T.P., and R.I., unpublished abservations). The presence of multiple messages for both the type 5 and 6 enzymes indicates that alternative splicing of these forms may occur. Comparison of clones from different laboratories and preliminary direct studies in our laboratory indicate that at least for the type 6 enzyme; there exist long and short variants which differ in the presence or absence of an alternative 14 amino acid sequence at the extreme N-terminal (R.T.P. and R.I., unpublished).

The type 8 enzyme appears to be a brain-specific species. cDNAs encoding this form have found in both in rat and human brain (PARMA at al. 1990; KUPRINSKI et al. 1992).

The tissue distribution for the mRNAs for the various froms of adenylyl cyclase are summarized in Table 2. It is clear that some tissues contain mRNAs encoding a great variety of adenylyl cyclase forms (brain, heart), while most tissues appear to have only a limited subset of the known subtypes. Studies to localize individual species within the brain and heart may reveal interesting clues as to the specalized function of individual adenylyl cyclase enzymes. The protein distributions and relative amounts of these various adenylyl cyclase forms in any tissue remains to be determined.

## D. G-Protein Regulation of Adenylyl Cyclases

### I. $G_s$-$\alpha$ Regulation

Expression of each of the six mammalian adenylyl cyclases for which full-length clones are available leads to enzymatic activity which is stimulated by $G_s$ or by ligands for receptors that function through $G_s$. There appear to be no discernable or bilogically significant differences in which $G_s$ form stimulates the various types of mammalian adenylyl cyclases. Currently there are no published data on the regulation of the various forms of adenylyl cyclases by the various splice variants of $G_s$-$\alpha$. Functional data from many tissues suggest that major differences are probably unlikely.

All of the cloned mammalian adenylyl cyclases that have been expressed are also stimulated by forskolin. Such stimulation appears to be independent of $G_s$ (Tang et al. 1991; Feinstein et al. 1991; Gao and Gilman 1991).

### II. $G_i$-$\alpha$ Regulation

Currently it is not known which, if any, of the cloned adenylyl cyclases is directly regulated by G-$\alpha_i$. Gilman and coworkers have reported that neither type 1 nor type 2 adenylyl cyclases are regulated by $\alpha_i$ (Linder et al. 1990; Tang and Gilman 1991). Since inhibition of adenylyl cyclase by $\alpha_i$ is readily observed in S49 *cyc*$^-$ cell membranes, and since the type 6 and 7 enzymes are found in S49 cells, it is possible that one or both of the type 6 or 7 enzymes might be inhibited by $\alpha_i$. This, however, is only speculation at this stage and needs to be experimentally verified. It is clear, however, that not all mammalian adenylyl cyclases are subject to inhibitory regulation by G-proteins.

### III. $\beta\gamma$ Regulation

The various cloned adenylyl cyclases have been shown to exhibit differential regulation by G-protein $\beta\gamma$ subunits.

The type 1 enzyme is directly inhibited by $\beta\gamma$ subunits in the range of 1–50 n*M*. Such inhibition, not due to $\beta\gamma$ subunit inactivation of $G_s$-$\alpha$, was first described by Katada et al. (1987). Studies on the recombinant type 1 enzyme have shown that such inhibition is not through the CaM regulation of this enzyme but rather likely to be a direct effect on the anzyme that occurs most prominently for CaM stimulation (Tang et al. 1991).

The type 2 and 4 enzymes are conditionally *stimulated* by $\beta\gamma$ subunits. By themselves $\beta\gamma$ subunits have no stimulatory or inhibitory effects on these enzymes. But in the presence of activated $G_s$-$\alpha$, the $\beta\gamma$ subunits greatly enhance the $G_s$-$\alpha$ stimulation (Tang and Gilman 1991; Gao and Gilman 1991). The $\beta\gamma$ effects are in the 1–100 n*M* range, with an $EC_{50}$ of 5–10 n*M*. This concentration range for the $\beta\gamma$ effect specifies that the biologically relevant effects of $\beta\gamma$ regulation of type 2 or 4 adenylyl cyclases can occur only when the more abundant G-proteins such as $G_i$ or $G_o$ are fully activated. The type 7 enzyme, which is also a member of the type 2 family, has not been expressed, but is expected to exhibit conditional regulation by $\beta\gamma$ subunits. Membranes from S49 lymphoma cells, which contain mRNA for the type 7 enzyme, show only modest enhancement by $\beta\gamma$ subunits. It is not yet clear, however, whether this is due to low abundance of this type of enzyme in S49 cells or lack of responsiveness of the enzyme itself.

The type 5 and 6 enzymes are not affected by $\beta\gamma$ subunits in the presence or absence of activated $G_s$-$\alpha$ (PREMONT et al. 1992a; J.C. and R.I., unpublished observations). This property, along with the widespread expression of these two forms of the enzyme, provides a reasonable explanation as to why in most tissues receptors that activate $G_i$ or other abundant G-proteins either have no effect on or inhibit adenylyl cyclase. The type 3 enzyme is also not susceptible to regulation by $\beta\gamma$ subunits (Tang and Gilman 1991).

## E. Type-Specific Regulation by Intracellular Ligands

### I. $Ca^{2+}$/CaM Regulation

The type 1 enzyme is stimulated by $Ca^{2+}$/CaM. Studies on the recombinant enzyme expressed in Sf9 cells show that this effect does not require the presence of $G_s$. By itself $Ca^{2+}$/CaM is able to stimulate the type 1 enzyme to close (60%–80%) to its maximal achievable stimulation (TANG et al. 1991). Expression of the type 1 enzyme in 293 cells show that increases in intracellular $Ca^{2+}$ by receptor-regulated pathways can stimulate the enzyme in vivo (CHOI et al. 1992a). The type 3 enzyme is also stimulated by $Ca^{2+}$/CaM (CHOI et al. 1992b). The other adenylyl cyclases tested thus far (types 2, 4, 5, 6) are not stimulated by $Ca^{2+}$/CaM (FEINSTEIN et al. 1991; GAO and GILMAN 1991; ISHIKAWA et al. 1992; KATSUSHIKA et al. 1992; YOSHIMURA and COOPER, 1992).

### II. Inhibition by Low Concentrations of $Ca^{2+}$

YOSHIMURA and COOPER (1992) have shown that the type 6 adenylyl cyclase is inhibited low (micromolar) concentrations of free $Ca^{2+}$. The extent of the inhibition is relatively modest (30%), as is seen in native systems. At the current time it is not known whether the type 5 enzyme is inhibited as well. None of the other forms appear to be negatively regulated by low concentrations of $Ca^{2+}$.

**Table 3.** Regulation of the various types of adenylyl cyclases

| Regulatory entity | Mode of regulation | Adenylyl cyclase types | Comments |
|---|---|---|---|
| $G_s$-$\alpha$ | Stimulatory | 1–6 | |
| $G_i$-$\alpha$ | Inhibitory | Currently unknown | Could be types 6 or 7 since S49 cells show good inhibition. |
| G-$\beta\gamma$ | Direct stimulation in the presence of $\alpha_s$ | 2, 4 | Type 7 may also be stimulated since it is like 2 and 4. |
| | Direct inhibition | 1 | |
| | No direct effect | 3, 5, 6 | |
| Forskolin | Stimulatory | 1–6 | Forskolin binds directly to types 1–6 adenylyl cyclases. |
| $Ca^{2+}$/CaM | Stimulatory | 1, 3 | CaM binds directly to the type 1 enzyme. |
| $Ca^{2+}$ | Inhibitory | 6 | Effect is not through CaM and probably directly on the enzyme. |
| Adenosine (P-site ligands) | Direct Inhibition | 1, 5, 6 | Though types 2, 3, and 4 have not been tested; they are also likely to be inhibited. |

Currently no functional properties of types 7 and 8 adenylyl cyclases are known because full-length functional clones have not been expressed.

### III. P-Site Inhibition

Types 1, 5, and 6 have been tested for inhibition by adenosine and its P-site analogs, and all three enzymes are inhibited (TANG et al. 1991; ISHIKAWA et al. 1992; KATSUSHIKA et al. 1992). The rank order of potency for several analogs is similar to that first described by JOHNSON et al. (1989). Currently, there is no published information on the effects of the P-site inhibitors on types 2, 3, and 4. However, given the divergence between the 5/6 family and the type 1 enzyme and the similarity of the characteristics of their P-site inhibition, it can be reasonably predicted that P-site inhibition will be a general property of all G protein-regulated adenylyl cyclases.

The regulation of the different types of adenylyl cyclases by G-protein subunits and by intracellular ligands is summarized in Table 3.

## F. Regulation by Protein Phosphorylation

### I. Regulation by Protein Kinase C

Studies on the S49 lymphoma cells have shown that treatment with phorbol esters results in enhanced adenylyl cyclase activities (BELL et al. 1985). This increase in activity is thought to result from protein kinase C dependent phosphorylation of adenylyl cyclase. Extensive stimulation of the frog erthrocyte adenylyl cyclase by phorbol esters has also been reported (SIBLEY et al. 1986). Partial purification of adenylyl cyclase from frog erythrocytes treated with phorbol esters indicated that adenylyl cyclase itself was phosphorylated in response to phorbol ester treatment (YOSHIMASA et al. 1987). An adenylyl cyclase preparation purified from bovine striatum was phosphorylated in vitro by protein kinase C (YOSHIMASA et al. 1987). Recently Pfeuffer and coworkers have demonstrated that phorbol ester treatment of human platelets leads to the phosphorylation of a 150 kDa adenylyl cyclase, but not G-protein subunits (SIMMONTEIT et al. 1991). The sensitizing effect of phorbol esters on adenylyl cyclase activity is not universal. For instance, neither rat nor chick hepatocytes show significant increases in adenylyl cyclase activities upon phorbol ester treatment (MURPHY et al. 1989; PREMONT and IYENGAR 1988).

All of the cloned adenylyl cyclases contain numerous potential protein kinase C phosphorylation sites. However, the functional data from various tissues suggests that certain types of adenylyl cyclase may be specifically susceptible to regulation by protein kinase C. Experiments in our laboratory show that the expressed type 2 enzyme can be stimulated by protein kinase C activation but types 5 and 6 are not. The stimulation is most prominent for basal and forskolin stimulated activity, and less so in $G_s$ mediated activities (O.J. and R.I., unpublished observations). S49 cells contain type 7 enzyme, which is closely related to the type 2 enzyme. Since S49 cells show extensive stimulation by phorbol esters it is likely that the type 7 enzyme may also be stimulated by protein kinase C-mediated phosphorylation. This prediction needs to be experimentally verified when the clone becomes available.

## II. Protein Kinase A Regulation: A Component of Heterologous Desensitization

During studies on the multiple loci of glucagon-induced desensitization in chick hepatocytes we had found that one component of desensitization was cAMP-dependent. Addition of excess purified $G_s$ to membranes from cells treated either with glucagon or 8-Br-cAMP did not result in adenylyl cyclase activities comparable to that seen in naive cells. Treatment of membranes from naive but not desensitized cells in vitro with protein kinase A resulted in decreased adenylyl cyclase activity (PREMONT et al. 1992b). Studies on the $kin^-$ variant of S49 cells also indicate that protein kinase A treatment results in decreased forskolin stimulated adenylyl cyclase activity (KUNKEL et al. 1989; PREMONT et al. 1992b). Since $G_s$ is not a target for protein kinase A regulation (PREMONT and LYENGAR 1989), it appears that adenylyl cyclase is the most likely target for the protein kinase A-mediated phosphorylation.

Examination of the sequences of the cloned adenylyl cyclases shows that the all of the enzymes except the type 4 contain one or two putative protein kinase A phosphorylation sites. An interesting observation is that the positions of putative protein kinase A phosphorylation sites are not conserved among the various adenylyl cyclases; the location of these sites for the various adenylyl cyclases are shown in Fig. 5. Recent studies in our laboratory have shown that chick hepatocytes contain the types 5 and 6

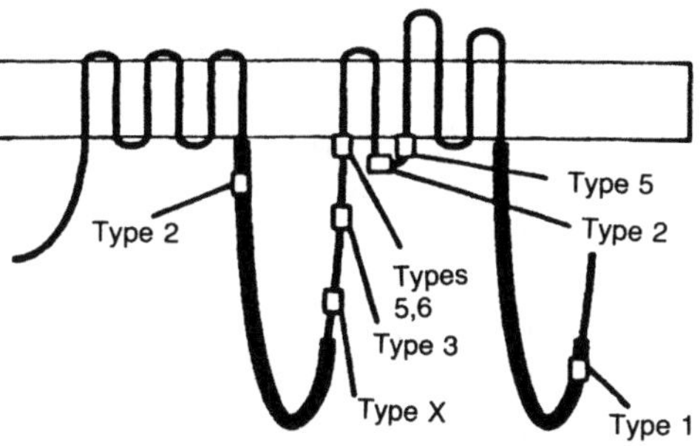


**Fig. 5.** Locations of the predicted protein kinase A phosphorylation sites for six mammalian adenylyl cyclases. The putative protein kinase A phosphorylation sites were detected by the PROSITE/PCGENE program. The type 4 enzyme has no consensus protein kinase A phosphorylation site sequence and hence is not represented. The putative protein kinase A site on the fourth cytoplasmic loop is seen for the type 5 rat enzyme (PREMONT et al. 1992) but not the dog enzyme (ISHIKAWA et al. 1992).

adenylyl cyclases (PREMONT et al. 1992b). We have also cloned the mouse homolog of the type 6 enzyme from S49 cells (PREMONT et al. 1992b). These observations indicate that the type 6 enzyme could be one form inhibited by protein kinase A dependent phosphorylation. Since one putative protein kinase A site is conserved between types 5 and 6, it is likely that type 5 adenylyl cyclase could also be inhibited by phosphorylation at this site. These conclusions remain to be directly verified. The varying locations of the protein kinase A sites for the other adenylyl cyclases also suggest that different types of adenylyl cyclases may be differentially regulated by protein kinase A dependent phosphorylation. This hypothesis also needs to be experimentally tested.

# G. Functional Consequences of Multiple Adenylyl Cyclases

## I. Integration of Multiple Signals

The multiplicity of adenylyl cyclases and their distinct patterns of responsiveness allow for a hitherto unrecognized locus where signals from receptors can be integrated. Such integration can be both positive and negative.

Positive integration can be achieved through $\beta\gamma$ stimulation of the $\alpha_s$-activated types 2 and 4 adenylyl cyclases. However, since the effective concentration range in which $\beta\gamma$ subunits affect the activity of adenylyl cyclases is in the 1–50 n$M$ range, it appears that only receptors that activate the abundant G-proteins are able to ultilize this pathway for signal integration. As has been noted by TANG and GILMAN (1991), in the brain, where both $G_i$ and $G_o$ are very abundant, receptors that activate these G-proteins could very easily produce sufficient $\beta\gamma$ subunits to enhance the stimulation of the type 2 enzyme. Indeed, over the years there have been reports of conditional stimulation of cAMP production in the brain by receptors that do not by themselves stimulate cAMP production (SATTIN et al. 1975; KARBON et al. 1985). It appears that one common condition in these reports is the presence of a second agonist that stimulates cAMP production (and presumably $G_s$) by itself. Indeed in one study (KARBON et al. 1985) the effects of the combination of the receptor agonists could not be mimicked by substituting the stimulatory ligand (isoprotererenol or other agonists that bind to $G_s$ coupled receptors) with a direct activator of adenylyl cyclase, forskolin. In light of our current understanding of the conditional nature of the regulation by $\beta\gamma$ subunits, it appears that these effects are most likely due to the stimulation of the type 2 or 4 adenylyl cyclase by $\alpha_s$ and $\beta\gamma$ subunits. Such integration could be the basis for synergism between dopamine $D_1$ and $D_2$ receptors in the regulation of the $Na^+$-$K^+$ ATPase in neostriatal neurons (BERTORELLO et al. 1990). In situations where full $\beta\gamma$ enhancement on $\alpha_s$-stimulated adenylyl cyclase activity is observed, the production of cAMP could reach such levels that the return of cellular cAMP levels to the resting state would take considerably longer than if $\alpha_s$ were alone activating the enzyme. Additional amplification could also occur through hormonal activation of protein kinase C to phosphorylate the type 2 adenylyl cyclase, leading to a three-pronged enhancement of the adenylyl cyclase activity. Thus the temporal and possibly the spatial characteristics of the cAMP signal from coincident signaling could be quite distinct from that observed when only $G_s$ is activated. Such differences may be particularly meaningful in neuronal cells.

Another aspect of the positive integration which is of potential importance is that at low levels of activation of $G_s$ the production of excess $\beta\gamma$ subunits by non-$G_s$-coupled receptors could well allow the cAMP signal to be suffieiently amplified that the increase in cAMP levels would be biologically meaningful. Thus due to simultaneous activation of a second receptor pathway, the cell would be able to detect and respond to a small stimulatory signal that it may otherwise have been incapable of detecting. Whether such detection is benificial to the cell, or whether such simultaneous signals can trigger altered responses that lead to pathological conditions remains to be determined.

Negative integration can be achieved by three mechanisms, depending on the type of adenylyl cyclase present. If certain types of adenylyl cyclases are uniquely suceptible to inhibition by the $\alpha_i$ subunits, the presence of these types of adenylyl cyclase would allow for receptors that activate $G_i$ to negatively modulate adenylyl cyclase directly. Alternatively, if types 2 and 4 are not present in substantial amounts, excess $\beta\gamma$ subunits can inhibit adenylyl cyclase by sequestering the $\alpha_s$ subunit (Northup et al. 1983). This mechanism should be operative for types 1, 3, 5, and 6 adenylyl cyclases at least, but requires substantial levels of $\beta\gamma$ release from G protein activation. Indeed, this mechanism had been proposed as the major mode of G-protein-dependent inhibition by Gilman and coworkers (Gilman 1984; Katada et al. 1984) and may well be operative in some tissues, such as brain. In addition, the type 1 adenylyl cyclase is directly inhibited by somewhat lower concentrations of $\beta\gamma$ subunits (Tang et al. 1991).

## II. Modulation of Signal Transmission

In addition to the signal integration by interactions of the G-protein subunits with the various forms of adenylyl cyclases, phosphorylation of specific types of adenylyl cyclases by protein kinase A and protein kinase C could be utilized to modulate both the duration as well as the amplitude of the signal through the cAMP pathway.

The possibility that the type 2 (and perhaps type 7) adenylyl cyclase can be uniquely stimulated by protein kinase C dependent pathways suggests that in cell types and tissues where these adenylyl cyclases constitute a significant portion of the total pool of adenylyl cyclase, activation of the phospholipase C pathway can substantially regulate the cAMP pathway. Similarly, in cells where the type 5 and 6 adenylyl cyclases comprise a significant portion of the total adenylyl cyclases, feedback inhibition of the adenylyl cyclase by protein kinase A could be an important component of desensitization. It is also noteworthy that one of the widely expressed forms of adenylyl cyclase, the type 4 enzyme, does not have any sequence that would be a consensus protein kinase A phosphorylation site and hence probably cannot be regulated by protein kinase A. Thus, ratios of the types 5/6 adenylyl cyclases to the type 4 could determine the extent to which heterologous desensitization at the level of adenylyl cyclase can affect signal transmission.

The multiplicity of $G_s$-regulated mammalian adenylyl cyclases, and their differing properties, add a new dimension to the complexity of signal sorting and transmission through this G-protein-coupled system. Differing ratios of these distinct adenylyl cyclases in a single cell type could endow each cell type with its own unique capabilities to receive and interpret signals. If the ratios of the different types of adenylyl cyclases can be altered in response to either altered physiology of the cell, or a changing environment, such alteration could provide the cell with the capability of having dynamic signal-transducing systems that could modulate and transmit signals in response to the previous history of the system as well as to its present state. It is possible that such signal-transducing systems themselves could well be the molecular corelates for memory and learning.

*Acknowledgements*. We thank Dr. J. Kuprinski for providing us the information to construct the tree in Fig. 2b and about the type 7 enzyme prior to publication. Research in our laboratories described here was supported by NIH grants CA-44998 and DK-38761. R.T.P. was supported by a NSF predoctoral fellowship. O.J. is supported by the MSTP trainining program and Endocrinology Training Grant DK-07645.

## References

Asano T, Pedersen SE, Scott CW, Ross EM (1984) Reconstitution of catecholamine-stimulated binding of guanosine-5′-O-(3-thiotriphosphate) to the stimulatory GTP-binding protein of adenylate cyclase. Biochemistry 23:5460–5467

Bakalyar HA, Reed RR (1990) Identification of a specialized adenylyl cyclase that may mediate odorant detection. Science 250:1403–1406

Bertorello AM, Hopfield JF, Aperia A, Greengard P (1990) Inhibition by dopamine of ($Na^+$+$K^+$) ATPase activity in neostriatal neurons through $D_1$ and $D_2$ dopamine receptor synergism. Nature 347:386–388

Bell JD, Buxton ILO, Brunton LL (1985) Enhancement of adenylate cyclase activity in S49 lymphoma cells by phorbol esters. Putative effects of C kinase on the $\alpha_s$-GTP-catalytic subunit interaction. J Biol Chem 260:2625–2628

Boyajian CL, Garritsen A, Cooper DMF (1991) Bradykinin stimulates [$Ca^{2+}$] mobilization in NCB-20 cells leading to direct inhibition of adenylyl cyclases: a novel mechanism for inhibition of cAMP production. J Biol Chem 266:4995–5003

Braun TD, Dods RF (1975) Devolopment of a $Mn^{2+}$ sensitive, soluble adenylate cyclase in rat testis. Proc Natl Acad Sci USA 72:1097–1101

Bray P, Carter A, Simons C, Guo V, Puckett C, Kamholz J, Speigel A, Nirenberg M (1986) Human cDNA clones for four species of $G\alpha_s$ signal transduction protein. Proc Natl Acad Sci USA 83:8893–8897

Brostrom CO, Huang Y, Breckenridge BM, Woolf DJ (1975) Identification of a calcium binding protein as the calcium dependent regulator of the brain adenylyl cyclase. Proc Natl Acad Sci USA 72:64–68

Caldwell KK, Boyajian CL, Cooper DMF (1992) The effects of $Ca^{2+}$ and calmodulin on adenylyl cyclase activity in plasma membranes of neuronal and non neuronal cells. Cell Calcium 13:107–121

Cassel D, Selinger Z (1978) Mechanism of adenylate cyclase stimulation through the $\beta$-adrenergic induced displacement of GTP for GDP. Proc Natl Acad Sci USA 75:4155–4159

Cheung WY, Bradham LD, Lynch TJ, Lin YM, Tallant EA (1975) Protein activator of cyclic 3′:5′ – nucleotide phosphodiesterase of bovine or rat brain also activates its adenylate cyclase. Biochem Biophys Res Commun 66:1055–1062

Choi E-J, Wong ST, Hinds TR, Storm DR (1992a) Calcium and muscarinic stimulation of type 1 adenylyl cyclase in whole cells. J Biol Chem 267:12440–12442

Choi E-J, Xia Z, Storm DR (1992b) Stimulation of the type III olfactory adenylyl cyclase by calcium and calmodulin. Biochemistry 31:6492–6498

Clark RB, Goka TJ, Green DA, Barber R, Butcher RW (1982) Differences in the forskolin activation of adenylate cyclases in wild typand variant lymphoma cells. Mol Pharmacol 22:609–613

Feinstein PG, Schrader KA, Bakalyar HA, Tang WJ, Kuprinski J, Gilman AG, Reed RR (1991) Molecular cloning and characterization of a $Ca^{2+}$/ calmodulin-insensitive adenylyl cyclase from rat brain. Proc Natl Acad Sci USA 88:10173–10177

Gao B, Gilman AG (1991) Cloning and expression of a widely distributed (type IV) adenylyl cyclase. Proc Natl Acad Sci USA 88:10178–10182

Gilman AG (1984) G proteins and dual control of adenylate cyclase. Cell 36:577–579

Graziano MP, Freissmuth M, Gilman AG (1987) Expression of $G_{s\alpha}$ in E. coli: purification and properties of two forms of the protein. J Biol Chem 264:409–418

Harrison JK, Hewlett HGK, Gnegy ME (1989) Regulation of the calmodulin sensitive adenylyl cyclase by the stimulatory G protein Gs. J Biol Chem 264:15880–15885

Hildebrandt JD, Sekura RD, Codina J, Iyengar R, Manclark CR, Birnbaumer L (1983) Stimulation and Inhibition of adenylyl cyclases is mediated by distinct regulatory proteins. Nature 302:702–709

Iyengar R, Birnbaumer L (1990) Overview. In: Iyengar R, Birnbaumer L (eds) G proteins. Academic, San Diego, pp 1–14

Ishikawa Y, Katsushika S, Chen L, Halnon NJ, Kawabe J, Homcy CJ (1992) Isolation and characterization of a novel cardiac adenylyl cyclase cDNA. J Biol Chem 267:13553–13557

Jakobs KH, Aktories K, Schultz G (1983) A nucleotide regulatory site for somatostatin inhibition of adenylate cyclase in S49 lymphoma cells. Nature 303:177–178

Jones DT, Reed RR (1987) Molecular cloning of five GTP binding protein cDNA species from rat olfactory neuroepithelium. J Biol Chem 262:14241–14249

Jones DT, Masters SB, Bourne HR, Reed RR (1990) Biochemical characterization of three stimulatory G proteins. The large and the small forms of $G_s$ and the olfactory specific G protein $G_{olf}$. J Biol Chem 262:2671–2676

Johnson RA, Yeung S-M H, Stubner D, Bushfield M, Shoshani I (1989) Cation and structural requirements for P-site mediated inhibition of adenylate cyclase. Molecular Pharmacology 35:681–688
Karbon EW, Enna SJ (1985) Characterization of the relationship between $\gamma$-aminobutyric acid-B agonists and transmitter coupled cyclic nucleotide generating systems in rat brain. Molecular Pharmacology 27:53–59
Katada T, Northup JK, Bokoch GM, Ui M, Gilman AG (1984) The inhibitory guanine nucleotide component of adenylyl cyclase: subunit dissociation and guanine nucleotide dependent hormonal inhibition. J Biol Chem 259:3578–3585
Katada T, Kasukabe K, Oinuma M, Ui M (1987) A novel mechanism for the inhibition of adenylate cyclase via inhibitory GTP binding proteins. Calmodulin dependent inhibition of the cyclase catalyst by the $\beta\gamma$-subunits of GTP binding proteins. J Biol Chem 262:11897–11900
Katsushika S, Chen L, Kawabe J-I, Nilakantan R, Halnon NJ, Homcy CJ, Ishikawa Y (1992) Cloning and characterization of a sixth adenylyl cyclase isoform: types V and VI constitute a subgroup within the mammalian adenylyl cyclase family. Proc Natl Acad Sci USA 89:8774–8778
Koesling D, Boheme E, Schultz G (1991) Guanylyl cyclases, a growing family of signal transducing enzymes. FASEB J 5:2785–2791
Kruprinski J, Coussen F, Bakalyar HA, Tang W-J, Feinstein PG, Orth K, Slaughter C, Reed RR, Gilman AG (1989) Adenylyl cyclase amino acid sequence: possible channel or transporter like structure. Science 244:1558–1564
Kruprinski J, Lehman PC, Frankenfield CD, Zwaagstra, Watson PA (1992) Molecular diversity in adenylyl cyclase family. J Biol Chem 267 (in press)
Kunkel MW, Freidman J, Shenolikar S, Clark RB (1989) Cell-free heterlogous desensitization of adenylyl cyclase in S49 cell membranes mediated by cAMP dependent kinase. FASEB J 3:2067–2074
Levin LR, Han P-Y, Hwang PM, Feinstein PG, Davis RL, Randall R Reed (1992) The Drosophila learning and memory gene rutabaga encodes a $Ca^{2+}$/calmodulin responsive adenylyl cyclase. Cell 68:479–489
Linder ME, Ewald DA, Miller RJ, Gilman AG (1990) Purification and characterization of $G_o$-$\alpha$ and three types of $G_i$-$\alpha$ after expression in E. coli. J Biol Chem 265:8243–8251
Livingstone MS (1985) Genetic dissection of Drosophila adenylate cyclase. Proc Natl Acad Sci USA 82:5992–5996
Livingstone MS, Sziber PP, Quinn WG (1984) Loss of calcium/calmodulin responsiveness in adenylate cyclase of rutabaga, a Drosophila learning mutant. Cell 37:205–215
Londos C, Salomon Y, Lin MC, Harwood JP, Schramm M, Wolff J, Rodbell M (1974) 5″-Guanylyl-imidodiphosphate, a potent activator of adenylyl cyclase systems in eukaryotic cells. Proc Natl Acad Sci USA 71:3087–3090
Mattera R, Graziano MP, Yatani A, Zhou Z, Graf R, Codina J, Birnbaumer L, Gilman AG, Brown AM (1989) Splice variants of the $\alpha$-subunits of the G protein $G_s$activate both adenylyl cyclases and calcium channels. Science 243:804–807
May DC, Ross EM, Gilman AG, Smigel MD (1985) Reconstitution of catecholamine stimulated adenylyl cyclase activity using three purified proteins. J Biol Chem 260:15829–15833
Murphy GJ, Gawler DJ, Milligan G, Wakelam MJO, Houslay MD (1989) Glucagon induced desensitization of adenylate cyclase and stimulation of inositol phospholipid metabolism does not involve the inhibitory guanine nucleotide regulatory protein $G_i$which is inactivated upon challenge of hepatocytes with glucagon. Biochemical J 259:191–197
Neer EJ (1978) Multiple forms of adenylyl cyclase. In: George WJ, Ignarro LJ (eds) Advances in cyclic nucleotide research. Raven, New York, vol 9, pp 69–83

Nelson CA, Seamon KB (1988) Binding of [$^3$H] forskolin to solubilized preparation of adenylate cyclase. Life Sciences 42:1375–1383

Northup JK, Smigel MD, Sternweis PC, Gilman AG (1983) The subunits of the stimulatory regulatory component of adenylyl cyclase. Resolution of the 45 000 Da ($\alpha$) subunit. J Biol Chem 258:11361–11368

Okamura N, Tajima Y, Onoe S, Sugita Y (1991) Purification of the bicarbonate sensitive sperm adenylyl cyclase by 4-acetamido-4″-isothiocyanostilbene-2,2″-disulfonic acid affinity chromatography. J Biol Chem 266:17754–17759

Parma J, Stengel D, Gannage M-H, Poyard M, Barouki R, Hanoune J (1991) Sequence of a human brain adenylyl cyclase partial cDNA. Evidence for a consensus cyclase specific domain. Biochem Biophys Res Comm 179:455–462

Pfeuffer T, Metzger H (1982) 7-O-Hemisuccinyl-deacetyl forskolin-Sepharose: a novel affinity support for purification of adenylate cyclase. FEBS Letts 146:369–375

Pfeuffer E, Dreher RM, Metzger H, Pfeuffer T (1985) Catalytic unit of adenylate cyclase; purification and identification by affinity cross linking. Proc Natl Acad Sci USA 82:3068–3090

Premont RT, Iyengar R (1989) Heterologous desensitization of the liver adenylyl cyclase: analysis of the role of G proteins. Endocrinology 125:1151–1160

Premont RT, Chen J, Ma H-W, Ponnapalli M, Iyengar R (1992) Two members of a widely expressed subfamily of hormone stimulated adenylyl cyclases. Proc Natl Acad Sci USA 89 (in press)

Premont RT, Jacobowitz O, Iyengar R (1992) Lowered responsiveness of the catalyst of adenylyl cyclase to stimulation by $G_s$ in heterologous desensitization: a role for cAMP dependent phosphorylation. Endocrinology 131 (in press)

Ross EM, Gilman AG (1977) Resolution of some components of adenylate cyclase necessary for catalytic activity. J Biol Chem 252:6966–6969

Sattin A, Rall TW, Zanella J (1975) Regulation of cyclic adenosine 3′: 5′ monophosphate levels in guniea pig cerebral cortex by interaction of alpha adrenergic and adenosine receptor activity. J Pharmacol Exp Ther 192:22–32

Schramm M, Rodbell M (1975) A persistent active state of the adenylate cyclase system produced by the combined actions of isoproterenol and guanylyl imidodiphosphate in frog erythrocyte membranes. J Biol Chem 250:2232–2237

Seamon KB, Daly JW (1981) Activation of adenylate cyclase by the diterpene forskolin does not require the guanine nucleotide regulatory protein. J Biol Chem 256:9799–9801

Seamon KB, Padgett W, Daly JW (1981) Forskolin: a unique diterpene activator of adenylate cyclase in membranes and intact cells. Proc Natl Acad Sci USA 78:3363–3367

Sibley DR, Jeffs RA, Daniel K, Nambi P, Lefkkowitz RJ (1986) Phorbol-ester treatment promotes enhanced adenylate cyclase activity in frog erythrocytes. Arch Biochem Biophys 244:373–381

Simon MI, Strathmann MP, Gautam N (1991) Diversity of G proteins in signal transduction. Science 252:802–808

Simmoteit RR, Schulzki HD, Palm D, Mollner S, Pfeuffer T (1991) Chemical and functional analysis of components of adenylyl cyclase from human platelets treated with phorbol esters. FEBS Lett 249:189–194

Smigel MD (1986) Purification of the catalyst of adenylate cyclase. J Biol Chem 261:1976–1982

Stengel D, Guenet L, Hanoune J (1982) Proteolytic solubilization of adenylyl cyclase from membranes deficient in regulatory component. J Biol Chem 257:10818–10826

Stengel D, Parma J, Gannage MH, Roeckel N, Mattaei M-G, Barouki R, Hanoune J (1992) Different chromosomal localization of two adenylyl cyclase genes expressed in human brain. Human Genetics (in press)

Sutherland EW, Rall TS (1958) Fractionation and characterization of a cyclic adenine ribonucleotide formed by tissue particles. J Biol Chem 232:1077–1091
Sutherland EW, Rall TS, Menon T (1962) Adenyl cyclase I distribution, preparation properties. J Biol Chem 237:1077–1091
Tang W-J, Gilman AG (1991) Type specific regulation of adenylyl cyclase by G protein $\beta\gamma$-subunits. Science 254:1500–1503
Tang W-J, Krupinski J, Gilman AG (1991) Expression and characterization of calmodulin activated adenylyl cyclase. J Biol Chem 266:8595–8603
Wong Y-H, Conklin BR, Bourne HR (1992) $G_z$ mediated cAMP accumulation. Science 255:339–342
Woolf J, Londos C, Cooper DMF (1981) Adenosine receptors and regulation of adenylate cyclase. In: Dumont JE, Greengard P, Robison AG (eds) Advances in cyclic nucleotide research. Raven Press N Y 14:199–214
Xia Z, Refsdal CD, Merchant KM, Dorsa DM, Storm DR (1991) Distribution of mRNA for cal modulin sensitive adenylyl cyclasein rat brain: expression in areas associated with learning and memory. Neuron 6:431–443
Yoshimasa T, Sibley DR, Bouvier M, Lefkowitz RJ, Caron MG (1987) Cross-talk between cellular signalling pathways suggested by phorbolester induced adenylated cyclase phosphorylation. Nature 327:67–70
Yoshimura M, Cooper DMF (1992) Cloning and expression of a $Ca^{2+}$ inhibitable adenylyl cyclase from NCB 20 cells. Proc Natl Acad Sci USA 89:6716–6720

CHAPTER 56

# The Light-Regulated cGMP Phosphodiesterase of Vertebrate Photoreceptors: Structure and Mechanism of Activation by $G_{t\alpha}$

T.G. WENSEL

## A. Physiological Role of cGMP Phosphodiesterase in Visual Signaling

The cGMP-specific phosphodiesterase (PDE) of vertebrate photoreceptor cells plays a central role in visual signal transduction. Activation of PDE by the G-protein transducin ($G_t$) is essential for the rapid reduction in cytoplasmic free cGMP concentration that is thought to be responsible for light-induced decreases in membrane current that lead to hyperpolarization and synaptic signaling. Mice with a mutation in a PDE subunit gene suffer from a hereditary retinal degeneration (*rd*; BOWES et al. 1990; PITTLER and BAEHR 1991).

A large number of reviews of this light-induced guanine nucleotide cascade have been published recently (e.g., PUGH and LAMB 1990; STRYER 1991; LOLLEY and LEE 1990; KAUPP and KOCH 1992; MCNAUGHTON 1990), including chapters in this work (see contributions by TING et al., HOFMANN), as has a review of cyclic nucleotide phosphodiesterases in general (BEAVO and REIFSNYDER 1990).

In large part because it is readily soluble and is present at relatively high concentration (on the order of $30\,\mu M$; BAEHR et al. 1979; HAMM and BOWNDS 1986) in rod outer segments (ROS), PDE was one of the first G-protein-coupled effectors to be purified (MIKI et al. 1975; BAEHR et al. 1979), and it has been one of the most extensively studied at both the structural and functional levels.

## B. Structure

### I. Subunit Composition

PDE from bovine rod cells is a heterotetramer of three different kinds of subunits (BAEHR et al. 1979): the large catalytic subunits $PDE_\alpha$ and $PDE_\beta$ and a smaller inhibitory subunit $PDE_\gamma$. It is generally agreed that the holoenzyme consists of one $\alpha$ and one $\beta$ subunit, and there is evidence that most rod PDE is in the form $PDE_{\alpha\beta\gamma\gamma}$ (DETERRE et al. 1988). Both the $\alpha$ and

$\beta$ subunits have acidic pI values of about 5.4 (BAEHR et al. 1979; YAMAZAKI et al. 1980), whereas $PDE_\gamma$ is basic.

Catalytic subunits free of $PDE_\gamma$ can be obtained by limited proteolytic digestion (MIKI et al. 1975; HURLEY and STRYER 1982) or by ion exchange chromatography following activation by transducin (DETERRE et al. 1986). $PDE_\gamma$ can be isolated by denaturation of $PDE_{\alpha\beta}$ with heat or extremes of pH (HURLEY and STRYER 1982; DUMLER and ETINGOF 1976; WENSEL and STRYER 1986) and purified by reverse-phase HPLC (BROWN and STRYER 1989). Separation of the catalytic subunits from one another without irreversible denaturation has been difficult, if not impossible, to achieve. Some 20%–30% of PDE in bovine rod outer segments is much more soluble than the rest and contains an additional 15-kDa subunit, $PDE_\delta$ (GILLESPIE et al. 1989). Cone cells contain a PDE (HURWITZ et al. 1985; GILLESPIE and BEAVO 1988) with similar properties and structure. It consists of two large identical catalytic subunits ($\alpha'$), with one or more $\gamma$-like subunits, as well as 13- and 15-kDa subunits.

## II. Size and Hydrodynamic Properties

The mass of PDE has been estimated to be 170 kDa by sedimentation velocity, with the apparent subunit $M_r$ values determined by SDS gel electrophoresis to be 88900 ($\alpha$), 84000 ($\beta$), and 13000 ($\gamma$; BAEHR et al. 1979). The apparent values vary with the gel system. More accurate values for bovine PDE, based on primary sequence data are 99261 ($\alpha$), 98308 ($\beta$), and 9700 ($\gamma$). Hydrodynamic studies suggest that the shape of PDE deviates greatly from spherical, with calculated axial ratios of 9:1 for a prolate ellipsoid and 12:1 for an oblate ellipsoid (GILLESPIE and BEAVO 1988).

## III. Primary Structure

The primary structures of all three subunits of bovine rod PDE (OVCHINNIKOV et al. 1986, 1987; LIPKIN et al. 1990; PITTLER et al. 1990), cone $PDE_{\alpha'}$, (LI et al. 1990; CHARBONNEAU et al. 1990), and a cone-specific $PDE_\gamma$ (HAMILTON and HURLEY 1990) have been determined by amino acid sequencing combined with cDNA sequencing, and closely related sequences from other species continue to appear. The rod $\alpha$ and $\beta$ subunits are 72% identical in amino acid sequence, and the cone $\alpha'$ sequence is 62% identical to rod $\beta$.

## IV. Posttranslational Modifications

The C termini of $\alpha$, $\beta$, and $\alpha'$ all have the CXXX motif that serves as a signal for *S*-isoprenylation of cysteine and proteolytic cleavage at the carboxylate side of cysteine, followed by methyl esterification (SWANSON and APPLEBURY 1983; ONG et al. 1989; CATTY and DETERRE 1991). Isoprenylation

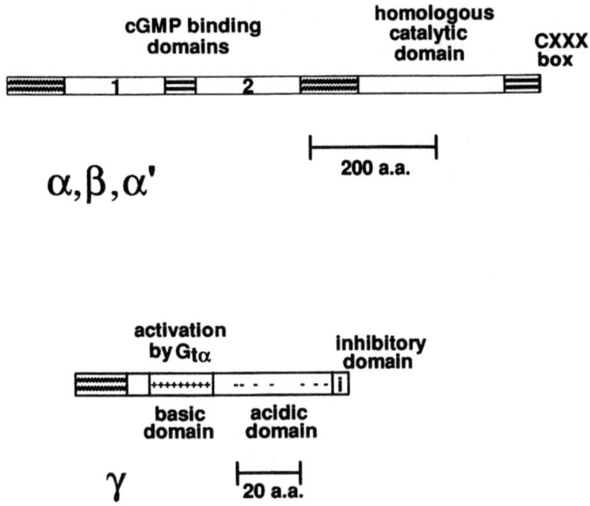


**Fig. 1.** Domain structure of PDE catalytic ($\alpha,\beta,\alpha'$), and inhibitory ($\gamma$) subunits. Further explanation and references are provided in the text. (The representation of the catalytic subunits is adapted from an unpublished drawing courtesy of Wolfgang Baehr)

is specific, with $PDE_\alpha$ receiving a farnesyl, and $PDE_\beta$ receiving a geranylgeranyl group (ANANT et al. 1992; QIN et al. 1992). The highly protease-sensitive C terminal peptide containing this group is required for PDE to bind membranes and be activated efficiently by $G_{t\alpha}$ (WENSEL and STRYER 1986; ONG et al. 1992; CATTY and DETERRE 1991; MALINSKI and WENSEL 1992). The amino termini of all three subunits are blocked, with the terminal groups reported to be acetylmethionine for $PDE_\gamma$ (OVCHINNIKOV et al. 1986) and acetyl-glycine for $PDE_\alpha$ (OVCHINNIKOV et al. 1987).

## V. Domain Structures of Subunits

### 1. Catalytic Subunits

The sequence of residues 555–790 in $\alpha$, $\beta$, and $\alpha'$ is homologous to domains found in all known cyclic nucleotide phosphodiesterase sequences and is thought to be the catalytic domain (reviewed by BEAVO and REIFSNYDER 1990). Each catalytic subunit contains two regions homologous to one another and to regions found in other cGMP-binding PDEs, spanning residues 89–251 and 295–464 (Fig. 1). These are proposed to form the noncatalytic cGMP binding domain(s) (CHARBONNEAU et al. 1990).

### 2. Inhibitory Subunit

The amino terminal half of $PDE_\gamma$ is rich in basic residues while the carboxyl terminal half is very acidic. Mutagenesis and peptide studies (LIPKIN et al.

1988; BROWN and STRYER 1989; BROWN 1992; CUNNICK et al. 1990) have implicated the carboxy terminal residues 82–87 as essential for inhibition of PDE. Lysine and arginine residues in the basic domain appear to be important in allowing activation by $G_{t\alpha}$. Residues involved in binding to $PDE_{\alpha\beta}$ appear to be dispersed throughout the sequence primary structure, but the amino terminal 17 residues do not appear to play an important role.

## C. Functional Properties

### I. Solubility

PDE from rods is a peripheral membrane protein with relatively high affinity for disc membranes in isotonic buffers. It can be eluted from the membranes if the ionic strength is lowered (KÜHN 1981). Cone PDE seems to be less tightly associated with membranes.

### II. Kinetic Properties

PDE is a highly efficient enzyme: maximum substrate turnover rates at saturating (cGMP) in the range of 4000–6900 $s^{-1}$ have been reported for both trypsin-activated and $G_{t\alpha}$-activated PDE (HURLEY and STRYER 1982; SITARAMAYYA et al. 1986; GILLESPIE and BEAVO 1988). The value of $K_m$ has been controversial. Values for tryspin-activated or unstimulated PDE have usually been reported in the range of 40–150 $\mu M$ (BAEHR et al. 1979; MIKI et al. 1975; HURLEY and STRYER 1982; SITARAMAYYA et al. 1986; WHALEN et al. 1990; ZIMMERMAN et al. 1986), with one report as low as 17 $\mu M$ (GILLESPIE and BEAVO 1988). PDE activated by $G_{t\alpha}$ in rod outer segments has a similar $K_m$ (YEE and LIEBMAN 1978; WHALEN et al. 1990) when assayed in room light. However the $K_m$ for PDE in ROS activated by a brief flash of light appears to be much higher, and values up to 1.4 m$M$ have been reported (ROBINSON et al. 1980; KAWAMURA and MURAKAMI 1986; SITARAMAYYA et al. 1986). This apparent "flash-induced" increase in $K_m$ is greatly diminished if the membranes are disrupted (KAWAMURA and MURAKAMI 1986; BOWNDS et al. 1992) and is probably an artifact caused by a radial cGMP concentration gradient in intact disc stacks.

### III. Noncatalytic cGMP Binding Sites

PDE binds cGMP at one or more (probably two) noncatalytic sites (YAMAZAKI et al. 1980, 1982; GILLESPIE and BEAVO 1988, 1989). Bovine cone PDE has a single class of high-affinity ($K_d$ = 11 n$M$) binding sites, while rod PDE has two classes of sites with estimated $K_d$ values of 16 and 234 n$M$. Dissociation from the rod high-affinity sites occurs with a half-life of 4 h. Amphibian rod PDE was reported to have two classes of low-affinity sites

($K_d$ values of 160 and 830 n$M$) that appear to exchange more rapidly. Occupation of these sites can account for most of the cGMP in rod cells.

## D. Regulation of Catalytic Activity

### I. Inhibition by $PDE_\gamma$

Purified intact PDE is essentially inactive but can be activated by treatment with proteases, polycations, or activated $G_{t\alpha}$ (MIKI et al. 1975; FUNG et al. 1981). Activation depends on relieving inhibition by $PDE_\gamma$ (HURLEY et al. 1982). Purified $PDE_\gamma$ can overcome activation by either trypsin (HURLEY and STRYER) or activated $G_{t\alpha}$ (WENSEL and STRYER 1986). $PDE_\gamma$ binds tightly to the catalytic subunits; $K_d$ is on the order of $10^{-11}$ $M$ (WENSEL and STRYER 1986, 1990; GILLESPIE et al. 1989; CLERC and BENNETT 1992), with reported values ranging from $10^{-13}$ $M$ (BENNETT and CLERC 1989) to 15 n$M$ (SITARMAYYA et al. 1986). Values of $10^{-9}$ $M$ and above probably reflect some inactive protein in the preparations. Binding is rapid ($\sim 10^8\,m^{-1}\text{-}s^{-1}$) while dissociation is slow [$k_{off} = (1\text{–}5) \times 10^{-3}\,s^{-1}$; WENSEL and STRYER 1986, 1990; BROWN 1992].

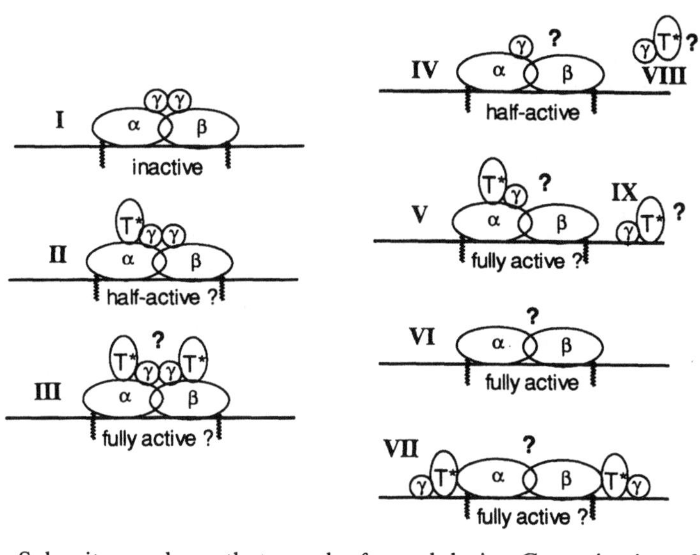


**Fig. 2.** Subunit complexes that may be formed during $G_{t\alpha}$ activation of PDE. *Greek letters*, PDE subunits; $T^*$, GTP-$G_{t\alpha}$. *Horizontal line*, the surface of the disc membrane; *tails* (on $\alpha$ and $\beta$), isoprenoid groups. It is not known whether they interact with membrane phospholipids as shown. All complexes shown have been proposed, but firm evidence for formation under physiological conditions exists only for those not marked with a *bold question mark*. *I* is the dark, fully inhibited state of PDE, and *IV*, *VI* can be formed by proteolytic digestion, so their activity levels are reasonably certain. Activities of other forms are unknown, and labels with *question marks* are based on an assumption that binding of one $G_{t\alpha}$ to one $PDE_\gamma$ produces half-activation of PDE. Alternative possibilities are discussed in the text

## II. Activation by G-Protein

### 1. Role of $G_{t\alpha}$

It is generally agreed that $G_{t\alpha}$-GTP activates PDE by altering the interactions between $PDE_{\alpha\beta}$ and $PDE_{\gamma}$. There is less agreement over whether $PDE_{\gamma}$ leaves the holoenzyme and/or the membrane when PDE is activated, and over whether two $G_{t\alpha}$-GTP molecules must act cooperatively or independently to activate fully one molecule of PDE.

### 2. Role of Membranes in PDE Activation by $G_{t\alpha}$

Activation of PDE by activated $G_{t\alpha}$ can be reconstituted using purified proteins, but membranes are necessary for efficient activation (FUNG and NASH 1983; HO and FUNG 1984; TYMINSKI and O'BRIEN 1984; WENSEL and STRYER 1986; MILLER et al. 1987; BENNETT and CLERC 1989; PHILLIPS et al. 1989; Fig. 2). Enhancement requires phospholipids but not integral membrane proteins (MALINSKI and WENSEL 1992). It has been argued (LIEBMAN et al. 1987 and references cited therein) that photoexcitation kinetics require two-dimensional diffusional encounter of membrane-bound species. A tightly membrane-bound subpopulation of $G_{t\alpha}$-GTP is much more potent at PDE activation than typical preparations of soluble $G_t$ and appears to account for much of the PDE activation observed in ROS (WENSEL and STRYER 1988).

### 3. Role of $PDE_{\gamma}$ in Activation by Transducin

When PDE in dark-adapted ROS membranes was activated by adding purified $G_{t\alpha}$-GppNHp (FUNG et al. 1981), activity close to the trypsin-induced level was observed, but $G_{t\alpha}$-GppNHp concentrations on the order of $10^{-6}\,M$ were required to activate PDE on the order of $10^{-9}\,M$. This result has been observed in a number of laboratories (e.g., TYMINSKI and O'BRIEN 1984; BENNETT 1982; MALINSKI and WENSEL 1992), and in general the titration curves are not hyperbolic. Rather, they are consistent with a strong ($K_d < 1\,nM$) interaction of a small fraction of the purified $G_{t\alpha}$ with PDE. While it may be that transducin simply interacts rather weakly with PDE ($K_d \sim 10^{-6}\,M$), it seems more likely that typical $G_t$ preparations are largely inactive, and thus mask a very strong activation by a minor subpopulation of $G_{t\alpha}$.

The simplest explanation for potent inhibition by $PDE_{\gamma}$ of $G_{t\alpha}$-activated PDE (WENSEL and STRYER 1986; BROWN and STRYER 1989) is that $G_{t\alpha}$-GTP causes dissociation of one or both $PDE_{\gamma}$ subunits, leaving behind active catalytic subunits. This interpretation is supported by reports that a complex of $G_{t\alpha}$ with $PDE_{\gamma}$ could be isolated by chromatography following release from frog ROS in isotonic buffers (YAMAZAKI et al. 1983, 1990) or from bovine ROS in hypotonic buffers (DETERRE et al. 1986) after activation of

$G_{t\alpha}$. $G_{t\alpha}$ antibodies immunoprecipitated a complex of activated $G_{t\alpha}$ with either $PDE_{\alpha\beta\gamma\gamma}$ (NAVON and FUNG 1987) or with $PDE_{\gamma}$ (FUNG and GRISWOLD-PENNER 1989) but did not immunoprecipitate $PDE_{\alpha\beta}$. Activation by $G_{t\alpha}$ accelerates exchange of labeled $PDE_{\gamma}$ for unlabeled endogenous $PDE_{\gamma}$ on PDE (WENSEL and STRYER 1990).

A difficulty with this mechanism is that in bovine ROS, a $G_{t\alpha}$-$PDE_{\gamma}$ complex does not dissociate from the membranes under the isotonic conditions in which PDE is activated, and added $PDE_{\gamma}$ is able to overcome the activation (WENSEL and STRYER 1986; DETERRE et al. 1986). Neither $PDE_{\gamma}$ nor activated $G_{t\alpha}$ has significant affinity for membranes, but it is conceivable that a complex of the two proteins may bind the membranes. However, despite attempts to measure membrane-binding of the $G_{t\alpha}$-$PDE_{\gamma}$ complex no evidence has been found so far for significant binding. In contrast, enhancement of PDE membrane affinity by activated $G_{t\alpha}$, or vice versa, can be readily demonstrated (CLERC and BENNETT 1992; MALINSKI and WENSEL 1992). The kinetics of activation and of $PDE_{\gamma}$ dissociation in the absence of $G_{t\alpha}$ argues strongly for formation of a complex of $G_{t\alpha}$ with $PDE_{\alpha\beta\gamma\gamma}$ as a necessary intermediate (HURLEY and STRYER 1982; WENSEL and STRYER 1990). It may be that much, and possibly most, PDE remains in this form throughout activation. Even if only a fraction of $PDE_{\gamma}$ dissociates, added $PDE_{\gamma}$ could have an inhibitory effect by pushing the equilibrium towards the inactive state (BENNETT and CLERC 1989); however, it is not clear whether the observed inhibition can be quantitatively explained with such a model (CLERC and BENNETT 1992).

### 4. Is There Cooperativity in the Action of $G_{t\alpha}$-GTP?

It has been suggested that each of the two $PDE_{\gamma}$ sites may exert a different influence on total activity (DETERRE et al. 1988), giving rise to a cooperative action of $G_{t\alpha}$ or to a GTPase-independent mechanism for rapid PDE inactivation. However, direct binding studies with trypsinized PDE indicated that each $\gamma$ binding site bound $PDE_{\gamma}$ with similar affinity, and parallel activity measurements revealed that each contributed approximately equally to inhibition (WENSEL and STRYER 1990). Moreover, most published data (e.g., FUNG et al. 1981; TYMINSKI and O'BRIEN 1984; PHILLIPS et al. 1989; BENNETT 1982; MALINSKI and WENSEL 1992) showing activation of PDE as a function of added $G_{t\alpha}$ shows the distinctly noncooperative dependence described above. Nonetheless, there have been reports of positive cooperativity in titrations of PDE with $G_{t\alpha}$ (BENNETT and CLERC 1989; WHALEN and BITENSKY 1989) and of both positive and negative cooperativity in titrations of $PDE_{\alpha\beta}$ with $PDE_{\gamma}$ (WHALEN and BITENSKY 1989) as well as of enhanced activation by $G_{t\alpha}$ bound to a bivalent antibody (PHILLIPS et al. 1989). Thus the issue remains controversial.

### 5. A Role for Noncatalytic cGMP Binding Sites?

It has recently been suggested that the presence or absence of cGMP bound to the noncatalytic sites of frog PDE may profoundly influence inactivation kinetics through a lowered affinity of the $G_{t\alpha}$-$PDE_\gamma$ complex for $PDE_{\alpha\beta}$ when cGMP dissociates from these sites, allowing acceleration of the $G_{t\alpha}$ GTPase rate by $PDE_\gamma$ (ARSHAVSKY et al. 1991; ARSHAVSKY and BOWNDS 1992). It remains to be determined whether similar effects can be observed with mammalian rod PDE, which has been found to release cGMP only on a time scale of hours (GILLESPIE and BEAVO 1989).

## References


Anant JS, Ong OC, Xie H, Clarke S, O'Brien PJ, Fung BK-K (1992) In vivo differential prenylation of retinal cyclic GMP phosphodiesterase catalytic subunits, J Biol Chem 267:687–690

Arshavsky VYu, Gray-Keller MP, Bownds MD (1991) cGMP suppresses GTPase activity of a portion of transducin equimolar to phosphodiesterase in frog rod outer segments. J Biol Chem 266:18350–18357

Arshavsky VYu, Bownds MD (1992) Regulation of deactivation of photoreceptor G protein by its target enzyme and cGMP. Nature 357:416–417

Baehr W, Devlin MJ, Applebury ML (1979) Isolation and characterization of cGMP phosphodiesterase from bovine rod outer segments. J Biol Chem 254:11669–11677

Beavo JA, Reifsnyder DH (1990) Primary sequence of cyclic nucleotide phosphodiesterase isozymes and the design of selective inhibitors. Trends Pharm Sci 11:150–155

Bennett N (1982) Light-induced interactions between rhodoopsin and the GTP-binding protein: relation with phosphodiesterase activation. Eur J Biochem 123:133–139

Bennett N, Clerc A (1989) Activation of cGMP phosphodiesterase in retinal rods: mechanism of interaction with the GTP-binding protein (transducin). Biochemistry 28:7418–7424

Bowes C, Li T, Danciger M, Baxter LC, Applebury ML, Farber DB (1990) Retinal degeneration in the *rd* mouse is caused by a defect in the β subunit of rod cGMP-phosphodiesterase. Nature 347:677–680

Bownds MD, Arshavsky VY, Calvert PD, Dumke CL (1992) Rod outer segment concentration and structure influence the apparent kinetics of cGMP phosphodiesterase. Invest Ophthal Vis Sci 33:873

Brown RL, Stryer L (1989) Expression in bacteria of functional inhibitory subunit of retinal rod cGMP phosphodiesterase. Proc Natl Acad Sci USA 86:4922–4926

Brown RL (1992) Functional regions of the inhibitory subunit of retinal rod cGMP phosphodiesterase identified by site-specific mutagenesis and fluorescence spectroscopy. Biochemistry (in press)

Catty P, Deterre P (1991) Activation and solubilization of the retinal cGMP-specific phosphodiesterase by limited proteolysis. Role of the C-terminal domain of the β-subunit. Eur J Biochem 199:263–269

Charbonneau H, Prusti RK, LeTrong H, Sonnenburg WK, Mullaney PJ, Walsh KA, Beavo JA (1990) Identification of a noncatalytic cGMP-binding domain conserved in both the cGMP-stimulated and photoreceptor cyclic nucleotide phosphodiesterases. Proc Natl Acad Sci USA 87:288–292

Clerc A, Bennett N (1992) Activated cGMP phosphodiesterase of retinal rods. A complex with transducin α subunit. J Biol Chem 267:6620–6627

Cunnick JM, Hurt D, Oppert B, Sakamoto K, Takemoto DJ (1990) Binding of the gamma-subunit of retinal rod outer segment phosphodiesterase with both transducin and the catalytic subunits of phosphodiesterase. Biochem J 271:721–727

Deterre P, Bigay J, Mylene R, Pfister C, Kühn H, Chabre M (1986) Activation of retinal rod cyclic GMP-phosphodiesterase by transducin: characterization of the complex formed by phosphodiesterase inhibitor and transducin $\alpha$ subunit. Proteins 1:188–193

Deterre P, Bigay J, Forquet F, Robert M, Chabre M (1988) cGMP phosphodiesterase of retinal rods is regulated by two inhibitory subunits. Proc Natl Acad Sci USA 85:2424–2428

Dumler IL, Etingof RN (1976) Protein inhibitor of cyclic adenosine 3′,5′-monophosphate phosphodiesterase in retina. Biochim Biophys Acta 429:474–484

Fung BK-K, Griswold-Prenner I (1989) G protein-effector coupling: binding of rod phosphodiesterase inhibitory subunit to transducin Biochemistry 28:3133–3137

Fung BK-K, Hurley JB, Stryer L (1981) Flow of information in the light-triggered cyclic nucleotide cascade of vision. Proc Natl Acad Sci USA 78:152–156

Fung BK-K, Nash CR (1983) Characterization of transducin from bovine retinal rod outer segments. II. Evidence for distinct binding sites and conformational changes revealed by limited proteolysis with trypsin. J Biol Chem 258:10503–10510

Gillespie PG, Beavo JA (1988) Characterization of a bovine cone photoreceptor phosphodiesterase purified by cyclic GMP-sepharose chromatography. J Biol Chem 263:8133–8141

Gillespie PG, Beavo JA (1989) cGMP is tightly bound to bovine retinal rod phosphodiesterase. Proc Natl Acad Sci USA 83:4311–4315

Gillespie PG, Prusti RK, Apel ED, Beavo JA (1989) A soluble form of bovine rod photoreceptor phosphodiesterase has a novel 15-kDa subunit. J Biol Chem 264:12187–12193

Hamilton SE, Hurley JB (1990) A phosphodiesterase inhibitor specific to a subset of bovine retinal cones. J Biol Chem 265:11259–11264

Hamm HE, Bownds MD (1986) Protein complement of rod outer segments of frog retina. Biochemistry 25:4512–4523

Ho Y-K, Fung BK-K (1984) Characterization to transducin from bovine retinal rod outer segments. The role of sulfhydryl groups. J Biol Chem 259:6694–6699

Hurley JB, Stryer L (1982) Purification and characterization of the $\gamma$ regulatory subunit of the cyclic GMP phosphodiesterase from retinal rod outer segments. J Biol Chem 257:11094–11099

Hurwitz RL, Bunt-Milam AH, Chang ML, Beavo JA (1985) cGMP phosphodiesterase in rod and cone outer segments of the retina. J Biol Chem 260:568–573

Kaupp UB, Koch K-W (1992) Role of cGMP and $Ca^{2+}$ in vertebrate photoreceptor excitation and adaptation. Annu Rev Physiol 54:153–175

Kawamura S, Murakami M (1986) Characterization of the light-induced increased in the Michaelis constant of the cGMP phosphodiesterase in frog rod outer segments. Biochim Biophys Acta 870:256–266

Kühn H (1981) Interactions of rod cell proteins with the disk membrane: influence of light, ionic strength, and nucleotides. Current Topics Memb Transp 15:171–201

Li T, Volpp K, Applebury ML (1990) Bovine cone photoreceptor cGMP phosphodiesterase structure deduced from a cDNA clone. Proc Natl Acad Sci USA 87:293–297

Liebman PA, Parker KR, Dratz EA (1987) The molecular mechanism of visual excitation and its relation to the structure and composition of the rod outer segment. Annu Rev Physiol 49:765–791

Lipkin VM, Dumler IL, Muradov KG, Artemyev NO, Etingof RN (1988) Active sites of the cGMP phosphodiesterase gamma-subunit of retinal rod outer segments. FEBS Lett 234:287–290
Lipkin VM, Khramtsov NV, Vasilevskaya IA, Atabekova NV, Muradov KG, Gubanov VV, Li T, Johnston JP, Volpp KJ, Applebury ML (1990) $\beta$-Subunit of bovine rod photoreceptor phosphodiesterase. J Biol Chem 265:12955–12959
Lolley RN, Lee RH (1990) Cyclic GMP and photoreceptor function. FASEB J 4:3001–3008
Malinski JA, Wensel TG (1992a) Effects of phospholipid composition and vesicle size on membrane binding and G protein-mediated activation of cGMP phosphodiesterase. Biophys J 61:A427
Malinski JA, Wensel TG (1992b) Membrane stimulation of cGMP phosphodiesterase activation by transducin: comparison of phospholipid bilayers to rod outer segment membranes. Biochemistry 31:9502–9512
McNaughton PA (1990) Light response of vertebrate photoreceptors. Phyiol Rev 70:847–843
Miki N, Baraban JM, Keirns JJ, Boyce JJ, Bitensky MW (1975) Purification and properties of the light-activated cyclic nucleotide phosphodiesterase of rod outer segments. J Biol Chem 250:6320–6327
Miller JL, Litman BJ, Dratz EA (1987) Binding and activation of rod outer segment phosphodiesterase and guanosine triphosphate binding protein by disc membranes: influence of reassociation method and divalent cations. Biochim Biophys Acta 898:81–89
Navon SE, Fung BK-K (1987) Characterization of transducin from bovine retinal rod outer segments. Participation of the amino-terminal of $T_\alpha$ in subunit interaction. J Biol Chem 262:15746–15751
Ong OC, Ota IM, Clarke S, Fung BK-K (1989) The membrane binding domain of rod cGMP phosphodiesterase is posttranslationally modified by methyl esterification at a C-terminal cysteine. Proc Natl Acad Sci USA 86:9238–9342
Ovchinnikov YuA, Lipkin VM, Kumarev VP, Gubanov VV, Khramtsov NV, Akmedov NB, Zagranichny VE, Muradov KG (1986) Cyclic GMP phosphodiesterase from cattle retina: amino acid sequence of the gamma-subunit and nucleotide sequence of the corresponding cDNA. FEBS Lett 204:288–292
Ovchinnikov YuA, Gubanov VV, Khramtsov NV, Ischenko KA, Zagranichny VE, Muradov KG, Shuvaeva TM, Lipkin VM (1987) Cyclic GMP phosphodiesterase from bovine retina: amino acid sequence of the alpha-subunit and the corresponding cDNA. FEBS Lett 223:2169–173
Phillips WJ, Trukawinski S, Cerione RA (1989) An antibody-induced enhancement of the transducin-stimulated cyclic GMP phosphodiesterase activity. J Biol Chem 264:16679–16688
Pittler SJ, Baehr W, Wasmuth JJ, Champagne M, vanTuinen P, Ledbetter D, Davis RL (1990) The primary structure of bovine and human rod cGMP phosphodiesterase $\alpha$-subunit and chromosomal localization of the human gene. Genomics 6:272–283
Pittler SJ, Baehr W (1991) Identification of a nonsense mutation in the rod photoreceptor cGMP phosphodiesterase $\beta$-subunit gene of the rd mouse. Proc Natl Acad Sci USA 88:8322–8326
Pugh EN, Lamb TD (1990) Cyclic GMP and calcium: the internal messengers of excitation and adaptation in vertebrate photoreceptors. Vision Res 30:1923–1945
Qin N, Pittler SJ, Baehr W (1992) In vitro isoprenylation and membrane association of mouse rod photoreceptor cGMP phosphodiesterase $\alpha$ and $\beta$ subunits expressed in bacteria. J Biol Chem 267:8458–8463
Robinson PR, Kawamura S, Abramson B, Bownds MD (1980) Control of the cyclic nucleotide phosphodiesterase of frog photoreceptor membranes. J Gen Physiol 76:631–645

Sitaramayya A, Harkness J, Parkes JH, Gonzales-Oliva C, Liebman PA (1986) Kinetic studies suggest that light-activated cyclic GMP phosphodiesterase is a complex with G-protein subunits. Biochemistry 25:651–656

Sitaramayya A, Casadevall C, Bennett N, Hakki SI (1988) Contribution of the guanosinetriphosphatase activity of G-protein to termination of light-activated guanosine cyclic 3′,5′-phosphate hydrolysis in retinal rod outer segments. Biochemistry 27:4880–4887

Stryer L (1991) Visual excitation and recovery. J Biol Chem 266:10711–10714

Swanson RJ, Applebury ML (1983) Methylation of proteins in photoreceptor outer segments. J Biol Chem 258:10599–10605

Tyminski PN, O'Brien DF (1984) Rod outer segment phosphodiesterase binding and activation in reconstituted membranes. Biochemistry 23:3986–3993

Wensel TG, Stryer L (1986) Reciprocal control of retinal rod cyclic GMP phosphodiesterase by its $\gamma$ subunit and transducin. Proteins Struct Funct Genet 1:90–99

Wensel TG, Stryer L (1988) Membrane-bound GTP transducin efficiently activates retinal cGMP phosphodiesterase. In: Chock PB et al. (eds) Enzyme dynamics and regulation. Springer, Berlin Heidelberg New York, pp 102–112

Wensel T, Stryer L (1990) Activation mechanism of retinal rod cGMP phosphodiesterase probed by fluorescein-labeled inhibitory subunit. Biochemistry 29:2155–2161

Whalen MM, Bitensky MW (1989) Comparison of the phosphodiesterase inhibitory subunit interactions of frog and bovine rod outer segments. Biochem J 259:13–19

Whalen MM, Bitensky MW, Takemoto DJ (1990) The effect of the $\gamma$-subunit of the cyclic GMP phosphodiesterase of bovine and frog (Rana catesbiana) retinal rod outer segments on the kinetic parameters of the enzyme. Biochem J 265:655–658

Yamazaki A, Sen I, Bitensky MW, Casnellie JE, Greengard P (1980) Cyclic GMP-specific, high affinity, non-catalytic binding sites on light-activated phosphodiesterase. J Biol Chem 255:11619–11624

Yamazaki A, Stein PJ, Chernoff N, Bitensky MW (1983) Activation mechanism of rod outer segment cyclic GMP phosphodiesterase. J Biol Chem 258:8188–8194

Yamazaki A, Hayashi F, Tatsumi M, Bitensky MW, George JS (1990) Interactions between the subunits of transducin and cyclic GMP phosphodiesterase in Rana catesbiana rod photoreceptors. J Biol Chem 265:11539–11548

Yee R, Liebman PA (1978) Light-activated phosphodiesterase of the rod outer segment. J Biol Chem 253:8902–8909

Zimmerman AL, Yamanaka GY, Eckstein F, Baylor DA, Stryer L (1985) Interaction of hydrolysis-resistant analogs of cyclic GMP with the phosphodiesterase and light-sensitive channel of retinal rod outer segments. Proc Natl Acad Sci USA 82:8813–8817

CHAPTER 57
# High-Voltage Activated $Ca^{2+}$ Channel

F. HOFMANN, M. BIEL, E. BOSSE, R. HULLIN, P. RUTH, A. WELLING, and V. FLOCKERZI

## A. Introduction

Calcium channels are part of the signal system which is vital for intercellular communication in higher multicellular organisms. They transduce electrical or hormonal signals into a chemical second messenger, namely calcium. The cytosolic calcium concentration controls numerous cellular functions by binding to distinct calcium receptor binding proteins such as calmodulin, troponin, or calcium-activated potassium channels. Voltage-dependent calcium channels are of particular interest since their opening or closing determinates the cellular calcium concentration of many cells. In the normal heart they are essential to the generation of normal cardiac rhythm, to impulse propagation through the atrioventricular node, and to contraction in atrial and ventricular muscle. In vascular smooth muscle calcium channels provide part of the calcium that controls smooth muscle contraction and vascular tone. In skeletal muscle they are an essential part of the tubular excitation-contraction coupling mechanism. In neuronal and neuroendocrine cells they are essential for neurotransmitter release (for recent reviews see BERTOLINO and LLINAS 1992; BROWN and BIRNBAUMER 1990; MILLER 1992; RIOS et al. 1992; TRAUTWEIN and HESCHELER 1990).

## B. Identified cDNAs of High-Voltage Activated Calcium Channels

High-voltage activated calcium channels are present in many tissues and are the major pathway for voltage-dependent calcium entry in excitable cells. They are activated at a high membrane potential, inactivate slowly (long lasting) and are readily blocked by different compounds. L-type calcium channels are blocked by the organic calcium channel blockers (CaCB) such as nifedipine and verapamil, N-type by $\omega$-conotoxin, and P-type by the funnel web spider toxin $\omega$-Age IVA (MINTZ et al. 1992). The principal channel-forming subunit of a high-voltage activated calcium channel is the $\alpha_1$ subunit. When purified from rabbit skeletal muscle, this protein (apparent molecular mass 165 kDa) is associated with a 55-kDa protein ($\beta$), a 32-kDa protein ($\gamma$), and a disulfide-linked dimer of 130/28 kDa ($\alpha_2/\delta$) (see HORMANN et al. 1990 and references cited there). The primary structure of

**Table 1.** Classification of cloned and expressed mammalian calcium channel cDNA's

| Gene | Snutch class | Source | Species | Functionally expressed | Sensitive to | Reference |
|---|---|---|---|---|---|---|
| $\alpha_1$ subunit | | | | | | |
| CaCh1 | – | Skeletal muscle | Rabbit | Yes | DHP | TANABE et al. 1987 |
| CaCh2a | C | Heart | Rabbit | Yes | DHP | MIKAMI et al. 1989 |
| | | Brain | Rat | – | | SNUTCH et al. 1991 |
| CaCh2b | C | Lung, smooth muscle | Rabbit | Yes | DHP | BIEL et al. 1990 |
| | | Brain | Rat | – | | SNUTCH et al. 1991 |
| | | Aorta | Rat | Yes | DHP | KOCH et al. 1990 |
| CaCh3 | D | Brain | Human | Yes | DHP, $\omega$-conotoxin | WILLIAMS et al. 1992a |
| | | Brain | Rat | – | | HUI et al. 1991 |
| | | Pancreatic islet | Human | – | | SEINO et al. 1992 |
| CaCh4 | A | Brain | Rabbit | Yes | Spider venom | MORI et al. 1991 |
| | | Brain | Rat | – | | STARR et al. 1991 |
| CaCh5 | B | Brain | Human | Yes | $\omega$-Conotoxin | WILLIAMS et al. 1992b |
| | | Brain | Rat | – | $\omega$-Conotoxin | BUBEL et al. 1992 |
| $\alpha_2/\delta$ subunit | | | | | | |
| $CaA_2$1a | – | Skeletal muscle | Rabbit | Yes | – | ELLIS et al. 1988; MIKAMI et al. 1989 |
| | | Brain | Human | Yes | | WILLIAMS et al. 1992a |
| $CaA_2$1b | – | Brain | Rat | – | | KIM et al. 1992 |
| $\beta$ subunit | | | | | | |
| CaB1 | – | Skeletal muscle | Rabbit | Yes | – | RUTH et al. 1989 |
| | | Brain | Rat | – | | PRAGNELL et al. 1991 |
| | | Brain | Human | Yes | – | WILLIAMS et al. 1992a |
| CaB2* | – | Heart | Rabbit | Yes | – | HULLIN et al. 1992 |
| | | Brain | Rat | Yes | – | PEREZ-REYES et al. 1992 |
| CaB3 | – | Heart | Rabbit | Yes | – | HULLIN et al. 1992 |
| $\gamma$ subunit | | | | | | |
| CaG1 | – | Skeletal muscle | Rabbit | Yes | – | JAY et al. 1990; BOSSE et al. 1990 |

Only full length clones have been included in this table. The nomenclature used for the $\alpha_1$ subunit is adapted from PEREZ-REYES et al. (1990). For the Snutch classes see SNUTCH et al. (1990). The references in the table refer to the first published sequence. In some cases functional expression of the particular clone has been reported in a different publication.

–, not reported; DHP, dihydropyridine; *, at least three different variants (a–c) of the same gene have been identified.

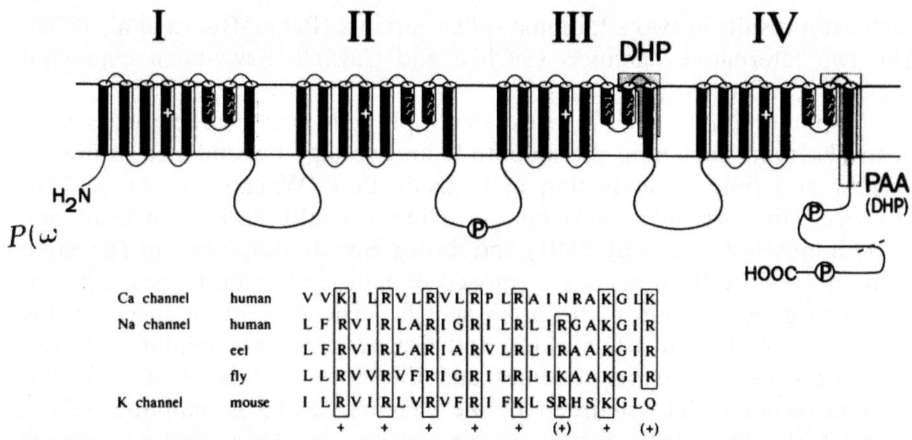


**Fig. 1.** Proposed topography of the $\alpha_1$ subunit of the skeletal muscle calcium channel gene CaCh1. *Shaded areas*, proposed binding site for dihydropyridines (*DHP*) and phenylalkylamines (*PAA*). *P*, the in vitro identified cAMP kinase phosphorylation sites, *dashes*, the proposed truncation of the carboxy terminus; +, the amphipathic helix S4. *Below*, the amphipathic helix S4 sequence is compared with the S4 helices of other voltage-dependent ion channels

these proteins has been deduced by cloning their cDNAs (see Table 1 for references).

## I. The $\alpha_1$ Subunit

Complete cDNA clones of four different $\alpha_1$ subunit gene products (CaCh 1–4) have been sequenced and shown to direct the synthesis of functional calcium channel after expression of their cRNA in *Xenopus oocytes* or cell culture cells (Table 1). The primary sequences of these different gene products are homologous to each other and predict a transmembrane topology which is similar to that of other voltage-dependent ion channels (Fig. 1). The primary sequences of the $\alpha_1$ subunits predict proteins of 212–273 kDa containing four homologous repeats, each of which is composed of five hydrophobic putative transmembrane $\alpha$ helices and one amphipathic segment (S4) (Tanabe et al. 1987; Fig. 1). The "extracellular" loop between the transmembrane helices 5 and 6 (SS1–SS2 region) is predicted to fold into the membrane and to form part of the pore of the channel (Guy and Conti 1990). The skeletal and the cardiac/smooth muscle calcium channels are encoded by two different genes, CaCh1 and CaCh2. Several splice variants of the CaCh2 gene have been identified (Biel et al. 1991). One major difference is the presence of two different exons at the transmembrane region IV S3 which alternate between the cardiac (CaCh2a) and smooth muscle (CaCh2b) isochannels. Polymerase chain reaction (PCR) amplification of the sequences around IV S3 suggests that a deletion within each exon results in two additional splice variants (PEREZ-REYES et al. 1990). The two alternative channels CaCh2a and CaCh2b have been expressed stably in CHO-cells.

No major differences in basic electrophysiological characteristics have been observed, including the amplitude and voltage dependence of inward current and time of activation and inactivation (WELLING et al. 1992b). However, the two splice variants are expressed differentially in heart and smooth muscle (BIEL et al. 1991) and during cardiac development (DIEBOLD et al. 1992). The third gene is expressed in neuroendocrine tissues, whereas the forth gene appears to be brain specific. The currents induced by the expression of the CaCh1, CaCh2, and CaCh3 genes are inhibited by low concentrations of dihydropyridines and therefore are classified as L-type calcium channels. The neuroendocrine channel CaCh3 is inhibited only at rather high concentration by $\omega$-conotoxin and is not an N-type calcium channel. The current through the CaCh4 gene product is not affected by dihydropyridines but is inhibited by low concentrations of a mixture of the funnel web spider toxines and has been classified as a neuronal P-type channel. The $\alpha_1$ subunit of CaCh5 has a high-affinity $\omega$-conotoxin binding site (DUBEL et al. 1992; WILLIAMS et al. 1992b) and is expressed as calcium channel only in the presence of the $\beta$ and $\alpha_2$ subunits. The current is inhibited at picomolar concentrations of $\omega$-conotoxin (WILLIAMS et al. 1992b), identifying the CaCh5 protein as a neuronal N-type channel.

## II. The $\alpha_2/\delta$ Subunit

The deduced amino acid sequence of the $\alpha_2/\delta$ protein is that of a membrane protein of 125 018 Da (ELLIS et al. 1988; Table 1). It contains three putative transmembrane segments and a large extracellular domain with several consensus sequences for glycosylation. The $\delta$ subunit sequence is identical with the carboxy terminal part of the deduced primary structure of the $\alpha_2$ cDNA starting at amino acid 935 of the predicted sequence (DE JONGH et al. 1990). Presumably, the mature $\alpha_2$ and $\delta$ proteins are the product of the same gene and arise by posttranslational processing. The mature $\alpha_2$ protein may be located completely extracellularly, linked by a disulfide bridge to the transmembrane $\delta$ protein (JAY et al. 1991). Immuno- (NORMAN et al. 1987) and northern blots (ELLIS et al. 1990; BIEL et al. 1991) show that the $\alpha_2/\delta$ protein is expressed in skeletal muscle, heart, brain, airway, vascular, and intestinal smooth muscle. Northern blots have identified a predominant 8-kB and a low-abundance 7-kB transkript with a skeletal muscle $\alpha_2$ probe at high and low stringency (BIEL et al. 1991). Recently an $\alpha_2/\delta$ cDNA has been cloned from human brain which is identical to the skeletal muscle $\alpha_2/\delta$ cDNA (WILLIAMS et al. 1992a). A splice variant of this $\alpha_2/\delta$ gene has been cloned from rat brain, which differs in part from the putative $\alpha_2$ proteins but contains an identical $\delta$ protein (KIM et al. 1992). These results suggest that heart, brain, and smooth muscle express a conserved $\alpha_2$ protein whereas brain contains an additional $\alpha_2/\delta$ protein.

### III. The $\beta$ Subunit

The deduced primary sequence of the skeletal muscle $\beta$ subunit is compatible with that of a peripheral membrane protein of 57 868 Da (Table 1; RUTH et al. 1989). It contains four $\alpha$-helical domains, each of which contains a homologous stretch of eight amino acids. Domains II, IIII, and IV contain a heptad repeat structure. Heptad repeats have been found in cytoskeletal proteins. This suggests that the $\beta$ subunit may be a cytoskeletal protein which anchors the $\alpha_1$ subunit to the cytoskeleton. The $\beta$-like proteins which are different from the skeletal muscle $\beta$ subunit exist in heart, aorta, and brain and are derived from two different genes (CaB2 and CaB3; HULLIN et al. 1992). The primary transcript of CaB2 is differentially spliced and leads to the expression of at least three different isoforms (CaB2a, CaB2b, and CaB2c). The overall homology between the növel $\beta$ subunits found in heart, aorta, and brain and the skeletal muscle $\beta$ subunit (CaB1) is 71% for CaB2a, 71.5% for CaB2b, and 66.6% for CaB3. Northern blot and PCR analyses show that CaB1 is present in large amounts in skeletal muscle and brain, CaB2 in heart and aorta, and CaB3 in brain and tissues which are rich in smooth muscle such as aorta, lung, and trachea.

### IV. The $\gamma$ Subunit

The deduced primary sequence of the skeletal muscle $\gamma$ subunit is in agreement with that of an integral membrane protein of 25 058 Da (Table 1; BOSSE et al. 1990; JAY et al. 1990). The deduced sequence contains four putative transmembrane domains and two glycosylation sites which are located extracellularly. A complete cDNA for the $\gamma$ subunit has been detected only in skeletal muscle. Northern and PCR analyses have not indicated that the same mRNA is present in higher concentrations in other tissues.

## C. Structure-Function of the Cloned Calcium Channel Proteins

### I. Expression and Function of the Channel Subunits

The cloned cDNA of the four calcium channel genes has been expressed in different cells (Table 2). The skeletal muscle $\alpha_1$ subunit (CaCh1) has been expressed in L cells (PEREZ-REYES et al. 1989; LACERDA et al. 1991; VARADI et al. 1991) and skeletal muscle myotubes from mice with the muscular dysgenesis mutation (TANABE et al. 1988). Neither cell type has a functional $\alpha_1$ subunit. The mice myotubes contained the $\alpha_2$ and the other subunits may also be present. Expression of the skeletal muscle $\alpha_1$ subunit (CaCh1) in L cells induces a barium current which activates extremely slowly ($\tau_{act.}$ $\approx$665 ms; PEREZ-REYES et al. 1989). Channel activation is accelerated 75-fold

**Table 2.** Functional effects of calcium channel subunits on currents induced by different $\alpha_1$ subunits

| $\alpha_1$ Gene | Subunits expressed | Cell line | DHP sites[a] | $I_{Ba}$[b] | Activation time | Voltage dependence[c] | Reference |
|---|---|---|---|---|---|---|---|
| Heterologous subunits | | | | | | | |
| CaCh2a | $\alpha_2$ | Oocyte | – | (↑) | – | – | MIKAMI et al. 1989 |
| CaCh2a | $\alpha_2$ | Oocyte | – | (↑) | ↓ | (↑) | SINGER et al. 1991 |
| CaCh2a | $\beta$ | Ooctye | – | (↑) | (↓) | (↑) | SINGER et al. 1991 |
| CaCh2a | $\gamma$ | Oocyte | – | (↑) | (↓) | ↑ ↑ | SINGER et al. 1991 |
| CaCh2a | $\alpha_2\beta$ | Oocyte | – | ↑ ↑ | ↓ | ↑ | SINGER et al. 1991 |
| CaCh2a | $\alpha_2\beta\gamma$ | Oocyte | – | ↑ ↑ | ↓ | ↑ ↑ | SINGER et al. 1991 |
| CaCh2a | $\beta$ | Oocyte | – | ↑ | ↓ | ≈ | WEI et al. 1991 |
| CaCh2a | $\beta\gamma$ | Oocyte | – | ↑ | (↓) | (↑) | WEI et al. 1991 |
| CaCh2b | $\beta$ | CHO | ↑ | ↑ | (↓) | (↑) | WELLING et al. 1992 |
| CaCh2a | $\beta$ | Oocyte | – | (↑) | – | – | ITAGAKI et al. 1992 |
| CaCh3 | $\beta$ | Ooctye | – | ↑ | – | – | WILLIAMS et al. 1992a |
| CaCh4 | $\alpha_2\beta$ | Ooctye | – | ↑ ↑ | – | – | MORI et al. 1991 |
| Homologous subunits | | | | | | | |
| CaCh1 | $\beta$ | L cell | ↑ | ≈ | ↓ | – | LACERDA et al. 1991 |
| CaCh1 | $\alpha_2$ | L cell | (↑) | ≈ | ≈ | – | VARADI et al. 1991 |
| CaCh1 | $\beta$ | L cell | ↑ | ↓ * | ≈ | – | VARADI et al. 1991 |
| CaCh1 | $\gamma$ | L cell | ↓ | ↓ * | ≈ | – | VARADI et al. 1991 |
| CaCh1 | $\beta\gamma$ | L cell | ↓ | ↓ * | (↓) | – | VARADI et al. 1991 |
| CaCh1 | $\alpha_2\beta\gamma$ | L cell | (↑) | (↓) | – | – | VARADI et al. 1991 |
| CaCh2a | $\beta_2$ | Oocyte | – | ↑ | ↓ | – | HULLIN et al. 1992 |
| CaCh2a | $\beta_3$ | Oocyte | – | ↑ | ↓ | – | HULLIN et al. 1992 |
| CaCh2a | $\beta_2$ | Oocyte | – | ↑ | ↓ | (↑) | PEREZ-REYES et al. 1992 |
| CaCh2a | $\beta_2$ | COS | ↑ | – | – | – | PEREZ-REYES et al. 1992 |
| CaCh5 | $\alpha_2\beta$ | HEK293 | ↑ ↑ § | ↑ ↑ | – | – | WILLIAMS et al. 1992b |

All effects are compared with that of cells expressing only the $\alpha_1$ subunit.
[a] The number of dihydropyridine binding sites per mg protein.
[b] Barium inward current.
[c] A shift in voltage dependence of the I/V curve or steady state activation or inactivation to more negative values.
$\alpha_2$, $\beta$ and $\gamma$ are identical with CaA1a, CaB1 and CaG1. –, not reported; ≈, similar to cells expressing $\alpha_1$ alone; *, $I_{Ba}$ not sensitive to BayK 8644. (↑), ↑, and ↑ ↑ small, moderate and large increase or shift; (↓) and ↓, small and moderate decrease; §, $\omega$-conotoxin binding sites

($\tau_{act.}$ ≈8 ms) by the coexpression of the skeletal muscle $\beta$ subunit (CaB1; LACERDA et al. 1991). Expression of the skeletal muscle $\alpha_1$ subunit in the dysgenic myotubes generates cells with a slowly activating calcium current and normal skeletal muscle excitation-contraction coupling, which does not depend on the influx of calcium (TANABE et al. 1988). Expression of the cardiac muscle $\alpha_1$ subunit (CaCh2a) produces myotubes with $Ca^{2+}$ currents and excitation contraction coupling as in cardiac muscle (TANABE et al. 1990). TANABE et al. (1990) constructed several chimeras by starting with the cardiac muscle $\alpha_1$ subunit and introducing skeletal musclelike intracellular loops. Changing the large intracellular loop that connects repeats II and III switched the mode of excitation-contraction coupling to that characteristic of skeletal muscle. Interestingly, however, the $Ca^{2+}$ current produced by this chimera remained characteristic of cardiac muscle, i.e., rapidly activating. Chimeras in which the four homologous repeats of the cardiac muscle protein were each switched to the equivalent skeletal muscle sequence showed that changing merely the first homologous repeat switched the characteristics of the $Ca^{2+}$ current from fast activating (cardiac type) to slowly activating (skeletal muscle type) whereas switching the other three repeats did not have this effect (TANABE et al. 1991).

Stable expression of the $\alpha_1$ subunits from smooth muscle (CaCh2b) in CHO cells induces dihydropyridine-sensitive barium currents, which have the physiological characteristics as a smooth muscle calcium channel (BOSSE et al. 1992). The single-channel conductance is 26 pSi in the presence of 80 m*M* $Ba^{2+}$. The channel has the same voltage dependence of activation and inactivation as reported for the naturally occurring smooth muscle calcium channel. The cardiac $\alpha_1$ subunit (CaCh2a) cDNA directs the expression of a channel with electrophysiological properties which are indistinguishable from those of the smooth muscle $\alpha_1$ subunit (WELLING et al. 1992b). Stable coexpression of the CaCh2b protein with the skeletal muscle $\beta$ gene (CaB1) increases in parallel the number of dihydropyridine binding sites and the amplitude of whole cell barium current, suggesting that the amplitude of inward current is directly related to the number of expressed $\alpha_1$ subunits of the protein (WELLING et al. 1993). In addition, the coexpression of the $\beta$ subunit decreases the activation time of the channel by a factor of two and shifts the voltage dependence of steady state inactivation by 18 mV to −13 mV (WELLING et al. 1993). Coexpression of the $\beta$ subunit does not influence the sensitivity of the expressed channel toward the dihydropyridine agonist Bay K 8644. Similar results were obtained by coexpression of the cardiac (CaCh2a), smooth muscle (CaCh2b), neuroendocrine (CaCh3), and neuronal (CaCh4) $\alpha_1$ subunit with the skeletal muscle $\beta$ subunit (CaB1) in *Xenopus* oocytes (Table 2). In each case the current density increased with coexpression of the $\beta$ subunit (CaB1). Expression of the N type $\alpha_1$ subunit CaCh5 in HEK239 cells requires the presence of a neuronal $\beta$ (CaB1) and $\alpha_2$ subunit to induce $\omega$-conotoxin binding sites and calcium current (WILLIAMS et al. 1992b).

The coexpression of the $\alpha_1$ (CaCH2a) and $\beta$ (CaB1–3) subunit together with the $\alpha_2$ subunit cRNA enhanced also the barium current in *Xenopus* oocytes. In *Xenopus* oocytes the $\alpha_2$ subunit (SINGER et al. 1991) and the $\beta$ subunit (WEI et al. 1991) decreased the activation time of the channel (CaCh2a). Identical results were observed when the cardiac $\alpha_1$ subunit (CaCh2a) was coexpressed with the cardiac (CaB2) or the smooth muscle/neuronal (CaB3) $\beta$ subunit and the $\alpha_2$ subunit (HULLIN et al. 1992) in *Xenopus* oocytes. The skeletal muscle $\gamma$ subunit (CaG1) shifted the voltage dependence of steady-state inactivation of the cardiac $\alpha_1$ subunit (CaCh2a) by 40 mV to a negative membrane potential as is observed in skeletal muscle. These results suggest that (a) the skeletal muscle $\beta$ subunit interacts with different $\alpha_1$ subunits, (b) the $\beta$ subunits increase barium currents by increasing the number of functional calcium channel proteins, and (c) the $\beta$ subunits affect the activation time of the channel and the voltage dependence of steady-state inactivation. These conclusions are not supported by the experiments of VARADI et al. (1991), who reported that homologous coexpression of skeletal muscle $\alpha_1$ and $\beta$, $\alpha_1$ and $\gamma$, $\alpha_1$, $\beta$ and $\gamma$, $\alpha_1$, $\alpha_2$, $\beta$ and $\gamma$ in L cells decreases the inward current and the stimulatory effect of the calcium channel agonist Bay K 8644. The latter results are difficult to reconcile with those from other laboratories. They could be caused by a nonstoichiometric expression of the channel subunits, i.e., a higher expression of the $\beta$ subunit than the $\alpha_1$ subunit (LORY et al. 1992).

## II. The Binding Sites for Calcium Channel Blockers

Photoaffinity labeling of the skeletal muscle $\alpha_1$ subunit and expression of CaCh1 and CaCH2b gene in L cells (KIM et al. 1990) or CHO cells (BOSSE et al. 1992) shows that the $\alpha_1$ subunit itself contains the binding sites for the known organic calcium channel blockers, the dihydropyridines, phenylalkylamines, and benzothiazepines. Binding of these drugs requires the binding of calcium to a high-affinity binding site (SCHNEIDER et al. 1991; STAUDINGER et al. 1991). The allosteric modulation of the dihydropyridine binding site by phenylalkylamine and benzothiazepine is preserved within each $\alpha_1$ subunit (KIM et al. 1990; BOSSE et al. 1992). The current induced in cell culture cells or *Xenopus* oocytes by the CaCh1, CaCh2a, and CaCh2b proteins is increased by Bay K 8644, a calcium channel agonist, and is inhibited by the known calcium channel blockers.

Photoaffinity labeling of the purified skeletal muscle $\alpha_1$ subunit by dihydrophyridines and phenylalkylamines suggests that the dihydropyridines bind to the SS1–SS2 region of repeat III (STRIESSNIG et al. 1991; NAKAYAMA et al. 1991) and apparently to a sequence following IVS6 (REGULLA et al. 1991) (Fig. 1). The extracellular location of the binding site at the SS1–SS2 region of repeat III is supported by the finding that dihydropyridines block the calcium channel from the extracellular space (KASS et al. 1991). The phenylalkylamines label a second putative intracellular site located directly after the IVS6 (Striessnig et al. 1990).

## III. Phosphorylation of the Channel Proteins

The L-type current of cardiac, smooth and skeletal muscle, neuroendocrine, and neuronal calcium channels is modulated by hormones through the $\alpha$ subunits of different G-proteins (Brown and Birnbaumer 1990). The open probability of the cardiac and skeletal muscle and of some neuroendocrine cells is increased by cAMP-dependent phosphorylation, suggesting that phosphorylation of the $\alpha_1$ subunit or a different subunit of the calcium channel is important for its hormonal control. In skeletal muscle, about 90% of the full-length $\alpha_1$ subunit (CaCh1) is apparently processed to a smaller protein with the carboxy terminus being close to amino acid residue 1690 (De Jongh et al. 1991). cAMP-kinase phosphorylates in vitro rapidly Ser-687 (Röhrkasten et al. 1988), which is located at the cytosolic loop between repeat II and III, and Ser-1854 (Rotman et al. 1992), which is present only in the full-length skeletal muscle $\alpha_1$ subunit, and slowly Ser-1617 (Röhrkasten et al. 1988). Phosphorylation of these sites may be significant, since the open probability of the reconstituted skeletal muscle CaCB-receptor/calcium channel is increased several-fold by cAMP-dependent phosphorylation (Flockerzi et al. 1986; Hymel et al. 1988; Nunoki et al. 1989; Mundina-Weilenmann et al. 1991). The $\alpha_1$ subunit is phosphorylated also in vivo at least at two sites in response to isoproterenol in isolated rat myocytes (Lai et al. 1990; Mundina-Weilenmann et al. 1991).

These in vivo phosphorylation sites may be identical with Ser-687 and Ser-1854. However, it is not clear which of these phosphorylation sites – one of which is present only in the unprocessed $\alpha_1$ subunit – affect the open probability of the skeletal muscle calcium channel. The in vitro identified phosphorylation sites of the CaCh1 gene are not conserved in the sequences of the other calcium channel genes and therefore are not important for the hormonal regulation of the calcium channel in heart and neuroendocrine cells. The hormonal control of the calcium channels may be exerted by tissue specific $\beta$ subunits. The deduced amino acid sequence of the skeletal $\beta$ subunit (CaB1) contains several potential phosphorylation sites. Two of these sites, Ser-182 and Thr-205, are phosphorylated in vitro by cAMP kinase (Ruth et al. 1988; De Jongh et al. 1989). The products of the CaB2 gene contain a cAMP-kinase phosphorylation site equivalent to Thr-205 of CaB1. This phosphorylation site is not present in the product of CaB3, which is expressed mainly in brain and smooth muscle (Hullin et al. 1992). This is interesting since in vivo whole cell calcium current is increased in heart (Kameyama et al. 1986) and skeletal muscle (Garcia et al. 1990) but not smooth muscle (Welling et al. 1992a) by cAMP-dependent phosphorylation. Expression of the cardiac $\alpha_1$ subunit with the $\alpha_2$ and $\beta$ subunit in *Xenopus* oocytes indicates that cAMP-dependent regulation of the cardiac calcium channel is mediated by phosphorylation of the $\beta$ subunit (KLÖCKNER et al. 1992; DASCAL et al. unpublished observation).

# D. Conclusion

High-voltage activated calcium channels are encoded by different genes. Their electrophysiological and hormonal regulation may depend on the coexpression of different subunits. The interaction of these channel subunits with additional proteins such as the $\alpha$ subunit of trimeric G-proteins may be required for basic and hormonal regulation of the channel (HAMILTON et al. 1991; CAVALIE et al. 1991; KLEUSS et al. 1991). The availability of the cloned cDNA of several channel proteins and channel regulators will facilitate understanding of the complexities of voltage-gated calcium channels.

*Acknowledgments*. The experimental work of the authors was supported by grants from DFG, Fond der chemischen Industrie, und Thyssen Stiftung. We thank Mrs B. Schatz for typing part of the manuscript.

## References

Bertolino M, Llinás R (1992) The central role of voltage-activated and receptor operated calcium channels in neuronal cells. Annu Rev Pharmacol Toxicol 32:399–421

Biel M, Hullin R, Freundner S, Singer D, Dascal N, Flockerzi V, Hofmann F (1991) Tissue-specific expression of high-voltage-activated dihydropyridine-sensitive L-type calcium channels. Eur J Biochem 200:81–88

Biel M, Ruth P, Bosse E, Hullin R, Stühmer W, Flockerzi V, Hofmann F (1990) Primary structure and functional expression of a high voltage activated calcium channel from rabbit lung. FEBS Lett 269:409–412

Bosse E, Bottlender R, Kleppisch T, Hescheler J, Welling A, Hofmann F, Flockerzi V (1992) Stable and functional expression of the calcium channel $\alpha_1$ subunit from smooth muscle in somatic cell lines. EMBO J 11 2033–2038

Bosse E, Regulla S, Biel M, Ruth P, Meyer HE, Flockerzi V, Hofmann F (1990) The cDNA and deduced amino acid sequence of the $\gamma$ subunit of the L-type calcium channel from rabbit skeletal muscle. FEBS Lett 267:153–156

Brown AM, Birnbaumer L (1990) Ionic channels and their regulation by G protein subunits. Annu Rev Physiol 52:197–213

Cavalié A, Allen TJA, Trautwein W (1991) Role of the GTP-binding protein Gs in the $\beta$-adrenergic modulation of cardiac Ca channels. Pflügers Arch 419:433–443

De Jongh KS, Merrick DK, Catterall WA (1989) Subunits of purified calcium channels: a 212-kDa form of $\alpha_1$ and partial amino acid sequence of a phosphorylation site of an independent $\beta$ subunit. Proc Natl Acad Sci USA 86:8585–8589

De Jongh KS, Warner C, Catterall WA (1990) Subunits of purified calcium channels; $\alpha_2$ and $\delta$ are encoded by the same gene. J Biol Chem 265:14738–14741

De Jongh KS, Warner C, Colvin AA, Catterall WA (1991) Characterization of the two size forms of the $\alpha_1$ subunit of skeletal muscle L-type calcium channels. Proc Natl Acad Sci USA 88:10778–10782

Diebold RJ, Koch WJ, Ellinor PT, Wang J-J, Muthuchamy M, Wieczorek DF, Schwartz A (1992) Mutually exclusive exon splicing of the cardiac calcium channel $\alpha_1$ subunit gene generates developmentally regulated isoforms in the rat heart. Proc Natl Acad Sci USA 89:1497–1501

Dubel SJ, Starr TVB, Hell J, Ahlijanian MA, Enyeart JJ, Catterall WA, Snutch TP (1992) Molecular cloning of the $\alpha$-1 subunit of an $\omega$-conotoxin-sensitive calcium channel. Proc Natl Acad Sci USA 89:5058–5062

Ellis SB, Williams ME, Ways NR, Brenner R, Sharp AH, Leung AT, Campbell KP, McKenna E, Koch WJ, Hui A, Schwartz A, Harpold MM (1988) Sequence and expression of mRNAs encoding the $\alpha_1$ and $\alpha_2$ subunits of a DHP-sensitive calcium channel. Science 241:1661–1664

Flockerzi V, Oeken HJ, Hofmann F, Pelzer D, Cavalié A, Trautwein W (1986) Purified dihydropyridine-binding site from skeletal muscle t-tubules is a functional calcium channel. Nature 323:66–68

Garcia J, Gamboa-Aldeco R, Stefani E (1990) Charge movement and calcium currents in skeletal muscle fibers are enhanced by GTP$\gamma$S. Pflügers Arch 417:114–116

Guy HR, Conti F (1990) Pursuing the structure and function of voltage-gated channels. TiNS 13:201–206

Hamilton S, Codina J, Hawkes MJ, Yatani A, Sawada T, Strickland FM, Froehner SC, Spiegel AM, Toro L, Stefani E, Birnbaumer L, Brown AM (1991) Evidence for direct interaction of $G_s\alpha$ with the $Ca^{2+}$ channel of skeletal muscle. J Biol Chem 266:19528–19535

Hofmann F, Flockerzi V, Nastainczyk W, Ruth P, Schneider T (1990) The molecular structure and regulation of muscular calcium channels. Curr Top Cell Regulation 31:223–239

Hui A, Ellinor PT, Krizanova O, Wang J-J, Diebold RJ, Schwartz A (1991) Molecular cloning of multiple subtypes of a novel rat brain isoform of the $\alpha_1$ subunit of the voltage-dependent calcium channel. Neuron 7:35–44

Hullin R, Singer-Lahat D, Freichel M, Biel M, Dascal N, Hofmann F, Flockerzi V (1992) Calcium channel $\beta$ subunit heterogeneity: functional expression of cloned cDNA from heart, aorta and brain. EMBO J 11:885–890

Hymel L, Striessnig J, Glossmann H, Schindler H (1988) Purified skeletal muscle 1,4-dihydropyridine receptor forms phosphorylation-dependent oligomeric calcium channels in planar bilayers. Proc Natl Acad Sci USA 85:4290–4294

Itagaki K, Koch WJ, Bodi I, Klöckner U, Slish DF, Schwartz A (1992) Native-type DHP-sensitive calcium channel currents are produced by cloned rat aortic smooth muscle and cardiac $\alpha_1$ subunits expressed in Xenopus laevis oocytes and are regulated by $\alpha_2$- and $\beta$-subunits. FEBS Lett 297:221–225

Jay SD, Ellis SB, McCue AF, Williams ME, Vedvick TS, Harpold MM, Campbell K (1990) Primary structure of the $\gamma$ subunit of the DHP-sensitive calcium channel from skeletal muscle. Science 248:490–492

Jay SD, Sharp AH, Kahl StD, Vedvick TS, Harpold MM, Campbell KP (1991) Structural characterization of the dihydropyridine-sensitive calcium channel $\alpha_2$-subunit and the associated $\delta$ peptides. J Biol Chem 266:3287–3293

Kameyama M, Hescheler J, Hofmann F, Trautwein W (1986) Modulation of Ca current during the phosphorylation cycle in the guinea pig heart. Pflügers Arch 407:123–128

Kass RS, Arena JP, Chin S (1991) Block of L-type calcium channels by charged dihydropyridines. J Gen Physiol 98:63–75

Kim HL, Kim H, Lee P, King RG, Chin HR (1992) Rat brain expresses an alternatively spliced form of the dihydropyridine-sensitive L-type calcium channel alpha-2 subunit. Proc Natl Acad Sci USA 89:3251–3255

Kim HS, Wei X, Ruth P, Perez-Reyes E, Flockerzi V, Hofmann F, Birnbaumer L (1990) Studies on the structural requirements for the activity of the skeletal muscle dihydropyridine receptor/slow $Ca^{2+}$ channel. J Biol Chem 265:11858–11863

Kleuss C, Hescheler J, Ewel C, Rosenthal W, Schultz G, Wittig B (1991) Assignment of G-protein subtypes to specific receptors inducing inhibition of calcium currents. Nature 353:43–49

Klöckner U, Itagaki K, Bodi I, Schwartz A (1992) β-Subunit expression is required for cAMP-dependent increase of cloned cardiac and vascular calcium channel currents. Pflügers Arch 420:413–415

Koch WJ, Ellinor PT, Schwartz A (1990) cDNA cloning of a dihydropyridine-sensitive calcium channel from rat aorta. J Biol Chem 265:17786–17791

Lacerda AE, Kim HS, Ruth P, Perez-Reyes E, Flockerzi V, Hofmann F, Birnbaumer L, Brown AM (1991) Normalization of current kinetics by interaction between the $\alpha_1$ and $\beta$ subunits of the skeletal muscle dihydropyridine-sensitive $Ca^{2+}$ channel. Nature 352:527–530

Lai Y, Seagar MJ, Takahashi M, Catterall W (1990) Cyclic AMP-dependent phosphorylation of two size forms of $\alpha_1$ subunits of L-type calium channels in rat skeletal muscle cells. J Biol Chem 34:20839–20848

Lory P, Varadi G, Schultz D, Schwartz A (1992) Subunit composition regulates the skeletal L-type Ca Channel. FASEB J 6:A406

Mikami A, Imoto K, Tanabe T, Niidome T, Mori Y, Takeshima H, Narumiya S, Numa S (1989) Primary structure and functional expression of the cardiac dihydropyridine-sensitive calcium channel. Nature 340:230–233

Miller RJ (1992) Voltage-sensitive $Ca^{2+}$ channels. J Biol Chem 267:1403–1406

Mintz IM, Venema VJ, Swiderek KM, Lee TD, Bean BP, Adams ME (1992) P-type calcium channels blocked by the spider toxin ω-Aga-IVA. Nature 355:827–830

Mori Y, Friedrich T, Kim M-S, Mikami A, Nakai J, Ruth P, Bosse E, Hofmann F, Flockerzi V, Furuichi T, Mikoshiba K, Imoto K, Tanabe T, Numa S (1991) Primary structure and functional expression from complementary DNA of a brain calcium channel. Nature 350:398–402

Mundina-Weilenmann C, Chang CF, Gutierrez LM, Hosey MM (1991) Demonstration of the phosphorylation of dihydropyridine-sensitive calcium channels in chick skeletal muscle and the resultant activation of the channels after reconstitution. J Biol Chem 266:4067–4073

Nakayama H, Taki M, Striessnig J, Glossmann H, Catterall WA, Kanaoka Y (1991) Identification of 1,4-dihydropyridine binding regions within the $\alpha_1$ subunit of skeletal muscle $Ca^{2+}$ channels by photoaffinity labeling with diazipine. Proc Natl Acad Sci USA 88:9203–9207

Norman RI, Burgess AJ, Allen E, Harrison TM (1987) Monoclonal antibodies against the 1,4-dihydropyridine receptor associated with voltage-sensitive $Ca^{2+}$ channels detect similar polypeptides from a variety of tissues and species. FEBS Letters 212:127–132

Nunoki K, Florio V, Catterall WA (1989) Activation of purified calcium channels by stoichiometric protein phosphorylation. Proc Natl Acad Sci 86:6816–6820

Perez-Reyes E, Castellano A, Kim HS, Bertrand P, Baggstrom E, Lacerda A, Wei X, Birnbaumer L (1992) Cloning and expression of cardiac/brain β subunit of the L-type calcium channel. J Biol Chem 267:1792–1797

Perez-Reyes E, Kim HS, Lacerda AE, Horne W, Wei X, Rampe D, Campbell KP, Brown AM, Birnbaumer L (1989) Induction of calcium currents by the expression of the $\alpha_1$ subunit of the dihydropyridine receptor from skeletal muscle. Nature 340:233–236

Perez-Reyes E, Wei X, Castellano A, Birnbaumer L (1990) Molecular diversity of L-type calcium channel. Evidence for alternative splicing of the transcripts of three non-allelic genes. J Biol Chem 265:20430–20436

Pragnell M, Sakamoto J, Jay SD, Campbell KP (1991) Cloning and tissue-specific expression of the brain calcium channel β-subunit. FEBS Lett 291:253–258

Regulla S, Schneider T, Nastainczyk W, Meyer HE, Hofmann F (1991) Identification of the site of interaction of the dihydropyridine channel blockers nitrendipine and azidopine with the calcium-channel $\alpha_1$ subunit. EMBO J 10:45–49

Rios E, Pizarro G, Stefani E (1992) Charge movement and the nature of signal transduction in skeletal muscle excitation-contraction coupling. Annu Rev Physiol 54:251–275

Röhrkasten A, Meyer HE, Nastainczyk W, Sieber M, Hofmann F (1988) cAMP-dependent protein kinase rapidly phosphorylates serine-687 of the skeletal muscle receptor for calcium channel blockers. J Biol Chem 263:15325–15329
Rotman EI, Florio V, Lai Y, De Jongh D, Catterall WA (1992) Specific phosphorylation of a C-terminal site on the 212kDa form of the $\alpha_1$ subunit of the skeletal muscle calcium channel by cAMP-dependent protein kinase. FASEB J 6:A246
Ruth P, Röhrkasten A, Biel M, Bosse E, Regulla S, Meyer HE, Flockerzi V, Hofmann F (1989) Primary structure of the $\beta$ subunit of the DHP-sensitive calcium channel from skeletal muscle. Science 245:1115–1118
Schneider T, Regulla S, Hofmann F (1991) The devapamil-binding site of the purified skeletal muscle receptor for organic-calcium channel blockers is modulated by micromolar and millimolar concentrations of $Ca^{2+}$. Eur J Biochem 200:245–253
Seino S, Chen L, Seino M, Blondel O, Takeda J, Johnson JH, Bell GI (1992) Cloning of the $\alpha_1$ subunit of a voltage-dependent calcium channel expressed in pancreatic $\beta$ cells. Proc Natl Acad Sci USA 89:584–588
Singer D, Biel M, Lotan I, Flockerzi V, Hofmann F, Dascal N (1991) The roles of the subunits in the function of the calcium channel. Science 253:1553–1557
Snutch TP, Leonard JP, Gilbert MM, Lester HA, Davidson N (1990) Rat brain expresses a heterogeneous family of calcium channels. Proc Natl Acad Sci USA 87:3391–3395
Snutch TP, Tomlinson WJ, Leonard JP, Gilbert MM (1991) Distinct calcium channels are generated by alternative splicing and are differentially expressed in the mammalian CNS. Neuron 7:45–57
Starr TVB, Prystay W, Snutch TP (1991) Primary structure of a calcium channel that is highly expressed in the rat cerebellum. Proc Natl Acad Sci USA 88:5621–5625
Staudinger R, Knaus H-G, Glossmann H (1991) Positive heterotopic allosteric regulators of dihydropyridine binding increase the $Ca^{2+}$ affinity of the L-type $Ca^{2+}$ channel. J Biol Chem 266:10787–10795
Striessnig J, Glossmann H, Catterall WA (1990) Identification of a phenylalkylamine binding region within the $\alpha_1$ subunit of skeletal muscle $Ca^{2+}$ channels. Proc Natl Acad Sci USA 87:9108–9112
Striessnig J, Murphy BJ, Catterall WA (1991) Dihydropyridine receptor of L-type $Ca^{2+}$ channels: identification of binding domains for [$^3$H] (+)-PN200-110 and [$^3$H]azidopine within the $\alpha_1$ subunit. Proc Natl Acad Sci USA 88:10769–10773
Tanabe T, Beam KG, Adams BA, Niidome T, Numa S (1990) Regions of the skeletal muscle dihydropyridine receptor critical for excitation-contraction coupling. Nature 346:567–569
Tanabe T, Beam KG, Powell JA, Numa S (1988) Restoration of excitation-contraction coupling and slow calcium current in dysgenic muscle by dihydropyridine receptor complementary DNA. Nature 336:134–139
Tanabe T, Brett AA, Numa S, Beam KG (1991) Repeat I of the dihydropyridine receptor is critical in determining calcium channel activation kinetics. Nature 352:800–803
Tanabe T, Takeshima H, Mikami A, Flockerzi V, Takahashi H, Kangawa K, Kojima M, Matsuo H, Hirose T, Numa S (1987) Primary structure of the receptor for calcium channel blockers from skeletal muscle. Nature 328:313–318
Trautwein W, Hescheler J (1990) Regulation of cardiac L-type calcium current by phosphorylation and G-proteins. Annu Rev Physiol 52:257–274
Varadi G, Lory P, Schultz D, Varadi M, Schwartz A (1991) Acceleration of activation and inactivation by the $\beta$ subunit of the skeletal muscle calcium channel. Nature 352:159–162
Wei X, Perez-Reyes E, Lacerda AE, Schuster G, Brown AM, Birnbaumer L (1991) Heterologous regulation of the cardiac $Ca^{2+}$ channel $\alpha_1$ subunit by skeletal muscle $\beta$ and $\gamma$ subunits. J Biol Chem 266:21943–21947

Welling A, Felbel J, Peper K, Hofmann F (1992a) Hormonal regulation of calcium current in freshly isolated airway smooth muscle cells. Am J Physiol 262:L351–L359

Welling A, Bosse E, Ruth P, Bottlender R, Flockerzi V, Hofmann F (1992b) Expression and regulation of cardiac and smooth muscle calcium channels. J J Pharmacol 58 Suppl-II 1258p–1262p

Welling A, Bosse E, Cavalie A, Bottlender R, Ludwig A, Nastainczyk W, Flockerzi V, Hofmann F (1993) Stable co-expression of Calcium Channel $\alpha_1$, $\beta$ and $\alpha_2$, $\delta$ Subunits in a Somatic Cell Line. J Physiol, in Press

Williams ME, Feldman DH, McCue AF, Brenner R, Velicelebi G, Ellis SB, Harpold MM (1992a) Structure and functional expression of $\alpha_1$, $\alpha_2$, and $\beta$ subunits of a novel human neuronal calcium channel subtype. Neuron 8:71–84

Williams ME, Brust PF, Feldman DH, Patti S, Simerson S, Maroufi A, McCue AF, Velicelebi G, Ellis SB, Harpold MM (1992b) Structure and functional expression of an $\omega$-conotoxin-sensitive human N-type calcium channel. Science 257:389–395

CHAPTER 58

# Phosphólipase C-$\beta$ Isozymes Activated by G$\alpha_q$ Members

D. PARK

A variety of agonists interact with surface receptors on their target cells to stimulate inositol phospholipid-specific phospholipase C (PLC; RANA and HOKIN 1990; MAJERUS 1992). Activated PLC catalyzes the hydrolysis of phosphatidylinositol 4,5-bisphosphate to generate two second messenger molecules, diacylglycerol and inositol 1,4,5-trisphosphate ($IP_3$). Diacylglycerol is the physiological activator of protein kinase C (PKC) and $IP_3$ induces the release of $Ca^{2+}$ from internal stores (RANA and HOKIN 1990).

Several distinct PLC enzymes have been purified from a variety of mammalian tissues (RHEE et al. 1989 and references therein), and a total of 16 amino acid sequences – 14 mammalian enzymes and 2 *Drosophila* enzymes – have been deduced from the nucleotide sequences of their corresponding cDNAs (RHEE and CHOI 1992a and references therein). Comparison of deduced amino acid sequences has indicated that the PLCs can be divided into three types – PLC-$\beta$, PLC-$\gamma$, and PLC-$\delta$ – and that each type contains more than one subtype; subtypes are designated by adding Arabic numerals after the Greek letters, as in PLC-$\beta$1 and PLC-$\beta$2.

Although the overall amino acid sequence similarity between the different PLC types is low, significant similarity is apparent in two regions – one of approximately 170 amino acids and the other of approximately 260 amino acids – which are designated the X and Y regions, respectively. Eighty amino acids are either invariant or conservatively changed in each of the X and Y regions of 14 or more PLC sequences (Fig. 1), whereas only eight amino acids are conserved outside these two regions. All PLCs contain an amino-terminal of about 300 amino acids that precede the X region. Whereas PLC-$\beta$ and PLC-$\delta$ contain short sequences of 50–70 amino acids separating the X and Y regions, PLC-$\gamma$ has a long sequence of some 400 amino acids, which contains the so-called *src* homology (SH2 and SH3) domains – domains first identified as noncatalytic regions common to a variety of *src* family tyrosine kinases (KOCH et al. 1991). Further, whereas the carboxyl-terminal sequence following the Y region is approximately 450 amino acids long in PLC-$\beta$, this region is almost nonexistent in PLC-$\delta$.

Three subtypes of mammalian PLC-$\beta$ are known. In addition to the two previously identified PLC-$\beta$1 (RYU et al. 1986; RYU et al. 1987; KATAN and PARKER 1987; SUH et al. 1988; KATAN et al. 1988) and -$\beta$2 (PARK et al. 1992),

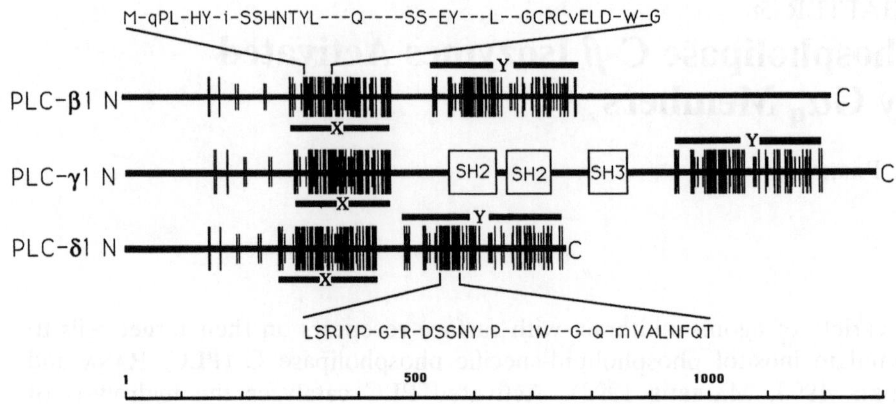


**Fig. 1.** Linear display of three types of PLCs (*β*, *γ*, and *δ* types) represented by PLC-*β*1, PLC-*γ*1, and PLC-*δ*1. Amino acid sequences of six PLC-*β* subtypes (rat *β*1, bovine *β*1, human *β*2, human *β*3, *Drosophila* norpA, and *Drosophila* plc-21), five PLC-*γ* subtypes (rat *γ*1, bovine *γ*1, human *γ*1, rat *γ*2, and human *γ*2), and five PLC-*δ* subtypes (rat *δ*1, human *δ*1, bovine *δ*1, bovine *δ*2, and human *δ*3) were aligned (RHEE and CHOI 1992a and references therein). Amino acids that are invariant or conservatively changed in 14 or more of the total of 16 PLC sequences were identified and are represented in PLC-*β*1, PLC-*γ*1, and PLC-*δ*1 by *long vertical lines* if invariant or by *short vertical lines* if conservatively changed. Conservative amino acids changes are changes that occur within one of the functionally similar groups: hydrophobic residues (I, V, L, and M), aromatic residues (F, W, and Y), acidic residues (E and D), basic residues (K and R), and uncharged polar residues (N and Q). *X*, *Y regions*, two regions with a high density of conserved amino acids. Amino acid sequences of two regions with especially high densities of conserved residues are shown in the standard one-letter code. The letter codes appear in *upper case* for invariant residues and in *lower case* for conservatively changed residues; *dashes* represent nonconserved residues. SH2 and SH3 regions are discussed in the text. Residue number is indicated at the *base*. (Reproduced from RHEE and CHOI 1992a with permission)

**Fig. 2.** Amino acid alignment of the members of PLC-*β* type. Amino acid sequences of five members of PLC-*β* type – bovine *β*1 (KATAN et al. 1988), rat *β*1 (SUH et al. 1988), human *β*2 (PARK et al. 1992), *Drosophila* norpA (BLOOMQUIST et al. 1988), and *Drosophila* plc-21 (SHORTRIDGE et al. 1991) – were compared and amino acid residues conserved in four or more of the five sequences are presented in *bold letter*. *Numbers* (left of each line), the amino-terminal residue numbers of each molecules. The amino acid residues conserved for all five members are presented at the *top of each set of sequences*. *Gaps* (*periods*) are introduced into the sequences to optimize the alignment. The single-letter code of an amino acid appears in *upper case* if the five amino acids in the same position are identical and in *lower case* if they are conservatively changed amino acids, as described in Fig. 1

```
                M          V   L  G  F   W        d           r
bov.β1     1    MAGAQPGVHA LQLKPVCVSD SLKKGTKFVK W......EDD STVVTPIILR
rat β1     1    MAGAQPGVHA LQLKPVCVSD SLKKGTKFVK W......DDD STIVTPIILR
hum.β2     1    M.....SLLN PVLLPPKVKA YLSQGERFIK W......DDE TTVASPVILR
   P21     1    M..MSAGGTY ISTASVEVPQ ALQDGEKFIR W......DDD SGTGTPVTMR
 norpA     1    M.....TKKY EFDWIIPVPP ELTTGCVFDR WFENEKETKE NDFERDALFK

                 D  Gff YW      e e l d     D R  G  AK P    Klr
bov.β1    45    TDPQGFFFYW TDQNKETELL DLSLVKDARC GKHAKAPKDP KLRELLDVGN
rat β1    45    TDPQGFFFYW TDQNKETELL DLSLVKDARC GKHAKAPKDP KLRELLDVGN
hum.β2    40    VDPKGYYLYW TYQSKEMEFL DITSIRDTRF GKFAKMPKSQ KLRDVFNMDF
   P21    43    VDAKGFFLYW VDQNNELDIL DIATIRDVRT GQYAKRPKDN KLRQIVTLG.
 norpA    46    VDEYGFFLYW KSEGRDGDVI ELCQVSDIRA GG...TPNDP KILDKVTKKN

                         r  iTv    D  v                  W     v  l
bov.β1    55    IGR...LEHR MITVVYGPDL VNISHLNLVA FQEEVAKEWT NEVFSLATNL
rat β1    55    IGH...LEQR MITVVYGPDL VNISHLNLVA FQEEVAKEWT NEVFSLATNL
hum.β2    90    PDN..SFLLK TLTVVSGPDM VDLTFHNFVS YKENVGKAWA EDVLALVKHP
   P21    92    PQD..TLEEK TVTVCHGSDF VNMTFVNFCC TRRDIAQLWT DGLIKLAYSL
 norpA    93    GTNIPELDKR SLTICSNTDY INITYHHVIC PDAATAKSWQ KNLRLITHNN

                   N       L K   kL    v   GrIPv K      F   K dRK V
bov.β1   142    LAQNMSRDAF LEKAYTKLKL QVTPEGRIPL KNIYRLFS.. .ADRKRVETA
rat β1   142    LAQNMSRDAF LEKAYTKLKL QVTPEGRIPL KNIYRLFS.. .ADRKRVETA
hum.β2   138    LTANASRSTF LDKILVKLKM QLNSEGKIPV KNFFQMFP.. .ADRKRVEAA
   P21   140    AQLNGSAIMF LQKAHTKLCL QVDKSGRIPV KNIIKLFAQN KEDRKRVEKA
 norpA   143    RATNVCPRVN LMKHWMRLSY CVEKSGKIPV KTLAKTFASG KTE.KLVYTC

                l    LP  k    i    F     y       l  R eid i F       K
bov.β1   189    LEACSLPSSR NDSIPQEDFT PEVYRVFLNN LCPRPEIDNI FSEFGAKSKP
rat β1   189    LEACSLPSSR NDSIPQEDFT PDVYRVFLNN LCPRPEIDNI FSEFGAKSKP
hum.β2   185    LSACHLPKGK NDAINPEDFP EPVYKSFLMS LCPRPEIDEI FTSYHAKAKP
   P21   190    LDVTGLPSGK VDSISVSKFQ FEDFYNLYKY LTQRSEVERL FDSIVGNSKR
 norpA   192    IKDAGLPDDK NATMTKEQFT FDKFYALYHK VCPRNDIEEL FTSI.TKGKQ

                  l        FiN  QRD R lN iLyP           lI   YE
bov.β1   239    .YLTVDQMMD FINLKQRDPR LNEILYPPLK QEQVQVLIEK YEPNNSLAKK
rat β1   239    .YLTVDQMMD FINLKQRDPR LNEILYPPLK QEQVQVLIEK YEPNSSLAKK
hum.β2   235    .YMTKEHLTK FINQKQRDSR LNSLLFPPAR PDQVQGLIDK YEPSGINAQR
   P21   240    KCMSIAQLVE FLNKTQRDPR LNEILYPYAN PARAKELIQQ YEPNKFNAQK
 norpA   241    DFISLEQFIQ FMNDKQRDPR MNEILYPLYE EKRCTEIIND YELDEEKKKN

                 Q S dG    yL   eN  v    kL    d M QPl HYfI NSSHNTYL
bov.β1   288    GQISVDGFMR YLSGEENGVV SPEKLDLNED MSQPLSHYFI NSSHNTYLTA
rat β1   288    GQMSVDGFMR YLSGEENGVV SPEKLDLNED MSQPLSHYFI NSSHNTYLTA
hum.β2   284    GQLSPEGMVW FLCGPENSVL AQDKLLLHHD MTQPLNHYFI NSSHNTYLTA
   P21   290    GQLSLDGFLR YLMGDDNPIM APSKLDLCDD MDQPMSHYFI NSSHNTYLTG
 norpA   291    VQMSLDGFKR YLMSDENAPV FLDRLDFYME MDQPLAHYYI NSSHNTYLSG

                 Q  G SS E mYRQ LL GC R VELD W G r  eEEPvi  HG     eI
bov.β1   338    GQLAGNSSVE MYRQVLLSGC RCVELDCWKG RTAEEEPVIT HGFTMTTEIS
rat β1   338    GQLAGNSSVE MYRQVLLSGC RCVELDCWKG RTAEEEPVIT HGFTMTTEIS
hum.β2   334    GQFSGLSSAE MYRQVLLSGC RCVELDCWKG KPPDEEPIIT HGFTMTTDIF
   P21   340    HQLTGKSSVE IYRQCLLAGC RSVELDFWNG RT..EEPVIV HGYTFVPEIF
 norpA   341    RQIGGKSSVE MYRQTLLAGC RCVELDCWNG KGEDEEPIVT HGHAYCTEIL

                 Ke i AIAe  AF  S yPi iLSFENH     kQQ KmA Y C   FGD LL
bov.β1   388    FKEVIEAIAE CAFKTSPFPI LLSFENHVDS PKQQAKMAEY CRLIFGDALL
rat β1   388    FKEVIEAIAE CAFKTSPFPI LLSFENHVDS PKQQAKMAEY CRLIFGDALL
hum.β2   384    FKEAIEAIAE SAFKTSPYPI ILSFENHVDS PRQQAKMAEY CRTIFGDMLL
   P21   388    AKDVLEAIAE SAFKTSEYPV ILSFENHCN. PRQQAKIANY CREIFGDMLL
 norpA   391    FKDCIQAIAD CAFVSSEYPV ILSFENHCNR A.QQYKLAKY CDDFFGDLLL
```

```
                PL   PL    v LP P    L  KIliKNK k
bov.β1   438 MEPLDKYPLE SGVPLPSPMD LMYKILVKNK KKSH...... ......KSSE
rat β1   438 MEPLEKYPLE SGVPLPSPMD LMYKILVKNK KKSH...... ......KSSE
hum.β2   434 TEPLEKFPLK PGVPLPSPED LRGKILIKNK KNQF...... ......SGPT
   P21   437 DKPLDSHPLE PNMDLPPPAM LRRKIIIKNK KKHHHHHHHH HHKKPAQVGT
 norpA   440 KEPLPDRPLD PGLPLPPPCK LKRKILIKNK RMK....... ..........

                         V             L    e              A
bov.β1   476 GSGKKKLS.. .......... .......EQA SNTYSDSSSV F......EPS
rat β1   476 GSGKKKLS.. .......... .......EQA SNTYSDSSSV F......EPS
hum.β2   472 SSSKDTGG.. .......... .......EAE GSSPPSAPAV WAGEEGTELE
   P21   487 PAANNKLTTA NSVDAKAAQQ VALSASHEDG GVTRSTANGD VATGTGTGSA
 norpA   473 .......PEV EKVELELWLK GELKTDDDPE EDASAGKPPE AAAAPAPAPE

bov.β1   501 SPGAGEADT. .....ESDDD DDDDDCKKSS MDEG...... ..........
rat β1   501 SPGAGEADT. .....ESDDD DDDDDCKKSS MDEG...... ..........
hum.β2   503 EEEVEEEEE. .....EESGN LDEEEIKKMQ SDEG...... ..........
   P21   537 AGTAGHAPPL QQIRQSSKDS TGSSDSDSSS EDESLPNTTP NLPSGNEPPP
 norpA   516 AAAAAEGAA. ........EG GGGAEAEAAA ANYS...... ..........

               A             mS lVNY  QP  F  Fe     kN     m SSF
bov.β1   529 .TAGSEAMAT EEMSNLVNYI QPVKFESFEI SKKRNRSFEM SSFVETKGLE
rat β1   529 .TAGSEAMAT EEMSNLVNYI QPVKFESFET SKKRNKSFEM SSFVETKGLE
hum.β2   531 .TAGLEVTAY EEMSSLVNYI QPTKFVSFEF SAQKNRSYVI SSFTELKAYD
   P21   587 EKAQKETEAG AEISALVNYV QPIHFSSFEN AEKKNRCYEM SSFDEKQATT
 norpA   541 ...GSTTNVH PWLSSMVNYA QPIKFQGFDK AIEKNIAHNM SSFAYLKQSS

                    v FV  YNK Q SRi YP GTR DSS NyMPQlFWNA GCQmV LNFQ
bov.β1   578 QLTKSPVEFV EYNKMQLSRI YPKGTRVDSS NYMPQLFWNA GCQMVALNFQ
rat β1   578 QLTKSPVEFV EYNKMQLSRI YPKGTRVDSS NYMPQLFWNA GCQMVALNFQ
hum.β2   580 LLSKASVQFV DYNKRQMSRI YPKGTRMDSS NYMPQMFWNA GCQMVALNFQ
   P21   637 LLKERPIEFV NYNKHQLSRV YPAGTRFDSS NFMPQLFWNA GCQLVALNFQ
 norpA   588 IDESAGMNFV NYNKRQMSRI YPKGTRADSS NYMPQVFWNA GCQMVSLNFQ

               DL MQ N     fEYN   G Y LK eFMRR  Dk   PF     vD ivA
bov.β1   628 TVDLAMQINM GMYEYNGKSG YRLKPEFMRR PDKHFDPFTE GIVDGIVANT
rat β1   628 TVDLAMQINM GMYEYNGKSG YRLKPEFMRR PDKHFDPFTE GIVDGIVANT
hum.β2   630 TMDLPMQQNM AVFEFNGQSG YLLKHEFMRR PDKQFNPFSV DRIDVVVATT
   P21   687 TLDLAMQLNL GIFEYNARSG YLLKPEFMRR SDRRLDPFEE STVDGIIAGT
 norpA   638 SSDLPMQLNQ GKFEYNGGCG YLLKPDFMRR ADKDFDPFAD EPVDGVIAAQ

              Sv vi GQF L dk   TyV EVdmfGLP D  rk frTk      N vNPv
bov.β1   678 LSVKIISGQF LSDKKVGTYV EVDMFGLPVD TRRKAFKTKT SQG.NAVNPI
rat β1   678 LSVKIISGQF LSDKKVGTYV EVDMFGLPVD TRRKAFKTKT SQG.NAVNPV
hum.β2   680 LSITVISGQF LSERSVRTYV EVELFGLPGD PKRR.YRTKL SPSTNSINPV
   P21   737 VSITVLSGQF LTDKRANTFV EVDMYGLPAD TVRKKFRTKT VRD.NGMNPL
 norpA   688 CSVKVIAGQF LSDKKVGTYV EVDMFGLPSD TVKKEFRTRL VAN.NGLNPV

               EeP VF K vvlP LA lR      EE  K  iG RilPv  l  GY  i L
bov.β1   727 WEEEPIVFKK VVLPSLACLR IAVYEEGGKF IGHRILPVQA IRPGYHYICL
rat β1   727 WEEEPIVFKK VVLPSLACLR IAAYEEGGKF IGHRILPVQA IRPGYHYICL
hum.β2   729 WKEEPFVFEK ILMPELASLR VAVMEEGNKF LGHRIIPINA LNSGYHHLCL
   P21   786 YDEEPFVFKK VVLPELASIR IAAYEEGGKL IGHRVLPVIG LCPGYRHVNL
 norpA   737 YNEDPFVFRK VVLPDLAVLR FGVYEESGKI LGQRILPLDG LQAGYRHVSL

               E   Pl l   lFv i vK  YvP          L  P  y          Q
bov.β1   777 RNERNQPLML PALFVYIEVK DYVPDTYADV IEALSNPIRY VNLMEQRAKQ
rat β1   777 RNERNQPLML PAVFVYIEVK DYVPDTYADV IEALSNPIRY VNLMEQRAKQ
hum.β2   779 HSESNMPLTM PALFIFLEMK DYIPGAWADL TVALANPIKF FS........
   P21   836 RSEVGQPIAL ASLFLCVVVK DYVPDDLSNF AEALANPIKY QSELEKRDIQ
 norpA   787 RTEANFPMSL PMLFVNIELK IYVPDGFEDF MAMLSDPRGF AGAAKQHNEQ
```

**Fig. 2.** (*Continued*)

```
               l L E      e D                  E                   l
bov.β1   827 LAALTLEDEE EVKKEAD... .PGETPSEAP SEARPTPAEN ..GVNHTTSL
rat β1   827 LAALTLEDEE EVKKEAD... .PGETSSEAP SETRTTPAEN ..GVNHTATL
hum.β2   821 .......... .......... .AHDTKSVKL KEAMGGLPEK ..PFPLASPV
   P21   886 LSVLTDEAEA LGSADEDLSK SCGQKKELRP VESLATSPKH RPSISAAAAM
 norpA   837 MKALGIEEQS .......... .......... .......... ..........

                                             l   V
bov.β1   871 TPKPPSQALH SQPAPGSVKA PAKT....ED LIQSVLTEVE ..........
rat β1   871 APKPPSQAPH SQPAPGSVKA PAKT....ED LIQSVLTEVE ..........
hum.β2   848 ASQVNGALAP TSNGSPAARA GARE....EA MKEA..AEPR ..........
   P21   936 SVDVTVDRTD GGRGEDSISI VAPSIQHQHS LDQSVSTSIR QVESSQFDVD
 norpA   847 .......... GGAARDAGKA KEEEKKEPPL VFEPV..... ..........

                   lE l      K           K l  L  kr  k      l
bov.β1   907 ...AQTIEEL KQQKSFVKLQ KKHYKEMKDL VKRHHKKTTD LIKEHTTKYN
rat β1   907 ...AQTIEEL KQQKSFVKLQ KKHYKEMKDL VKRHHKKTTE LIKEHTTKYN
hum.β2   882 ...TASLEEL RELKGVVKLQ RRHEKELREL ERRGARRWEE LLQRGAAQLA
   P21   986 LVLAEPLEKI LDHKSVKEKR LEMEKKLESL RKKHDKEKIK IAGQKSSPLE
 norpA   872 .....TLESL RQEKGFQKVG KKQIKELDTL RKKHAKERTS VQKTQ.....

              i
bov.β1   954 EIQNDYLRRR AALEKTAKKD NKKKSEPSSP DHVSSTIEQD LAALDA..EM
rat β1   954 EIQNDYLRRR AALEKSAKKD SKKKSEPSSP DHGSSAIEQD LAALDA..EM
hum.β2   929 ELGPPGVGGV GACKLGPGKG SRKKRSLPRE ESAGAAPGEG PEGVDG....
   P21  1036 GKKPKFAITN KLVKRLSNKS LEPGVEIPAC PLDGGDSSEE SAAADAGEDL
 norpA   912 .........N AAIDKLIKGK SK........ ....DDIRND ANIKNSINDQ

                   l r              Y                        v  Q
bov.β1  1002 TQKLVDLKDK QQQQLLNLRQ EQYYSEKYQK REHIKLLIQK LTD.VAEECQ
rat β1  1002 TQKLIDLKDK QQQQLLNLRQ EQYYSEKYQK REHIKLLIQK LTD.VAEECQ
hum.β2   975 ..RVRELKDR LELELLRQGE EQYECVLKRK EQHVAEQISK MME.LAREKQ
   P21  1086 AGGSSSLDGR TQESRLRSAC REYTSQYREL QEKYHEAIYS AAEKVLKTSQ
 norpA   941 TKQWTDMIAR HRKEEWDM.. .........L RQHVQD.SQD AMKALMLTVQ

                 K L           el                           R
bov.β1  1051 NNQLKKLKEI CEKEKKELKK KMDKKRQEKI TEAKSKD..K SQMEEE..KT
rat β1  1051 NNQLKKLKEI CEKEKKELKK KMDKKRQEKI TEAKSKD..K SQMEEE..KT
hum.β2  1022 AAELKALKET SENDTKEMKK KLETKRLERI QGMTKVT..T DKMAQERLKR
   P21  1136 TGQTKQLKAS LDKVTGEVMH QLQEARRNEV KNLATVH..R DRDELIRMKR
 norpA   979 AAQIKQLEDR HARDIKDLNA KQAKMSADTA KEVQNDKTLK TKNEKDRRLR

             E      i   v     l      r  e L
bov.β1  1097 EMIRSYIQEV VQYIKRLEEA QSKRQEKLVE KHKEIRQQIL DEKPKLQVEL
rat β1  1097 EMIRSYIQEV VQYIKRLEEA QSKRQEKLVE KHKEIRQQIL DEKPKLQMEL
hum.β2  1070 EINNSHIQEV VQVIKQMTEN LERHQEKLEE KQAACLEQIR EMEKQFQKEA
   P21  1184 EVASSVVERG VAERVRLKQT FDRRTDELQK QHDSVRNALA EHRSKARQIL
 norpA  1029 EKRQNNVKRF MEEKKQIGVK QGRAMEKLKL AHSKQIEEFS TDVQKLMDMY

               E
bov.β1  1147 EQEYQDKFKR LPLEILEFVQ EAMK....GK ISEDSNHSSA PPLMTSDSGK
rat β1  1147 EQEYQDKFKR LPLEILEFVQ EAMK....GK VSEDSNHGSA PPSLASDPAK
hum.β2  1120 LAEYEARMKG LEAEVKESVR ACLRTCFPSE AKDKPERACE CPPELCEQDP
   P21  1234 DKEAESRSCV SSNGFLVLFH GP........ .HHHGCTGSG SSALSGNNLT
 norpA  1079 KIEEEAYKTQ GKTEFCAK 1096

             l                  k
bov.β1  1193 LNQKPPSSEE LEGENPGKEF DTPL 1216
rat β1  1193 VNLKSPSSEE VQGENAGREF DTPL 1216
hum.β2  1170 LIAKADAQES RL 1181
   P21  1275 LNLDAGAAGS HSAISPAKSH NSIAAAAEMK T 1305
```

**Fig. 2.** (*Continued*)

the presence of a third member of mammalian PLC-$\beta$ isozyme, PLC-$\beta$3, was identified both by purification and cloning of partial cDNAs from human fibroblast and rat brain library (CAROZZI et al. 1992; JHON et al. 1993). Two *Drosophila* PLC genes, norpA (BLOOMQUIST et al. 1988) and plc-21 (SHORTRIDGE et al. 1991), are also classified as PLC-$\beta$ type based on their primary structure and high sequence homology with mammalian PLC-$\beta$ isozymes. The amino acid sequences of five PLC-$\beta$ isozymes; bovine $\beta$1, rat $\beta$1, human $\beta$2, *Drosophila* norpA and plc-21 (P21) are aligned in Fig. 2, and their comparisons are schematically represented in Fig. 3. All five enzymes have a long carboxyl-terminal tail region which is a characteristic of the PLC-$\beta$ type. Sequence identities in these carboxyl-terminal regions are considerably lower than those in amino-terminal region, as well as X and Y regions. The carboxyl-terminal regions of all PLC-$\beta$ isozymes contain an unusually high number of charged residues (about 40%) and many of these charged residues are well conserved in all five PLC-$\beta$ isozymes aligned here.

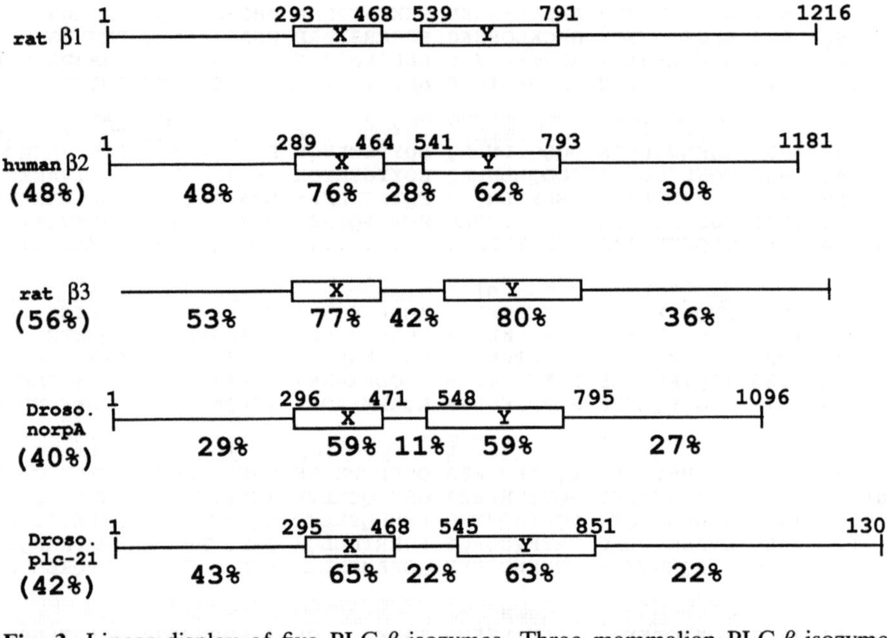


**Fig. 3.** Linear display of five PLC-$\beta$ isozymes. Three mammalian PLC-$\beta$ isozymes (rat $\beta$1, human $\beta$2, and rat $\beta$3) and two *Drosophila* PLCs (norpA, and plc-21) are shown. *Open boxes X, Y* denote the highly conserved regions of PLCs, as described in Fig. 1. The beginning and the end of these regions are indicated *at the top* of the representations by numbers of amino acid residue in each isozyme. The beginning and the end of each molecule are also indicated by numbers of amino acid residue. Since the amino-terminal sequence of rat $\beta$3 is not obtained yet, the total numbers of amino acid residues and the locations of X and Y regions cannot be assigned by the numbers of amino acid residues. *Percentages* (under the names of each isozyme), percentage identity with rat $\beta$1; *percentages* (under the various domains of each isozyme), percentage identity with the corresponding domains in rat $\beta$1

These charged residues might be important in G-protein–PLC interaction. In addition to the PLC-$\beta$ isozymes described above, another 140-kDa PLC-$\beta$1-like protein has been purified from particulate fractions of bovine brain (LEE et al. 1987). Unlike 150-kDa PLC-$\beta$1 which distributed in both cytosol and particulate fractions, the 140-kDa PLC protein was located primarily in particulate fractions. Polyclonal antisera and monoclonal antibodies to five different epitopes of 150-kDa PLC-$\beta$1 cross-reacted with 140-kDa PLC protein. The cDNA for 140-kDa PLC-$\beta$1-like protein has not been isolated yet.

Studies over the past several years have revealed the existence of distinct mechanisms for the activation of different PLC isozymes. PLC-$\gamma$1 and PLC-$\gamma$2 are activated when specific tyrosine residues on PLC-$\gamma$s become phosphorylated. This phosphorylation is known to be mediated by the intrinsic tyrosine kinase activity of the receptors for certain peptide growth factors such as platelet-derived growth factor and epidermae growth factor or by nonreceptor tyrosine kinases coupled to immune system receptors such as membrane IgM, the T cell antigen receptor, the high-affinity IgE receptor, and the IgG receptor (RHEE and CHOI 1992b; RHEE et al. 1992). Activation of PLC-$\beta$s is achieved by a completely different mechanism in which the $\alpha$ subunits of $G_q$ family are involved, as described below. The mechanism of PLC-$\delta$ activation has not yet been elucidated.

In this chapter I summarize our current knowledge on the $G\alpha_q$-activatable PLC-$\beta$ isozyme with emphasis on the primary structure of PLC-$\beta$ subtypes and on the specificity in the interaction between the subtypes of $G\alpha_q$ and PLC-$\beta$. Several lines of evidence have implicated a mandatory role for a G-protein in transduction of the signal from certain receptors to PLC and that there might be two distinct types – pertussis toxin sensitive and pertussis toxin insensitive types – of such G-proteins. However, the nature of these PLC-activating G-proteins remained elusive until recently, when the G protein subfamily $G_q$ was characterized independently in several laboratories.

SIMON and coworkers (1991) obtained and sequenced a number of cDNAs corresponding to previously uncharacterized $\alpha$ subunits of the $G_q$ subfamily. There are four distinct members – $G\alpha_q$, $G\alpha_{11}$, $G\alpha_{14}$, and $G\alpha_{16}$ – in the $G_q$ subfamily. The amino acid sequences of $G\alpha_q$ and $G\alpha_{11}$ are 88% identical, whereas $G\alpha_{14}$ and $G\alpha_{16}$ are more distantly related to $G\alpha_q$, with amino acid identities of 55%–60%. None of the four $G_q$ members contain a site for pertussis toxin modification. Both $G\alpha_q$ and $G\alpha_{11}$ are widely distributed while $G\alpha_{14}$ and $G\alpha_{16}$ are found primarily in cells derived from hematopoietic lineage.

Sternweis and coworkers (SMRCKA et al. 1991) purified a mixture of $G\alpha_q$ and $G\alpha_{11}$ from rat brain with the use of an affinity matrix containing immobilized G-protein $\beta\gamma$ subunits, and they subsequently demonstrated that these brain G-proteins activated partially purified PLC. PLC activation was observed only in the presence of $AlF_4^-$, and not with the nonhydrolyzable GTP analog GTP$\gamma$S; the purified G$\alpha_q$ and G$\alpha_{11}$ contained tightly associated GDP, which exchanges only very slowly with GTP$\gamma$S. Concurrently, Exton and coworkers (TAYLOR et al. 1991) purified a mixture of G$\alpha_q$ and G$\alpha_{11}$, on the basis of its ability to activate partially purified PLC, from bovine liver membranes that had been incubated with GTP$\gamma$S. When reconstituted in the presence of GTP$\gamma$S with isozymes of PLC, the mixture of G$\alpha_q$ and G$\alpha_{11}$ specifically activated PLC-$\beta$1 but not PLC-$\gamma$1 and PLC-$\delta$1 (TAYLOR et al. 1991). Half-maximal activation of PLC-$\beta$1 required 4 $\mu M$ GTP$\gamma$S, suggesting that the affinity of these G-proteins for GTP analogs is low (BLANK et al. 1991). $AlF_4$u2– also activated PLC-$\beta$1 in the presence of the G-proteins. Subsequently, G$\alpha_q$ and G$\alpha_{11}$ were resolved and both were shown to have PLC-stimulatory activity.

Our recent experiments show that the carboxyl-terminal truncated 100-kDa form of PLC-$\beta$1 retained catalytic activity similar to that of untruncated form but unlike 150-kDa PLC-$\beta$1 it cannot be activated by G$\alpha_q$ in reconstitution assay (PARK et al. 1993). This result suggests that the carboxyl-terminal tail portion behind Y domain of the PLC-$\beta$1 contains the domains interacting with G$\alpha_q$.

Avian homologs of mammalian PLC-$\beta$1 and G$\alpha_q$ (or G$\alpha_{11}$), which are associated with $P_{2y}$ purinergic receptor-stimulated phosphatidylinositol metabolism in turkey erythrocytes, have also been purified and reconstituted (WALDO et al. 1991). Although other studies had suggested that G-protein-mediated stimulation of PLC activity results in a decreased $Ca^{2+}$ requirement of the enzyme, the extent of PLC-$\beta$1 activation by G$\alpha_q$ and G$\alpha_{11}$ was shown to be similar over the range of $Ca^{2+}$ concentration examined; activation thus appeared to be achieved mainly by increasing the intrinsic activity of PLS (SMRCKA et al. 1991; TAYLOR et al. 1991; WALDO et al. 1991).

The receptors that activate PLC via G$\alpha_q$ or G$\alpha_{11}$ include those for thromboxane $A_2$, bradykinin, angiotensin, histamine, vasopressin, and acetylcholine (muscarinic) (SHENKER et al. 1991; WANGE et al. 1991; GUTOWSKI et al. 1991). Interaction with the ligand-occupied receptor causes dissociation of the heterotrimeric GDP-bound $G_q$ to yield GTP-bound G$\alpha_q$, which remains in the membrane. PLC-$\beta$1 then binds the GTP-bound G$\alpha_q$, probably via the carboxyl-terminal regions of both proteins, which results in the activation of PLC-$\beta$1.

The specificity of the interaction between different G$\alpha$ subunits and PLC has been further assessed by introducing cDNAs corresponding to various G$\alpha$ subunits into Cos-7 cells and measuring inositol phosphates formed after stimulation with $AlF_4^-$ (WU et al. 1992). Transfection with G$\alpha_q$ or G$\alpha_{11}$ cDNA resulted in a marked increase in inositol phosphate formation. Cotransfection of G$\alpha_q$ (or G$\alpha_{11}$) cDNA and PLC-$\beta$1 cDNA resulted in even higher levels of inositol phosphate formation. The introduction of mutations (Gln-209 → Leu) that constitutively activate G$\alpha_q$ and G$\alpha_{11}$ resulted in persistent activation of PLC and high basal levels of inositol phosphates. On the other hand, transfection with a variety of other G$\alpha$ subunit cDNAs – G$\alpha_{i2}$, G$\alpha_{OA}$, G$\alpha_{OB}$, transducin, and the constitutively activated Gln-205 $\rightarrow$ Leu mutants of G$\alpha_Z$ and G$\alpha_{OA}$ – did not increase inositol phosphate formation. These results are consistent with the conclusion that G$\alpha_q$ and G$\alpha_{11}$ cDNAs encode proteins that specifically activate PLC.

The relative abilities of members of the $G_q$ subfamily to activate PLC-$\beta$ isozymes have also been determined. Whereas G$\alpha_q$ purified from a $\beta\gamma$-affinity gel failed to stimulate PLC-$\beta$2, PLC-$\beta$1 activity was enhanced markedly (Park et al. 1992). Membranes from Cos-7 cells transfected with cDNAs corresponding to G$\alpha_q$, G$\alpha_{11}$, G$\alpha_{14}$, or G$\alpha_{16}$ have also been reconstituted with purified PLC-$\beta$1 or PLC-$\beta$2 in the presence of GTP$\gamma$S (Lee et al. 1992). All four members of the $G_q$ subfamily were found to stimulate PLC-$\beta$1, with G$\alpha_q$ and G$\alpha_{11}$ most efficient. On the other hand, G$\alpha_{16}$ was found to activate most effectively PLC-$\beta$2 while G$\alpha_q$, G$\alpha_{11}$, and G$\alpha_{14}$ showed much less stimulation. Therefore, there appears to be specificity in the interaction of different members of the $G_q$ subfamily with different PLC-$\beta$ effectors. This specificity may be important in generating tissue-specific or receptor-specific responses in vivo.

Mammalian cells have multiple subtypes of PLC-$\beta$ and G$\alpha_q$. These subtypes are expressed differently among various tissues. In addition, there appear, to be certain degree of specificities in the interaction between the subtypes of PLC-$\beta$ and G$\alpha_q$. Considering the number of receptors utilizing the G-protein–PLC pathway, the delineation of corresponding receptor–G-protein–PLC pathway will be one of the main task remaining to be clarified in future. Another important problem that remains unsolved is the identities of pertussis toxin-sensitive PLC-activating G-protein(s) and PLC isozyme(s). An intriguing hypothesis is that activation of PLC by G$\beta\gamma$ subunits may put pertussis toxin-sensitive G protein with unknown function such as $G_O$ in this place.

Evidence suggests that activation of PKC or protein kinase A attenuates receptor-coupled PLC activity in certain types of cell, thus providing a negative feedback signal to limit the magnitude and duration of receptor signaling. PLC-$\beta$1 appears to be a target for such regulation by PKC in certain cells. Prior treatment with phorbol myristate acetate of several cell lines containing PLC-$\beta$1, PLC-$\gamma$1, and PLC-$\delta$1 elicits a large increase in the phosphorylation of serine residues in PLC-$\beta$1, with only a small increase observed for PLC-$\gamma$1 and no effect on PLC-$\delta$1 (Ryu et al. 1990). It was suggested that, rather than a direct effect on enzyme activity, phosphorylation of PLC-$\beta$1 by PKC may alter its interaction with $G_q$.

Most recently, Ross and collegues (Berstein et al. 1992) demonstrated that addition of purified PLC-$\beta$1 to $M_1$ muscarinic receptor and G$\alpha$q/11 coreconstituted lipid vesicle stimulates the receptor-promoted steady-state GTPase activity of G$\alpha$q/11 up to 20-fold. The result suggests that PLC-$\beta$1 is not only the effector molecule for G$\alpha$q/11 but also function as a GTPase activating protein (GAP) for the same G proteins. Together with the phosphodiesterase $\gamma$-subunit, which also functions as a GAP for its own activator G protein, transducin (Arshavsky and Bownds 1992), PLC-$\beta$1 belongs to a novel class of enzymes that regulate the activity of their own activators.

## References

Arshavsky VY, Bownds MD (1992) Regulation of deactivation of photoreceptor G protein by its target enzyme and cGMP. Nature 357:416–417

Berstein G, Blank JL, John DY, Exton JH, Rhee SG, Ross EM (1992) Phospholipase C-$\beta$1 is a GTPase-activating protein for Gq/11, its physiologic regulator. Cell 70:411–418

Blank JL, Ross AH, Exton JH (1991) Purification and characterization of two G-proteins that activate the $\beta$1 isozyme of phosphoinositide-specific phospholipase C. J Biol Chem 266:18206–18216

Bloomquist BT, Shortridge RD, Schneuwly S, Perdew M, Montell C, Steller H, Rubin G, Pak WL (1988) Isolation of a putative phospholipase C gene of Drosophila, norpA, and its role in phototransduction. Cell 54:723–733

Carozzi AJ, Kriz RW, Webster C, Parker PJ (1992) Identification, purification and characterization of a novel phosphatidylinositol-specific phospholipase C, a third member of the $\beta$ subfamily. Eur J Biochem 210:521–529

Gutowski S, Smrcka A, Nowak L, Wu D, Simon M, Sternweis PC (1991) Antibodies to the $\alpha$q subfamily of guanine nucleotide-binding regulatory protein $\alpha$ subunits attenuate activation of phosphatidylinositol 4,5-bisphosphate hydrolysis by hormons. J Biol Chem 266:20519–20524

Jhon DY, Lee HH, Park D, Lee CW, Lee KH, Rhee SG (to be published, 1993) Cloning, sequencing, purification and Gq-dependent activation of phospholipase C-$\beta$3. J Biol Chem

Katan M, Parker PJ (1987) Purification of phosphoinositide-specific phospholipase C from a particulate fraction of bovine brain. Eur J Biochem 168:413–418

Katan M, Kriz RW, Totty M, Philip R, Meldrum E, Aldape RA, Knopf JL, Parker PJ (1988) Determination of the primary structure of PLC-154 demonstrates diversity of phosphoinositide-specific phospholipase C activities. Cell 54:171–177

Koch AC, Anderson DA, Moran MF, Ellis C, Pawson T (1991) SH2 and SH3 domains: elements that control interactions of cytoplasmic signaling proteins. Science 252:668–674

Lee CH, Park D, Wu D, Rhee SG, Simon MI (1992) Members of the Gq $\alpha$ subunit gene family activate phospholipase C $\beta$ isozymes. J Biol Chem 267:16044–16047

Lee KY, Ryu SH, Suh PG, Choi WC, Rhee SG (1987) Phospholipase C associated with particulate fractions of bovine brain. Proc Natl Acad Sci USA 84:5540–5544

Majerus PW (1992) Inositol phosphate biochemistry. Annu Rev Biochem 61:225–250

Park D, Jhon DY, Kriz R, Knopf J, Rhee SG (1992) Cloning, sequencing, expression, and Gq-independent activation of phospholipase C-$\beta$2. J Biol Chem 267:16048–16055

Park D, Jhon DY, Lee CW, Ryu SH, Rhee SG (to be published 1993) Removal of the carboxyl-terminal region of phospholipase C-$\beta$1 by calpain abolishes activation by G$\alpha$q. J Biol Chem

Rana RS, Hokin LE (1990) Role of phosphoinositides in transmembrane signaling. Physiol Rev 70:115–164

Rhee SG, Suh PG, Ryu SH, Lee SY (1989) Studies of inositol phospholipide-specific phospholipase C. Science 244:546–550

Rhee SG, Choi KD (1992a) Multiple forms of phospholipase C isozymes and their activation mechnisms. In: Putney JW Jr (ed) Advances in second messenger and phosphoprotein research. Raven, New York, 26:35–61

Rhee SG, Choi KD (1992b) Regulation of inositol phospholipide-specific phospholipase C isozymes. J Biol Chem 267:12393–12396

Rhee SG, Park DJ, Park D (1992) Regulation of phospholipase C isozymes. In: Cochrane CG, Gimbrone MA (eds) Cellular and molecular mechanisms of inflammation: signal transduction. Academic, New York, 3:57–88

Ryu SH, Cho KY, Lee KY, Suh PG, Rhee SG (1986) Two forms of phosphatidylinositol-specific phospholipase C from bovine brain. Biochem Biophys Res Comm 141:137–144

Ryu SH, Cho KS, Lee KY, Suh PG, Rhee SG (1987) Purification and characterization of two immunologically distinct phosphoinositide-specific phospholipase C from bovine brain. J Biol Chem 262:12511–12518

Ryu SH, Kim UH, Wahl MI, Brown AB, Carpenter G, Huang KP, Rhee SG (1990) Feedback regulation of phospholipase C-$\beta$ by protein kinase C. J Biol Chem 265:17941–17945

Shenker A, Goldsmith P, Unson CG, Spiegel AM (1991) The G protein coupled to the thromboxan A2 receptor in human platelets is a member of the novel Gq family. J Biol Chem 266:9309–9313

Shortridge RD, Yoon J, Lending CR, Bloomquist BT, Perdew MH, Pak WL (1991) A Drosophila phospholipase C gene that is expressed in central nervous system. J Biol Chem 266:12474–12480

Simon MI, Strathmann MP, Gautam N (1991) Diversity of G proteins in signal transduction. Science 252:802–808

Smrcka AV, Hepler JR, Brown KO, Sternweis PC (1991) Regulation of polyphosphoinositide-specific phospholipase C activity by purified Gq. Science 251:804–807

Suh PG, Ryu SH, Moon KH, Suh HW, Rhee SG (1988) Cloning and sequence of multiple forms of phospholipase C. Cell 54:161–169

Taylor SJ, Chae HZ, Rhee SG, Exton JH (1991) Activation of the $\beta$1 isozyme of phospholipase C by $\alpha$ subunits of the Gq class of G proteins. Nature 350:516–518

Waldo GL, Boyer JW, Morris AJ, Harden TK (1991) Purification of an $AlF_4^-$ and G-protein $\beta\gamma$-subunit-regulated phospholipase C-activating protein. J Biol Chem 266:14217–14225

Wange RL, Smrcka AV, Sternweis PC, Exton JH (1991) Photoaffinity labeling of two rat liver plasma membrane proteins with [$^{32}$P]$\gamma$-azidoanilido GTP in responce to vasopressin. Immunologic identification as $\alpha$ subunits of the Gq class of G proteins. J Biol Chem 266:11409–11412

Wu D, Lee CH, Rhee SG, Simon MI (1992) Activation of phospholipase C by the $\alpha$ subunits of the Gq and G11 proteins in transfected Cos-7 cells. J Biol Chem 267:1811–1817

CHAPTER 59
# Stimulation of Phospholipase C by G-Protein $\beta\gamma$ Subunits

P. GIERSCHIK and M. CAMPS

## A. Introduction

The hydrolysis by phospholipase C of phosphatidylinositol 4,5-bisphosphate (PtdIns$P_2$) to inositol 1,4,5-trisphosphate (Ins$P_3$) and diacylglycerol is a key mechanism by which many extracellular signaling molecules (e.g., hormones, growth factors, and neurotransmitters) regulate functions of their target cells (BERRIDGE and IRVINE 1989; NISHIZUKA 1988). There is ample evidence to suggest that many of the receptors interacting with these mediators stimulate phospholipase C via a guanine nucleotide binding protein (G-protein) (COCKCROFT and STUTCHFIELD 1988; DE VIVO and GERSHENGORN 1990). In certain cell types, e.g., neutrophils and cultured granulocytes like HL-60 cells (OHTA et al. 1985; KRAUSE et al. 1985), mast cells (NAKAMURA and UI 1985), or renal mesangial cells (PFEILSCHIFTER and BAUER 1986), the relevant G-protein appears to be a substrate for ADP-ribosylation by pertussis toxin. In other cells, e.g., cultured chick heart cells (MASTERS et al. 1985), 1321N1 astrocytoma cells (HEPLER and HARDEN 1986), $GH_3$ pituitary cells (MARTIN et al. 1986), or hepatocytes (UHING et al. 1986) stimulation of phospholipase C is mediated by G-protein(s) resistant to modification by pertussis toxin. Recently, evidence has been presented suggesting that, in the latter systems, PLC$\beta_1$, but not PLC$\gamma_1$ or PLC$\delta_1$, is activated by members of the newly discovered $\alpha_q$ subfamily of the G-protein $\alpha$ subunits (see Chaps. 58 and 67 by EXTON and PARK, this volume, for review and references). With regard to stimulation of phospholipase C by pertussis-toxin-sensitive G-proteins, however, little information has been available until recently on the nature the G-protein involved.

We have previously reported that HL-60 granulocytes contain a cytosolic phospholipase C, which is stimulated by the poorly hydrolyzable GTP analog GTP[S] (CAMPS et al. 1990). To investigate whether this stimulation proceeded via a soluble $\alpha$ subunit of heterotrimeric G-proteins or a soluble small GTPase, we examined the effect of exogeneous G-protein $\beta\gamma$ subunits on the activity of this phospholipase C. G-protein $\beta\gamma$ subunits have previously been shown to accelerate the deactivation of GTP[S]-activated $\alpha_s$ (NORTHUP et al. 1983), increase the rate of dissociation of GTP[S] from $\alpha_o$ and $\alpha_i$, and increase the affinity of $\alpha_o$ and $\alpha_i$ for GDP (HIGASHIJIMA et al. 1987). G-protein $\beta\gamma$ subunits are therefore expected to alter the rate and/or extent of the activation of free $\alpha$ subunits by GTP[S]. In contrast, there is no evidence available to suggest that $\beta\gamma$ subunits interfere with the activation of small GTPases by GTP[S]. Contrary to all expectations, we found that $\beta\gamma$ subunits markedly and specifically stimulated phospholipase C (CAMPS et al. 1992a).

The purpose of this chapter is to summarize what we currently know about the regulation of phospholipase C by G-protein $\beta\gamma$ subunits and to discuss the role of $\beta\gamma$ subunits in mediating the stimulation of inositol phosphate formation by G-protein-coupled receptors.

## B. Stimulation of Soluble Phospholipase C of HL-60 Granulocytes by G-Protein $\beta\gamma$ Subunits

To avoid potential artifacts due to detergents present in $\beta\gamma$ subunit preparations purified from other sources (YATANI et al. 1990), we initiated our studies using $\beta\gamma$ subunits of retinal transducin ($\beta\gamma_t$) which can be purified and remain soluble in the absence of detergents (FUNG 1983). $\beta_t$ is identical in structure to the 36-kDa $\beta_1$ subunit found in nonretinal cells. $\gamma_t$ is related to the nonretinal $\gamma$ subunits, but structurally distinct (see Chap. 10 ITOH and KAZIRO, vol 108 I).

Much to our surprise, $\beta\gamma_t$ led to a marked stimulation of inositol phosphates by a soluble phospholipase C present in the cytosol of myeloid-differentiated HL-60 cells (Fig. 1). Half-maximal and maximal ($\approx$ sixfold) stimulation were observed at $\approx 0.3\ \mu M$ and $2\ \mu M$ $\beta\gamma_t$, respectively. In additional experiments, we found that Ins$P_3$ was the main hydrolysis product of PtdIns$P_2$ hydrolysis in the absence and in the presence of $\beta\gamma_t$. $\beta\gamma_t$ did not promote the hydrolysis of PtdIns$P_2$ in the absence of HL-60 cytosol, and heat-treated $\beta\gamma_t$ was without effect (CAMPS et al. 1992a). Taken together, these results clearly demonstrated that $\beta\gamma_t$ stimulated a phospholipase C present in the cytosol of HL-60 granulocytes.

G-protein $\beta\gamma$ subunits have previously been suggested to promote formation of arachidonic acid and/or its metabolites by stimulating phospholipase $A_2$ (JELSEMA and AXELROD 1987; KIM et al. 1989). As arachidonic acid has been shown to stimulate inositol phospholipid breakdown (IRVINE et al. 1979; ZEITLER and HANDWERGER 1985), we examined the possibility that the effect of $\beta\gamma_t$ on PtdIns$P_2$ hydrolysis by HL-60 cytosol was secondary to a stimulation of phospholipase $A_2$. Under the conditions used to analyze the regulation of phospholipase C, $\beta\gamma_t$ did not cause formation of arachidonic acid from the phospholipid substrate. These findings demonstrated that stimulation of endogeneous phospholipase $A_2$ was unlikely to be the mechanism by which $\beta\gamma_t$ stimulated phospholipase C. An immunochemical analysis of our $\beta\gamma_t$ preparations using antisera reactive against $\alpha_q$ and $\alpha_{11}$ showed that our $\beta\gamma_t$ preparations either did not contain $\alpha_q$ and/or $\alpha_{11}$ or contained these $\alpha$ subunits at level lower than 0.1%. These

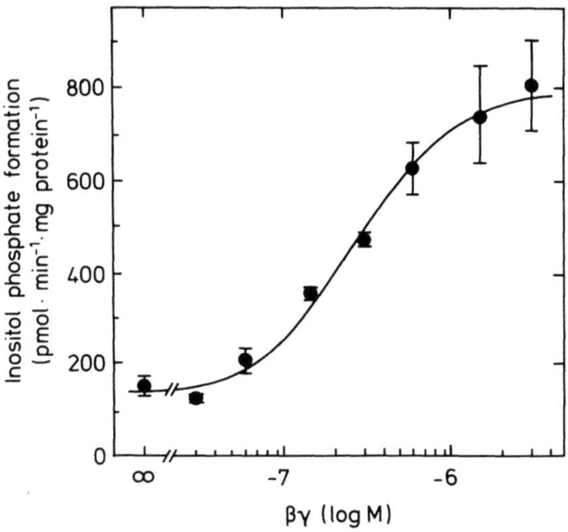


**Fig. 1.** Stimulation of a soluble PLC of HL-60 cells by $\beta\gamma_t$. Cytosol from myeloid differentiated HL-60 cells was incubated in the presence of increasing concentrations of $\beta\gamma_t$ with phospholipid vesicles containing PtdIns$P_2$. The reaction was terminated by addition of chloroform/methanol/HCl and analyzed for inositol phosphates. See Camps et al. (1990, 1992a) for experimental details

results, together with the fact that $\beta\gamma$ stimulation of phospholipase C was observed in the absence of activating guanine nucleotides or $AlF_4^-$ (Fig. 1), argued strongly against activation of phospholipase C by contaminating $\alpha_q$ and/or $\alpha_{11}$.

To investigate whether G-protein dissociation was required for $\beta\gamma$ subunit stimulation of phospholipase C, the effect of GDP-liganded $\alpha_t$ was examined on the stimulation of PtdIns$P_2$ hydrolysis by $\beta\gamma_t$. Figure 2 illustrates that $\alpha_t$ completely prevented $\beta\gamma_t$ stimulation of phospholipase C, but did not affect basal PtdIns$P_2$ hydrolysis. GTP[S]-liganded $\alpha_t$ did not affect the ability of $\beta\gamma_t$ to stimulate phospholipase C (results not shown). Taken together, these results not only provide further strong evidence for the specificity of the effect of $\beta\gamma_t$, but also clearly demonstrate that phospholipase C is activated by free $\beta\gamma$ subunits, but not by the $\alpha\beta\gamma$ heterotrimer.

Stimulation of phospholipase C by $\beta\gamma_t$ was not restricted to cytosol prerared from differentiated HL-60 cells, but was also observed with cytosol from undifferentiated HL-60 cells as well as bovine and human peripheral blood neutrophils (Camps et al. 1992a). Soluble preparations from bovine cerebral and cerebellar cortex, striatum, lung, kidney, heart, liver, and spleen also contain $\beta\gamma_t$-sensitive phospholipases C (Camps et al., unpublished results). $\beta\gamma_t$ also led to a marked stimulation of PtdIns$P_2$ hydrolysis when HL-60 cell membranes rather than cytosol were used as a source of phospholipase C. A comparison of the effect of $\beta\gamma$ subunits

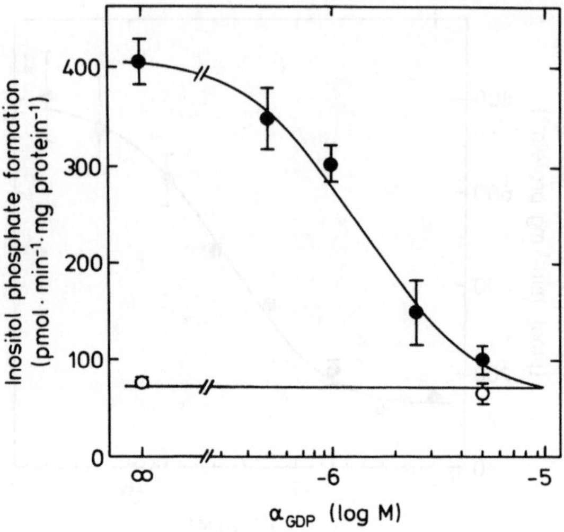


**Fig. 2.** Reversal of the $\beta\gamma_t$-dependent stimulation of phospholipase C by GDP-liganded $\alpha_t$. Cytosol from myeloid-differentiated HL-60 cells was incubated in the presence (*closed circles*) or absence (*open circles*) of $0.7\,\mu M$ $\beta\gamma_t$ with increasing concentrations of GDP-liganded $\alpha_t$ and phospholipid vesicles containing PtdIns$P_2$. See CAMPS et al. (1992a) for experimental details

purified from bovine brain membranes ($\beta\gamma_G$) and $\beta\gamma_t$ on soluble phospholipase C activity of differentiated HL-60 cells showed that both $\beta\gamma$ subunit preparations were stimulatory, although $\beta\gamma_G$ was clearly less efficient than $\beta\gamma_t$ (CAMPS et al. 1992a).

## C. Identification of the $\beta\gamma$-Sensitive Phospholipase C of HL-60 Granulocytes as PLC$\beta_2$

Phosphoinositide-specific phospholipases C are members of a large gene family. To date, at least eight distinct isozymes of phospholipase C, designated $\beta_1$, $\beta_2$, $\beta_3$, $\gamma_1$, $\gamma_2$, $\delta_1$, $\delta_2$, and $\delta_3$, are recognized in mammalian cells (see Chap. 58 by PARK, this volume, for review and references). The recent identification of the cDNA of PLC$\beta_2$ in an HL-60 cell cDNA library (PARK et al. 1992) prompted us to examine whether the phospholipase C encoded by this cDNA corresponded to the $\beta\gamma$-subunit-sensitive phospholipase C isozyme identified in these cells. To this end, we examined and compared the effect of purified $\beta\gamma_t$ on the activities of PLC$\beta_1$ and PLC$\beta_2$ transiently expressed in COS-1 cells (CAMPS et al. 1992b). Figure. 3 shows that $\beta\gamma_t$ had no significant effect on the inositol phosphate formation by lysates of cells transfected with vector containing no insert. In contrast, $\beta\gamma_t$ clearly stimulated the formation of inositol phosphates by lysates of cells transfected with pMT2-PLC$\beta_1$ and pMT2-PLC$\beta_2$. Most interestingly,

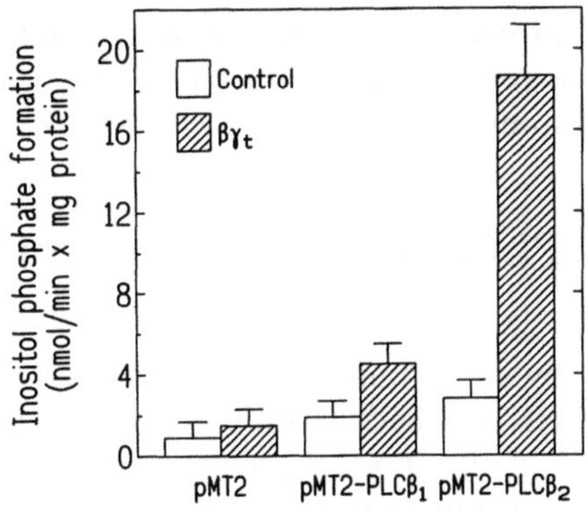


**Fig. 3.** Effect of $\beta\gamma_t$ on inositol phosphate formation by PLC$\beta_1$ and PLC$\beta_2$ expressed in cultured mammalian cells. COS-1 cells were transfected as indicated with the mammalian expression vector pMT2 without insert, with pMT2-PLC$\beta_1$, or with pMT2-PLC$\beta_2$. Cell lysates were prepared from transfected cells by treatment with detergent-containing buffer and incubated in the absence (*open bars*) or presence (*hatched bars*) of 2 $\mu M$ $\beta\gamma_t$ with phospholipid vesicles containing PtdIns$P_2$. See CAMPS et al. (1992b) for experimental details

however, the degree of stimulation by $\beta\gamma_t$ was considerably higher for the cells transfected with pMT2-PLC$\beta_2$ than for those transfected with pMT2-PLC$\beta_1$. Additional experiments showed that the differential sensitivity of the two phospholipase C preparations was not due to differences in transfection efficiencies, since the two preparations displayed similar maximal activities when stimulated by $Ca^{2+}$ and deoxycholate. To examine the specificity of the stimulation of PLC$\beta_2$ by $\beta\gamma_t$, we determined the effect of GDP-liganded $\alpha_t$ on $\beta\gamma_t$-stimulated PtdIns$P_2$ hydrolysis by PLC$\beta_2$. $\alpha_{t.GDP}$ completely eliminated the stimulatory effect of $\beta\gamma_t$, but had no effect in the absence of $\beta\gamma_t$. Thus, PLC$\beta_2$ is specifically stimulated by the free $\beta\gamma$ subunit, but not by the $\alpha\beta\gamma$ heterotrimer.

The results presented in Fig. 3 demonstrate that both PLC$\beta_1$ and PLC$\beta_2$ are stimulated by G-protein $\beta\gamma$ subunits and identify PLC$\beta_2$ as a major target of $\beta\gamma$ subunit stimulation. These findings, together with our previous observation of an unidentified $\beta\gamma_t$-resistant phospholipase C in HL-60 granulocytes (CAMPS et al. 1992a) and the observation that purified PLC$\delta_1$ showed no response to $\beta\gamma_t$ (CAROZZI et al., unpublished results), strongly suggest that the stimulation of PLC$\beta_2$ by $\beta\gamma_t$ is specific for members of the PLC$\beta$ subfamily and is isozyme selective within this group of phospholipases C. A third phospholipase C-$\beta$ isozyme (PLC$\beta_3$) was recently identified in WI-38 fibroblasts and purified from HeLa S3 cells (KRIZ et al. 1990; CAROZZI et al. 1992). When purified PLC$\beta_3$ and PLC$\beta_1$ were reconstituted with purified $\beta\gamma_t$, both isozymes were stimulated by $\beta\gamma_t$. However, PLC$\beta_3$ was stimulated by $\beta\gamma_t$ with much greater efficacy than PLC$\beta_1$ (CAROZZI et al. 1993). These results not only suggest that $\beta\gamma$ subunits stimulate phospholipase C-$\beta$ isozymes by direct protein–protein interaction, but also allow us to tentatively define the relative sensitivity of the three phospholipase C-$\beta$ isozymes to $\beta\gamma$ subunit stimulation as follows: $\beta_2 \approx \beta_3 \gg \beta_1$.

## D. Stimulation of PLC$\beta_2$ by G-Protein $\beta\gamma$ Subunits in Intact Cells

Stimulation of PLC$\beta_2$ by $\beta\gamma$ subunits is not restricted to in vitro systems composed of cell extracts, purified $\beta\gamma$ subunits, and exogeneous PtdIns$P_2$, but is also observed when the inositol phosphate formation by intact cells expressing PLC$\beta_2$ and $\beta\gamma$ subunits is analyzed in vivo (Fig. 4). Transfection of COS-1 cells with the cDNAs of either PLC$\beta_2$ or the $\beta_1\gamma_2$ subunit followed by radiolabeling of the cells with [$^3$H]inositol led to only small (≈35%) increases in the formation of [$^3$H]Ins$P_3$. In contrast, cotransfection of the cells with the cDNAs of both PLC$\beta_2$ and the $\beta_1\gamma_2$ dimer caused a much higher (≈200%) increase in [$^3$H]Ins$P_3$ formation. KATZ et al. (1992) recently extended these findings and showed that both $\beta_1\gamma_1$ and $\beta_1\gamma_2$ dimers specifically stimulated PLC$\beta_2$, but not PLC$\beta_1$, upon coexpression in COS-7 cells. Stimulation of PLC$\beta_2$ was dependent on coproduction of $\beta$ and $\gamma$ subunits and was not observed when the cDNA of a mutant $\gamma_1$ subunit lacking the consensus sequence for $\gamma$ subunit isoprenylation, proteolysis, and carboxyl methylation was used for transfection. Coexpression of $\alpha_{i2}$ in cells expressing $\beta_1\gamma_1$ and PLC$\beta_2$ eliminated the activation of phospholipase C.

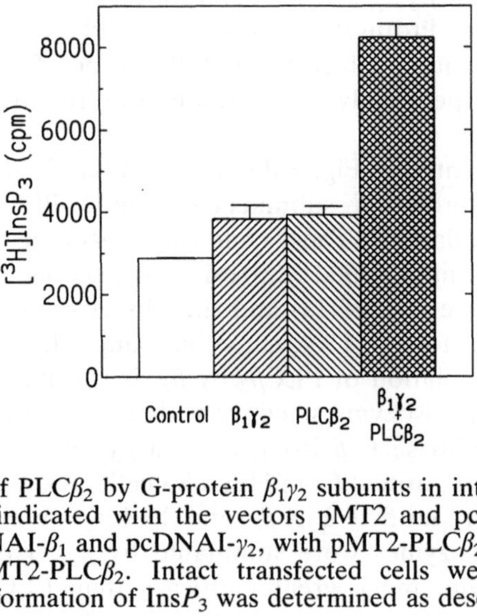


**Fig. 4.** Stimulation of PLC$\beta_2$ by G-protein $\beta_1\gamma_2$ subunits in intact cells. COS-1 cells were transfected as indicated with the vectors pMT2 and pcDNAI without insert (*Control*), with pcDNAI-$\beta_1$ and pcDNAI-$\gamma_2$, with pMT2-PLC$\beta_2$, or with pcDNAI-$\beta_1$, pcDNAI-$\gamma_2$, and pMT2-PLC$\beta_2$. Intact transfected cells were radiolabeled with [$^3$H]inositol and the formation of Ins$P_3$ was determined as described in SCHNABEL et al. (1992)

Taken together, these findings not only confirm the results of CAMPS et al. (1992a, 1992b) for intact cells, but also suggest that $\beta\gamma$ subunits need to be associated with the plasma membrane in order to be capable of stimulating phospholipase C.

# E. Role of $\beta\gamma$ Subunits in Mediating Receptor Stimulation of Phospholipase C

An important issue to be discussed at this point is the relevance of $\beta\gamma$ subunits in mediating the stimulation of inositol phosphate formation by agonist-activated cell surface receptors. Of interest in this regard, BOYER et al. (1989) reported that $\beta\gamma$ subunits purified from a variety of sources inhibited $AlF_4^-$-stimulated phospholipase C activity, but stimulated $P_{2Y}$-purinergic receptor-stimulated inositol phosphate formation when reconstituted into turkey erythrocyte membranes. The authors suggested that $\beta\gamma$ subunits may stimulate PtdIns$P_2$ hydrolysis indirectly by facilitating the interaction of a phospholipase-C-stimulating $\alpha$ subunit with the receptor. It is equally possible, however, that the latter finding reflects a direct stimulatory effect of $\beta\gamma$ subunits on a $\beta\gamma$-sensitive phospholipase C present in these membranes. Also pertinent to this issue, injection of $\beta\gamma_t$ into *Xenopus laevis* oocytes has been reported to lead to a marked enhancement of serotonin receptor-dependent whole cell $Cl^-$ currents (DASCAL et al. 1986), a process that is triggered by receptor-mediated generation of Ins$P_3$ followed by an increase in $[Ca^{2+}]_i$ (TAKAHASHI et al. 1987).

It is clear from the findings described above that stimulation of phospholipase C by $\beta\gamma$ subunits is much more likely to occur in cells expressing predominantly PLC$\beta_2$ and/or PLC$\beta_3$ rather than PLC$\beta_1$. Membranes of peripheral blood neutrophils and HL-60 granulocytes contain significant quantities of PLC$\beta_2$, but lack PLC$\beta_1$ (CAMPS et al., unpublished results). The calculated concentration of $\beta\gamma$ subunits in granulocyte plasma membranes is approximately 70 $\mu M$ (CAMPS et al. 1992a). This concentration is by far in excess of the concentration required for maximal stimulation of phospholipase C (cf. Fig. 1) and is, therefore, high enough to account for G-protein stimulation of PtdIns$P_2$ hydrolysis in this compartment. Granulocyte membranes also contain high numbers of receptors for *N*-formyl methionine containing chemotactic peptides (PIKE 1990), which are coupled to stimulation of phospholipase C via the pertussis-toxin-sensitive G-proteins $G_{i2}$ and/or $G_{i3}$ (GIERSCHIK et al. 1989). $G_{i2}$ and $G_{i3}$ are highly abundant in granulocytes (GIERSCHIK et al. 1986; UHING et al. 1987; BOKOCH et al. 1988), and it is likely that a considerable proportion of the cellular $\beta\gamma$ subunits is complexed to $\alpha_{i2}$ and $\alpha_{i3}$. Agonist occupancy of formyl peptide receptors leads to a catalytic activation of $G_{i2}$ and $G_{i3}$ (GIERSCHIK et al. 1991), and may very well cause a transient, pertussis-toxin-sensitive increase in the concentration of free $\beta\gamma$ subunits that is sufficiently high to stimulate

PLC$\beta_2$. Thus, at least in granulocytes, stimulation of PLC$\beta_2$ by the free $\beta\gamma$ subunits of the pertussis-toxin-sensitive G-proteins $G_{i2}$ and/or $G_{i3}$ is likely to be a mechanism by which receptors stimulate phospholipase C.

If this hypothesis is correct, one would, of course, predict that other G-protein-coupled receptors capable of inducing the release of a similarly high concentration of $\beta\gamma$ subunits are also capable of stimulating inositol phosphate formation, at least in cells expressing phospholipase C isozymes sensitive to $\beta\gamma$ stimulation. Recent results obtained by expressing a variety of cloned G-protein-coupled receptors in cultured mammalian cells suggest that this may indeed be the case. Thus, several receptors well known to inhibit adenylyl cyclase via $G_i$ have been shown to also stimulate phospholipase C, at least in certain cells and under certain conditions. Recombinant m2 and m4 muscarinic receptors (ASHKENAZI et al. 1987, 1989; PERALTA et al. 1988), the 5-$HT_{1A}$ receptor (FARGIN et al. 1989, 1991; LIU and ALBERT 1991; ABDEL-BASET et al. 1992), two types of $\alpha_2$ adrenergic receptors (COTTECCHIA et al. 1990; SEUWEN et al. 1990), and the D2 dopaminergic receptor (VALLAR et al. 1990) are examples of this kind of receptors. It is interesting to note that, at least in some instances, rather high levels of receptors (ASHKENAZI et al. 1987, 1989; SEUWEN et al. 1990) or high concentrations of the stimulatory agonist (ASHKENAZI et al. 1987; PERALTA et al. 1988; FARGIN et al. 1989; SEUWEN et al. 1990) were required to observe receptor-mediated stimulation of phospholipase C. In all cases, pertussis toxin treatment of the transfected cells suppressed both adenylyl cyclase inhibition and phospholipase C stimulation, attesting to the fact that either $G_i$ or $G_i$-like proteins are involved in both processes. Using synthetic peptide antibodies recognizing the $\alpha$ subunits of $G_{i1}$, $G_{i2}$, and $G_{i3}$, FARGIN et al. (1991) demonstrated that the coupling of the cloned 5-$HT_{1A}$ receptor to both adenylyl cyclase inhibition and phospholipase C stimulation is mediated via the same population of $G_i$ proteins. KATZ et al. (1992) recently reported that stimulation of inositol phosphate formation by a recombinant m2 muscarinic receptor in COS-7 cells was markedly enhanced in cells expressing PLC$\beta_2$. A further increase in inositol phosphate formation was observed in cells expressing recombinant $\beta_1\gamma_1$ or $\beta_1\gamma_2$ dimers in addition to the receptor and PLC$\beta_2$. In contrast, coexpression of $\alpha_{i2}$ inhibited m2 receptor activation of PLC$\beta_2$.

Very recent evidence suggests that several cloned adenylyl cyclase stimulatory receptors, e.g., cloned thyrotropin (TSH), luteinizing hormone (LH), calcitonin, and parathyroid hormone (PTH) receptors, are also capable of stimulating phospholipase C when expressed in cultured cells and activated by the appropriate ligands (VAN SANDE 1990; GUDERMANN et al. 1992; CHABRE et al. 1992; FORCE et al. 1992; ABOU-SAMRA et al. 1992). In some cases, ≈10- to 25-fold higher agonist concentrations were required for stimulation of phospholipase C than for activation of the bona fide effector adenylyl cyclase (VAN SANDE et al. 1990; CHABRE et al. 1992; FORCE et al. 1992). Notably, pertussis toxin did not suppress the stimulation of phospholipase C triggered by LH and calcitonin receptors (GUDERMANN et al. 1992; CHABRE et al. 1992).

The activation of two signal transduction pathways by a single cloned receptor is frequently believed to reflect the promiscuous interaction of this receptor with several G-proteins, which might occur when the receptor is expressed at high density in a nonphysiologic environment. The observation, however, that receptor regulation of adenylyl cyclase and phospholipase C was similarly sensitive or insensitive to pertussis toxin in many of the above cases argues against a receptor–G-protein interaction at random. In addition, several of the receptors specified above are well known to regulate more than one signal transduction pathway when expressed in their natural cellular environments (for references, see VAN SANDE et al. 1990; GUDERMANN et al. 1992; CHABRE et al. 1992; ABOU-SAMRA 1992; FORCE et al. 1992). It thus appears likely that these receptors retain their G-protein specificity even upon heterologous expression and that their ability to interact with more than one effector is an intrinsic feature of the G-protein they couple to. In light of the findings presented above, we suggest that agonist occupancy of all aforementioned receptors liberates two active G-protein subunits, $\alpha_{GTP}$ and $\beta\gamma$, and that both of these subunits are capable of regulating effector activities on their own. Depending on the relative affinities of $\alpha_{GTP}$ and $\beta\gamma$ for their respective effectors and the relative abundance of receptors, G-proteins, and effectors in a particular cellular system, receptor activation may be followed by an alteration of adenylyl cyclase activity, activation of phospholipase C, or a concomittant or consecutive regulation of both signal transduction pathways. If this is true, this model would not only imply that $\beta\gamma$ subunits of $G_i$ and/or $G_o$ correspond to the long-sought pertussis-toxin-sensitive form of "$G_p$", but would also raise the possibility that $\beta\gamma$ subunits of other G-proteins, e.g., $G_s$, may act as pertussis-toxin-insensitive stimulators of phospholipase C.

# F. Perspectives

The findings summarized in this chapter raise a number of important questions to be clarified by future experimentation. First, it will be important to examine the specificity of the phospholipase C stimulation with regard to the subunit composition of the $\beta\gamma$ dimer. Differences have been noted in the relative ability of various $\beta\gamma$ dimers to stimulate atrial phospholipase $A_2$ (SCHMIDT and NEER 1991) or adenylyl cyclase (TANG and GILMAN 1991). It will, therefore, be interesting to see whether different species of $\beta\gamma$ dimers will be more potent stimulators of phospholipase C than $\beta\gamma_t$. A second important issue raised by our findings is that members of the PLC$\beta$ family are apparently regulated by both G-protein $\alpha$ and $\beta\gamma$ subunits. Thus, it is clear from a variety of studies that PLC$\beta_1$ is activated by all four members of the $\alpha_q$ family. PLC$\beta_2$ appears to be sensitive to stimulation by

$\alpha_{16}$ and is only poorly stimulated by the other $\alpha_q$ subunits (see Chaps. 58 and 67 by PARK and EXTON, this volume; LEE et al. 1992; PARK et al. 1992). It will, therefore, be important to find out how G-protein $\alpha$ and $\beta\gamma$ subunits cooperate at the level of these effector enzymes and to identify the structural elements of the phospholipase C-$\beta$ isozymes responsible for $\alpha$ and $\beta\gamma$ subunit interaction. Finally, it will be an important task to identify the parameter(s) which determine whether or not a given receptor is coupled to stimulation of phospholipase C in a given cellular environment. The LH receptor clearly stimulated phospholipase C in Ltk− fibroblasts (GUDERMANN et al. 1992), but barely activated this effector in embryonal kidney (293) cells (CHABRE et al. 1992). Both the dopamine D2 and 5-$HT_{1A}$ receptor stimulated inositol phosphate formation in Ltk− cells, but not in $GH_4C_1$ pituitary cells (VALLAR et al. 1990; LIU et al. 1991). In contrast, even high levels of vasopressin V2 receptors failed to stimulate phospholipase C in $Ltk^-$ cells (GUDERMANN et al. 1992). Cell-specific expression of signal transduction components specifically required for $\beta\gamma$ subunit stimulation of phospholipase C or selective interaction of receptors with only certain $\alpha\beta\gamma$ combinations (cf. KLEUSS et al. 1993) are intriguing possibilities to be considered.

*Acknowledgements*. Studies performed in the authors' laboratory reported herein were supported by grants from the Deutsche Forschungsgemeinschaft and the Fritz Thyssen Stiftung. M.C. received a grant from the Commission of the European Communities.

## References

Abdel-Baset H, Bozovic V, Szyt M, Albert PR (1992) Conditional transformation mediated via a pertussis toxin-sensitive receptor signaling pathway. Mol Endocrinol 6:730–740

Abou-Samra A-B, Jüppner H, Force T, Freeman MW, Kong Y-F, Schipani E, Urena P, Richards J, Bonventre JV, Potts JT, Kronenberg HM, Segre GV (1992) Expression cloning of a common receptor for parathyroid hormone and parathyroid hormone-related peptide from rat osteoblast-like cells: a single receptor stimulates intracellular accumulation of both cAMP and inositol trisphosphates and increases intracellular free calcium. Proc Natl Acad Sci USA 89:2732–2736

Ashkenazi A, Peralta EG, Winslow JW, Ramachandran J, Capon DJ (1989) Functionally distinct G proteins selectively couple different receptors to PI hydrolysis in the same cell. Cell 56:487–493

Ashkenazi A, Winslow JW, Peralta EG, Peterson GL, Schimerlik MI, Capon DJ, Ramachandran J (1987) An M2 muscarinic receptor subtype coupled to both adenylyl cyclase and phosphoinositide turnover. Science 238:672–675

Berridge MJ, Irvine RF (1989) Inositol phosphates and cell signalling. Nature 341:197–205

Bokoch GM, Bickford K, Bohl BP (1988) Subcellular localization and quantitation of the major neutrophil pertussis toxin substrate, $G_n$. J Cell Biol 106:1927–1936

Boyer JL, Waldo GL, Evans T, Northup JK, Downes CP, Harden TK (1989) Modification of $ALF_4^-$- and receptor-stimulated phospholipase C activity by G protein $\beta\gamma$ subunits. J Biol Chem 264:13917–13922

Camps M, Hou C, Jakobs KH, Gierschik P (1990) Guanosine 5′-[$\gamma$-thio]triphosphate-stimulated hydrolysis of phosphatidylinositol 4,5-bisphosphate in HL-60 granulocytes: evidence that the guanine nucleotide acts by relieving phospholipase C from an inhibitory constraint. Biochem J 271:743–748

Camps M, Hou C, Sidiropoulos D, Stock JB, Jakobs KH, Gierschik P (1992a) Stimulation of phospholipase C by G protein $\beta\gamma$-subunits. Eur J Biochem 206:821–831

Camps M, Carozzi A, Schnabel P, Scheer A, Parker PJ, Gierschik P (1992b) Isozyme-selective stimulation of phospholipase C-$\beta_2$ by G protein $\beta\gamma$-subunits. Nature 360:684–686

Carozzi A, Kriz RW, Parker PJ (1992) Identification, purification and characterization of a novel phosphatidylinositol-specific phospholipase C: a third member of the $\beta$ subfamily. Eur J Biochem 210:521–529

Carozzi A, Camps M, Gierschik P, Parker PJ (1993) Activation of phosphatidylinositol lipid-specific phospholipase C-$\beta$3 by G-protein $\beta\gamma$ subunits. FEBS Lett 315:340–342

Chabre O, Conklin BR, Lin HY, Lodish HF, Wilson E, Ives HE, Catanzariti L, Hemmings BA, Bourne HR (1992) A recombinant calcitonin receptor independently stimulates 3′,5′-cyclic adenosine monophosphate and $Ca^{2+}$/inositol phosphate signaling pathways. Mol Endocrinol 6:551–556

Cockcroft S, Stutchfield J (1988) G proteins, the inositol lipid signaling pathway, and secretion. Philos Trans R Soc London Ser B 320:247–265

Cotecchia S, Kobilka BK, Daniel KW, Nolan RD, Lapetina EG, Caron MG, Lefkowitz RJ, Regan JW (1990) Multiple second messenger pathways of $\alpha$-adrenergic receptor subtypes expressed in eukaryotic cells. J Biol Chem 265:63–69

Dascal N, Ifune C, Hopkins R, Snutch TP, Lübbert H, Davidson N, Simon MI, Lester HA (1986) Involvement of a GTP-binding protein in mediation of serotonin and acetylcholine responses in *Xenopus* oocytes injected with rat brain messenger RNA. Molecular Brain Research 1:201–209

De Vivo M, Gershengorn MC (1990) G proteins and phospholipid turnover. In: Moss J, Vaughan M (eds) ADP-ribosylating toxins and G proteins: insights into signal transduction. American Society for Microbiology, Washington, D.C., p 267

Fargin A, Raymond JR, Regan JW, Cotecchia S, Lefkowitz RJ, Caron MG (1989) Effector coupling mechanisms of the cloned 5-HT1A receptor. J Biol Chem 264:14848–14852

Fargin A, Yamamoto K, Cotecchia S, Goldsmith PK, Spiegel AM, Lapetina EG, Caron MG, Lefkowitz RJ (1991) Dual coupling of the cloned 5-$HT_{1A}$ receptor to both adenylyl cyclase and phospholipase C is mediated via the same $G_i$ protein. Cell Signal 3:547–557

Force T, Bonventre JV, Flannery MR, Gorn AH, Yamin M, Goldring SR (1992) A cloned porcine renal calcitonin receptor couples to adenylyl cyclase and phospholipase C. Am J Physiol 262:F1110–F1115

Fung BK-K (1983) Characterization of transducin from bovine retinal rod outer segments: separation and reconstitution of the subunits. J Biol Chem 258:10495–10502

Gierschik P, Falloon J, Milligan G, Pines M, Gallin JI, Spiegel AM (1986) Immunochemical evidence for a novel pertussis toxin substrate in human neutrophils. J Biol Chem 261:8058–8062

Gierschik P, Sidiropoulos D, Jakobs KH (1989) Two distinct $G_i$-proteins mediate formyl peptide receptor signal transduction in human leukemia (HL-60) cells. J Biol Chem 264:21470–21473

Gierschik P, Moghtader R, Straub C, Dieterich K, Jakobs KH (1991) Signal amplification in HL-60 granulocytes: evidence that the chemotactic peptide receptor catalytically activates G proteins in native plasma membranes. Eur J Biochem 197:725–732

Gudermann T, Birnbaumer M, Birnbaumer L (1992) Evidence for dual coupling of the murine luteinizing hormone receptor to adenylyl cyclase and

phosphoinositide breakdown and $Ca^{2+}$ mobilization: studies with the cloned murine luteinizing hormone receptor expressed in L cells. J Biol Chem 267:4479–4488
Hepler JR, Harden TK (1986) Guanine nucleotide-dependent pertussis-toxin-insensitive stimulation of inositol phosphate formation by carbachol in a membrane preparation from human astrocytoma cells. Biochem J 239:141–146
Higashijima T, Ferguson KM, Sternweis PC, Smigel MD, Gilman AG (1987) Effects of $Mg^{2+}$ and the $\beta\gamma$-subunit complex on the interactions of guanine nucleotides with G proteins. J Biol Chem 262:762–766
Irvine RF, Letcher AJ, Dawson RMC (1979) Fatty acid stimulation of membrane phosphatidylinositol hydrolysis by brain phosphatidylinositol phosphodiesterase. Biochem J 178:497–500
Jelsema CL, Axelrod J (1987) Stimulation of phospholipase $A_2$ activity in bovine rod outer segments by the $\beta\gamma$ subunits of transducin and its inhibition by the $\alpha$ subunit. Proc Natl Acad Sci USA 84:3623–3627
Katz A, Wu D, Simon MI (1992) $\beta\gamma$ Subunits of heterotrimeric G proteins activate an isoform of phospholipase C. Nature 360:686–688
Kim D, Lewis DL, Graziadei L, Neer EJ, Bar-Sagi D, Clapham DE (1989) G-protein $\beta\gamma$-subunits activate the cardiac muscarinic $K^+$-channel via phospholipase $A_2$. Nature 337:557–560
Kleuss C, Scherübl H, Hescheler J, Schultz G, Wittig B (1993) Selectivity in signal transduction is determined by each of the subunits of heterotrimeric G proteins Science 259:832–834
Krause K-H, Schlegel W, Wollheim CB, Andersson T, Waldvogel FA, Lew PD (1985) Chemotactic peptide activation of human neutrophils and HL-60 cells: pertussis toxin reveals correlation between inositol trisphosphate generation, calcium ion transients, and cellular activation. J Clin Invest 76:1348–1354
Kriz RW, Lin L-L, Sultzman L, Ellis C, Heldin C-H, Pawson T, Knopf JL (1990) Phospholipase C isozymes: structural and functional similarities. In: Bock G, Marsh J (eds) Proto-oncogenes in cell development. Ciba Foundation Symposium 150, John Wiley & Sons, Chichester, p 112
Lee CH, Park DJ, Wu D, Rhee SG, Simon MI (1992) Members of the $G_q$ $\alpha$ subunit gene family activate phospholipase C $\beta$ isozymes. J Biol Chem 267:16044–16047
Liu YP, Albert PR (1991) Cell-specific signaling of the 5–HT1A receptor: modulation by protein kinases C and A. J Biol Chem 266:23689–23697
Martin TFJ, Bajjalieh SM, Lucas DO, Kowalchyk JA (1986) Thyrotropin-releasing hormone stimulation of polyphosphoinositide hydrolysis in $GH_3$ membranes is GTP dependent but insensitive to cholera or pertussis toxin. J Biol Chem 261:10041–10049
Masters S, Martin MW, Harden K, Brown JH (1985) Pertussis toxin does not inhibit muscarinic-receptor-mediated phosphoinositide hydrolysis or calcium mobilization. Biochem J 227:933–937
Nakamura T, Ui M (1985) Simultaneous inhibitions of inositol phospholipid breakdown, arachidonic acid release, and histamine secretion in mast cells by islet-activating protein, pertussis toxin: a possible involvement of the toxin-specific substrate in the $Ca^{2+}$-mobilizing receptor-mediated biosignaling system. J Biol Chem 260:3584–3593
Nishizuka Y (1988) The molecular heterogeneity of protein kinase C and its implications for cellular regulation. Nature 334:661–665
Northup JK, Smigel MD, Sternweis PC, Gilman AG (1983) The subunits of the stimulatory regulatory component of adenylate cyclase: resolution of the activated 45,000-dalton ($\alpha$) subunit. J Biol Chem 258:11369–11376
Ohta H, Okajima F, Ui M (1985) Inhibition by islet-activating protein of a chemotactic peptide-induced early breakdown of inositol phospholipids and $Ca^{2+}$ mobilization in guinea pig neutrophils. J Biol Chem 260:15771–15780

Park DJ, Jhon D-Y, Kriz RW, Knopf JL, Rhee SG (1992) Cloning, sequencing, expression, and $G_q$-independent activation of phospholipase C-$\beta$2. J Biol Chem 2670:16048–16055
Peralta EG, Ashkenazi A, Winslow JW, Ramachandran J, Cappon DJ (1988) Differential regulation of PI hydrolysis and adenylyl cyclase by muscarinic receptor subtypes. Nature 334:434–437
Pfeilschifter J, Bauer C (1986) Pertussis toxin abolishes angiotensin II-induced phosphoinositide hydrolysis and prostaglandin synthesis in rat renal mesangial cells. Biochem J 236:289–294
Pike MC (1990) Chemoattractant receptors as regulators of phagocytic cell function. Curr Top Member Transp 35:19–43
Schmidt CJ, Neer EJ (1991) *In vitro* synthesis of G protein $\beta\gamma$ dimers. J Biol Chem 266:4538–4533
Schnabel P, Schreck R, Schiller DL, Camps M, Gierschik P (1992) Stimulation of phospholipase C by a mutationally activated G protein $\alpha_{16}$ subunit. Biochem Biophys Res Commun 188:1018–1023
Seuwen K, Magnaldo I, Kobilka BK, Caron MG, Regan JW, Lefkowitz RJ, Pouyssegur J (1990) $\alpha_2$-Adrenergic agonists stimulate DNA synthesis in Chinese hamster lung fibroblasts transfected with $\alpha_2$-adrenergic receptor gene. Cell Regul 1:445–451
Takahashi T, Neher E, Sakmann B (1987) Rat brain serotonin receptors in *Xenopus* oocytes are coupled by intracellular calcium to endogenous channels. Proc Natl Acad Sci USA 84:5063–5067
Tang W-J, Gilman AG (1991) Type-specific regulation of adenylyl cyclase by G protein $\beta\gamma$ subunits. Science 254:1500–1503
Uhing RJ, Prpic V, Jiang H, Exton JH (1986) Hormone-stimulated polyphosphoinositide breakdown in rat liver plasma membranes: roles of guanine nucleotides and calcium. J Biol Chem 261:2140–2146
Uhing RJ, Polskis PG, Snyderman R (1987) Isolation of GTP-binding proteins from myeloid HL-60 cells: identification of two pertussis toxin substrates. J Biol Chem 262:15575–15579
Vallar L, Muca C, Magni M, Albert P, Bunzow J, Meldolesi J, Civelli O (1990) Differential coupling of dopaminergic $D_2$ receptors expressed in different cell types: stimulation of phosphatidylinositol 4,5-bisphosphate hydrolysis in $LtK^-$ fibroblasts, hyperpolarizytion, and cytosolic-free $Ca^{2+}$ concentration decrease in $GH_4C_1$ cells. J Biol Chem 265:10320–10326
Van Sande J, Raspé E, Perret J, Lejeune C, Maenhaut C, Vassart G, Dumont JE (1990) Thyrotropin activates both the cyclic AMP and the $PIP_2$ cascades in CHO cells expressing the human cDNA of TSH receptor. Mol Cell Endocrinol 74:R1–R6
Yatani A, Okabe K, Birnbaumer L, Brown AM (1990) Detergents, dimeric $G\beta\gamma$, and eicosanoid pathways to muscarinic atrial $K^+$ channels. Am J Physiol 258:H1507–H1514
Zeitler P, Handwerger S (1985) Arachidonic acid stimulates phosphoinositide hydrolysis and human placental lactogen release in an enriched fraction of placental cells. Mol Pharmacol 28:549–554

## *Note Added in Proof*

Following our original report on $\beta\gamma$ subunit stimulation of phospholipase C in HL-60 granulocytes (CAMPS et al. 1992a), several papers have appeared confirming our results for the same and other systems:

Blank JL, Brattain KA, Exton JH (1992) Activation of cytosolic phosphoinositide phospholipase C by G-protein $\beta\gamma$ subunits. J Biol Chem 267:23069–23075
Boyer JL, Waldo GL, Harden TK (1992) $\beta\gamma$-Subunit activation of G-protein-regulated phospholipase C. J Biol Chem 267:25451–25456
Park D, Jhon D-Y, Lee C-W, Lee K-H, Rhee SG (1993) Activation of phospholipase C by G protein $\beta\gamma$ subunits. J Biol Chem 268:4573–4576
Smrcka AV, Sternweis PC (1993) Regulation of purified subtypes of phosphatidylinositol-specific phospholipase C $\beta$ by G protein $\alpha$ and $\beta\gamma$ subunits. J Biol Chem 268:9667–9674

# *E. Specialized Systems*

CHAPTER 60

# Rhodopsin/G-Protein Interaction

K.P. HOFMANN

## A. Introduction

Rhodopsin, the visual pigment of the rod cell, governs early steps of the visual process by its concerted interactions with the G-protein transducin, rhodopsin kinase, and arrestin. The precise interaction with transducin is crucial for single photon detection by rod cells. The rate of transducin turnover catalyzed by a single rhodopsin molecule in the dark is lower than $10^{-6}$ per second and increases to $10^3$ after light activation. The visual receptor/G-protein system, designed for low noise-signal amplification, may operate near to the upper physical limit set to biological signal transduction. On the other hand, rhodopsin and transducin are fundamentally similar to other receptors and G-proteins, and understanding this specialized system may help to understand such systems in general. Studies of the visual system are supported by specific physical properties, as the activation by light, the colored receptor intermediates, and the variable membrane binding of the G-protein. Part of this chapter is dedicated to techniques of physical biochemistry and how to employ them in the study of rhodopsin/G-protein interaction. Some of the insights may also apply to the related systems. Other recent reviews deal with rhodopsin and transducin from different perspectives, including the role of rod outer segment structure (LIEBMAN et al. 1987), molecular details of the interaction (HAMM 1991), structure and function of rhodopsin (HARGRAVE and MCDOWELL 1992; HARGRAVE et al. 1993) and of seven-helix receptors in general (HARGRAVE 1991), molecular biology (KHORANA 1992) and spectroscopy (SIEBERT 1992) of rhodopsin, and the role of transducin in the visual process (HOFMANN and KAHLERT 1992; LAMB and PUGH 1992).

## B. Interactions of Rhodopsin in the Visual Cascade

Rhodopsin is embedded in disc-shaped intracellular membrane vesicles, which form long stacks in the outer segments of rod cells. A pair of discs is schematically shown in Fig. 1A. G-protein (transducin, $G_t$) and effector (a cyclic GMP phosphodiesterase, PDE) are peripheral membrane proteins. Activation of the PDE effector via receptor and $G_t$ leads to the hydrolysis of

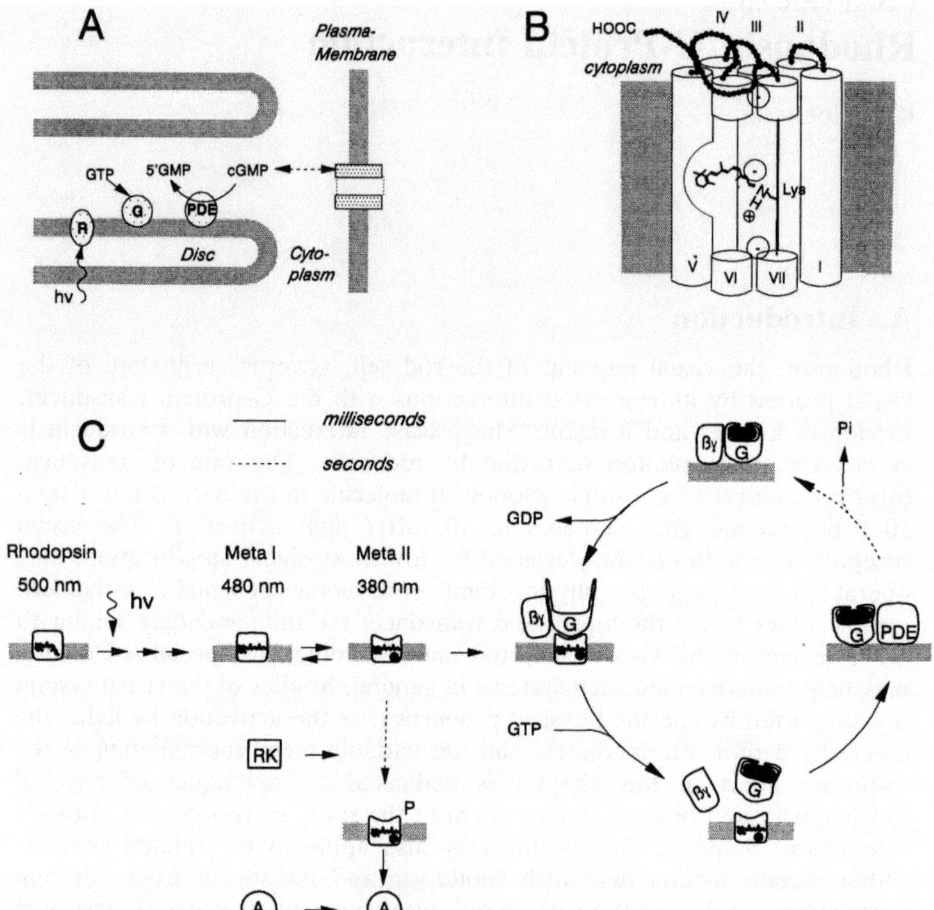


**Fig. 1A–C.** Rhodopsin and the visual cascade. **A** Components of the visual cascade. Rhodopsin (*R*), G-protein (*G*), and cGMP phosphodiesterase (*PDE*) are located in and on the disc membrane. Bovine disc membranes are ca. 1.5 $\mu$m in diameter and form long stacks in the outer segment. The very fluid disc membrane contains R at a density of 25 000 $\mu m^{-2}$; the relative abundance of R, G and PDE is 100:10:1. See text and LIEBMAN et al. (1987) for details. **B** Rhodopsin and its chromophore, 11-*cis* retinal. The schematic picture shows the disposition of the seven transmembrane helical stretches, the cytoplasmic loops and the negative charges along the third helix. The intradiscal loops are omitted. **C** The visual cascade. The scheme shows the light-induced formation of the active photoproduct, MII, its deactivation by phosphorylation via a rhodopsin kinase (*RK*) and subsequent binding of arrestin (*A*), and the G-protein cycle. See text for details

cyclic GMP in the cytoplasm and to the closure of some 1000 cGMP-dependent channels around the excited disc.

Bovine rhodopsin is an intrinsic membrane protein of 40 kDa molecular weight. Numerous biochemical and biophysical approaches have contributed to elucidate the disposition of rhodopsin in the disc membrane (for reviews see OVCHINNIKOV 1982; DRATZ and HARGRAVE 1983; APPLEBURY and HARGRAVE 1986; HARGRAVE and McDOWELL 1992). The proposed seven-transmembrane helix structure was based on the original analysis for bacteriorhodopsin (HENDERSON and UNWIN 1975). Rhodopsin has half of its mass embedded in the lipid bilayer and exposes the remaining half in equal portions to the cytoplasmic and intradiscal surfaces. Since part of the carboxyl terminus is anchored to the bilayer by palmitoyl-cysteines, four "loop" regions and the carboxyl terminus face the rod cytoplasm. Figure 1B shows schematically the helical regions as columns, the cytoplasmic loops, and the chromophore, 11-*cis* retinal, which is bound as a protonated Schiff base to a lysine residue (Lys-296). Negative charges along the third helix, at carboxylic acids E113, E112, and E134, are also shown. E113 has been proposed to provide the counterion of the retinal Schiff base (SAKMAR et al. 1989), E122 may be involved in proton transfer from the Schiff base (GANTER et al. 1989), and the E134/R135 charge pair is required for proper interaction with the G-protein (FRANKE et al. 1990).

Figure 1C shows the protein interactions that are involved in the visual cascade, with emphasis on the rhodopsin/G-protein interaction. The interaction of $G_t$ with PDE is simplified, because it is dealt with elsewhere in this volume (T.G. WENSEL, Chap. 56). The light-activated form of rhodopsin, metarhodopsin II (Meta II, MII) eventually arises from the very rapid isomerization of the chromophore from the 11-*cis* into a distorted *trans* form. Via intermediates MII is formed by relaxation of the photochemically activated state. During its lifetime MII remains in equilibrium with its predecessor MI (and its successor, MIII, not shown). Only MII couples with measurable affinity to the G-protein (EMEIS and HOFMANN 1981) and to arrestin (SCHLEICHER et al. 1989). The functional equivalence of MII and the $G_t$-activating form of rhodopsin is supported by kinetic analysis in urea-stripped, purified disc membranes (KIBELBEK et al. 1991) and in intact photoreceptors (KAHIERT and HOFMANN 1991).

Transducin, rhodopsin kinase and arrestin interact with MII within seconds after its formation (Fig. 1C). First, transducin and rhodopsin kinase compete for binding to free, unphosphorylated MII (PFISTER et al. 1983). Phosphorylation of rhodopsin at multiple sites increasingly reduces the affinity for the G-protein and switches to tight binding of arrestin. Bound arrestin strongly inhibits further interaction with transducin (WILDEN et al. 1986; PALCZEWSKI et al. 1992). Thus MII is shut off by this biochemical mechanism long before its spontaneous decay to MIII, which takes minutes (CHABRE and BRETON 1979). The overall deactivation reaction of rhodopsin as an activator of transducin takes 2–3 s, depending on species, as determined in situ from the recovery of the light-scattering amplified transient retina (ATR) signal (see Sect. C.I; PEPPERBERG et al. 1988) or from the onset of the falling phase of the rod photoresponse (PEPPERBERG et al. 1991). Since the binding of arrestin to phosphorylated rhodopsin is much faster, in the order of 200 ms at 13°C (SCHLEICHER et al. 1989), rhodopsin phosphorylation is the rate-limiting step for the overall shut-off reaction. In the presence of GTP, the MII-$G_t$ complex dissociates, MII is released and can thus sequentially interact with many copies of $G_t$. With the kinetic parameters of activation and deactivation, the lifetime of active MII of 2 s, the MII-$G_t$ interaction time of 0.3 ms, and the deactivation of $G_t$ (GTPase reaction time) of 1 s (VUONG and CHABRE 1991), $G_t$ transiently accumulates in its GTP binding active form.

# C. Biophysical Monitors of G-Protein Activation

## I. Description of the Monitors

Biophysical effects accompany the formation and dissociation of the MII-$G_t$ complex. They can be used as intrinsic reporters, which monitor the different stages of rhodopsin–transducin interaction.

*Intrinsic $G_\alpha$ Fluorescence.* The fluorescence assay is based on a change of the tryptophane fluorescence of the $G_\alpha$ subunit of heterotrimeric G-proteins, arising from the change of conformation accompanying their activation (HIGASHIJIMA et al. 1987; GUY et al. 1990). As Fig. 2A shows, the addition of GTP$\gamma$S to a mixture of photoactivated disc membranes and $G_t$ leads to a pronounced increase in the fluorescence. Recent development of the method allows records within 10–20 s, which is fast enough to overcome the relatively fast decay of MII in detergent solution (ca. 2 min at 20°C).

*Extra-MII.* The extra-MII effect (EMEIS and HOFMANN 1981; EMEIS et al. 1982) arises from the stabilization by G-protein (and also by arrestin and anti-rhodopsin antibodies; SCHLEICHER et al. 1989; B. KÖNIG, A. ARENDT, J.H. MCDOWELL, M. KAHLERT, P.A. HARGRAVE, K.P. HOFMANN, unpublished) of the 380-nm MII form at the expense of its tautomeric forms, MI (480 nm) and MIII (470 nm). The effect can either be measured as an enhanced absorption change after a light flash (as discussed in the examples below) or by static spectrophotometry (BENNETT et al. 1982; HOFMANN et al. 1992). The amount of MII formed in excess of the normal MII is a stoichiometric measure of the complexes formed with the interacting protein (EMEIS et al. 1982; SCHLEICHER et al. 1989). In the absence of nucleotides, the enhancement of MII by $G_t$ is stable, reflecting the stability of the complex. When GTP is present and the complex dissociates, extra-MII is transient in its time course (Fig. 2B), reflecting the formation and decay of the MII-$G_t$ complex (HOFMANN 1985; KOHL and HOFMANN 1987). Extra-MII can only work under conditions where the normal equilibrium is not on the MII side. For rhodopsin in the host of its native or egg lecithin bilayer, this is the case at 4°C and pH 7.5–8. The presence of noninteractive proteins does not disturb the assay. The necessary accumulation of MII-$G_t$ demands low GTP concentration. For a review see HOFMANN 1986.

*Near-Infrared Light Scattering.* Light-scattering changes ("signals") in the 800 nm range were first observed upon flash activation of rod outer segment preparations (HOFMANN et al. 1976; BIGNETTI et al. 1980). In 1981 KÜHN et al. demonstrated a stoichiometric relation between the binding and dissociation of the rhodopsin/G-protein complex and light-scattering signals. Figure 2C shows an example of a dissociation signal. It is based on the solubility of the active $G_t$ (KÜHN 1980). Since the diffusion of active $G_t$ from the membrane is very fast, it measures in real time the dissociation of the MII-$G_t$ complex. This interpretation (SCHLEICHER and HOFMANN 1987) was experimentally confirmed by a rapid centrifugation assay, using flash excitation and silver stain of the supernatant (M. HECK, K.P. HOFMANN, unpublished). Fast light-scattering signals work best with particles of the size of discs (ca. 1 $\mu$m), which provide sufficiently high basic scattering (HOFMANN and EMEIS 1981) in the standard configuration (scattering angle $15° < \theta < 20°$). All light scattering signals can also be measured as transmission changes (with inverse polarity, see for example KÜHN et al. 1981) in a normal spectrophotometer, with a loss of resolution. Note that the signal in Fig. 2C reflects in its delay the time it takes to form the active rhodopsin and to reach a steady state of transducin turnover.

*Light Scattering In Situ.* $G_t$ activation can also be measured in situ, on intact or permeabilized rod outer segments, in which the activated $\alpha$ subunit does not leave the interdisc space within the time domain of G activation (VUONG et al. 1984; BRUCKERT et al. 1988) or may even remain membrane-bound (WAGNER et al. 1987). The interpretation of these light-scattering signals is based on a stoichiometric and kinetic comparison to the signals from membrane preparations. In the intact retina, the time course of active $G_t$ is directly seen in the ATR signal (KAHLERT and HOFMANN 1991; Fig. 2D). This signal is best measured in the angular range $6° < \theta < 10°$ due to the diffraction of the retinal rod lattice. Simultaneously measured with the electroretinogram, the ATR signal allows the study of $G_t$ activation in intact functioning photoreceptors (PEPPERBERG et al. 1988; KAHLERT et al. 1990b).

## II. Instrumentation

The biophysical techniques employ specific optics and electronic devices. These are described elsewhere (HOFMANN and EMEIS 1979, 1981; KREUTZ et al. 1984). For absorption studies, the modified UV 300 Shimadzu two-wavelength spectrophotometer offers low preactivation by the measuring light (at the expense of resolution; examples are Figs. 2B and 3A).

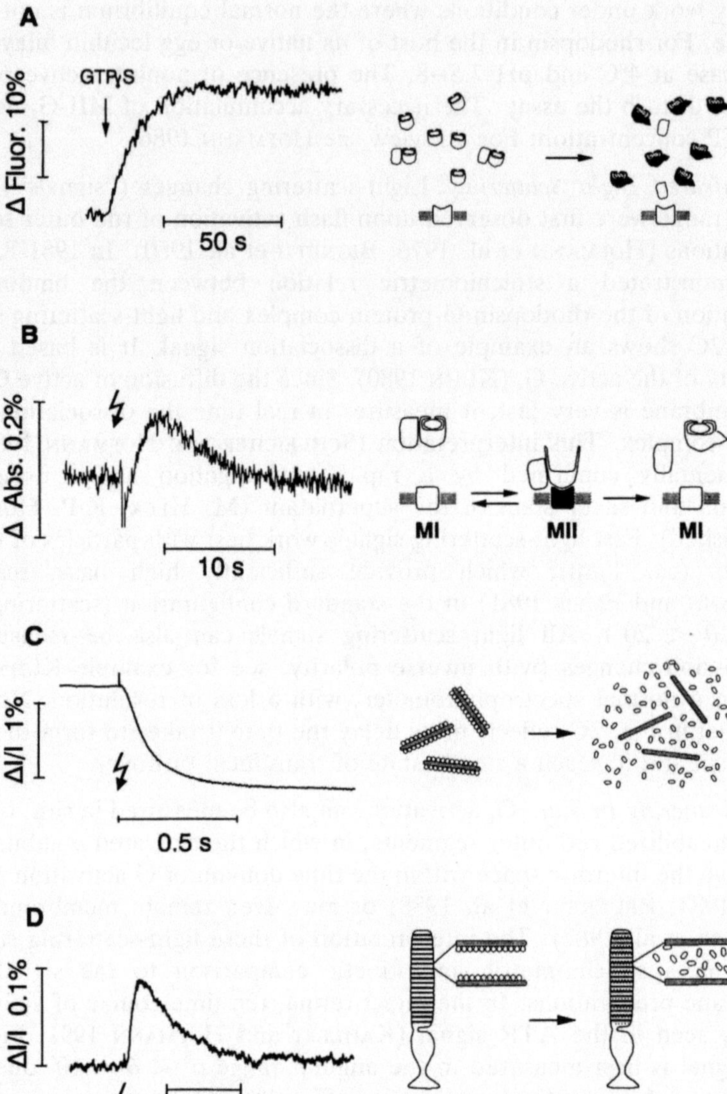


**Fig. 2A–D.** Monitors of transducin activation. **A** Fluorescence of $G_\alpha$. The positive fluorescence change indicates the transition of the G-protein into its active conformation. Washed disc membranes (Sect. C.IV) were recombined with purified $G_t$ and photoactivated by green light prior to the measurement. The sample (800 $\mu$l) contained 400 n*M* $G_t$ and 60 n*M* rhodopsin in washed disc membranes; 1 $\mu M$ GTP$\gamma$S added, as indicated. T = 20°C, pH 7.2, isotonic saline (120 m*M* NaCl, 2 m*M* $MgCl_2$, 20 m*M* HEPES). **B** Metarhodopsin II absorption. The MII conformation of photoactivated rhodopsin absorbs at 380 nm; its formation is measured as the change of the absorption difference between 380 nm (maximal absorption of MII) and 417 nm (isosbestic point of MI/MII). The record on the left shows a measuring example. The initial negative jump is due to photoproducts earlier than MII,

### III. Application to the Analysis of R*-$G_t$ Interaction

*Reaction Sequence In Vitro*. Each reaction monitored by the techniques above must reflect at least one step in the following reaction sequence (we denote the active MII by the more general term R*):

$$\begin{aligned}
&1.\ G_s\text{-GDP} \rightleftharpoons G_m\text{-GDP} \\
&2.\ R^* + G_m\text{-GDP} \rightleftharpoons R^*\text{-}G_{empty} + \text{GDP} \\
&3.\ R^*\text{-}G_{empty} + \text{GTP} \rightarrow R^* + G_m\text{-GTP}\ (\rightarrow G_s\text{-GTP}) \\
&4.\ G_s\text{-GTP} \rightarrow G_s\text{-GDP}
\end{aligned} \tag{1}$$

Step 1 is the equilibration between soluble ($G_s$) and membrane-bound ($G_m$) $G_t$ in its inactive, GDP-binding form (Sect. E.I). Steps 2 and 3 are the formation and dissociation of the R*-$G_t$ complex, respectively, and step 4 is the return into the G-GDP form by GTP hydrolysis. For the following, it will be important to evaluate how much the steps 1 and/or 4, which are slow under conditions in vitro, contribute to the measured overall rates of activation. The on-reaction of step 1 takes seconds at $\mu$M concentration, with an apparent $K_D$ of $10^{-5}$–$10^{-6}\,M$, depending on the preparation (Schleicher and Hofmann 1987). Similar rates are measured in micellar solutions of rhodopsin where the rate of R*-$G_t$ formation may depend on the transition of G-protein from detergent micelles to the solubilized rhodopsin molecules

←

including MI. As illustrated on the right, binding of $G_t$ stabilizes MII at the cost of other, noninteractive and spectrophotometrically different forms of photolyzed rhodopsin (extra-MII). With GTP, MII·$G_t$ forms only transiently due to the GTP induced MII·$G_t$ dissociation, the time course of which is seen in the decay of the enhanced 380 nm absorption. The sample (400 $\mu$l) contained 1.25 $\mu M$ GTP, 1.5 $\mu M$ $G_t$, and 5 $\mu M$ rhodopsin in washed membranes (Sect. C.IV), isotonic saline as in **A**. The flash, applied at the indicated time, photolyzed 12% of the rhodopsin. T = 6°C, pH 7.2, optical path length 2 mm. The noise of the record in this example results from the small absorbance (0.04). **C** Light scattering. The activation of transducin is accompanied by its rapid release/mainly the $\alpha$ subunit from the membrane. In a suspension of disc vesicles (*right*), this results in a decrease in light scattering (measured in the near infrared, at 800–850 nm) due to the loss of scattering mass from the discs. The time course of this scattering change (dissociation signal, measuring example with 500 $\mu M$ GTP) is therefore a stoichiometric monitor of the formation of activated transducin (*G**). The flash, applied as indicated, photolyzed 0.2% of the rhodopsin. The sample (400 $\mu$l) contained 3 $\mu M$ rhodopsin in washed disc membranes, 0.3 $\mu M$ $G_t$, T = 18°C, pH 7.2, optical path length 10 mm. Isotonic saline 120 m$M$ KCl, 5 m$M$ $MgCl_2$, 20 m$M$ HEPES. To enhance the stability of the measurements, different salines were tested, without significant difference in the results. **D** Retina light scattering. In intact photoreceptors, the plasma-membrane prohibits the loss of activated $G_t$ from the scattering object, the rods; however, the release of $G_t$ is seen by the scattering effect of the more random distribution of the active form. The measuring example is from a 5 mm diameter patch of isolated bovine retina (ATR signal). The signal is transient (and repeatable) due to the functioning G* and R* deactivation reactions. The flash photolyzed 0.02% of the rodopsin, T = 18°C

(SCHLEICHER et al. 1987). Step 4, the hydrolysis of GTP, takes minutes at room temperature.

*Use of Extra-MII and Light Scattering.* Under the conditions of extra-MII the formation of the $R^*$-$G_t$ complex is limited by the rate of MII formation (note that the initial slope of MII and of extra-MII signals is the same, Figs. 3, 5). The dissociation of the complex, however, is at low GTP concentration slow enough to make the decay of extra-MII a real-time monitor of $R^*$-$G_t$ dissociation, i.e., $G_t$ activation kinetics (HOFMANN 1985; KOHL and HOFMANN 1987). Typical working conditions for investigating properties of the MII-$G_t$ complex are 3–15 $\mu M$ rhodopsin and 1–5 $\mu M$ $G_t$. Soluble $G_t$ is only seen with slow kinetics, when the amount of activated rhodopsin exceeds the membrane-bound $G_t$ (SCHLEICHER and HOFMANN 1987).

The conditions for measuring dissociation signals are similar, with the exception that the photoexcitation can be much smaller (typically a $10^{-3}$ mol fraction of rhodopsin) since this monitor measures the catalytic product, $G_t$-GTP. The final level of the signal is a measure of the membrane-bound $G_t$ pool, and the initial slope is a measure of the overall activation speed. Since the dissociation from the membrane is not rate limiting, the rise of the signal follows with high fidelity the production of active $G_t$. For sufficiently high GTP concentration, the $R^*$-$G_t$ complex formation rate limits the overall rate (see below). Soluble $G_t$ can influence the final level inasmuch as the loss of active $G_t$ (mainly the $\alpha$ subunit) does not exactly balance the uptake of $G_t$ holoprotein to the membrane. One uses high protein concentrations to enhance membrane binding (with small optical path length).

*The Problem of Intrinsically Slow Assays.* For both extra-MII and light scattering in vitro, conditions can be adjusted so that step 1 in Eq. 1 does not cause kinetic artifacts. Unfortunately, light scattering needs membraneous particles, and extra-MII needs a spectrophotometrically identified receptor form with high affinity to the binding protein. In many interesting cases, these conditions are not met [e.g., for constitutively active rhodopsin (ROBINSON et al. 1992) or other dark active forms, e.g., the pseudo-photoproducts (HOFMANN et al. 1992); see Sect. D.3]. In the intrinsically slow assays applied in this case (fluorescence, GTPase, GTP uptake filtration assay), the reaction is usually limited by $R^*$, added to the assay in the form of fully photolyzed or otherwise activated membranes (FERRETTI et al. 1986; SAKMAR et al. 1989; GANTER et al. 1989; GUY et al. 1990). The kinetic analysis is based on

$$dG^*/dt = k \cdot [R^*] \cdot [G]_o \qquad (2)$$

where $[G]_o$ denotes the total concentration of activatable G-protein in the GDP-binding form. In the absence of slow reaction steps, the initial rate of $G^*(t)$ is a measure of the catalytic efficiency of the $R^*$-$G_t$ interaction. However, when the amount of photolyzed membranes is limiting, almost all

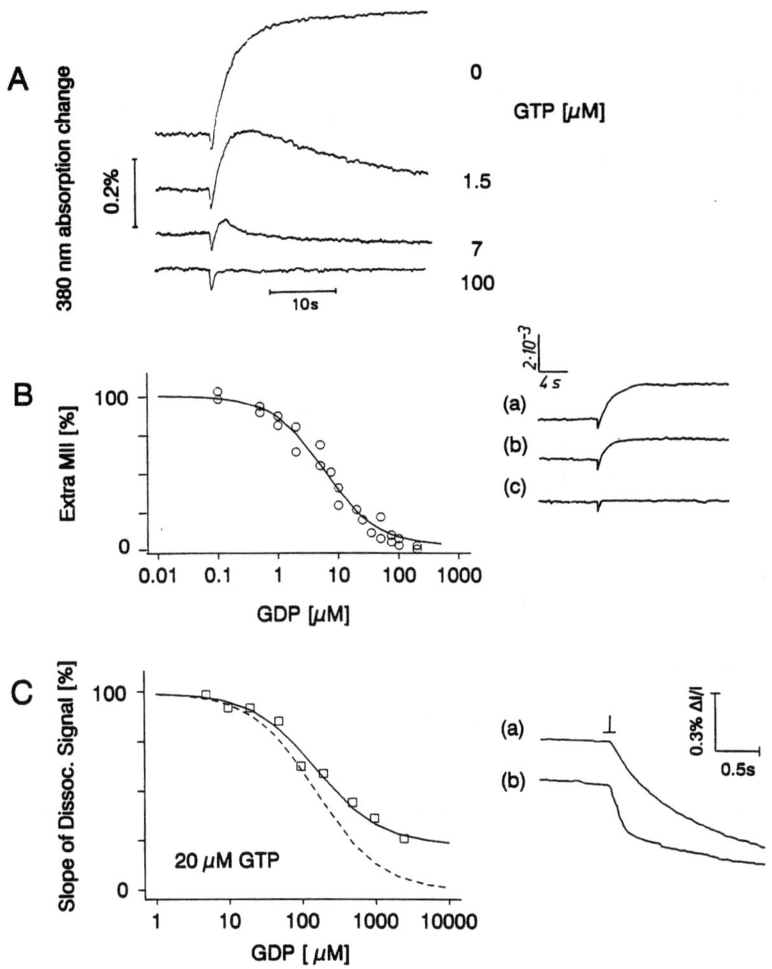


**Fig. 3A–C.** Effect of nucleotides on Meta II stabilization and $G_t$ activation rate. **A** Effect of GTP. The flash-induced extra-MII absorption change (Fig. 2B) rises with the kinetics of MII formation, which limits the rate of formation of the MII-$G_t$ complex. The lower the GTP concentration the more complex can accumulate and the slower is the decay of extra-MII (Hofmann 1985; Kohl and Hofmann 1987); $3\,\mu M$ rhodopsin, $0.4\,\mu M$ low ionic strength extract containing $G_t$ (Kühn 1980), T = 4°C. The flash photolyzed 4% of the rhodopsin. Isotonic saline 130 m*M* KCl, 0.5 mM $MgCl_2$, 1 m*M* $CaCl_2$, 0.5 m*M* EDTA, 10 m*M* PIPES. **B** Effect of GDP. The level of formation of $G_t$ induced extra-MII, set to 100% in the absence of exogenous nucelotide, is gradually suppressed with increasing GDP concentration (*a*, no exogenous GDP; *b*, $5\,\mu M$; *c*, $200\,\mu M$). The data points are the level of extra-MII (MII with $G_t$ minus control). *Solid line*, a linear hyperbolic fit to the data, yielding half suppression of extra-MII at $6\,\mu M$ GDP, $1.5\,\mu M$ rhodopsin, $0.6\,\mu M$ $G_t$, flash photolysis 5%; isotonic saline as in Fig. 1B, pH 7.6, 4°C; effective optical path length in the flash photolysis apparatus is 7 mm (see Hofmann and Emeis 1981). **C** Competition between GDP and GTP. The dissociation signals (Fig. 2C) on the right were measured with different GDP concentrations (*a*, $5\,\mu M$; *b*, $500\,\mu M$). The data points plot the initial slope of the dissociation signal as a measure of the $G_t$ activation rate. *Solid line*, a linear hyperbolic fit to the data; *Dotted line*, corrected for the GTP impurity (see Kahlert et al. 1990a); $1.3\,\mu M$ rhodopsin, $1\,\mu M$ $G_t$, flash photolysis 1.2%, isotonic saline as in Fig. 1C, room temperature

the G-protein is free, or it may be, in solubilized systems, associated with micelles that do not contain the receptor. Therefore, the reaction may be limited by the efficiency of collision of G-protein with membranes or micelles rather than by the actual formation of the $R^*$-$G_t$ complex. In terms of Eq. 1, step 1 remains rate limiting as long as the intrinsically fast steps 2 and/or 3 do not take over. A practical example is the following: the dissociation signal rises the faster the more GTP is present and only saturates in its rise for [GTP] $> 100\,\mu M$ (Sect. F.II.2). This shows that GTP at $1\,\mu M$ concentration cannot occupy more than 1% of the $G_t$ nucleotide sites during the turnover of G-protein seen in the signal. The fluorescence change, however, saturates for a GTP concentration in the $1\,\mu M$ range (S. JÄGER, M. HECK, K.P. HOFMANN, unpublished). In this case, the formation of the empty site complexes (Fig. 1C, 7) is limited by the additional association step 1, and accelerating step 3 by more GTP cannot accelerate the turnover. This example shows that one must be careful in quantitative determinations of receptor or G interactive efficiency. It is true that *if* the slope of $G^*(t)$ is smaller than in the native system, $R^*$ or G-protein must be impaired. However, a receptor can be impaired in its activity without any indication in the assay.

*Bimolecular Rate Constants.* It is instructive to compare the activation of $G_t$ by photolyzed membranes to the activation of G-protein by peptides from a $\beta_2$-adrenergic receptor. Photolyzed disc membranes in $60\,nM$ concentration activate the first 20% of the total transducin pool ($[G]_o = 400\,nM$) in 3 s (Fig. 2A; S. JÄGER and K.P. HOFMANN, unpublished). Equation 2 yields

$$(0.2 \times [G]_o)/3\,s = k \times [G]_o \times 0.06\,\mu M. \qquad (3)$$

The resulting bimolecular rate constant is $k = 1\,\mu M^{-1}\,s^{-1}$. OKAMOTO et al. (1991) show (Fig. 6) that 20% of $G_s$ ($[G]_o \times 10\,nM$) is activated by $1\,\mu M$ peptide in 1 min. The resulting bimolecular rate constant is $0.003\,\mu M^{-1}\,s^{-1}$ for the peptide, revealing the activation of soluble $G_t$ via the membrane-bound dark state (Sect. E.I) as quite effective. The activation of membrane-bound $G_t$ is much faster for conditions in vitro, due to the low concentrations of soluble $G_t$. However, we may consider the efficiency for the intact retina where the concentration of $G_t$ is very high ($300\,\mu M$; KÜHN 1984). Figure 6 shows that 20% of the total $G_t$ is activated within 0.07 s by a mole fraction of $3 \times 10^{-4}$ of the rhodopsin ($[R]_o = 3\,mM$); Eqs. 2 and 3 yield $k = 3\,\mu M^{-1}\,s^{-1}$. The higher value may be understood as a first approximation to the enhancement of collisional coupling by the membrane (KAHLERT and HOFMANN 1991).

## IV. Preparations

From rod outer segments, greater quantities of rhodopsin and transducin can be prepared than for other receptor/G-protein pairs. This has enhanced the application of biochemical and biophysical techniques. Preparations suited to investigate rhodopsin/$G_t$ include the perfused retina (PEPPERBERG et al. 1988; KAHLERT et al. 1990b), isolated rod outer segments of different integrity from bovine (KÜHN 1984; EMEIS and HOFMANN 1981; UHL et al. 1987) and frog retinae (VUONG et al. 1984; BRUCKERT et al. 1988), washed membrane preparations (SMITH et al. 1975; SCHLEICHER and HOFMANN 1987), vesicle preparations (KIBELBEK et al. 1991), rhodopsin in pure detergent (SCHLEICHER et al. 1987 and references therein) and in "doped" micelle systems (KÖNIG et al. 1989a). Kinetics and extent of MII formation are sensitive monitors of the rhodopsin hydrophobic environment (KÖNIG et al. 1989a; FRANKE et al. 1990).

The doped micelle system is best suited to the study of rhodopsin mutants (FRANKE et al. 1990; see Sect. F.I and Fig. 5). They can now be solubilized directly from the COS cell plasma membranes into the detergent and doped subsequently with egg lecithin (O. ERNST, T.P. SAKMAR, K.P. HOFMANN, unpublished). The cytoplasmic loop peptide competition assay (Fig. 4) has so far been applied only to membranes, but it should work as well with the micellar systems. Most peptide preparations of the effective loops CII–CIV compete, with $IC_{50}$ values between $50\,\mu M$ (Fig. 4) and $200\,\mu M$. There are cases of considerably reduced competitive potency and/or solubility, even with peptides that are to more than 80% pure (M. HECK, A. ARENDT, J.H. McDOWELL, P.A. HARGRAVE, K.P. HOFMANN, unpublished). The reasons for this variability are unknown. The buffer appears to be critical; best results were obtained with HEPES buffer, while Tris buffer reduces the competitive potency of the CIII peptide (H. HAMM, personal communication).

# D. Interactive States of Rhodopsin

## I. Molecular Nature of Metarhodopsin II

Rhodopsin shares with other receptors the existence of an active state in which the "agonist" retinal maintains a structural state of the apoprotein, which is compatible with the G-protein. In the normal sequence of events (Fig. 1C), the active form of rhodopsin, which is able to bind and to activate transducin, can develop only when MII is formed. In MII the retinal chromophore is still bound to the apoprotein but already has the absorption of the free chromophore. For a more detailed description of the physicochemical properties of MII, for example, its positive reaction volume, the electrical effects accompanying its formation, and the characteristic changes in the infrared spectrum, see HOFMANN (1986) and SIEBERT (1992). As in other retinal proteins, the crucial event for the biological function of the protein is the deprotonation of the Schiff base binding to the apoprotein. When the Schiff base nitrogen is methylated instead of protonated, transducin activation is blocked (LONGSTAFF et al. 1986), and the modified rhodopsin does not form MII but decays directly from MI to a MIII-like species (GANTER et al. 1991). The accepting group for the Schiff base proton is still under discussion (SIEBERT 1992), but it is interesting to note that the removal by site-directed mutagenesis of the counterion of the Schiff base, Glu-113, does not affect formation of a 380-nm active species (SAKMAR et al. 1989). A crucial role of histidine residues in the formation of MII has recently been described (WEITZ and NATHANS 1992). Molecular spectroscopy has suggested that considerable changes occur in the apoprotein moiety during the transition to MII (reviewed by DEGRIP 1988; SIEBERT 1992) and, in combination with partial digestion, a specific contribution of the cytoplasmic loop regions (GANTER et al. 1992; see also Sect. F.I).

## II. Active Forms of Rhodopsin from Alternative Light-Induced Pathways

When opsin is regenerated with 9-demethyl retinal, the structural changes normally seen with the formation of MI are already absent and the pigment fails to form a 380-nm photoproduct. The product that it does form is active toward transducin only to some percentage; thus its conformation is not fully compatible with the G-protein (GANTER et al. 1989).

Substitution of the active site lysine by glycine (mutant K296G) prevents covalent binding of retinal. However, this mutant can bind the protonated retinyl Schiff base 11-*cis*-retinyl-n-propyl amine noncovalently and activates transducin with light (ZHUKOVSKY et al. 1991).

## III. Activation of Rhodopsin in the Dark

The recently discovered rhodopsin forms that are active in the dark underscore the functional similarity to other receptors. In native rhodopsin, a key for the neutralization of the protonated Schiff base in its hydrophobic environment is the residue Glu-113 (SAKMAR et al. 1989). When a mutant, E113Q, is reconstituted with 11-*cis* retinal, it can activate transducin after absorption of light. The mutant can bind all-*trans* retinal and active transducin in the dark.

It has recently been demonstrated (HOFMANN et al. 1992) that even native opsin apoprotein can bind all-*trans* retinal. Purified opsin forms with all-*trans* retinal a receptor-agonist complex (380-nm "pseudo-photoproduct") which interacts tightly with arrestin and rhodopsin kinase. Like the normal, light-induced MII, the pseudo-MII is in equilibrium with one or several forms absorbing maximally in the 470 nm range, indicating the presence of protonated retinal Schiff bases. Mutants E113Q and K296G are both constitutively active, in dark or light, and without any binding of retinal (ROBINSON et al. 1992; D. OPRIAN, personal communication). Thus available data agree with the following scheme: (a) when the protonated Lys-296 and the negatively charged group at position 113 are both present, the protein cannot activate transducin; (b) *trans*-retinal can bind to Lys-296 via a deprotonated Schiff base and lead to selectively active rhodopsin; (c) when the retinal binding site is occupied by 11-*cis*-retinal, any interactive conformation of rhodopsin is blocked.

# E. Interactive States of Transducin

## I. Dark Binding

Binding of $G_t$ to the surface of the disc membrane is dependent on ionic strength, with an optimum between 70 and 400 m*M*. For low and high ionic strength, $G_t$ is released from the membrane, indicating a complex binding mechanism, presumably involving both ionic and hydrophobic interactions. At normal ionic strength, $G_t$ in vitro is in equilibrium between a membrane-bound ($G_m$) and a soluble form ($G_s$). By binding of a fraction of the $G_m$ to MII, the equilibrium is shifted towards $G_m$. The analysis is consistent with saturable binding sites for $G_t$ in the membrane (SCHLEICHER and HOFMANN 1987). The related light-scattering binding signal reveals a bimolecular rate constant in the order of $10^6 M^{-1} s^{-1}$ (in agreement with the fluorescence assay, see Sect. C.III) and an activation energy of 44 kJ/mol. The surface of rhodopsin may contribute to dark binding (HAMM et al. 1987). Potential roles of the different $G_t$ subunits are discussed by HAMM (1991).

## II. Stable Light Binding with Empty Nucleotide Site

Specific binding of $G_t$ to MII rhodopsin is much tighter than dark binding, with an apparent $K_D$ in the order of $10^{-7} M$ (HOFMANN 1984; BENNETT and DUPONT 1985). In the absence of any GDP, the actual "empty site" interaction is so strong that $G_t$ cannot be removed from MII without denaturation (BORNANCIN et al. 1989). In the absence of nucleotide MII-$G_t$ is stabilized during the whole lifetime of MII; when the chromophore is removed by formation of the retinaloxime, $G_t$ is released (HOFMANN et al. 1983). The $G_\alpha$ or $G_{\beta\tau}$ subunits alone (50 $\mu M$) cannot stabilize MII, suggesting that the holoprotein is required for this type of interaction (KÖNIG 1989). However, both subunits alone can inhibit the access of rhodopsin kinase (KELLEHER and JOHNSON 1988).

Two regions near the C-terminal of $G_\alpha$ are involved in the interaction with MII according to peptide competition (HAMM et al. 1988). Two synthetic peptides from $G_\alpha$, 340–350 and acetyl-311–329-amide, competed against $G_t$-induced extra-MII at lower concentrations (<100 $\mu M$) and exhibited a stabilizing effect at concentrations above 300 $\mu M$. The synergism of the latter effect (simultaneous presence of both peptides; KÖNIG 1989) may suggest that the peptides can mimic the interaction domain of $G_\alpha$ to some degree (HAMM 1991). The competition of bovine serum albumin against $G_t$ in its binding to MII has been interpreted to suggest a contribution of hydrophobic regions of the $G_t$ surface (BUZDYGON and LIEBMAN 1984). However, the interaction also depends on ionic interactions as is seen by the effect of mutagenic charge pair reversal (Glu-134/Arg-135) in the rhodopsin sequence, at the cytoplasmic border of the third cytoplasmic loop. This mutant forms normal MII but does not show any MII stabilization when $G_t$ is added (FRANKE et al. 1990). Bound transducin slows the hydroxylamine-induced decay of MII by a factor of 20 (HOFMANN et al. 1983; HOFMANN 1986), suggesting that a region of the $G_t$ holoprotein may inhibit the access to the retinal binding pocket.

### III. Rhodopsin/G-Protein Interaction with Bound Nucleotides

As shown in Fig. 3A, GTP accelerates the decay of extra-MII, which monitors the dissociation of the complex into free metarhodopsin and active $G_t$ (Sect. C.I) in a dose-dependent manner. There is a linear relationship between GTP or its analogs and the rate of dissociation (KOHL and HOFMANN 1987). For sufficiently high GTP, extra-MII is apparently abolished, providing a simple criterion for MII-$G_t$ interaction leading to dissociated, active $G_t$ (see Sect. F.I).

Collisional interaction of $G_t$ with MII occurs when the nucleotide site is occupied with GDP (Fig. 1C). In this form, $G_t$ does not stabilize MII, as seen from the suppression by GDP of extra-MII to the control MII level (Fig. 3B). It is interesting to note that the analog GDP$\beta$S does not exert this effect while it binds to MII-$G_t$ with similar affinity (KAHLERT et al. 1990b). The kinetics of extra-MII at intermediate GDP concentration is consistent with a rapid equilibration between GDP- and MII-binding of $G_t$, which leads to an intermediate level of MII-stabilization with identical initial rate (see also PANICO et al. 1990).

Figure 3C shows dissociation signals at $20\,\mu M$ GTP, i.e., in the range where the rate is linearly related to [GTP]. When GDP (or GDP$\beta$S, not shown) is added, the rate of $G_t$ dissociation decreases according to a hyperbolic competition curve (the dotted line is corrected for GTP contamination). Figure 3C shows tant GDP and GTP compete for the same binding site (for detail see KAHLERT et al. 1989a).

## F. Mechanism of Transducin Activation

### I. Role of Rhodopsin's Cytoplasmic Loops

#### 1. Three Loops Contribute to MII-$G_t$ Interaction

Chemical probing of MII-$G_t$ interaction, employing biophysical monitors of the interaction, was first applied with sulfhydryl modification of R and/or $G_t$

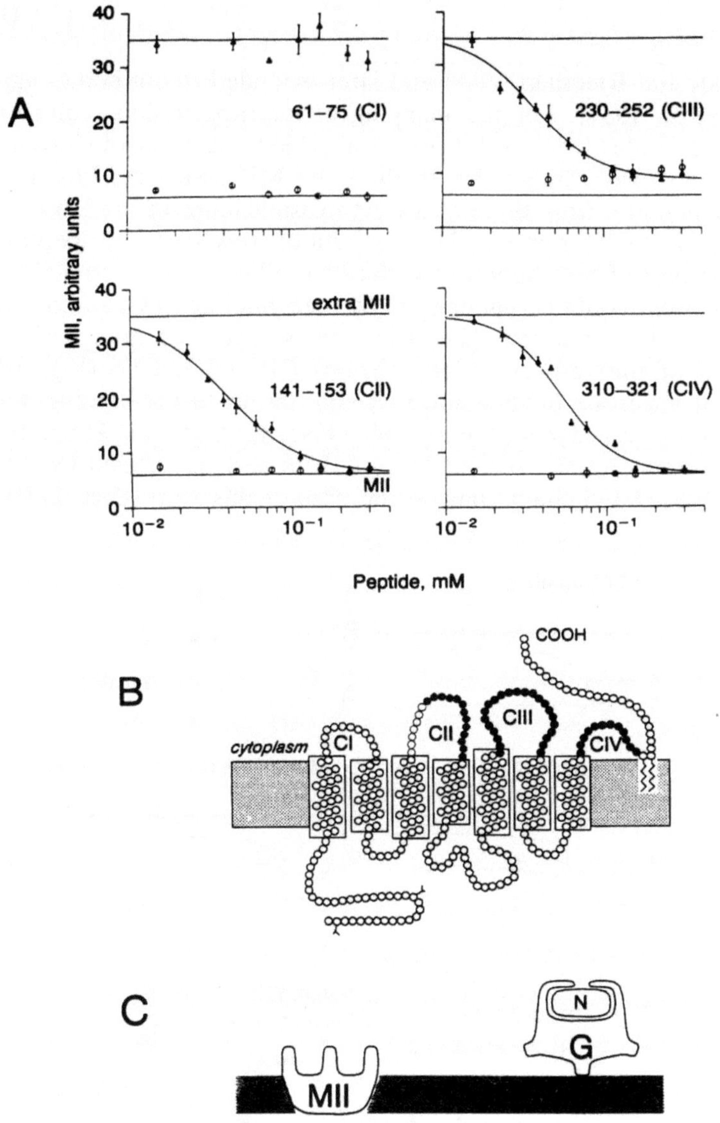


**Fig. 4A–C.** Interaction sites of Meta II and $G_t$. **A** Competition of rhodopsin cytoplasmic loop peptides against MII-$G_t$ formation. Data points [mean ± SEM, $n$ = 8 or 4 (CI)] are from measurements of MII-formation (see Sect. C.I), in the presence (*filled symbols*) or absence (*open symbols*) of $G_t$, and in absence of nucleotide. *Solid lines*, the control levels measured in the absence of peptides. The curves are hyperbolic fits to the data, with exponents 1.83, 2.11, and 2.46 for CII, CIII, and CIV, respectively (KÖNIG et al. 1989b). Competition curves as shown in this figure were obtained with different membrane and peptide preparations (KÖNIG 1989; see Sect. C.IV). **B** Topographic model of rhodopsin. *Filled circles*, peptide sequences that were found to compete against MII-$G_t$ binding. **C** Model of Meta II and $G_t$. The MII state of rhodopsin presents the cytoplasmic loops CII-CIV in a special relative position and/or geometry. Only the MII conformation is compatible with all corresponding $G_t$ binding sites (see also KÖNIG et al. 1989; HOFMANN and KAHLERT, 1992). The model of the interaction sites is used in Fig. 7 to symbolize the different interactive states of rhodopsin and $G_t$

(HOFMANN and REICHERT 1985) and later extended to antibodies against $G_t$ (HAMM et al. 1987). For the mapping of interaction sites, competition of synthetic peptides from $G_t$ (HAMM et al. 1988) and from rhodopsin (KÖNIG et al. 1989b) was combined with the extra-MII assay. As Fig. 4A shows, synthetic peptides from three of the cytoplasmic loops of rhodopsin compete against extra-MII formation. No competition was seen with peptides from any other part of the rhodopsin surface. In all cases, the competitive effect of the peptides could be understood by their binding to $G_t$ and a reversal of $G_t$ induced MII formation. The observations have been interpreted to show that each of the respective loop regions CII, CIII, CIV (Fig. 4B) must contain at least one binding site that contributes to the binding domain of rhodopsin for transducin (KÖNIG et al. 1989b; KAHLERT and HOFMANN 1992). The importance of the second loop, CII, is also underlined by the crucial role of the 134/135 charge pair at the base of this loop (Sect. E.II) and by

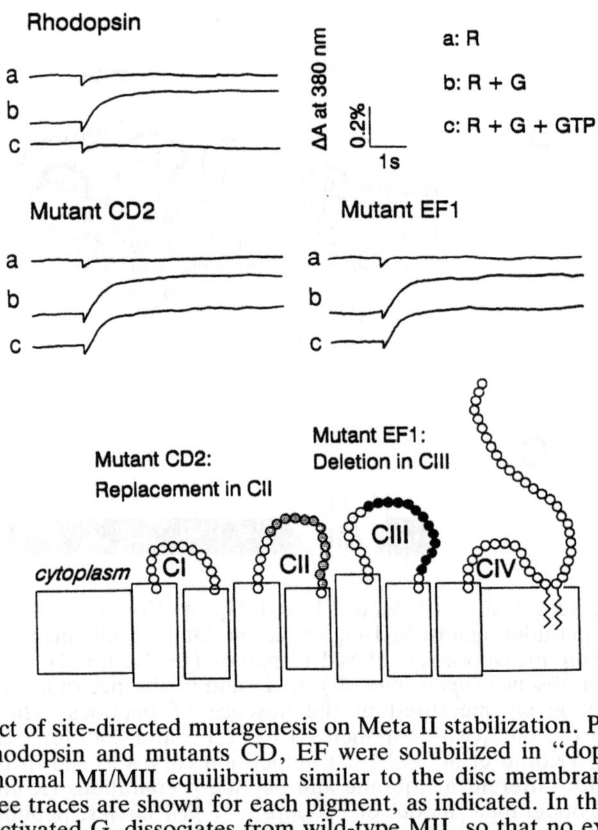


**Fig. 5.** Effect of site-directed mutagenesis on Meta II stabilization. Purified COS cell wild-type rhodopsin and mutants CD, EF were solubilized in "doped micelles" to adjust the normal MI/MII equilibrium similar to the disc membrane (KÖNIG et al. 1989a). Three traces are shown for each pigment, as indicated. In the presence of $G_t$ and GTP, activated $G_t$ dissociates from wild-type MII, so that no extra-MII is seen. However, for the mutants CD2 and EF1, a stable extra-MII level is observed. The sample contained 500 n$M$ pigment, 40 n$M$ $G_t$, and 20 $\mu M$ GTP in HEPES-buffered saline, pH 8 (see FRANKE et al. 1990). The flash photolyzed about 10% of the rhodopsin; T = 4°C

the loss of activity when a part of the sequence is replaced by an inactive sequence (Sect. F.I.2). The involvement of the large CIII loop in rhodopsin-$G_t$ interaction became evident since KÜHN and HARGRAVE (1981) had studied the effect of partial digestion on $G_t$ binding and activation. For the significance of the CIV loop, independent evidence comes from studies of the related $\beta$-AR system (O'DOWD et al. 1988). The loop regions of seven helix receptors in general are discussed in a recent review (HARGRAVE 1991). Interaction sites of rhodopsin and $G_t$ are schematically shown in Fig. 4C.

### 2. Loop Mutants: Binding and Activation in MII-$G_t$ Interaction

To further test the function of loop sites by site-directed mutagenesis, a segment of the second cytoplasmic loop (residues 140–152) was replaced and a segment of the third loop (residues 237–249) was deleted (FRANKE et al. 1990). These mutant opsins, regenerated with 11-*cis*-retinal to rhodopsin with a normal absorption spectrum and formed MII normally on flash excitation. In "doped micelles" (KÖNIG et al. 1989a) both loop mutants form extra-MII indistinguishable from wild-type rhodopsin (Fig. 5). This result shows, in agreement with peptide competition (KÖNIG et al. 1989b) and the effect of GDP$\beta$S (KAHLERT et al. 1990b), that the MII conformation does not need more than two loops interacting with $G_t$ to be stabilized. In neither loop mutant does GTP cause dissociation of the MII-$G_t$ complex (Fig. 5). Thus the $G_t$ activation process is impaired after formation of the complex with MII, as also shown by the absence of GTPase activity (FRANKE et al. 1990). Interaction at the second and third cytoplasmic loops is obligatory for $G_t$ activation, and each of them needs in addition the interaction at the other two effective loops. At least three loops of rhodopsin must interact to active $G_t$ in the fast and effective mode tested in these experiments (see also Sect. G). Alteration of the loops stops the transduction chain after complex formation, as does the 134/135 charge pair reversal after MII formation (Sect. E.II). These findings indicate complex and delicate structural requirements for the different steps of $G_t$ activation. In this regard, it is interesting to note that purified $G_t$ does not take up GTP$\gamma$S over hours (MATESIC and LIEBMAN 1992) and also remains susceptible, in the presence of GTP$\gamma$S, to subsequent rhodopsin-catalyzed change of its fluorescence (S. JÄGER, K.P. HOFMANN, unpublished). The complexity of the MII-$G_t$ interaction is underlined by the observation that depalmitylation of rhodopsin enhances the GTPase activity of $G_t$ but does not affect its light-dependent binding to rhodopsin (MORRISON et al. 1991).

## II. Dissection of Reaction Steps

### 1. The GDP/MII Switch

Taking into account that GDP-bound $G_t$ does not stabilize MII (Sect. E.III), and that binding via two loops is sufficient to stabilize (Sect. F.II), it follows that GDP-bound $G_t$ cannot be compatible with MII on more than one loop (KÖNIG et al. 1989b; FRANKE et al. 1990). Only the empty site $G_t$ is fully compatible, while GDP (and, differently, its analog GDPβS) distort the interactive domain (see Sect. E.III). It is obvious, on the other hand, that the GDP-binding form of $G_t$ must be able to interact with photoactivated R (R*), because it is the form which R* encounters (Fig. 1C). In this type of interaction, R* must induce the opening of the nucleotied binding site. It can only be transient because it fails to cause a kinetic delay in the formation of the stable empty site $G_t$ complex with MII rhodopsin (Sect. C.I). MII and GDP displace each other at $G_t$ (BOURNE et al. 1988) via the transient state.

## 2. The MII/GTP Switch

GTP binds very strongly to $G_t$ after the G-protein has dissociated from MII (BENNETT and DUPONT 1985). This is in contrast to the binding of GTP to the MII-$G_t$ complex. It has been stated above (Sect. C.III) that the $G_t$ activation rate in vitro does not saturate up to at least 100 $\mu M$ GTP, leading to the conclusion that the nucleotide site cannot be occupied completely at this concentration. This applies even more to measurements of the ATR signal in situ (Fig. 6). Dissociation rates measured at micromolar GTP (KOHL and HOFMANN 1987) must be extrapolated linearly up to the millimolar range to explain the rapid transducin turnover in situ (30, 120, 800, 2500, and 4000 activated transducins per photoactivated rhodopsin and second at 5°, 10°, 20°, 30°, and 37°C, respectively; KAHLERT and HOFMANN 1991). Thus the interactions of GTP which it has in the active $G_\alpha$ conformation cannot be built up while $G_t$ is bound to MII. GTP binds with other or less interactions to the complex than it does to the active conformation. Our interpretation is that, analogous to collisional coupling of MII and $G_t$-GDP, GTP binds weakly to the MII-$G_t$ complex to induce the transition into stable $G_t$-GTP (Fig. 7).

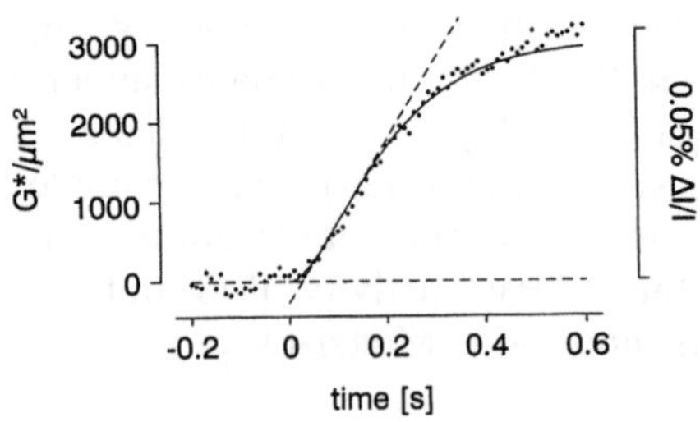


**Fig. 6.** ATR signal and model calculation of G* generation after a flash. The ATR signal (*points*) was recorded from an intact retina at T = 21°C and with flash excitation of rhodopsin R*/R = $2.7 \cdot 10^{-4}$ at time 0. *Solid curve*, obtained from numerical solution of the differential equations for $G^*(t)$ (KAHLERT and HOFMANN 1991) with $[G_o]$ = 3000/$\mu m^2$, $[R^*_0]$ = 10/$\mu m^2$. Only with an assumed GTP concentration of 500 $\mu M$ can the observed speed of the reaction be fitted

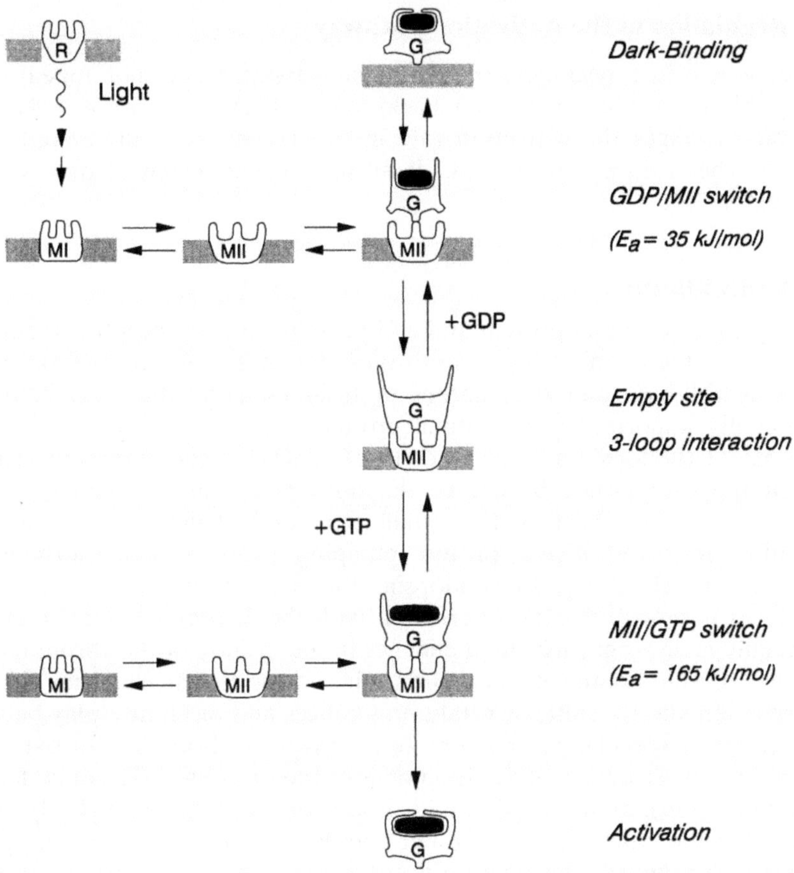


**Fig. 7.** Reaction model of transducin activation. The reaction pathway begins with the collision of active MII and $G_t$ in the GDP-binding state (for symbolization of the binding sites see Fig. 4). Formation of the catalytic MII-stabilizing three-loop "empty-site" interaction must pass through a transient state, in which rhodopsin induces the opening of the nucleotide site while the three-loop interaction is not yet established. Stable light-binding is necessarily connected with the release of GDP. This can be viewed as an "induced fit" mechanism, where MII induces the fitting conformation of $G_t$ in that it enables the loss of GDP (GDP/MII switch). Binding of GDP to the empty site complex leads inversely to the transient state and dissociation. The action of GTP can be understood by analogy: weak binding of the triphosphate allows the transition into another stable state in which MII is replaced by GTP (MII/GTP switch)

When reading the reaction scheme in Fig. 7 from the empty site interaction upwards and downwards, the two switch processes disclose their similarity: $G_t$ can switch from MII binding to the binding of GDP or GTP only via transient states of higher enthalpy. In this regard, it is of interest that the activation energy for MII-$G_t$ dissociation is very high ($E_a$ = 165 kJ/mol; KOHL and HOFMANN 1987; KAHLERT and HOFMANN 1991).

### III. Regulation of the Activation Pathway

The double-switch machinery of transducin activation does not appear to be susceptible to regulation: neither background (PEPPERBERG et al. 1988) nor bleaching changes the activation rate in situ (KAHLERT et al. 1990b). This point has been discussed in more detail in a recent review (HOFMANN and KAHLERT 1991).

## G. Conclusion

Rhodopsin, as a retinal protein and a G-protein coupled receptor, reflects in its functions the membership of both these families. The protonation of the retinal Schiff base and the charges of residues along the third helix are functionally important in all retinal proteins and are also involved in the functions of rhodopsin as a receptor (Sect. D.I). Crucial here is the charge pair at the cytoplasmic border of the third helix, an invariant of all G-coupled receptors (Sect. E.II). Small structural changes accompany the transition into the active proton pumping state of bacteriorhodopsin (DENCHER et al. 1991). In rhodopsin, the presentation of multiple loop structures necessary for the interaction with the G-protein (Sect. F.I) may eventually arise from such small changes in the vicinity of the chromophore. At the present state of our knowledge, possible mechanisms include transmission via tilt, shift, or rotation of helices and/or an interplay between charges and molecular interactions (e.g., hydrogen bridges; ZUNDEL 1992) controlling the structure of the loop region. Importantly, MII can be formed in mutants which are blocked in their interaction with $G_t$ (Sect. E.II), which shows that the coupling between chromophore-protein and protein-protein interactions cannot be entirely rigid.

The rhodopsin/G-protein interaction is notable by its sharp distinction between the active and the inactive receptor and by its high catalytic rates. It would be of interest to compare the catalytic efficiency of light-activated rhodopsin and also the dark-active receptor–all-*trans* retinal complex (Sect. D.III), to other receptor systems. Unfortunately, the available data are not directly comparable. However, we have seen (Sect. C.III) that a peptide from the $\beta_2$-adrenergic receptor, which is comparable in its efficiency to the receptor (OKAMOTO et al. 1991), is a much less efficient catalyst of nucleotide exchange than light-activated rhodopsin. The same has been experienced with mastoparan (M. HECK, K.P. HOFMANN, unpublished). It seems that, in agreement with the complex loop interactions required for $G_t$ activation (Sect. F.I.2), simple structures are not competent to activate transducin in the normal fast mode. It remains to be investigated how the dark active rhodopsin forms compare to other receptors. Conceivably, the rhodopsin/transducin system is optimized toward fidelity and speed, while other systems may be tuned to other tasks, for example, to the sorting of different G-protein pathways.

*Acknowledgments.* The work carried out in my laboratory was supported by grants from the Deutsche Forschungsgemeinschaft (SFB 60 and SFB 325). I wish to thank Martin Heck, Stefan Jäger, and Oliver Ernst for critical reading of the manuscript and for providing me with data prior to publication.

## References

Applebury ML, Hargrave PA (1986) Molecular biology of the visual pigments. Vision Res 26:1881–1885

Bennett N, Michel-Villaz M, Kühn H (1982) Light-induced interaction between rhodopsin and the GTP-binding protein – metarhodopsin II is the major photoproduct involved. Eur J Biochem 127:97–103

Bennett N, Dupont Y (1985) The G-protein of retinal rod outer segments (transducin) – mechanism of interaction with rhodopsin and nucleotides. J Biol Chem 260:4156–4168

Bignetti E, Cavaggioni A, Fasella P, Ottonello S, Rossi GL (1980) Light and GTP effects on the turbidity of frog visual membrane suspensions. Mol Cell Biochem 30:93–99

Bornancin F, Pfister C, Chabre M (1989) The transitory complex between photoexcited rhodopsin and transducin – reciprocal interaction between the retinal site in rhodopsin and the nucleotide site in transducin. Eur J Biochem 184:687–698

Bourne HR, Masters SB, Miller RT, Sullivan KA, Heidemann W (1988) Mutations probe structure and function of G-protein $\alpha$ chains. Cold Spring Harbor Symp Quant Biol LIII:221–228

Bruckert F, Vuong TM, Chabre M (1988) Light and GTP dependence of transducin solubility in retinal rods. Eur Biophys J 16:207–218

Buzdygon BE, Liebman PA (1984) Albumin inhibits light activation of cGMP phosphodiesterase on rod disc membranes. J Biol Chem 259:14567–14571

Chabre M, Breton J (1979) The orientation of the chromophore of vertebrate rhodopsin in the "meta" intermediate states and the reversibility of the Meta II–Meta III transition. Vision Res 19:1005–1018

DeGrip (1988) Recent chemical studies related to vision. Photochem Photobiol 48:799–810

Dencher NA, Heberle J, Bark C, Koch MHJ, Rapp G, Oesterhelt D, Bartels K, Büldt G (1991) Proton translocation and conformational changes during the bacteriorhodopsin photocycle: time-resolved studies with membrane-bound optical probes and X-ray diffraction. Photochem Photobiol 54:881–887

Dratz EA, Hargrave PA (1983) The structure of rhodopsin and the rod outer segment disc membrane. Trends Biochem Sci 8:128–131

Emeis D, Hofmann KP (1981) Shift in the relation between flash-induced metarhodopsin I and metarhodopsin II within the first 10% rhodopsin bleaching in bovine disc membranes. FEBS Lett 136:201–207

Emeis D, Kühn H, Reichert J, Hofmann KP (1982) Complex formation between metarhodopsin II and GTP-binding protein in bovine photoreceptor membranes leads to a shift of the photoproduct equilibrium. FEBS Lett 143:29–34

Ferretti L, Karnik SS, Khorana HG, Nassal M, Oprian DD (1986) Total synthesis of a gene for bovine rhodopsin. Proc Natl Acad Sci USA 83:599–603

Franke RR, König B, Sakmar TP, Khorana HG, Hofmann KP (1990) Rhodopsin mutants that bind but fail to activate transducin. Science 250:123–125

Ganter UM, Schmid ED, Perez-Sala D, Rando RR, Siebert F (1989) Removal of the 9-methyl group of retinal inhibits signal transduction in the visual process – a Fourier transform infrared and biochemical investigation. Biochemistry 28:5954–5962

Ganter UM, Longstaff C, Pajares MA, Rando RR, Siebert F (1991) Fourier-transform Infrared studies of active-site methylated rhodopsin – implications for

chromophore-protein interaction, transducin activation, and the reaction pathway. Biophys J 59:640–644
Ganter UM, Charitopoulos T, Virmaux N, Siebert F (to be published) Conformational changes of cytosolic loops of bovine rhodopsin during the transition to metarhodopsin. II. An investigation by Fourier transform infrared difference spectroscopy. Photochem Photobiol
Guy PM, Koland JG, Cerione RA (1990) Rhodopsin-stimulated activation-deactivation cycle of transducin: kinetics of the intrinsic fluorescence response of the $\alpha$-subunit. Biochemistry 29:6954–6964
Hamm HE (1991) Molecular interactions between the photoreceptor G protein and rhodopsin. Cell Mol Neurobiol 11:563–578
Hamm HE, Deretic D, Hofmann KP, Schleicher A, Kohl B (1987) Mechanism of action of monoclonal antibodies that block the light-activation of the guanyl nucleotide binding protein, transducin. J Biol Chem 262:10831–10838
Hamm HE, Deretic D, Arendt A, Hargrave PA, König B, Hofmann KP (1988) Site of G protein binding to rhodopsin mapped with synthetic peptides from the $\alpha$ subunit. Science 241:832–835
Hargrave PA (1991) Seven-helix receptors. Curr Opinion Struct Biol 1:575–581
Hargrave PA, McDowell JM (to be published) Rhodopsin and phototransduction: a model system for G-protein linked receptors. FASEB J
Hargrave PA, Hamm HE, Hofmann KP (1993) Interaction of rhodopsin with the G-protein, transducin. Bioessays 15:43–50
Henderson R, Unwin PNT (1975) Three-dimensional model of purple membrane obtained by electron microscopy. Nature 257:28–32
Higashijima T, Ferguson KM, Sternweis PC, Ross EM, Smigel MD, Gilman AG (1987) The effect of activating ligands on the intrinsic fluorescence of guanine nuleotide-binding regulatory proteins. J Biol Chem 262:752–756
Hofmann KP, Uhl R, Hoffmann W, Kreutz W (1976) Measurements of fast light-induced light-scattering and absorption changes in outer segments of vertebrate light sensitive rod cells. Biophys Struct Mech 2:61–77
Hofmann KP, Emeis D (1979) Differential light detector. Rev Sci Instr 50:249–252
Hofmann KP, Emeis D (1981) Comparative kinetic light-scattering and -absorption photometry. Biophys Struct Mech 8:23–34
Hofmann KP, Emeis D, Schnetkamp PPM (1983) Interplay between hydroxylamine, metarhodopsin II and GTP-binding protein in bovine photoreceptor-membranes. Biochim Biophys Acta 725:60–70
Hofmann KP (1984) Light-scattering signal of G-binding in rod outer segments depends on osmolarity. Invest Ophthal Vis Sci 25a:56
Hofmann KP (1985) Effect of GTP on the rhodopsin G-protein complex by transient formation of extra metarhodopsin II. Biochim Biophys Acata 810:278–281
Hofmann KP, Reichert J (1985) Chemical probing of the light-induced interaction between rhodopsin and G-protein – near infrared light scattering and sulfhydryl modifications. J Biol Chem 260:7990–7995
Hofmann KP (1986) Photoproducts of rhodopsin in the disc membrane. Photobiochem Photobiophys 13:309–338
Hofmann KP, Kahlert M (1992) The activation of transducin: studies on its mechanism and modulation. In: Hargrave PA, Hofmann KP, Kaupp UB (eds) Signal transduction in photoreceptor cells. Springer, Heidelberg, New York
Hofmann KP, Pulvermüller A, Buczylko J, Hooser PV, Palczewski K (1992) The role of arrestin and retinoids in the regeneration pathway of rhodopsin. J Biol Chem 267:15701–15706
Kahlert M, König B, Hofmann KP (1990a) Displacement of rhodopsin by GDP from three-loop interaction with transducin depends critically on the diphosphate $\beta$-position. J Biol Chem 265:18928–18932
Kahlert M, Pepperberg DR, Hofmann KP (1990b) Effect of bleached rhodopsin on signal amplification in rod visual receptors. Nature 345:537–539

Kahlert M, Hofmann KP (1991) Reaction rate and collisional efficiency of the rhodopsin-transducin system in intact retinal rods. Biophys J 59:375–386
Khorana HG (1992) Rhodopsin, photoreceptor of the rod cell. J Biol Chem 267:1–4
Kelleher DJ, Johnson GL (1988) Transducin inhibition of light-dependent rhodopsin phosphorylation: evidence for $\beta\tau$ subunit interaction with rhodopsin. Mol Pharm 34:452–460
Kibelbek J, Mitchell DC, Beach JM, Litman BJ (1991) Functional equivalence of metarhodopsin II and the $G_t$-activating form of photolyzed bovine rhodopsin. Biochemistry 30:6761–6768
König B (1989) Blitzlichtphotometrische Untersuchung der Interaktion des Photorezeptors Rhodopsin mit dem G-Protein Transducin: Identifizierung der relevanten Bereiche mit Hilfe monoklonaler Antikörper und synthetischer Peptide. Thesis, University of Freiburg
König B, Welte W, Hofmann KP (1989a) Photoactivation of rhodopsin and interaction with transducin in detergent micelles: effect of "doping" with steroid molecules. FEBS Lett 257:163–166
König B, Arendt A, McDowell JH, Kahlert M, Hargrave PA, Hofmann KP (1989b) Three cytoplasmic loops of rhodopsin interact with transducin. Proc Natl Acad Sci USA 86:6878–6882
Kohl B, Hofmann KP (1987) Temperature dependence of G-protein activation in photoreceptor membranes – transient extra metarhodopsin II on bovine disc membranes. Biophys J 52:271–277
Kreutz W, Siebert F, Hofmann KP (1984) Kinetic infrared spectorscopy and kinetic light-scattering – two new methods for studying fast trigger processes. In: Chapman D (ed) Biological membranes V. Academic, New York, pp 240–277
Kühn H (1980) Light- and GTP-regulated interaction of GTPase and other proteins with bovine photoreceptor membranes. Nature 283:587–589
Kühn H, Hargrave PA (1981) Light-induced binding of GTPase to bovine photoreceptor membranes: effects of limited proteolysis of the membranes. Biochemistry 20:2410–2417
Kühn H, Bennett N, Michel-Villaz M, Chabre M (1981) Interactions between photoexcited rhodopsin and GTP-binding protein: kinetic and stoichiometric analysis from light scattering changes. Proc Natl Acad Sci USA 78:6873–6877
Kühn H (1984) Interactions between photoexcited rhodopsin and light-activated enzymes in rods. In: Osborne N, Chader J (eds) Progress in retinal research. Pergamon, New York, pp 123–153
Lamb TD, Pugh EN Jr (1992) A quantitative account of the activation steps involved in phototransduction in amphibian photoreceptors. J Physiol 449:719–758
Liebman PA, Parker KR, Dratz EA (1987) The molecular mechanism of visual excitation and its relation to the structure and composition of the rod outer segment. Ann Rev Physiol 49:765–791
Longstaff C, Calhoon RD, Rando RR (1986) Deprotonation of the Schiff base of rhodopsin is obligate in the activation of the G protein. Proc Natl Acad Sci USA 83:4209–4213
Matesic D, Liebman PA (1992) Spontaneous G-protein nucleotide exchange rate: implications for rod dark noise. Invest Ophthalm Vis Sci: ARVO abstract 2049
Morrison DF, O'Brien PJ, Pepperberg DR (1991) Depalmitylation with hydroxylamine alters the functional properties of rhodopsin. J Biol Chem 266:20118–20123
O'Dowd DF, Hnatovich, Regan JW, Leader WM, Caron MG, Lefkowitz RJ (1988) Site-directed mutagenesis of cytoplasmic domains of the human $\beta_2$-adrenergic receptor. Localization of regions involved in G protein-receptor coupling. J Biol Chem 262:15985–15992
Okamoto, T, Murayama, Y, Hayashi, Y, Inagaki M, Ogata E, Nishimoto I (1991) Identification of a $G_s$ activator region of the $\beta_2$-adrenergic receptor that is autoregulated via protein kinase A-dependent phosphorylation. Cell 67:723–730

Ovchinnikov YA (1982) Rhodopsin and bacteriorhodopsin: structure-function relationships. FEBS Lett 148:179–191

Palczewski K, Rispoli G, Detwiler PB (1992) The influence of arrestin (48K protein) and rhodopsin kinase on visual transduction. Neuron 8:117–126

Pepperberg DR, Kahlert M, Krause A, Hofmann KP (1988) Photic modulation of a highly sensitive, near-infrared light-scattering signal recorded from intact retinal photoreceptors. Proc Natl Acad Sci USA 85:5531–5535

Panico J, Parkes JH, Liebman PA (1990) The effect of GDP on rod outer segment G-protein interactions. J Biol Chem 265:18922–18927

Pepperberg DR, Cornwall MC, Kahlert M, Hofmann KP, Jin J, Jones GJ, Ripps H (1991) Light-dependent delay in the falling phase of the retinal rod photoresponse. Vis Neuroscience 8:9–18

Pfister C, Kühn H, Chabre M (1983) Interaction between photoexcited rhodopsin and peripheral enzymes in frog retinal rods. Influence on the postmetarhodopsin II decay and phosphorylation rate of rhodopsin. Eur J Biochem 136:489–499

Robinson PR, Cohen GB, Zhukovsky EA, Oprian D (1992) Constitutive activation of rhodopsin by mutation of $Lys^{296}$. FASEB J 6:A46

Sakmar TP, Franke RR, Khorana HG (1989) Glutamic acid-113 serves as the retinylidene Schiff base counterion in bovine rhodopsin. Proc Natl Acad Sci USA 86:8309–8313

Schleicher A, Hofmann KP (1987) Kinetic study on the equilibrium between membrane-bound and free photoreceptor G-protein. J Membr Biol 95:269–279

Schleicher A, Franke R, Hofmann KP, Finkelmann H, Welte W (1987) Deoxylysolecithin and a new biphenyl detergent as solutilizing agents for bovine rhodopsin – functional test by formation of metarhodopsin II and binding of G-protein. Biochemistry 26:5908–5916

Schleicher A, Kühn H, Hofmann KP (1989) Kinetics, binding constant, and activation energy of the 48-KDa protein-rhodopsin complex by extra-metarhodopsin II. Biochemistry 28:1770–1775

Siebert F (1992) Infrared spectroscopic investigation of retinal proteins. In: Clark RJH, Hester RE (eds) Biomolecular spectroscopy. Advances in spectroscopy. Wiley, Chichester

Smith HG, Stubbs GW, Litman BJ (1975) The isolation and purification of osmotically intact discs from retinal rod outer segments. Exp Eye Res 20:211–217

Uhl R, Desel H, Ryba N, Wagner R (1987) J Biochem Biophys Meth 14:127–138

Vuong TM, Chabre M, Stryer L (1984) Millisecond activation of transducin in the cyclic nucleotide cascade of vision. Nature 311:659–661

Vuong TM, Chabre M (1991) Deactivation kinetics of the transduction cascade of vision. Proc Natl Acad Sci USA 88:9813–9817

Wagner R, Ryba NJP, Uhl R (1987) Rapid transducin deactivation in intact stacks of bovine rod outer segment disks as studied by light scattering techniques – arrestin requires additional soluble proteins for rapid quenching of rhodopsin catalytic activity. FEBS Lett 235:103–108

Weitz CJ, Nathans J (1992) Histidine residues regulate the transition of photoexcited rhodopsin to its active conformation, metarhodopsin II. Neuron 8:465–472

Wilden U, Hall SW, Kühn H (1986) Phosphodiesterase activation by photoexcited rhodopsin is quenched when rhodopsin is phosphorylated and binds the intrinsic 48-kDa protein of rod outer segments. Proc Natl Acad Sci USA 83:1174–1178

Zhukovsky EA, Robinson PR, Oprian DD (1991) Transducin activation by rhodopsin without a covalent bond to the 11-cis-retinal chromophore. Science 251:558–560

Zundel G (to be published) Proton polarizability and proton transfer processes in hydrogen bonds and cation polarizabilities of other cation bonds – their importance to understand molecular processes in electrochemistry and biology. Trends Phys Chem

CHAPTER 61

# Fast Kinetics of G-Protein Function In Vivo

G. SZABO, Y. LI, and A.S. OTERO

## A. Introduction

The muscarinic receptor-mediated activation of the potassium current $I_{K(ACh)}$ in atrial myocytes is the most extensively studied prototype of the direct activation of effectors by G-proteins (reviewed in SZABO and OTERO 1990; BIRNBAUMER et al. 1990). Electrophysiological methods applied both at the level of whole cell and single channel recording have shown that this is a membrane-delimited process that operates without diffusible cytosolic intermediates (SOEJIMA and NOMA 1984). The high sensitivity and time resolution inherent in the patch clamp technique (HAMILL et al. 1981) make it possible to follow the time course of activation and deactivation of the effectors, in this case a specific class of $K^+$-permeable channels, in response to rapid application and removal of agonist. Coupled with maneuvers that specifically alter receptor or G-protein function, these methods can be used to examine in a quantitative manner the kinetics of the processes interposed between receptor activation and channel opening.

In particular, monitoring of whole cell $I_{K(ACh)}$ reveals the functioning of G-protein-mediated signal transduction in a native environment that preserves spatial relationships between the basic components of the system as well as potentially important interactions with other cellular elements. The differences between this and biochemically pure, reconstituted systems may be significant if the specificity and kinetics of G-protein-mediated signal transduction processes depend on their arrangement in the membrane and their immediate environment in the cell.

This chapter concentrates on the kinetics of muscarinic acetylcholine receptor-dependent activation of $I_{K(ACh)}$ mediated by the G-protein $G_k$. While the quantitative details are presumably specific for this system, the general aspects of the mechanisms of G-protein function that emerge from these studies are likely to be applicable to membrane-delimited G-protein-coupled processes in general.

## B. Kinetics of Muscarinic $K^+$ Channel Activation

Rapid application of muscarinic agonists to atrial myocytes activates $I_{K(ACh)}$ with a time course that depends on the extracellular agonist concentration

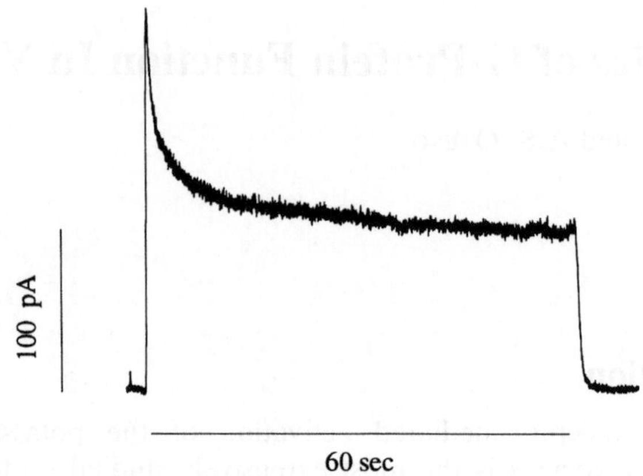


**Fig. 1.** Typical time course of the muscarinic potassium channel activation in frog atrial myocytes. Single cells were obtained by enzymatic dissociation of bullfrog (Rana catesbeiana) atria. Detailed procedures have been described elsewhere (BREITWIESER and SZABO 1988; OTERO et al. 1991). The whole-cell configuration of the patch-clamp technique was used to hold the intracellular potential at −35 mV. Membrane currents were recorded as a function of time while ACh (1 $\mu M$) was applied to the cell by a rapid change in the solution bathing the cell. Upward deflections correspond to outward currents. Control experiments show that the ACh containing solution could be applied and washed out within 20 ms

and the intracellular levels of guanine nucleotides. Figure 1 shows the typical response of an atrial cell to 1 $\mu M$ ACh as a function of time. The membrane potential was held at −35 mV, which is more positive than the potassium reversal potential $E_K$, so that $I_{K(ACh)}$ appears as an outward current (upward deflection). ACh was applied for approximately 60 s. Initially, the muscarinic $K^+$ currents rise rapidly to a peak value ($I_p$) and then decrease spontaneously to what is essentially a steady-state value ($I_\infty$). On removal of agonist, the currents return to basal levels with a rapid but measurable time course (BREITWIESER and SZABO 1988).

Detailed analysis of this response shows that the onset of $I_{K(ACh)}$ induced by rapid application of agonist has a sigmoidal time course. This is illustrated in Fig. 2A for two different concentrations of ACh with the aid of an expanded time scale. For these experiments a change in $E_K$ was used to mark the arrival of agonist-containing solution at the cell membrane (see below). It is obvious that $I_{K(ACh)}$, seen as a downward deflection in these experiments, develops only after a considerable delay and reaches a maximum exponentially. Thus, the overall process of channel opening by ACh displays two temporally distinct stages. The first is a latency period that can be defined as the time elapsed from the initial contact of the cell with the agonist to the point where $I_{K(ACh)}$ can be distinguished from the background noise. The second phase marks the period during which channel activation takes place and $I_{K(ACh)}$ rises quasi-exponentially towards a peak

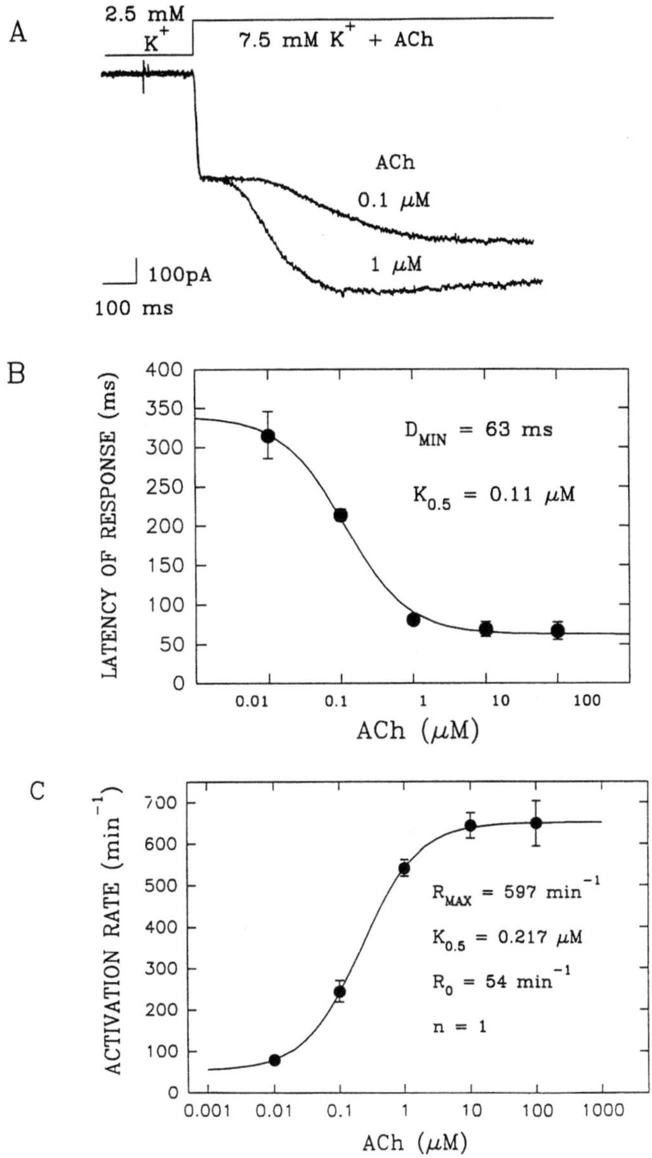


**Fig. 2.** Time course of $I_{K(ACh)}$ following the application of 0.1 and 1 $\mu M$ ACh using an expanded time scale. The holding potential was −85 mV, near $E_K$ for the normal physiological solution containing 2.5 m$M$ $K^+$. The rapidly applied ACh-containing solutions also had a higher $K^+$ concentrations (7.5 m$M$) in order to monitor the precise moment and rapidity of the exchange of external solution by way of an altered potassium reversal potential $E_K$. The sudden change in $E_K$ induced an inward current through the background $K^+$ channel, thereby providing a practically instantaneous sensor of solution exchange at the cell membrane. A typical time course of this "marker current" and the subsequent rise in $I_{K(ACh)}$ is shown in **A**. The rise of $I_{K(ACh)}$ manifests itself as an inward current (downward deflections) since the holding potential is more negative than $E_K$. The ACh concentration dependence of latencies is plotted in **B**, while that of the activation rates, calculated as the inverse of the exponential part of the rise of $I_{K(ACh)}$, is plotted in **C**. Procedures used to draw the *solid lines* are given in the text

value. In principle, separate analysis of each process can yield information regarding the kinetics not only of the channel activation process itself but also the events preceding channel opening.

An accurate measurement of the latency of the response requires fast solution exchange, and a clear marker to indicate the time at which the exchange takes place. Using a rotary valve to switch solutions and placing the cell near the opening of a short tube that connects the valve to the recording chamber, exchange of the solution bathing the cell is usually complete within 20 ms. To determine the exact time when solution exchange occurs, one can take advantage of the background $K^+$ conductance of the cell membrane responding instantaneously (on our time scales) to a change in the potassium reversal potential produced by a larger $K^+$ concentration in the agonist-containing solutions. This method was used to determine the agonist dependence of response latencies. Figure 2B shows for ACh that the latency of response D decreases as the concentration of agonist is raised,

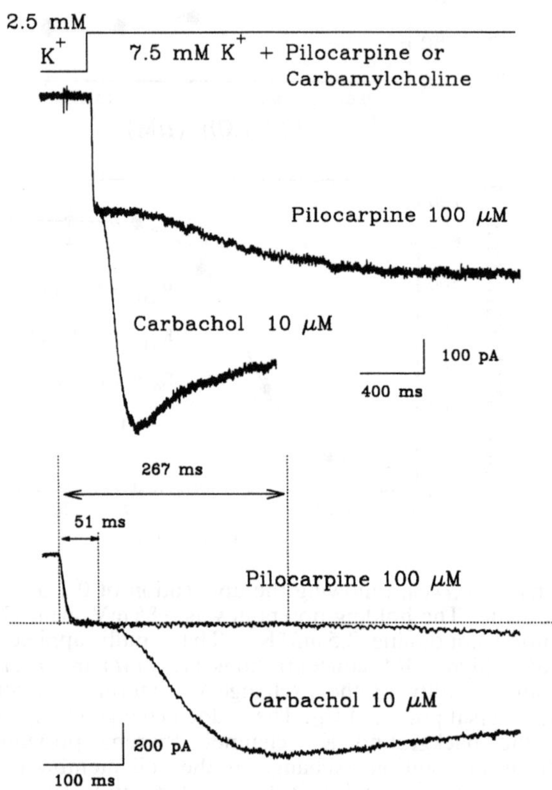


**Fig. 3.** Time course of $I_{K(ACh)}$ activation by saturating concentrations of carbamylcholine (carbachol, 10 $\mu M$) and pilocarpine (100 $\mu M$). Methods were similar to those of Fig. 2

from a maximal value $D_{MAX}$ at low [ACh] down to a limiting value $D_{MIN}$ at saturating agonist. The dependence on agonist concentration is hyperbolic, as shown by fitting the Hill equation to the data (solid line)

$$D = D_{MAX}/(1 + ([ACh]/K_{50})^n + D_{MIN}$$

with $K_{50} = 0.11\,\mu M$, $D_{MIN} = 63\,ms$, $D_{MAX} = 275\,ms$, and $n = 1$.

The actual rate of activation of $I_{K(ACh)}$ (R) can be determined by fitting a single exponential to the rising phase of the response and is calculated as the inverse of the time constant of activation. R increases hyperbolically with the concentration of agonist, and for ACh the concentration dependence profile is well fitted by the Hill equation

$$R = R_{MAX}/(1 + (K_{50}/[ACh])^n) + R_{MIN}$$

with parameters $K_{50} = 0.22\,\mu M$, $R_{MIN} = 54\,min^{-1}$, $R_{MAX} = 597\,min^{-1}$, and $n = 1$.

In general, the kinetics of $I_{K(ACh)}$ activation by other muscarinic cholinergic agonists are similar to those seen with ACh. Figure 3 shows time courses of $I_{K(ACh)}$ activation for carbamylcholine and pilocarpine applied at nearly saturating concentrations, 10 and $100\,\mu M$, respectively. Among full agonists the differences lie mainly in relative potency for $I_{K(ACh)}$ activation ($K_{50}$) but not in kinetic features. Thus, ACh and carbamylcholine exhibit practically identical kinetics when applied at equipotent concentrations: as illustrated in Table 1, the latencies and activation rates for these two agonists are indistinguishable. In contrast, partial agonists such as pilocarpine activate $I_{K(ACh)}$ with significantly slower kinetics even at saturating concentrations (Table 1). These results parallel observations made in reconstituted systems (TOTA and SCHIMERLIK 1990) and can be understood in terms of a less frequent attainment of the activating conformation of the receptor even at full occupancy by partial agonists.

## C. Rapid Desensitization

The rapid, agonist-induced rise of $I_{K(ACh)}$ is invariably followed by a decrease in channel activity, despite the continuing presence of agonist (Fig.

**Table 1.** Kinetic parameters of $I_{K(ACh)}$ activation for different muscarinic cholinergic agonists

| Agonist | Concentration ($\mu M$) | Latency (ms) | Activation rate ($min^{-1}$) |
|---|---|---|---|
| Acetylcholine | 10 | 70 ± 7.7 | 582 ± 34 |
| Carbamylcholine | 10 | 63 ± 13 | 568 ± 37 |
| Pilocarpine | 100 | 288 ± 30 | 112 ± 15 |

Number of determinations: ACh, $n = 5$; carbamylcholine, $n = 3$; pilocarpine, $n = 5$.

1). This spontaneous decay of $I_{K(ACh)}$ is often referred to as "rapid desensitization".

The ratio of the peak value of $I_{K(ACh)}$ ($I_p$) to the steady state response ($I_\infty$) is a practical measure of rapid desensitization. Figure 4 shows that both the peak response (open circles) and the steady state response (filled circles) follow a saturating agonist concentration dependence. Both data sets were fit by the Hill equation:

$$I_{K(ACh)} = I_{MAX}/(1 + (K/[ACh])^n)$$

For the peak response the best fitting parameters were: $I_{MAX} = 257\,pA$, $K = 0.097\,\mu M$, and $n = 1.34$, while for the steady state response $I_{MAX} = 102\,pA$, $K = 0.067\,\mu M$, and $n = 1.12$. Note that while the Hill coefficients for these fits are larger than unity, a simple linear dependence can not be ruled out on statistical grounds and in fact results in a reasonably good fit, with $I_{MAX}$ and K remaining essentially the same ($I_{MAX} = 260\,pA$, $K = 0.1\,\mu M$ and $I_{MAX} = 103\,pA$, $K = 0.069\,\mu M$, respectively, for the peak and steady-state responses). It has been previously reported (KURACHI et al. 1987) that the peak response increases linearly with ACh concentrations up to 1 m*M*. The present study shows that when agonist is applied rapidly so that agonist concentrations stabilize well before the appearance of the $I_{K(ACh)}$ transient, the peak $I_{K(ACh)}$ response does reach a limiting value. Consequently, the relaxation amplitude ($I_p/I_\infty$) rises from a value of approximately 1 at low agonist concentrations to a maximum of 2.5 at saturating [ACh].

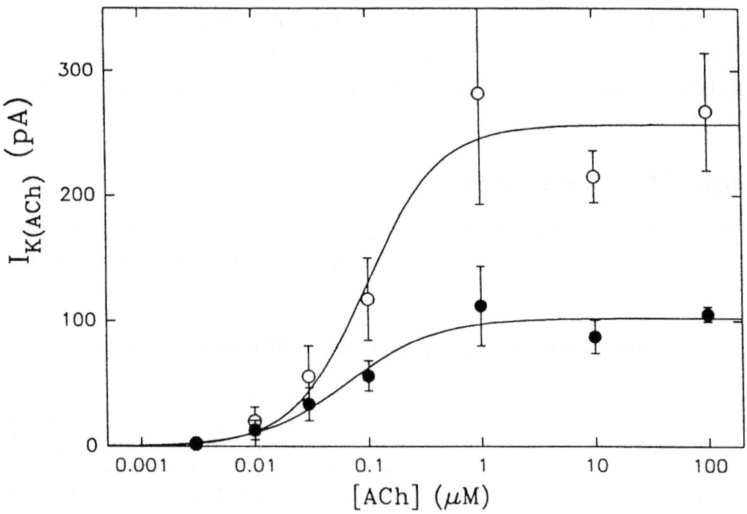


**Fig. 4.** Dependence of the peak and steady state $I_{K(ACh)}$ response on the ACh concentration. The points were fitted to the appropriate Hill equation using parameters given in the text

It is likely that the initial peak of $I_{K(ACh)}$ activation represents a distinct and transient state of activation of the muscarinic $K^+$ channel and/or the G-protein ($G_k$) that regulates it. This is suggested by experiments in which the magnitude of a control, muscarinic agonist-induced $I_{K(ACh)}$ response is compared to that produced by a subsequent intracellular injection of a hydrolysis-resistant GTP analog, GppNHp or GTP$\gamma$S. An experiment of this kind is shown in Fig. 5. ACh was applied at a concentration large enough (1 $\mu M$) to produce a well-defined initial peak and was continued to be applied until steady state was reached. The agonist was then washed out and GppNHp (1 m$M$) was applied intracellularly through the patch pipette. The agonist-independent response is similar in magnitude to the steady-state value of the ACh-induced response, but obviously and reproducibly smaller than $I_p$. This result runs counter to expectations since hydrolysis – resistant trinucleotides are believed to activate $G_k$ irreversibly and completely, and therefore should give rise to the largest attainable channel activation. This is clearly not the case. The presence of agonist-bound receptors or the possibility of guanine nucleotide turnover may cause the observed initial "hyperactivation" of $I_{K(ACh)}$. Alternatively, it is possible that upon activation by muscarinic agonists the system has transient access to a state that induces channel activation more effectively. GTP analogs might eliminate this state, for instance by activating G-proteins that are not linked to muscarinic receptors but affect the function of $G_k$ or the channel through

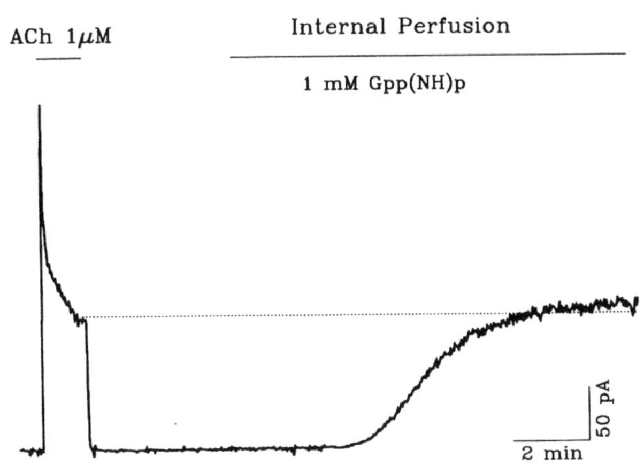


**Fig. 5.** The $I_{K(ACh)}$ response induced by the hydrolysis resistant GTP analog GppNHp is similar in magnitude to the steady-state $I_{K(ACh)}$ response induced by 1 $\mu M$ ACh, and it is clearly smaller than the peak ACh response. $I_{K(ACh)}$ was sampled at 750-ms intervals at the end of a 250-ms pulse of the membrane potential to −5 mV from a holding potential of −85 mV, and plotted on the figure as a function of time. ACh was applied extracellularly during the time indicated by the *bar*. GppNHp (1 m$M$) was applied by intracellular dialysis through of perfusion of the patch pipette, as described elsewhere (Lapointe and Szabo 1987)

modulatory processes such as phosphorylation. Single channel experiments lending support to this possibility have been reported recently (Kim 1991).

Additional possibilities are that activated $G_k$ subunits are taken up by other effectors, adenylyl cyclase for example, or that $G_k$ activation is transiently more effective in the presence of GTP due to a membrane-associated nucleotide diphosphate kinase (Otero 1990; Kaibara et al. 1991). Whether these mechanisms, individually or collectively, are responsible for rapid desensitization remains to be established.

## D. Kinetics of $I_{K(ACh)}$ Deactivation

When agonist is removed from the bathing medium, $I_{K(ACh)}$ decays with a simple exponential time course without any observable delay (see Fig. 1). The time constant of $I_{K(ACh)}$ decay is typically 0.5 s. Since solution exchange and agonist diffusion away from the membrane are likely to be complete well within 50 ms (Hess et al. 1986), the decay of $I_{K(ACh)}$ upon agonist washout reflects the actual deactivation process of $I_{K(ACh)}$ when stimulation by agonist ceases. The rate of deactivation $k_w$ seen upon washout of agonist, calculated as the inverse of the time constant of $I_{K(ACh)}$ decay, is independent of the ACh concentration but it is somewhat affected by the duration of ACh application. In experiments where ACh is washed out at varying times of ACh application, $k_w$ was smaller (1/s) following short (near 2 s) exposures to ACh, and reached a limiting value of 2/s following long exposures to ACh. The increase of $k_w$ as a function of the duration ACh exposure is gradual. It can be described by an exponential function having an initial value of 1.0, a steady-state value of 2.0, and a time constant near 17 s. It is interesting to note that the increase of $k_w$ following ACh application has approximately the same time course as the slow component of the fast desensitization process (14 s for the response shown in Fig. 1). Therefore, it is likely that the slow component of rapid desensitization reflects an increased deactivation rate during the course of a prolonged agonist application.

The deactivation rate of 2/s is much faster than commonly observed rates of GTP hydrolysis by purified G-proteins, in the range of 0.1/s (Breitwieser and Szabo 1988). The significantly larger rate of $I_{K(ACh)}$ deactivation implies either that the rate of $G_k$ deactivation through GTP breakdown is much faster in vivo, possibly due to GTPase activity enhanced by interactions with effector or some other regulatory factor (Arshavsky and Bownds 1992; Berstein et al. 1992), or that $G_k$ deactivation is not an absolute requirement for $I_{K(ACh)}$ deactivation. It is possible, for example, that channel activation is terminated by dissociation of GTP-bound $G_k$ preceding GTP hydrolysis (Erickson et al. 1992).

Although the rate of channel deactivation must be at least 2/s, one cannot rule out the possibility that the process is actually more rapid. It is possible, for example, that the decay of $I_{K(ACh)}$ following ACh washout reflects the rate of agonist dissociation from receptor. If this were the case, then the "off" rate of ACh from the cardiac (m2) muscarinic receptor would have to be near 2/s. This possibility is not inconsistent with the activation rates measured for the system. Specifically, one would expect that at the concentration of ACh that produces half maximal activation of $I_{K(ACh)}$ ($0.1\,\mu M$), agonist "off" and "on" rates should be equal. From the data of Fig. 2C, one calculates a half-maximal $I_{K(ACh)}$ activation rate of 5/s, a value that is of the same magnitude as the deactivation rate of 2/s. It is therefore possible that the kinetics of both the activation and deactivation of $I_{K(ACh)}$ are limited by the kinetics of agonist-receptor interactions. This would imply very rapid kinetics for receptor–$G_k$ interactions as well as activated $G_k$–channel interactions. Further experiments, using a variety of agonists, will be required to examine this possibility.

The rapidity of membrane-delimited, G-protein-mediated effector activation may imply a proximity or linkage of the components of the system in the membrane. There are a number of reports suggesting this possibility (LOGOTHETIS et al. 1987; KARSCHIN et al. 1991; COULTER and RODBELL 1992). Our observations of the effect of pertussis toxin, which disables $G_k$ by ADP-ribosylation, on the kinetics of $I_{K(ACh)}$ activation may support this idea. Briefly, we find that pertussis toxin treatment decreases the magnitude of the currents, but not the rate of $I_{K(ACh)}$ activation (LI et al. 1992). This result implies that unless there is a large excess of channel protein in the membrane relative to $G_k$, an unlikely possibility, channel activation does not procede via a reaction of activated G-protein with channel protein but rather by activation of a preexisting G-protein–channel complex. This would explain the stoichiometric rather than catalytic action of pertussis toxin on $I_{K(ACh)}$ activation.

## E. Basic Kinetic Model for Membrane-Delimited Effector Activation by G-Protein

It is possible to summarize the salient features of a membrane-delimited signal transduction process by considering a simplified kinetic scheme in which the effector is either in a resting, nonactivated state, or in a stimulated, activated state. Conversion between these states is governed by trinucleotide-dependent cycling of the appropriate G-protein between active (GTP-bound) and inactive (GDP-bound) forms. The rate of guanine trinucleotide dependent activation is taken to be proportional to the fraction of receptor molecules present in an activating conformation which, in turn, is enhanced by agonist. In addition, the scheme allows for a small fraction of receptors being in the activating conformation even in the absence of agonist (OKABE et al. 1991; HANF et al. 1993).

This simplified but practical kinetic model is summarized in Fig. 6.

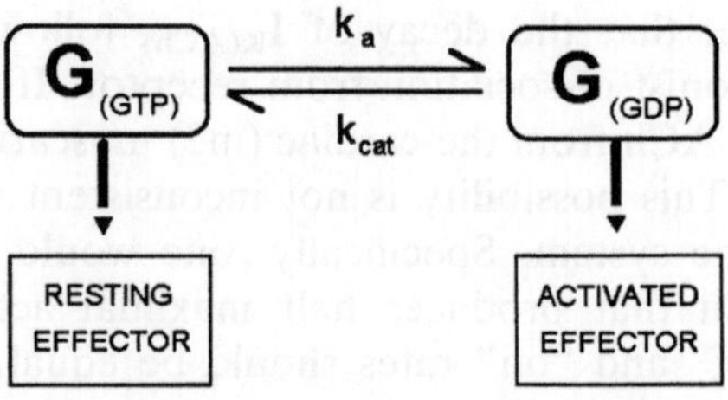


**Fig. 6.** Simple kinetic scheme for membrane-delimited effector activation by G-protein

The overall activation rate $k_a$ is taken to be proportional to the concentration of receptors in activating form, [R*]

$$k_a = \alpha[R^*]$$

[R*] being directly related to the number of agonist bound receptors:

$$[R^*] = R_{TOT}/(1 + K_{ACh}/[ACh])$$

where $R_{TOT}$ represents the total receptor concentration in the membrane and $K_{ACh}$ is the equlibrium constant for the agonist–receptor interaction. This leads to a simple expression for the agonist dependence of the activation rate $k_a$:

$$k_a = \alpha R_{TOT}/(1 + K_{ACh}/[ACh])$$

The data of Fig. 2C together with previously obtained results for low ACh concentrations (BREITWIESER and SZABO 1988) yields the following experimentally determined overall activation rate for ACh:

$$k_a = 9.95/(1 + 0.12/[ACh]) + 0.01$$

where the rates are expressed in units of $s^{-1}$ and [ACh] is the micromolar ACh concentration.

The kinetic scheme developed above accounts for some of the main features of $I_{K(ACh)}$ channel activation and allows one to assign a numerical value to $\alpha R_{TOT}$ and $K_{ACh}$. However, it has also a number of obvious shortcomings. In particular, it does not account for the experimentally observed latency of response and it does not incorporate rapid desensitization, largely because the mechanism of these processes is unclear. Also the model does not predict a higher than first-order dependence of the activation rates as a function of intracellular guanine trinucleotides (LI et al. 1991; ITO et al. 1991). Clearly, further studies will be required to relate the kinetic constants to the detailed molecular events taking place during $I_{K(ACh)}$ activation.

## F. Conclusions

The muscarinic-activated potassium channel in cardiac atrial myocytes provides a well-defined and readily accessible system in which, using the whole-cell patch-clamp technique, it is possible to characterize a membrane-delimited, G-protein-coupled signal transduction system in a quantitative manner in vivo. Thus, we have used the $I_{K(ACh)}$ channel as a rapid, sensitive, and specific indicator of the activation of a specific G-protein, $G_k$, during its interaction with agonist-bound receptor. This approach, coupled with specific interventions to alter selected components of the system, is likely to provide a description of this signal transduction process in terms of chemical kinetic processes that can be related to a sequence of interactions between well-defined components in the membrane as a function of their molecular structures.

## References


Arshavsky VY, Bownds MD (1992) Regulation of deactivation of photoreceptor G proteins by its target enzyme and cGMP. Nature 357:416–417

Berstein G, Blank JL, Jhon DK, Exton JH, Rhee SG Ross EM (1992) Phospholipase C-$\beta$1 is a GTPase-activating protein for $G_{q/11}$, its physiologic regulator. Cell 70:411–418

Birnbaumer L, Abramowitz J, Brown AM (1990) Receptor-effector coupling by G proteins. Biochim Biophys Acta 1031:163–224

Breitwieser GE, Szabo G (1985) Uncoupling of cardiac muscarinic and -adrenergic receptors from ion channels by a guanine nucleotide analogue. Nature 317:538–540

Breitwieser GE, Szabo G (1988) Mechanism of muscarinic receptor-induced $K^+$ channel activation as revealed by hydrolysis resistant GTP analogues. J Gen Physiol 91:469–463

Coulter S, Rodbell M (1992) Heterotrimeric G proteins in synaptoneurosome membranes are crosslinked by p-phenylenedimaleimide, yielding structures comparable in size to crosslinked tubulin and F-actin. Proc Natl Acad Sci USA 89:5842–5846

Erickson MA, Robinson P, Lisman J (1992) Deactivation of visual transduction without guanosine triphosphate hydrolysis by G protein. Science 257:1255–1258

Hanf R, Li Y, Szabo G, Fischmeister R (to be published) Atropine reduces agonist-independent activation of G proteins mediating muscarinic regulation of $Ca^{2+}$ and $K^+$ currents in frog and rat cardiac cells. J Physiol

Hamill OP, Martin A, Neher E, Sakmann B, Sigworth FJ (1981) Improved patch -clamp techniques for high resolution current recording from cells and cell-free membrane patches. Pflugers Arch 391:85–100

Hess GP, Udgaonkar JB, Olbricht WL (1986) Chemical kinetic measurements of transmembrane processes using rapid reaction techniques: acetylcholine receptor. Annu Rev Biophys Biophys Chem 16:507–534

Ito H, Sugimoto T, Kobayashi I, Takahasi K, Katada T, Ui M, Kurachi Y (1991) On the mechanism of basal and agonist-induced activation of the G protein-gated muscarinic $K^+$ channel in atrial myocytes of guinea pig heart. J Gen Physiol 98:517–533

Kaibara M, Nakajima T, Irisawa T, Giles W (1991) Regulation of spontaneous opening of muscarinic $K^+$ channels in rabbit atrium. J Physiol (Lond) 433:589–613

Karschin A, Ho BY, Labarca C, Elroystein O, Moss B, Davidson N, Lester HA (1991) Heterologously expressed serotonin 1A receptors couple to muscarinic K channels in heart. Proc Natl Acad Sci USA 88:5694–5698

Kim D (1991) Modulation of acetylcholine-activated $K^+$ channel function in rat atrial cells by phosphorylation. J Physiol (Lond) 437:133–155

Kurachi Y, Nakajima T, Sugimoto T (1987) Short-term desensitization of muscarinic K channel current in isolated atrial myocytes and possible role of GTP binding proteins. Pflugers Arch 410:227–233

Lapointe JY, Szabo G (1987) A novel holder allowing internal perfusion of patch clamp pipettes. Pflugers Arch 410:212–216

Li Y, Otero AS, Szabo G (1991) Kinetics of muscarinic agonist dependent activation and deactivation of the G-protein coupled $K^+$ current $I_{K(ACh)}$ in atrial myocytes. Biophys J 59:349a

Li Y, Otero AS, Hanf R, Fischmeister R, Szabo G (1992) Different sensitivities to pertussis toxin of muscarinic activation of $I_{K(ACh)}$ and inhibition of $I_{Ca}$ in cardiac myocytes. Biophys J 61:A145

Logothetis DE, Kurachi Y, Galper J, Neer EJ, Clapham DE (1987) The bc subunits of GTP binding proteins activate the muscarinic K channel in heart. Nature 325:321–326

Okabe K, Yatani A, Brown A (1991) The nature and origin of spontaneous noise in G protein-gated ion channels. J Gen Physiol 97:1279–1293

Otero AS (1990) Transphosphorylation and G protein activation. Biochem Pharmacol 39:1399–1405

Otero AS, Li Y, Szabo G (1991) Receptor-mediated deactivation of Gk in cardiac myocytes. Pflugers Arch 417:543–545

Soejima M, Noma A (1984) Mode of regulation of the ACh sensitive K channel by the muscarinic receptor in rabbit atrial cells. Pflugers Arch 400:424–431

Szabo G, Otero AS (1990) G protein mediated regulation of $K^+$ channels in heart. Annu Rev Physiol 52:293–305

Tota MR, Schimerlik MI (1990) Partial agonist effects on the interaction between the atrial muscarinic receptor and the inhibitory guanine nucleotide binding protein in reconstituted system. Mol Pharmacol 37:996–1004

CHAPTER 62

# The Yeast Pheromone Response G-Protein

K. Clark and M. Whiteway

## A. Introduction

The initial steps of the pheromone response pathway of the unicellular baker's or brewer's yeast *Saccharomyces cerevisiae* involve a heterotrimeric G-protein, and this response pathway is necessary for the sexual phase of the yeast life cycle. Yeast cells can grow as either haploid or diploid cells, and the haploid yeast cells can exist in one of two sexes, or mating types, designated **a** and $\alpha$ (Herskowitz 1988). The two haploid cell types communicate by means of small diffusible molecules, termed pheromones, which interact with cell-type-specific receptor proteins on the surface of the two types of haploid cells (Marsh et al. 1991; Kurjan and Whiteway 1990). These receptor proteins are members of the rhodopsin-like seven-transmembrane-domain, or serpentine, class of receptors. As is characteristic of such proteins, the pheromone receptors communicate to the intracellular machinery through a G-protein (Simon et al. 1991). The availability of a G-protein system in an experimentally manipulatable unicellular eukaryotic cell provides a useful model system for the genetic investigation of G-proteins.

There are several reasons why the yeast pheromone response system is convenient for the analysis of G-protein function. *S. cerevisiae* is unrivaled as a eukaryotic organism for molecular genetic studies. Genes can be readily isolated from yeast cells, modified in vitro and reintroduced into yeast. A variety of tools, such as stable low-copy vectors, stable high-copy vectors, strong regulated promoters, and highly representative genomic libraries make the identification and manipulation of cloned sequences very efficient (Guthrie and Fink 1991). These general properties have allowed for the cloning and mutagenesis both of the G-protein components involved in mating signal transduction as well as a variety of elements which interact with this G-protein.

In addition, there are specific characteristics of the yeast pheromone response that make it very suitable for analysis of G-protein function. Mating is dispensable for the vegetative growth of yeast cells, so that even mutants that are completely defective in the mating response pathway can be grown and studied. The mating process is also easily measured by simple assays, and this can provide a sensitive assessment of the function of components of this pathway. In addition, the yeast mating response pathway involves a single G-protein, and there is no evidence for informational cross-talk between signaling pathways. Thus the specific function of the G-protein components can be assessed in the absence of interference from other signaling pathways. Finally, the same G-protein interacts with a structurally distinct receptor protein in each of the two haploid cell types, and this allows for the analysis of the specificity of interaction between receptors and G-proteins.

In this paper we review the recent molecular genetic studies on the G-protein components. This review does not discuss the overall pheromone response pathway, which has been reviewed extensively recently (MARSH et al. 1991; SPRAGUE 1991; WHITEWAY and KURJAN 1990). We put the G-protein system into context with a brief overview of the G-protein's role in the signaling pathway and then present the information obtained from molecular genetic studies on the three subunits.

## B. Overview

The heterotrimeric G-protein in the yeast pheromone response pathway is encoded by three unlinked genes. The $\alpha$ subunit is encoded by the *GPA1* (*SCG1*) gene found on chromosome 8 (MIYAJIMA et al. 1987), the $\beta$ subunit is encoded by the *STE4* gene on chromosome 15 (SHUSTER 1982), and the $\gamma$ subunit is encoded by *STE18* on chromosome 10 (WHITEWAY et al. 1988b). The identification of the products of these genes as G-protein subunits was initially based on the sequence similarity between the predicted amino acid sequences and the sequences of mammalian G-protein subunits. In the case of the $\alpha$ and $\beta$ subunits these sequence similarities are striking given the evolutionary divergence of mammals and yeast. The *GPA1* gene encodes a protein that has 40% identity with the $G_o$ and $G_i$ $\alpha$ subunits (NAKAFUKU et al. 1987; DEITZEL and KURJAN 1987a), and the *STE4* gene encodes a protein with similar sequence identity to the $G\beta1$ and $G\beta2$ proteins (WHITEWAY et al. 1989). The *STE18* gene product has less similarity to the $G\gamma$ subunits (30% identity to transducin $\gamma$) but shares with the $G\gamma$ subunits and Ras a carboxyl-terminal motif termed a CAAX box that is required for posttranslational isoprenylation (WHITEWAY et al. 1988a; FINEGOLD et al. 1990).

Subsequent studies have revealed that the yeast gene products are functionally as well as structurally similar to mammalian G-proteins. In mammalian cells, G-protein-coupled receptors have `a higher affinity for their ligands than do receptors which are not coupled to G-proteins (LEFKOWITZ et al. 1983). In the yeast system, deletion of either the *GPA1*, *STE4*, or *STE18* genes prevents the formation of a high-affinity receptor for $\alpha$-factor (BLUMER and THORNER 1990). In mammalian cells the $\alpha$ and $\beta\gamma$ elements provide opposing functions. In many of the mammalian model systems, the $\alpha$ subunit serves to activate the pathway when it is dissociated from the $\beta\gamma$ dimer, while the signaling response is inactivated when the subunits are all associated (GILMAN 1987; STRYER and BOURNE 1986). A similar situation holds true in the case of the yeast G-protein, with the important distinction that it is the $\beta\gamma$ element that serves as the activator of the mating response. Deletion of the *GPA1* gene, encoding the $\alpha$ subunit, activates the response pathway (DEITZEL and KURJAN 1987a; JAHNG et al. 1988), and this activation is prevented by deletion of either or both of the $\beta$ and $\gamma$ subunits (WHITEWAY et al. 1989). In addition, the response pathway can be activated by overproduction of the *STE4* gene product, and this activation can be prevented by a concomitant overproduction of the *GPA1* gene product (WHITEWAY et al. 1990; COLE et al. 1990; NOMOTO et al. 1990). Thus the $\alpha$ and $\beta\gamma$ elements play opposing roles in all the G-protein systems from yeast to man.

The mammalian $\beta$ and $\gamma$ subunits have been shown to form a tight physical association (FUNG 1983; NEER and CLAPHAM 1988), and association of the yeast *STE4* and *STE18* gene products has been detected using an in vivo protein association assay (CLARK et al., submitted). Finally, in the case of the G$\alpha$ subunit, expression of certain mammalian genes in yeast has allowed partial complementation of null mutations of the yeast gene (DIETZEL and KURJAN 1987a). Thus the yeast genes encode the functional and structural homologs of the classical G-protein subunits defined from mammalian cells.

## C. Gpa1, the $G_\alpha$ Subunit

The *GPA1* gene was identified by five different approaches. NAKAFUKU et al. (1987) isolated *GPA1* by virtue of its homology to rat $G_{\alpha i}$ and $G_{\alpha o}$ cDNAs. The same gene was isolated by DEITZEL and KURJAN (1987a) as a high-copy suppressor that conferred pheromone resistance to a strain supersensitive to pheromone. They named the gene *SCG1*. JAHNG et al. (1988) identified recessive temperature-sensitive mutations (*cdc70*) that allowed a receptorless strain to mate at the restrictive temperature. The gene was cloned by complementation of the temperature-sensitive phenotype and found to be *GPA1*. Furthermore, in a screen for mutations that would activate the pheromone response pathway, BLINDER et al. (1989) isolated several mutations of *GPA1* that were identified because they were complemented by *GPA1*, and they mapped to the *gpa1* locus. Finally, FUJIMURA (1989) identified *GPA1* in a selection for pheromone-resistant mutants; this mutant gene *dac1*, which was a recessive mutant, was allelic to *GPA1*. The sequence of Gpa1 is 40% identical to rat $G_{\alpha i}$ and $G_{\alpha o}$, and the greatest identity is present in the regions involved in guanine nucleotide binding and GTPase activity (NAKAFUKU et al. 1987; DEITZEL and KURJAN 1987a). Null mutants of *GPA1* are very sick or lethal, and their phenotype is consistent with a constitutive activation of the pheromone response signal; cells are very slow growing and morphologically aberrant (DEITZEL and KURJAN 1987a; MIYAJIMA et al. 1987).

## I. Random Mutagenesis

Since loss-of-function alleles of *GPA1* activate the pathway, presumably because free $\beta\gamma$ is available to initiate the signaling, STONE and REED (1990) were interested in identifying mutant alleles of *GPA1* that were unresponsive to pheromone (*phun* alleles). The *dac1* mutation identified by FUJIMURA (1989) is an example of one such allele. By analogy with the mammalian system, one would predict that such mutations would create mutant proteins that were unable to dissociate from $\beta\gamma$, either because they can not couple to the receptor, have a higher affinity for $\beta\gamma$, or can not change conformations upon GDP-GTP exchange. Another type of mutant would be one that could turn off the pheromone-induced signal (STONE and REED 1990).

Six different mutations were identified from a library of hydroxylamine-mutagenized *GPA1* (Fig. 1), and two classes of unresponsive alleles were identified: inactivating and activating. Three inactivating alleles were identified: G322E, G322R, and G470D. These alleles complemented a *gpa1-null* mutation but showed little or no response to pheromone. The mutations are thought to affect the $\alpha$ subunit's ability to dissociate from the $\beta\gamma$ subunit, but for different reasons. Residue G322 is within the consensus sequence DXX<u>G</u> in the hinge region of the $\alpha$ subunit believed to be required for a

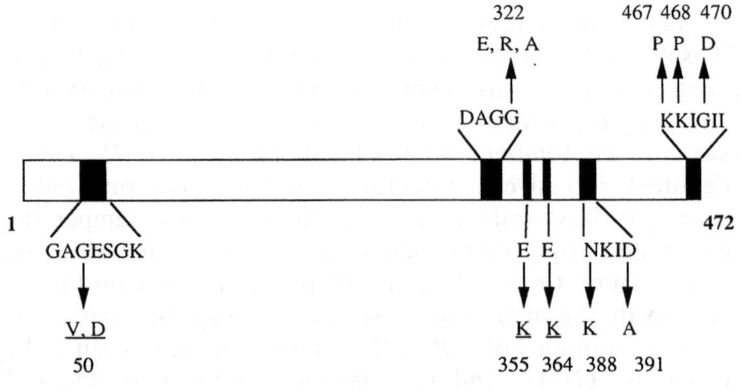


**Fig. 1.** The yeast G$\alpha$ subunit, Gpa1. The Gpa1 protein from amino acid 1 to 472 is shown. Mutations shown above the box are inactivating alleles; they complement a *gpa1-null* allele but confer a mating defect. Mutations presented below are of two types. One type consists of mutants N388K and D391A which cannot complement a *gpa1-null* allele. The other type is the underlined mutants which can complement a *gpa1-null* allele, but there is controversy about their phenotypes and interpretation of these phenotypes

conformational change associated with activation of the protein upon guanine nucleotide exchange. The corresponding mutation in $G_{\alpha s}$ causes a defect in dissociation of the mutant $\alpha$ subunit from the $\beta\gamma$ subunit (MILLER et al. 1988). In contrast, the C-terminal G470 mutation may affect dissociation because it is within the putative receptor interaction region, and so would be less sensitive to signaling from the receptor. In both cases the mutant Gpa1 can associate with $\beta\gamma$ and so suppress the lethality of a *gpa1-null* mutation, but the strain is sterile since the mutant Gpa1 cannot dissociate from $\beta\gamma$.

The activating *phun* mutations were identified as G50D, E355K, and E364K. G50D is within the phosphate box which is important for guanine nucleotide binding and GTPase activity. The E355K and E364K mutations are in conserved regions of unknown function. All three alleles formed growth inhibition zones (halos) in response to lower concentrations of $\alpha$-factor than wild-type cells did, but the halos filled in (the cells recovered from the arrest by pheromone and resumed growth) much faster than was the case for wild-type cells. An interpretation of their results is presented in the following section.

## II. Site-Directed Mutagenesis

Essentially two classes of mutants have been identified by site-directed mutagenesis- recessive mutations that activate the pathway, and dominant or partially dominant mutations that complement the growth defect of *gpa1-null* but have a mating defect themselves (Fig. 1). The targets of mutagenesis were chosen because of their homology with mammalian G$\alpha$ subunits and Ras, another GTP-binding protein. Recessive, signal-activating mutations include deletion of the carboxyl-terminal 22 residues, as well as mutations of asparagine or aspartic acid in the consensus NKXD sequence involved in guanine nucleotide binding in EF-Tu and Ras (KURJAN et al. 1991; HIRSCH et al. 1991). The phenotype of these mutants could be due to an increase in the guanine nucleotide dissociation rate, which is the phenotype of the analogous Ras mutations. In *S. cerevisiae* such an increase would result in the presence of more GTP-bound Gpa1 protein. Another possibility is that the mutant Gpa1 proteins are unstable.

These mutants fit with the model of the G-protein's activity in mating. Lack of the $\alpha$ subunit or a shift toward the GTP-bound form would create free $\beta\gamma$, which would initiate the pheromone response signal. In mating-competent diploids (**a**/**a** or $\alpha/\alpha$) only part of the full signal would be produced in the absence of pheromone since wild-type Gpa1 protein would be present to associate with $\beta\gamma$, so the cell would be viable and the mutant phenotype would be recessive.

Several different mutations created a mutant Gpa1 protein that complemented the *gpa1-null* allele but conferred a mating defect in a partially dominant manner. This class of mutants corresponds to the inactivating class obtained in the random mutagenesis. Among these mutations was G322A, a mutation of the same residue mutated in the random mutagenesis. As discussed in that section, the G322A mutant may be unable to dissociate from the $\beta\gamma$ subunits because this mutation within the hinge region affects the protein's ability to adopt the correct conformation upon guanine nucleotide exchange (KURJAN et al. 1991).

Three mutations at the carboxyl-terminus of Gpa1 are also in this second class. Deletion of the five carboxyl-terminal residues created a mutant protein that complemented the *gpa1-null* allele but was sterile in both haploid cell types (HIRSCH et al. 1991). Mutations of two adjacent lysines positioned five and six residues from the carboxyl-terminus of Gpa1 were viable but had different mating abilities in the two haploid cell types (HIRSCH et al. 1991). Mutant K467P mated better in **a** than in $\alpha$ cells, whereas K468P mated much better in $\alpha$ cells than in **a** cells. The mutation in $G_{\alpha s}$ corresponding to K467P causes uncoupling from the receptor (HAGA et al. 1977; RALL and HARRIS 1987; SULLIVAN et al. 1987). Thus these mutations, together with G470D from the random mutagenesis, suggest that this region of Gpa1 is important for receptor interaction and perhaps differentiation between the two pheromone receptors.

The phenotype of these mutants also fits with the model of G-protein function. If these mutant Gpa1 proteins were uncoupled from the receptor, they would be unable to be stimulated by pheromone-bound receptor to exchange GTP for GDP and to dissociate from $\beta\gamma$. Hence, mating-competent diploids would have a mating defect because only the wild-type Gpa1 would be able to interact with the receptor and so initiate a signal.

Glycine 50 of the consensus sequence GXGXXGK was mutated to valine by two groups, and was mutated to aspartic acid in the random mutagenesis described above. The G50V mutation corresponds to the activating G12V mutation of Ras (for review, see BARBACID 1987) and the activating G19V mutation of *S. cerevisiae RAS2* (BROEK et al. 1985). In both of these cases, the mutants have decreased GTPase activity. $RAS^{G12V}$ causes cellular transformation and $RAS2^{G19V}$ causes increased activation of its effector, adenylyl cyclase. However, although the analogous $G_{\alpha s}{}^{G49V}$ mutant has decreased GTPase activity in vitro and in vivo, it causes an increase in effector activation (adenylyl cyclase activity) only in vitro, not in vivo (GRAZIANO and GILMAN 1989; MASTERS et al. 1989).

The two groups assayed their $Gpa1^{G50V}$ mutants differently and obtained different results. KURJAN et al. (1991) assayed the phenotype when the mutant $GPA1^{G50V}$ gene was present on the chromosome, whereas MIYAJIMA et al. (1989) assayed the phenotype when the mutant gene was present on a low-copy plasmid. The G50D, E355K, and E364K mutations of STONE and REED (1990) were assayed on plasmids and yielded results similar to those of MIYAJIMA et al. (1989).

KURJAN et al. (1991) interpreted their results as indicating that the $Gpa1^{G50V}$ mutant has an altered equilibrium towards GTP-bound Gpa1, causing a high basal level of the signaling pathway. The mutant then has a defect in pheromone response, either because the addition of pheromone has little effect on the amount of GTP-Gpa1 present, or because the high level of GTP-Gpa1 causes desensitization to pheromone. In contrast, MIYAJIMA et al. (1989) and STONE and REED (1990) interpreted their results as indicating that Gpa1$^{G50V}$ and Gpa1$^{G50D}$, as well as Gpa1$^{E355K}$ and Gpa1$^{E364K}$, are present predominantly in a GTP-bound form and that this activated Gpa1 causes the cells to be supersensitive to pheromone and also activates an adaptation pathway.

Obviously, the presence of the mutant gene on a plasmid or in its native chromosomal location affects its phenotype. More investigation of this mutant under various conditions will be necessary to understand its phenotype. Since the $G_{\alpha s}{}^{G49V}$ has a different phenotype in vivo than the analogous ras$^{G12V}$ or RAS2$^{G19V}$ mutants, it is clear that this residue may have different roles in the various GTP-binding proteins.

## D. Ste4, the Gβ Subunit

The *STE4* mutation was identified in the initial characterization of nonmating mutants (MACKAY and MANNEY 1974). It was also identified in a selection for temperature-sensitive sterile mutants that were resistant to the cell-cycle arrest caused by α-factor (HARTWELL 1980). Finally, *STE4* mutant alleles which gave a modified response to α-factor were detected in a selection for mutant cells which were insensitive to an arrest loop created by expression of α-factor in *MAT***a** cells responsive to the α-factor pheromone (WHITEWAY et al. 1988a, 1988b). The mating defect in these strains allowed the cloning of the *STE4* gene, and sequencing of the gene revealed the strong sequence similarity to the β subunits of the mammalian G-proteins (WHITEWAY et al. 1989).

### I. Random Mutagenesis

Several distinct classes of *STE4* mutant alleles have been detected by random mutagenesis of the *STE4* gene, either by mutagenesis of chromosomal copies of the gene, or by mutagenesis of the cloned *STE4* gene. The first detected were nulls leading to sterility, and temperature-sensitive alleles leading to sterility at the restrictive temperature. Currently, only one of these essentially null alleles has been sequenced, and this was found to be a change of asparagine 81 to tyrosine (CLARK et al., submitted). However, mutations causing loss of function could be found throughout the *STE4* gene, as the entire gene was mutagenized using 17 sets of randomly substituted oligonucleotides, and each of the 17 regions was found to generate null mutations (LEBERER et al., in preparation). Such loss-of-function alleles are not very useful in defining the molecular functioning of the *STE4* gene product. A wide variety of defects such as nonsense mutations, frame shift mutations, and amino acid substitutions which disrupt overall protein folding would cause loss of function by significantly perturbing the protein structure. Such mutations tell little about the *STE4* gene product other than, as proteins in general, it fails to function when its structure is altered too greatly.

It is much more useful to identify mutations which only partially disturb function, or create a new phenotype, or interfere with functioning even in the presence of a wild type *STE4* protein, because these mutations are less likely to result simply from an uninformative loss of function. One very interesting class of mutations leads to activation of the pheromone response pathway, and thus these mutant proteins act as the opposite of *STE4* null mutations. This class of mutations was initially identified in a screen for mutations which lead to haploid-specific lethality (Blinder et al. 1989). Constitutive activation of the pheromone response pathway leads to irreversible cell cycle arrest, but this lethality is limited to haploid cells because all the elements necessary for pheromone-mediated cell cycle arrest are not expressed in diploid cells (Cross et al. 1988). Because activation of the response pathway also occurs when the *GPA1* gene is deleted, it was suggested that the *STE4*$^{HPL}$ allele was unable to be shut off by the *GPA1* product (Blinder 1989). A direct search for mutants of *STE4* that still activate the response pathway in the presence of a high level of the *GPA1* gene product identified a number of residues that could mutate to create this phenotype (Whiteway et al., in preparation), and these mutations, together with the initial Hpl allele, are clustered in a small region of the *STE4* gene (Fig. 2). This region, around amino acids 124–140 was also identified as a region critical for the proper association of mammalian $\alpha$ and $\beta$ G-protein subunits, because antibodies directed against a peptide close to this region could interfere with the proper association of the transducin $\alpha$ and $\beta$ subunits (Murakami et al. 1992). Thus genetic evidence from yeast and biochemical evidence from mammalian cells point to the region around amino acid 130 as being critical for the interaction of the $\alpha$ and $\beta$ G-protein subunits.

A second class of mutations that was identified by screening the population of oligodirected mutants contained dominant negative alleles of the *STE4* gene. Dominant negative alleles interfere with normal cellular function even in the presence of the wild type allele of the gene (Herskowitz 1987). In the case of the *STE4* gene, the dominant negative alleles were identified because they interfered with normal pheromone-mediated cell cycle arrest when they were overexpressed (Leberer et al., in preparation). This was the opposite phenotype to that of overexpression of the wild type *STE4* gene, which caused cell cycle arrest in the absence of any pheromone (Whiteway et al. 1990; Cole et al. 1990; Nomoto et al. 1990). As with the *STE4*$^{HPL}$ alleles, the dominant negative alleles were clustered in the *STE4* protein; only two small stretches of the gene could mutate to create the dominant negative phenotype (Fig. 2). When these

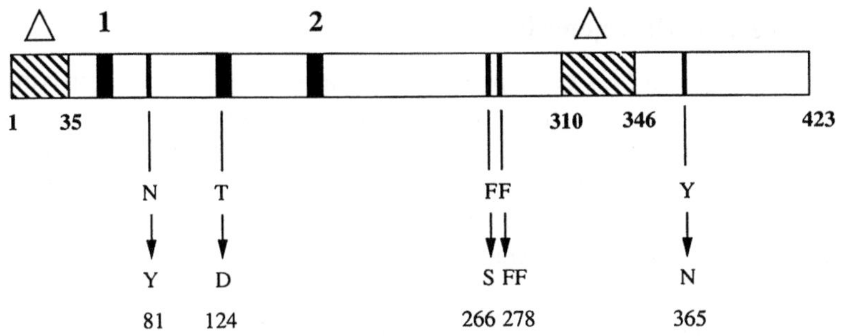


**Fig. 2.** The yeast Gβ subunit, Ste4. The Ste4 protein from amino acid 1 to 423 is shown. *Hatched areas*, the two blocks of nonhomology to mammalian Gβ subunits. These regions can be deleted (*Δ*) without dramatic effects on Ste4 function. Two regions of the protein can be mutated to give dominant negative phenotypes, one block is around amino acid 60 and the other around amino acid 175. These regions are designated region *1* and region *2* above the box representing the Ste4 protein. Mutations shown below the box affect Ste4 function. The change at position 81 leads to an almost total defect in Ste4 function. Mutations which cause an Hpl phenotype by interfering with association between the α and β G-protein subunits cluster around amino acid 130. A representative Hpl mutation is shown at position 124. The three mutations in the C-terminal half of the protein partially reduce Ste4 function, and may affect the association of the β and γ subunits of the G-protein

mutant proteins were expressed in cells which contained a deletion of the wild type *STE4* gene, expression of the dominant negative mutant allele alone was not sufficient to allow normal levels of mating. In fact, the majority of the dominant negative alleles were highly defective in their ability to transmit the mating signal.

A third class of *STE4* alleles included those which created a partially functional protein. One set of such partially functional proteins was identified from among the survivors of a self-arrest selection involving pheromone-supersensitive *MAT***a** cells that were both producing and responding to α factor (Whiteway et al. 1988b). These alleles caused cells to be less responsive to high levels of pheromone than to lower levels, and thus a ring of responding cells was formed at a distance from a source of pheromone (Whiteway et al. 1988a, 1989). This response required that the responding cells also be defective in the *SST2* gene function which is required for proper adaptation to pheromone (Deitzel and Kurjan 1987b). Four of these "ring-forming" alleles were sequenced (Fig. 2). The three alleles which had the least effect on overall mating ability were found to modify closely situated phenylalanine residues (F266 and F278), while the fourth allele modified a more distant tyrosine residue (Y365; Clark et al., submitted). The F278FF mutation disrupted the association of the *STE4* gene product with the *STE18* gene product, and thus these alleles may define amino acids that are important in the formation of the βγ heterodimer (Clark et al., submitted)

## II. Site-Directed Mutagenesis

The structure of the *STE4* gene product contains 7 repeats of approximately 40 amino acids that were initially described for the transducin $\beta$ subunit (FONG et al. 1986). This repeat structure characterizes all G$\beta$ subunits that have been described, but it is also found in a number of proteins that do not appear to be members of heterotrimeric G-proteins, such as Cdc4 (YOCHEM and BYERS 1987) and Prp4 (DALRYMPLE et al. 1989) of yeast, and Enhancer of Split in *Drosophila* (HARTLEY et al. 1987). The yeast Ste4 protein has two blocks of amino acids that are not part of this repeat structure, one at the amino terminus and a second between repeats 5 and 6 (WHITEWAY et al. 1989; COLE and REED 1991). Because transmission of the yeast pheromone response signal involves the $\beta\gamma$ element acting in a positive manner, rather than the $\alpha$ subunit as was found in most mammalian systems, these two unique regions were logical targets for site-directed mutagenesis.

Intriguingly, deletion of these two blocks of nonhomology did not dramatically perturb Ste4 function, showing that the positive signaling capability of the Ste4 protein did not depend on these unique blocks of amino acids (COLE and REED 1991). Indeed, the deletion of the internal block of nonhomology increased the cellular responsiveness to pheromone in a manner that suggested that this region was involved in the process of adaptation to the mating signal, and in fact, this deletion removed the region of Ste4 that is phosphorylated in response to pheromone (COLE and REED 1991). Phosphorylation of the carboxyl-terminus of the pheromone receptors is a component of the adaptation response, and deletion of the receptors' C-termini increases cellular sensitivity to pheromones (RENEKE et al. 1988; KONOPKA et al. 1988; N. DAVIS and G. SPRAGUE, personal communication). Thus it is possible that the internal nonhomologous block in the Ste4 protein is a target of a signal-modulating phosphorylation event (COLE and REED 1991). However, similar but nonidentical deletions which include the potential phosphorylation site do not enhance the sensitivity to pheromone (E. LEBERER, unpublished). Thus the relationship between adaptation, Ste4 phosphorylation, and the Ste4-specific block of amino acids may be complex.

# E. Ste18, the G$\gamma$ Subunit

The *STE18* gene was identified by a mutation in a group of mating-competent survivors of the autocrine selection (WHITEWAY et al. 1988b). This mutation led to a unique "ring" phenotype in pheromone response halo assays, similar to that previously described for certain *STE4* mutations. However, cloning of the gene by complementing this phenotype established that the gene encoded a novel protein, distinct from *STE4*, that was also essential for pheromone responsiveness (WHITEWAY et al. 1989).

## I. Random Mutagenesis

The cloned *STE18* gene was subjected to the same oligodirected random mutagenesis strategy that was applied to the *STE4* gene. The region encompassing amino acids 24–92 of the 110 amino acid protein was targeted in this mutagenesis. Two classes of mutants were screened for: (a) those which enhanced the mating of the essentially null *STE4* allele which contained a tyrosine in place of the normal asparagine (N81Y) and (b) dominant negative alleles which created resistance to pheromone-mediated cell cycle arrest.

Overproduction of the *STE18* gene product partially supressed the mating defect of strains containing the N81Y allele of the *STE4* gene. A further enhancement of mating was achieved by mutations of the *STE18* gene; these mutations may have increased protein stability, or increased the affinity of the *STE18* product for the mutant *STE4* product (WHITEWAY et al., in press). Those *STE18* mutations that were sequenced were not clustered – this result may be more compatible with a general effect on protein stability rather than a change in the relative affinities of the Ste4 and Ste18 proteins (Fig. 3).

Dominant negative alleles were also found throughout the gene, but appeared most prevalent at the carboxyl-terminus. Two C-terminal alleles were sequenced, and both contained nonsense codons at position 83

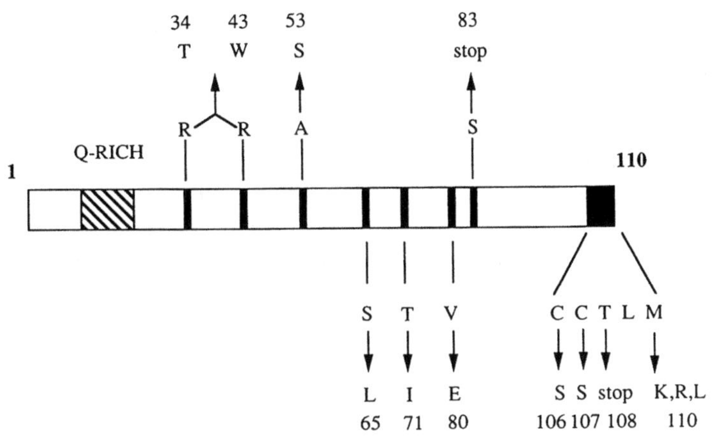


**Fig. 3.** The yeast Gγ subunit, Ste18. The Ste18 protein from amino acid 1 to 110 is shown. *Hatched*, the glutamine rich region. Mutations which lead to a dominant negative phenotype are shown above the box. The mutations at the carboxyl-terminal CAAX box affect Ste18 function; mutations of the cysteine residues significantly reduce function, while those of the other CAAX box residues have a dramatic effect only in cells which lack the Gα subunit encoded by *GPA1*. The other mutations noted below the box improve the mating ability of cells with a specific defective Gβ subunit; these mutations otherwise do not create a detectable phenotype

(WHITEWAY et al., in press). These mutations deleted the C-terminal CAAX box. As described below, other mutations which modified the CAAX box residues also created dominant negative phenotypes. Two other mutations were identified; one was a double mutation that changed arginines 34 and 43 to threonine and trytophan, respectively, and the second changed alanine 53 to serine (Fig. 3). The R34T R43W double mutant differed from the other dominant negative alleles because it did not interfere with the signaling created by overproduction of Ste4 (WHITEWAY et al., in press). This mutant protein therefore must sequester Ste4 in an inactive complex, and overexpression of Ste4 can alleviate the interference caused by this sequestration.

## II. Site-Directed Mutagenesis

Ste18 has two distinctive motifs, a glutamine-rich tract from amino acid 11 to 22, and a carboxyl terminal CAAX box (WHITEWAY et al., 1989). The carboxyl-terminal CAAX boxes of a variety of proteins serve as a signal for a complex posttranslational modification (SINENSKY and LUTZ 1992). This modification involves the addition of a 15- or 20-carbon isoprenoid lipid to the cysteine of the CAAX box, the proteolytic cleavage of the last three amino acids, leaving this prenylated cysteine residue as the carboxyl-terminus, and finally a carboxyl-methylation of this C-terminal cysteine. Site-directed mutagenesis has been applied to investigate the roles of the CAAX box residues in G$\gamma$ function.

The carboxyl-terminus of the *STE18* protein is modified by the same enzymatic machinery that modifies Ras and **a** factor in *S. cerevisiae* (FINEGOLD et al., 1990), and thus is almost certainly modified by a 15-carbon farnesyl group. Inactivation of the farnesyl transferase $\beta$ subunit, the product of the *DPR1* (*RAM1*, *STE16*) gene, does not prevent G protein-mediated mating, since *MAT*$\alpha$ cells containing a *dpr1-null* mutation are not sterile. However, *dpr1* mutant cells do have a modified pheromone response. *MAT***a** *dpr1* mutant cells which are also *sst1 sst2* mutants are less responsive to high levels of pheromone than they are to lower levels, and thus they exhibit the "ring" phenotype that was noted for some *STE4* alleles (WHITEWAY et al., in preparation). An identical phenotype is found for *STE18* alleles which change the ultimate residue of the CAAX box from a methionine to an arginine (WHITEWAY et al., 1988a) or to a lysine (WHITEWAY et al., in preparation). This result implies that the prenylation machinery is unable to recognize CAAX boxes that end in arginine or lysine, and in fact, these residues have never been found in naturally occurring CAAX boxes.

Mutations which affect the CAAX box cysteine residues have a much more profound effect on the mating ability of yeast cells. Changing the prenylation target cysteine 107 to a serine (Fig. 3) totally abolished the function of Ste18 in both mating types (FINEGOLD et al., 1990). Because elimination of the prenylation machinery created a relatively minor effect on mating of *MATα* cells, it is clear that C107 serves as more than simply a site of farnesylation. When the adjacent cysteine (C106) was modified to a serine, the activity of Ste18 was greatly reduced but not eliminated (FINEGOLD et al., 1990). Cysteines located in proximity to the prenylated CAAX box cysteine are often targets of palmitylation (DESCHENES et al., 1990), but there is no current evidence that the C106S effect on Ste18 function is related to a defect in palmitylation.

The M110K and M110R mutations, as well as M110L and T108 replacement by a termination codon have minor effects on *STE18* function in otherwise wild type cells, but have profound effects on cells that lack *GPA1* function (WHITEWAY et al., in preparation). These mutations suppress the constitutive signaling usually found in *GPA1* null mutant strains suggesting that there may be an overlap in function between the Ste18 C-terminus and the Gpa1-protein.

## F. Conclusions

Genetic tools have been used to investigate the roles of the three subunits of the heterotrimeric G-protein involved in the *S. cerevisiae* pheromone response pathway. These results show that the yeast pathway uses the $\beta\gamma$ subunit as the positive signaling component, and that this function is antagonized by the $\alpha$ subunit. Mutations of the G-protein subunit genes have further refined our understanding of G-protein function. Random and site-directed mutagenesis of the G$\alpha$ subunit have allowed an assessment of the functional role of guanine nucleotide binding and hydrolysis. C-terminal mutations have pointed out residues that are important in the differential interaction of G-proteins and receptors. Mutagenesis of the G$\gamma$ CAAX box has shown a complex involvement of this region in G-protein function, while mutations of the G$\beta$ subunit have identified residues that are implicated in the association of the $\alpha$ and $\beta$ subunits. Future work, involving approaches such as the isolation of suppressors of defined mutants, should lead to a detailed understanding of the functional interactions of the yeast pheromone response G protein. Thus the yeast system should provide a valuable tool for the continuing analysis of G-protein structure and function.

*Acknowledgments.* This is NRC publication number 36826.

## References

Barbacid M (1987) Ras genes. Ann Rev Biochem 56:779–827

Blinder D, Bouvier S, Jenness DD (1989) Constitutive mutants in the yeast pheromone response: ordered function of the gene products. Cell 56:479–486

Blumer KJ, Thorner J (1990) $\beta$ and $\gamma$ subunits of a yeast guanine nucleotide-binding protein are not essential for membrane association of the $\alpha$ subunit but are required for receptor coupling. Proc Natl Acad Sci 87:4363–4367

Broek D, Nasrollah S, Fasano O, Fujiyama A, Tamanoi F, Northup J, Wigler M (1985) Differential activation of yeast adenylate cyclase by wild-type and mutant RAS proteins. Cell 41:763–769
Clark KL, Dignard D, Thomas DY, Whiteway M (submitted) The STE4 and STE18 gene products of Saccharomyces cerevisiae physically interact. Mol Cell Biol
Cole GM, Reed SI (1991) Pheromone-induced phosphorylation of a G protein $\beta$ subunit in S. cerevisiae is associated with an adaptive response to mating pheromone. Cell 64:703–716
Cole GM, Stone DE, Reed SI (1990) Stoichiometry of G protein subunits affects the Saccharomyces cerevisiae mating pheromone signal transduction pathway. Mol Cell Biol 10:510–517
Cross F, Hartwell LH, Jackson C, Konopka JB (1988) Conjugation in Saccharomyces cerevisiae. Ann Rev Cell Biol 4:429–457
Dalrymple MA, Petersen-Bjorn S, Friessen JD, Beggs JD (1989) The product of the PRP4 gene of S. cerevisiae shows homology to $\beta$ subunits of G proteins. Cell 58:811–812
Deschenes RJ, Resh MD, Broach JR (1990) Acylation and prenylation of proteins. Curr Opinion Cell Biol 2:1108–1113
Dietzel C, Kurjan J (1987a) The yeast SCG1 gene: a $G_\alpha$-like protein implicated in the a- and $\alpha$-factor response pathway. Cell 50:1001–1010
Dietzel C, Kurjan J (1987b) Pheromonal regulation and sequence of the Saccharomyces cerevisiae SST2 gene: a model for desensitization to pheromone. Mol Cell Biol 7:4169–4177
Finegold AA, Schafer WR, Rine J, Whiteway M, Tamanoi F (1990) Common modifications of trimeric G proteins and ras protein: involvement of polyisoprenylation. Science 249:165–169
Fong HKW, Hurley JB, Hopkins RS, Miake-Lye R, Johnson MS, Doolittle RF, Simon M (1986) Repetitive segmental structure of the transducin $\beta$ subunit: homology with the CDC4 gene and identification of related mRNAs. Proc Natl Acad Sci USA 84:3792–3796
Fujimura H (1989) The yeast G-protein homolog is involved in the mating pheromone signal transduction system. Mol Cell Biol 9:152–158
Fung BK (1983) Characterization of transducin from bovine retinal rod outer segments. I. Separation and reconstitution of the subunits. J Biol Chem 258:10495–10502
Gilman A (1987) G proteins: transducers of receptor-generated signals. Ann Rev Biochem 56:615–649
Graziano MP, Gilman AG (1989) Synthesis in Escherichia coli of GTPase-deficient mutants of Gs$\alpha$. J Biol Chem 264:15475–15482
Guthrie C, Fink GR (1991) Guide to yeast genetics and molecular biology. Methods in enzymology, vol 194. Academic Press, San Diego New York Boston
Haga T, Ross EM, Anderson HJ, Gilman AG (1977) Adenylate cyclase permanently uncoupled from hormone receptors in a novel variant of S49 mouse lymphoma cells. Proc Natl Acad Sci 74:2016–2020
Hartwell LH (1980) Mutants of Saccharomyces cerevisiae unresponsive to cell division control by polypeptide mating pheromone. J Cell Biol 85:811–822
Hartley DA, Preiss A, Artavanis-Tsakonas S (1988) A deduced gene product from the Drosophila neurogenic locus, Enhancer of Split, shows homology to mammalian G-protein $\beta$ subunit. Cell 55:785–795
Herskowitz I (1987) Functional inactivation of genes by dominant negative mutations. Nature 329:219–222
Herskowtiz I (1988) Life cycle of the budding yeast Saccharomyces cerevisiae. Microbiol Rev 52:536–553
Hirsch JP, Dietzel C, Kurjan J (1991) The carboxy terminus of Scg1, the G subunit involved in yeast mating, is implicated in interactions with the pheromone receptors. Genes Dev 5:467–474

Jahng K-Y, Ferguson J, Reed SI (1988) Mutations in a gene encoding the $\alpha$ subunit of a Saccharomyces cerevisiae G protein indicate a role in mating pheromone signaling. Mol Cell Biol 8:2484–2493

Konopka JB, Jenness DD, Hartwell LH (1988) The C terminus of the S. cerevisiae $\alpha$ pheromone receptor mediates an adaptive response to pheromone. Cell 54:609–620

Kurjan J, Hirsch JP, Dietzel C (1991) Mutations in the guanine nucleotide binding domains of a yeast G$\alpha$ protein confer a constitutive or uninducible state to the pheromone response pathway. Genes Dev 5:475–483

Kurjan J, Whiteway M (1990) The Saccharomyces cerevisiae pheromone response pathway. Sem Dev Biol 1:151–158

Lefkowitz RJ, Stadel JM, Caron, MG (1983) Adenylate cyclase-coupled $\beta$-adrenergic receptors: structure and mechanism of activation and desensitization. Ann Rev Biochem 52:159–186

MacKay VL, Manney TR (1974) Mutations affecting sexual conjugation and related processes in Saccharomyces cerevisiae. II. Genetic analysis of non-mating mutants. Genetics 76:273–288

Marsh L, Neiman AM, Herskowitz I (1991) Signal transduction during pheromone response in yeast. Ann Rev Cell Biol 7:699–728

Masters SB, Miller RT, Chi M-H, Chang F-H, Beiderman B, Lopez NG, Bourne HR (1989) Mutations in the GTP-binding site of Gs$\alpha$ alter stimulation of adenylyl cyclase. J Biol Chem 264:15467–15474

Miller RT, Masters SB, Sullivan KA, Beiderman B, Bourne HR (1988) A mutation that prevents GTP-dependent activation of the $\alpha$ chain of $G_s$. Nature 334, 712–715

Miyajima I, Arai K, Matsumoto K (1989) GPA1$^{Val-50}$ mutation in the mating-factor signaling pathway in Saccharomyces cerevisiae. Mol Cell Biol 9:2289–2297

Miyajima I, Nakafuku M, Nakayama N, Brenner C, Miyajima A, Kaibuchi K, Arai K, Kaziro Y, Matsumoto K (1987) GPA1, a haploid essential gene, encodes a yeast homolog of mammalian G protein which may be involved in mating factor signal transduction. Cell 50:1011–1019

Murakami T, Simonds WF, Spiegel AM (1992) Site-specific antibodies directed against G protein $\beta$ and $\gamma$ subunits: effects on $\alpha$ and $\beta\gamma$ subunit interaction. Biochemistry 31:2905–2911

Nakafuku M, Itoh H, Nakamura S, Kaziro Y (1987) Occurrence in Saccharomyces cerevisiae of a gene homologous to the cDNA coding for the $\alpha$ subunit of mammalian G proteins. Proc Natl Acad Sci USA 84:2140–2144

Neer EJ, Clapham DE (1988) Roles of G protein subunits in transmembrane signalling. Nature 333:129–134

Nomoto S, Nakayama N, Arai K-I, Matsumoto K (1990) Regulation of the yeast pheromone response pathway by G protein subunits. EMBO J 9:691–696

Rall T, Harris BA (1987) Identification of the lesion in the stimulatory GTP-binding protein of the uncoupled S49 lymphoma. FEBS Lett 224:365–371

Reneke JE, Blumer KJ, Courchesne WE, Thorner J (1988) The carboxy-terminal segment of the yeast $\alpha$-factor receptor is a regulatory domain. Cell 55:221–234

Shuster JR (1982) Mating-defective ste mutations are suppressed by cell division cycle start mutations in Saccharomyces cerevisiae. Mol Cell Biol 2:1052–1063

Simon MI, Strathmann MP, Gautam N (1991) Diversity of G proteins in signal transduction. Science 252:802–808

Sinensky M, Lutz RJ (1992) The prenylation of proteins. BioEssays 14:25–31

Sprague GF Jr (1991) Signal transduction in yeast mating. Trends in Genetics 7:393–398

Stone DE, Reed SI (1990) G protein mutants that alter the pheromone response in Saccharomyces cerevisiae. Mol Cell Biol 10:4439–4446

Stryer L, Bourne HR (1986) G proteins: a family of signal transducers. Ann Rev Cell Biol 2:391–419

Sullivan KA, Miller RT, Masters SB, Beiderman B, Heideman W, Bourne HR (1987) Identification of receptor contact site involved in receptor-G protein coupling. Nature 330:758–760
Whiteway M, Dignard D, Thomas DY (in press) Mutagenesis of Ste18, a putative Gγ subunit in the S. cerevisiae pheromone response pathway. Biochem Cell Biol
Whiteway M, Hougan L, Dignard D, Bell L, Saari G, Grant F, O'Hara P, MacKay VL, Thomas DY (1988a) Function of the STE4 and STE18 genes in mating pheromone signal transduction in Saccharomyces cerevisiae. Cold Spring Harbor Symp Quant Biol 53:585–590
Whiteway M, Hougan L, Dignard D, Thomas DY, Bell L, Saari G, Grant F, O'Hara P, MacKay VL (1989) The STE4 and STE18 genes of yeast encode potential β and γ subunits of the mating factor receptor-coupled G protein. Cell 56:467–477
Whiteway M, Hougan L, Thomas DY (1988b) Expression of MFα1 in MATa cells supersensitive to α-factor leads to self-arrest. Mol Gen Genet 214:85–88
Whiteway M, Hougan L, Thomas DY (1990) Overexpression of the STE4 gene leads to mating response in haploid Saccharomyces cerevisiae. Mol Cell Biol 10:217–222
Yochem J, Byers B (1987) Structural comparison of the yeast cell division cycle gene CDC4 and a related pseudogene. J Mol Biol 195:233–245

CHAPTER 63

# G$\alpha$ Proteins in *Drosophila*: Structure and Developmental Expression

M. FORTE, F. QUAN, D. HYDE, and W. WOLFGANG

## A. Introduction

### I. G-Protein-Coupled Signaling in Development

Many of the complex cellular interactions involved in regulating cellular proliferation, determination, and differentiation during development depend on signal transduction processes that allow target cells to alter the expression of specific genes in response to extracellular signals. Given the importance of these inductive interactions, it is likely that many of the genes expressed throughout development and required for accurate specification of cell fate encode molecules that are components of individual signal transduction pathways. Consistent with this notion, a variety of studies have suggested that a number of receptor-mediated signal transduction processes are employed to relay extracellular signals of developmental consequence. Of these signaling mechanisms, the most ancient and diverse involves the coupling of structurally related extracellular receptors to intracellular effectors by means of guanyl nucleotide-binding proteins, or G-proteins. Studies in simple organisms such as yeast (HERSKOWITZ 1989) and *Dictyostelium* (FIRTEL et al. 1989) clearly point out the potential significance of this form of sensory transduction in mediating fundamental developmental processes. In both of these systems, the receptors involved possess the typical seven-transmembrane-domain architecture and the G-protein pathways they activate are the exclusive mechanism for conveying developmentally significant signals. In metazoan organisms, the developmental role of G-protein-coupled receptor systems has been more difficult to elucidate, primarily because of the diversity of ligand/receptor systems that utilize this signaling pathway. It has been impossible to predict then, a priori, ligand/receptor interactions that convey developmentally important information. Thus, G-protein-coupled signaling pathways involved in determining cell fate during development remain to be identified systematically.

### II. The *Drosophila* System

Clearly, complex events that regulate cellular proliferation and differentiation events cannot be completely understood at the single-cell level and can only be studied meaningfully in intact animals. One approach to examining the role of G-protein-activated pathways that participate in specific developmental events has been to identify and characterize such systems in *Drosophila*. In this experimental system, the tools of classical and molecular genetics can be applied systematically in an organism in which each of the developmental stages is well-described and easily accessible (RUBIN 1988). For example, in *Drosophila*, all development takes place outside the mother in a well-characterized temporal sequence. Thus, preparation of large quantities of biological material from distinct developmental stages is relatively straightforward. Further, once a gene has been isolated in flies, its position in the genome can be mapped by in situ hybridization to polytene chromosomes found in salivary glands. From this map position, over 80 years of classical and molecular genetic manipulation of the *Drosophila* genome can be used to identify or create strains in which the gene of interest has been deleted, mutated, or disrupted by the insertion of a single-copy transposable element. If such alterations are lethal, they can be maintained as heterozygotes, and the homozygous lethal phenotype can be examined following appropriate genetic crosses. Clones of cells homozygous for the lethal mutation can also be produced in a background of heterozygous cells. The ease of creating transgenic flies is another distinct advantage of the *Drosophila* system. Specific changes that result in proteins with defined biochemical phenotypes can be introduced into cloned genes, and such modified genes can be reintroduced into the genome. The effects of these changes on the development of transformed flies in the heterozygous or homozygous condition can then be assessed systematically.

In *Drosophila*, the studies of the role of specific G-protein pathways has focused on the identification and characterization of individual *Drosophila* $G\alpha$ genes and the use of specific antibodies to follow the expression of each protein. These studies indicate that in adults, the expression of individual $G\alpha$-proteins within the *Drosophila* nervous system is anatomically restricted, suggesting defined roles for specific pathways in mediating distinct sensory transduction events or complex behavioral processes. In addition, as the common component of this signaling pathway, the spatial and temporal expression of individual $G\alpha$-proteins should define specific developmental processes that involve the participation of pathways activated by these signaling molecules. Studies on the developmental expression of individual $G\alpha$-proteins suggest that G-protein pathways may mediate interactions both at early embryonic stages and in specific developing tissues later in embryogenesis. These results predict specific roles for each G-protein activated pathway both in the adult organism and during development; predictions that can be tested using the molecular and genetic approachs outlined above within the context of the well-developed *Drosophila* experimental system.

**Table 1.** Molecular characteristics of identified *Drsophila* Gα-proteins

| | Adult transpripts | Maternal transpripts | Proteins | Vertebrate homology (>70%) | Functional characteristics | Toxin sensitivity | Map position |
|---|---|---|---|---|---|---|---|
| $DG_{s\alpha}$ | Two, each 1.9 kb | None | 51 kDa ($DG_{sl}$)<br>48 kDa ($DG_{ss}$) | $G_{s\alpha}$ | ↑ cAMP | CTX | 60A |
| $DG_{i\alpha}$ | 2.3 kb | 2.3 kb (major)<br>1.7 kb (minor) | 41 kDa | $G_{i\alpha 1}$ | ? | – | 65D |
| $DG_{o\alpha}$ | 6.0 kb (head specific)<br>4.2 kb<br>3.5 kb (body specific) | 3.5 kb | Two, each 40 kDa | $G_{o\alpha}$ | ↓ $Ca^{2+}$ channels? | PTX | 47A |
| $DG_{q\alpha}$ | Two, each 1.8 kb (eye specific) | – | 41.3 kDa<br>37.2 kDa | $G_{q\alpha}$ | Activate phospholipase C? | – | 49B |
| *concertina* | 2.8 kb<br>2.2 kb | 2.8 kb<br>2.2 kb | 52.7 kDa | $G\alpha_{12,13}$ | ? | – | 40A |

CTX, Cholera toxin; PTX, pertussin toxin.
References: $DG_{s\alpha}$, QUAN et al. 1989 (MSI), QUAN and FORTE 1990 (MS2); $DG_{i\alpha}$, PROVOST et al. 1988, QUAN and FORTE, unpublished; $DG_{o\alpha}$, THAMBI et al. 1989 (MS4), YOON et al. 1989, DE SOUSA et al. 1989; $DG_{q\alpha}$, LEE et al. 1990, STRATHMANN and SIMON 1990; *concertina*, PARKS and WIESCHAUS 1991.

## B. G$\alpha$-Proteins in *Drosophila*

Table 1 summarizes current molecular and biochemical information about identified *Drosophila* G-protein $\alpha$ subunits. *Drosophila* cDNAs encoding homologs (DG$\alpha$) of mammalian $G_{s\alpha}$, $G_{i\alpha}$, and $G_{o\alpha}$ were identified in screens of head libraries at reduced stringency with probes generated from appropriate mammalian homologs. Homologs of mammalian $G_{q\alpha}$ have been identified in screens for transcripts specifically expressed in the adult eye. The *concertina* gene has been identified by the phenotype resulting from null mutations in this gene. The expression of each DG$\alpha$ has been assessed at the RNA level by Northern and in situ analysis and at the protein level in many cases using antibodies to peptide sequences uniquely found in each DG$\alpha$-protein. In addition to these $\alpha$ subunit genes, two *Drosophila* $\beta$ subunit genes have been identified and characterized; one specifically expressed in the CNS (YARFITZ et al. 1988) and another specifically found in the adult eye (YARFITZ et al. 1991). These are not discussed further in this review.

### I. $DG_{s\alpha}$

#### 1. Gene Structure

Northern blot and in situ hybridization studies indicate that a single 1.9-kb transcript encoding the $DG_{s\alpha}$-protein is preferentially expressed in neuronal cell bodies of the adult CNS (QUAN et al. 1989). The gene encoding $DG_{s\alpha}$ is approximately 4.5-kb in length and is divided into 9 exons. The intron/exon structure of the $DG_{s\alpha}$ gene shows substantial similarity to the human gene for $G_{s\alpha}$ (QUAN and FORTE 1990). Immunoblot analysis indicates that membranes prepared from adult *Drosophila* heads contain two forms of $DG_{s\alpha}$ protein (WOLFGANG et al. 1990). Analysis of the $DG_{s\alpha}$ gene indicates that these two forms are generated by alternate splicing of intron 7 involving

**Fig. 1A–E.** DG$\alpha$ expression in the adult CNS. **A–C** Horizontal sections through fly heads stained with antibodies specific for $DG_{o\alpha}$ (**A**) $DG_{s\alpha}$ (**B**), and $DG_{i\alpha}$ (**C**) subunits. $DG_{s\alpha}$ and $DG_{o\alpha}$ show elevated expression in the central neuropil regions relative to the cortical cell body regions except in the first-order optic neuropil (the lamina) where they are both low (*between the pairs of arrows*). $DG_{i\alpha}$ is at low levels throughout the CNS except in the lamina, outer medulla, and antennal glomerulae, all of which contain the terminals of primary sensory axons (see text). All three show only background levels of staining in the eyes. **D** A sagital, **E** a frontal section of a fly head showing elevated staining of $DG_{s\alpha}$ in all neuropil regions of the mushroom bodies. **D** The calyx (*CA*) and peduncle (*P*) are shown. **E** The $\alpha$, $\beta$, and $\gamma$ lobes are shown. This pattern is remarkably similiar to that of the PDE coded by the *dunce* locus (see text) even to the extent that the lateral portions of the peduncle stain more intensely than the medial portion. *C*, Cell bodies in cortex of brain; *N*, neuropil; *AG*, antenal glomerulae; *bar* = 100 $\mu$m

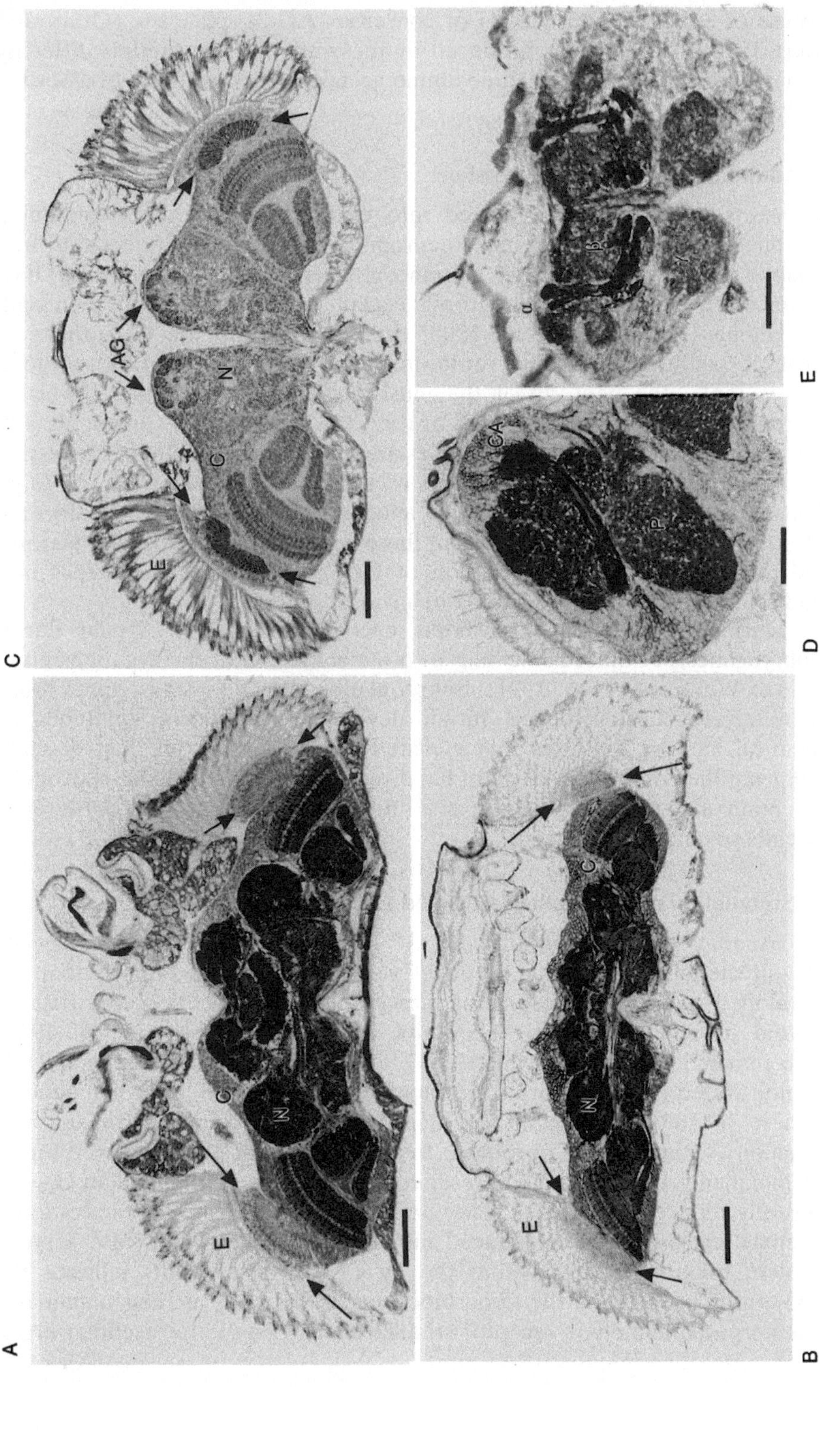
A
B
C
D
E
AG
N
C
E
CA
P
α
β

the use of either an unusual TG or consensus AG 3′ splice site (QUAN and FORTE 1990). The proteins produced by these alternate transcripts differ by the inclulsion or deletion of three amino acids and the substitute of a Ser for a Gly.

## 2. Adult and Embryonic Expression

In insects, the CNS is organized into outer cortical regions containing neuronal cell bodies and an inner neuropil containing nerve fibers and their synaptic terminals. Light level immunocytochemistry has shown that $DG_{s\alpha}$ proteins are elevated in all neuropils except the first-order optic neuropil, the lamina (WOLFGANG et al. 1990; Fig. 1B). Within the CNS there is, however, additional regional variation in the expression of $DG_{s\alpha}$-protein. For example, the neuropil of the mushroom bodies has higher levels of expression than surrounding areas of the CNS (Fig. 1D,E). Consistent with this observation, the mushroom bodies are the preferential site of expression of genes that encode proteins which modulate cellular cAMP levels (e.g., *dunce*, encoding a phosphodiesterase and *rutabagga*, encoding an isoform of adenylyl cyclase). Since mutations in these genes result in deficits in learning and memory, selective manipultion of $DG_{s\alpha}$ pathways in these cells may lead to alterations in this complex behavioral process.

During development, transcripts encoding $DG_{s\alpha}$ first appear during embryogenesis at times coinciding with the activation of the zygotic genome (2–6 h; WOLFGANG et al. 1991). Subsequently, the $DG_{s\alpha}$ transcript is found at relatively constant levels in all developmental stages. Immunocytochemical analysis indicates that elevated levels of $DG_{s\alpha}$ are first detected following the completion of germ band retraction in the forming neuropil of the brain and ventral ganglion; a pattern which persists for the duration of embryogenesis.

## 3. Stimulation of Mammalian Adenylyl Cyclase Through $DG_{s\alpha}$

To examine the function of both fly $DG_{s\alpha}$ isoforms as well as specific site-directed $DG_{s\alpha}$ mutants, a novel vaccina virus expression system was developed to allow expression of these proteins in S49 *cyc*$^-$ cells that are devoid of endogenous $G_{s\alpha}$ transcript and protein (QUAN et al. 1991). Receptor-independent activation (GTP$\gamma$S, $AlF_3$) of each wild-type $DG_{s\alpha}$ demonstrated that both forms are capable of activating mammalian adenylyl cyclase to the same extent as their mammalian counterparts and thus function as stimulatory G-proteins. In addition, as has been observed for the mammalian proteins, specific site-directed mutations of mammalian $G_{s\alpha}$ that severely reduce $\alpha$ subunit GTPase activity lead to constitutive activation of adenylyl cyclase when introduced into $DG_{s\alpha}$. $DG_{s\alpha}$-proteins also serve as cholera toxin substrates within the fly CNS. These results indicate that $DG_{s\alpha}$-proteins activate the same biochemical pathways as their mammalian counterparts, namely in receptor-stimulated increases in intracellular cAMP levels. Surprisingly, $DG_{s\alpha}$ interact poorly with mammalian $G_{s\alpha}$-coupled receptors despite the fact that $DG_{s\alpha}$- and rat $G_{s\alpha}$-proteins differ by only three conservative amino acid replacements over C-terminal regions responsible for interaction with receptors. Based on this observation, sequence comparisons of *Drosophila* and rat proteins have helped to define a region of high variability in $G_{s\alpha}$-proteins that is likely to be important for efficient receptor–G-protein interaction.

## II. $DG_{o\alpha}$

### 1. Gene Structure

Northern blots probed with $DG_{o\alpha}$ cDNAs indicate that three different $DG_{o\alpha}$ transcripts are expressed in adults; a 6.0-kb message is expressed primarily in adult heads, a 3.5-kb message is present primarily in adult bodies, and a 4.2-kb message present in both heads and bodies (THAMBI et al. 1989; YOON et al. 1989; DE SOUSA et al. 1989; SCHMIDT et al. 1989). For the most part, the intron/exon structure of the $DG_{o\alpha}$ gene remains to be determined. Partial characterization of the $DG_{o\alpha}$ gene indicates that there are eight coding exons (YOON et al. 1989; DE SOUSA et al. 1989). Alternate usage of 5′ exons result in the production of two $DG_{o\alpha}$ transcripts that differ in 5′ untranslated regions and first coding exon but share remain coding exons. These cDNAs result in the production of two $DG_{o\alpha}$ proteins that differ in 7 of 21 N terminal amino acids. These results suggest that the three transcripts observed on Northern blots differ by the inclusion of exons resulting in extension of 5′ and/or 3′ untranslated regions.

### 2. Adult and Embryonic Expression

In situ hybridization to head sections indicate that $DG_{o\alpha}$ transcripts are expressed predominantly in neuronal cell bodies in the brain, optic lobe, and thoracic ganglion (THAMBI et al. 1989; YOON et al. 1989; DE SOUSA et al. 1989; SCHMIDT et al. 1989). Immunolocalization studies revealed that, similar to $DG_{s\alpha}$, $DG_{o\alpha}$ is present at highest levels in neuropils (except the lamina, where it is low) and at intermediate levels in the cortex of all brain and thoracic ganglion regions (WOLFGANG et al. 1990; Fig. 1A). Northern blot analysis indicates that the three $DG_{o\alpha}$ transcripts are differentially expressed during development (DE SOUSA et al. 1989; WOLFGANG et al. 1991). The 3.5-kb transcript specific to adult bodies appears to be primarily maternal in origin since it is present only in 0- to 2-h embryos and in female ovaries. The 4.2-kb transcript is present at extremely low levels in early embryos and reaches peak expression later (10–18 h) and in pupae. The 6.0-kb transcript, specific to adult heads, first appears as the product of zygotic transcription in 10- to 14-h embryos, coincident with the highest levels of expression of $DG_{o\alpha}$-proteins in developing neurons (see below).

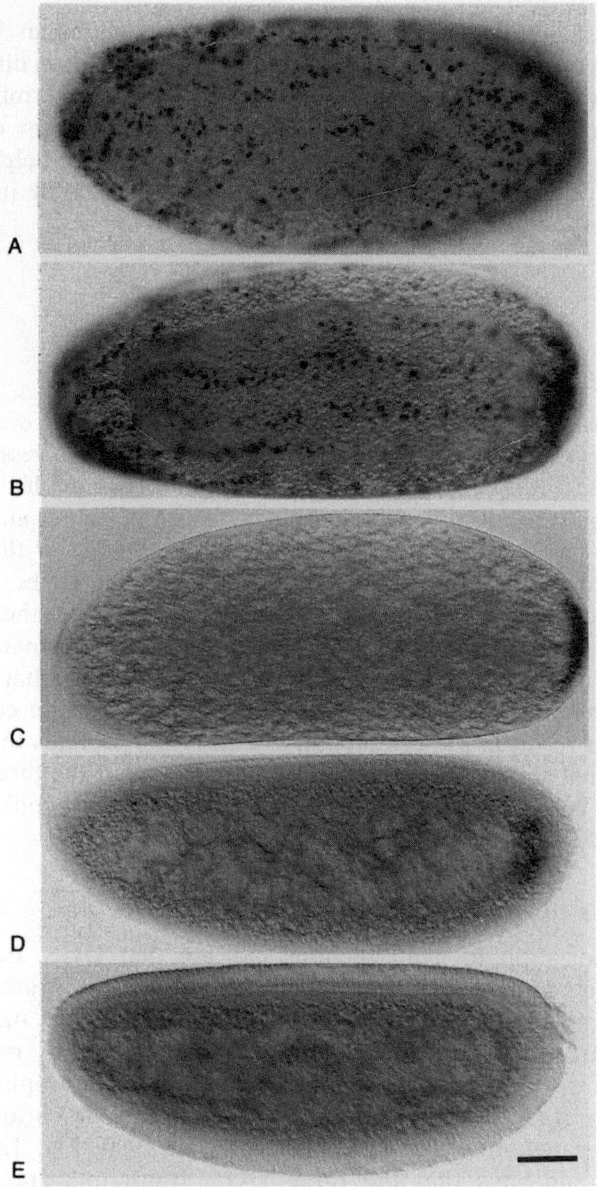


**Fig. 2A–E.** $DG_{i\alpha}$-protein distribution in whole mounts of cleavage and blastoderm stage embryos. **A–C** Early cleavage stage embryos. **A** Uniform peripheral distribution of $DG_{i\alpha}$ granules in a newly laid egg. **B** Granules organized into longitudinal stripes in a slightly older embryo. **C** Granules restricted to the postrior pole after about 45 min of embryonic development. **D** Syncytial blastoderm stage embryo about 2 h after fertilization showing that only a few granules remain at the posterior pole. **E** Blastoderm stage embryo about 2.5 h after fetilization, showing that the granules are lost. **A, B** Montages. Posterior is to the *right*; Bar = 50 $\mu$m

Maternal expression of the $DG_{o\alpha}$ gene has been confirmed by in situ hybridization studies which demonstrate the presence of $DG_{o\alpha}$ transcripts in female abdomens in nurse cells and oocytes (DE SOUSA et al. 1989; WOLFGANG et al. 1991). Immunocytochemical studies failed to demonstrate significant levels of $DG_{o\alpha}$ protein in nurse cells, oocytes, or early embryos, suggesting that expression of $DG_{o\alpha}$ proteins may be regulated at the translational level. As with $DG_{s\alpha}$, elevated levels of $DG_{o\alpha}$-protein first become evident after germ band retraction in the forming neuropil of the brain and ventral ganglion (WOLFGANG et al. 1991). $DG_{o\alpha}$ neuropil staining persists through adult life.

Immunological and molecular characterization indicates that $DG_{o\alpha}$ is the major, if not exclusive pertussis toxin substrate in the *Drosophila* CNS (HOPKINS et al. 1988; THAMBI et al. 1989; GUILLEN et al. 1990). Recently, the pertussis toxin gene has been introduced into the *Drosophila* genome under the control of a heat shock promoter (DE SOUSA S. and HURLEY J., unpublished). When heat-shocked, tranformed flies produce active toxin which ADP-ribosylates endogenous $G_{o\alpha}$. Pertussis toxin expression results in alterations in physiological responses of photoreceptor cells, block of embryonic development, shortened life span of adult flies and cessation of feeding. Since high energy metabolites are significantly depleted shortly after heat shock, it is not yet clear that the phenotypes observed following pertussis expression are due directly to a modification of $G_{o\alpha}$ function.

## III. $DG_{i\alpha}$

### 1. Gene Structure

Northern blot analysis indicates that $DG_{i\alpha}$ is expressed primarily as a 2.3-kb transcript. The $DG_{i\alpha}$ gene is composed of five exons; the most 5′ encoding untranslated sequences and the remaining representing $DG_{i\alpha}$ coding sequences (PROVOST et al. 1988). Despite intensive searches for additional homologs of mammalian $G_{i\alpha}$ by low stringency screens of genomic and cDNA libraries and by polymerase chain reaction amplification strategies (QUAN and FORTE, unpublished), $DG_{i\alpha}$ remains the only homolog of this abundant family of mammalian G-proteins identified in *Drosophila*.

### 2. Adult and Embryonic Expression

$DG_{i\alpha}$ transcripts are present at very low levels in adult heads but are abundantly expressed in female bodies (PROVOST et al. 1988; WOLFGANG et al. 1991). The $DG_{i\alpha}$ protein could not be detected in head membranes by immunoblotting using antibodies to peptide sequences specifically found in $DG_{i\alpha}$, consistent with the negligible levels of $DG_{i\alpha}$ message in adult heads. These antibodies however, did detect $DG_{i\alpha}$ fusion proteins expressed in *Escherichia coli*. Immunocytochemical studies demonstrate that $DG_{i\alpha}$ has a surprisingly restricted distribution in the adult CNS (WOLFGANG et al. 1990; Fig. 1C). It is present at significant levels only in photoreceptor cell terminations, glomerulae of the antennal lobes, and the ocellar retina. Little or no $DG_{i\alpha}$-protein was detected in other brain regions or the thoracic ganglion. Thus, $DG_{i\alpha}$ appears to be uniquely associated with most primary sensory afferents and their terminations. This suggests the presence of specific receptor and/or effector systems that mediate the transmission of primary sensory information in *Drosophila*.

Northern blot analysis indicates that $DG_{i\alpha}$ transcripts are primarily expressed maternally (PROVOST et al. 1988; WOLFGANG et al. 1991). Peak levels are found in early embryos (0–2h) and decline thereafter. Maternal expression of $DG_{i\alpha}$ has also been confirmed by in situ hybridization to tissue sections prepared from adult female abdomens. Abundant $DG_{i\alpha}$ message is uniformly distributed in nurse cells and oocytes. This general pattern of expression has been confirmed by immunoblotting. Immunocytochemical analysis of mature oocytes and early embryos indicates that $DG_{i\alpha}$-protein is initially found in uniformly distributed granules (Fig. 2). As development proceeds, $DG_{i\alpha}$ immunoreactivity is redistributed so that by the third or fourth syncytial nuclear division (i.e., 30–40 min after fertilization) it becomes exclusively associated with the posterior pole. Posterior localization of $DG_{i\alpha}$ persists throughout the formation of the syncytial blastoderm and disappears upon the formation of the cellular blastoderm. During these early nuclear cleavage stages, embryonic axes (i.e., anterior/posterior, dorsal/ventral), determined primarily by the polarized localization of maternal determinants within the egg are actively interpreted by the developing embryo. The exclusive posterior localization of $DG_{i\alpha}$ during these stages suggests that signal transduction as mediated by this G-protein may play a role in the interpretation of this positional information. At later stages of embryogenesis, $DG_{i\alpha}$-protein reappears in a specific set of developing tissues. For example, high levels are specifically associated with advancing cardioblasts during dorsal closure and later, in a subset of peripheral neurons during the formation of a specific sensory organ, the chordotonal organ. These observation suggest the presence of specific $DG_{i\alpha}$-coupled sensory transduction systems that mediate signals critical to the formation of these specific tissues. The restricted expression pattern of $DG_{i\alpha}$ during early development, in the elaboration of specific tissue systems, and in adults is likely to reflect the association of *Drosophila* $DG_{i\alpha}$ with a restricted set of receptor/second-messenger systems.

## IV. $DG_{q\alpha}$

### 1. Gene Structure

Drosophila homologs of mammalian $G_{q\alpha}$ have been identified either as a transcript with expression restricted to the visual system (LEE et al. 1990) or by homology to the mammalian forms (STRATHMANN and SIMON 1990). Complete characterization of the gene indicates that it contains eight exons and seven introns. Two transcripts are expressed; the first containing all eight exons and the second lacking the seventh exon. Because this exon is within the coding sequence, its absence removes 36 amino acids from one form of the protein and results in the production of long ($DG_{q\alpha 1}$) and short ($DG_{q\alpha 2}$) forms of $DG_{q\alpha}$.

### 2. Adult Expression

By both Northern blot and in situ hybridization, the expression of $DG_{q\alpha}$ is restricted to the retina and occellus of the adult head, the two components of the adult visual system (LEE et al. 1990). Recently, immunocytochemical studies have indicated that $DG_{q\alpha 1}$ and $DG_{q\alpha 2}$ may be differentially localized within photoreceptor cells (HYDE D., unpublished). Polyclonal antiserum directed against amino acid sequences encoded by exon 7, and therefore specific for $DG_{q\alpha 1}$, stained only the rhabdomere of the photoreceptor cell. The rhabdomere is a microvillar specialization of cells in the visual system that is the site of accumulation of photoreceptor complexes containing rhodopsin. Antisera generated against either a common C terminal region or a large common region in the middle of the protein, and therefore identifying both $DG_{q\alpha 1}$ and $DG_{q\alpha 2}$, stain both the rhabdomere and cell body of photoreceptor cells. This observation suggests that while $DG_{q\alpha 1}$ is specifically localized to the rhabdomere and may therefore play a role exclusively in phototransduction; $DG_{q\alpha 2}$ may be localized to both the rhabdomere and cell body, suggesting a diverse set of functions for this isoform.

### 3. Role in Phototransduction

Several observations suggest that $DG_{q\alpha}$ encodes the *Drosophila* transducin analog, i.e., the G-protein required for phototransduction. First, $DG_{q\alpha}$ is the only identified $G\alpha$ protein that is expressed specifically in adult photoreceptor cells. Second, characterization of the *Drosophila norpA* mutation demonstrates that phospholipase C is a key effector molecule in the *Drosophila* phototransduction pathway. Mutants in *norpA* are defective in phototransduction and lack phosopholipase C activity relative to wild-type flies. Molecular characterization indicates that the conceptual translation product of the *norpA* gene is a molecule with high homology to vertebrate phospholipase C. In recent studies, the vertebrate $G_{q\alpha}$ isoforms have been shown to couple receptors to the stimulation of a specific isoform of vertebrate phosopholipase C, phospholipase $C\beta$ (TAYLOR et al. 1991; WU et al. 1992). Given these observations, the identification of the major $G_\alpha$ in *Drosophila* photoreceptor cells as a member of the $G_{q\alpha}$ family is consistent with predictions of the functional interactions that must be mediated by the *Drosophila* transducin analog. Direct demonstration of the participation of

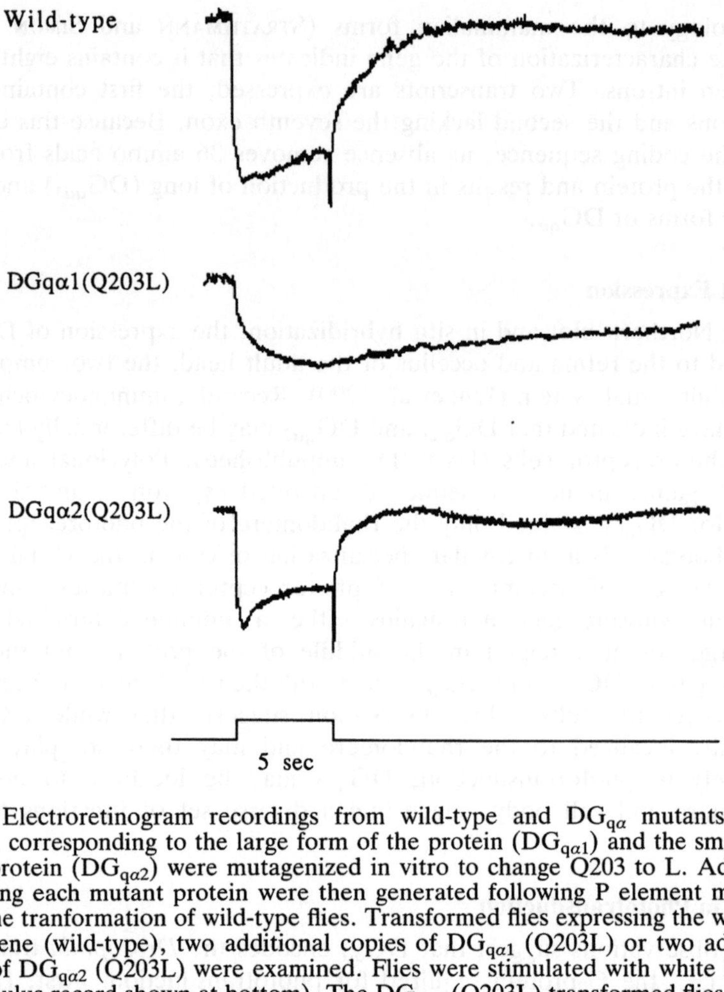


**Fig. 3.** Electroretinogram recordings from wild-type and $DG_{q\alpha}$ mutants. $DG_{q\alpha}$ cDNAs, corresponding to the large form of the protein ($DG_{q\alpha 1}$) and the small form of the protein ($DG_{q\alpha 2}$) were mutagenized in vitro to change Q203 to L. Adult flies expressing each mutant protein were then generated following P element mediated germ line tranformation of wild-type flies. Transformed flies expressing the wild-type $DG_{q\alpha}$ gene (wild-type), two additional copies of $DG_{q\alpha 1}$ (Q203L) or two additional copies of $DG_{q\alpha 2}$ (Q203L) were examined. Flies were stimulated with white light for 5 s (stimulus record shown at bottom). The $DG_{q\alpha 2}$ (Q203L) transformed flies showed typical rates for depolarizationd and repolarization, levels of depolarization, and on and off transients. The $DG_{q\alpha 1}$ transformants showed a slower rate of depolarization, much delayed repolarization, a reduced level of depolarization, and the absence of any transients

$DG_{q\alpha 1}$ in phototransduction has been obtained by the expression of a constitutively activated, site-directed mutant form (Q203L) following germ line transformation (D. HYDE, unpublished). The visual defect in the transformed fly, as measured by electroretiongram (ERG), is slightly slowed on response, a smaller level of depolarization, and a greatly decreased cessation of physiological responses following termination of a light stimulation (Fig. 3), as expected if $DG_{q\alpha 1}$ plays a primary role in phototransduction. In addition, these transformants require almost a 1000 times brighter light stimulus than wild-type to generate an equivalent physiolgical response. Interestingly, transformants expressing an equivalent, constitutively activated form of $DG_{q\alpha s}$ exhibit a normal ERG response to a 100- to 1000-fold lower light stimulus than wild-type and are generally nonviable and show high levels of sterility. Thus, $DG_{q\alpha s}$ may play a more general role in signal transduction in cells other than those in the adult visual system.

## V. *concertina*

### 1. Mutant Phenotype

Mutations in the *concertina* (*cta*) gene were initially isolated in screens for female sterile mutations of the second chromosome (SCHUPBACH and WIESCHAUS 1989). All *cta* mutations produce exclusively a maternal effect; females homozygous for these mutations survive but are sterile. By detailed morphological analysis, embryos from homozygous *cta* mutant mothers have been shown to gastrulate abnormally. In *Drosophila*, normal gastrulation begins immediately upon cellularization of the blastoderm stage embryo with the formation of the ventral furrow and posterior midgut. Cells that form both of these invaginations become tightly apposed and change their shape via apical constriction. Embryos from mothers homozygous for mutations in the *cta* gene begin furrow formation by forming a zone of tightly apposed cells. Some cells constrict but not in sufficient numbers to form an organized groove (PARKS and WIESCHAUS 1991). Thus, gastrulation is disorganized, leading to the generation of embryos with multiple invaginations (*concertina-like*) rather than the production of a uniform ventral furrow.

### 2. Cloning and Gene Structure

The gene encoding the *cta* protein was identified by the generation of *cta* alleles that result from transposon (i.e., P element) insertion (PARKS and WIESCHAUS 1991). Each of four P element induced *cta* mutations were shown by genetic methods to be allelic (i.e., within the same gene) to EMS (Ethyl Methanesulfonate) induced *cta* mutations. A genomic library constructed using DNA from one P element strain was screened with a P element probe, allowing the isolation of the genomic region containing the *cta* gene. This genomic region has been used to identified partial length *cta* cDNAs. The partial cDNAs appear to contain the complete coding region for the *cta* protein, suggesting that the remaining information in the two *cta* transcripts identified on northern blots corresponds to 5′ and 3′ untranslated regions. The *cta* transcripts encode a unique $G\alpha$-protein that shows the highest similarity over guanine nucleotide binding domains to the recently described $G\alpha_{12,13}$ family of vertebrate $G_{s\alpha}$ (STRATHMANN et al. 1989) whose function remains to be identified. In addition, the *cta* product extends an additional 100 amino acids upstream of the translational initiation site of other G$\alpha$-proteins. This region shows no homology to other G$\alpha$-proteins nor significant homology to other known proteins.

### 3. Expression of *cta*

By Northern blot analysis, both *cta* transcripts are present throughout development and are found even in adult males, although the relative levels of the two transcripts may vary at different developmental stages. By in situ hybridization, *cta* transcripts first appear in the germarium (the site of oocyte formation) of the female ovary, as expected from the maternal effect of mutations in this gene. Throughout oogenesis, the accumulation of *cta* RNA is restricted to the germline and is observed in both nurse cells and ooctyes. In early embryos, *cta* transcripts are found throughout the cytoplasm and appear to be concentrated in the region of the cortex below the syncytial nuclei. Later, high levels of *cta* RNA accumulate in the mesoderm, presumably from zygotic expression of the gene.

While the genetics of *cta* shows that is a maternal-effect gene whose expression is required only during oogenesis for viability, expression studies show that *cta* RNA is present not only during oogenesis and early embryogenesis but is found throughout embryonic, larval, and adult life in both males and females. The function that the *cta* protein plays in these later stages of development is unclear, given that females homozygous for *cta* mutations survive, provided their mothers had one copy of a functional *cta* gene. Possible explanations for this observation may be that the receptors or effectors that interact with *cta* may be nonessential at later stages of development and in adults or that another G$\alpha$-protein may functionally compensate for the absense of *cta* protein during these stages. The exact role played by the *cta* protein in mediating the shape changes required for normal gastrulation also remains to be defined.

## C. Summary

Over the past 3 years, individual *Drosophila* G$\alpha$ proteins have been identified and characterized. In addition, tools such as antibodies that can be used to specifically follow expression of each protein have been developed. These tools as well as detailed molecular characterization have been used to describe and analyze the spacial and temporal expression pattern of each G$\alpha$-protein, both in the adult organism and during development. These results indicate that G-protein pathways may mediated developmental interactions both at early embryonic stages and in specific developing tissues later in embryogenesis. In addition, these studies have identified G$\alpha$ that mediate visual transduction in this organism, as well as a novel G$\alpha$ (*cta*) that is specifically required for the changes in cell shape associated with invagination during gastrulation. These observations predict specific roles for each Gα activated pathway within the context of a carefully characterized and easily manipulated whole animal system. Critical and defining tests of hypotheses on the role of individual G protein-activated pathways during development and in adults will in the future rest on the use of the classical and molecular genetic approaches availible in *Drosophila*. These approaches may also lead to the discovery of novel interactions between signaling pathways and the identification of a range of previously unknown molecules that participate in G-protein-coupled signaling events. The *Drosophila* system will allow then, not only the identification and characterization of G-protein pathways expressed at various stages of development but also an experimental approach to test and confirm predictions of the specific role of each pathway during critical stages of development, in the formation of specific tissues, and in complex signal transduction events in the adult organism.

## References


de Sousa S, Hoveland L, Yarfitz S, Hurley J (1989) The Drosophila Goα-like G protein gene produces multiple transcripts and is expressed in the nervous system and in ovaries. J Biol Chem 264:18544–18551

Firtel R, van Haastert P, Kimmel A, Devreotes P (1989) G protein linked signal transduction pathways in development: Dictyostelium as an experimental system. Cell 58:235–239

Guillen A, Jallon J, Fehrentz J, Pantaloni C, Bockaert J, Homburger V (1990) A Go-like protein in Drosophila melanogaster and its expression in memory mutants. EMBO J 9:1449–1455

Herskowitz I (1989) A regulatory hierarchy for cell specialization in yeast. Nature 342:749–757

Hopkins R, Stamnes M, Simon M, Hurley J (1988) Cholera and pertussis toxin substrates and endogenous ADP-ribosyltransferase activity in Drosophila melanogaster. Biochim Biophys Acta 970:355–362

Lee Y, Dobbs M, Verardi M, Hyde D (1990) dgq: A Drosophila gene encoding a visual system-specific Gα molecule. Neuron 5:889–898

Parks S, Wieschaus E (1991) The Drosophila gastrulation gene concertina encodes a Gα-like protein. Cell 64:447–458

Provost N, Somers D, Hurley J (1988) A Drosophila melanogaster G protein α subunit gene is expressed primarily in embryos and pupae. J Biol Chem 263:12070–12076

Quan F, Wolfgang W, Forte M (1989) The Drosophila gene coding for the α subunit of a stimulatory G protein is preferentially expressed in the nervous system. Proc Natl Acad Sci USA 86:4321–4325

Quan F, Forte M (1990) Two forms of Drosophila melanogaster Gsα are produced by alternate splicing involving and unusual splice site. Mol Cell Biology 10:910–917

Quan F, Thomas L, Forte M (1991) Drosophila stimulatory G protein α subunit activates mammalian adenylyl cyclase but interacts poorly with mammalian receptors: implications for receptor-G protein interaction. Proc Natl Acad Sci USA 88:1898–1902

Rubin G (1988) Drosophila melanogaster as an experimental system. Science 240:1453–1459

Schmidt C, Garen-Fazio S, Chow Y, Neer E (1989) Neuronal expression or a newly identified Drosophila melanogaster G protein αo subunit. Cell Reg 1:125–134

Schupbach G, Wieschaus E (1989) Female sterile mutations on the second chromosome of Drosophila melanogaster. I. Maternal effect mutations. Genetics 121:101–117

Strathmann M, Wilkie T, Simon M (1989) Diversity of the G-protein family: sequences from five additional α subunits in the mouse. Proc Natl Acad Sci USA 86:7407–7409

Strathmann M, Simon M (1990) G protein diversity: a distinct class of α subunits is present in vertebrates and invertebrates. Proc Natl Acad Sci USA 87:9113–9117

Taylor S, Chae HZ, Rhee SG, Exton J (1991) Activation of the β1 isozyme of phospholipase C by α subunits of the Gq class of G proteins. Nature 350:516–518

Thambi N, Quan F, Wolfgang W, Spiegel A, Forte M (1989) Immunological and molecular characterization of Goα-like proteins in the Drosophila central nervous system. J Biol Chem 264:18552–18560

Wolfgang W, Quan F, Goldsmith P, Unson C, Spiegel A, Forte M (1990) Immunolocalization of G protein α subunits in the Drosophila CNS. J Neuroscience 10:1014–1024

Wolfgang W, Quan F, Thambi N, Forte M (1991) Restricted spatial and temporal expression of G-protein α subunits during Drosophila embryogenesis. Development 113:527–538

Wu D, Lee CH, Rhee SG, Simon M (1992) Activation of phospholipase C by the α subunits of Gq and G11 proteins in transfected Cos-7 cells. J Biol Chem 267:1811–1817

Yarfitz S, Provost N, Hurley J (1988) Cloning of a Drosophila melanogaster guranine nucleotide regulatory protein β subunit gene and characterization of its expression during development. Proc Natl Acad Sci USA 85:7134–7138

Yarfitz S, Niemi G, McConnell J, Fitch C, Hurley J (1991) A Gβ protein in the Drosophila compound eye is different from that in the brain. Neuron 7:429–438

Yoon J, Shortridge R, Bloomquist B, Schneuwly S, Perdew M, Pak W (1989) Molecular characterization of the Drosophila gene encoding Goα subunit homolog. J Biol Chem 264:18536–18543

CHAPTER 64

# Signal Transduction by G-Proteins in *Dictyostelium discoideum*

L. Wu, C. Gaskins, R. Gundersen, J.A. Hadwiger, R.L. Johnson, G.S. Pitt, R.A. Firtel, and P.N. Devreotes

## A. Introduction

G-protein-linked signal transduction pathways play essential roles during the differentiation process of *Dictyostelium discoideum*, a simple developing eucaryotic organism. These transmembrane signaling systems are essentially the same as those in mammalian cells, and there are simple methods to disrupt genes by homologous recombination and to create cell lines expressing mutant genes. In addition, *Dictyostelium* is easy to grow, and development is synchronous, allowing one to readily obtain $10^{11}$ cells for biochemical studies. Thus, *Dictyostelium* provides a model system to study G-protein-linked signal transduction.

## B. Signal Transduction in *Dictyostelium*

The life cycle of *Dictyostelium* consists of distinct growth and developmental phases. In the developmental phase, triggered by starvation, about $10^5$ individual amoebae aggregate to form a multicellular structure. This process is organized by extracellular adenosine 3′,5′-monophosphate (cAMP) that is secreted by cells at aggregation centers. Surrounding cells respond by moving chemotactically toward the signaling cells and by relaying the signal to cells further from the center. The resulting multicellular aggregate undergoes further morphogenesis, in which the signaling system continues to play a role. Cells in the aggregate differentiate into prestalk and prespore cells which eventually form the stalk and spore mass of a fruiting body (Fig. 1). This cell-cell signaling process occurs via cAMP binding to cell surface receptors, which in turn triggers numerous responses (Devreotes 1989; Firtel 1991).

Genes encoding four surface cAMP receptors (cARs), which comprise a family highly related by sequence, have been identified (Klein et al. 1988; Saxe et al. 1991a,b). Each gene is expressed at a different time in development (Fig. 1). cAR1 mRNA is present mainly during early aggregation, although two additional transcripts are induced later in development at much lower levels (Saxe et al. 1991a). cAR3 is expressed next, being induced at late aggregation with maximal expression occurring at the mound stage and continuing through later development at reduced

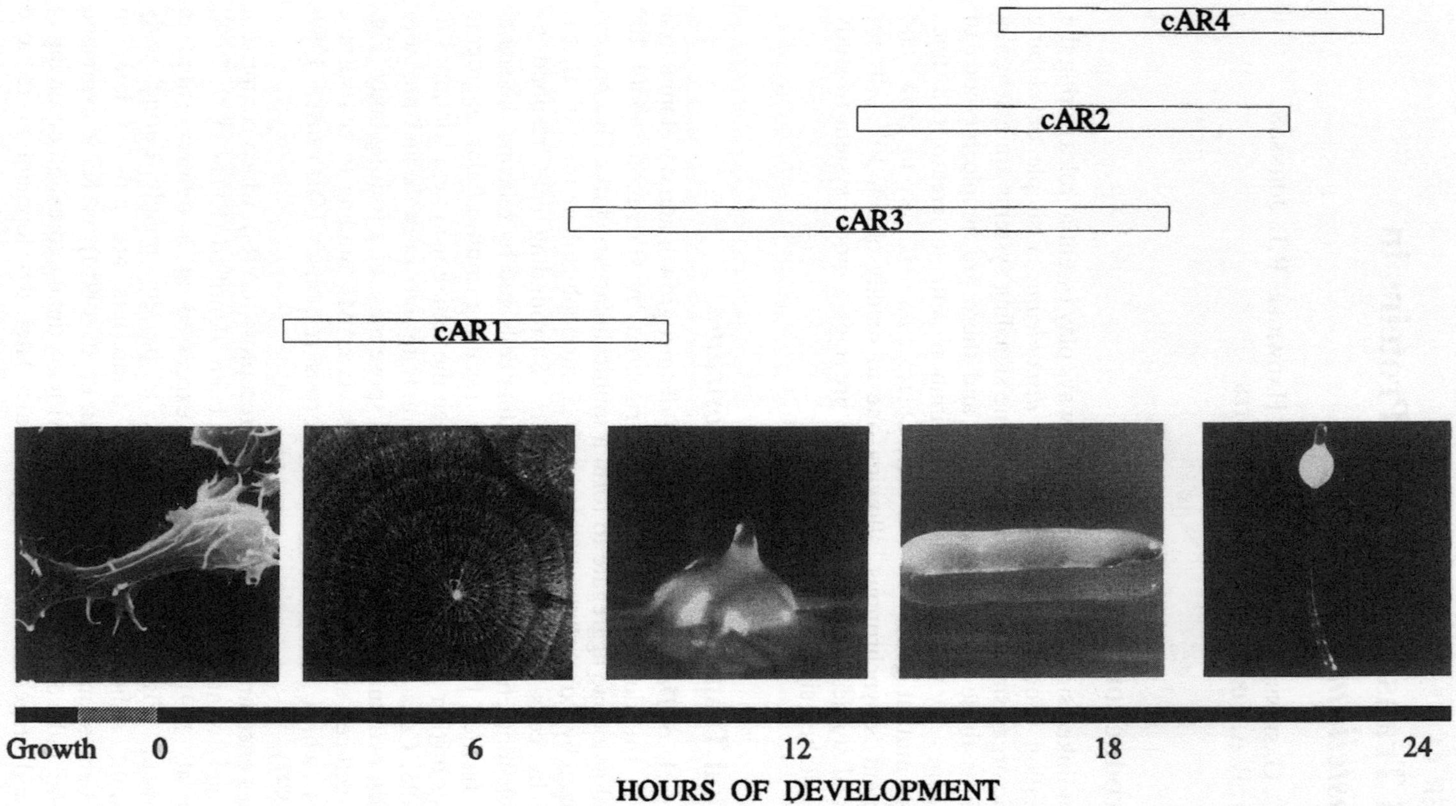


**Fig. 1.** The prominent stages of *Dictyostelium* development and the expression of cAMP receptors (*cARs*). Five stages of *Dictyostelium* development are shown (*bottom, left to right*: growth, aggregation, mounds, slugs, and fruiting bodies. *Open bars*, developmental expression of the major RNA transcript for each cAR. (Saxe et al. 1991a; Kimmel, personal communications)

levels. The cAR2 transcript is enriched in prestalk cells and is expressed initially at the mound stage but is present predominantly at the slug stage (SAXE et al. 1991a). cAR4 is expressed lastly, and it appears during the culmination (A. KIMMEL, personal communication).

All of these receptors contain seven putative transmembrane domains, a characteristic of receptors that are linked to G-proteins, such as the $\beta$-adrenergic receptor and rhodopsin. It has been shown that cAR1 is needed for normal development (SUN et al. 1990; SUN and DEVREOTES 1991). Cells that lack cAR1 do not aggregate and have almost no detectable surface cAMP binding sites. The expression of early genes in the cAR1-null cells is delayed, and late gene expression is blocked. Preliminary results suggest that the other cARs also serve critical functions in the signaling processes and in controlling development in this organism (JOHNSON et al. 1993; SAXE et al. 1993).

## C. Diversity of G-Proteins in *Dictyostelium*

By using oligonucleotides based on the sequences of the conserved GTP-binding domains of G-protein $\alpha$ subunits to screen a *Dictyostelium* cDNA libary or to perform polymerase chain reactions (PCR), genes encoding eight G-protein $\alpha$ subunits have been cloned (PUPILLO et al. 1989; HADWIGER et al. 1991; WU and DEVREOTES 1991; PUPILLO and DEVREOTES, in preparation; WU et al., in preparation; CUBITT et al. 1992). Comparison of the predicted amino acid sequences indicates that the eight G-protein $\alpha$ subunits share 30%–50% identity to each other and to mammalian G-protein $\alpha$ subunits. The eight $\alpha$subunits do not fall into any obvious subtypes related to the four $\alpha$ subunit classes, $G_s$, $G_i$, $G_q$, and $G_{12}$ found in higher eucaryotes.

Despite the relatively low degree of identity among these G-proteins overall, some regions are highly conserved. Figure 2 shows the sequence comparison of the most conserved regions between the *Dictyostelium* G$\alpha$ subunits and several mammalian G$\alpha$ subunit subtypes. These sequence motifs are believed to be important for G-protein function. G$\alpha$4 and G$\alpha$7 have unusual amino acids in region A (. . . GAGESG . . .), which is involved in $\beta\gamma$ release and GTP hydrolysis (SIMON et al. 1991). G$\alpha$8, moreover, possesses some very interesting and unusual features. The amino acid sequences of the N-terminal portion (about 75% of the molecule) of G$\alpha$8 is similar to other G$\alpha$ subunits, but its C-terminus portion has an additional 50 amino acids consisting of long stretches of Asn and Ser. Such stretches of repeated sequence have been observed for several cAMP receptors, adenylyl cyclase genes, and the catalytic subunit of cAMP-dependent protein kinase in *Dictyostelium* (JOHNSON et al. 1993; PITT et al. 1992; MANN and FIRTEL 1991), but to our knowledge G$\alpha$8 is the first G-protein $\alpha$ subunit identified possessing this motif. Moreover, the

| | Region A | Region C | Region G | Region T |
|---|---|---|---|---|
| Consensus: | KLLLLGAGESGKSTIXKQMK | DVGGQR | LFLNKXD | TCATDT |
| Gα1: | KLLLLGAGESGKSTIAKQMK | DVGGQR | LFLNKRD | TCATDT |
| Gα2: | KLLLLGAGESGKSTISKQMK | DVGGQR | LFLNKSD | TCATDT |
| Gα4: | KLLLLGPGESGKSTIFKQMK | DVGGQR | LFLNKKD | TCAVDT |
| Gα5: | KLLLLGAGESGKSTIFKQMK | DVGGQR | YFLNKVD | TCAIDT |
| Gα6: | GAGESGKSTIFKQLK | DVGGQR | | |
| Gα7: | KLLLLGTGDSGKSTVVKQMK | DVAGQR | LFLNKRD | TTATDT |
| Gα8: | RILLLGAGESGKSTVVKQLK | DVGGQR | LVLNKKD | IAARYK |
| Gs: | RLLLLGAGESGKSTIVKQMR | DVGGQR | LFLNKQD | TCAVDT |
| Gi: | KLLLLGAGESGKSTIVKQMK | DVGGQR | LFLNKKD | TCATDT |
| Gq: | KLLLLGTGESGKSTFIKQMR | DVGGQR | LFLNKKD | TCATDT |
| G12: | KILLLGAGESGKSTFLKQMR | DVGGQR | LFLNKKD | TTAIDT |

**Fig. 2.** Amino acid sequence comparison of *Dictyostelium* Gα1–Gα8 and mammalian Gα subunits in the most conserved regions. Gα3 sequence is not shown and is cloned by PUPILLO and DEVREOTES (in preparation). The complete sequence of Gα6 has not been determined. The sequences of Gα1 and Gα2 are taken from PUPILLO et al., the sequences of Gα4 and Gα5 are taken from HADWIGER et al. (1991) and HADWIGER and FIRTEL (in preparation), and the sequences of Gs, Gi, Gq, and G12 are taken from SIMON et al. (1991)

well-conserved TCATDT motif of Gα subunits (SIMON et al. 1991) is totally missing in Gα8. It has been suggested that the C-terminal region of the G protein is involved in receptor interactions (SIMON et al. 1991). This suggestion is supported by the observation that modification of the α subunit of the $G_i$ class by pertussis toxin blocks its interaction with receptor, and antibodies or peptides that specifically interact with C-terminal regions, including the TCATDT region, of some of the Gα proteins also block interaction with receptor (DERETIC and HAMM 1987; SULLIVAN et al. 1987; MASTERS et al. 1988). The unusual structure of Gα8 at the C-terminal region may suggest that Gα8 interacts with a structurally different receptor and thus represents a very different class of G-protein superfamily.

Northern blot analyses indicate that each of these genes has a distinct pattern of expression during development of *Dictyostelium* (Fig. 3). Most of these genes hybridize to multiple RNA species that are presumably driven by different promoters. Gα6 is expressed primarily in vegetative cells. Upon starvation, the level of Gα6 mRNA declines rapidly. Gα3 mRNA is detected mainly in growing and very early aggregation stages. Gα1 is expressed at moderate levels in vegetative cells and increases to a maximal level at 10–12 h. Gα2 is expressed at very low levels in vegetative cells. Upon initiation of development, Gα2 RNA levels increase, reaching a maximum level during aggregation and then declining. A second transcript of Gα2 is preferentially expressed late in development in the anterior prestalk region as determined by *lacZ* expression studies (CARREL and FIRTEL, in preparation). Gα8 has a similar expression pattern as Gα2. The expression time course of both Gα2 and Gα8 parallels that of cAR1 during

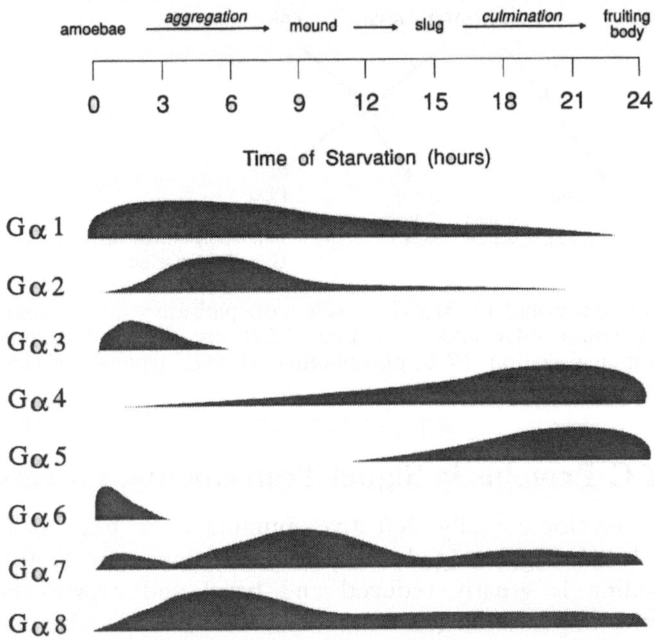


**Fig. 3.** Developmental expression of G$\alpha$ subunits during *Dictyostelium* development. The relative *width of the bar* indicates the relative level of mRNA detected at the time indicated. *Above*, the time course of the development. The expression pattern of G$\alpha$1 and G$\alpha$2 is from Pupillo et al. (1989), of G$\alpha$3 from Pupillo and Devreotes (in preparation), of G$\alpha$4 and G$\alpha$5 from Hadwiger et al. (1991), of G$\alpha$6, G$\alpha$7, and G$\alpha$8 from Wu and Devreotes (1991) and Cubitt et al. (1992). See text for a detailed explanation

early development. The level of G$\alpha$7 peaks in late aggregation and early mound stages, and declines thereafter. Finally, G$\alpha$4 and G$\alpha$5 are synthesized predominantly in late development when the multicellular structure is undergoing differentiation, although G$\alpha$4 is also expressed at low levels in vegetative cells.

The presence of at least eight G-protein subtypes during development is intriguing. It is unclear why there is such a diversity of G-proteins in the slime mold, and whether they are functionally redundant. The distinct time course of expression of these G-proteins in combination with the fact that the functions of G$\alpha$2 and G$\alpha$4 cannot be replaced by other G-proteins (see below) suggest that each of these proteins is probably involved in a different signal transduction pathway and thus plays a distinct role.

cDNA encoding for one G-protein $\beta$ subunit has been isolated (Lilly et al. 1993). The predicted amino acid sequences of G$\beta$ share an extensive degree of identity to its mammalian counterpart, and it is expressed throughout the growth and developmental stages.

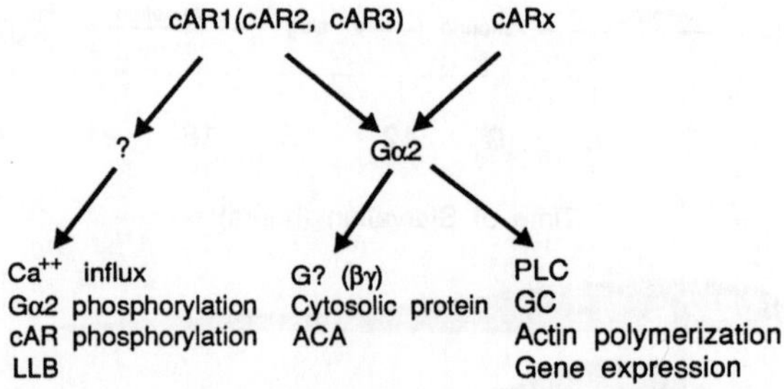


**Fig. 4.** The proposed model for signal transduction pathways during aggregation. See text for explanation. *cAR*, cAMP receptor; *LLB*, loss of ligand binding; *ACA*, adenylyl cyclase in aggregation; *PLC*, phospholipase C; *GC*, guanylyl cyclase

## D. Roles of G-Proteins in Signal Transduction Processes

A number of developmentally defective mutants have been isolated in *Dictyostelium*. In the frigid A (*fgdA*) mutants, the guanine nucleotide effect on cAMP binding is greatly reduced and basal and cAMP-stimulated GTPase activities are lowered (KESBEKE et al. 1988). Molecular cloning of the G$\alpha$2 gene indicates that the defective alleles in *fgdA* mutants reside in G$\alpha$2 (KUMAGAI et al. 1989). A gene-targeting experiment has generated *g$\alpha$2*-null cells that display same phenotypes as *fgdA* (KUMAGAI et al. 1991). The studies with these *g$\alpha$2*-null mutant cells have shown clearly that G$\alpha$2, in coupling to a cAMP receptor, plays an important role in signaling and development. A proposed pathway in early aggregation stage of the *Dictyostelium* development is shown in Fig. 4.

The *g$\alpha$2*-null cells do not aggregate and lack cAMP-mediated activation of adenylyl cyclase, guanylyl cyclase, phosphatidylinositol (PI)-specific phospholipase C (PLC), and regulation of gene expression (KESBEKE et al. 1988; SNAAR-JAGALSKA et al. 1988; KAMAGAI et al. 1991; OKAICHI et al. 1992). They also display a loss of GTP-mediated decrease in receptor affinity for cAMP but have no effect on chemotaxis to folate or folate activation of guanylyl cyclase (KUMAGAI et al. 1991), suggesting that G$\alpha$2 is coupled to a cAMP receptor but not to folate receptors. These phenotypes can be rescued by transformation with a vector expressing G$\alpha$2, indicating that the defects are caused by the absence of G$\alpha$2. It has also been demonstrated that G$\alpha$2 is required for actin polymerization (HALL et al. 1989). There are several cAMP receptor-mediated responses, however, that appear to be independent of G$\alpha$2 and will be discussed later (see below).

On stimulation of cells with cAMP, G$\alpha$2 is phosphorylated on one or more serine residues, resulting in an alteration of its electrophoretic mobility (GUNDERSEN and DEVREOTES 1990). Figure 5 shows a cAMP dose response of the G$\alpha$2 mobility shift. Triggered by increased occupancy of the surface

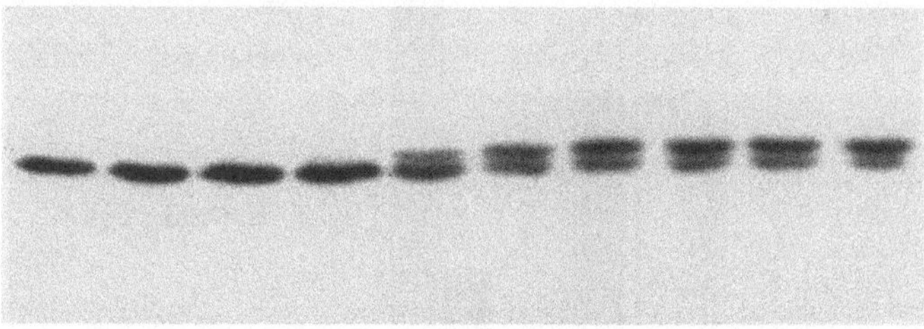


**Fig. 5.** The cAMP dose response of the Gα2 mobility shift (phosphorylation) on SDS-PAGE. Aggregation-competent cells were stimulated with various concentrations of cAMP for 1 min, and proteins were isolated and subjected to immunoblot for Gα2

cAMP receptor, phosphorylation of Gα2 is rapid and transient, coinciding with the time course of activation of physiological responses. The cAMP receptor is essential for Gα2 phosphorylation since cells that do not express the receptor do not phosphorylate Gα2. The site of phosphorylation has been mapped thus far to the N-terminal region of Gα2 (GUNDERSEN and DEVREOTES, unpublished results), which contains 12 serine residues, 4 of which are highly conserved among α subunits. It is unclear what role the phosphorylation of Gα2 plays. Its transient kinetics suggest that it might be involved in the activation of the protein, yet phosphorylation of Gα2 still occurs in certain *fgdA* mutants (R. GUNDERSEN, unpublished results). On the other hand, phosphorylation of Gα2 may affect inherent α subunit functions, such as GTP hydrolysis or binding to the βγ complex, or it may be important in receptor and/or effector recognition.

Further analysis of the functions of Gα2 has been obtained by expressing Gα2 containing amino acid substitutions in the highly conserved GTP-binding domains (OKAICHI et al. 1992). Two of the mutants analyzed are a G40V change in the GAGES domain and a Q208L change in GGQRS region. The equivalent mutations in *ras* and mammalian Gα subunit Gαs have been shown to substantially reduce the intrinsic GTPase activity of these proteins. The Q227L or R201C in Gαs results in a constitutive, dominant activating phenotype presumably because the protein is "locked" in the on or activating configuration (LANDIS et al. 1989). Expression of Gα2 proteins carrying these mutations in wild-type cells results in an aggregation-deficient phenotype, and the activation of guanylyl cyclase and phospholipase C is almost completely blocked and the activation of adenylyl cyclase is substantially inhibited (Fig. 6). Neither of the mutant proteins is capable of complementing *gα2*-null cells. Overexpression of wild-type Gα2 results in a cAMP-dependent stimulation of a maximum level of guanylyl cyclase activation and an inhibition of the adenylyl cyclase activation (OKAICHI et al. 1992).

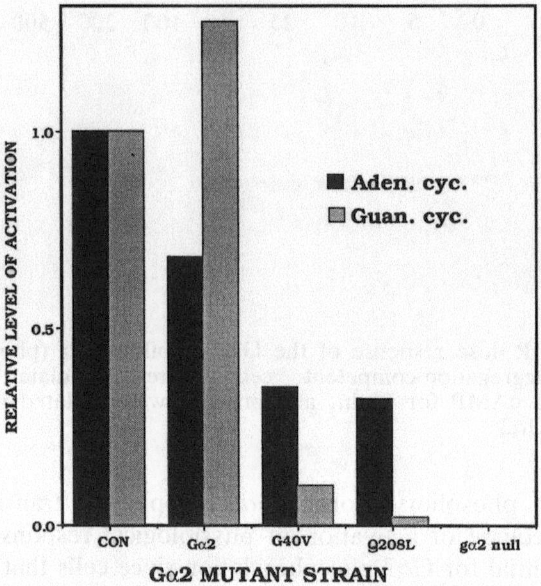


**Fig. 6.** Effect of amino acid substitutions in G$\alpha$2 on the activation of adenylyl cyclase and guanylyl cyclase. Wild-type KAx-3 cells containing expression vectors expressing wild-type G$\alpha$2 or G$\alpha$2 substitution mutations (G40V and Q208L) were plated on nonnutrient agar. Cells were harvested, and cAMP activation of adenylyl cyclase and guanylyl cyclase was assayed. The maximum levels of activation, as determined by the maximum level of cAMP or cGMP produced, were compared. The value for control KAX-3 cells (CON) was given a value of 1.0. *G$\alpha$2*, Cells transformed with wild-type G$\alpha$2; *g$\alpha$2*, *g$\alpha$2* null cells. In all G$\alpha$2 transformants G$\alpha$2 protein was over-expressed tenfolds compared to the level found in control cells. See Okaichi et al. (1992) for details

These results suggest that both the G40V and Q208L mutations have a dominant negative phenotype in vivo, in contrast to the expected dominant activating phenotype seen with similar G$\alpha$ subunit mutations in other cells. The cAMP receptor-mediated effector pathways that require G$\alpha$2, such as adenylyl and guanylyl cyclase, adapt rapidly during persistent cAMP stimulation. Perhaps the activated G$\alpha$ proteins cause a low-level constitutive activation of these pathways that in turn results in the pathway being constitutively adapted or down-regulated.

As in mammalian cells, adenylyl cyclase activity in *Dictyostelium* is regulated by G-proteins. This suggestion was initially demonstrated by the ability of GTP, Gpp(NH)p (guanyl-5′-yl imidodiphosphate), and GTP$\gamma$S to activate and GDP$\beta$S to inactivate adenylyl cyclase in lysates of aggregation-competent cells (Theibert et al. 1986). Thus, it was predicted that the adenylyl cyclase present during aggregation would closely resemble mammalian adenylyl cyclase. Recently, the gene encoding this adenylyl cyclase, ACA, has been isolated (Pitt et al. 1992). Analysis of the predicted

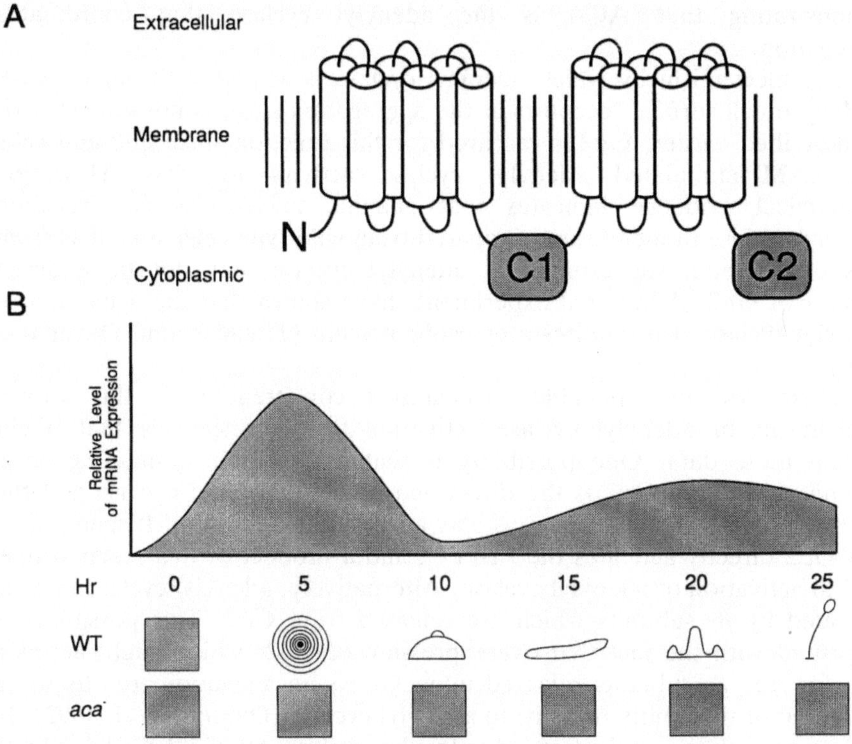


**Fig. 7A,B.** Topological model and expression of ACA and phenotype of *aca*-null cells. **A** Model of ACA. *Vertical bars*, the plasma membrane; *N*, the amino termini. **B** *Top*, the developmentally regulated expression of ACA, showing the relative amount of mRNA expressed at time indicated; *bottom*, schematic developmental phenotype of wild-type and *aca*-null cells to show that the disruption of ACA locus results in the loss of the capability to aggregate

amino acid sequence demonstrates that the ACA gene product shares significant topological and sequence homology with its mammalian counterparts. These molecules have two large hydrophobic domains, each of which contains six transmembrane spanning domains and two homologous hydrophilic domains (Fig. 7). Sequence homology among mammalian adenylyl cyclases and between mammalian adenylyl cyclase and ACA is highest in the hydrophilic domains. The two hydrophilic domains in ACA are about 50% similar at the amino acid level to each other and to their counterparts within the mammalian adenylyl cyclases (Pitt et al. 1992). ACA is expressed maximumly during the aggregation stage, decreases after aggregation, and is induced again in the later stage of development (Fig. 7). An *aca*-null cell line has been created by gene disruption. These cells have little detectable adenylyl cyclase activity and fail to aggregate (Fig. 7), demonstrating that ACA is the adenylyl cyclase that coordinates aggregation.

The mechanism by which adenylyl cyclase is activated through cAMP binding to cell surface receptors at the aggregation stage is not understood. As described earlier, G$\alpha$2 is required for this function since *g$\alpha$2*-null cells lack cAMP-stimulated adenylyl cyclase activity in vivo. However, biochemical evidence indicates that guanine nucleotides can regulate adenylyl cyclase in membranes prepared from wild-type cells as well as from cells of *g$\alpha$2*-null, suggesting that adenylyl cyclase may not be a direct effector of G$\alpha$2. Additional experiments have shown that the activation of adenylyl cyclase also requires a cytosolic protein (THEIBERT and DEVREOTES 1986).

There are two possible mechanisms consistent with the known mechanisms of adenylyl cyclase activation in other systems that might explain these data. One possibility is that a G-protein containing an $\alpha$ subunit other than G$\alpha$2 is the direct activator of adenylyl cyclase and the G$\alpha$2-mediated signaling pathways play a role in its activation. It seems clear that G$\alpha$2 directly activates the PI-PLC, and a product of this reaction may lead to activation of adenylyl cyclase. Alternatively, adenylyl cyclase may be activated by $\beta\gamma$ subunits which are released from G$\alpha$2. This possibility is consistent with the fact that overexpression of G$\alpha$2, which might act as a sink for free $\beta\gamma$ subunits released upon G-protein activation, results in an inhibition of the ability to activate adenylyl cyclase (OKAICHI et al. 1992). In such a model the activation of adenylyl cyclase by GTP$\gamma$S in *g$\alpha$2*-null membranes would be mediated through the release of $\beta\gamma$ subunits from other G-proteins, such as those containing G$\alpha$1, G$\alpha$7, and G$\alpha$8, which are known to be preferentially expressed at this time during development. Further in vitro analysis is required to distinguish between these two possibilities as well as other mechanisms.

G$\alpha$2 appears to be important for many key transmembrane signaling processes. However, it is not required for response to folic acid (see above) and several cAMP-stimulated responses occurring in its absence (Fig. 4), suggesting that other G-proteins might mediate these functions. It has been shown that cAMP receptor-mediated $Ca^{2+}$ uptake is independent of G$\alpha$2 (MILNE and COUKELL 1991). Phosphorylation of G$\alpha$2, which is mediated by cAR1, seems to be independent of functional G$\alpha$2 since the protein can still be phosphorylated in certain *fgdA* alleles. The cAMP-induced phosphorylation of cAMP receptors and the loss of ligand binding, which are components of the desensitization process, are not affected in cells that lack G$\alpha$2 protein (VAN HAASTERT et al. 1992).

The discovery of at least eight G-protein $\alpha$ subunits in *Dictyostelium* has provided candidates for the G-proteins that might fulfill these specific roles. The gene-targeting technique provides a powerful tool to investigate the functions of individual G-proteins. Cell lines lacking each subunit have recently been generated and are being analyzed. In the case of G$\alpha$1, loss of

G$\alpha$1 expression results in no visible growth or developmental defects (KUMAGAI et al. 1991), suggesting that some of them are functionally redundant or have subtle effects. Thus, it may be necessary to construct cells that are deficient in multiple G-protein $\alpha$ subunits to address the possibility of redundancy.

## E. Roles of G-Proteins in Morphogenesis and Differentiation

As shown in Fig. 3, both G$\alpha$4 and G$\alpha$5 are expressed at high levels at the time of mound and tip formation following aggregation, implying that they might play a role during this stage of development. The role of G$\alpha$4 in development was investigated by creating and examining *g$\alpha$4*-null mutants and cells that overexpress the G$\alpha$4 gene (HADWIGER and FIRTEL 1992). As might be expected from the temporal pattern of expression, the early stages of development appear normal in *g$\alpha$4*-null cells. They aggregate and form mounds that differentiate into an erect finger morphology similar to wild-types cells (Fig. 8). In contrast to wild-type development in which the finger falls over forming a migrating slug, the apical portion of the tip in the *g$\alpha$4*-null cells continues to elongate while the basal region remains more rounded. In many cases, the apical projection becomes thinner, while in other cases it falls back on itself, producing a "knotted" structure (Fig. 8). Similarly, the overexpressing strain shows normal mound formation and then produces a very abnormal "fruiting-body-like" structure (Fig. 8). The *g$\alpha$4*-null cells do not produce mature spores, while the overexpressor cells show a 25-fold reduction in total number of spores produced. The *g$\alpha$4*-null cells can be complemented with a low copy number G$\alpha$4-expression vector, restoring the normal morphological differentiation and the production of mature spores as wild-type cells.

Further insight into the possible role of G$\alpha$4 comes from the analysis of the temporal and spatial late gene expression in *g$\alpha$4*-null cells. Northern blot analysis indicates that the prestalk-specific *ras* gene *DdrasD* is expressed at 50% of wild-type levels, and the prespore-specific protein *SP60* mRNA is reduced to very low levels. Expression of *DdrasD* and *SP60* are preferentially localized to the anterior and more basal regions, respectively, as in wild-type cells, suggesting that the initial spatial patterning of both of these genes is not affected in *g$\alpha$4*-null cells.

Both cAMP-induced prestalk and prespore-specific genes can also be induced in a shaken suspension of cells, a method that enables one to bypass the effects of a mutation on morphogenesis (MCHDY and FIRTEL 1985). Under these conditions, the induction of the cAMP-inducible prestalk gene *DdrasD* is reduced but that of another cAMP-inducible prestalk gene, *pst-cath/CP2*, is not affected, while the expression of *SP60* is still only induced to a low level. To distinguish cell-autonomous and non-cell-

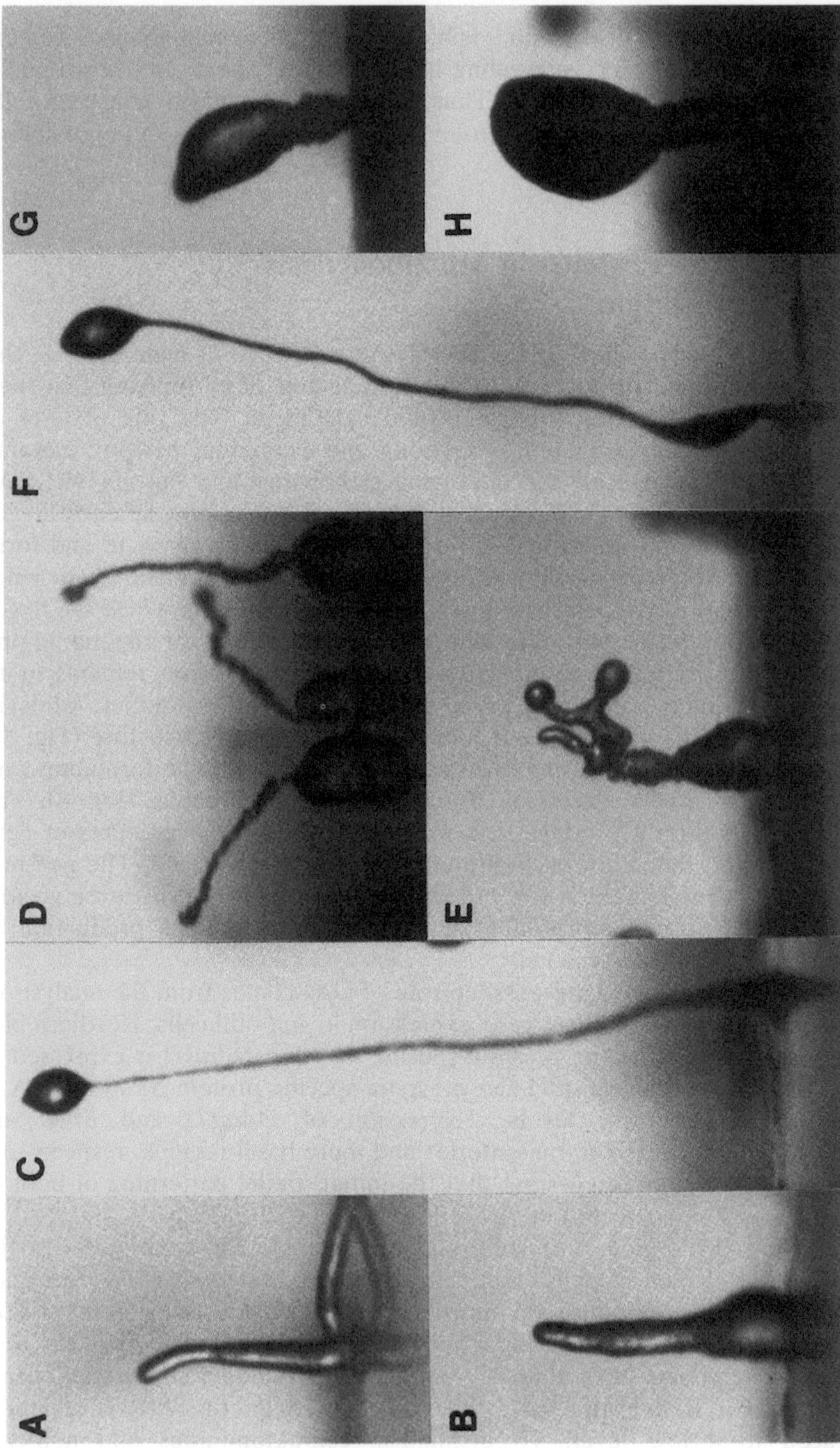
A
B
C
D
E
F
G
H

autonomous function of G$\alpha$4, wild-type and mutant strains or two mutant strains are mixed in various proportions and the developmental potential of both cell types examined. When the *g$\alpha$4*-null cells or the G$\alpha$4-overexpressing cells are mixed with wild-type cells, or the *g$\alpha$4*-null cells are mixed with the overexpressor cells, at a 50:50 ratio, morphologically normal fruiting bodies are observed. The G$\alpha$4-overexpressing cells in the chimera with either wild-type cells or *g$\alpha$4*-null cells produced spores at a level similar to that of wild-type cells. The *g$\alpha$4*-null cells also produce spores in either chimera, although the level of spore formation was only approximately 3%–4% of that of wild-type cells, indicating that the wild-type and G$\alpha$4-overexpressing cells can partially complement *g$\alpha$4*-null cells. In these chimeras, *g$\alpha$4*-null cells are found in all of the cell types of the mature fruiting body except that there appear to be a slightly lower level in the spore mass and a slightly higher level in the stalk.

These combined results suggest that G$\alpha$4 is essential for proper development during culmination and spore production. The function of G$\alpha$4 appears to be non-cell-autonomous because *g$\alpha$4*-null cells can participate in the formation of a wild-type fruiting body in chimeras with either wild-type or overexpressing cells, and the wild-type cells can partially complement the spore production of *g$\alpha$4*-null cells. Thus G$\alpha$4 may be involved in an extracellular signaling process in which G$\alpha$4-producing cells are required for producing either an intercellular soluble signal or a cell-cell surface molecule that directly interacts with downstream cells. Studies with G$\alpha$4 promoter/*lacZ* reporter gene indicate that G$\alpha$4 is expressed at a detectable level in only a small subpopulation of cells within the multicellular aggregate that are known as anteriorlike cells. Although we cannot exclude the fact that G$\alpha$4 may also be expressed at a very low level in prespore cells and possibly also in the general prestalk population as a whole, the expression of G$\alpha$4 in anterior-like cells is probably important in controlling both the spatial patterning in *Dictyostelium* multicellular morphogenesis and prespore differentiation.

# F. Conclusions and Perspectives

In *Dictyostelium*, genes encoding four surface receptors, eight G-protein $\alpha$ subunits and one $\beta$ subunit, and an adenylyl cyclase have been identified.

**Fig. 8A–H.** Developmental morphology of *g$\alpha$4*-null, G$\alpha$4-overexpressor, and wild-type cells. Logarithmically grown cells were washed and plated for development on nonnutrient agar. **A** Wild-type cells at the finger stage (14 h after starvation); **B** *g$\alpha$4* null cells at finger stage (15 h after starvation); **C** wild-type cells at fruiting-body stage (>26 h after starvation); **D, E** *g$\alpha$4* cells at final morphological stage (photo taken at 36 h); **F** *g$\alpha$4* null cells complemented with a low copy of the G$\alpha$4 expression vector at the fruiting-body stage (photo taken at 36 h); **G, H** G$\alpha$4-overexpressing cells at the final morphological stage (photo taken at 36 h). See HADWIGER and FIRTEL (1992) for details

The rather large diversity of the $\alpha$ subunits indicates that they may be involved in a variety of signal transduction pathways. The characterization of several of these genes, cAR1, G$\alpha$2, G$\alpha$4, and adenylyl cyclase, has shown that they are essential for the proper development in *Dictyostelium*. Gene-targeting and other genetic and molecular techniques have provided powerful tools to investigate the functions of these proteins. Since the mechanisms of signaling processes in *Dictyostelium* are very similar to those in mammalian cells, the molecular and genetic dissection of these processes will elucidate their possible roles not only in *Dictyostelium* but also in other developmental systems.

*Acknowledgements*. We would like to thank D. Hereld for critical reading of this manuscript. This work was supported by NIH grants GM28007 to P.N.D. and GM37830 to R.A.F.

## References

Cubitt AB, Carrel F, Dharmawardhane S, Gaskins C, Hadwiger JA, Howard PH, Mann SKO, Okaichi K, Zhou K, Firtel RA (1992) Molecular genetic analysis of signal transduction pathways controlling multicellular development in Dictyostelium. Cold Spring Harbor Symposium for Quant. Biology 177–192

Deretic D, Hamm HE (1987) Topographic analysis of antigentic determinants recognized by monoclonal antibodies to the photoreceptor guanyl nucleotide-binding protein, transducin. J Biol Chem 262:10839–10847

Devreotes PN (1989) Dictyostelium discoideum: a model system for cell-cell interactions in development. Science 245:1054–1058

Firtel RA (1991) Signal transduction pathways controlling multicellular development in Dictyostelium. Trends in Genetics 7:381–388

Gundersen RE, Devreotes PN (1990) In vivo receptor-mediated phosphorylation of a G-protein in Dictyostelium. Science 248:591–593

Hadwiger JA, Wilkie TM, Stratmann M, Firtel RA (1991) Identification of Dictyostelium G$\alpha$ genes expressed during multicellular development. Proc Natl Acad Sci USA 88:8213–8217

Hadwiger JA, Firtel RA (1992) Analysis of G$\alpha$4, a G protein subunit required for multicellular development in Dictyostelium. Genes Dev 6:38–49

Hall AL, Warren V, Cordellis J (1989) Transduction of the chemotactic signal to the actin cytoskeleton of Dictyostelium discoideum. Dev Biol 136:517–525

Johnson RL, Saxe CL, Gollop R, Kimmel AR, Devreotes PN (1993) Identification and targeted gene disruption of cAR3, a cAMP receptor subtype expressed during multicellular stages of Dictyostelium development. Genes Dev 7:273–282

Kesbeke F, Snaar-Jagalska BE, van Haastert (1988) Signal transduction in Dictyostelium fgdA mutants with a defective interaction between surface cAMP receptor and a GTP-binding regulatory protein. J Cell Biol 107:521–528

Klein PS, Sun TJ, Saxe CL, Kimmel AR, Johnson RL, Devreotes P (1988) A chemoattractant receptor controls development in Dictyostelium discoideum. Science 241:1467–1472

Kumagai A, Pupillo M, Gundersen R, Miake-lye R, Devreotes PN, Firtel RA (1989) Regulation and function of G$\alpha$ protein subunits in Dictyostelium. Cell 57:265–275

Kumagai A, Hadwiger JA, Pupillo M, Firtel A (1991) Molecular genetic analysis of two G$\alpha$ protein subunits in Dictyostelium. J Biol Chem 266:1220–1228

Landis CA, Masters SB, Spada A, Pace AM, Bourne HR, Vallar L (1989) GTPase inhibiting mutations activate the $\alpha$ chain of Gs and stimulate adenylyl cyclase in human pituitary tumours. Nature 340:692–696

Lilly P, Wu L, Welker DL, Devreotes PN (1993) A G-protein $\beta$-subunit is essential for Dictyostelium development. Genes Dev 7:786–995

Mann SKO, Firtel RA (1991) A developmentally regulated, putative sereine/threonine protein kinase is essential for development in Dictyostelium. Mech of Develop 35:89–101

Masters SB, Sullivan KA, Miller BB, Lopez NG, Ramachandran J, Bourne HR (1988) Carboxyl terminal domain of Gs$\alpha$ specifies coupling of receptors to stimulation of adenyl cyclase. Science 241:448–451

Mchdy MC, Firtel RA (1985) A secreted factor and cyclic AMP jointly regulate cell-type-specific gene expression in Dictyostelium discoideum. Mol Cell Biol 5: 705–713

Miline JA, Coukell B (1991) A $Ca^{++}$ transport system associated with the plasma membrane of Dictyostelium discoideum is activated by different chemoattractant receptors. J Cell Biol 112:103–110

Okaichi K, Cubitt AB, Pitt GS, Firtel RA (1992) Amino acid substitutions in the Dictyostelium G$\alpha$ subunit G$\alpha$2 produce dominant negative phenotypes and inhibit the activation of adenylyl cyclase, guanylyl cyclase, and phospholipase C. Mol Biol Cell 735–747

Pitt G, Milona N, Borleis J, Lin K, Reed R, Devreotes PN (1991) Structurally distinct and stage-specific adenylyl cyclase genes play different role in Dictyostelium development. Cell 69:305–315

Pupillo M, Kumagai A, Pitt GS, Firtel RA, Devreotes PN (1989) Multiple $\alpha$ subunits of guanine nucleotide-binding proteins in Dictyostelium. Proc Natl Acad Sci USA 86:4892–4896

Saxe CL, Johnson RL, Devreotes PN, Kimmel AR (1991a) Multiple genes for cell surface cAMP receptors in Dictyostelium discoideum. Dev Gen 12:6–13

Saxe CL, Johnson RL, Devreotes PN, Kimmel AR (1991b) Expression of a cAMP receptor gene of Dictyostelium and evidence for a multigene family. Genes Dev 5:1–8

Saxe CL, Ginsburg GT, Louis JM, Johnson RL, Devreotes PN, Kimmel AR (1993) CAR2, a prestalk cAMP receptor required for normal tip formation and late development of Dictyostelium discoideum. Genes Dev 7:262–272

Simon MI, Strathmann P, Gautam N (1991) Diversity of G proteins in signal transduction. Science 252:802–808

Snaar-Jagalska BE, van Haastert PJM (1988) G-proteins in the signal-transduction pathways of Dictyostelium discoideum. Dev Genet 9:215–226

Sullivan K, Miller RT, Masters SB, Beiderman B, Heideman W, Bourne HR (1987) Identification of receptor contact site involved in receptor-G protein coupling. Nature 330:758–760

Sun TJ, VanHaastert PJM, Devreotes PN (1990) Surface cAMP receptors mediate multiple responses during development in Dictyostelium: evidenced by antisense mutagenesis. J Cell Biol 110:1549–1554

Sun TJ, Devreotes PN (1991) Gene targeting of the aggregation stage cAMP receptor cAR1 in Dictyostelium. Genes Dev 5:572–582

Theibert A, Devreotes PN (1986) Surface receptor-mediated activation of adenylate cyclase in Dictyostelium. J Biol Chem 261:15121–15125

Van Haastert PJM, Wang M, Bominaar AA, Devreotes PN, Schaap P (1992) cAMP-induced desensitization of surface cAMP receptors in Dictyostelium: different second messengers mediate receptor phosphorylation, loss of ligand binding, degradation of receptor, and reduction of receptor mRNA levels. Mol Biol Cell 3:603–612

Wu L, Devreotes PN (1991) Dictyostelium transiently expresses eight distinct G-protein $\alpha$-subunits during its developmental program. Biochem Biophys Res Commun 179:1141–1147

CHAPTER 65

# Functional Expression of Mammalian Receptors and G-Proteins in Yeast

K.J. BLUMER

## A. Introduction

Biologists studying G-protein-linked signal transduction pathways have used heterologous expression systems to make many important discoveries, including the molecular description of pharmacologically distinct subtypes of receptors, and the specificity with which receptors, G-proteins, and their effectors interact with one another. However, commonly used expression systems, including those employing *Xenopus* oocytes or transfected mammalian cell lines, pose limitations for basic and applied scientists alike. Such systems are not usually amenable to rigorous formal genetic analysis of signal transduction pathways, and they are costly to use as biological screening tools in drug discovery programs.

Accordingly, a potentially attractive alternative to classicial expression systems involves use of the yeast *Saccharomyces cerevisiae*. Yeast cells can be manipulated genetically with ease, and they possess a well-characterized G-protein-linked signaling pathway (Fig. 1) that controls expression of specific and easily scored phenotypes (BLUMER and THORNER 1991; MARSH et al. 1991). Thus if mammalian receptors, G-proteins, and their effectors can be expressed in yeast such that they substitute for their yeast counterparts, their functional properties can be elucidated rapidly through detailed molecular genetic and biochemical analyses. Such studies might reveal, for example, how receptors are activated by agonists and blocked by antagonists, and uncover fundamental cell biological principles regarding the synthesis, assembly, targeting, and posttranslational modifications of receptors, G-protein subunits, and their effector molecules. Moreover, yeast cells expressing receptors and G-proteins of interest may be well suited as rapid and economical biological drug screening tools.

In this review I discuss the current status of efforts to express functional mammalian receptors and G-protein subunits in yeast. Highlighted are methods that have been or could be used to improve the likelihood that functional interaction will occur between mammalian receptors and G-protein subunits. Although components of other types of signal transduction pathways have been expressed in yeast, including ion channels and transporters (YELLEN and MIGEON 1990; RAYMOND et al. 1992; ANDERSON et al. 1992; SENTENAC et al. 1992), steroid hormone receptors (SCHENA and

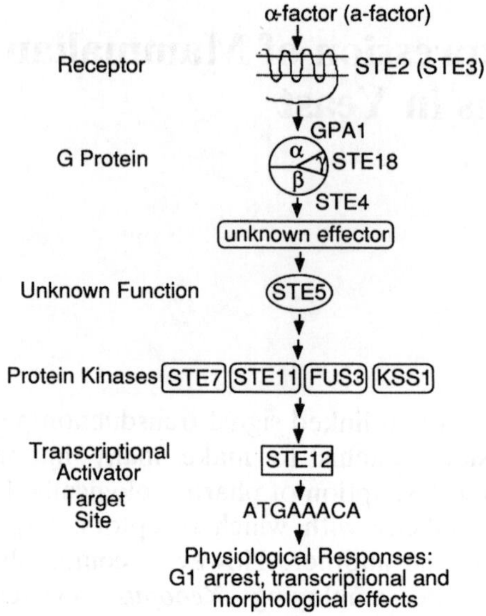


**Fig. 1.** Mating pheromone response pathway in the yeast *S. cerevisiae*. Current evidence indicates that the gene products shown have the indicated biochemical activities, and function in the order depicted. A hypothetical effector molecule has been interposed between the G-protein and the *STE5* gene product solely to indicate that the mechanism of signaling between these molecules is unknown. The precise functional order for the protein kinase homologs encoded by the *STE7*, *STE11*, *FUS3*, and *KSS1* genes has yet to be determined; they are therefore shown functioning at a similar step. *Single arrows*, direct interaction between signaling components; *double arrows*, direct interactions have yet to be established

YAMAMOTO, 1989), *ras* protein and its regultors (BALLESTER et al. 1989; XU et al. 1990), protein kinases, and transcription factors (GILL and TJIAN 1991), discussion of these areas is beyond the scope of this review. Also omitted are detailed discussions of the biogenesis of integral membrane proteins, and the G-protein-linked signaling pathway in yeast that mediates response to oligopeptide mating pheromones; appropriate reviews are available (SINGER 1990; BLUMER and THORNER 1991; MARSH et al. 1991), including one by CLARKE et al. in this volume.

## B. Expression of Mammalian G-Protein-Coupled Receptors

Several kinds of G-protein-coupled receptors have been expressed in yeast, including human $\beta_2$-adrenergic receptor (KING et al. 1990), human $M_1$ muscarinic acetylcholine receptor (PAYETTE et al. 1990), and rat $M_5$ muscarinic acetylcholine receptor (HUANG et al. 1992). Table 1 summarizes the properties of these receptors when they are expressed in yeast and

**Table 1.**

| Receptor[a] | Cell type | $B_{max}$ (pmol/mg)[b] | $K_d$ (n$M$)[c] | $K_i$ (n$M$)[d] | Reference |
|---|---|---|---|---|---|
| $\alpha$ Factor receptor | Yeast | 0.65 | 2 ($\alpha$ factor) | NA | Blumer et al. (1988) |
| Human $\beta_2$-AR | Yeast | 115 | 0.093 (ICYP) | 103 (−ISO), 3670 (+ISO), 664 (−EPI), 6000 (−NOR) | King et al. (1990) |
| | COS-7 cells | 24 | 0.110 (ICYP) | 130 (−ISO), 4000 (+ISO), 360 (−EPI), 5800 (−NOR) | King et al. (1990) |
| Human $M_1$ mAChR | Yeast | 0.001–0.02 | 0.179 (NMS) | 0.235 (ATR), 7.26 (PZP), 230 (METH) | Payette et al. (1990) |
| | CHO cells | 0.180 | 0.065 (NMS) | 0.21 (ATR), 16 (PZP), 16 (METH) | Buckley et al. (1989) |
| Rat $M_5$ mAChR | Yeast | ND | 23 (NMS) | ND | Huang et al. (1992) |
| | L cells | ND | 0.41 (NMS) | 2.0 (ATR), 150 (PZP) | Liao et al. (1989) |

[a] $\alpha$ Factor receptor, *STE2* gene product; $\beta_2$-AR, $\beta_2$-adrenergic receptor; mAChR, muscarinic acetylcholine receptor.
[b] $B_{max}$, Ligand binding sites expressed per milligram of protein in total cellular membrane preparations.
[c] ICYP, Iodocyanopindolol; NMS, *N*-methylscopolamine.
[d] ISO, Isoproterenol; EPI, epinephrine; NOR, norepinephrine; ATR, atropine; METH, methoctramine; PZP, pirenzepine.

mammalian cell lines. These findings indicate that human $\beta_2$-adrenergic receptors display essentially identical ligand-binding properties when expressed in yeast and mammalian cells (KING et al. 1990). Similar evidence indicates that three subtypes of $\alpha_2$-adrenergic receptors (C2, C4, and C10-2) and $\alpha_{1B}$-adrenergic receptors can be expressed in yeast without significant perturbation of their subtype-specific ligand-binding properties (LOMASNEY and LEFKOWITZ, personal communication; PETERSEN et al., personal communication).

Muscarinic acetylcholine receptors seem to display somewhat different ligand-binding characteristics when they are expressed in yeast and mammalian cells. Differences occur principally in the absolute affinities for certain muscarinic antagonists (BUCKLEY et al. 1989; HUANG et al., 1992; LIAO et al. 1989; PAYETTE et al. 1990; PERALTA et al. 1987). However, because quantitative differences in antagonist binding properties are also observed when muscarinic acetylcholine receptors are expressed in different mammalian cell types (BUCKLEY et al. 1989; PERALTA et al. 1987), these effects may not be due simply to expression of receptors in yeast. Despite these quantitative differences, the rank order for binding of various muscarinic antagonists seems to be fairly characteristic of each receptor subtype, when they are expressed in yeast. However, it remains to be seen whether mammalian receptors always retain their pharmacologically distinct properties when expressed in yeast.

Inspection of Table 1 also reveals that mammalian receptors are expressed in functional form at levels ($B_{max}$) that span at least a 10000-fold range. Because it appears that functional receptors must be expressed above a certain threshold level in order to display signaling activity in vivo (see Sect. D), it is important to establish whether transcriptional, translational, or posttranslational events control formation of active receptors in yeast. Each of these areas is discussed below.

Differences in transcriptional efficiency or mRNA stability are unlikely to account for wide variations in expression levels since strong inducible (*GAL* or *CUP1*) or constitutive (*ADH*) promoters are used nearly invariably to drive expression of receptor cDNAs. However, because receptor mRNA levels have not generally been measured, rapid degradation of receptor mRNAs could limit expression in some cases.

Differences in translational efficiency might be more likely to limit receptor expression, particularly if the 5′ leader region of a receptor mRNA forms a stable secondary structure, or the initiation codon is placed in an unfavorable sequence context (CIGAN and DONAHUE 1987). Translational efficiency could also be reduced by an abundance of rarely used codons in a receptor coding sequence; however, rare codons limit translation only under special circumstances (ANDERSSON and KURLAND, 1990).

By contrast, there is good evidence that posttranslational events control biogenesis of receptors in yeast. For example, human $M_1$ muscarinic acetylcholine receptor and $D_1$ dopamine receptor polypeptides are expressed efficiently, yet they display little ligand binding activity (PAYETTE et al. 1990; DENNIS, personal communication; DIN et al., personal communication). In principle, nonfunctional receptors could accumulate because they insert improperly into the endoplasmic reticulum (ER), fail to fold properly, or are mistargeted to inappropriate intracellular destinations, such as the vacuole. Clear resolution of these issues, however, is difficult due to a paucity of information regarding the factors that specifically control the biogenesis of polytopic membrane proteins in yeast. As a consequence, most investigators have used various kinds of manipulations, as outlined below, that might be expected to improve expression of functional receptors.

One possible means of enhancing expression of functional receptors has been the use protease-deficient strains of yeast. In one instance, $\alpha_{2C2}$-adrenengic receptor is expressed at least 100-fold more efficiently in a strain deficient in three major vacuolar proteases (encoded by *PEP4*, *PRB1*, and *PRC1*; PETERSEN et al., personal communication). However neither wild-type nor protease-deficient strains express significant levels of functional human $D_1$ dopamine receptors (DIN et al., personal communication). Thus it remains uncertain whether protease deficient strains increase expression of functional receptors in any systematic way.

As a second means of increasing expression of functional receptors, amino acid sequences have been fused to the N-terminal regions of mammalian receptors such that translocation into the ER may occur more efficiently. One variation on this theme involves the use of signal sequences derived from precursor forms of secreted yeast proteins, such as the mating pheromone $\alpha$ factor. Although classical signal sequences are normally absent from G-protein-coupled receptors, including the yeast mating pheromone receptors, heterologous chimeric receptors containing signal sequences could nonetheless be imported more efficiently into the ER of yeast. Indeed, a hybrid protein containing the $\alpha$ factor signal sequence enhances expression of functional rat $M_5$ muscarinic acetylcholine receptors (HUANG et al. 1992). However, signal sequences do not always have a positive effect because $\alpha_{2C2}$-adrenergic receptors are expressed at similar levels whether or not a signal sequence is employed (PETERSEN et al., personal communication). These findings are not easily squared with one another because neither study addressed whether signal sequences actually enhance import of receptors into the ER. However, since signal sequences do improve expression of nicotinic acetylcholine receptor subunits in yeast (YELLEN et al. 1990), the use of signal sequences may be a generally productive strategy if native receptors are poorly expressed.

Attempts to improve import of mammalian receptors into the ER have also involved exchanging the N-terminal extracellular domain of mammalian receptors with the corresponding region of the yeast $\alpha$ factor receptor. Here the logic is that the N-terminal domain of yeast receptors might be translocated more efficiently across the ER membrane and assume a lumenal (extracellular) orientation. Whereas chimeras of this kind do seem to improve expression of functional $\beta_2$-adrenergic receptors (King et al. 1990; Lomasney and Lefkowitz, personal communication), a similar hybrid involving $\alpha_{2C2}$-adrenergic receptor actually decreases expression of functional receptors (Petersen et al., personal communication). Again, none of these studies have addressed at the molecular or cell biological level how these constructions exert their apparently opposite effects on expression of functional receptors. However, lack of a generally positive effect might not be surprising because the extracellular domain of the $\alpha$ factor receptor fails to direct translocation and secretion of a soluble heterologous protein (Cartwright and Tipper 1991). Although it is conceivable that the N-terminal extracellular domains of receptors have passive or secondary roles in the insertion and translocation processes, the first two transmembrane segments are probably of key importance (Singer 1990).

Even if mammalian receptors are imported efficiently into the ER in yeast, they could be retained in an intracellular compartment because they are improperly folded or assembled. Indeed an important role for folding or assembly events in the biogenesis of G-protein-coupled receptors has been obtained from genetic studies of *Drosophila melanogaster*. Here, folding or assembly of certain rhodopsin isoforms requires a proline cis-trans isomerase encoded by the *ninaA* gene (Stammes et al. 1991). Similarly, emerging data from yeast suggest that a family of genes (*STP22*, *STP24*, *STP26*, and *STP27*) controls the biogenesis of plasma membrane proteins (Cartwright and Tipper 1991; Jenness, personal communication). Mutations in these genes dramatically increase (up to 100-fold) cell surface expression of mutant forms of plasma membrane proteins, including the $\alpha$ factor receptor and arginine permease (Jenness, personal communication). Although the means by which *STP* genes exert their effects is uncertain (Jenness, personal communication), it is known that *stp22* mutations elevate expression of $\beta_2$-adrenergic receptor two- to fourfold (Lee, personal communication). This effect is modest, but it will nonetheless be useful to determine whether mutations in other *STP* genes affect expression of a variety of mammalian receptors.

As a possible means of assisting the folding or assembly of mammalian receptors, hydrophobic antagonists have been included in the growth medium (King et al. 1990). By diffusing across membranes such molecules presumably can bind to and stabilize intracellular receptors that are correctly folded (Waldo et al. 1983; Dohlman et al. 1990). Correctly folded receptors may therefore be transported efficiently through the secretory pathway. Although the general utility of this technique remains to be established, it may prove to be the simplest means of improving functional expression of receptors for which hydrophobic antagonists are available.

# C. Expression of Mammalian G-Protein Subunits

## I. Physiological Roles of Yeast G-protein Subunits

Yeast cells possess two known G-protein $\alpha$ subunit homologs, encoded by the *GPA1* (*SCG1*) and *GPA2* genes (BLUMER and THORNER 1991). The function of Gpa2 protein is uncertain. Overexpression of *GPA2* suppresses *ras2* temperature-sensitive mutations and elevates cyclic AMP levels under certain conditions (NAKAFUKU et al. 1988). The *gpa2* mutants, however, display no obvious phenotypes. Currently there is no indication whether Gpa2 protein functions in association with $G_\beta$ and $G_\gamma$ subunit homologs. For these reasons, functional replacement of Gpa2 protein by mammalian $G_\alpha$ subunits has not been evaluated.

The *GPA1* gene encodes the $\alpha$ subunit of a heterotrimeric G-protein that couples mating pheromone receptors with downstream elements of the pheromone response pathway. The $\beta$ and $\gamma$ subunits of this G-protein are encoded by the *STE4* and *STE18* genes, respectively. Gpa1 protein represses the response pathway in the absence of mating pheromone by sequestering $\beta\gamma$ complexes. Activation of pheromone receptors, or deletion of *GPA1* releases $\beta\gamma$ subunits, and triggers the mating response pathway. Because this response includes arrest in the $G_1$ phase of the cell division cycle, *gpa1* deletion mutations are lethal in haploid cells.

Functional expression of mammalian $G_\alpha$ subunits or chimeric yeast/mammalian $G_\alpha$ subunits can be easily assessed based on their ability to suppress the lethal phenotype of *gpa1* deletion mutations, interfere with the function of wild-type Gpa1 protein (inhibition of pheromone response or mating), or interact with mammalian receptors that are coexpressed in yeast. Presumably, these tests all require that native mammalian or chimeric yeast-mammalian $G_\alpha$ subunits form complexes with yeast $G_{\beta\gamma}$ subunits.

## II. Mammalian $G_\alpha$ Subunits

### 1. Intact $G_\alpha$ Subunits

Four intact $G_\alpha$ subunits have been expressed in yeast: $G_{\alpha i2}$, $G_{\alpha o}$, and the short and long forms of rat $G_{\alpha s}$ produced by alternative mRNA splicing (DIETZEL and KURJAN, 1988; KANG et al. 1990). Of these, the short and long forms of $G_{\alpha s}$ replace Gpa1 protein most effectively. However, high level expression of $G_{\alpha s}$ is needed to suppress the growth defect of *gpa1* mutants, or inhibit mating of wild-type cells.

Expression of a $G_{\alpha i2}$ cDNA from the strong *PGK* and *CUP1* promoters only weakly suppresses a *gpa1* mutation. This result is somewhat surprising since of the mammalian $G_\alpha$ subunits that have been expressed in yeast, $G_{\alpha i2}$ is most similar in sequence to Gpa1 protein. $G_{\alpha o}$ appears to be even less active, since expression of its cDNA from the *CUP1* promoter fails to suppress a *gpa1* mutation to a detectable degree.

Several factors could explain why mammalian $G_\alpha$ subunits differ in their ability to substitute for or interfere with Gpa1 protein, including differences in protein levels, subcellular localization, posttranslational modification, or affinity for yeast $G_{\beta\gamma}$ subunits. Whereas the first two possibilities have not been rigorously ruled out, lack of posttranslational modification is an unlikely explanation because mammalian $G_\alpha$ subunits are known to be substrates for the yeast *N*-myristoyltransferase (LINDER et al. 1991). The relative activities of mammalian $G_\alpha$ subunits in yeast, however, is consistent with known differences in their affinities for mammalian $G_{\beta\gamma}$ subunits ($G_{\alpha s} > G_{\alpha i2} > G_{\alpha o}$; FREISSMUTH et al. 1989; KANG et al. 1990). It remains to be determined whether mammalian $G_\alpha$ subunits actually display a similar order of binding affinity toward the yeast $G_{\beta\gamma}$ subunits.

### 2. Chimeric Yeast/Mammalian $G_\alpha$ Subunits

Because most intact mammalian $G_\alpha$ subunits seem to have rather low affinity for yeast $G_{\beta\gamma}$ subunits, hybrid yeast/mammalian $G_\alpha$ subunits have been expressed. This approach takes advantage of the observation that the N-terminal portions of mammalian $G_\alpha$ subunits are required for binding $G_{\beta\gamma}$ subunits (FREISSMUTH et al. 1989). Thus, if one assumes that the functional domains of yeast and mammalian $G_\alpha$ subunits are organized similarly, chimeric $G_\alpha$ subunits consisting of the N-terminal domain of Gpa1 protein and the C-terminal domains of various mammalian $G_\alpha$ subunits may have high affinity for yeast $G_{\beta\gamma}$ subunits. Moreover, since the C-terminal domains of $G_\alpha$ subunits mediate interaction with specific receptors (FREISSMUTH et al. 1989), such chimeric $G_\alpha$ subunits might also couple to mammalian receptors that are coexpressed in yeast. To increase the likelihood that chimeric proteins retain their native properties, most chimeric yeast/mammalian $G_\alpha$ subunits contain fusions within a highly conserved 32 amino acid sequence that, by analogy with *ras* proteins, forms part of the nucleotide-binding pocket.

Chimeric $G_\alpha$ subunits consisting approximately of the N-terminal two-thirds of Gpa1 protein and the C-terminal one-third of $G_{\alpha s}$, $G_{\alpha i2}$, or $G_{\alpha o}$ are all functional (KANG et al. 1990). Expression of each chimera suppresses the growth defect of *gpa1* mutants. Similarly, chimeras constructed between Gpa1 protein and $G_{\alpha i1}$, $G_{\alpha i3}$, or $G_q$ are also active (LOMASNEY and LEFKOWITZ, personal communication), although the precise structures of these chimeras has not been described.

Chimeric $G_\alpha$ subunits do differ in activity, however. Based on their ability to inhibit pheromone response and mating in wild type cells, chimeras involving $G_{\alpha o}$ and $G_{\alpha i2}$ are more active than one involving $G_{\alpha s}$ (KANG et al. 1990). It is curious that the relative activities of these chimeric proteins and their corresponding native subunits are inverted. In the absence of further data, this observation remains puzzling.

### III. Mammalian $G_\beta$ and $G_\gamma$ Subunits

Reports have yet to appear describing attempts to express mammalian $G_\beta$ and $G_\gamma$ subunits in yeast. This area may warrant study since expression of intact $G_\beta$ or $G_\gamma$ subunits or appropriate chimeric molecules may allow mammalian receptors and native $G_\alpha$ subunits to interact more effectively with one another and with components of the yeast pheromone response pathway.

## D. Signaling Between Mammalian Receptors and G-Proteins

Convincing evidence for signaling between a mammalian receptor and G-protein in yeast has been described (King et al. 1990). In this study, human $\beta_2$-adrenergic receptor was expressed from the inducible *GAL* promoter, rat $G_{\alpha s}$ was expressed from the inducible *CUP1* promoter, and a *gpal* mutant was used as a host. Because $G_{\alpha s}$ subunits interact with $\beta_2$-adrenergic receptors and associate with yeast $G_{\beta\gamma}$ subunits, cells coexpressing these mammalian signaling components should respond to $\beta$-adrenergic agonists by triggering the pheromone response pathway. Indeed, the $\beta$-adrenergic agonist (−)isoproterenol induces the expression of a pheromone-responsive gene (*FUS1-lacZ*) three- to fourfold, and causes morphological changes similar to those elicited in wild-type cells exposed to mating pheromone. Appropriate controls showed that a $\beta$-adrenergic antagonist, alprenolol, completely blocks the effects of (−)isoproterenol, and cells treated with (+)isoproterenol fail to respond.

Productive signaling between mammalian receptors and G-proteins in yeast seems to be influenced by at least two factors (Lomasney and Lefkowitz, personal communication). First, if active receptors constitute less than about 0.5 pmol/mg membrane protein, reproducible induction of a reporter gene is difficult to observe. Second, it is crucial to induce expression of the chimeric $G_\alpha$ subunits to a level that is just sufficient to suppress the growth defect of *gpal* mutant cells. Expressing $G_\alpha$ subunits beyond this level inhibits agonist-induced signaling, presumably because excess $G_\alpha$ subunits sequester yeast $G_{\beta\gamma}$ subunits and prevent them from interacting with their downstream targets.

## E. Perspectives

Functional analysis of mammalian receptors and G-proteins expressed in yeast is in its infancy. Current data suggest that two advances must be made before yeast can be considered a generally useful expression system. First, reliable, high-level expression of various receptors must be achieved.

Second, chimeric G-protein subunits must be designed such that the pheromone response pathway is repressed effectively in the absence of agonist, and strongly induced in its presence. How might these goals be best achieved?

First we need to know where mammalian receptors are localized in yeast cells when they are not efficiently expressed at the cell surface. An elegant means of addressing this issue has been described by Cartwright and Tipper (1991). In this system, if a receptor–$\beta$-lactamase fusion is capable of reaching the Golgi apparatus, it will be cleaved by a resident protease (*KEX2* gene product), resulting in efficient secretion of soluble $\beta$-lactamase activity. This technique, coupled with classical methods of immunofluorescence, cell fractionation, and analysis of protein glycosylation should establish the intracellular fates of mammalian receptors when they are expressed in yeast.

Assuming that receptor biogenesis is controlled principally by the rate of protein import into or folding within the ER, then it might be useful to overexpress yeast gene products that participate in these processes. For example, receptor expression might be improved in cells overexpressing the *KAR2*, *SEC63*, and *PDI1* gene products, which encode homologs of mammalian BiP, *Escherichia coli* DnaJ, and protein disulfide isomerase, respectively (Rose et al. 1989; Sadler et al. 1989; Lamantia et al. 1991). Overexpression of *PDI1* might be particularly useful. Formation of disulfide bonds in the extracellular (lumenal) loops of $\beta_2$-adrenergic receptor is required for normal receptor conformation and ligand binding activity (Dohlman et al. 1990), and most mammalian receptors contain cysteine residues that could form disulfide bonds in extracellular loops.

To address the problems posed by existing chimeric G protein subunits, it will be useful to identify domains or sequences in yeast $G_\beta$ and $G_\gamma$ subunits that are required for interaction with downstream signaling components. If the minimal effector interaction regions of the yeast $\beta$ and $\gamma$ subunits can be defined, perhaps they could be transplanted into mammalian $G_{\beta\gamma}$ subunits. Chimeric $\beta$ and $\gamma$ subunits may associate tightly with native mammalian $G_\alpha$ subunits, thereby reducing basal expression of the pheromone response pathway, yet permitting robust induction by an appropriate agonist. Coupled with efficient expression of receptors, the use of chimeric $G_\beta$ and $G_\gamma$ subunits may allow the potential of a yeast expression system to be realized fully.

## References


Anderson JA, Huprikar SS, Kochian LV, Lucas WJ, Gaber RF (1992) Functional expression of a probable *Arabadopsis thaliana* potassium channel in *Saccharomyces cerevisiae*. Proc Natl Acad Sci USA 89:3736–3740

Andersson SGE, Kurland CG (1990) Codon preferences in free-living microorganisms. Microbiol Rev 54:198–210

Ballester R, Michaeli T, Ferguson K, Xu HP, McCormick F, Wigler M (1989) Genetic analysis of mammalian GAP expressed in yeast. Cell 594:681–686

Blumer KJ, Reneke JE, Thorner J (1988) The *STE2* gene product is the ligand binding component of the $\alpha$-factor receptor of *Saccharomyces cerevisiae*. J Biol Chem 263:10836–10842
Blumer KJ, Thorner J (1991) Receptor-G protein signaling in yeast. Annu Rev Physiol 53:37–57
Buckley NJ, Bonner TI, Buckley CM, Brann MR (1989) Antagonist binding proterties of five cloned muscarinic receptors expressed in CHO-K1 cells. Mol Pharm 35:469–476
Cartwright CP, Tipper DJ (1991) In vivo topological analysis of Ste2, a yeast plasma membrane protein, by using $\beta$-lactamase gene fusions. Mol Cell Biol 11:2620–2628
Cigan AM, Donahue TF (1987) Sequence and structural features associated with translational initiator regions in yeast – a review. Gene 59:1–18
Dietzel C, Kurjan J (1987) The yeast *SCG1* gene: a $G_\alpha$-like protein implicated in the a- and $\alpha$-factor response pathway. Cell 50:1001–1010
Dohlman HG, Caron MG, DeBlasi A, Frielle T, Lefkowitz RJ (1990) Role of extracellular disulfide-bonded cysteines in the ligand binding function of the beta 2-adrenergic receptor. Biochemistry 29:2335–2342
Freissmuth M, Casey PJ, Gilman AG (1989) G proteins control diverse pathways of transmembrane signalling. FASEB J 3:2125–2131
Gill G, Tjian R (1991) A highly conserved domain of TFIID displays species specificity in vivo. Cell: 65:333–340
Huang H-J, Liao C-F, Yang B-C, Kuo T-T (1992) Functional expression of rat M5 muscarinic acetylcholine receptor in yeast. Biochem Biophys Res Comm 182:1180–1186
Kang Y-S, Kane J, Kurjan J, Stadel JM, Tipper DJ (1990) Effects of expression of mammalian $G_\alpha$ and hybrid mammalian-yeast $G_\alpha$ proteins on the yeast pheromone response signal transduction pathway. Mol Cell Biol 10:2582–2590
King K, Dohlman HG, Thorner J, Caron MG, Lefkowitz RJ (1990) Control of yeast mating signal transduction by a mammalian $\beta_2$-adrenergic receptor and $G_s$ $\alpha$ subunit. Science 250:121–123
LaMantia M, Miura T, Tachikawa H, Kaplan HA, Lennarz WJ, Mizunaga T (1991) Glycosylation site binding protein and protein disulfide isomerase are identical and essential for cell viability in yeast. Proc Natl Acad Sci USA 88:4453–4457
Liao C-F, Themmen APN, Joho R, Barberis C, Birnbaumer M, Birnbaumer L (1989) Molecular cloning and expression of a fifth muscarinic acetylcholine receptor. J Biol Chem 264:7328–7337
Linder ME, Pang IH, Duronio RJ, Gordon JI, Sternweiss PC, Gilman AG (1991) Lipid modifications of G protein subunits. myristoylation of $G_o$ alpha increases its affinity for beta gamma. J Biol Chem 266:4654–4659
Marsh L, Neiman AM, Herskowitz I (1991) Signal transduction during pheromone response in yeast. Annu Rev Cell Biol 7:699–728
Nakafuku M, Obara K, Kaibuchi K, Miyajima I, Miyajima A, Itoh H, Nakamura S, Arai K-I, Matsumoto K, Kaziro Y (1988) Isolation of a second yeast *Saccharomyces cerevisiae* gene (*GPA2*) coding for guanine nucleotide-binding regulatory protein: studies on its structure and possible functions. Proc Natl Acad Sci USA 85:1374–1378
Payette P, Gossard F, Whiteway M, Dennis M (1990) Expression and pharmacological characterization of the human M1 muscarinic receptor in *Saccharomyces cerevisiae*. FEBS Lett 266:21–25
Peralta EG, Ashkenazi A, Winslow JW, Smith DH, Ramachandran J, Capon DJ (1987) Distinct primary structures, ligand-binding properties and tissue-specific expression of four human muscarinic acetylcholine receptors. EMBO J 6:3923–3929
Raymond M, Gros P, Whiteway M, Thomas DJ (1992) Functional complementation of yeast *ste6* by a mammalian multidrug resistance *mdr* gene. Science 256:232–234

Rose MD, Misra L, Vogel JP (1989) *KAR2*, a karyogamy gene, is the yeast homolog of the mammalian BiP/GRP78 gene. Cell 48:1211–1221
Sadler I, Chiang A, Kurihara T, Rothblatt J, Way J, Silver P (1989) A yeast gene important for protein assembly into the endoplasmic reticulum and the nucleus has homology to DnaJ, an Escherichia coli heat shock protein. J Cell Biol 109:2665–2675
Schena M, Yamamoto KR (1989) Mammalian glucocortocoid receptor derivatives enhance transcription in yeast. Science 241:965–967
Sentenac H, Bonneaud N, Minet M, Lacroute F, Salmon J-M, Gaymard F, Grignon C (1992) Cloning and expression in yeast of a plant potassium ion transport system. Science 256:663–665
Singer SJ (1990) The structure and insertion of integral proteins in membranes. Annu Rev Cell Biol 6:247–296
Stammes MA, Shieh BH, Chuman L, Harris GL, Zucker CS (1991) The cyclophilin homolog ninaA is a tissue-specific integral membrane protein required for the proper synthesis of a subset of drosophila rhodopsins. Cell 65:219–227
Waldo GL, Northup JK, Perkins JP, Harden TK (1983) Characterization of an altered membrane form of the β-adrenergic receptor produced during agonist-induced desensitization. J Biol Chem 258:13900–13908
Xu GF, Lin B, Tanaka K, Dunn D, Wood D, Gesteland R, White R, Weiss R, Tamanoi F (1990) The catalytic domain of the neurofibromatosis type 1 gene product stimulates ras GTPase and complements ira mutants of S. cerevisiae. Cell 63:835–841
Yellen G, Migeon JC (1990) Expression of Torpedo nicotinic acetylcholine receptor subunits in yeast is enhanced by use of yeast signal sequences. Gene 86:145–152

CHAPTER 66

# G-Proteins in the Signal Transduction of the Neutrophil

K.R. McLeish and K.H. Jakobs

## A. Introduction

Heterotrimeric G protein-mediated transmembrane signaling has been extensively studied in polymorphonuclear leukocytes (PMNs), due in part to the ability to obtain large numbers of these cells from peripheral blood and their importance to health. Studies using PMNs and the human cell line HL-60 have produced significant advances in understanding the molecular mechanisms of PMN responses to infection and inflammation. Additionally, these cells provide a useful model for studying the coupling of multiple receptors to multiple effectors by G-proteins.

## B. Receptor-Mediated PMN Functions

The migration of PMNs from the circulation to a site of infection or inflammation and subsequent phagocytosis and killing of bacteria require orchestration of a complex series of events. This orchestration is accomplished through functional regulation by multiple receptors.

### I. Adherence

PMNs exist in the circulation in two states, either rapidly moving in the center of the blood stream or rolling along vascular endothelial cells. Rolling requires PMN expression of a lectin receptor, leukocyte cell adhesion molecule-1, which is rapidly shed upon chemoattractant stimulation (Smith et al. 1991). At sites of inflammation PMNs cease rolling by adhering tightly to venular endothelial cells via adhesion receptors termed integrins, including Mac-1 (CR3, CD11b/CD18, $\alpha_M\beta_2$) and leukocyte function antigen-1 (LFA-1, CD11a/CD18, $\alpha_L\beta_2$; Hynes 1992). The endothelial cell ligands for these adhesion receptors include intracellular adhesion molecule-1 and endothelial-leukocyte adhesion molecule-1. The expression of both PMN adhesion receptors and endothelial adhesion molecules is regulated by endotoxins, chemoattractants, and proinflammatory cytokines, including tumor necrosis factor-$\alpha$ (TNF) and interleukin-1. Subsequent migration of adherent PMNs across the vascular wall depends on weakening of the PMN-endothelial cell attachment by endothelial cell release of the chemoattractant cytokine interleukin-8 (IL-8; Huber et al. 1991).

### II. Chemotaxis

Following adherence to vascular endothelium, PMNs migrate toward a chemoattractant gradient using agonist-specific plasma membrane receptors. In addition to IL-8, other chemoattractants include formylated peptides, C5a, leukotriene $B_4$ ($LTB_4$), and platelet activating factor (PAF). $LTB_4$, IL-8, and PAF are produced by a number of different cells, including PMNs. Formyl peptides are components of bacterial cell walls and mitochondrial membranes. C5a is produced from the complement cascade, usually in circulating blood. Chemotaxis is dependent on recognition of differences in chemoattractant concentration as small as 0.1% along the length of cells. At high chemoattractant conentrations, receptors no longer detect a gradient and migration ceases.

### III. Phagocytosis and Bactericidal Activity

Immunoglobulin receptors (Fc$\gamma$R) and receptors for the complement cleavage product C3bi (CR3 receptors) mediate phagocytosis of opsonized bacteria. The major mechanisms by which PMNs kill bacteria are through generation of toxic oxygen radicals by NADPH oxidase, a process called respiratory burst activity, or through formation of lysosomes containing proteolytic enzymes. Respiratory burst activity and lysosomal enzyme release are stimulated by phagocytosis and by chemoattractant concentrations that exceed those which stimulate chemotaxis.

### IV. Regulatory Receptors

PMN function is not static, but can be enhanced or impaired by other extracellular messengers. The cytokines TNF, interferon-$\gamma$ (IFN), and granulocyte-macrophage colony-stimulating factor (GM-CSF) enhance PMN responses by receptor-mediated signals. Activation of purinergic receptors modulates several PMN functions, including respiratory burst activity, adherence, chemotaxis, and granular secretion.

## C. G-Protein-Coupled Receptors

Many of the receptors that coordinate PMN activity couple to G proteins. PMNs and HL-60 granulocytes express a number of signal transducing G proteins, including $G_s$, $G_i2$, and $G_i3$ (Murphy et al. 1987). About 60% of $G_i$ proteins are localized to PMN plasma membranes, while another 35% exists as a translocatable pool in specific granules (Rotrosen et al. 1988).

### I. Chemoattractant Receptors

Chemoattractant receptors remain the most extensively studied G-protein-coupled receptors in PMNs. PMNs express specific receptors for formyl peptides, C5a, IL-8, $LTB_4$, and PAF; while differentiated HL-60 granulocytes express receptors for all chemoattractants except IL-8. Receptors for formyl peptides, C5a, PAF, and IL-8 have been cloned and sequenced. Although sequence homology among these receptors is relatively low, less than 30%, all sequences predict seven transmembrane spanning regions (BOULAY et al. 1990, 1991; HOLMES et al. 1991; NAKAMURA et al. 1991). Functional activity varies among individual chemoattractants. For example, $LTB_4$ and PAF are less effective than C5a or formyl peptides in stimulating chemotaxis, respiratory burst activity, and lysosomal enzyme release (PALMBLAD et al. 1988; MCLEISH et al. 1989a).

The ability of guanine nucleotides to reduce receptor affinity for agonists, considered characteristic of G-protein-coupled receptors, is readily observed with several chemoattractant receptors. For example, expression of high-affinity formyl peptide or C5a receptors in HL-60 membranes is efficiently inhibited by guanine nucleoside tri- and diphosphates, with GTP$\gamma$S being most potent and GDP the least potent nucleotide (GIERSCHIK et al. 1989b,c; WIELAND and JAKOBS, unpublished observations). In isolated membranes, a small percentage (5%–15%) of formyl peptide receptors retain a high agonist affinity in the presence of guanine nucleotides or following uncoupling of receptors from G-proteins by pertussis toxin treatment (GIERSCHIK et al. 1989c). In permeabilized PMNs formyl peptide receptors are completely interconvertible between high- and low-affinity states by addition of guanine nucleotides (SKLAR et al. 1987), suggesting that all formyl peptide receptors expressed on PMNs are coupled to G-proteins.

Pertussis toxin catalyzes ADP-ribosylation of a carboxy terminal cysteine of $G\alpha_i2$ ($M_r$ 40 kDa) and $G\alpha_i3$ ($M_r$ 41 kDa) in PMN and HL-60 membranes (GIERSCHIK et al. 1989a). Functional responses to chemoattractants are inhibited by pretreatment with pertussis toxin. However, PMN responses to fMet-Leu-Phe are not equally sensitive to inhibition by pertussis toxin (OMANN et al. 1991). Incomplete pertussis toxin treatment resulting in a 90% reduction in respiratory burst activity produces only a 50% reduction in actin polymerization. The requirement for exhaustive treatment to completely inhibit actin polymerization indicates that this function may be the most sensitive indicator of complete pertussis toxin-catalyzed ADP-ribosylation in PMNs.

In the absence of added guanine nucleotides, cholera toxin ADP-ribosylates $G\alpha_i2$ and $G\alpha_i3$, as well as $G\alpha_s$ in PMN and HL-60 membranes (GIERSCHIK and JAKOBS 1987; GIERSCHIK et al. 1989a; MCLEISH et al. 1989a). GDP and GppNHp inhibit cholera toxin labeling of $G\alpha_i$, but ADP-ribosylation of both $G\alpha_i2$ and $G\alpha_i3$ is reestablished by addition of fMet-Leu-Phe. This receptor-specific labeling indicates that formyl peptide receptors are coupled to both $G_i$ proteins present in these membranes (GIERSCHIK et al. 1989a). The coupling of chemoattractant receptors to $G_i$ is also supported by the ability of purified formyl peptide receptors and $G_i$ proteins to reconstitute agonist binding in liposomes and by the coisolation of C5a and formyl peptide receptors with $G_i$ (Williamson et al. 1988; Polakis et al. 1989; Rollings et al. 1991). Receptor-specific cholera toxin labeling, however, is not uniformly stimulated by chemoattractants. In PMN and HL-60 membranes, fMet-Leu-Phe and C5a induce labeling of $G\alpha_i$, while $LTB_4$ and PAF do not (McLeish et al. 1989a; Scherpers and McLeish, unpublished observations).

Functional activation of G-proteins by chemoattractant receptors has been demonstrated in plasma membranes isolated from PMNs and HL-60 granulocytes by measuring GTP$\gamma$S binding and GTP hydrolysis. Although GTP hydrolysis is the molecular off switch for G-proteins, it can be used to measure receptor-induced GTP binding to G-proteins because the rate of hydrolysis exceeds the rate of guanine nucleotide exchange. A two- to sixfold increase in GTP hydrolysis is stimulated by fMet-Leu-Phe, C5a, IL-8, and $LTB_4$ (McLeish et al. 1989a; Kupper et al. 1992; Schepers et al. 1992). The rate of GTP hydrolysis stimulated by chemoattractants is linear in the presence of adequate substrate, although the rate differs for different chemoattractants (Schepers et al. 1992).

Maximal chemoattractant-stimulated GTP hydrolysis and GTP$\gamma$S binding by G-proteins requires the presence of sodium ions, which apparently interfer with receptor–G-protein coupling via a binding site on receptors (Gierschik et al. 1989b). The ability to measure chemoattractant-stimulated GTP$\gamma$S binding to $G_i$-proteins is also dependent on the presence of GDP (Gierschik et al. 1991), as GDP appears to be only loosely bound to $G_i$-proteins. However, at low temperatures, for example 0°C, GDP is not spontaneously released from HL-60 membranes, and addition of GDP is thus not required to measure agonist-stimulated GTP$\gamma$S binding (Wieland et al. 1992). Finally, $Mg^{2+}$ is absolutely and specifically required to detect chemoattractant-stimulated G-protein activation. Two distinct binding sites for $Mg^{2+}$ have been described, a divalent cation binding site on receptors and a $Mg^{2+}$-specific binding site on G-proteins (Gierschik et al. 1989c). Whether or not $Na^+$, GDP, and $Mg^{2+}$ regulate receptor–G-protein coupling in intact phagocytes remains to be determined.

Significant differences between basal and agonist-stimulated GTP$\gamma$S binding are seen within a few seconds, and binding equilibrium is reached by about 60 min at 30°C (Gierschik et al. 1991; Schepers et al. 1992). Using this assay, we have found that all chemoattractants stimulate GTP$\gamma$S binding to PMN membranes, while only fMet-Leu-Phe, $LTB_4$, and C5a stimulate binding to HL-60 membranes. fMet-Leu-Phe and C5a are more efficient agonists than $LTB_4$, and the initial rate of GTP$\gamma$S binding stimulated by fMet-Leu-Phe is twice that stimulated by $LTB_4$ (Schepers et al. 1992). At GTP$\gamma$S binding equilibrium, the number of G-proteins coupled to chemoattractant receptors and their affinity for GTP$\gamma$S can be estimated (Gierschik et al. 1991). Using this assay, formyl peptide and $LTB_4$ receptors were shown to share a common pool of G-proteins representing 30%–40%

of total membrane G-proteins (SCHEPERS et al. 1992). The apparent affinity for GTPγS was significantly lower following $LTB_4$ activation compared to that for fMet-Leu-Phe, possibly due to different capacities of the two agonists to release GDP from G proteins and thus to allow GTPγS binding. The ability of formyl peptide receptor-specific cholera toxin-catalyzed ADP-ribosylation to inhibit both fMet-Leu-Phe- and $LTB_4$-stimulated GTP hydrolysis also supports a common pool of G-proteins for these receptors (MCLEISH et al. 1989a). Recent studies showing that fMet-Leu-Phe and C5a inhibit receptor binding of the other ligand in the presence of poorly hydrolyzable GTP analogs suggest that formyl peptide and C5a receptors also share a common pool of G-proteins (WIELAND and JAKOBS, unpublished observations). The size of the receptor-coupled G-protein pool indicates that chemoattractant receptors catalytically activate G-proteins.

ADP-ribosylation by pertussis toxin completely inhibits fMet-Leu-Phe and $LTB_4$-induced G-protein activation (GIERSCHIK et al. 1989b, 1991; MCLEISH et al. 1989a), suggesting that these receptors do not couple to G-proteins other than $G_i2$ and $G_i3$. Although cholera toxin acts at a site distinct from that for pertussis toxin, pretreatment of HL-60 membranes with cholera toxin under conditions causing ADP-ribosylation of $G\alpha_i$ inhibits chemoattractant receptor-stimulated GTP hydrolysis (MCLEISH et al. 1989a). Furthermore, agonist-stimulated GTPγS binding, measured at low nucleotide concentration, is also inhibited. The molecular basis for this cholera toxin action is presently unknown. In addition to preventing agonist activation of G-proteins, pertussis toxin treatment reduces basal GTPγS binding and GTP hydrolysis. This finding, as well as the inhibition of basal activity by $Na^+$, suggests that nonliganded chemoattractant receptors induce low-level stimulation of G-proteins. No data are available to indicate whether agonist-free receptors are active in intact cells.

Chemoattractant receptors have been shown to activate several effector enzymes in PMNs and HL-60 granulocytes, including phospholipases C, D, and $A_2$, a tyrosine kinase, and a phosphatidylinositol kinase (PAI et al. 1988; COCKCROFT and STRUTCHFIELD 1989a; GOMEZ-CAMBRONERO et al. 1989; PIKE et al. 1990; KANAHO et al. 1991). Chemoattractant stimulation of all of these effectors is prevented by pertussis toxin, suggesting a $G_i$-protein(s) intermediary in the signaling pathway. Demonstration of direct activation of effectors by chemoattractant receptor-coupled G-proteins in membrane preparations has been shown only for the phospholipases. As with G-protein activation, there are different patterns of effector enzyme activation for the different chemoattractants. While fMet-Leu-Phe and C5a induce rapid and sustained activation of phospholipases, generation of second messengers following $LTB_4$ and PAF stimulation is more transient and quantitatively reduced (REIBMAN et al. 1988; TRAYNOR-KAPLAN et al. 1989; KANAHO et al. 1991). The question of which G-protein subunit interacts with these phospholipases remains unresolved. $G\alpha_q$ has been shown to activate phospholipase C-$\beta_1$ (Wu et al. 1992), but activation of phospholipase C by $G\alpha_i$ has not been reported. Data supporting a role for G protein $\beta\gamma$ subunit activation of a phospholipase C in HL-60 granulocytes, most likely of the $\beta_2$ subtype, has recently been reported (Camps et al. 1992).

Both low molecular weight GTP binding proteins and $\gamma$ subunits of heterotrimeric G-proteins are isoprenylated and carboxylmethylated (Spiegel et al. 1991). The functional role of these posttranslational modifications has recently been examined in HL-60 granulocytes. Inhibition of isoprenylation by lovastatin inhibited respiratory burst activity stimulated by fMet-Leu-Phe or phorbol diesters (Bokoch and Prossnitz 1992). The authors interpreted these findings to indicate that an isoprenylated low molecular weight G-protein participates in NADPH oxidase activation. The specific role of carboxylmethylation has been examined by cultivation of HL-60 cells with the AdoMet-dependent methylation inhibitor, periodate oxidized adenosine (Adox; Lederer and McLeish, unpublished observations). Adox pretreatment for 24 h resulted in a 30%–40% reduction in superoxide release stimulated by fMet-Leu-Phe or NaF, but no inhibition of phorbol diester stimulation. Membranes prepared from Adox-treated cells showed a significant reduction in fMet-Leu-Phe and $LTB_4$ stimulation of GTP$\gamma$S binding and GTP hydrolysis, however, GTP$\gamma$S was able to inhibit fMet-Leu-Phe binding with equal potency to control membranes. Immunoblotting showed expression of $\alpha_i$ and $\beta$ subunits to be unchanged by Adox treatment, while pertussis toxin labeling was reduced in membranes from Adox-treated cells. These preliminary results suggest that carboxylmethylation participates in $\alpha$-$\beta\gamma$ subunit association, and consequently in receptor–G-protein coupling.

## II. Purinergic Receptors

PMNs and HL-60 granulocytes express several purinergic receptor subtypes, including $A_1$ and $A_2$ adenosine receptors and $P_2$ purinergic receptors (Cowen et al. 1990; Cronstein et al. 1990, 1992; Xie et al. 1991). Acting through $A_1$ receptors, adenosine promotes PMN adherence to endothelial cells and chemotaxis by a pertussis toxin-inhibitible mechanism (Cronstein et al. 1990, 1992). Adenosine inhibits chemoattractant-stimulated respiratory burst activity through occupancy of $A_2$ receptors (Cronstein et al. 1990, 1992). This activity probably results from $G_s$-mediated $A_2$ receptor stimulation of adenylyl cyclase, as adenosine stimulates cAMP formation in PMNs (Iannone et al. 1989) and cAMP inhibits PMN responses to chemoattractants. $P_2$ purinergic agonists (ATP, UTP) activate phospholipases C, D, and $A_2$ by a mechanism at least partially inhibited by pertussis toxin (Cockroft and Stutchfield 1989b; Cowen et al. 1990; Xie et al. 1991). However, $P_2$ agonists stimulate PMN and HL-60 granulocyte respiratory burst activity and granule secretion poorly compared to fMet-Leu-Phe (Cockcroft and Stutchfield 1989b).

### III. Other PMN Receptors

The three classes of Fc$\gamma$R belong to the immunoglobulin supergene family. Little is known about the early signaling events intitiated by Fc$\gamma$R, but they appear to be distinct from those used by chemoattractant receptors. Pertussis toxin fails to inhibit the calcium transient induced by aggregated IgG (WALKER et al. 1991). Variable effects of pertussis toxin on the respiratory burst stimulated by Fc$\gamma$R have been reported (BLACKBURN and HECK 1989; WALKER et al. 1991). Although IgG has been reported to stimulate GTP hydrolysis by and GTP$\gamma$S binding to PMN plasma membranes, only a 20%–40% increase in these activities was seen (BLACKBURN and HECK 1989). Thus a role for G proteins in Fc$\gamma$R activity remains speculative.

## D. Regulation of Neutrophil Responses

PMN responses to chemoattractants are not fixed but can be both up- and down-regulated by processes that alter G-protein-mediated signaling.

### I. Priming

TNF, GM-CSF, and IFN are released during host infection or inflammation and enhance PMN responses to chemoattractants, a process termed priming. Cytokines may induce priming by multiple mechanisms, but only the regulation of chemoattractant–G-protein signaling is discussed here.

TNF priming of PMNs can be detected within 10 min and is maximal by 15–20 min (KLEIN et al. 1992a). This priming is associated with a 20%–25% increase in formyl peptide receptors, and a marked increase in plasma membrane expression of $G\alpha_s$ and $G\alpha_i$. On the other hand, onset of TNF priming of HL-60 granulocytes requires 6 h and is maximal at 24 h (KLEIN et al. 1991a; MCLEISH et al. 1991). TNF priming of HL-60 cells results in a 50% increase in membrane expression of formyl peptide receptors but no difference in $G_i$ expression. Basal and fMet-Leu-Phe-stimulated GTP hydrolysis and GTP$\gamma$S binding are enhanced in membranes from both TNF-primed PMNs and HL-60 granulocytes. The signal transduction pathway used by TNF receptors to initiate priming has been suggested to include G-proteins. IMAMURA and colleagues (1988) reported that TNF stimulated GTP$\gamma$S binding to HL-60 membranes; however, we have been unable to duplicate these results. More recently, TNF receptors on an osteoblast cell line have been shown to couple to a pertussis toxin-sensitive G protein (YANAGA et al. 1992).

IFN enhances HL-60 granulocyte respiratory burst to fMet-Leu-Phe and NaF, enhances expression of formyl peptide receptors and $G_i$-proteins, and results in increased basal and fMet-Leu-Phe-stimulated GTP hydrolysis and GTP$\gamma$S binding (KLEIN et al. 1991a, 1992b). GM-CSF increases formyl peptide receptor expression and enhances respiratory burst activity stimulated by NaF and GTPγS in PMNs (McColl et al. 1990). Membranes from GM-CSF treated PMNs demonstrate an increased basal and fMet-Leu-Phe-stimulated GTP hydrolysis (Gomez-Cambronera et al. 1989b).

## II. Desensitization

PMN activation by chemoattractants is quite transient, as the generation of second messengers and the production of oxygen radicals ceases within seconds to minutes despite continued exposure to agonists. These transient responses are due to the loss of responsiveness, or desensitization, of receptors. Both homologous and receptor class-specific desensitization of chemoattractant receptors on PMNs has been reported (Wilde et al. 1989; Didsbury et al. 1991), while in HL-60 granulocytes only homologous desensitization has been found (McLeish et al. 1989b; Lederer and McLeish, unpublished observations). Formyl peptide receptor desensitization is associated with reduced membrane receptor expression and uncoupling of the remaining receptors from their G-proteins (McLeish et al. 1989b; Wilde et al. 1989). The mechanism by which formyl peptide receptors are uncoupled from G-proteins has not been determined. Activation of protein kinase C does not reproduce either reduced receptor expression or receptor–G-protein uncoupling (Lederer and McLeish, unpublished observations). Based on the similarities in receptor loss and uncoupling following desensitization of formyl peptide receptors and other G-protein-coupled receptors, for example, rhodopsin and $\beta_2$-adrenoceptors, one can speculate that receptor phosphorylation by a receptor-specific kinase and subsequent binding of an arrestinlike protein is involved in the rapid desensitization of chemoattractant receptors.

*Acknowledgements.* The authors' studies reported herein were supported by the Deutsche Forschungsgemeinschaft, the Fonds der Chemischen Industrie, the Alexander von Humboldt Foundation, and the Department of Veterans Affairs.

## References

Blackburn WD Jr, Heck LW (1989) Neutrophil activation by surface bound IgG is via a pertussis toxin insensitive G protein. Biochem Biophys Res Commun 164:983–989

Bokoch GM, Prossnitz V (1992) Isoprenoid metabolism is required for stimulation of the respiratory burst oxidase of HL-60 cells. J Clin Invest 89:402–408

Boulay F, Tardif M, Brouchon L, Vignais P (1990) The human N-formylpeptide receptor. Characterization of two cDNA isolates and evidence for a new subfamily of G-protein-coupled receptors. Biochemistry 29:11124–11133

Boulay F, Mery L, Tardif M, Brouchon L, Vignais P (1991) Expression cloning of a receptor for C5a anaphylatoxin on differentiated HL-60 cells. Biochemistry 30:2993–2999

Camps M, Hou C, Sidiropoulos D, Stock JB, Jakobs KH, Gierschik P (1992) Stimulation of phospholipase C by guanine-nucleotide-binding protein βγ subunits. Eur J Biochem 68:1–11

Cockcroft S, Stutchfield J (1989a) The receptors for ATP and fMetLeuPhe are independently coupled to phospholipases C and $A_2$ via G-proteins(s). Biochem J 263:715–723
Cockcroft S, Stutchfield J (1989b) ATP stimulates secretion in human neutrophils and HL60 cells via a pertussis toxin-sensitive guanine nucleotide-binding protein coupled to phospholipase C. FEBS Lett 245:25–29
Cowen DS, Baker B, Dubyak GR (1990) Pertussis toxin produces differential inhibitory effects on basal, $P_2$-purinergic, and chemotactic peptide-stimulated inositol phospholipid breakdown in HL-60 cells and HL-60 cell membranes. J Biol Chem 265:16181–16189
Cronstein BN, Daguma L, Nichols D, Hutchison AJ, Williams M (1990) The adenosine/neutrophil paradox resolved: human neutrophils possess both $A_1$ and $A_2$ receptors that promote chemotaxis and inhibit $O_2$ generation, respectively. J Clin Invest 85:1150–1157
Cronstein BN, Levin RI, Philips M, Hirschhorn R, Abramson SB, Weissmann G (1992) Neutrophil adherence to endothelium is enhanced via adenosine $A_1$ receptors and inhibited via adenosine $A_2$ receptors. J Immunol 148:2201–2206
Didsbury JR, Uhing RJ, Tomhave E, Gerard C, Gerard N, Snyderman R (1991) Receptor class desensitization of leukocyte chemoattractant receptors. Proc Natl Acad Sci USA 88:11564–11568
Gierschik P, Jakobs KH (1987) Receptor-mediated ADP-ribosylation of a phospholipase C-stimulating G protein. FEBS Lett 224:219–223
Gierschik P, Sidiropoulos D, Jakobs KH (1989a) Two distinct Gi-proteins mediate formyl peptide receptor signal transduction in human leukemia (HL-60) cells. J Biol Chem 264:21470–21473
Gierschik P, Sidiropoulos D, Steisslinger M, Jakobs KH (1989b) $Na^+$ regulation of formyl peptide receptor-mediated signal transduction in HL 60 cells. Evidence that the cation prevents activation of the G-protein by unoccupied receptors. Eur J Pharmacol 172:481–492
Gierschik P, Steisslinger M, Sidiropoulos D, Herrmann E, Jakobs KH (1989c) Dual $Mg^{2+}$ control of formyl-peptide-receptor-G-protein interaction in HL 60 cells. Eur J Biochem 183:97–105
Gierschik P, Moghtader R, Straub C, Dieterich K, Jakobs KH (1991) Signal amplification in HL-60 granulocytes. Evidence that the chemotactic peptide receptor catalytically activates guanine-nucleotide-binding regulatory proteins in native plasma membranes. Eur J Biochem 197:725–732
Gomez-Cambronero J, Huang CK, Bonak VA, Wang E, Casnellie JE, Shiraishi T, Sha'afi RI (1989a) Tyrosine phosphorylation in human neutrophil. Biochem Biophys Res Commun 162:1478–1485
Gomez-Cambronero J, Yamazaki M, Metwally F, Molski TFP, Bonak VA, Huang C-K, Becker EL, Sha'afi RI (1989b) Granulocyte-macrophage colony-stimulating factor and human neutrophils: role of guanine nucleotide regulatory proteins. Proc Natl Acad Sci USA 86:3569–3573
Holmes WE, Lee J, Kuang WJ, Rice GC, Wood WI (1991) Structure and functional expression of a human interleukin-8 receptor. Science 253:1278–1280
Huber AR, Kunkel SL, Todd FR III, Weiss SJ (1991) Regulation of transendothelial neutrophil migration by endogenous interleukin-8. Science 25:99–102
Hynes RO (1992) Integrins: versatility, modulation, and signaling in cell adhesion. Cell 69:11–25
Iannone MA, Wolberg G, Zimmerman TP (1989) Chemotactic peptide induces cAMP elevation in human neutrophils by amplification of the adenylate cyclase response to endogenously produced adenosine. J Biol Chem 264:20177–20180
Imamura K, Sherman ML, Spriggs D, Kufe D (1988) Effect of tumor necrosis factor on GTP binding and GTPase activity in HL-60 and L929 cells. J Biol Chem 263:10247–10253

Kanaho Y, Kanoh H, Saitoh K, Nozawa Y (1991) Phospholipase D activation by platelet-activating factor, leukotriene B4, and formyl-methionyl-leucyl-phenylalanine in rabbit neutrophils. J Immunol 146:3536–3541

Klein JB, Scherzer JA, McLeish KR (1991a) Interferon-$\gamma$ enhances superoxide production by HL-60 cells stimulated with multiple agonists. J Interferon Res 11:69–74

Klein JB, Sonnenfeld G, McLeish KR (1991b) Priming of the HL-60 cell respiratory burst response by tumor necrosis factor-$\alpha$. Lymphokine Res 10:173–176

Klein JB, Scherzer JA, McLeish KR (1992a) Rapid modulation of neutrophil transmembrane signaling by tumor necrosis factor-$\alpha$. Clin Res 40:241A

Klein JB, Scherzer JA, McLeish KR (1992b) IFN-$\gamma$ enhances expression of formyl peptide receptors and guanine nucleotide-binding proteins by HL-60 granulocytes. J Immunol 148:2483–2488

McColl SR, Beauseigle D, Gilbert C, Naccache PH (1990) Priming of the human neutrophil respiratory burst by granulocyte-macrophage colony-stimulating factor and tumor necrosis factor-$\alpha$ involves regulation at a post-cell surface receptor level. J Immunol 145:3047–3053

McLeish KR, Gierschik P, Schepers T, Sidiropoulos D, Jakobs KH (1989a) Evidence that activation of a common G-protein by receptors for leukotriene $B_4$ and N-formylmethionyl-leucyl-phenylalanine in HL-60 cells occurs by different mechanisms. Biochem J 260:427–434

McLeish KR, Gierschik P, Jakobs KH (1989b) Desensitization uncouples the formyl peptide receptor-guanine nucleotide-binding protein interaction in HL60 cells. Mol Pharmacol 36:384–390

McLeish KR, Klein JB, Schepers T, Sonnenfeld G (1991) Modulation of transmembrane signalling in HL-60 granulocytes by tumour necrosis factor-$\alpha$. Biochem J 279:455–460

Murphy PM, Eide B, Goldsmith P, Brann M, Gierschik P, Spiegel A, Malech HL (1987) Detection of multiple forms of $G_{i\alpha}$ in HL60 cells. FEBS Lett 221:81–86

Nakamura M, Honda Z, Izumi T, Sakanaka C, Mutoh H, Minami M, Bito H, Seyama Y, Matsumoto T, Noma M, Shimizu T (1991) Molecular cloning and expression of platelet-activating factor receptor from human leukocytes. J Biol Chem 266:20400–20405

Omann GM, Porasik-Lowes MM (1991) Graded G-protein uncoupling by pertussis toxin treatment of human polymorphonuclear leukocytes. J Immunol 146:1303–1308

Palmblad J, Gyllenhammar H, Ringertz B, Nilsson E, Cottell B (1988) Leukotriene B4 triggers highly characteristic and specific functional responses in neutrophils: studies of stimulus specific mechanisms. Biochim Biophys Acta 970:92–102

Pike MC, Bruck ME, Arndt C, Lee CS (1990) Chemoattractants stimulate phosphatidylinositol-4-phosphate kinase in human polymorphonuclear leukocytes. J Biol Chem 265:1866–1873

Polakis PG, Evans T, Snyderman R (1989) Multiple chromatographic forms of the formylpeptide chemoattractant receptor and their relationship to GTP-binding proteins. Biochem Biophys Res Commun 161:276–283

Rollins TE, Siciliano S, Kobayashi S, Cianciarulo DN, Bonilla-Argudo V, Collier K, Springer MS (1991) Purification of the active C5a receptor from human polymorphonuclear leukocytes as a receptor-$G_i$ complex. Proc Natl Acad Sci USA 88:971–975

Rotrosen D, Gallin JI, Spiegel AM, Malech HL (1988) Subcellular localization of $G_{i\alpha}$ in human neutrophils. J Biol Chem 263:10958–10964

Schepers TM, Brier ME, McLeish KR (1992) Quantitative and qualitative differences in guanine nucleotide binding protein activation by formyl peptide and leukotriene $B_4$ receptors. J Biol Chem 267:159–165

Sklar LA, Bokoch GM, Button D, Smolen JE (1987) Regulation of ligand-receptor dynamics by guanine nucleotides. J Biol Chem 262:135–139

Smith CW, Kishimoto TK, Abbass O, Hughes B, Rothlein R, McIntire LV, Butcher E, Anderson DC (1991) Chemotactic factors regulate lectin adhesion molecule 1 (LECAM-1)-dependent neutrophil adhesion to cytokine-stimulated endothelial cells in vitro. J Clin Invest 87:609–618

Spiegel AM, Backlund PS Jr, Butrynski JE, Jones TLZ, Simonds WF (1991) The G protein connection: molecular basis of membrane association. TIBS 16:338–341

Traynor-Kaplan AE, Thompson BL, Harris AL, Taylor P, Omann GM, Sklar LA (1989) Transient increase in phosphatidylinositol 3,4-bisphosphate and phosphatidylinositol trisphosphate during activation of human neutrophils. J Biol Chem 264:15668–15673

Walker BAM, Hagenlocker BE, Stubbs EB Jr, Sandborg RR, Agranoff BW, Ward PA (1991) Signal transduction events and Fc$\gamma$R engagement in human neutrophils stimulated with immune complexes. J Immunol 146:735–741

Wieland T, Kreiss J, Gierschik P, Jakobs KH (1992) Role of GDP in formyl-peptide-receptor-induced activation of guanine-nucleotide-binding proteins in membranes of HL 60 cells. Eur J Biochem 205:1202–1206

Wilde MW, Carlson KE, Manning DR, Zigmond SH (1989) Chemoattractant-stimulated GTPase activity is decreased on membranes from polymorphonuclear leukocytes incubated in chemoattractant. J Biol Chem 264:190–196

Williamson K, Dickey BF, Pyun HY, Navarro J (1988) Solubilization and reconstitution of the formylmethionylleucylphenylalanine receptor coupled to guanine nucleotide regulatory protein. Biochemistry 27:5371–5377

Wu D, Lee CH, Rhee SG, Simon MI (1992) Activation of phospholipase C by the $\alpha$-subunits of the $G_q$ and $G_{11}$ proteins in transfected Cos-7 cells. J Biol Chem 267:1811–1817

Xie MS, Jacobs LS, Dubyak GR (1991) Regulation of phospholipase D and primary granule secretion by P2-purinergic- and chemotactic peptide-receptor agonists is induced during granulocytic differentiation of HL-60 cells. J Clin Invest 88:45–54

Yanaga F, Abe M, Koga T, Hirata M (1992) Signal transduction by tumor necrosis factor-$\alpha$ is mediated through a guanine nucleotide-binding protein in osteoblast-like cell line, MC3T3-E1. J Biol Chem 267:5114–5121

CHAPTER 67

# Hormonal Regulation of Phospholipid Metabolism via G-Proteins: Phosphoinositide Phospholipase C and Phosphatidylcholine Phospholipase D

J.H. EXTON

## A. Introduction

Many hormones, neurotransmitters, and growth factors produce their effects by inducing the hydrolysis of phosphatidylinositol 4,5-bisphosphate ($PtdInsP_2$) in the plasma membranes of their target cells by activating a phospholipase C. This produces *myo*inositol 1,4,5-trisphosphate ($InsP_3$), which releases $Ca^{2+}$ from intracellular stores, and 1,2-diacylglycerol (DAG), which activates protein kinase C (BERRIDGE 1987). There are two basic mechanisms by which agonists activate $PtdInsP_2$ hydrolysis. In the case of growth factors, the tyrosine kinase activity of their receptors phosphorylates and activates the $\gamma$ isozymes of phosphoinositide phospholipase C (MELDRUM et al. 1991; RHEE and CHOI 1992). In the case of hormones, neurotransmitters, and certain other agonists, G-proteins transduce the signal from the receptor to phospholipase C $\beta$ isozymes.

$PtdInsP_2$ is not the only phospholipid hydrolyzed in response to agonists. Studies of the chemical composition and molecular species of DAG produced in response to a variety of agonists, and labeling studies using [$^3$H]choline, have indicated that phosphatidylcholine (PtdCho) is also hydrolyzed by phospholipases of the C and D types (EXTON 1990; BILLAH and ANTHES 1990). In some cell types, direct control of PtdCho phospholipases by G-proteins appears to occur, but in most cells the major mechanism involves protein kinase C (EXTON 1990; BILLAH and ANTHES 1990). In contrast to $PtdInsP_2$ hydrolysis, which occurs within a few seconds but is transient, PtdCho breakdown is slower and more sustained and may be secondary to $PtdInsP_2$ breakdown in some cells (EXTON 1990).

## B. Identification of the G-Proteins Regulating $PtdInsP_2$ Phospholipase C

Beginning in 1983 G-proteins were implicated in the regulation of $PtdInsP_2$ hydrolysis by several lines of evidence (EXTON 1988). However, identification of the specific G-proteins involved remained elusive until TAYLOR et al. (1990) demonstrated that a 42-kDa G-protein $\alpha$ subunit purified from GTP$\gamma$S-treated liver plasma membranes activated PtdInsP$_2$ phospholipase C (PI-PLC). Western blotting showed that this protein was different from any $\alpha$ subunit known at that time. Later, SRMCKA et al. (1991) showed that a 42-kDa $\alpha$ subunit prepared from rat brain by affinity chromatography also activated PI-PLC in the presence of $ALF_4^-$. It had been shown earlier (PANG and STERNWEIS 1990) that tryptic peptides from the 42-kDa $\alpha$ subunit had sequences identical to those deduced from two novel $\alpha$ subunit cDNAs cloned from a mouse brain cDNA library using polymerase chain reaction (STRATHMANN and SIMON 1990). The cloned $\alpha$ subunits were named $\alpha_q$ and $\alpha_{11}$ and were designated members of a new family of G-proteins ($G_q$) since they showed less than 50% sequence identity with other $\alpha$ subunits (SIMON et al. 1991). The $\alpha$ subunits of $G_q$ and $G_{11}$ show close (88%) sequence identity, but those of other members of the $G_q$ family ($G_{14-16}$) are less closely related (SIMON et al. 1991). All lack the cysteine residue that is ADP-ribosylated in other $\alpha$ subunits by pertussis toxin.

STERNWEIS and associates (PANG and STERNWEIS 1990; SMRCKA et al. 1991) developed polyclonal antibodies to peptide sequences in $\alpha_q$ and $\alpha_{11}$. One of them recognized $\alpha_q$ uniquely, whereas two others recognized both $\alpha_q$ and $\alpha_{11}$. Using these antisera, TAYLOR et al. (1991) and TAYLOR and EXTON (1991) showed that their $\alpha$ subunit preparations contained a 42-kDa protein recognized by antisera to $\alpha_q$ and a 43-kDa protein recognized by antisera to sequences common to $\alpha_q$ and $\alpha_{11}$ but not by an antiserum selective for $\alpha_q$. Western blotting with an antiserum raised to a sequence unique to $\alpha_{11}$ later confirmed that the 43-kDa protein was $\alpha_{11}$ (TAYLOR and EXTON 1991). Recently, WALDO et al. (1991) purified a 43-kDa protein from turkey erythrocyte membranes that activated PI-PLC in the presence of $AlF_4^-$. Western blotting indicated that it was a member of the $\alpha_q$ family.

$\alpha_q$ and $\alpha_{11}$ are present very widely in mammalian tissues (PANG and STERNWEIS 1990; STRATHMANN and SIMON 1990). A homologue of $\alpha_q$ ($DG_q$) is present in the eye of *Drosophila* and probably couples rhodopsin to PI-PLC (LEE et al. 1990). A cDNA for an $\alpha$ subunit that is 98% identical to $\alpha_{11}$ has been identified in bovine liver (NAKAMURA et al. 1991).

BLANK et al. (1991) purified $G_q$ and $G_{11}$ in the heterotrimeric ($\alpha\beta\gamma$) form from bovine liver. These G-proteins markedly activated purified bovine brain PI-PLC $\beta$1 in the presence of $AlF_4^-$ and analogues of GTP, and consisted of 42- and 43-kDa proteins which cross-reacted with antisera to peptide sequences in $\alpha_q$ and $\alpha_{11}$, respectively. They also contained 35- and 36-kDa proteins recognized by an antiserum specific for transducin $\beta$ subunit, and an 8-kDa protein presumed to be a $\gamma$ subunit. The stoichiometry between $\alpha$ and $\beta$ subunits was approximately 1:1. Activation of hepatic $G_{q/11}$ by GTP$\gamma$S was very rapid, but required high concentrations of the nucleotide ($>1\,\mu M$). Activation was blocked by GDP$\beta$S and $\beta\gamma$ subunits. Significant activation of PI-PLC was achieved with sub-stoichiometric amounts of G protein, and equimolar amounts stimulated enzyme activity six fold. As expected from previous studies, neither the 42- nor 43-kDa protein was ADP-ribosylated or inactivated by pertussis toxin.

Since pertussis toxin partly or completely inhibits the effects of certain $Ca^{2+}$-mobilizing agonists on $PtdInsP_2$ hydrolysis in some cell types (MELDRUM et al. 1991; ASHKENAZI et al. 1989; COTECCHIA et al. 1990), pertussis toxin-sensitive G proteins must also be involved. In HL60 cells the chemotactic peptide f-Met-Leu-Phe (FMLP) stimulated the cholera toxin-dependent ADP-ribosylation of 40- and 41-kDa pertussis toxin substrates that comigrated on gels with $\alpha_{i2}$ and $\alpha_{i3}$ (GIERSCHIK et al. 1989; IIRI et al. 1989). The 41-kDa protein was also photolabeled with [$\alpha$-$^{32}$P]GTP azidoanilide in HL60 membranes incubated with FMLP (SCHÄFER et al. 1990). A complex between the FMLP receptor and a 40-kDa pertussis toxin substrate was partially purified from HL60 membranes (POLAKIS et al. 1988) and binding of FMLP to the complex was inhibited by GTP$\gamma$S. These data suggest that $G_{i2}$ or $G_{i3}$ may couple the FMLP receptor to PI-PLC in HL60 cells. However, the coupling of these G-proteins to PI-PLC has not been shown directly.

There was an early report that $G_o$ and $G_i$ restore FMLP-stimulated formation of $InsP_3$ in pertussis toxin-treated HL60 membranes (KIKUCHI et al. 1986). However, both G-proteins were equally effective and their purity was not established. $G_o$ has also been implicated in the regulation of PI-PLC because its injection into *Xenopus* oocytes enhanced the muscarinic receptor-stimulated $Cl^-$ current, which is due to $InsP_3$-mediated $Ca^{2+}$ mobilization (MORIARTY et al. 1990). The current was also stimulated by two forms of $G_o$ isolated from bovine brain and by recombinant $\alpha_o$ (PADRELL et al. 1991). However, the specificity of the effect was not determined by testing the potency of $G_o$ against other G proteins, for example, $G_q$. Irrespective of this, the limited tissue distribution of $G_o$ (PRICE et al. 1989; MUMBY et al. 1988; STRATHMANN et al. 1990) means that it cannot be the pertussis-sensitive G protein that controls PI-PLC in most cells. In some cells there is evidence that some receptors are linked to PI-PLC by pertussis toxin-sensitive G proteins, whereas others are linked via toxin-insensitive G proteins (PERNEY and MILLER 1989; COWEN et al. 1990; RAYMOND et al. 1991).

## C. Coupling of G-Proteins to $Ca^{2+}$-Mobilizing Receptors

Recently, WANGE et al. (1991) showed that vasopressin and other $Ca^{2+}$-mobilizing agonists induce the labeling of 42- and 43-kDa proteins with [$^{32}$P]GTP-azidoanilide in liver plasma membranes. The proteins comigrated on gels with proteins recognized by antisera to $\alpha_q$ and $\alpha_{11}$, respectively, and were selectively immunoprecipitated with an antiserum against the C-terminal dodecapeptide common to $\alpha_q$ and $\alpha_{11}$. In another study, SHENKER et al. (1991) found that a similar antiserum inhibited the stimulation of

GTPase by thromboxane $A_2$ in human platelets and recognized a 42-kDa protein, presumably $\alpha_q$.

A mixture of $G_q$ and $G_{11}$ ($G_{q/11}$) has been reconstituted with the $M_1$-muscarinic cholinergic receptor in phospholipid vesicles (BERSTEIN et al. 1992). In this system, GTPase activity and [$^{35}$S]GTP$\gamma$S binding were markedly stimulated by carbachol, but weakly by atropine. Furthermore, negligible stimulation of GTP$\gamma$S binding was observed when the $M_2$ receptor replaced the $M_1$ receptor and $G_s$, $G_i$, $G_o$, $G_z$ were used in place of $G_{q/11}$. These observations are consistent with previous findings on the effects of muscarinic receptor subtypes on PtdIns$P_2$ hydrolysis (PERALTA et al. 1988).

Activation of $G_{q/11}$ by carbachol in phospholipid vesicles containing $M_1$ muscarinic receptors was also measured using PI-PLC$\beta_1$ in a two-step assay (BERSTEIN et al. 1992). The time course of binding of [$^{35}$S]GTP$\gamma$S followed closely the activation of the phospholipase, and the concentrations of carbachol for half-maximal stimulation of GTP$\gamma$S binding and phospholipase activation were similar. Coreconstitution of $M_1$ receptor, $G_{q/11}$, and PI-PLC$\beta$1 allowed the demonstration of GTP$\gamma$S-dependent, carbachol-stimulated hydrolysis of PtdIns$P_2$. These data show that the three components are sufficient for reconstitution of agonist-stimulated Ins$P_3$ formation in vitro.

There is other evidence that $G_q$ and $G_{11}$ are the transducing G-proteins for PtdIns$P_2$ hydrolysis. GUTOWSKI et al. (1991) showed that an antiserum to the C-terminal dodecapeptide of $\alpha_q$ and $\alpha_{11}$ abrogates the stimulation of PtdIns$P_2$ hydrolysis by various agonists in membranes from NG108-15, rat liver, and 1321N1 cells. Activation of the enzyme by GTP$\gamma$S alone was also blocked, and the inhibition was attenuated by addition of the C-terminal peptide.

WU et al. (1992) showed that transfection of cDNAs coding for $\alpha_q$ and $\alpha_{11}$ into COS-7 cells increased the stimulation of Ins$P_3$ formation by $AlF_4^-$ and enhanced the activation of PI-PLC$\beta$1 by GTP$\gamma$S in membranes isolated from the cells. On the other hand, transfection with cDNAs for $\alpha_{oa}$, $\alpha_{ob}$, $\alpha_t$, $\alpha_z$, or $\alpha_{12}$ did not result in enhanced formation of Ins$P_3$. Transfection with cDNAs for mutant forms of $\alpha_{11}$ or $\alpha_q$ in which Gln-209 was changed to Leu or Arg-183 was changed to Cys resulted in high levels of Ins$P_3$ in the absence of $AlF_4^-$, consistent with these mutations rendering the $\alpha$-subunits constitutively active.

CONKLIN et al. (1992) have also examined the effects of mutations in $\alpha_q$ on agonist-stimulated Ins$P_3$ production. In agreement with WU et al. (1992), transient expression of $\alpha_q$ with mutation of Arg-183 to Cys in COS-7 or HEK-293 cells resulted in consitutive activation of PI-PLC. To examine the coupling of $\alpha_q$ to receptors, CONKLIN et al. (1992) cotransfected cDNAs for $\alpha_q$- and the $\alpha_2$-adrenergic receptor into COS-7 or HEK-293 cells. In these cells, the Ins$P_3$ response to the $\alpha_2$-adrenergic agonist UK-14304 was enhanced.

The domains on the $\alpha_1$-adrenergic and $M_1$-muscarinic receptors that interact with the G-proteins which regulate PI-PLC, presumably $G_q$ and $G_{11}$, have been mapped by mutational analysis and construction of chimerae with $\beta$-adrenergic receptors (LOMASNEY et al. 1991; WONG et al. 1990). A major region defining interaction with the G-proteins is a small segment in the third cytoplasmic loop, but other regions are involved in fully determining selectivity. For example, the coupling region of the $\alpha_{1b}$-adrenergic receptor resides in a 27 amino acid sequence in the N-terminal part of the third cytoplasmic loop (COTECCHIA et al. 1991), but a site in the C-terminal part greatly affects the efficiency of the coupling, and the substitution of a single residue (Ala-293) leads to constitutive activation (KJELSBERG et al. 1992).

## D. Specificity of Phosphoinositide Phospholipase C Linked to $G_q$ and $G_{11}$

There are at least eight isozymes of PI-PLC, which are of four types ($\alpha$, $\beta$, $\gamma$, $\delta$; MELDRUM et al. 1991; RHEE and CHOI 1992). The reported sequence for the $\alpha$ isozyme differs from the others and may be that of thiol:protein disulfide oxidoreductase (SRIVASTAVA et al. 1991). The $\beta$, $\gamma$, and $\delta$ isozymes have two domains, designated X and Y, which have high (40%–60%) sequence identity, and the $\gamma$ isozymes contain regions ($SH_2$ and $SH_3$) which are homologous to conserved regions in nonreceptor tyrosine kinases.

The $\gamma$-isozyme exists in two forms of high sequence identity. PI-PLC$\gamma_1$ is widely distributed, and is the isozyme coupled to growth factor receptor tyrosine kinase. Binding of platelet-derived growth factor and epidermal growth factor to their receptors leads to activation of the tyrosine kinase activity and phosphorylation of several cytosolic proteins, including PI-PLC$\gamma_1$ (MELDRUM et al. 1991; RHEE and CHOI 1992). In contrast, agonists acting through $G_q$ and $G_{11}$ activate PI-PLC$\beta_1$. The evidence comes from reconstitution studies with GTP$\gamma$S-activated $\alpha_q$ and $\alpha_{11}$ and various PI-PLC isozymes (TAYLOR et al. 1991) which showed unequivocally that the $\beta_1$ isozyme was stimulated but not the $\gamma_1$ and $\delta_1$ isozyme. Further evidence comes from the observation that antibodies to the $\beta_1$ isozyme selectively blocked the stimulation of PI-PLC in liver plasma membranes by GTP$\gamma$S-activated $\alpha_{q/11}$ (TAYLOR et al. 1991).

Support for PI-PLC$\beta$1 as the PLC isozyme responding to $\alpha_q$ and $\alpha_{11}$ comes from some cell transfection studies in which coexpression of these $\alpha$ subunits and PI-PLC$\beta$1 in COS-7 cells led to enhanced accumulation of $InsP_3$ in response to $AlF_4^-$ (WU et al. 1992). Furthermore, the PI-PLC target of GTP$\gamma$S-activated $\alpha_{q/11}$ has been partially purified from bovine liver plasma membranes and identified immunologically as the $\beta$1 isozyme (SHAW and EXTON 1992).

The PI-PLC isozyme(s) regulated by the pertussis toxin-sensitive G-protein(s) has not been defined. However, in a mutant of the CCL39 fibroblast line that is defective in pertussis toxin-sensitive, thrombin-induced mitogenesis, a deficiency in PI-PLC$\delta$ was observed (RATH et al. 1990).

## E. Mechanisms of Agonist-Stimulated Phosphatidylcholine Breakdown

The hydrolysis of PtdCho is a widespread cellular response to hormones, neurotransmitters, and growth factors (EXTON 1990; BILLAH and ANTHES 1990). In some cell types there is evidence that G-proteins are involved. For example, addition of GTP$\gamma$S (0.1–10 $\mu M$) to liver plasma membranes stimulated the $Mg^{2+}$-dependent release of choline, PtdCho, DAG, and PtdOH (IRVING and EXTON 1987; BOCCKINO et al. 1987a; HURST et al. 1990), and of PtdEtOH (Ptd ethyl alcohol) in the presence of EtOH (BOCCKINO et al. 1987b). Other hydrolysis-resistant analogues of GTP were effective, but not the corresponding ATP analogues (IRVING and EXTON 1987; HURST et al. 1990). The effect of GTP$\gamma$S was inhibited by millimolar GDP$\beta$S (IRVING and EXTON 1987; HURST et al. 1990). Based on the products of PtdCho hydrolysis, it appears that both phospholipase C and D activities are stimulated.

GTP$\gamma$S effects on PtdCho phospholipases C and D have also been reported in permeabilized Swiss 3T3 fibroblasts (DIAZ-MECO et al. 1989; QUILLIAM et al. 1990), HL-60 cells (ANTHES et al. 1989), pulmonary artery endothelial cells (MARTIN and MICHAELIS 1989), and cerebral cortical membranes (QUIAN and DREWES 1989; QUIAN et al. 1990). In addition, stimulation of phospholipase D (without identification of the phospholipid substrate) has been reported in permeabilized HL-60 (XIE and DUBYAK 1991) and NG108-15 cells (LISCOVITCH and ELI 1991) and in platelet membranes (VAN DER MUELEN and HASLAM 1990). In some cases other GTP analogues were effective and GDP$\beta$S reversed the effect.

Evidence for G-protein-mediated agonist effects on PtdCho hydrolysis is limited. In liver plasma membranes, PtdCho hydrolysis was induced by $P_{2y}$-purinergic agonists, but not vasopressin, angiotensin II or epinephrine (IRVING and EXTON 1987; BOCCKINO et al. 1987a). The effects of the purinergic agonists were dependent on GTP analogues. Mediation by $P_{2y}$-purinergic receptors was indicated by the potency order of the adenine nucleotides and the inhibitory effects of $\alpha$, $\beta$-methylene ATP and 2,2′-pyridylisatogen (IRVING and EXTON 1987). GTP$\gamma$S-dependent effects of ATP and its analogues on phopholipase D activity have been observed in other broken cell preparations (XIE and DIBYAK 1991; LISCOVITCH and ELI 1991), although a phosphorylation mechanism may also be involved. Carbachol, GTP$\gamma$S, NaF, and cholera toxin have been reported to stimulate PtCho breakdown in brain synaptosomes (QUIAN and DREWES 1989). Effects of pertussis toxin on agonist-induced PtdCho hydrolysis have been noted in some cell lines (XIE et al. 1991; KANAHO et al. 1991; MACNULTY et al. 1991) but not others. This implies that more than one type of G-protein regulates PtdCho hydrolysis.

The nature of the G-proteins that regulate the PtdCho phospholipases remains elusive and, because the phospholipases have not been purified, it is unclear that the action of the G-proteins is direct. In an interesting study using HL-60 cells, ANTHES et al. (1991) found that GTP$\gamma$S stimulation of PtdCho phospholipase D activity was not observed with membranes alone, but required a heat-labile factor in the cytosol. The nature of this factor remains unclear.

## F. Summary

The pertussis toxin-insensitive G proteins that stimulate PI-PLC consist of 42- and 43-kDa $\alpha$ subunits, 35- and 36-kDa $\beta$ subunits and 8-kDa $\gamma$ subunits. The $\alpha$ subunits are members of the $G_q$ class of G-proteins, namely, $\alpha_q$ and $\alpha_{11}$, and both specifically stimulate the $\beta_1$ isozyme of PI-PLC. Agonist activation of $Ca^{2+}$-mobilizing receptors in liver plasma membranes selectively induces the photoaffinity labeling of $\alpha_q$ and $\alpha_{11}$ by [$\alpha$-$^{32}$P]GTP-azidoanilide. Reconstitution of $G_q$ and $G_{11}$ with recombinant $M_1$ muscarinic receptors in phospholipid vesicles permits carbachol stimulation of [$^{35}$S]GTP$\gamma$S binding and GTPase activity. Substituting atropine, $M_2$ receptors, or other G-proteins results in little change in GTP$\gamma$S binding or GTPase. Reconstitution of $G_q$ and $G_{11}$ with the $M_1$ muscarinic receptor and PI-PLC$\beta_1$ allows GTP$\gamma$S-dependent stimulation of $PtdInsP_2$ hydrolysis by carbachol. These data establish unequivocally that $G_q$ and $G_{11}$ function as the transducing G-proteins between $Ca^{2+}$-mobilizing receptors and the $\beta_1$ isozyme of PI-PLC.

The hydrolysis of PtdCho by hormones, neurotransmitters and growth factors is a widespread cellular response. It involves activation of phospholipases of the C and D types, but these have not been identified, and the physiological functions of the DAG and PtdOH produced are not defined. Control of the phospholipases may involve G-proteins, $Ca^{2+}$ and protein kinase C. However, the specific details of the control mechanisms remain to be determined.

### References

Anthes JC, Eckel S, Siegel MI, Egan RW, Billah MM (1989) Phospholipase D in homogenates from HL-60 granulocytes: implications of calcium and G protein control. Biochem Biophys Res Commun 163:657–664

Ashkenazi A, Peralta EG, Winslow JW, Ramachandran J, Capon DJ (1989) Functionally distinct G proteins selectively couple different receptors to PI hydrolysis in the same cell. Cell 56:487–493

Berridge MJ (1987) Inositol trisphosphate and diacylglycerol: two interacting second messengers. Ann Rev Biochem 56:159–193
Berstein G, Blank JL, Smrcka AV, Higashima T, Sternweis PC, Exton JH, Ross EM (1992) Reconstitution of agonist-stimulated phosphatidylinositol 4,5-bisphosphate hydrolysis using purified ml muscarinic receptor, $G_{q/11}$ and phospholipase C-$\beta_1$. J Biol Chem 267:8081–8088
Billah MM, Anthes JC (1990) The regulation and cellular functions of phosphatidylcholine hydrolysis. Biochem J 269:281–291
Blank JL, Ross AH, Exton JH (1991) Purification and characterization of two G-proteins which activate the $\beta$1 isozyme of phosphoinositide-specific phospholipase C. Identification as members of the $G_q$ class. J Biol Chem 266:18206–18216
Bocckino SB, Blackmore PF, Wilson PB, Exton JH (1987a) Phosphatidate accumulation in hormone-treated hepatocytes via a phospholipase D mechanism. J Biol Chem 262:15309–15315
Bocckino SB, Wilson PB, Exton JH (1987b) $Ca^{2+}$-mobilizing hormones elicit phosphatidylethanol accumulation via phospholipase D activation. FEBS Lett 225:201–204
Conklin BR, Chabre O, Wong YH, Federman AD, Bourne HR (1992) Recombinant $G_q\alpha$. Mutational activation and coupling to receptors and phospholipase C. J Biol Chem 267:31–32
Cotecchia S, Kobilka BK, Daniel KW, Nolan RD, Lapetina EY, Caron MG (1990) Multiple second messenger pathways of $\alpha$-adrenergic receptor subtypes expressed in eukaryotic cells. J Biol Chem 265:63–69
Cotecchia S, Ostrowski J, Kjelsberg MA, Caron MC, Lefkowitz RJ (1992) Discrete amino acid sequences of the $\alpha_1$-adrenergic receptor determine the selectivity of coupling to phosphatidylinositol hydrolysis. J Biol Chem 267:1633–1639
Cowen DS, Baker B, Dubyak GR (1990) Pertussis toxin produces differential inhibitory effects on basal, $P_2$-purinergic, and chemotactic peptide-stimulated inositol phospholipid breakdown in HL-60 cells and HL-60 cell membranes. J Biol Chem 265:16181–16189
Diaz-Meco MT, Larrodera P, Lopez-Barahona M, Cornet ME, Barreno PG, Moscat J (1989) Phospholipase C-mediated hydrolysis of phosphatidylcholine is activated by muscarinic agonists. Biochem J 263:115–120
Exton JH (1988) The roles of calcium and phosphoinositides in the mechanisms of $\alpha_1$-adrenergic and other agonists. Rev Physiol Biochem Pharmacol 111:118–224
Exton JH (1990) Signaling through phosphatidylcholine breakdown. J Biol Chem 265:1–4
Gierschik P, Sidiropoulos D, Jakobs KH (1989) Two distinct $G_i$-proteins mediate formyl peptide receptor signal transduction in human leukemia (HL-60) cells. J Biol Chem 264:21470–21473
Gutowski S, Smrcka A, Nowak L, Wu D, Simon M, Sternweis PC (1991) Antibodies to the $\alpha_q$ subfamily of guanine nucleotide-binding regulatory protein $\alpha$ subunits attenuate activation of phosphatidylinositol 4,5-bisphosphate hydrolysis by hormones. J Biol Chem 266:20519–20524
Hurst KM, Hughes BP, Barritt GJ (1990) The roles of phospholipase D and a GTP-binding protein in guanosine 5′-[$\gamma$-thio]triphosphate-stimulated hydrolysis of phosphatidylcholine in rat liver plasma membranes. Biochem J 272:749–753
Iiri T, Tohkin M, Morishima N, Ohoka Y, Ui M, Katada T (1989) Chemotactic peptide receptor-supported ADP-ribosylation of a pertussis toxin substrate GTP-binding protein by cholera toxin in neutrophil-type HL-60 cells. J Biol Chem 264:21394–21400
Irving H, Exton JH (1987) Phosphatidylcholine breakdown in rat liver plasma membranes: roles of guanine nucleotides and $P_2$-purinergic agonists. J Biol Chem 262:3440–3443

Kanaho Y, Kanoh H, Nozawa Y (1991) Activation of phospholipase D in rabbit neutrophils by fMet-Leu-Phe is mediated by a pertussis toxin-sensitive GTP-binding protein that may be distinct from a phospholipase C-regulating protein, FEBS Lett 279:249–252

Kikuchi A, Kozawa O, Kaibuchi K, Katada T, Ui M, Takai Y (1986) Direct evidence for involvement of a guanine nucleotide-binding protein in chemotactic peptide-stimulated formation of inositol bisphosphate and trisphosphate in differentiated human leukemic (HL-60) cells. J Biol Chem 261:11558–11562

Kjelsberg MA, Cotecchia S, Ostrowki J, Caron MG, Lefkowitz RJ (1992) Constitutive activatism of the $\alpha_{1B}$-adrenergic receptor by all amino acid substitutions at a single site. J Biol Chem 267:1430–1433

Lee Y-J, Dobbs MB, Verardi ML, Hyde DR (1990) dgq: Drosophila gene encoding a visual system-specific $G_\alpha$ molecule. Neuron 5:889–898

Liscovitch M, Elil Y (1991) $Ca^{2+}$ inhibits guanine nucleotide-activated phospholipase D in neural-derived NG108-15 cells. Cell Regulation 2:1011–1019

Lomasney JW, Cotecchia S, Lefkowitz RJ, Caron MG (1991) Molecular biology of adrenergic receptors: implications for receptor classification and for structure-function relationships. Biochim Biophys Acta 1095:127–139

MacNulty EE, McClue SJ, Carr IC, Jess T, Wakelam MJO, Milligan G (1992) $\alpha_2$-C10 adrenergic receptors expressed in rat 1 fibroblasts can regulate both adenylylcyclase and phospholipase D-mediated hydrolysis of phosphatidylcholine by interacting with pertussis toxin-sensitive guanine nucleotide-binding proteins. J Biol Chem 267:2149–2156

Martin TW, Michaelis K (1989) $P_2$-purinergic agonists stimulate phosphodiesteratic cleavage of phosphatidylcholine in endothelial cells. Evidence for activation of phospholipase D. J Biol Chem 264:8847–8856

Meldrum E, Parker PJ, Carozzi A (1991) The PtdIns-PLC superfamily and signal transduction. Biochim Biophys Acta 1092:49–71

Moriarty TM, Padrell E, Carty DJ, Omri G, Landau EM, Iyengar R (1990) $G_o$ protein as signal transducer in the pertussis toxin-sensitive phosphatidylinositol pathway. Nature 343:79–82

Mumby S, Pang I-H, Gilman AG, Sternweis PC (1988) Chromatographic resolution and immunologic identification of the $\alpha_{40}$ and $\alpha_{41}$ subunits of guanine nucleotide-binding regulatory proteins from bovine brain. J Biol Chem 263:2020–2026

Nakamura F, Ogata K, Shiozaki K, Kameyama K, Oharak K, Haga T, Nukada T (1991) Identification of two novel GTP-binding protein $\alpha$-subunits that lack apparent ADP-ribosylation sites for pertussis toxin. J Biol Chem 266:12676–12681

Padrell E, Carty DJ, Moriarty TM, Hildebrandt JD, Landau EM (1991) Two forms of the bovine brain $G_o$ that stimulate the inositol trisphosphate-mediated $Cl^-$ currents in Xenopus oocytes. J Biol Chem 266:9771–9777

Pang I-H, Sternweis PC (1990) Purification of unique $\alpha$ subunits of GTP-binding regulatory proteins (G proteins) by affinity chromatography with immobilized $\beta\gamma$ subunits. J Biol Chem 265:18707–18712

Peralta EG, Ashkenazi A, Winslow JW, Ramachandran J, Capon DJ (1988) Differential regulation of PI hydrolysis and adenylate cyclase by muscarinic receptor subtypes. Nature 334:434–437

Perney TM, Miller RJ (1989) Two different G-proteins mediate neuropeptide Y and bradykinin-stimulated phospholipid breakdown in cultured rat sensory neurons. J Biol Chem 264:7317–7327

Polakis PG, Uhing RJ, Snyderman R (1988) The formylpeptide chemoattractant receptor copurifies with a GTP-binding protein containing a distinct 40-kDa pertussis toxin substrate. J Biol Chem 263:4969–4976

Price SR, Tsai S-C, Adamik R, Angus CW, Serventi IM, Tsuchiya M, Moss J, Vaughan M (1989) Expression of $G_{o\alpha}$ mRNA and protein in bovine tissues. Biochemistry 28:3803–3807
Qian Z, Drewes LR (1989) Muscarinic acetylcholine receptor regulates phosphatidylcholine phospholipase D in canine brain. J Biol Chem 264:21720–21724
Qian Z, Reddy PV, Drewes LR (1990) Guanine nucleotide-binding protein regulation of microsomal phospholipase D activity of canine cerebral cortex. J Neurochem 54:1632–1638
Quilliam LA, Der CJ, Brown JH (1990) Binding protein-stimulated phospholipase C and phospholipas D activities in ras-transformed NIH 3T3 fibroblasts. Second Messengers and Phosphoproteins 13:59–67
Rath HM, Fee JA, Rhee SG, Silbert DF (1990) Characterization of phosphatidylinositol-specific phospholipase C defects associated with thrombin-induced mitogenesis. J Biol Chem 265:3080–3087
Raymond JR, Albers FJ, Middleton JP, Middleton JP, Lefkowitz RJ, Caron MG, Obeid LM, Dennis VW (1991) 5-$HT_{1A}$ and histamine $H_1$ receptors in HeLa cells stimulate phosphoinositide hydrolysis and phosphate uptake via distinct G protein pools. J Biol Chem 266:372–379
Rhee SG, Choi WC (1992) Regulation of inositol phospholipid-specific phospholipase C isozymes. J Biol Chem 267:12393–12396
Schäfer SO, Hoffmann B, Bombien E, et al. (1990) Agonist-sensitive binding of a photoreactive GTP analog to a G-protein $\alpha$-subunit in membranes of HL-60 cells. FEBS Lett 260:14–18
Shaw K, Exton JH (1991) Identification in bovine liver plasma membranes of a $G_q$-activatable phosphoinositide phospholipase C. Biochemistry In Press
Shenker A, Goldsmith P, Unson CG, Spiegel AM (1991) The G protein coupled to the thromboxane $A_2$ receptor in human platelets is a member of the novel $G_q$ family. J Biol Chem 266:9309–9313
Simon MI, Strathmann MP, Gautam N (1991) Diversity of G proteins in signal transduction. Science 252:802–808
Smrcka AV, Hepler JR, Brown KO, Sternweis PC (1991) Regulation of polyphosphoinositide-specific phospholipase C activity by purified $G_q$. Science 251:804–807
Strathmann M, Simon MI (1990a) G protein diversity: a distinct class of $\alpha$ subunits is present in vertebrates and invertebrates. Proc Natl Acad Sci USA 87:9113–9117
Stathmann M, Wilkie TM, Simon MI (1990b) Alternative splicing produces transcripts encoding two forms of the $\alpha$ subunit of GTP-binding protein $G_o$. Proc Natl Acad Sci USA 87:6477–6481
Taylor SJ, Smith JA, Exton JH (1990) Purification from bovine liver membranes of a guanine nucleotide-dependent activator of phosphoinositide-specific phospholipase C. Immunologic identification as a novel G-protein $\alpha$ subunit. J Biol Chem 265:17150–17156
Taylor SJ, Chae HZ, Rhee SG, Exton JH (1991) Activation of the $\beta$1 isozyme of phospholipase C by $\alpha$ subunits of the $G_q$ class of G proteins. Nature 350:516–518
Taylor SJ, Exton JH (1991) Two $\alpha$ subunits of the $G_q$ class of G proteins stimulate phosphoinositide phospholipase C-$\beta$1 activity. FEBS Lett 286:214–216
Van Der Meulen J, Haslam RJ (1990) Phorbol ester treatment of intact rabbit platelets greatly enhances both the basal and guanosine 5′-[$\gamma$-thio]triphosphate-stimulated phospholipase D activities of isolated platelet membranes. Biochem J 271:693–700
Waldo GL, Boyer JL, Morris AJ, Harden TK (1991) Purification of an $AlF_4^-$ and G-protein $\beta\gamma$-subunit-regulated phospholipase C-activating protein. J Biol Chem 266:14217–14225

Wange RL, Smrcka AV, Sternweis PC, Exton JH (1991) Photoaffinity labeling of two rat liver plasma membrane proteins with [$^{32}$P]$\gamma$-azidoanilido GTP in response to vasopressin. J Biol Chem 266:11409–11412
Wong SK-F, Parker EM, Ross EM (1990) Chimeric muscarinic cholinergic: $\beta$-adrenergic receptors that activate $G_s$ in response to muscarinic agonists. J Biol Chem 265:6219–6224
Wu D, Lee CH, Rhee SG, Simon MI (1992) Activation of phospholipase C by the $\alpha$ subunits of the $G_q$ and $G_{11}$ proteins in transfected Cos-7 cells. J Biol Chem 267:1811–1817
Xie M, Dubyak GR (1991) Guanine-nucleotide- and adenine-nucleotide-dependent regulation of phospholipase D in electropermeabilized HL-60 granulocytes. Biochem J 278:81–89
Xie M, Jacobs LS, Dubyak GR (1991) Regulation of phospholipase D and primary granule secretion by $P_2$-purinergic- and chemotactic peptide-receptor agonists is induced during granulocytic differentiation of HL-60 cells. J Clin Invest 88:45–54

CHAPTER 68

# Hormonal Regulation of Phospholipid Metabolism via G-proteins II: $PLA_2$ and Inhibitory Regulation of PLC

D. CORDA

## A. Introduction

Lipid derivatives play a major role as second messengers. A large body of evidence has accumulated on the mechanisms leading to their formation and on their role in cell regulation (reviewed in DENNIS et al. 1991). They are formed by the action of different phospholipases ($A_2$, C, D) acting on membrane phospholipids. Of particular physiological relevance has been the elucidation of the hormonal regulation of these enzymes. A large number of receptors for hormones and neurotransmitters are coupled to cellular phospholipases and regulate the cytosolic levels of second messengers such as $Ca^{2+}$, diacylglycerol, inositol trisphosphate, and arachidonic acid. The activation of phospholipases can be indirectly induced by a receptor-mediated increase in cytosolic $Ca^{2+}$, or it can be due to the activation of a heterotrimeric GTP-binding (G) protein directly coupled to the enzyme.

This chapter discusses the role of G-proteins in the modulation of phospholipase $A_2$ ($PLA_2$) and in the inhibitory regulation of phospholipase C (PLC). Information on other aspects of phospholipase regulation ($Ca^{2+}$ and protein kinase C dependence, subtypes, structure, etc.) are extensively discussed in recent reviews (DENNIS et al. 1991; BEREZIAT et al. 1990; RHEE et al. 1989; FAIN 1990) and elsewhere in this volume (see Chaps. 58, 59, 67, 69).

## B. Modulation of $PLA_2$

$PLA_2$ releases a fatty acid from the sn-2 position of phospholipids. In mammalian cells, the sn-2 position of phosphatidylcholine, phosphatidylethanolamine, and phosphatidylinositol is preferentially occupied by arachidonic acid (BURCH 1989). Alternative mechanisms of arachidonic acid formation have been described (BURGOYNE and MORGAN 1990); one of these involves the hormonal activation of PLC, which produces inositol trisphosphate and diacylglycerol; the latter is substrate of diacylglycerol lipase that liberates arachidonic acid (AXELROD et al. 1988). In several cell systems, the hormone-dependent generation of arachidonic acid has been attributed to the activation of $PLA_2$ rather than of PLC (either by the use of specific inhibitors of these enzymes or by their coupling to different

G-proteins, see below). The cellular systems in which the generation of arachidonic acid has been clearly related to the activation only of $PLA_2$ have been used as models to analyze the molecular mechanism involved in the modulation of this enzyme (AXELROD 1988, 1990; BURCH 1989, 1990).

## I. Molecular Forms of $PLA_2$

There are two families of identified $PLA_2$ in mammalian cells: one comprises the approximately 14-kDa secretory forms, prevalent in digestive organs and homologous to the $PLA_2$ of snake venoms. These enzymes have been extensively characterized and structurally defined; they require millimolar $Ca^{2+}$ concentrations for activation and do not show selectivity for the fatty acid in position sn-2 (DAVIDSON and DENNIS 1990). The second family comprises the cytosolic form of $PLA_2$, which is selective for arachidonic acid and requires submicromolar $Ca^{2+}$ concentration for activation (BEREZIAT et al. 1990; CHANNON and LESLIE 1990). A cytosolic $PLA_2$ has recently been purified from a monocytic tumor cell line (U937), a macrophage-like cell line (RAW 264.7), and rat kidney as a 110-kDa protein (CLARK et al. 1990; KRAMER et al. 1991; LESLIE et al. 1988; GRONICH et al. 1990). The enzyme has also been cloned; the predicted molecular mass from the 749 amino acid sequence, which does not share any homology with the secretory $PLA_2$, is approximately 85 kDa (CLARK et al. 1991; SHARP et al. 1991). By radiation inactivation analysis it has been confirmed that the active form of the enzyme is a monomer of about 80 kDa (TREMBLAY et al. 1992). The cytosolic form of $PLA_2$ is thought to be the one that responds to hormonal stimuli and is activated by G-proteins (BEREZIAT et al. 1990). The latter event requires translocation of the enzyme to the plasma membrane. In RAW 264.7 cells a cytosolic $PLA_2$ was found associated with the membrane when cytosolic $Ca^{2+}$ increased to 230–450 n$M$, levels that are usually evoked in response to hormone stimulation (CHANNON and LESLIE 1990). The amino-terminal portion of the cytosolic $PLA_2$ shares significant sequence homology with the $Ca^{2+}$-dependent forms of protein kinase C in the region believed to be involved in protein translocation and membrane phospholipid binding (CLARK et al. 1991). Moreover, an amino-terminal fragment of 138 amino acids which includes the homologous region has been shown to associate with the plasma membrane in a $Ca^{2+}$-dependent manner (CLARK et al. 1991). This sequence has thus been proposed to be involved in the translocation of different cytosolic proteins to the plasma membrane (CLARK et al. 1991). Receptor activation of $PLA_2$ might therefore proceed in two steps: a rapid rise in cytosolic $Ca^{2+}$ promoting the translocation to the plasma membrane of the cytosolic $PLA_2$, followed by interaction of the enzyme with the G-protein subunit responsible for its activation.

## II. G-Protein-Mediated Activation of $PLA_2$

Early evidence indicating that $PLA_2$ could be under the control of G-proteins has come from the observation that arachidonic acid release was

**Table 1.** Cellular systems in which G-proteins sensitive to pertussis toxin have been shown to modulate $PLA_2$ activity

| Cell type | Agent | Reference |
|---|---|---|
| Stimulation[a] | | |
| Thyroid FRTL5 | Norepinephrine/GTP$\gamma$S | Burch et al. 1986 |
| Thyroid FRTL5 | Carbachol | Di Girolamo et al. 1991 |
| Thyroid FRTL5 | Thyroid-stimulating hormone | Corda et al. 1989 |
| Retina | $\beta\gamma$ | Jelsema and Axelrod 1987 |
| Fibroblast | Bradykinin | Huang et al. 1991 |
| Neutrophils | fMet-Leu-Phe | Cockcroft et al. 1991 |
| Keratinocytes | Bradykinin | Kast et al. 1991 |
| RBL (Basophils) | IgE | Narasimhan et al. 1990 |
| Kidney | Epidermal growth factor | Teitelbaum 1990 |
| Inhibition[b] | | |
| Retina | GTP$\gamma$S | Jelsema 1987 |
| Macrophages | Pertussis toxin | Burch et al. 1988 |
| Pancreas | GTP$\gamma$S | Rubin et al. 1991 |

[a] Presented are a few representative examples of systems in which the direct activation of $PLA_2$ by G-proteins has been demonstrated. Additional studies have been reviewed by Burch (1989, 1990), Axelrod (1988, 1990), and Cockcroft et al. (1991).
[b] This mechanism has been deduced from a GTP$\gamma$S-dependent stimulation or pertussis toxin-mediated inhibition of a putative $G_{pi}$ specific for $PLA_2$.

inhibited in cells pretreated with pertussis toxin (i.e., under conditions where G-proteins sensitive to the toxin are ADP-ribosylated and therefore unable to dissociate into the active $\alpha$ and $\beta\gamma$ subunits; see Chap. 8, this volume, part I, Bokach and Gilman 1984). Moreover, activators of G-proteins such as guanine nucleotides or fluoride were able to stimulate arachidonic acid release in several systems (Table 1; reviewed in Burch 1989, 1990). These experiments however, did not clarify whether G-proteins interacts directly with $PLA_2$, or whether an indirect effect, mediated by $Ca^{2+}$ or other second messengers, induced the activation of this enzyme. The demonstration that a direct interaction could be the mechanism used by hormones was provided by the work of Axelrod and coworkers, first in a thyroid cell line (FRTL5) and then in several other systems (for comprehensive reviews, see Axelrod 1988, 1990; Burch 1989, 1990). The authors demonstrated that thyroid cells released arachidonic acid upon adrenergic stimulation in a manner that was independent of the hormonal stimulation of PLC, based on the evidence that the arachidonic acid release was not affected by the addition of the diacylglycerol lipase inhibitor, RHC 80267 and was potently inhibited by pertussis toxin. The G-protein-mediated activation of PLC was much less sensitive to the toxin, indicating that different G-proteins couple the adrenergic receptor to the two enzymes (Burch et al. 1986; Corda and Kohn 1986). Moreover, in permeabilized cells, the arachidonic acid release was directly activated by GTP$\gamma$S in the presence of the PLC inhibitor, neomycin (Burch et al. 1986). Similar mechanisms of $PLA_2$ activation have subsequently been described in several other systems, leading to the conclusion that the G-protein-mediated activation of $PLA_2$ is a rather general and ubiquitous phenomenon (Table 1). It is noteworthy that the G-protein-dependent mechanism is not only coupled to the activation of classical seven transmembrane domain receptors (adrenergic, muscarinic, purinergic, thyroid-stimulating hormone, etc.; see Table 1), but also involved in the action of epidermal growth factor and IgE receptors (TEITELBAUM 1990; NARASIMHAN et al. 1990).

## III. Molecular Aspects

The molecular mechanism by which G-proteins activate $PLA_2$ has not been completely elucidated. It has been suggested that both $\beta\gamma$ and $\alpha$ subunits might be direct activators of this enzyme. When free $\beta\gamma$ subunit obtained from $G_t$ was added to dark-adapted $G_t$-depleted rod outer segment membranes, it activated $PLA_2$ (JELSEMA and AXELROD 1987). This activation was inhibited by addition of equimolar amounts of $\alpha_t$, which associates with the available $\beta\gamma$, but not by the addition of $\alpha_t$ bound to GTP$\gamma$S, which prevents subunit reassociation (JELSEMA and AXELROD 1987). Addition of equivalent amounts of $\alpha_t$ alone had a slight stimulatory activity. Another $\alpha$ subunit, the $\alpha_o$ from brain, was also able to activate $PLA_2$ in the retina, and its effect was additive to the effect of $\beta\gamma$ (JELSEMA et al. 1989). The role of $\alpha$ and $\beta\gamma$ subunits has been studied in other cellular systems. Additional evidence for a function of the $\beta\gamma$ subunits came from the work of KIM et al. (1989), who in patch-clamp experiments, demonstrated that the addition of this subunit activates the atrial $K^+$ channel; this effect was mediated by the activation of $PLA_2$ and the production of arachidonic acid metabolites (KIM et al. 1989; KURACHI et al. 1989). In similar experiments, YATANI et al. (1988) showed that the $\alpha_i$ subunits and, more specifically, the three forms $\alpha_{i1}$, $\alpha_{i2}$, and $\alpha_{i3}$ were able to activate the $K^+$ channel; in their hands the $\beta\gamma$ subunit had an inhibitory effect (OKABE et al. 1990). These apparently contrasting pieces of evidence could be explained in part by the different experimental conditions used in the two studies (OKABE et al. 1990). Alternatively, it has been proposed that two mechanisms may exist: a direct activation of the $K^+$ channel mediated by an $\alpha$ subunit and an indirect modulation due to the interaction of the $\beta\gamma$ subunit with $PLA_2$ (BROWN 1991). In A6 cells (*Xenopus laevis* renal tubular cells) a similar arachidonic acid-mediated mechanism of $Na^+$ channel regulation has been proposed by CANTIELLO et al. (1990). The $\alpha_{i3}$ subunit added to the cytosolic side of excised patches increased the open-time and number of active $Na^+$ channels in pertussis toxin-treated cells by stimulating $PLA_2$ and the formation of lipo-oxygenase metabolites. In chromaffin cells the stimulation of $PLA_2$ by GTP$\gamma$S and the formation of lipo-oxygenase metabolites activated a $Cl^-$ current (DOROSHENKO 1991).

GUPTA et al. (1990) have proposed that $\alpha_{i2}$ is the subunit involved in the modulation of $PLA_2$ by thrombin and $P_2$ purinergic receptors. Chimeras of

$G_2$- and $G_{i2}$-proteins (in which the C-terminal of the $\alpha_s$ protein had been replaced with the 38 C-terminal amino acids of the $\alpha_{i2}$) when overexpressed in Chinese hamster ovary cells, prevented the hormonal activation of $PLA_2$; the same effect was seen after a pertussis toxin pretreatment. The authors propose that these experiments indicate the relevance of the C-terminal region of $\alpha_{i2}$ in the modulation of $PLA_2$ activity. This role of $\alpha_{i2}$ is also supported by a different approach in the same cell system. LOWNDES et al. (1991) showed that expression of the oncogene *gip2* (a mutated $\alpha_{i2}$ subunit, lacking GTPase activity) attenuates the receptor-induced stimulation of $PLA_2$; surprisingly, however, no effect was seen on the basal activity of $PLA_2$. The molecular mechanism by which *gip2* acts on $PLA_2$ has not been completely determined; the authors propose that the expression of this oncogene alters the G-protein-controlled sensitivity of $PLA_2$ to $Ca^{2+}$-dependent activation (the $Ca^{2+}$ requirement of $PLA_2$ has been shown to be affected by the interaction with G-proteins; see NAKASHIMA et al. 1988; BURCH 1990).

In summary, there is no evidence for a single molecular mechanism by which G-proteins modulate $PLA_2$ activity. Rather, it appears that both $\beta\gamma$ and $\alpha$ subunits are involved, possibly depending on the cellular system and on the G-protein distribution and relative abundance. The recent cloning of the cytosolic form of $PLA_2$ might help clarify this matter since the molecular reconstitution of the hormonal activation of $PLA_2$ can now be attempted using the purified components.

## IV. Inhibitory Regulation of $PLA_2$

Based on indirect evidence, G-proteins appear to be involved also in the inhibitory regulation of $PLA_2$. JELSEMA (1987) has shown that the addition of GTP$\gamma$S to rod outer segments inhibits the activity of light-stimulated $PLA_2$, and that basal $PLA_2$ is inhibited by GTP$\gamma$S in rod outer segments depleted of $G_t$. Both cholera and pertussis toxin decreased the basal and light-stimulated activity of $PLA_2$ (JELSEMA 1987). The molecular mechanism has been proposed to involve stabilization of the inactive form of $G_t$ by both pertussis and cholera toxin-induced ADP-ribosylation (JELSEMA 1987). A mechanism of dual regulation of $PLA_2$ by G-proteins has been proposed on the basis of the cholera and pertussis toxin action in macrophages, RAW 264.7 (BURCH et al. 1988). In this system both toxins increased $PLA_2$ activity; the authors proposed a mechanism similar to the one described for the regulation of adenylyl cyclase, where the cholera toxin-dependent ADP-ribosylation of $G_S$ leads to cyclase activation whereas the pertussis toxin-dependent ADP-ribosylation of $G_i$ prevents the inhibition of the enzyme. More recently, RUBIN et al. (1991) have shown that also in pancreatic membranes $PLA_2$ may be under a dual G-protein-dependent regulation. GTP$\gamma$S evoked an increase in arachidonic acid release under resting conditions, whereas it inhibited $Ca^{2+}$-stimulated $PLA_2$.

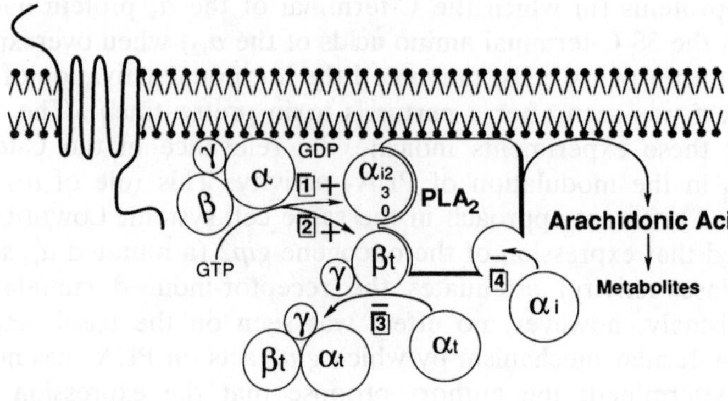


**Fig. 1.** Scheme of the molecular mechanisms involved in the dual regulation of $PLA_2$ by G-proteins. The stimulatory regulation may involve the direct interaction of (*1*) the $\alpha_{i2}$, $\alpha_{i3}$, and $\alpha_o$ or (*2*) the $\beta\gamma_t$ subunits. The inhibitory regulation might either be due to (*3*) the interaction of the $\alpha_t$ subunit made available by the inhibitory stimulus with the stimulatory $\beta\gamma_t$ or (*4*) to the direct interaction of an inhibitory $\alpha$ subunit with the enzyme. +, −, the stimulatory and inhibitory pathways, respectively. See text for details and relevant references

In summary, the available evidence on $PLA_2$ inhibition is indirect and does not address the molecular identity of the inhibitory G-protein involved in the regulation of $PLA_2$. It appears that the mechanism of dual regulation described for the adenylyl cyclase could apply to $PLA_2$ as well (a scheme that summarizes the molecular mechanisms proposed is given in Fig. 1).

## C. Activity of $PLA_2$ in *ras*-Transformed Cells

A *ras*-mediated transformation has been proposed to alter the metabolism of membrane phospholipids (ALONSO et al. 1988; Chap. 20, this volume, part I). The link between *ras* transformation and $PLA_2$ activity was first addressed in fibroblasts by BAR-SAGI and FERAMISCO (1986) who showed that microinjection of p21 (the small G-protein product of the *ras* oncogene; see Chaps. 17, 18, 20, this volume, part I) induced morphological changes correlating in time with the production of lysolipids. As these compounds are produced by $PLA_2$, the authors proposed that this enzyme is activated by the *ras* protein.

In cell systems transformed by the *ras* oncogene, ALONSO et al. (1988; in fibrobasts) and VALITUTTI et al. (1991; in epithelial thyroid cells) found a correlation between the expression of the p21 and a pronounced increase in glycerophosphoinositol, a phosphoinositide derivative formed by the action of $PLA_1$ and $PLA_2$. This was associated with increased levels of arachidonic acid, confirming that *ras*-transformed cells are indeed characterized by an enhanced basal activity of $PLA_2$ (VALITUTTI et al. 1991). Interestingly, phosphoinositides seem to be preferential substrates of this enzyme (VALITUTTI et al. 1991). Increased levels of glycerophosphoinositol were also present in cell lines transformed by membrane-associated or cytosolic oncogenes (such as *src*, *met*, *trk*, *mos* and *raf*). This led to the proposal that the biochemical events initiated by p21 are a common step in the transformation induced by several oncogenes (ALONSO et al. 1988; ALONSO and SANTOS 1990).

The mechanism linking *ras*-induced transformation to $PLA_2$ activation is unclear. PRICE et al (1989) have proposed that the activation of $PLA_2$ (evaluated as increase in arachidonic acid release) induced by oncogenic p21, in scrape-loading experiments, is mediated by protein kinase C. Another factor that could be involved is cytosolic $Ca^{2+}$; however, there is no evidence that basal $Ca^{2+}$ levels are increased in *ras*-transformed cells, at least to such an extent as to justify a persistent elevation of basal $PLA_2$ activity (VALITUTTI et al. 1991). An alternative possibility is that the *ras* GTPase-activating protein (see Chap. 23, this volume, part I), could modulate $PLA_2$ activity, as it has been shown to inhibit the coupling between muscarinic receptors and atrial $K^+$ channels (YATANI et al. 1990).

The altered activity of $PLA_2$ in *ras*-transformed cells could also result from defects in its coupling to G-proteins. Indeed, a partial decrease (about 30%) in glycerophosphoinositol levels in *ras*-transformed cells has been observed upon pretreatment with pertussis toxin (VALITUTTI and CORDA, unpublished observations). This would indicate that the phosphoinositide-specific $PLA_2$ of *ras*-transformed cells is controlled by a G-protein sensitive to pertussis toxin. In the same cell system it has been reported that the pertussis toxin-sensitive G-proteins are twice as abundant as in normal cells. Moreover, the regulation of their toxin-dependent ADP-ribosylation is altered, since an inhibitor of the ADP-ribosylation reaction appears to be inactivated after *ras*-transformation (DI GIROLAMO et al. 1992). Other evidence of G-protein dysfunction in these cells is that pertussis toxin and cholera toxin, which in normal thyroid cells markedly increase cAMP levels, were much less effective in transformed cells (COLLETTA et al. 1988). Direct activation of G protein by $AlF_4^-$ in fibroblasts transformed by *ras* and other membrane associated or cytosolic oncogenes, have uncovered an altered coupling to PLC stimulation (ALONSO et al. 1990).

From this information it can be hypothesized that in transformed cells the alteration of G-proteins results in changes in the activity of $PLA_2$ and, possibly, of other transducing enzymes.

## D. Inhibitory Regulation of PLC

PLCs are a family of hormonally regulated enzymes that hydrolyze phosphoinositide-4,5-bisphosphate to form inositol trisphosphate and diacylglycerol (RHEE et al. 1989; see Chaps. 58, 59, 67, 69, this volume, part I). Whereas the molecular mechanisms of PLC activation have been partially elucidated, very little is known about the hormonal inhibition of

**Table 2.** Cellular systems in which the inhibition of PLC has been proposed to be mediated by a $G_{pi}$ protein

| Cell type | Agent | Reference |
|---|---|---|
| Pituitary | Dopamine | ENJALBERT et al. 1986, 1990 |
| Cerebral cortex | NaF | GODFREY and WATSON 1988 |
| Pituitary GH3 | Adenosine | DELAHUNTY et al. 1988 |
| Cerebral cortex | GTP$\gamma$S | LITOSCH 1989 |
| Pituitary | GTP$\gamma$S, GDP$\beta$S | LIMOR et al. 1989 |
| Thyroid FRTL5 | Carbachol | BIZZARRI et al. 1990 |
| Parotid acini | Carbachol | HORN et al. 1990 |
| Cerebellum | Opioids | MISAWA et al. 1990 |
| Astrocytoma | Adenosine | NAKAHATA et al. 1991 |

PLC. The possibility that inhibitory G-proteins intervene in this modulation is supported by a series of studies. Table 2 contains a list of systems in which a dual mechanism of PLC regulation by stimulatory and inhibitory G-proteins ($G_{ps}$ and $G_{pi}$) has been proposed.

Several receptors have been shown to be able to inhibit PLC activity (for a review see LINDEN and DELAHUNTY 1989). The first proposal that G-proteins mediate this inhibition came from experiments in pituitary cells in which the stimulation of inositol trisphosphate accumulation by angiotensin II was enhanced by pertussis toxin. Moreover, dopamine inhibited the angiotensin II effect in a toxin-dependent manner (ENJALBERT et al. 1986). These results led to the suggestion of a dopamine-induced inhibitory modulation of PLC mediated by a $G_{pi}$; this study, however, did not conclusively prove that the effect of dopamine consisted of a direct inhibition of the enzyme (ENJALBERT et al. 1986), as it was later shown that the dopamine inhibition was mediated largely by a decrease in cytosolic $Ca^{2+}$ due to dopamine-induced closure of a $Ca^{2+}$ channel (VALLAR et al. 1988). In a rat pituitary cell line (GH3) the inhibitory effect of adenosine on PLC was proposed to be mediated by a pertussis toxin-sensitive G-protein acting on the enzyme (DELAHUNTY et al. 1988). Their claim was based on the observation that no other second messenger could duplicate the inhibitory effect of adenosine. However, it is difficult in experiments performed in intact cells to exclude completely the intervention of second messengers in the inhibition of an intracellular enzyme such as PLC. This type of conclusion can be reached more safely using permeabilized cells or cell-free systems where second messengers can be effectively "clamped" at the desired concentrations. These experiments were performed in permeabilized thyroid cells; it was shown that the muscarinic agonist carbachol inhibited both the basal and stimulated level of inositol trisphosphate (BIZZARRI et al. 1990). The effect was dependent on the presence of GTP and was inhibited by pertussis toxin. The role of other signals such as cAMP, $Ca^{2+}$, diacylglycerol, and kinases was excluded. It could therefore be proposed

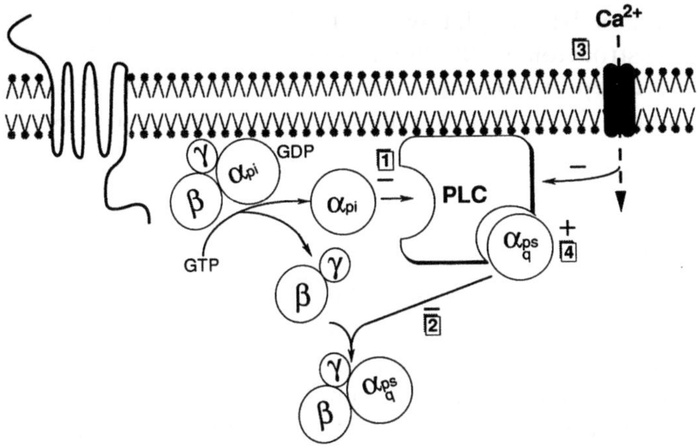


**Fig. 2.** Scheme of the molecular mechanisms involved in the inhibitory regulation of PLC. The inhibitory regulation might involve (*1*) the direct interaction of an $\alpha_{pi}$ subunit with the enzyme. The indirect mechanism involves (*2*) the reduced effect of the stimulatory $\alpha_{ps}$ or $\alpha_q$ subunits, upon interaction with the $\beta\gamma$ subunit made available by the inhibitory stimulus. Alternatively, (*3*) a decrease in the intracellular levels of $Ca^{2+}$ due to the closure of a $Ca^{2+}$ channel might inhibit the enzyme. The stimulation due to (*4*) the direct interaction of $\alpha_{ps}$ or $\alpha_q$ with PLC is also indicated. +, − the stimulatory and inhibitory pathways, respectively. See text for details and relevant references

that the mechanism leading to the inhibition of PLC in thyroid cells involves the direct coupling between a pertussis toxin-sensitive G protein and the enzyme (Bizzarri et al. 1990). Using astrocytoma cell membranes, Nakahata et al. (1991) reached the same conclusion showing that adenosine is able to inhibit the GTP$\gamma$S-dependent accumulation of inositol phosphates; this effect is mediated by a pertussis toxin-sensitive G-protein. These authors also reconstituted the inhibition of GTP$\gamma$S-stimulated PLC in astrocytoma cell membranes by adding a purified preparation of $G_i/G_o$ from brain.

## I. Molecular Aspects

PLC could, in theory, be inhibited directly by the interaction with the $\alpha$ or $\beta\gamma$ subunits of $G_{pi}$. Alternatively, the $\beta\gamma$ subunit of $G_{pi}$ could decrease the activity of the stimulatory $\alpha_p$ by associating with it (see the scheme in Fig. 2).

A role of $\beta\gamma$ has been demonstrated by Moriarty et al. (1988) and Boyer et al. (1989), who showed that PLC could be inhibited by adding exogenous $\beta\gamma$ subunit in *Xenopus* oocytes stimulated by acetylcholine and turkey erythrocytes stimulated by $AlF_4^-$. Although this effect has been attributed to a decrease in enzyme activity by association of the $\alpha_p$ stimulatory subunit with $\beta\gamma$, a direct inhibitory action of the $\beta\gamma$ subunit on the enzyme cannot be completely excluded. NAKAHATA et al. (1991) have shown that in astrocytoma cells the $\beta\gamma$ subunit released by the activation of an adrenergic receptor does not affect PLC activity, suggesting that the adenosine-induced inhibition is unlikely to be due to release of the $\beta\gamma$ subunit.

## E. Conclusion

From the available data on the regulation of PLC and $PLA_2$ it is possible to suggest a general scheme of interaction with G-proteins as shown in Figs. 1 and 2. The stimulation of these enzymes would be direct, i.e., it would involve the interaction of $\alpha_{ps}$ with the enzyme. In some cases the $\beta\gamma$ subunit might also play a direct stimulatory role. The inhibitory regulation remains less well understood. It is likely that more than one inhibitory mechanism exists. Similarly to the situation of adenylyl cyclase, a likely mechanism for PLC and $PLA_2$ might be the reassociation of the $\beta\gamma$ subunit with the stimulatory $\alpha_{ps}$, which decreases the $\alpha_{ps}$ activity. A role of $\beta\gamma$ as direct inhibitor of $PLA_2$ and PLC has not been described. As $\beta\gamma$ under certain conditions is able to inhibit adenylyl cyclase, we can foresee that a similar mechanism might be uncovered in the future also for $PLA_2$ and PLC. This would confirm the familiar scheme of dual regulation, originally described for adenylyl cyclase, as more generally valid for G-protein-controlled transducing enzymes.

*Acknowledgements*. I wish to thank Ms. D. Spadano and Ms. C. Di Sebastiano for help with the references, Ms. S. Falcone for typing the manuscript, and Ms. R. Bertazzi for preparing the figures. Work in the author's laboratory was supported in part by the Italian Association for Cancer Research (AIRC), the Agenzia per la Promozione e lo Sviluppo del Mezzogiorno (PR-3), and the Italian National Research Council (Progetto Finalizzato ACRO).

### References

Alonso T, Morgan RO, Marvizon JC, Zarbl H, Santos E (1988) Malignant transformation by *ras* and other oncogenes produces common alterations in inositol phospholipid signaling pathways. Proc Natl Acad Sci USA 85:4271–4275
Alonso T, Santos E (1990) Increased intracellular glycerophosphoinositol is a biochemical marker for transformation by membrane-associated and cytoplasmic oncogenes. Biochem Biophys Res Comm 171:14–19
Alonso T, Srivastava S, Santos E (1990) Alterations of G-protein coupling function in phosphoinositide signaling pathways of cells transformed by *ras* and other membrane-associated and cytoplasmic oncogenes. Mol Cell Biol 10:3117–3124
Axelrod J, Burch RM, Jelsema CL (1988) Receptor-mediated activation of phospholipase A2 via GTP-binding proteins: arachidonic acid and its metabolites as second messengers. Trends Neurosci 11:117–123
Axelrod J (1990) Receptor-mediated activation of phospholipase A2 and arachidonic acid release in signal transduction. Biochem Soc Trans 18:503–507
Bar-Sagi D, Feramisco JR (1986) Induction of membrane ruffling and fluid-phase pinocytosis in quiescent fibroblasts by *ras* proteins. Science 233:1061–1068

Béréziat G, Etienne J, Kokkinidis M, Olivier JL, Pernas P (1990) New trends in mammalian non-pancreatic phospholipase A2 research. J Lipid Mediat 2:159–172

Bizzarri C, Di Girolamo M, D'Orazio MC, Corda D (1990) Evidence that a guanine nucleotide-binding protein linked to a muscarinic receptor inhibits directly phospholipase C. Proc Natl Acad Sci USA 87:4889–4893

Bokach GM, Gilman AG (1984) Inhibition of receptor-mediated release of arachidonic acid by pertussis toxin. Cell 39:301–308

Boyer JL, Waldo GL, Evans T, Northup JK, Downes CP, Harden TK (1989) Modification of $AlF_4^-$-and receptor-stimulated phospholipase C activity by G-protein $\beta\gamma$ subunits. J Biol Chem 264:13917–13922

Brown AM (1991) A cellular logic for G protein-coupled ion channel pathways. FASEB J 5:2175–2179

Burch RM, Luini A, Axelrod J (1986) Phospholipase A2 and phospholipase C are activated by distinct GTP-binding proteins in response to $\alpha_1$-adrenergic stimulation in FRTL5 thyroid cells. Proc Natl Acad Sci USA 83:7201–7205

Burch RM, Jelsema C, Axelrod J (1988) Cholera toxin and pertussis toxin stimulate prostaglandin E2 synthesis in a murine macrophage cell line. J Pharmacol Exp Ther 244:765–773

Burch RM (1989) G Protein regulation of phospholipase A2. Mol Neurobiol 3:155–171

Burch RM (1990) G protein regulation of phospholipase A2: partial reconstitution of the system in cells. In: Mukherjee AB (ed) Biochemistry, molecular biology, and physiology of phospholipase A2 and its regulatory factors. Plenum Press. New York and London, pp 185–195

Burgoyne RD, Morgan A (1990) The control of free arachidonic acid levels. Trends Biochem Sci 15:365–366

Cantiello HF, Patenaude CR, Codina J, Birnbaumer L, Ausiello DA (1990) Gαi-3 regulates epithelial $Na^+$ channels by activation of phospholipase A2 and lipoxygenase pathways. J Biol Chem 265:21624–21628

Channon JY, Leslie CC (1990) A calcium-dependent mechanism for associating a soluble arachidonoyl-hydrolyzing phospholipase A2 with membrane in the macrophage cell line RAW 264.7. J Biol Chem 265:5409–5413

Clark JD, Milona N, Knopf JL (1990) Purification of a 110-kilodalton cytosolic phospholipase A2 from the human monocytic cell line U937. Proc Natl Acad Sci USA 87:7708–7712

Clark JD, Lin LL, Kriz RW, Ramesha CS, Sultzman LA, Lin AY, Milona N, Knopf JL (1991) A novel arachidonic acid-selective cytosolic PLA2 contains a $Ca^{2+}$ dependent translocation domain with homology to PKC and GAP. Cell 65:1043–1051

Cockcroft S, Nielson CP, Stutchfield J (1991) Is phospholipase A2 activation regulated by G-proteins? Biochem Soc Trans 19:333–336

Colletta G, Corda D, Schettini G, Cirafici AM, Kohn LD, Cosinglio E (1988) Adenylate cyclase activity of v-ras-k transformed rat epithelial thyroid cells. FEBS Lett 228:37–41

Corda D, Kohn LD (1986) Role of pertussis toxin sensitive G proteins in the alpha1 adrenergic receptor but not in the thyrotropin receptor mediated activation of membrane phospholipases and iodide fluxes in FRTL-5 thyroid cells. Biochem Biophys Res Comm 141:1000–1006

Corda D, Bizzarri C, Di Girolamo M, Valitutti S, Luini A (1989) G protein-linked receptors in the thyroid. Adv Exp Med Biol 261:245–269

Delahunty TM, Cronin MJ, Linden J (1988) Regulation of GH3-cell function via adenosine A1 receptors. Biochem J 255:69–77

Davidson FF, Dennis EA (1990) Evolutionary relationships and implications for the regulation of phospholipase A2 from snake venom to human secreted forms. J Mol Evol 31:228–238

Dennis EA, Rhee SG, Billah MM, Hannun YA (1991) Role of phospholipases in generating lipid second messengers in signal transduction. FASEB J 5:2068–2077

Di Girolamo M, D'Arcangelo D, Bizzarri C, Corda D (1991) Muscarinic regulation of phospholipase A2 and iodide fluxes in FRTL-5 thyroid cells. Acta Endocrinol 125:192–200

Di Girolamo M, D'Arcangelo D, Cacciamani T, Gierschik P, Corda D (1992) K-ras transformation greatly increases the toxin-dependent ADP-ribosylation of GTP binding proteins in thyroid cells. Involvement of an inhibitor of the ADP-ribosylation reaction. J Biol Chem 267:17397–17403

Doroshenko P (1991) Second messengers mediating activation of chloride current by intracellular GTP$\gamma$S in bovine chromaffin cells. J Physiol 436:725–738

Enjalbert A, Sladeczek F, Guillon G, Bertrand P, Shu C, Epelbaum J, Garcia-Sainz A, Jard S, Lombard C, Kordon C, Bockaert J (1986) Angiotensin II and dopamine modulate both cAMP and inositol phosphate productions in anterior pituitary cells. Involvement in prolactin secretion. J Biol Chem 261:4071–4075

Enjalbert A, Guillon G, Mouillac B, Audinot V, Rasolonjanahary R, Kordon C, Bockaert J (1990) Dual mechanisms of inhibition by dopamine of basal and thyrotropin-releasing hormone-stimulated inositol phosphate production in anterior pituitary cells. Evidence for an inhibition not mediated by voltage-dependent $Ca^{2+}$ channels. J Biol Chem 265:18816–18822

Fain JN (1990) Regulation of phosphoinositide-specific phospholipase C. Biochim Biophys Acta 1053:81–88

Godfrey PP, Watson SP (1988) Fluoride inhibits agonist induced formation of inositol phospates in rat cortex. Biochem Biophys Res Comm 155:664–669

Gronich JH, Bonventre JV, Nemenoff RA (1990) Purification of a high-molecular-mass-form of phospholipase A2 from rat kidney activated at physiological calcium concentrations. Biochem J 271:37–43

Guillon G, Mouillac B, Savage AL (1992) Modulation of hormone-sensitive phospholipase C. Cell Signal 4:11–23

Gupta SK, Diez E, Heasley LE, Osawa S, Johnson GL (1990) A G protein mutant that inhibits thrombin and purinergic receptor activation of phospholipase A2. Science 249:662–666

Horn VJ, Baum BJ, Ambudkar IS (1990) Attenuation of inositol trisphosphate generation and cytosolic $Ca^{2+}$ elevation in dispersed rat parotid acini stimulated simultaneously at muscarinic and $\alpha_1$-adrenergic receptors. Biochem Biophys Res Comm 166:967–972

Huang NN, Wang DJ, Gonzalez F, Heppel LA (1991) Multiple signal transduction pathways lead to extracellular ATP-stimulated mitogenesis in mammalian cells. II. A pathway involving arachidonic acid release, prostaglandin synthesis, and cyclic AMP accumulation. J Cell Physiol 146:483–494

Jelsema CL (1987) Light activation of phospholipase A2 in rod outer segments of bovine retina and its modulation by GTP-binding proteins. J Biol Chem 262:163–168

Jelsema CL, Axelrod J (1987) Stimulation of phospholipase A2 activity in bovine rod outer segments by the $\beta\gamma$ subunits of transducin and its inhibition by the $\alpha$ subunit. Proc Natl Acad Sci USA 84:3623–3627

Jelsema CL, Burch RM, Jaken S, Ma AD, Axelrod J (1989) Modulation of phospholipase A2 activity in rod outer segments of bovine retina by G protein subunits, guanine nucleotides, protein kinases, and calpactin. In Redburn DA, Pasantes-Morales H (eds) Extracellular and intracellular second messengers in the vertebrate retina, vol 49, Neurology and neurobiology, Alan R. Liss, New York, pp 25–46

Kast R, Fürstenberger G, Marks F (1991) Activation of a keratinocyte phospholipase A2 by bradykinin and 4$\beta$-phorbol 12-myristate 13-acetate. Eur J Biochem 202:941–950

Kim D, Lewis DL, Graziadei L, Neer EJ, Bar-Sagi D, Clapham DE (1989) G-protein $\beta\gamma$-subunits activate the cardiac muscarinic $K^+$-channel via phospholipase A2. Nature 337:557–560

Kramer RM, Roberts EF, Manetta J, Putnam JE (1991) The $Ca^{2+}$-sensitive cytosolic phospholipase A2 is a 100-kDa protein in human monoblast U937 cells. J Biol Chem 266:5268–5272

Kurachi Y, Ito H, Sugimoto T, Shimizu T, Miki I, Ui M (1989) Arachidonic acid metabolites as intracellular modulators of the G protein-gated cardiac $K^+$ channel. Nature 337:555–557

Leslie CC, Voelker DR, Channon JY, Wall MM, Zelarney PT (1988) Properties and purification of an arachidonoyl-hydrolyzing phospholipase A2 from a macrophage cell line, RAW 264.7. Biochim Biophys Acta 963:476–492

Limor R, Schvartz I, Hazum E, Ayalon D, Naor Z (1989) Effect of guanine nucleotides on phospholipase C activity in permeabilized pituitary cells: possible involvement of an inhibitory GTP-binding protein. Biochem Biophys Res Comm 159:209–215

Linden J, Delahunty TM (1989) Receptors that inhibit phosphoinositide breakdown. Trends Pharmacol Sci 10:114–120

Litosch I (1989) Guanine nucleotides mediate stimulatory and inhibitory effects on cerebral-cortical membrane phospholipase C activity. Biochem J 261:245–251

Lowndes JM, Gupta SK, Osawa S, Johnson GL (1991) GTPase-deficient G$\alpha$i2 oncogene gip2 inhibits adenylyl cyclase and attenuates receptor-stimulated phospholipase A2 activity. J Biol Chem 266:14193–14197

Misawa H, Ueda H, Satoh M (1990) $\kappa$-Opioid agonist inhibits phospholipase C, possibly via an inhibition of G-protein activity. Neurosci Lett 112:324–327

Moriarty TM, Gillo B, Carty DJ, Premont RT, Landau EM, Iyengar R (1988) $\beta\gamma$ Subunits of GTP-binding proteins inhibit muscarinic receptor stimulation of phospholipase C. Proc Natl Acad Sci USA 85:8865–8869

Nakahata N, Abe MT, Matsuoka I, Ono T, Nakanishi H (1991) Adenosine inhibits histamine-induced phosphoinositide hydrolysis mediated via pertussis toxin-sensitive G-protein in human astrocytoma cells. J Neurochem 57:963–969

Nakashima S, Nagata KI, Ueeda K, Nozawa Y (1988) Stimulation of arachidonic acid release by guanine nucleotide in saponin-permeabilized neutrophils: evidence for involvement of GTP-binding protein in phospholipase A2 activation. Arch Biochem Biophys 261:375–383

Narasimhan V, Holowka D, Baird B (1990) A guanine nucleotide-binding protein participates in IgE receptor-mediated activation of endogenous and reconstituted phospholipase A2 in a permeabilized cell system. J Biol Chem 264:1459–1464

Okabe K, Yatani A, Evans T, Ho YK, Codina J, Birnbaumer L, Brown AM (1990) $\beta\gamma$ Dimers of G proteins inhibit atrial muscarinic $K^+$ channels. J Biol Chem 265:12854–12858

Price BD, Morris JDH, Marshall CJ, Hall A (1989) Stimulation of phosphatidylcholine hydrolysis, diacylglycerol release, and arachidonic acid production by oncogenic *ras* is a consequence of protein kinase C activation. J Biol Chem 264:16638–16643

Rehfeldt W, Hass R, Goppelt-Struebe M (1991) Characterization of phospholipase A2 in monocytic cell lines. Functional and biochemical aspects of membrane association. Biochem J 276:631–636

Rhee SG, Suh P-G, Ryu S-H, Lee SY (1989) Studies of inositol phospholipid-specific phospholipase C. Science 244:546–550

Rubin RP, Withiam-Leitch M, Laychock SG (1991) Modulation of phospholipase A2 activity in zymogen granule membranes by GTP$\gamma$(S); evidence for GTP-binding protein regulation. Biochem Biophys Res Comm 177:22–26

Sharp JD, White DL, Chiou XG, Goodson T, Gamboa GC, McClure D, Burgett S, Hoskins J, Skatrud PL, Sprtsman JR, Becker GW, Kang LH, Roberts EF,

Kramer RM (1991) Molecular cloning and expression of human $Ca^{2+}$-sensitive cytosolic phospholipase $A_2$. J Biol Chem 266:14850–14853
Teitelbaum I (1990) The epidermal growth factor receptor is coupled to a phospholipase A2-specific pertussis toxin-inhibitable guanine nucleotide-binding regulatory protein in cultured rat inner medullary collecting tubule cells. J Biol Chem 265:4218–4222
Tremblay NM, Nicholson D, Potier M, Weech PK (1992) Cytosolic phospholipase $A_2$ from U937 cells: size of the functional enzyme by radiation inactivation. Biochem Biophys Res Comm 183:121–127
Valitutti S, Cucchi P, Colletta G, Di Filippo C, Corda D (1991) Transformation by the K-*ras* oncogene correlates with increases in phospholipase A2 activity, glycerophosphoinositol production and phosphoinositide synthesis in thyroid cells. Cell Signal 3:321–332
Vallar L, Vicentini LM, Meldolesi J (1988) Inhibition of inositol phosphate production is a late, $Ca^{2+}$-dependent effect of D2 dopaminergic receptor activation in rat lactotroph cells. J Biol Chem 263:10127–10134
Yatani A, Mattera R, Codina J, Graf R, Okabe K, Padrell E, Iyengar R, Brown AM, Birnbaumer L (1988) The G protein-gated atrial $K^+$ channel is stimulated by three distinct Gi$\alpha$-subunits. Nature 336:680–682
Yatani A, Okabe K, Polakis P, Halenbeck R, McCormick F, Brown AM (1990) *ras* p21 and GAP inhibit coupling of muscarinic receptor to atrial $K^+$ channels. Cell 61:769–776

CHAPTER 69

# G-Protein Regulation of Phospholipase C in the Turkey Erythrocyte

A.J. MORRIS, D.H. MAURICE, G.L. WALDO, J.L. BOYER, and T.K. HARDEN

## A. Introduction

A wide variety of stimuli send signals into target cells by stimulating the hydrolysis of a minor phospholipid component of the inner leaflet of the plasma membrane, phosphatidylinositol 4,5-bisphosphate [PtdIns(4,5)$P_2$]. This reaction is catalyzed by an inositol lipid-specific phospholipase C (PLC) and produces two compounds each with intracellular second messenger functions. The lipid product of this reaction is 1,2-diacylglycerol which activates a $Ca^{2+}$ and phospholipid-dependent protein kinase, protein kinase C. D-Myoinositol 1,4,5,-trisphosphate [Ins(1,4,5)$P_3$] is a water-soluble product of this reaction that releases $Ca^{2+}$ ions from stores within the cell (BERRIDGE and IRVINE 1987; KIKKAWA et al. 1989).

Understanding the molecular mechanisms by which cell surface receptors increase the activity of PLC in target cells has been the subject of intense experimental investigation in recent years. PLC-coupled receptors fall into two broad classes. The growth factors epidermal growth factor (EGF) and platelet-derived growth factor (PDGF) both increase inositol lipid hydrolysis in target cells. Receptors for these growth factors are single polypeptides that span the plasma membrane once. Their intracellular domain expresses a protein tyrosine kinase activity that is stimulated by ligand binding to the extracellular domain (YARDEN and ULLRICH 1988), and both in vitro and in intact cells these growth factor receptors associate with and catalyze the phosphorylation of a specific isoenzyme of PLC, PLC-$\gamma$ (WAHL et al. 1989; MEISENHELDER et al. 1989). Several lines of evidence including studies with site-directed mutants of PLC-$\gamma$ suggest that the tyrosine phosphorylation of PLC-$\gamma$ is responsible for the increased inositol lipid hydrolysis observed in growth factor-stimulated cells (KIM et al. 1991). Additionally, under certain assay conditions, the in vitro catalytic activity of PLC-$\gamma$ has been shown to be increased by tyrosine phosphorylation (NISHIBE et al. 1990), suggesting that this covalent modification may be necessary and sufficient for activation of PLC-$\gamma$ by growth factor receptors (see RHEE 1991a for review). Some evidence for a role of G-proteins in EGF receptor action has also been reported. For example, treatment of primary cultures of rat hepatocytes with pertussis toxin inhibits EGF receptor stimulation of PLC (JOHNSON et al. 1989). In these cells, evidence has been presented for a physical association between EGF receptors, PLC-$\gamma$, and $G\alpha_{i2}$ (Yang et al. 1992), although pertussis toxin does not appear to block the small amount of EGF-promoted tyrosine phosphorylation of PLC-$\gamma$ that occurs (Liang and Garrison 1992). These results raise the possibility that a G-protein $\alpha$ subunit plays some role in growth factor receptor regulation of PLC.

The majority of cell surface receptors known to couple to PLC belong to the family of membrane proteins characterized by their rhodopsinlike arrangement of seven transmembrane-spanning $\alpha$-helices. These receptors employ G-proteins as intermediaries in their regulation of intracellular effectors (O'Dowd et al. 1989). Detailed studies of two prototypical systems, hormone receptor-regulated adenylyl cyclase and rhodopsin-regulated retinal cyclic GMP-phosphodiesterase, have led to a general model for the function of G-proteins as transducers in transmembrane signaling. G-proteins are heterotrimers with an $\alpha$, $\beta$, $\gamma$ subunit structure. The $\alpha$ subunit binds GTP in a manner that is promoted by association with an appropriate activated hormone receptor. The $\alpha$ subunit has an intrinsic GTPase activity and is thought to be responsible for most of the effector functions of these proteins (Gilman 1987).

In 1985 two groups published results that suggested a role for G-proteins in receptor-mediated regulation of PLC (Litosch et al. 1985; Cockcroft and Gomperts 1985). These workers observed that hormonal increases in PLC activity in broken cell systems required GTP. Nonhydrolyzable analogues of GTP could activate PLC in the absence of hormones, suggesting the involvement of a GTP hydrolysis step terminating the catalytic cycle of PLC. Subsequent work from many laboratories using membranes from different target tissues established a general phenomenology of G-protein-dependent regulation of PLC for a broad range of receptors for extracellular stimuli. These results suggested that a specific G-protein or family of such proteins with properties analagous to those of the stimulatory G-protein, $G\alpha_s$, of the adenylyl cyclase system or to the retinal G-protein transducin served to couple hormone receptors to PLC (see Harden 1992 for review). Identification of these G-protein $\alpha$ subunits and the PLC isoenzymes that they regulate has only recently been accomplished.

Turkey erythrocytes provided an excellent system in which to investigate G-protein-dependent regulation of adenylyl cyclase by the $\beta$-adrenergic receptor. Our laboratory also has found turkey erythrocytes to be a useful model in which to investigate G-protein-dependent regulation of PLC. Turkey erythrocytes express a PLC activity that is regulated by G-protein-coupled $P_{2Y}$ purinergic receptors, and ghosts prepared by hypotonic lysis of these cells retain the capacity for $P_{2Y}$ purinergic receptor and G-protein regulation of PLC. Turkey erythrocytes possess several other properties that make them ideal for studies of receptor regulation of inositol lipid hydrolysis. These cells are relatively simple, containing a large nucleus, some mitochondria and few other intracellular organelles visible under the electron microscope and de novo synthesis of phosphatidylinositol, and its phosphorylation of PtdIns(4,5)$P_2$ actively occurs in these cells. Thus, [$^3$H]inositol can be used as a tracer with which to label PLC substrates in intact cells. The simplicity of these cells allows them to be readily fractionated by hypotonic lysis and centrifugation into ghosts which under the electron microscope appear as a "cytosol-less cell" with the plasma membrane collapsed around the nucleus. Cytosolic, membrane, and nuclear fractions can be prepared by nitrogen cavitation of intact turkey erythrocytes followed by differential centrifugation. The quantities in which these cells can be obtained in a homogeneous state provide an ideal starting material from which to purify the relatively rare protein components of the inositol lipid-dependent signaling system. Over the past 2 years we have isolated both the G-protein-regulated PLC and its regulatory G-protein from turkey erythrocytes in sufficient quantities to permit detailed characterizations of their biochemical properties and partial determinations of their amino acid sequences. This review considers the biochemical and structural properties of the protein components of the receptor-regulated inositol lipid signaling system in turkey erythrocytes and reviews what we have learned of their interactions in both native membranes and artificial assay systems.

## B. Properties of $P_{2Y}$ Purinergic Receptor and G-Protein-Regulated PLC in Turkey Erythrocytes

The ease with which PLC substrates can be labeled to high specific activity in turkey erythrocytes has allowed us to make a detailed description of the properties of this receptor and G-protein-regulated PLC of these cells. These studies include investigations of the $Ca^{2+}$ dependence, substrate specificity, and kinetics of activation of this PLC by $P_{2Y}$ purinergic receptor agonists and guanine nucleotides.

### I. Initial Observations

Our initial descriptions of G-protein-mediated inositol lipid hydrolysis in turkey erythrocyte ghosts employed [$^{32}$P]$PO_4$ to label polyphosphoinositide substrates in intact cells. In contrast to ghosts prepared by hypotonic lysis of human and rabbit erythrocytes, turkey erythrocyte ghosts expressed a PLC activity that was stimulated by aluminum fluoride and GTP$\gamma$S. These studies employed a PLC assay mixture that contained ATP in an attempt to sustain phosphorylation of phosphatidylinositol to phosphatidylinositol 4-phosphate (PtdIns4P) and PtdIns(4,5)$P_2$, hopefully producing an amplified PLC activity that was not limited by substrate supply. Using this assay system, we observed concentration-dependent activation of PLC by GTP$\gamma$S, GppNHp, and to a lesser extent GTP. Stimulation by guanine nucleotides was blocked by GDP$\beta$S (HARDEN et al. 1987). We subsequently observed that ATP and certain of its analogues stimulated PLC in intact turkey erythrocytes. This effect was apparently mediated by a $P_{2Y}$ purinergic receptor (BERRIE et al. 1989). A reassessment of the effects of ATP on guanine nucleotide-stimulated PLC activity in [$^3$H]inositol-labeled turkey erythrocyte ghosts supported the idea that these cells expressed a $P_{2Y}$ purinergic receptor that coupled to PLC in a guanine nucleotide-dependent manner. Thus, ATP and a $P_{2Y}$ purinergic receptor-selective agonist, 2-methylthioadenosine triphosphate, greatly increased PLC activity observed in the presence of low concentrations of GTP$\gamma$S, assessed by both measurements of inositol phosphate release and inositol lipid breakdown. Receptor- and guanine nucleotide-dependent stimulation of PLC in turkey erythrocyte ghosts was unaltered by prior treatment of intact cells with cholera and pertussis toxins. We also established that, although dependent on $Ca^{2+}$ for activity, the turkey erythrocyte PLC was not a $Ca^{2+}$-controlled enzyme. PLC activity measured under both basal conditions, and with submaximal or maximal stimulation by GTP$\gamma$S occurred with similar $Ca^{2+}$ dependence (HARDEN et al. 1988). We attempted to assess the selectivity of the receptor and G-protein-regulated PLC for the three inositol lipids. Inositol phosphates generated by the PLC were subjected to anion-exchange HPLC analysis. No D-myoinositol 1-monophosphate (Ins1P) was detected, indicating that this G-protein-regulated PLC does not hydrolyze phosphatidylinositol. Determination of the selectivity of the PLC for the polyphosphoinositides PtdIns4P and PtdIns(4,5)$P_2$ was complicated by the observation that the turkey erythrocyte ghost preparation contained both Ins(1,4,5)$P_3$ 3-kinase and Ins(1,4,5)$P_3$ 5-phosphatase activities (MORRIS et al. 1987). High concentrations of unlabeled Ins(1,4,5)$P_3$ were used to isotopically dilute [$^3$H]Ins(1,4,5)$P_3$ released during activation of PLC and to demonstrate that the primary substrate of this enzyme was PtdIns(4,5)$P_2$ although a small amount of D-myoinositol 1,4-bisphosphate [Ins(1,4)$P_2$] appeared to be produced directly by PLC-catalyzed hydrolysis of PtdIns4P (HARDEN et al. 1988).

## II. Kinetics of Activation of PLC by $P_{2Y}$ Purinergic Receptor Agonists and Guanine Nucleotides

BOYER et al. (1989) undertook a detailed examination of the kinetics of activation of PLC by $P_{2Y}$ purinergic receptor agonists and guanine nucleotides. The capacity of agonists to stimulate PLC activity was completely dependent on the presence of guanine nucleotide. GTP$\gamma$S-mediated stimulation of PLC occurred with a considerable time lag, and the rate of activation of PLC by GTP$\gamma$S was increased by $P_{2Y}$ purinergic receptor agonists in a concentration-dependent manner. At a fixed agonist concentration, the rate of activation was independent of guanine nucleotide concentration. GDP$\beta$S competitively inhibited activation of PLC by $P_{2Y}$ receptor agonists and GTP$\gamma$S. Addition of GDP$\beta$S to ghosts in which PLC was preactivated by $P_{2Y}$ receptor agonists and GTP restored PLC activity to basal levels. These observations were similar to those made for G-protein-dependent regulation of adenylyl cyclase in both turkey erythrocyte membranes and other tissues.

The turkey erythrocyte ghost system was also used to examine the effects of G-protein $\beta\gamma$ subunits on receptor and G-protein-regulated PLC activity. Boyer et al. (1990) observed that reconstitution of purified $\beta\gamma$ subunits with turkey erythrocyte ghosts caused a $\beta\gamma$ subunit concentration-dependent inhibition of both aluminum fluoride stimulation of PLC activity and of adenylyl cyclase. These results were taken to support the idea that a heterotrimeric G-protein was involved in the activation of PLC.

## C. Identification, Purification, and Primary Structure of the Protein Components of the Turkey Erythrocyte Inositol Lipid-Dependent Signaling System

The receptor-coupled isoenzyme of PLC and its regulatory G-protein have been purified from turkey erythrocytes and their primary structures determined by peptide sequencing and cDNA cloning. The following summarizes our findings and makes a particular attempt to relate them to the recent advances in identification of PLC isoenzymes and their regulatory G-proteins in mammalian systems.

### I. G-Protein-Regulated PLC

Mammalian PLCs are a diverse group of proteins ranging in molecular weight from 56 to 150 kDa. Four defined classes of PLC isoenzymes exist, and current terminology identifies these as PLCs $\alpha$, $\beta$, $\gamma$, and $\delta$. Multiple isoenzymes of PLC appear to exist within each class (reviewed by Rhee et al. 1989; Rhee 1991a). Reports of the isolation of these proteins appeared over a period of years from several different groups working with a variety of tissues as sources and using PLC assays that employed artificial substrate preparations (most commonly detergent and phospholipid micelles). Although their biochemical properties and, to some extent, structural relationships were examined in detail, none of these purified PLC isoenzymes was demonstrated to participate in cell surface receptor-regulated inositol lipid hydrolysis. Consequently, two intimately related questions faced workers seeking to understand the biochemical basis of G-protein regulation of PLC: which G-protein(s) regulates PLC, and which PLC isoenzyme(s) is subject to regulation? We wanted to devise an assay in which the capacity of G-protein-containing detergent extracts of turkey erythrocyte membrane proteins to activate PLC could be determined in the hope that such an assay could be employed to purify the G-protein to homogeneity. Attempts to achieve selective functional inactivation of the G-protein in the turkey erythrocyte membrane preparation, for example, by bacterial toxin treatment or by gentle alkylation, were not successful. Therefore, we focused our attention on the PLC component of the signaling system with the idea that identification of this protein would provide a necessary reagent for identification of its regulatory G-protein.

## 1. Purification and Properties of a G-Protein-Regulated PLC from Turkey Erythrocytes

Washed packed turkey erythrocytes were disrupted by nitrogen cavitation and cytosolic, plasma membrane, and nuclear fractions prepared by differential centrifugation. These fractions were assayed for PLC activity using a substrate preparation of PtdIns(4,5)$P_2$ presented to the enzyme as a component of mixed phospholipid and cholate micelles. Almost all of the cellular PLC activity measured in this way was associated with the cytosolic fraction, and by repeated washing it was possible to prepare a membrane fraction with less than 0.1% of total PLC activity. We purified the cytosolic PLC of turkey erythrocytes to homogeneity using a procedure that employed sequential ammonium sulfate precipitation and chromatography on Q-Sepharose, hydroxylapatite, heparin-Sepharose, Sephacryl S-300 gel filtration, and Mono-Q FPLC. The purified protein was a single polypeptide of molecular mass 150 kDa. The PLC was dependent on $Ca^{2+}$ for activity and very selective for the inositol lipids. In assays employing pure inositol lipid substrates the turkey erythrocyte PLC showed a marked preference for the polyphosphoinositides PtdIns4P and PtdIns(4,5)$P_2$ over phosphatidylinositol (MORRIS et al. 1990a).

PLC-$\beta_1$ and PLC-$\gamma_1$ are two mammalian PLC isoenzymes of similar molecular mass to the turkey erythrocyte PLC. However, both polyclonal and monoclonal antisera raised against these mammalian proteins failed to react with the turkey erythrocyte PLC on western blots. Of four preparations of polyclonal antisera raised against the turkey erythrocyte PLC three did not react with PLC-$\beta_1$ and PLC-$\gamma_1$ on western blots, but one weakly recognized PLC-$\beta_1$. Antipeptide antisera raised against sequences present in the so-called X and Y domains common to PLC-$\beta$, -$\gamma$, and -$\delta$ also reacted with the turkey erythrocyte PLC on western blots. Direct determination of the amino acid sequence of peptides derived from a tryptic digest supported these findings and suggested that the turkey erythrocyte PLC shared regions of homology with mammalian PLC-$\beta$, -$\gamma$, and -$\delta$, and that it belonged to the PLC-$\beta$ class of PLC isoenzymes. Thus, sequences were obtained with pronounced homology to regions within the X and Y domains, and one peptide was sequenced that was homologous to a sequence present in the COOH-terminus of PLC-$\beta_1$. No cognate sequence is present in the PLC-$\gamma$ and PLC-$\delta$ isoenzymes (WALDO et al. 1991). KRIZ et al. (1989) have discussed the cDNA cloning of two additional isoenzymes of the PLC-$\beta$ class, PLC-$\beta_2$ and PLC-$\beta_3$. Unfortunately, the limited sequence data available for the turkey erythrocyte PLC do not allow definitive subclassification of the turkey erythrocyte PLC within the PLC-$\beta$ family of isoenzymes although, based on the restricted tissue distribution of PLC-$\beta_1$ (RHEE et al. 1991b), it would seem unlikely that the turkey erythrocyte PLC is closely related to this PLC isoenzyme. We have recently cloned and partially sequenced a cDNA from a turkey brain cDNA library that encodes a protein with 96% sequence similarity to bovine brain PLC-$\beta_1$ and that presumably represents the avian homologue of PLC-$\beta_1$ isoenzyme (A.J. MORRIS, unpublished data). The cDNA sequence of this protein does not encode any of the peptides sequenced from the turkey erythrocyte PLC. Certain, but not all of the peptides sequenced show a provocative similarity to the cDNA sequence of human PLC-$\beta_2$ (J. KNOPF and R. KRIZ, personnal communication). Our finding that the 150-kDa turkey erythrocyte PLC is a G-protein-regulated form of PLC (MORRIS et al. 1990b) and the subsequent observation that in mammalian systems PLC-$\beta_1$ is specifically regulated by members of the $G_q$ class of G-protein $\alpha$ subunits (TAYLOR et al. 1991) further supports the idea that the turkey erythrocyte PLC is a member of the PLC-$\beta$ subfamily.

## 2. Receptor and G-Protein Regulation of the Purified Turkey Erythrocyte PLC

We have demonstrated that the 150-kDa turkey erythrocyte PLC is a G-protein-regulated enzyme by recombination of the purified protein with preparations functionally deficient in PLC activity. These preparations have included turkey erythrocyte ghosts exposed to low concentrations of $Mg^{2+}$ during their preparation, EDTA-washed turkey erythrocyte plasma membranes, and artificially prepared phospholipid vesicles into which the regulatory G-protein was introduced from detergent solution (see Sect. C.2).

Although the turkey erythrocyte ghosts exhibit a highly responsive G-protein-regulated PLC activity, exposure of these ghosts to low concentrations of $Mg^{2+}$ during preparation results in much reduced PLC responsiveness to guanine nucleotides and $P_{2Y}$ receptor agonists. The biochemical basis of this lesion appears to involve a physical dissociation of PLC from the membranes although the reqirement for $Mg^{2+}$ to maintain stability of the nuclei in these ghosts makes it difficult to examine this phenomenon in detail. Addition of native but not heat-denatured PLC to these unresponsive ghosts restored agonist and guanine nucleotide-stimulated inositol lipid hydrolysis. This effect of the PLC was concentration dependent and saturable. With 10 $\mu$g ghost protein a half-maximal effect was observed at around 5 ng PLC, and maximal effects were attained with 100 ng PLC. Reconstitution of the purified PLC resulted in a marked increase in the maximal rate obtained but did not alter the rate of activation of inositol lipid hydrolysis by $P_{2Y}$ purinergic receptor agonists in the presence of nonhydrolyzable guanine nucleotides. The $Ca^{2+}$ dependence and substrate selectivity of the purified reconstituted PLC was identical to that of the native PLC activity of turkey erythrocyte ghosts. During the latter stages of our purification scheme we observed that PLC activity determined using an exogenous substrate assay cochromatographed with G-protein-regulated PLC activity measured using the reconstitution assay. From these results we conclude that the 150-kDa PLC is the major and perhaps the sole G-protein-regulated form of PLC expressed in turkey erythrocytes (Morris et al. 1990b).

We have also observed receptor and G-protein regulation of the purified PLC in reconstitution systems employing turkey erythrocyte plasma membranes prepared by nitrogen cavitation and differential centrifugation of [$^3$H]inositol-prelabeled turkey erythrocytes. The results obtained were qualitatively similar to those found for regulation of the PLC in the ghost reconstitution system; PLC activity was stimulated by $P_{2Y}$ purinergic agonists in combination with guanine nucleotides, by nonhydrolyzable GTP analogues, and by aluminum fluoride (Waldo et al. 1991).

## II. G-Protein Activators of PLC

Although the involvement of G-proteins in receptor regulation of phospholipase C was generally accepted in 1985, the identification of these G-protein(s) was not achieved until 1991. The major reason for this difficulty was an inability to design appropriate assays that could be used to follow activity of PLC-activating proteins during chromatographic purification from detergent extracts of plasma membranes. Ultimately, the identification of PLC-activating G-proteins was achieved by several laboratories using a combination of fortuitously effective and more directed protein purification strategies. These proteins were shown to be encoded by cDNAs encoding G-protein $\alpha$ subunits isolated without knowledge of their function.

Simon and coworkers reported the sequences of four members of a new class of G-protein $\alpha$ subunits collectively termed $G_q$ by amplification of mouse brain cDNA using the polymerase chain reaction and primers based on sequences conserved in previously cloned G-protein $\alpha$ subunits (Strathmann et al. 1989; Strathmann and Simon 1990; Wilkie et al. 1991; Amatruda et al. 1991). These $\alpha$ subunits lack sites for pertussis toxin-catalyzed ADP-ribosylation. Members of the $G_q$ family of G-protein $\alpha$ subunits were expressed in a wide variety of tissues although there were some specific differences in patterns of expression of the individual members of this class (see Simon et al. 1991 for review). Nakamura et al. (1991) reported the isolation of cDNAs encoding proteins which are apparently $\alpha_{11}$ and $\alpha_{14}$ from bovine liver cDNA libraries.

Taylor et al. (1990) described the purification of a 42-kDa protein from GTP$\gamma$S-pretreated bovine liver membranes that increased the activity of a partially purified PLC isolated from the same tissue preparation. Reaction with an antiserum raised against a sequence conserved amongst G-protein $\alpha$ subunits suggested that this protein was a G-protein $\alpha$ subunit although it failed to react with a battery of G-protein $\alpha$ subunit selective antisera. PANG and STERNWEIS (1990) devised an affinity matrix composed of agarose-immobilized G-protein $\beta\gamma$ subunits. Passage of detergent extracts of bovine brain membranes through a column of this resin resulted in selective retention of a 42-kDa protein that could be eluted with the G-protein activator, aluminum fluoride. Direct sequencing of peptides isolated from a tryptic digest of this protein preparation suggested that it was primarily composed of $\alpha_q$ (with small amounts of $\alpha_{11}$). Reactivity of the purified protein with an antiserum X-384 raised against sequences common to $\alpha_q$ and $\alpha_{11}$ supported this idea. The $\alpha_q$ purified in this manner did not bind GTP$\gamma$S and did not exhibit measurable GTPase activity. SMRCKA et al. (1991) reported that this preparation of $\alpha_q$ increased the activity of a partially purified preparation of bovine brain phospholipase C. The bovine liver PLC activator preparation was subsequently reported to contain proteins of molecular weights 42 and 43 kDa, which based on their immunoreactivity with selective antipeptide antisera were proposed to represent $\alpha_q$ and $\alpha_{11}$. Both resolved and unresolved preparations of these two proteins activated purified bovine brain PLC-$\beta_1$ with apparently equal effectiveness while having no effect on the activity of PLC-$\gamma$ and PLC-$\delta$ (TAYLOR and EXTON 1991; TAYLOR et al. 1991). BLANK et al. (1991) purified $\alpha_q$ and $\alpha_{11}$ from bovine liver membranes in their heterotrimeric form using nonactivating conditions. Aluminum fluoride-dependent stimulation of PLC by these $\alpha_q$ and $\alpha_{11}$ preparations was inhibited by excesses of G-protein $\beta\gamma$ subunit. Recombinantly expressed $\alpha_q$ and $\alpha_{11}$ have also been shown to activate PLC-$\beta_1$ (WU et al. 1992).

The interaction of members of the $G_q$ family with hormone receptors has been studied in less detail. WANGE et al. (1991) found that angiotensin specifically stimulated the incorporation of a photoreactive analogue of GTP into proteins of molecular weight 42 and 43 kDa in rat liver plasma membranes. These proteins were immunoprecipitated by antiserum X-384 raised against the common amino terminus of $\alpha_q$ and $\alpha_{11}$. BERNSTEIN et al. (1992) found that upon reconstitution with recombinant muscarinic cholinergic receptors in phospholipid vesicles, carbachol promoted GTP$\gamma$S binding to $\alpha_q$ and $\alpha_{11}$. Upon extraction from the phospholipid vesicle preparation GTP$\gamma$S-liganded $\alpha_q$ and $\alpha_{11}$ increased the activity of PLC-$\beta_1$.

### 1. Purification and Properties of the Turkey Erythrocyte PLC-Activating G-Protein

Isolation of the G-protein-regulated form of PLC from turkey erythrocytes placed us in a position to use this enzyme in a reconstitution assay to measure activity of its regulatory G-protein. Proteins were extracted from turkey erythrocyte plasma membranes with Na cholate, and this extract was combined with phosphatidylserine, phosphatidylethanolamine, and lesser amounts of [$^3$H]PtdIns4P or [$^3$H]PtdIns(4,5)$P_2$. This mixture was passed over a small Sephadex G-50 gel filtration column, and material excluded from the resin was collected. This fraction which appears to contain small unilamellar phospholipid vesicles was combined with the purified PLC in an appropriate assay mixture, and PLC activity directed against the phosphoinositide substrates was determined. The PLC exhibited a basal activity under these conditions. GTP$\gamma$S consistently produced a small increase in PLC activity that was inhibited by GDP$\beta$S. Aluminum fluoride markedly stimulated PLC activity when added to the assay mixture. This effect was concentration dependent and was mediated by a heat-labile component of the turkey erythrocyte plasma membrane extract. Using the capacity to confer aluminum fluoride sensitivity to PLC as an assay, WALDO et al. (1991) purified the protein responsible for this activity. The preparation consists primarily of a 43-kDa protein with lesser amounts of a 35-kDa protein that, based on its mobility on SDS polyacrlylamide gel electrophoresis and its immunoreactivity with selective antisera is a G-protein $\beta$ subunit. The capacity of this purified preparation to confer aluminum fluoride sensitivity to the purified PLC was inhibited by G-protein $\beta\gamma$ subunits in a concentration-dependent manner. The 43-kDa PLC-activator-containing preparation increased the activity of purified PLC-$\beta_1$ in an aluminum fluoride-dependent manner and with a similar potency to that with which it stimulated the turkey erythrocyte PLC. Higher concentrations of $\beta\gamma$ subunits increase the activity of the turkey erythrocyte PLC while having little effect on the activity of PLC-$\beta_1$. This observation strongly suggests that direct activation of PLC by G-protein $\beta\gamma$-subunits is also an important regulatory mechanism (BOYER et al. 1992).

### 2. cDNA Sequence of the Turkey Erythrocyte PLC-Activating G-protein and Its Relationship to Mammalian G-Protein $\alpha$ Subunits

The 43-kDa protein has been identified as a member of the $G_q$ family of G-protein $\alpha$ subunits by its reactivity with certain sequence-directed antisera, by determination of portions of its internal amino acid sequence, and by cloning a cDNA that precisely encodes a protein containing these sequences (MAURICE et al. 1992). The purified protein reacted strongly with two preparations of antisera raised against the carboxyl terminal 12 amino acids conserved between the sequences of $\alpha_q$ and $\alpha_{11}$. An antiserum W082 generated against a sequence in $\alpha_q$ that is not well conserved in $\alpha_{11}$, and which selectively recognizes $\alpha_q$ (PANG and STERNWEIS 1990; GUTOWSKI et al. 1991) reacted only weakly with the 43-kDa protein. By contrast, another antiserum E976 raised against a sequence of $\alpha_{11}$ that is not well conserved in $\alpha_q$ (TAYLOR and EXTON 1991) reacted strongly with the 43-kDa protein. These results suggested a relationship of the turkey erythrocyte PLC-activating G-protein to the $G_q$ family of G-protein $\alpha$ subunits. Amino acid sequences determined from peptides isolated from a tryptic digest of the

43-kDa protein supported this idea. Thus, sequences were obtained that exhibited an obvious similarity to sequences deduced from the $\alpha_q$, $\alpha_{11}$ and $\alpha_{14}$ cDNAs isolated from mouse brain by Simon and colleagues (STRATHMANN et al. 1990) and from bovine liver (NAKAMURA et al. 1991). Unambiguous identification of the 43-kDa protein was achieved by cloning a cDNA that precisely encodes eight of these peptide sequences. The cDNA contains an open reading frame that encodes a protein of 359 amino acids. This protein displays the greatest sequence similarity to $\alpha_{11}$ (96%–98%) and we concluded that the 43-kDa PLC activator purified from turkey erythrocytes represents a turkey homologue of mammalian $\alpha_{11}$. The turkey $\alpha_{11}$ cDNA has been expressed recombinantly in sf9 insect cells using a baculovirus vector. Detergent extracts of membranes from cells infected with recombinant baculovirus contained a protein with strong immunoreactivity against the antisera 118 and E976 discussed above, and these extracts, but not those from uninfected cells or cells infected with virus containing an irrelevant construct, conferred aluminum fluoride sensitivity to the turkey erythrocyte PLC in a reconstitution assay.

## D. Concluding Comments

The identification of the component proteins of the G-protein-regulated PLC raises some important questions for further investigation. Of the identified PLC isoenzymes only PLC-$\beta_1$ has been demonstrated to be G-protein regulated. The tissue distribution of this PLC isoenzyme is restricted to the nervous system yet G-protein-regulation of PLC is a widespread phenomenon. This raises the possibility that additional G-protein-regulated forms of PLC exist, and the additional members of the PLC-$\beta$ family, PLCs-$\beta_2$ and -$\beta_3$, are obviously good candidates for this function. Similarly, of the $G_q$ family, only $\alpha_q$ and $\alpha_{11}$ have been directly implicated in the regulation of PLC. Current concepts of structure and function in G-protein $\alpha$ subunits suggest that specificity in effector coupling resides in the carboxyl terminal portions of the molecule, making it inviting to speculate that the remaining members of the $G_q$ family are also involved in receptor regulation of PLC. It will be important to determine the specificity of interaction between these G-protein-regulated PLCs and the members of the $G_q$ family of G-protein $\alpha$ subunits.

In certain cell types, particularly those of myeloid origin, G-protein regulation of PLC is inhibited to some extent by prior exposure to pertussis toxin. Since the members of the $G_q$ family are not substrates for ADP-ribosylation catalyzed by this toxin, they would not appear to be involved in this process. Possibly one of the known pertussis toxin G-protein $\alpha$ subunits, i.e., $\alpha_o$ and $\alpha_i$, are involved directly in regulation of an as yet unidentified PLC isoenzyme. It should be noted that these two pertussis toxin substrate G-protein $\alpha$ subunits do not increase the activity of PLC-$\beta_1$ under conditions where $\alpha_q$ was found to stimulate activity of this protein (SMRCKA et al. 1991;

Taylor et al. 1991). Our recent work strongly suggests that G-protein $\beta\gamma$ subunits mediate pertussis toxin-sensitive stimulation of PLC (Boyer et al. 1992).

As demonstrated by Bernstein et al. (1992), the $G_q$/PLC system offers some unique opportunities to study G-protein effector coupling. It will be particularly interesting to determine the mechanisms by which these $G_q$ $\alpha$ subunits increase PLC activity. PLC is only one of a number of lipolytic enzymes involved in cellular signal transduction; thus understanding of the biochemical basis for receptor regulation of this enzyme may provide important information about the ways in which arachidonic acid-releasing phospholipase $A_2$ and phosphatidylcholine-specific phospholipases C and D are controlled.

Finally, the interactions between the protein components of the G-protein-regulated PLC system are impinged upon by other signal transduction processes. One such example is the process of agonist-induced desensitization, which in the case of the adenylyl cyclase system appears to be mediated primarily by phosphorylation of its component proteins. Turkey erythrocytes offer a readily available and homogeneous source from which to obtain defined preparations of these signaling proteins. These cells will continue to be of value in studies of the inositol lipid-dependent signaling system in the future.

*Acknowledgements*. Work in our laboratory has been supported by United States Public Health Service Grants GM29536 and GM38213, American Heart Association Grant-In-Aid 91015240, a Patrick Mitchell Fellowship Award from the American Heart Association North Carolina Affiliate (A.J.M), and a Medical Research Council of Canada Fellowship (D.H.M).

## References

Amatruda T, Steele D, Slepak V, Simon M (1991) $G\alpha_{16}$, a G-protein $\alpha$-subunit specifically expressed in hematopoietic cells. Proc Natl Acad Sci USA 88:5587–5591

Berridge MJ, Irvine RF (1987) Inositol trisphosphate and diacylglycerol: two interacting second messengers. Ann Rev Biochem 56:159–193

Berrie CP, Hawkins PT, Stephens LR, Harden TK, Downes CP (1989) Phosphatidylinositol 4,5-bisphosphate hydrolysis in turkey erythrocytes is regulated by $P_{2y}$-purinoceptors. Mol Pharmacol 35:526–532

Berstein G, Blank JL, Smrcka AV, Higashijima T, Sternweis PC, Exton JH, Ross EM (1992) Reconstitution of agonist-stimulated phosphatidylinositol 4,5-bisphosphate hydrolysis using purified M1 muscarinic receptors, $G_q/_{11}$, and phospholipase C-$\beta_1$. J Biol Chem 267:8081–8088

Blank J, Ross A, Exton JH (1991) Purification and characterization of two G-proteins that activate the $\beta_1$ isozyme of phosphoinositide-specific phospholipase C. J Biol Chem 266:18206–18216

Boyer J-L, Downes CP, Harden TK (1989) Kinetics of activation of phospholipase C by $P_2$ purinergic receptor agonists and guanine nucleotides. J Biol Chem 264:884–890

Boyer J-L, Waldo GL, Evans T, Northup JK, Downes CP, Harden TK (1989) Modification of $AlF_4^-$- and receptor-stimulated phospholipase C by G-protein $\beta\gamma$-subunits. J Biol Chem 264:13917–13922

Boyer J-L, Waldo GL, Harden TK (1992) $\beta\gamma$-Subunit activation of G-protein-regulated phospholipase C. J Biol Chem 267:25451–25456

Cockcroft S, Gomperts BD (1985) Role of guanine nucleotide binding protein in the activation of polyphosphonositide phosphodiesterase. Nature 314:534–536

Gilman AG (1987) G proteins: transducers of receptor-generated signals. Ann Rev Biochem 56:615–649

Gutowski S, Smrcka A, Nowak L, Wu D, Simon M, Sternweis P (1991) Antibodies to the $\alpha_q$ subfamily of G-protein $\alpha$-subunits attenuate activation of phosphatidylinositol 4,5-bisphosphate hydrolysis by hormones. J Biol Chem 266:20519–20524

Harden TK, Stephens L, Hawkins PT, Downes CP (1987) Turkey erythrocyte membranes as a model for regulation of phospholipase C by guanine nucleotides. J Biol Chem 262:9057–9061

Harden TK, Hawkins PT, Stephens L, Boyer J-L, Downes CP (1988) Phosphoinositide hydrolysis by guanosine 5′-0-(3-thiotriphosphate)-activated phospholipase C of turkey erythrocyte membranes. Biochem J 252:583–593

Harden TK (1992) G-protein-regulated phospholipase C: identification of component proteins. In: Putney JW Jr (ed) Advances in Second Messenger and Phosphoprotein Research. Raven, New York, pp 11–33

Johnson RM, Connelly PA, Sisk RB, Pobiner BF, Hewlett EL, Garrison JC (1986) Pertussis toxin or phorbol 12-myristate 13-acetate can distinguish between epidermal growth factor- and angiotensin-stimulated signals in hepatocytes. Proc Natl Acad Sci USA 83:2032–2036

Kikkawa U, Kishimoto A, Nishizuka Y (1989) The protein kinase C family: heterogeneity and its implications. Ann Rev Biochem 58:31–44

Kim HK, Kim JW, Zilberstein A, Margolis B, Kim JG, Schlessinger J, Rhee SG (1991) PDGF stimulation of inositol phospholipid hydrolysis requires PLC-$\gamma_1$ phosphorylation on tyrosine residues 783 and 1254. Cell 65:435–441

Kriz R, Lin L-L, Sultzman L, Ellis C, Heldin C-H, Pawson T, Knopf J (1989) Phospholipase C isozymes: structural and functional similarities. CIBA Found Symp 150:112–127

Liang M, Garrison J (1991) The epidermal growth factor receptor is coupled to a pertussis toxin-sensitive guanine nucleotide regulatory protein in rat hepatocytes. J Biol Chem 266:13342–13349

Litosch I, Wallis C, Fain JN (1985) 5-Hydroxytryptamine stimulates inositol phosphate production in a cell-free system from blowfly salivary glands. Evidence for a role of GTP in coupling receptor activation to phosphoinositide breakdown. J Biol Chem 260:5464–5471

Maurice DH, Waldo GL, Morris AJ, Nicholas RA, Harden TK (1992) Identification of $G\alpha_{11}$ as the phospholipase C-activating G-protein of turkey erythrocytes. Biochem J 290:765–770

Meisenhelder J, Suh P-G, Rhee SG, Hunter T (1989) Phospholipase C-$\gamma$ is a substrate for the PDGF and EGF receptor protein tyrosine kinases in vivo and in vitro. Cell 57:1099–1122

Morris AJ, Downes CP, Harden TK, Michell RH (1987) Turkey erythrocytes possess a membrane-associated inositol 1,4,5-trisphosphate 3-kinase that is activated by $Ca^{2+}$ in the presence of calmodulin. Biochem J 248:489–493

Morris AJ, Waldo GL, Downes CP, Harden TK (1990a) A receptor and G-protein-regulated polyphosphoinositide specific phospholipase C. I. Purification and properties. J Biol Chem 265:13501–13507

Morris AJ, Waldo GL, Downes CP, Harden TK (1990b) A receptor and G-protein-regulated polyphosphoinositide specific phospholipase C. II. $P_{2Y}$-purinergic receptor and G-protein-mediated regulation of the purified enzyme. J Biol Chem 265:13508–13514

Nakamura F, Ogata K, Shiozaki Kameyama K, Ohara K, Haga T, Nukada T (1991) Identification of two novel GTP-binding protein $\alpha$-subunits that lack apparent ADP-ribosylation sites for pertussis toxin. J Biol Chem 266:12676–12681
Nishibe S, Wahl MI, Hernandez-Sotamayor SMT, Tonks NK, Rhee SG, Carpenter G (1990) Increase of the catalytic activity of phospholipase $C_{\gamma\text{-}1}$, by tyrosine phosphorylation. Science 250:1253–1256
O'Dowd BF, Lefkowitz RJ, Caron MG (1989) Structure of the adrenergic and related receptors. Ann Rev Neurosci 12:67–83
Pang IH, Sternweis PC (1989) Purification of unique $\alpha$-subunits of GTP-binding regulatory proteins by affinity chromatography with immobilized $\beta\gamma$-subunits. J Biol Chem 265:18707–18712
Rhee SG, Suh P-G, Ryu S-H, Lee SY (1989) Studies of inositol-specific phospholipase C. Science 244:546–550
Rhee SG (1991a) Inositol phospholipid-specific phospholipase C: interaction of the $\gamma_1$ isoform with tyrosine kinase. Trends Biochem Sci 16:297–301
Rhee SG, Kim H, Suh P-G, Choi WC (1991b) Multiple forms of phosphoinositide specific phospholipase C and different modes of activation. Biochem Soc Trans 19:337–341
Simon MI, Strathmann MP, Gautam N (1991) Diversity of G proteins in signal transduction. Science 252:802–808
Smrcka AV, Hepler JR, Brown KO, Sternweis PC (1991) Regulation of polyphosphoinositide-specific phospholipase C by purified $G_q$. Science 251:804–807
Strathmann M, Simon MI (1990) G-protein diversity: a distinct class of $\alpha$-subunits is present in vertebrates and invertebrates. Proc Natl Acad Sci USA 87:9113–9117
Strathmann M, Wilkie TM, Simon MI (1989) Diversity of the G-protein family: sequences from five additional $\alpha$-subunits in the mouse. Proc Natl Acad Sci USA 86:7407–7409
Taylor S, Exton JH (1991) Two $\alpha$-subunits of the $G_q$ class of G-protein stimulate phosphoinositide phospholipase C-$\beta_1$ activity. FEBS Lett 286:214–216
Taylor SJ, Smith JA, Exton JH (1990) Purification from bovine liver membranes of a guanine nucleotide-dependent activator of phosphoinositide-specific phospholipase C: immunological identification as a novel G-protein $\alpha$-subunit. J Biol Chem 265:17150–17157
Taylor SJ, Chae HZ, Rhee SG, Exton JH (1991) Activation of the $\beta_1$ isozyme of phospholipase C by $\alpha$-subunits of the $G_q$ class of G-proteins. Nature 350:516–518
Wahl MI, Nishibe S, Suh P-G, Rhee SG, Carpenter G (1989) Epidermal growth factor stimulates tyrosine phosphorylation of phospholipase C-II independently of receptor internalization and extracellular calcium. Proc Natl Acad Sci USA 86:1568–1572
Waldo GL, Boyer J-L, Morris AJ, Harden TK (1991a) Purification of an $AlF_4^-$ and G-protein $\beta\gamma$-subunit-regulated phospholipase C-activating protein. J Biol Chem 261:14217–14225
Waldo GL, Morris AJ, Klapper DG, Harden TK (1991b) Receptor and G-protein regulated 150 kDa avian phospholipase C: inhibition of enzyme activity by isoenzyme specific antisera and nonidentity with mammalian phospholipase C isoenzymes established by immunoreactivity and peptide sequence. Mol Pharmacol 40:480–489
Wange R, Smrcka AV, Sternweis PC, Exton JH (1991) Photoaffinity labeling of two rat liver plasma membrane proteins with [$^{32}$P]-azidoanilido GTP in response to vasopressin. J Biol Chem 266:11409–11412
Wilkie TM, Scherle PA, Strathmann MP, Slepak VZ, Simon MI (1991) Characterization of G-protein $\alpha$-subunits in the $G_q$ class: expression in murine tissues and in stromal and hematopoietic cell lines. Proc Natl Acad Sci USA 88:10043–10049

Wu D, Lee CH, Rhee SG, Simon MI (1992) Activation of phospholipase C by the $\alpha$-subunits of the $G_q$ and $G_{11}$ proteins in transfected Cos-7 cells. J Biol Chem 267:1811–1817

Yang L, Baffy G, Rhee SG, Manning D, Hansen CA, Williamson JR (1991) Pertussis toxin-sensitive $G_i$ protein involvement in epidermal growth factor-induced activation of phospholipase G-$\gamma$ in rat hepatocytes. J Biol Chem 266:22451–22458

Yarden Y, Ullrich A (1988) Growth factor receptor tyrosine kinases. Ann Rev Biochem 57:443–478

CHAPTER 70

# Hormonal Inhibition of Adenylyl Cyclase by $\alpha_i$ and $\beta\gamma$, $\alpha_i$ or $\beta\gamma$, $\alpha_i$ and/or $\beta\gamma$

J.D. HILDEBRANDT

## A. Introduction

All eukaryotic cells use heterotrimeric GTP-binding proteins, G-proteins, in their responses to extracellular signals. Our current understanding of the transduction processes mediated by these proteins began with the discovery that GTP regulates glucagon binding to its receptor (RODBELL et al. 1971b), and that the nucleotide is required for hormone stimulation of adenylyl cyclase (RODBELL et al. 1971a). Those observations led to the discovery of the family of G-proteins and their widespread involvement in many different signal transduction pathways (BIRNBAUMER 1990; GILMAN 1984). The responses of cells to individual signals using G-proteins are probably complex and seem to depend upon a large number of factors related to the diversity of G-proteins within the cell and their mechanism(s) of activation. The identification of these mechanisms has often involved studies of receptor regulation of adenylyl cyclase activity.

Early in the study of the regulation of adenylyl cyclase it was recognized that, while many hormones stimulate the enzyme to synthesis cAMP, other hormones inhibit its catalytic activity (MURAD et al. 1962). The realization that both of these events require GTP (RODBELL 1980) ultimately resulted in the isolation of $G_i$ (BOKOCH et al. 1984; CODINA et al. 1983), and with it the establishment of the G-proteins as a family of signal transduction proteins (GILMAN 1984; BIRNBAUMER et al. 1985). Struggling to understand how G-proteins mediate inhibition of adenylyl cyclase has provided the basis for our understanding of many of the fundamental mechanisms by which G-proteins mediate hormonal effects. Whereas $\alpha_s$ stimulation of adenylyl cyclase has proven to be relatively straightforward (at least so far), hormone inhibition of the enzyme has turned out to be complex, ambiguous, and controversial. Nevertheless, the proposed mechanisms of hormone-mediated inhibition of adenylyl cyclase are so intimately related to the structure, diversity, and mechanism of activation of G-proteins that there are few mechanistic problems more central to our understanding of how these proteins transmit signals from extracellular receptors to intracellular second messenger processes. Understanding how hormones inhibit adenylyl cyclase has provided and promises to continue to provide paradigms that describe how the G-proteins in general interact with one another, with upstream receptors and with downstream effectors.

## B. Mechanism(s) Mediating Inhibition of Adenylyl Cyclase

Almost immediately after the conclusive identification and isolation of $G_i$ there arose a controversy over the molecular mechanism by which the G-proteins inhibit adenylyl cyclase (HILDEBRANDT et al. 1985b). Initially two mechanisms (HILDEBRANDT et al. 1983; KATADA et al. 1984) and subsequently a third (KATADA et al. 1987) were proposed for the G-protein. When, how, and even whether all these mechanisms are relevant is still uncertain. If anything, the physiologically relevant mechanisms for inhibition of adenylyl cyclase are less clear now than they were when originally proposed. It is possible that all of these mechanisms play a role in receptor-mediated inhibition of adenylyl cyclase, and that the challenge is to determine under what circumstances each is important.

### I. Direct Inhibition of Adenylyl Cyclase by $\alpha_i$

The idea that GTP-activated $\alpha_i$ directly inhibits adenylyl cyclase was based upon analogy to the demonstrated functions of the $\alpha$ subunits of $G_s$ and $G_t$ (FUNG et al. 1981; NORTHUP et al. 1983). This hypothesis rested on the conclusive identification of $G_i$ as a protein separate and distinct from $G_s$ and responsible for guanine nucleotide (HILDEBRANDT et al. 1982, 1983) and hormone inhibition (JAKOBS et al. 1983) of adenylyl cyclase activity in $\alpha_s$-deficient *cyc*$^-$S49 cell membranes. This activity appeared to have the same requirements and sensitivities, as to pertussis toxin, for example, as inhibitory regulation in wild-type S49 cells, which contain a fully functional $G_s$ protein. These observations seemed to preclude any obligatory role for $G_s$ in receptor-mediated inhibition of adenylyl cyclase. Since there was no obvious difference in either the $\beta$ (MANNING and GILMAN 1983) or $\gamma$ (HILDEBRANDT et al. 1985a) subunits associated with $G_s$ and $G_i$, it seemed reasonable to assume that the specific $\alpha$ subunit of $G_i$ was responsible for its regulation of downstream effectors. In agreement with this idea, GTP$\gamma$S-liganded $\alpha_i$ inhibited adenylyl cyclase activity in both wild-type and *cyc*$^-$ cell membranes (KATADA et al. 1984), and this effect was specific for $\alpha_i$ and not reproduced by $\alpha_o$ or $\alpha_t$ (ROOF et al. 1985).

Although the original case for $\alpha_i$-mediated inhibition of adenylyl cyclase might seem quite strong, it was not universally accepted. Others believed that it was in fact the $\beta\gamma$ complex from $G_i$ that was responsible for its inhibition of adenylyl cyclase (KATADA et al. 1984). At least in wild-type S49 cell membranes, and under the right conditions, $\beta\gamma$ was actually more potent at inhibiting enzyme activity than was $\alpha_i^{GTP\gamma S}$. In addition, attempts by several laboratories failed to demonstrate the ability of activated $\alpha_i$ to inhibit the purified or partially purified adenylyl cyclase enzyme (CERIONE et al. 1985; SMIGEL 1986). This is certainly not proof of the inability of $\alpha_i$ to inhibit the enzyme (BIRNBAUMER 1987), and the recent realization of the diversity of adenylyl cyclase isoforms (see TANG and GILMAN 1991) suggests that the right isoform(s) or preparations of the enzyme have not yet been identified. Nevertheless, the inability of activated $\alpha_i$ to inhibit purified or partially purified enzyme preparations and the relatively low potency for its inhibitory effects in membranes have often been cited as reasons for doubting its physiological role in mediating inhibition of adenylyl cyclase.

## II. Indirect Inhibition of Adenylyl Cyclase by $\beta\gamma$ Suppression of $\alpha_s$ Activation

This idea was a logical extension of the notion that subunit dissociation is required for G-protein activation, as described in Eq. 1. Here the isolated $\alpha$ subunit with GTP

$$\alpha^{GDP}\beta\gamma + GTP \xrightarrow[Mg^{2+}]{HR\ Complex} \alpha^{GTP} + \beta\gamma + GDP \quad (1)$$

bound is thought to be the active complex for regulation of downstream effector enzymes (FUNG et al. 1981; NORTHUP et al. 1983). This mechanism of $G_i$-mediated inhibition was based upon evidence for $\beta\gamma$ inhibition of $\alpha$ subunit ($\alpha_s$) activation and the observation that $\beta\gamma$ reconstituted into intact membranes inhibited adenylyl cyclase activity in a manner remarkably similar to hormone-mediated inhibition (KATADA et al. 1984). This mechanism suggested that $\beta\gamma$ liberated upon the activation of $G_i$ would feed back by mass action and block formation of the activated form of $G_s$ ($\alpha_s{}^{GTP}$). In other words, $\beta\gamma$ would drive Eq. 1 backwards. Perhaps the most striking piece of evidence supporting this idea was the fact that hormone (somatostatin in S49 cell membranes) and $\beta\gamma$ were both only able to inhibit forskolin-stimulated adenylyl cyclase activity by a maximum of 50%, and these effects were not additive. This seemed so coincidental, and so consistent with the subunit dissociation idea, that it was hard to believe that $\beta\gamma$ did not mediate the effects of somatostatin. There were also other indirect supporting observations that seemed to favor this idea, such as the stoichiometry of $G_i$ to $G_s$ in membranes (~5–10:1) that would seemingly favor mass action inhibition of $\alpha_s$ activation by $\beta\gamma$ liberated from the activation of $G_i$. An extension of this hypothesis was that $\alpha_i$ would therefore have some other function, i.e., regulate some effector other than adenylyl cyclase, a prediction that turned out to be true (YATANI et al. 1987; BOKOCH and GILMAN 1984; KIKUCHI et al. 1986). Further, reconstitution experiments with purified components seem to confirm that $\beta\gamma$ could block activation of $\alpha_s$, whereas $\alpha_i$ was ineffective (CERIONE et al. 1985; SMIGEL 1986).

As elegant as the subunit dissociation idea was as an explanation for $G_i$-mediated inhibition of adenylyl cyclase, it was not without its problems too as a simple explanation of the data. Even as it was being proposed it could not explain how $G_i$ would inhibit enzyme activity in *cyc*-S49 cell membranes which do not contain $\alpha_s$. Although it was suggested that this was an aberration of a mutant cell line, S49 has generated a succession of mutants extremely successful at identifying components and mechanisms involved in the regulation and production of cAMP. Why should it have suddenly been misleading? Ultimately, however, the subunit dissociation mechanism led to other inherent inconsistencies as well. For example, even though cholera toxin activation of $\alpha_s$ substantially decreases its apparent affinity for $\beta\gamma$ (KAHN and GILMAN 1984), hormone inhibition of adenylyl cyclase activity in membranes made from cholera toxin-treated cells is unaltered (TORO et al. 1987). Likewise, although $\beta\gamma$ inhibition of adenylyl cyclase resembles closely hormonal inhibition in some cases, as in the case of basal activity or in the presence of forskolin (KATADA et al. 1984), it does not in others, as when adenylyl cyclase is stimulated by a hormone that activates $G_s$ (HILDEBRANDT and KOHNKEN 1990). Whereas somatostatin inhibits adenylyl cyclase in S49 cell membranes at least as well in the presence as in the absence of isoproterenol, inhibition by $\beta\gamma$ is much more effective in the absence of isoproterenol (Fig. 1). Although $\beta\gamma$-mediated inhibition of adenylyl cyclase cannot be unequivocally ruled out as a relevant physiological mechanism, it apparently cannot on its own explain the way hormones inhibit the enzyme either.

## III. Direct Inhibition of Adenylyl Cyclase by $\beta\gamma$

As if inhibitory regulation of adenylyl cyclase were not complex enough, instead of two possible mechanisms, there was soon a third as well, direct inhibition of the adenylyl cyclase enzyme itself by isolated $\beta\gamma$ (KATADA et al. 1987). The realization that there are multiple isoforms of adenylyl cyclase, each with unique regulatory properties, has made this even more complex

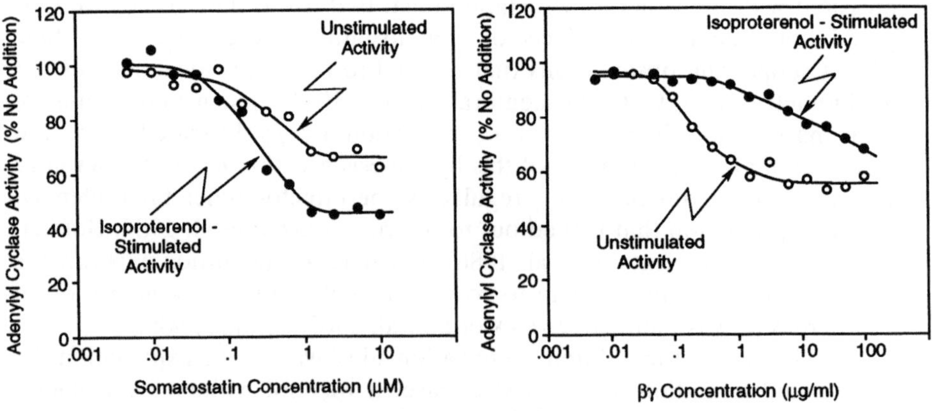


**Fig. 1.** The effect of somatostatin or $\beta\gamma$ concentration on the adenylyl cyclase activity of S49 cell membranes in the absence and presence of $10\,\mu M$ isoproterenol. (With permission, from HILDEBRANDT and KOHNKEN 1990)

(TANG and GILMAN 1991). Thus, although the calmodulin-sensitive type I enzyme is inhibited by $\beta\gamma$, the calmodulin-insensitive type II and type IV enzymes are stimulated by $\beta\gamma$, and another calmodulin-sensitive, olfactory-specific form (type III) is unaffected. In addition, there are probably several additional forms of the enzyme. Since type I is expressed primarily in the nervous system, its regulation by $\beta\gamma$ cannot account for hormone inhibition of adenylyl cyclase in most tissues. It is possible though that there are other forms of the enzyme in peripheral tissues where direct inhibition by $\beta\gamma$ could account for hormone inhibition.

The mechanism by which $\beta\gamma$ directly affects adenylyl cyclase, either stimulation or inhibition, is unclear. Effects have been described primarily on crude (and more heterogeneous than first realized) preparations of enzyme (KATADA et al. 1987), or on material expressed in insect cells with the baculovirus vector (TANG and GILMAN 1991; TANG et al. 1991). Inhibitory regulation of the enzyme is totally lost upon purification (TAGN et al. 1991), and although stimulation seems to be retained, purification of a viable enzyme preparation is apparently quite difficult (TANG and GILMAN 1991). Such results may indicate indirect effects of $\beta\gamma$ on adenylyl cyclase in one or both of these cases. Such may also be the case for other effector systems suggested to be regulated by $\beta\gamma$. For example, it was at one time suggested that $\beta\gamma$ rather than $\alpha$ mediates muscarinic receptor activation of the cardiac, inward-rectifying potassium channel (LOGOTHETIS et al. 1987). This was contested (BIRNBAUMER 1987), and finally revised (KIM et al. 1989) to suggest that the original $\beta\gamma$ effects were in fact indirectly due to its stimulation of phospholipase $A_2$ (JELSEMA and AXELROD 1987) and the production of arachidonic acid metabolites. It has not actually been conclusively demonstrated that the regulation of phospholipase $A_2$ is a direct effect of $\beta\gamma$, but if it and similar effects are, it is possible that analogous effects could mediate $\beta\gamma$ regulation of adenylyl cyclase, at least under specific conditions.

## C. Current View of Inhibition of Adenylyl Cyclase

There are at least three possible mechanisms by which hormones could inhibit adenylyl cyclase activity. Are all of these physiologically relevant? At present it is difficult to say, but it is quite probable that each can play a role in receptor regulation of adenylyl cyclase under specific circumstances. It is even possible that the mechanisms involved in hormone inhibition of the enzyme vary with cell type, receptor system within a cell, and possibly physiological state.

Previous arguments to discount each of the possible mechanisms do not seem to make a lot of sense. The observation that activated $\alpha_i$ can specifically inhibit the S49 adenylyl cyclase (ROOF et al. 1985), even if at lower apparent affinity than $\beta\gamma$, still argues quite strongly for the existence of an $\alpha_i$-dependent mechanism (HILDEBRANDT et al. 1985b). It just seems unlikely that the protein would do exactly what it is expected to do (i.e., inhibit adenylyl cyclase), and other G-protein $\alpha$ subunits not do so, if this was not a physiologically relevant effect. On the other hand, the striking similarity and nonadditivity of the effects of $\beta\gamma$ and somatostatin in S49 cell membranes (KATADA et al. 1984), even if only under a defined set of conditions (HILDEBRANDT and KOHNKEN 1990), would also seem to be more than just coincidence. Likewise, although direct inhibition of the adenylyl cyclase enzyme by $\beta\gamma$ (KATADA et al. 1987) was initially difficult to document (SMIGEL 1986), it now seems easy to accept that this reflected the relative purity or the specific isoforms of the enzyme(s) being studied. Although this may make inhibitory regulation of adenylyl cyclase potentially complex, it may nevertheless be possible to propose general rules about how and when each of these mechanisms is important.

## I. The Mechanism of Inhibition of Adenylyl Cyclase in S49 Cells

One of the most studied systems for determining the mechanism of hormone inhibition of adenylyl cyclase is somatostatin inhibition of the enzyme in the S49 mouse lymphoma cell line (JAKOBS et al. 1983). It is in membranes prepared from these cells that $\beta\gamma$ and somatostatin were shown to have equivalent and nonadditive effects, at least in the presence of forskolin (KATADA et al. 1984). When the enzyme is stimulated by an agonist such as isoproterenol that activates $G_s$, however, somatostatin and $\beta\gamma$ no longer have similar effects (HILDEBRANDT and KOHNKEN 1990). Somatostatin, if anything, has greater efficacy in the presence of isoproterenol, whereas $\beta\gamma$ is much less effective (Fig. 1). This seems to rule out the possibility that $\beta\gamma$-mediated inhibition alone can account for how somatostatin inhibits the enzyme. This would be true regardless of the mechanism of inhibition by $\beta\gamma$. Although this could indicate that the inhibitory effects of $\beta\gamma$ are not relevant to how somatostatin inhibits adenylyl cyclase, this ignores the striking relationship between the effects of $\beta\gamma$ and somatostatin in the absence of isoproterenol.

An alternative idea suggests that both $\beta\gamma$ and $\alpha_i$ mediate the effects of somatostatin, but under different conditions (HILDEBRANDT and KOHNKEN 1990). In the absence of stimulation by isoproterenol $\beta\gamma$ would mediate inhibition, whereas in the presence of isoproterenol $\alpha_i$ would mediate inhibition. Since S49 cells do not seem to contain the type I adenylyl cyclase that is inhibited by $\beta\gamma$ (cited from TANG and GILMAN 1991), presumably the mechanism of inhibition must involve its inhibition of the activation of $G_s$. The explanation of why different mechanisms are required in the presence and absence of isoproterenol is based upon the subunit dissociation idea for the activation of $G_s$ (Fig. 2).

If activation of $G_s$ requires both binding of an activating nucleotide (GTP) and dissociation of $\alpha$ from $\beta\gamma$, there are two possible ways to

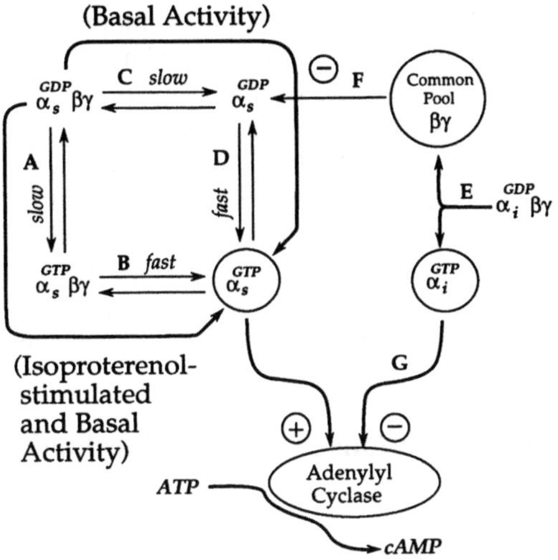


**Fig. 2.** A model describing the role of multiple mechanisms mediating somatostatin inhibition of adenylyl cyclase activity (HILDEBRANDT and KOHNKEN 1990). *A–G*, specific steps referred to in the text

generate an active protein ($\alpha_s^{GTP}$). The receptor-stimulated pathway appears to be one in which receptors promote exchange of GDP for GTP (Fig. 2, A), a process that requires the intact trimer (FUNG 1983; FLORIO and STERNWEIS 1989), followed by the subsequent dissociation of the protein (Fig. 2, B). On the other hand, it is also possible for the intact trimer, even with GDP bound, to dissociate into an isolated $\alpha$ subunit with GDP bound and a $\beta\gamma$ complex (Fig. 2, C). Since isolated $\alpha$ subunits have decreased affinity for GDP and increased affinity for GTP (or actually its analogs; HIGASHIJIMA et al. 1987), this spontaneous dissociation process should promote the subsequent rapid activation of the G protein (i.e., formation of $\alpha_s^{GTP}$) (Fig2, D). The liberation of $\beta\gamma$ from the activation of $G_i$ (Fig. 2, E) could readily inhibit this later process by binding to the isolated $\alpha^{GDP}$ complex before it has a chance to exchange for GTP and become activated (Fig. 2, F). One of the most interesting properties of inhibition of adenylyl cyclase in the S49 cell system is that inhibition by somatostatin is only partial (about 50%). In the absence of isoproterenol this would be explainable based upon what fraction of basal adenylyl cyclase activity is derived from subunit dissociation prior to nucleotide exchange (Fig. 2, C and D), compared to the fraction derived from spontaneous nucleotide exchange prior to subunit dissociation (Fig. 2, A and B). Under normal circumstances in S49 cells, but not necessarily in other cells, these must contribute approximately equivalently to basal activation of $G_s$.

Inhibition of adenylyl cyclase in S49 cell membranes stimulated by isoproterenol would have to be by another mechanism. $G_s$ activation by a hormone would facilitate nucleotide exchange prior to subunit dissociation (Fig. 2, A and B), and this would preclude inhibition by $\beta\gamma$ liberated from the activation of $G_i$. This is because hormone activation of $G_s$ bypasses the $\beta\gamma$-sensitive form of $\alpha_s$ (the $\alpha_s{}^{GDP}$ complex) and results directly in a $\beta\gamma$-insensitive (relatively) complex ($\alpha_s{}^{GTP}$). Under these conditions, $\alpha_i$ must be able to account for inhibition by somatostatin (Fig. 2, G). Presumably, the amount of inhibition of adenylyl cyclase activity in the presence of isoproterenol is determined by the amount of $G_i$ present in the membranes.

## II. Significance and Predications of Multiple Mechanism for Inhibition

The idea that $\beta\gamma$ preferentially inhibits basal adenylyl cyclase activity is similar to the idea that one of the functions of the $\beta\gamma$ complex is to suppress basal activity, making G protein activation hormone-dependent (CERIONE et al. 1985). This idea may define a general mechanism by which all G proteins at least potentially interact with one another. In fact, a similar phenomenon has been described for $G_i$ itself, where exogenous $\beta\gamma$ preferentially inhibits atrial potassium channel opening in the absence rather than in the presence of a muscarinic receptor agonist (OKABE et al. 1990). If G-protein activation contributes free $\beta\gamma$ to a pool accessible to all G-proteins in the cell, this may provide a mechanism for cross-regulation of G-protein function. In the case of inhibition of adenylyl cyclase, this may indicate that activation of any G-protein that dissociates and provides a sufficient $\beta\gamma$ concentration into a generally accessible pool of cellular $\beta\gamma$ would inhibit adenylyl cyclase in the absence of hormones that activates $G_s$. On the other hand, only hormones that specifically activate $G_i$ would inhibit adenylyl cyclase when $G_s$-activating hormones are present. Potentially, similar mechanisms could contribute to the regulation of all G-proteins in a cell. The relative sensitivity of a G-protein $\alpha$ subunit to this kind of mechanism would depend upon its relative concentration and its affinity for $\beta\gamma$.

This proposed scheme suggests a number of predictions that are at least potentially testable. For example, in order to explain why inhibition by somatostatin is not greater in the absence of isoproterenol, when both mechanisms could operate, than in its presence, the adenylyl cyclase enzyme should be more sensitive to inhibition by $\alpha_i$ when it is also activated by $G_s$. Conversely, if $\beta\gamma$ inhibits $\alpha_s$ activation in the absence of isoproterenol, but not in its presence, then somatostatin inhibition should be differentially sensitive to agents or conditions that interfere with G-protein subunit interactions. Blocking $\beta\gamma$ binding to $\alpha_s$ should prevent somatostatin inhibition of adenylyl cyclase in the absence of isoproterenol, but not in its presence. Such predictions are at least potentially testable.

## III. Unresolved Structural and Functional Issues About G-proteins Affecting the Mechanism(s) Mediating Hormone Inhibition of Adenylyl Cyclase

There are many as yet unresolved issues about the mechanism(s) mediating hormone inhibition of adenylyl cyclase. For example, it remains to be determined how the existence of multiple mechanisms relates to the variable patterns of regulation of cAMP levels in intact cells. Also, the possibility should be considered that compartmentalization of G-proteins within cells could limit their access to a general pool of $\beta\gamma$, thus limiting the interactions of some proteins. At the moment, however, probably the most important issue affecting our ideas about the mechanism(s) mediating hormone inhibition of adenylyl cyclase, is the origin and nature of the diversity of the G-proteins and their subunits.

The most obvious question is the significance of the multiple forms of $\alpha_i$. In part, the tissue specificity and hormone receptor specificity of these isoforms may provide a partial explanation for their diversity. However, there may be other more subtle differences between them that relate more closely to the mechanistic focus of the work summarized here. The observation that these proteins have different guanine nucleotide binding properties that may give rise to different kinetic patterns of activation (CARTY et al. 1990) may indicate that these proteins differ also in their contribution to $\beta\gamma$ effects on inhibition of adenylyl cyclase. Thermodynamic arguments would suggest that the different guanine mucleotide binding properties of these proteins should also reflect differences in the affinity or regulation of their interactions with $\beta\gamma$. Differences in subunit interactions of these proteins remain to be established.

Perhaps the biggest issue to be resolved is the significance and extent of variation in $\beta\gamma$ isoforms. There are at least four $\beta$ subunits and four to six $\gamma$ subunits (see SIMON et al. 1991). Post-translational modification may give rise to additional diversity, including functionally different $\beta\gamma$ complexes (FUKADA et al. 1990). How many forms of $\beta\gamma$ exist is not certain. It is possible that a given $\gamma$ subunit associates with only a specific form(s) of $\beta$, and that there are a minimum number of different complexes, perhaps four to six. It is also possible though that there is random association of $\beta$ and $\gamma$ subunits, and that the number of complexes is determined by the regulatory mechanisms that control their coexpression in different tissues. There could be as many different $\beta\gamma$ complexes as there are $\alpha$ subunits (20 or more).

It is unclear under what circumstances and whether all $\alpha$ subunits interact with all possible $\beta\gamma$ complexes. Nor is it clear what exactly the role of these multiple $\beta\gamma$ complexes might play in the function of the trimer. It is possible that $\alpha$ subunits have absolute or relative preferences for different $\beta\gamma$ complexes. It is also possible, however, that $\alpha$ subunits can and do interact with all possible $\beta\gamma$ complexes, and that there are literally hundreds of possible G-protein heterotrimers. In all likelihood the composition of the $\beta\gamma$

complexes in association with a given $\alpha$ subunit is either selective and/or effects the functional properties of the protein. If either of these is true, the role of $\beta\gamma$ in mediating inhibition of adenylyl cyclase may be quite fascinating, even if it turns out to be complex. As has often been the case before, such mechanisms affecting inhibition of adenylyl cyclase may reflect and describe not just the behavior of this system, but that of the entire family of G-proteins.

# D. Conclusion

The mechanism(s) mediating hormone inhibition of adenylyl cyclase have been difficult, if not painful, to define. In trying to determine what these mechanisms are, and which ones are physiologically relevant, many of the properties of the G-proteins in general have been elucidated and characterized. The regulation of adenylyl cyclase by multiple G proteins has provided the best available model system for investigating at the molecular level the kinds of interactions that must be taking place inside a cell, but pertaining to all of those different processes that are regulated by G-proteins. Although a complete understanding of how hormones inhibit adenylyl cyclase still eludes us, and may turn out to be more complex than we might ever have imagined, the continued investigation of this phenomenon with open minds promises to uncover additional attributes of the signal transduction processes mediated by the larger family of heterotrimeric G-proteins. In particular, our full understanding of the mechanisms involved in hormone inhibition of adenylyl cyclase will undoubtedly require our uncovering the significance and range of the diversity of both $\beta\gamma$ complexes and heterotrimer structure itself, and may require that we re-evaluate our current ideas about G-protein activation as well.

*Acknowledgements.* This work was supported in part by NIH grant DK37219.

## References

Birnbaumer L, Codina J, Mattera R, Cerione RA, Hildebrandt JD, Sunyer T, Rojas FJ, Caron MG, Lefkowitz RJ, Iyengar R (1985) Regulation of hormone receptors and adenylyl cyclases by guanine nucleotide binding N proteins. Recent Prog Horm Res 41:41–99

Birnhaumer L (1987) Which G protein subunits are the active mediators in signal transduction. Trends Pharmacol Sci 8:209–211

Birnbaumer L (1990) Transduction of receptor signal into modulation of effector activity by G proteins: the first 20 years or so. FASEB J 4:3178–3188

Bokoch GM, Gilman AG (1984) Inhibition of receptor-mediated release of arachidonic acid by pertussis toxin. Cell 39:301–308

Bokoch GM, Katada T, Northup JK, Ui M, Gilman AG (1984) Purification and properties of the inhibitory guanine nucleotide-binding regulatory component of adenylate cyclase. J Biol Chem 259:3560–3567

Carty DJ, Padrell E, Codina J, Birnbaumer L, Hildebrandt JD, Iyengar R (1990) Distinct guanine nucleotide binding and release properties of the three Gi proteins. J Biol Chem 265:6268–6273
Cerione RA, Staniszewski C, Caron MG, Lefkowitz RJ, Codina J, Birnbaumer L (1985) A role for Ni in the hormonal stimulation of adenylate cyclase. Nature 318:293–295
Codina J, Hildebrandt J, Iyengar R, Birnbaumer L, Sekura RD, Manclark CR (1983) Pertussis toxin substrate, the putative Ni component of adenylyl cyclase, is an alpha beta heterodimer regulated by guanine nucleotide and magnesium. J Biol Chem 80:4276–4280
Florio VA, Sternweis PC (1989) Mechanisms of muscarinic receptor action on Go in reconstituted phospholipid vesicles. J Biol Chem 264:3909–3915
Fukada Y, Takao T, Ohguro H, Yoshizawa T, Akino T, Shimonishi Y (1990) Farnesylated gamma-subunit of photoreceptor G protein indispensable for GTP-binding. Nature 346:658–660
Fung BK, Hurley JB, Stryer L (1981) Flow of information in the light-triggered cyclic nucleotide cascade of vision. Proc Natl Acad Sci USA 78:152–156
Fung BK (1983) Characterization of transducin from bovine retinal rod outer segments. I. Separation and reconstitution of the subunits. J Biol Chem 258:10495–10502
Gilman AG (1984) G proteins and dual control of adenylate cyclase. Cell 36:577–579
Higashijima T, Ferguson KM, Sternweis PC, Smigel MD, Gilman AG (1987) Effects of Mg2+ and the beta gamma-subunit complex on the interactions of guanine nucleotides with G proteins. J Biol Chem 262:762–766
Hildebrandt JD, Kohnken RE (1990) Hormone inhibition of adenylyl cyclase. Differences in the mechanisms for inhibition by hormones and G protein beta/gamma. J Biol Chem 265:9825–9830
Hildebrandt JD, Hanoune J, Birnbaumer L (1982) Guanine nucleotide inhibition of $cyc^-$ S49 mouse lymphoma cell membrane adenylyl cyclase. J Biol Chem 257:14723–14725
Hildebrandt JD, Sekura RD, Codina J, Iyengar R, Manclark CR, Birnbaumer L (1983) Stimulation and inhibition of adenylyl cyclases mediated by distinct regulatory proteins. Nature 302:706–709
Hildebrandt JD, Codina J, Rosenthal W, Birnbaumer L, Neer EJ, Yamazaki A, Bitensky MW (1985a) Characterization by two-dimensional peptide mapping of the gamma subunits of Ns and Ni, the regulatory proteins of adenylyl cyclase, and of transducin, the guanine nucleotide-binding protein of rod outer segments of the eye. J Biol Chem 260:14867–14872
Hildebrandt JD, Codina J, Rosenthal W, Sunyer T, Iyengar R, Birnbaumer L (1985b) Properties of human erythrocyte Ns and Ni, the regulatory components of adenylate cyclase, as purified without regulatory ligands. Adv Cyclic Nucleotide Protein Phosphorylation Res 19:87–101
Jakobs KH, Aktories K, Schultz G (1983) A nucleotide regulatory site for somatostatin inhibition of adenylate cyclase in S49 lymphoma cells. Nature 303:177–178
Jelsema CL, Axelrod J (1987) Stimulation of phospholipase A2 activity in bovine rod outer segments by the beta/gamma subunits of transducin and its inhibition by the alpha subunit. Proc Natl Acad Sci USA 84:3623–3627
Kahn RA, Gilman AG (1984) ADP-ribosylation of Gs promotes the dissociation of its alpha and beta subunits. J Biol Chem 259:6235–6240
Katada T, Bokoch GM, Smigel MD, Ui M, Gilman AG (1984) The inhibitory guanine nucleotide-binding regulatory component of adenylate cyclase. Subunit dissociation and the inhibition of adenylate cyclase in S49 lymphoma $cyc^-$ and wild type membranes. J Biol Chem 259:3586–3595

Katada T, Kusakabe K, Oinuma M, Ui M (1987) A novel mechanism for the inhibition of adenylate cyclase via inhibitory GTP-binding proteins. Calmodulin-dependent inhibition of the cyclase catalyst by the beta gamma-subunits of GTP-binding proteins. J Biol Chem 262:11897–11900

Kikuchi A, Kozawa O, Kaibuchi K, Katada T, Ui M, Takai Y (1986) Direct evidence for involvement of a guanine nucleotide-binding protein in chemotactic peptide-stimulated formation of inositol bisphosphate and trisphosphate in differentiated human leukemic (HL-60) cells. Reconstitution with Gi or Go of the plasma membranes ADP-ribosylated by pertussis toxin. J Biol Chem 261:11558–11562

Kim D, Lewis DL, Graziadei L, Neer EJ, Bar Sagi D, Clapham DE (1989) G-protein beta gamma-subunits activate the cardiac muscarinic $K^+$-channel via phospholipase A2. Nature 337:557–560

Logothetis DE, Kurachi Y, Galper J, Neer EJ, Clapham DE (1987) The beta gamma subunits of GTP-binding proteins activate the muscarinic $K^+$ channel in heart. Nature 325:321–326

Manning DR, Gilman AG (1983) The regulatory components of adenylate cyclase and transducin. A family of structurally homologous guanine nucleotide-binding proteins. J Biol Chem 258:7059–7063

Murad F, Chi Y-M, Rall TW, Sutherland EW (1962) Adenyl cyclase III. The effect of catecholamines and choline esters on the formation of adenosine 3′-5′-phosphate by preparations of cardiac muscle and liver. J Biol Chem 237:1233–1238

Northup JK, Smigel MD, Sternweis PC, Gilman AG (1983) The subunits of the stimulatory regulatory component of adenylate cyclase. Resolution of the activated 45000-dalton (alpha) subunit. J Biol Chem 258:11369–11376

Okabe K, Yatani A, Evans T, Ho YK, Codina J, Birnbaumer L, Brown AM (1990) Beta gamma dimers of G proteins inhibit atrial muscarinic $K^+$ channels. J Biol Chem 265:12854–12858

Rodbell M (1980) The role of hormone receptors and GTP-regulatory proteins in membrane transduction. Nature 284:17–22

Rodbell M, Birnbaumer L, Pohl SL, Krans HMJ (1971a) The glucagon-sensitive adenyl cyclase system in plasma membranes of rat liver. V. An obligatory role of guanyl nucleotides in glucagon action. J Biol Chem 246:1877–1882

Rodbell M, Krans HMJ, Pohl SL, Birnbaumer L (1971b) The glucagon-sensitive adenyl cyclase system in plasma membranes of rat liver. IV. Effects of guanyl nucleotides on binding of $^{125}$I-glucagon. J Biol Chem 246:1872–1876

Roof DJ, Applebury ML, Sternweis PC (1985) Relationships within the family of GTP-binding proteins isolated from bovine central nervous system. J Biol Chem 260:16242–16249

Simon MI, Strathmann MP, Gautam N (1991) Diversity of G-proteins in signal transduction. Science 252:802–808

Smigel MD (1986) Purification of the catalyst of adenylate cyclase. J Biol Chem 261:1976–1982

Tang W, Gilman AG (1991) Type-specific regulation of adenylyl cyclase by G-protein beta/gamma subunits. Science 254:1500–1503

Tang W, Krupinski J, Gilman AG (1991) Expression and characterization of calmodulin-activated (type I) adenylycyclase. J Biol Chem 266:8595–8603

Toro MJ, Montoya E, Birnbaumer L (1987) Inhibitory regulation of adenylyl cyclases. Evidence inconsistent with beta/gamma-complexes of Gi proteins mediating hormonal effects by interfering with activation of Gs. Mol Endocrinol 1:669–676

Yatani A, Codina J, Brown AM, Birnbaumer L (1987) Direct activation of mammalian atrial muscarinic potassium channels by GTP regulatory protein GK. Science 235:207–211

CHAPTER 71

# Neurobiology of $G_o$

P. BRABET, V. HOMBURGER, and J. BOCKAERT

## A. Introduction

$G_o$, the subscript standing for other (other than $G_s$ and $G_i$), has several interesting features (for a review see BOCKAERT et al. 1990) (1) it is highly concentrated in brain (0.5 to 1% of membrane bound proteins), (2) it is present in most excitable cells including endocrine cells and skeletal muscle (TOUTANT et al. 1990), (3) some $G_o$ protein epitopes and the gene organization are almost similar in vertebrates and invertebrates (HOMBURGER et al. 1987; YOON et al. 1989), (4) one of the roles of $G_o\alpha$ is to negatively couple some receptors to voltage-sensitive $Ca^{++}$ channels (see Table 1), (5) in view of (a) the high content of $G_o\alpha$ in brain tissues, (b) the close associations of $G_o\alpha$ with cytoskeletal elements, including microtubules, actin fibers, cytoplasmic matrices (see chapter A, I, II), and growth cone structures (STRITTMATTER et al. 1990), (c) the differential expression of two splice variants of $G_o$ ($G_{o1}\alpha$ and $G_{o2}\alpha$) during neuronal differentiation and neurite outgrowth (BRABET et al. 1990; ROUOT et al. 1992), it is likely that $G_o\alpha$ plays also a key structural role in neuronal cell architecture.

## B. Gene Structure of $G_o\alpha$ in Vertebrates and Invertebrates

### I. Gene Structure and Transcription in Vertebrates

Both the cDNA nucleotide and the deduced amino acid sequences are remarkably conserved among mammals, with 94% and 98% identity (ITOH et al. 1986; JONES and REED 1987; LAVU et al. 1988; OLATE et al. 1989). The deduced protein amino acid sequences show that $G_o\alpha$ consists of 354 amino acids, giving a calculated molecular weight of 40–40.5 kDa. Moreover, the alignment of cDNA sequences shows that an even higher degree of interspecies conservation of both 5′ and 3′ untranslated regions. Therefore, regulatory elements at the transcriptional and posttranscriptional levels, which undergo selective pressure throughout mammalian evolution, have presumably important roles in the expression of the $G_o\alpha$-protein.

Three classes of cDNA variants have been cloned from insulin-secreting tumor cells (HSU et al. 1990; BERTRAND et al. 1990) and mouse brain (STRATHMANN et al. 1990). They are derived from mRNAs produced by alternative splicing from the same $G_o\alpha$ gene. Two of these transcripts encode $G_o\alpha$-proteins with the same length (354 amino acids), which differ in 26 of the last 112 carboxy-terminal amino acids. These products were named independently by HSU et al. and STRATHMANN et al. as $G_{o1}\alpha$ and $G_{o2}\alpha$, and $G_oA\alpha$ and $G_oB\alpha$, respectively. The third transcript is identical to $G_{o1}\alpha$ (designated $G_{o1}B\alpha$, as opposed to $G_{o1}A\alpha$) in its coding region but differs in the 3′ untranslated region (3′UTR) 28 nucleotides downstream the termination codon (PRICE et al. 1990; BERTRAND et al. 1990). Concerning the size of these mRNAs, BERTRAND et al. described selective hybridization of 3.2-, 4.2-, and 5.7-kb hamster brain mRNAs with probes specific for $G_{o1}A\alpha$, $G_{o1}B\alpha$, and $G_{o2}\alpha$, respectively. PRICE et al. (1990) found an additional mRNA of approximately 2.0 kb which hybridized with an oligonucleotide complementary to a sequence of the 3′UTR of $G_{o1}B\alpha$. Three recent studies have contributed to establish the structural organization of the human gene encoding $G_o\alpha$ (Fig. 1). LAVU et al. (1988) first published a partial sequence containing exons 1 and 2. Recently, TSUKAMOTO et al. (1991) reported that the open reading frame of $G_o\alpha$ is encoded by 11 exons distributed over more than 100 kb of chromosomal DNA. Restriction mapping and nucleotide sequence analysis show that exons 1–6 are common to all the transcripts identified so far. Then follow two alternate exons $7_2$ and $8_2$ which code for amino acid residues 293–354 as well as a 3′ untranslated region (3′UTR 2) specific for $G_{o2}\alpha$. Next are the exons $7_1$ and $8_1$ which code for $G_{o1}\alpha$. MURTAGH et al. (1991) amplified by polymerase chain reaction and sequenced the 3′ untranslated regions of the human, bovine, and mouse chromosomal DNA. They inferred that the two splice variants of $G_{o1}\alpha$ ($G_{o1}A\alpha$ and $G_{o1}B\alpha$) derived from alternative 3′ untranslated regions which are located relatively close to one another within the genome (depicted by $9A_1$ and $9B_1$ in Fig. 1). Actually, $9B_1$ is directly adjacent to the 3′ end of exon $8_1$ while splicing out of the former plus 783 nucleotides brings $9A_1$ to the same position.

The 5′ untranslated region of the $G_o\alpha$ gene possesses common features of $G\alpha$ genes such as an extremely high G+C content, the presence of several GC boxes but no typical TATA box. Interestingly, it also contains the cAMP-responsive element (TSUKAMOTO et al. 1991).

## II. Gene Structure and Transcription in Invertebrates

In invertebrates a $G_o\alpha$-like protein has been identified in nervous tissue by pertussis toxin-catalyzed ADP-ribosylation and polyclonal antibodies generated to vertebrate $G_o\alpha$ (HOMBURGER et al. 1987; HOPKINS et al. 1988). Most of the studies concerning this $G_o\alpha$-like protein have early referred to the fruit fly *Drosophila melanogaster* because it is amenable to genetic manipulation and to study behavior and sensory transductions (SCHMIDT et al. 1989; YOON et al. 1989; DE SOUSA et al. 1989; THAMBI et al. 1989; QUAN et al. 1989; GUILLEN et al. 1990, 1991).

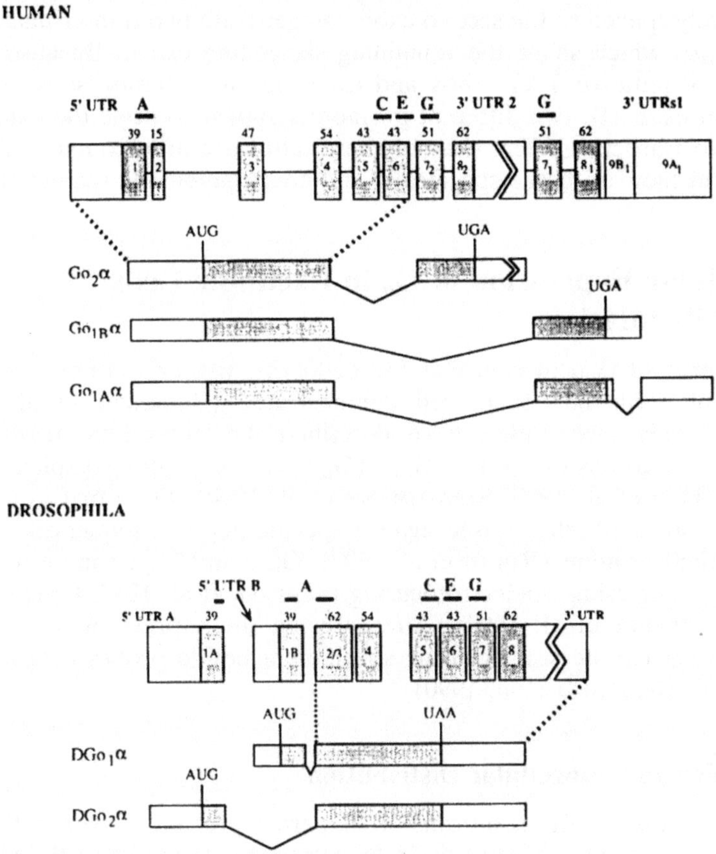


**Fig. 1.** Comparison of the intron-exon organization and transcription of human and *Drosophila* $G_o\alpha$ genes. In each human and *Drosophila panel*, the *top* depicts the organization of the gene. *Closed boxes*, coding regions of the exons; *open boxes*, 5'- and 3'-untranslated regions; *interrupted line*, introns. *A*, *C*, *E*, *G*, regions that are highly conserved in all G-proteins and contribute to binding and hydrolysis of GTP. *Numbers above boxes*, the number of amino acids encoded by each coding exons. They point out the conservation of intron-exon splice junctions. The bottom depicts the alternative splice variants. *Open and closed boxes*, the transcribed genomic sequences; *downstream lines*, nontranscribed genomic sequences. The initiation (AUG) and termination (UGA and UAA) codons limit the translated sequences

The *Drosophila* $G_o\alpha$ ($DG_o\alpha$) gene likely spans at least 40 kb which encompasses eight coding exons (Fig. 1). It is interesting to note that the intron-exon organization of this $DG_o\alpha$ gene and human $G_o\alpha$ and $G_i\alpha$ genes (Kaziro et al. 1990) are identical except that $DG_o\alpha$ lacks an intron corresponding to intron 2 of $G_o\alpha/G_i\alpha$ genes. Two first exons, 1A and 1B, encoding for amino-terminal stretches of 39 amino acids which differ in seven positions, has been found at different genomic locations. A comparison of genomic and cDNA sequences revealed that these exons 1 are alternatively spliced to the second exon and generate two transcripts, $DG_{o1}\alpha$ and $DG_{o2}\alpha$, which share the remaining six coding exons. Because of the presence of putative TATA box and transcription initiation sites in the 5′ UTR B of exon 1B, two different promoters appear to drive the expression of the two transcripts. The sizes of these latter are approximately 3.8 and 5.3 kb, and they are tissue-specifically and developmentally regulated.

## C. Cellular Expression of $G_o$ in Excitable Cells and Its Regulation

The presence of $G_o\alpha$ in neuronal and endocrine tissues and more generally in excitable cells has been well demonstrated (BOCKAERT et al. 1990). However, only few studies have described the differential tissular and cellular distributions of the two $G_{o1}\alpha$ ($G_oA$) and $G_{o2}\alpha$ ($G_oB$) splice variant proteins (HSU et al. 1990; STRATHMANN et al. 1990). $G_{o1}\alpha$ or $G_{o2}\alpha$ can be detected with antibodies raised against specific peptidic sequences of these splice variant proteins (ROUOT et al. 1992). $G_{o1}\alpha$ and $G_{o2}\alpha$ can be separated on urea gels or using isoelectrofocusing (SCHERER et al. 1987; GOLDSMITH et al. 1988; BRABET et al. 1990). Alternatively, the expression of $G_{o1}\alpha$ and $G_{o2}\alpha$ mRNA can be studied using specific nucleotide probes (STRATHMANN et al. 1990; BERTRAND et al. 1990).

### I. Cellular and Subcellular Distribution

In neuronal tissues, the two forms of $G_o\alpha$ are expressed; however, the $G_{o1}\alpha$ form is predominant in adult brain (STRATHMANN et al. 1990; BERTRAND et al. 1990; ROUOT et al. 1992). The two forms of $G_o\alpha$ proteins are present in both glial cells and neurons (BRABET et al. 1990; GRANNEMAN et al. 1990). However, glial cells exhibit a higher proportion of $G_{o2}\alpha$ as compared to neurons (BRABET et al. 1990).

In endocrine cells and more generally in excitable nonneuronal cells, the amount of $G_o\alpha$ is smaller (ASANO et al. 1988a). Fewer studies have been reported to discriminate $G_{o1}\alpha$ and $G_{o2}\alpha$. However, in these tissues or cell lines, expression of the $G_{o2}\alpha$ protein or detection of $G_{o2}\alpha$ transcripts seem to be major, if not exclusive (SPICHER et al. 1991; ASANO et al. 1991; BERTRAND et al. 1990).

#### 1. Neurons

Immunocytochemical studies in vertebrate brain indicate that the $G_o\alpha$ protein is abundant in synapse-rich neuropil (TERASHIMA et al. 1987; WORLEY et al. 1986; CHANG et al. 1988). In contrast, in situ hybridization experiments indicated that most of the staining is observed in neuronal cell bodies (LARGENT et al. 1988; VINCENT et al. 1990).

Subcellular localization in primary culture of neuronal cells from mouse with anti-$G_o\alpha$ antibodies indicated a positive reaction at the periphery of the cells, but also in the cytoplasm and neurite arborization with reinforcement at cell-cell contacts (BRABET et al. 1988). In clonal cells, such as NIE-115, the same pattern is observed in differentiated cells, but in the undifferentiated state, only a positive reaction was seen in the cytoplasmic matrix, indicating a relocalization of the $G_o\alpha$ protein during differentiation (BRABET et al. 1990; Fig. 2).

In amphibians, such as pleurodeles, the $G_o$ protein is detected in the nervous system and in well-differentiated neurons; localization exhibited the

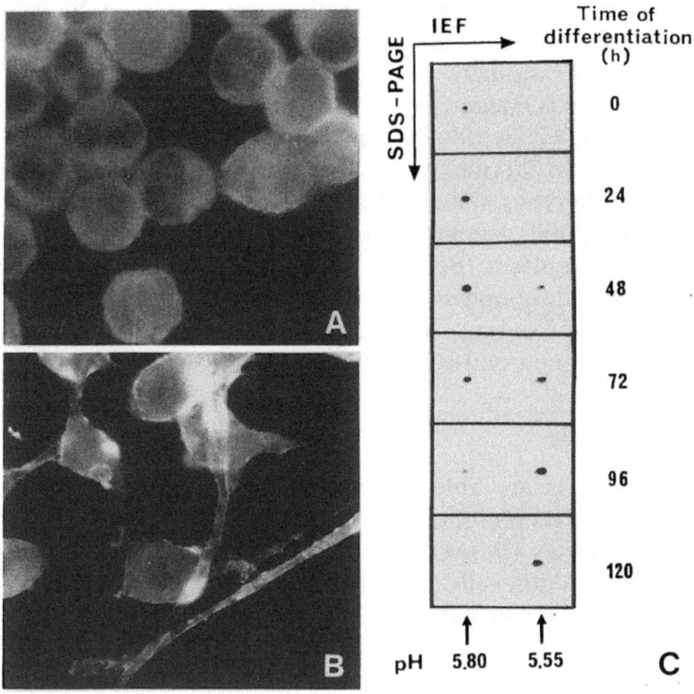


**Fig. 2A–C.** Expression and localization of $G_o\alpha$ during differentiation of NIE-115 neuroblastoma cells. **A**, **B** Indirect immunofluorescent localization of $G_o\alpha$ in undifferentiated (**A**) and 72-h differentiated NIE-115 cells (**B**). Cells cultured on glass coverslips were fixed, permeabilized with Triton X 100 and incubated with affinity purified anti-$G_o\alpha$ and rhodamin-conjugated goat anti-rabbit IgG. Note the diffuse immunolabeling over the cytoplasm in undifferentiated cells (**A**) and the marked staining at the periphery of cell bodies, neurites and cell–cell contact areas in differentiated cells (**B**). **C** Expression of $G_{o2}\alpha$ and $G_{o1}\alpha$ during differentiation of NIE-115 neuroblastoma cells. Undifferentiated cells (time 0) or at different stages of differentiation were homogeneized and particulate fraction were prepared. Proteins from particulate fractions (50 $\mu$g protein) were subjected to two-dimensional analysis, transferred to nitrocellulose, and incubated with anti-$G_o\alpha$ antiserum and $^{125}$I-labeled protein A. Two spots migrated at pH 5.80 ($G_{o2}\alpha$) and 5.55 ($G_{o1}\alpha$) were revealed by anti-$G_o\alpha$ antibodies

same pattern of staining as in well-differentiated mammalian neurons (PITUELLO et al. 1991). However, in a 3-day old culture of neuroectodermal cells, neuritic processes but also growth cones appeared strongly fluorescent using anti-$G_o\alpha$ antibodies (PITUELLO et al. 1991). Such a localization is similar to the one described by Neer's group using differentiated PC12 cells and showed that $G_o\alpha$ is concentrated at the distal tips of the cellular processes (STRITTMATTER et al. 1990).

Ultrastructural studies using anti-$G_o\alpha$ antibodies indicated two types of labeling in neuronal cells. Firstly, a membranous staining at the inner cytoplasmic face of the plasma membrane, at a distance from synaptic content, but mainly in contacting membranes. Secondly, a positive reaction in the cytoplasmic matrix. Some labeling was observed between rough endoplasmic reticulum and Golgi cisternae and at the cytoplasmic face of coated vesicles, occasionally an intense labeling was observed around centriolar structures (GABRION et al. 1989).

In *Drosophila*, $G_o$ is detected in the brain with anti-$G_o\alpha$ antibodies (GUILLEN et al. 1990; THAMBI et al. 1989; SCHMIDT et al. 1989; WOLFGANG et al. 1990). In the nervous system of adult flies, the $G_o\alpha$ protein is detected mainly in the neuropil, but in cell bodies it appears uniformly distributed throughout the cytoplasm (SCHMIDT et al. 1989; WOLFGANG et al. 1991). During embryonic development the $G_o$ protein is present at all stages but markedly increases during axonal tract development and is highly concentrated in the neuropil (GUILLEN et al. 1991; WOLFGANG et al. 1991).

## 2. Nonneuronal Cells

Other excitable cells are able to express the $G_o$ protein. To this group belong astrocytes (BRABET et al. 1988), oligodendrocytes (BRAUN et al. 1990), ependymocytes (PERALDI et al. 1989), and endocrine cells such as lactotroph or chromaffin cells (JOURNOT et al. 1987; ASANO et al. 1988a).

Few studies have been devoted to the subcellular and ultrastructural localization of the $G_o$ protein in these cells. In astrocytes not all the cells express the $G_o$ protein. Astrocytes derived from cerebellum are poorly reactive, while those derived from the striatum display specific staining which appears relatively diffuse in the cytoplasmic matrix. $G_o$ is never observed to be associated with the plasma membrane or at the cell-cell contacts (BRABET et al. 1988). In choroidal ependymocytes both immunofluorescence and ultrastructural studies have indicated that $G_o\alpha$ is localized at the apical pole of the cell in association with the cytoplasmic matrix rich in actin microfilaments. Some coated pits and vesicles appeared to be decorated with $G_o\alpha$ (PÉRALDI et al. 1987). In ciliated ependymocytes, positive labeling was distributed as spots along the axonemes of kinocilia (PERALDI 1987). This is reminiscent of the labeling of some centriolar structures of neurons (GABRION et al. 1989).

In outer hair cells, immunocytochemical localization of the $G_o$ protein indicated a positive reaction mainly in the region of the cuticular plate, at the base of the cell, along the cell membrane, and in some cells an infracuticular network was fluorescent (CANLON et al. 1991). These localizations of $G_o\alpha$ are similar to those of actin filaments. In ependymocytes, and outer hair cells, immunocytolocalization did suggest a possible interaction of the $G_o$ protein with some cytoskeletal elements (GABRION et al. 1989).

## II. Control of $G_o$, $G_{o1}$, and $G_{o2}$ Expression During Neuronal Differentiation

In primary culture of neurons or in neuronal cell lines, the level of $G_o$ increases during neuronal differentiation, i.e., growth of neurites and synaptogenesis (GRANNEMAN et al. 1990; BRABET et al. 1990; MULLANEY et al. 1988; ASANO et al. 1989). Brain $G_o$ content increases during embryogenesis and postnatal period reaching a maximal value 30 days after birth in rodents (ROUOT et al. 1992; ASANO et al. 1988b; CHANG et al. 1988; MILLIGAN et al. 1987). $G_o$ mRNA content also increases during neuronal cell differentiation (GARRIBAY et al. 1991). Interestingly the two splice variant $G_o\alpha$ proteins ($G_{o1}\alpha$ and $G_{o2}\alpha$) are not expressed similary during neuronal differentiation. In NIE-115 neuroblastoma cells, cellular differentiation can be obtained following treatment with DMSO. This leads to inhibition of cell growth and extension of long neurites (BRABET et al. 1990). In differentiated cells, two forms of $G_o\alpha$, identified with specific antibodies as $G_{o2}\alpha$ and $G_{o1}\alpha$, can be focalized in two-dimensional analysis at respective pI levels of 5.80 and 5.55 (BRABET et al. 1990; ROUOT et al. 1992; Fig. 2). In contrast, in undifferentiated neuroblastoma cells, only $G_{o2}\alpha$ was detected (Fig. 2). Using two-dimensional analysis coupled to metabolic cell labeling and immunoprecipitation of $G_o\alpha$, it was possible to show that differentiation was accompanied by a decrease in the degradation of both $G_o\alpha$ splice variants. However, cell differentiation induces a decrease in the synthesis of $G_{o2}\alpha$ and an increase in the synthesis of $G_{o1}\alpha$ (BRABET et al. 1991). This suggests that neuronal differentiation is responsible for the on/off switch of the expression of the $G_o\alpha$ forms (ROUOT et al. 1992). Interestingly, $G_{o2}\alpha$ is predominant in different brain areas of embryonic mice. Thereafter, there is a progressive decline of the $G_{o2}\alpha$ with a concomitant increase in the $G_{o1}\alpha$ which is temporally correlated to the formation of neurites (ROUOT et al. 1992). This latter observation, the fact that $G_o$ is a major protein in growth cones (STRITTMATTER et al. 1990; EDMONDS et al. 1990), and the immunolocalization of $G_o$ at the tips of neurites during differentiation of PC12 with nerve growth factor, suggest that $G_{o1}\alpha$ may play a physiological structural role during the growth of neurites.

# D. Neurotransmitter Receptors Coupled to $G_o$ and Their Inhibitory Effects on Voltage-Sensitive $Ca^{2+}$ Channels

## I. Nature of Receptors

### 1. Reconstitution of Resolved Receptors and $G_o$ Proteins

Only a few biochemical studies using purified receptors and $G_o$ proteins have been reported. FLORIO and STERNWEIS (1989) and KUROSE et al. (1986) and RICHARDSON et al. (1992) reported that both $G_i$ and $G_o$-proteins can restore the GTP effect on binding of agonists to resolved muscarinic receptors. Similar results were obtained by UEDA et al. (1988) with opiates. Recently two studies have demonstrated the coupling of purified adenosine $A_1$ receptors with either $G_{i1}$, $G_{i2}$, and $G_o$ purified from bovine brain (MUNSHI et al. 1991) or synthesized from *Escherichia coli*. (FREISSMUTH et al. 1991). All $G_i/G_o$ proteins (but not $G_s$ or $G_z$) were able to interact with adenosine $A_1$ receptors with the following order of specificity: $G_{i2} > G_o \geq G_{i1}$ in the study by MUNSHI et al. and $G_{i3} > G_{i1} > G_{i2} > G_o$ in the report of FREISSMUTH et al.

### 2. Reconstitution of Receptor Coupling to VSCC with $G_o$ Protein in PTX-Treated Cells

Several investigations have used purified $G_o$ and $G_i$ proteins to reconstitute the negative coupling of receptors to voltage-sensitive $Ca^{2+}$ channels (VSCC) of cells in which $G_o/G_i$ proteins have previously been inactivated with pertussis toxin (PTX). These proteins were perfused into the cell with a patch pipette (HESCHELER et al. 1987; EWALD et al. 1988, 1989; TOSELLI et al. 1989). In all reports, $G_o$ was able to reconstitute totally dopamine, $\delta$-opioid, neuropeptide Y, muscarinic receptor induced inhibition of VSCC, and only partially the negative coupling of bradykinin receptor to VSCC. When tested (Table 1), (HESCHELER et al. 1987; EWALD et al. 1989; TOSELLI et al. 1989), $G_o$ (or $G_o\alpha$) was more potent than $G_i$ (or $G_i\alpha$). In rat dorsal root ganglion neurons, part of the bradykinin coupling to VSCC could be reconstituted only with $G_{i1}\alpha$ and $G_{i2}\alpha$ (EWALD et al. 1989).

An original approach was made by TAUSSIG et al. (1992). They constructed an NG 108-15 cell line that stably expressed a mutated PTX-resistant $G_{o1}\alpha$. After treatment with PTX, the $G_{o1}\alpha$ mutant rescued Leu-enkephaline and norepinephrine coupling to VSCC but not the coupling of somatostatin to these channels. Since $G_o$ seems to be coupled to somatostatin receptors (OFFERMANNS et al. 1991a,b), it would seem that $G_{o2}\alpha$ is involved rather than $G_{o1}\alpha$ (see also KLEUSS et al. 1991).

Reconstitution experiments demonstrate the ability of some receptors to couple to exogeneous $G_o$; they do not necessarily imply that coupling to

**Table 1.** Receptors coupled to $G_o$ proteins

| Methods used for determination | Receptor | Effector | Cell | References |
|---|---|---|---|---|
| Reconstitution from purified proteins | Muscarinic | – | – | FLORIO and STERNWEIS 1985<br>KUROSE et al. 1986<br>RICHARDSON et al. 1992 |
| | Mu-opioid | – | – | UEDA et al. 1988 |
| | Adenosine $A_1$ | – | – | MUNSHI et al. 1991<br>FREISSMUTH et al. 1991 |
| Reconstitution in intact cell or isolated membranes (after PTX treatment) | Delta-opioid | VSCC | NG 108-15 hybrid cells | HESCHELER et al. 1988; TAUSSIG et al. 1992 |
| | NPY, bradykinin | VSCC | Rat dorsal root ganglion neurons | EWALD et al. 1988, 1989 |
| | Muscarinic | VSCC | Hippocampal neurons | TOSELLI et al. 1989 |
| | Alpha-2-adrenergic | VSCC (N) | NG 108-15 hybrid cells | TAUSSIG et al. 1992 |
| Stimulation of Go photolabeling with [$\alpha$-$^{32}$P]azido-GTP | Delta-opioid, somatostatin, alpha-2-adrenergic, bradykinin | – | NG 108-15 hybrid cells | OFFERMANNS et al. 1991 a,b |
| | Alpha-2-adrenergic | – | RINm 5F cells | SCHMIDT et al. 1991 |
| Intracellular injection of $G_o$-specific antibodies | Dopamine | VSCC | Snail neurons | HARRIS-WARRICK et al. 1988 |
| | Dopamine $D_2$ | VSCC (T and L) | Lactotroph cells | LLÉDO et al. 1992 |
| | Alpha-2-adrenergic | VSCC | NG 108-15 hybrid cells | MACFAZDEAN et al. 1989 |
| Intracellular injection of antisense oligonucleotides | Muscarinic (coupling to $G_{o1}$) | VSCC | GH3 | KLEUSS et al. 1991 |
| | Somatostatin (coupling to $G_{o2}$) | VSCC | GH3 | KLEUSS et al. 1991 |
| Immunoprecipitation of receptor $G_o$ complexes with anti-$G_o$ anti-receptor antibodies | Somatostatin | – | AtT 20 | LAW et al. 1991 |
| | Muscarinic | – | Heart | MATESIC et al. 1989 |

these proteins occurs in the native plasma membranes and even less in intact cell systems.

### 3. Stimulation of $G_o$ Photolabeling with [$\alpha^{32}$P]GTP Azidoanilide by Neurotransmitters

Photolabeling G-proteins with ([$\alpha^{32}$P]GTP)-azidoanilide is increased by the agonists of the receptors to which these G-proteins are coupled. This method respects the receptor–G-protein organization of the membrane. This provided a way of showing a coupling of $G_o$ but also $G_i$ proteins with (a) $\delta$-opioid, somatostatin, adrenaline, and bradykinin receptors in NG 108-15 hybrid cells (OFFERMANNS et al. 1991a,b) and (b) $\alpha_2$-adrenergic receptors in an insulin-secreting cell line (RINm5F; SCHMIDT et al. 1991; Table 1).

### 4. Intracellular Injections of G-Protein Antibodies and of Antisense Oligonucleotides Complementary to G-Protein DNA Sequences To Demonstrate the Specificity of the Negative Coupling Between Receptors and VSCC via $G_o$

To reduce the lack of specificity which could occur in reconstituted systems, we first used injection of specific anti-$G_o\alpha$ antibodies to demonstrate that dopamine receptor of snail neurons inhibit VSCC via the unique G-protein recognized by $G_o\alpha$ antibodies in snail neurons (HARRIS-WARRICK et al. 1988). More recently, we demonstrated in pituitary lactotroph cells that $G_o\alpha$ antibodies inhibit the negative coupling of dopamine $D_2$ receptors to VSCC but not the positive coupling of these receptors to $K^+$ channels (Fig. 3). $G_{i3}\alpha$ antibodies had a symmetric profile of action, they block the stimulation of $K^+$ channel activity by dopamine $D_2$ receptors, but not the negative coupling of $D_2$ dopamine receptors to VSCC (Fig. 3). $G_{i2}\alpha$ antibodies were inactive on both coupling. We verified the specificity of the antibodies by showing that they inhibit specifically the PTX-induced ADP-ribosylation of the $G\alpha$ protein against which they were directed (LLÉDO et al. 1992).

We do not know whether it is the same type of $D_2$ receptors which are coupled to both VSCC and $K^+$ channels through distinct G-proteins, or whether different $D_2$ receptors are implied. Similarly, MCFADZEAN et al. (1989) showed that in NG 108-15 cells, $G_o\alpha$ but not $G_i\alpha$ antibodies inhibit $\alpha_2$ adrenergic-induced inhibition of VSCC.

Using intranuclear injection of antisense DNA, KLEUSS et al. (1991) demonstrated that $G_o\alpha$- but not $G_i\alpha$-proteins are responsible for the negative coupling of muscarinic and somatostatin receptors to VSCC in $GH_3$ cells.

In addition, they showed that $G_{o1}\alpha$ and $G_{o2}\alpha$ mediate inhibition through muscarinic and somatostatin receptors, respectively. All these results obtained with the antisense approach indicate a higher specificity of coupling between receptors and G-proteins that was expected from reconstitution

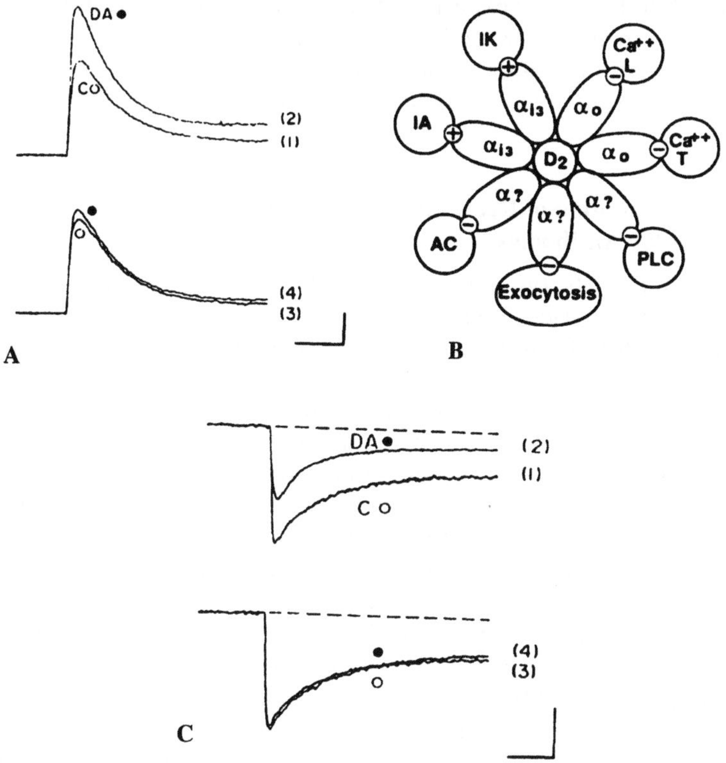


**Fig. 3A–C.** Differential G-protein-mediated coupling of $D_2$ dopamine receptors in rat anterior pituitary cells. Cells were dialyzed with pipette solutions containing affinity-purified antibodies against synthetic peptides corresponding to the C-termini of $\alpha_{i3}$ (anti-$G_{i3}\alpha c$ in A) and $\alpha_o$ (anti-$G_o\alpha c$ in C). Antibodies were dissolved in internal solutions to the final concentration of 180 $\mu$g/ml. Lactotrophs were voltage-clamped in the whole-cell patch-clamp configuration at room temperature. Currents were elicited every 10 s from −70 mV to +10 mV and from −80 mV to 0 mV for potassium and calcium. Bath applications of dopamine (10 n*M*) at 1 min (**A**, **C**, *upper recordings*) and 15 min (**A**, **C**, *lower recordings*) after beginning the whole-cell recording and dialyses with anti-$G_{i3}\alpha c$ (**A**) and anti-$G_o\alpha c$ (**C**). **A** Effects of intracellular dialysis with anti-$G_{i3}\alpha c$ on the dopamine-induced increase in potassium currents. At 1 min, dopamine induced a 54% increase in the amplitude of the peak potassium current (*2*), as compared to the control current (*1*). *Below*, superimposed potassium currents before (*3*) and during (*4*) the second application of dopamine 15 min later dialysis with anti-$G_{i3}\alpha c$. *Vertical calibration*, 100 pA; *horizontal*, 40 ms. **B** Coupling of $D_2$ dopaminergic receptors. $D_2$ dopamine receptors were coupled to potassium channels ($I_A$, $I_K$) via $C_{i3}\alpha$ ($\alpha_{i3}$) and to voltage-sensitive calcium channels (*T*, *L*), via $G_o\alpha$ ($\alpha_o$). Unidentified G proteins ($\alpha_?$) coupled $D_2$ dopaminergic receptors to adenylyl cyclase (*AC*), phospholipase C (*PLC*) and exocytosis. **C** Effect of intracellular dialysis with anti-$G_o\alpha c$ on the dopamine-induced decrease in calcium currents. At 1 min, dopamine induced a 46% decrease in the amplitude of the peak calcium current (*2*) as compared to the control current (*1*). *Below*, superimposed calcium currents before (*3*) and during (*4*) the second application of dopamine 15 min later dialysis with anti-$G_o\alpha c$. *Vertical calibration*, 60 pA; *horizontal*, 35 ms. (Results from Llédo et al. 1992)

experiments. Similarly, our experiments using lactotroph cells indicated that $G_{i3}\alpha$, but not $G_{i2}\alpha$ nor $G_{i1}\alpha$, activated $K^+$ channels (LLÉDO et al. 1992). In contrast, using reconstitution experiments, it has been shown that $G_{i1}\alpha$, $G_{i2}\alpha$, and $G_{i3}\alpha$ are all able to activate $K^+$ channels in atrial cells (YATANI et al. 1988).

### 5. Immunoprecipitation of Receptor-$G_o$ Complexes with Anti-$G_o$ Antibodies and Anti-receptor Antibodies

Recently, two reports on a new approach to study receptor–G-protein coupling has been used. After solubilization, immunoprecipitation of muscarinic receptor–G-protein complexes with anti-muscarinic antibodies was performed. Results indicated that $G_o$ ($G_\alpha$ plus $\beta\gamma$) as well as $G_i$ ($G_i\alpha$ plus $\beta$) were coimmunoprecipitated with muscarinic receptors (MATESIC et al. 1989). LAW et al. (1991) extensively studied the immunoprecipitation of the somatostatin receptor–G-protein complexes with different anti-$G\alpha$ and anti-$G\beta$ antibodies. They showed that $G_i\alpha_1$, $G_i\alpha_3$ and $G_o\alpha$ but not $G_i\alpha_2$ coimmunoprecipitate somatostatin receptors of AtT20 cells (LAW et al. 1991). The surprising finding was that brain somatostatin receptors were not immunoprecipitated by anti-$G_o\alpha$ antibodies. This indicates that either brain somatostatin receptors are not coupled to $G_o\alpha$, or that the antiserum epitope is not accessible when $G_o\alpha$ is associated with the brain somatostatin receptor (LAW et al. 1991). In addition, antiserum against $G\beta 36$ but not against $G\beta 35$ immunoprecipitate brain somatostatin receptors (LAW et al. 1991). This is the first indication that a receptor can be coupled to a specific $G\beta\gamma$ complex.

## II. Nature of VSCC Inhibited by $G_o$

We have demonstrated that both T- and L-type VSCC are inhibited by $D_2$ dopamine receptors via $G_o$ (LLÉDO et al. 1992). Other data indicate that G proteins inhibited N- and T-type VSCC (DOLPHIN 1990). Since N-type channels play a dominant role in neurosecretion, the inhibition of N-type channels by neurotransmitters may be relevant for inhibition of neurosecretion via presynaptic receptors. Indeed, the neurotransmitter receptors coupled to $G_o$ protein (see Table 1) are generally considered as inhibitory presynaptic receptors. Furthermore, the role of $G_o$ in controlling negatively neurosecretion is suggested in the report by Collu et al. (1988) showing a correlation between the expression of $G_o$ and the inhibition of prolactin secretion by dopamine.

## III. Colocalization of $G_o$ and L-Type VSCC in T-Tubule

We have investigated (TOUTANT et al. 1990) the cellular distribution of G-protein subunits in comparison with that of the voltage-dependent $Ca^{2+}$

channels by immunofluorescence on transversal and longitudinal sections of fast and slow muscles. With affinity purified antibodies against $G\beta$ subunits, a fluorescent labeling underlined the myofibrils and sarcolemma, whereas a strong immunoreaction in a dotted pattern evoked the presence of the subunit in repetitive triadic structures. With anti-$G_o\alpha$ antibodies, the immunofluorescence was more clearly focused on a dotted pattern, and the colocation with the voltage-dependent $Ca^{2+}$ channel immunoreactivity indicates that both proteins were located in very close subcellular structures.

### IV. Conclusions

Numerous types of studies using different methodological approaches indicate a rather specific, but probably not an exclusive role for $G_o\alpha$ in inhibiting VSCC. Only two studies provide evidence for a possible role of $G_i\alpha$ protein in inhibiting these channels (EWALD et al. 1989; TAUSSIG et al. 1992), whereas another study shows that $G_o\alpha$ can also activate $K^+$ channels of hippocampal neurons (VAN DONGEN et al. 1988). $K^+$ channels of atrial cells are not activated by $G_o\alpha$ but rather by $G_i\alpha$ (YATANI et al. 1987). When the mechanism by which $G_o\alpha$ inhibits VSCC has been analyzed, it has been demonstrated that it does not imply known second messengers. Therefore, a direct interaction between $G_o$ and VSCC is likely but remains to be demonstrated. The indication that some receptors control VSCC via $G_{o1}\alpha$ while others (such as somatostatin) may control these channels via $G_{o2}\alpha$ (KLEUSS et al. 1991; TAUSSIG et al. 1992) is intriguing since the amino acid domain which is the most divergent between these two $G_o$ proteins is not generally considered as being a domain of interaction with receptors.

## E. General Conclusion

It is now clearly established that $G_o\alpha$ has a key role in the negative coupling of some receptors (which are generally considered as being presynaptic inhibitory receptors) to VSCC. It is also likely that $G_{o1}\alpha$ has, at least in some cells, a stimulatory role on phospholipase C (see S.D. KROLL and R. IYENGAR, this volume). The possibility that the interactions of $G_o\alpha$ with some cytoskeletal elements (e.g., microtubules, actin) and growth cone associated proteins (GAP43) has some role to play in the architectural organization of neurons, remains to be demonstrated.

### References

Asano T, Semba R, Kamiya N, Ogasawara N, Kato K (1988a) $G_o$, a GTP binding protein: immunochemical and immunohistochemical localization in the rat. J Neurochem 50:1164–1169

Asano T, Kamiya N, Semba R, Kato K (1988b) Ontogeny of the GTP-binding protein $G_o$ in rat brain and heart. J Neurochem 51:1711–1716

Asano T, Morishita R, Sano M, Kato K (1989) The GTP binding proteins $G_o$ and $G_{i2}$, of neural cloned cells and their changes during differentiation. J Neurochem 53:1145–1198

Asano T, Morishita R, Kato K (1991) Immuno-affinity purification and characterization of the $\alpha$ subunits of $G_o$ type G proteins from various bovine tissues. J Biochem 110:571–574

Bertrand P, Sanford J, Rudolph U, Codina J, Birnbaumer L (1990) At least three alternatively spliced mRNAs encoding two $\alpha$ subunits of the $G_o$ GTP binding protein can be expressed in a single tissue. J Biol Chem 265:18576–18580

Bockaert J, Brabet Ph, Gabrion J, Homburger V, Rouot B, Toutant M (1990) Structural, immunological and functional characterization of guanine nucleotide binding protein $G_o$. In: Iyengar R, Birnbaumer L (eds) G proteins, Acad Press Inc, San Diego, pp 81–103

Brabet P, Dumuis A, Sebben M, Pantaloni C, Bockaert J, Homburger V (1988) Immunocytochemical localization of the guanine nucleotide binding protein $G_o$ in primary cultures of neuronal and glial cells. J Neurosci 8:701–708

Brabet P, Pantaloni C, Rodriguez M, Martinez J, Bockaert J, Homburger V (1990) Neuroblastoma differentiation involves the expression of two isoforms of the $\alpha$ subunit of $G_o$. J Neurochem 54:1310–1320

Brabet P, Pantaloni C, Bockaert J, Homburger V (1991) Metabolism of two $Go_\alpha$ isoforms in neuronal cells during differentiation. J Biol Chem 266: 12825–12828

Braun PE, Horvath E, Yong VW, Bernier L (1990) Identification of GTP binding proteins in myelin and oligodendrocyte membranes. J Neurosci Res 26:16–23

Canlon B, Homburger V, Bockaert J (1991) The identification and localization of the guanine nucleotide binding protein $G_o$ in the auditory system. Eur J Neurosci 3:1388–1342

Chang KJ, Pugh N, Blanchard SG, McDermed J, Tam JP (1988) Antibody specific to the $\alpha$ subunit of the guanine nucleotide binding regulatory protein $G_o$. Developmental appearance and immunocytochemical localization in brain. Proc Natl Acad Sci USA 85:4929–4933

Collu R, Bouvier C, Lagacé G, Unson CG, Milligan G, Goldsmith P, Spiegel A (1988) Selective deficiency of the guanine nucleotide binding protein $G_o$ in two dopamine-resistant pituitary tumors. Endocrinology 122:1176–1178

De Sousa SM, Hoveland LL, Yarfitz S, Hurley JB (1989) The Drosophila $G_o\alpha$-like G protein gene produces multiple transcripts and is expressed in the nervous system and ovaries. J Biol Chem 264:18544–18551

Dolphin A (1990) G protein modulation of calcium currents in neurons. Ann Rev Physiol 52:243–255

Edmonds BT, Moomaw CR, Hsu JT, Slaughter C, Ellis L (1990) The p38 and p34 polypeptides of growth cone particle membranes are the $\alpha$- and $\beta$-subunits of G proteins. Dev Brain Res 56:131–136

Ewald DA, Sternweis PC, Miller RJ (1988) Guanine nucleotide binding protein $G_o$ induced coupling of neuropeptide Y receptors to $Ca^{++}$ channels in sensory neurons. Proc Natl Acad Sci USA 85:3633–3637

Ewald DA, Pang IH, Sternweis PC, Miller EJ (1989) Differential G protein mediated coupling of neurotransmitter receptors to $Ca^{++}$ channels in rat dorsal root ganglion neurons in vitro. Neuron 2:1185–1193

Florio VA, Sternweis PC (1989) Mechanisms of muscarinic receptor action on $G_o$ in reconstituted phospholipid vesicles. J Biol Chem 264:3905–3915

Freissmuth M, Schütz W, Linder ME (1991) Interaction of bovine brain $A_1$-adenosine receptors with recombinant G protein $\alpha$ subunits. J Biol Chem 266:17778–17783

Gabrion J, Brabet P, Nguyen Than Dao B, Homburger V, Dumuis A, Sebben M, Rouot B, Bockaert J (1989) Ultrastructural localization of the GTP binding protein $G_o$ in neurons. Cell Signalling 1:107–132

Garribay JLR, Kozasa T, Itoh H, Tsukamoto T, Matsuoka M, Kaziro Y (1991) Analysis by mRNA levels of the expression of six G protein $\alpha$ subunit genes in mammalian cells and tissues. Biochem Biophys Acta 1094:193–199

Goldsmith P, Backlund PS, Rossiter K, Carter A, Milligan G, Unson CG, Spiegel A (1988) Purification of heterotrimeric GTP binding proteins from brain: identification of a novel form of $G_o$. Biochemistry 27:7085–7090

Granneman J, Kapatos G (1990) Developmental expression of $G_o$ in neuronal cultures from rat mesencephalon and hypothalamus. J Neurochem 54:1995–2001

Guillen A, Jallon JM, Fehrentz JA, Pantaloni C, Bockaert J, Homburger V (1990) A $G_o$-like protein in Drosophila melanogaster and its expression in memory mutants. EMBO J 9:1449–1455

Guillen A, Sémériva M, Bockaert J, Homburger V (1991) The transduction signalling protein $G_o$ during embryonic development of Drosophila melanogaster: Its possible role in axonogenesis. Cellular Signalling 3:341–352

Harris-Warrick R, Hammond C, Paupardin-Tritsch D, Homburger V, Rouot B, Bockaert J, Gerschenfeld HM (1988) The $\alpha$ subunit of a GTP binding protein homologous to mammalian $G_o\alpha$ mediates a dopamine-induced decrease of calcium current in snail neurons. Neuron 1:27–32

Hescheler J, Rosenthal W, Trautwein W, Schultz G (1987) The GTP binding protein $G_o$ regulated neuronal calcium channels. Nature 325:445–447

Homburger V, Brabet P, Audigier Y, Pantaloni C, Bockaert J, Rouot B (1987) Immunological localization of the GTP binding protein $G_o$ in different tissues of vertebrates and invertebrates. Mol Pharmacol 31:313–319

Hopkins RS, Stamnes MA, Simon MI, Hurley JB (1988) Cholera-toxin and pertussis-toxin substrates and endogeneous ADP-ribosyltransferase activity in Drosophila melanogaster. Biochim Biophys Acta 970:355–362

Hsu WH, Rudolph U, Sanford J, Bertrand P, Olate J, Nelson C, Moss LG, Boyd AE III, Codina J, Birnbaumer L (1990) Molecular cloning of a novel splice variant of the $\alpha$ subunit of the mammalian $G_o$ protein. J Biol Chem 265:11220–11226

Itoh H, Kozasa T, Nagata S, Nakamura SI, Katada T, Ui M, Iwai S, Ohtsuka E, Kawasaki H, Suzuki K, Kaziro Y (1986) Molecular cloning and sequence determination of cDNAs for $\alpha$ subunits of the guanine nucleotide binding protein $G_s$, $G_i$ and $G_o$ from brain. Proc Natl Acad Sci USA 83:3776–3780

Jones DT, Reed RR (1987) Molecular cloning of five GTP binding protein cDNA species from rat olfactory neuroepithelium. J Biol Chem 262:14241–14249

Journot L, Homburger V, Pantaloni C, Priam M, Bockaert J, Enjalbert A (1987) An IAP-sensitive G protein is involved in dopamine inhibition of angiotensin- and TRH-stimulated inositol phosphate production in anterior pituitary cells. J Biol Chem 262:15106–15110

Kaziro Y, Itoh H, Nakafuku M (1990) Organization of genes coding for G protein $\alpha$ subunits in higher and lower eukaryotes. In: Iyengar R, Birnbaumer L (eds) G proteins. Acadomic, San Diego, p 63

Kleuss C, Hescheler J, Ewel C, Rosenthal W, Schultz G, Wittig B (1991) Assignment of G-protein subtypes to specific receptors inducing inhibition of calcium currents. Nature 353:43–48

Kurose H, Katada T, Haga T, Haga K, Ichiyama A, Ui M (1986) Functional interaction of purified receptors with purified inhibitory guanine nucleotide regulatory proteins reconstituted in phospholipid vesicles. J Biol Chem 261:6423–6428

Largent BL, Jones DT, Reed RR, Pearson RCA, Snyder SH (1988) G protein mRNA mapped in rat brain in situ hybridization. Proc Natl Acad Sci USA 85:2864–2868

Lavu S, Clark J, Swarup R, Matsushima K, Paturu K, Moss J, Kung H (1988) Molecular cloning and DNA sequence analysis of the human guanine nucleotide binding protein $G_o\alpha$. Biochim Biophys Acta 150:811–815

Law SF, Manning D, Reisine T (1991) Identification of the subunits of GTP binding proteins coupled to somatostatin receptors. J Biol Chem 266:17885–17897
Llédo PM, Homburger V, Bockaert J, Vincent J-D (1992) Differential G protein-mediated coupling of $D_2$ dopamine receptors to $K^+$ and $Ca^{++}$ channels in rat anterior pituitary cells. Neuron 8:455–463
Matesic DF, Manning DR, Wolfe BB, Luthin GR (1989) Pharmacological and biochemical characterization of complexes of muscarinic acetylcholine receptors and guanine nucleotide binding proteins. J Biol Chem 264:21638–21645
MacFazdean I, Mullaney I, Brown DA, Milligan G (1989) Antibodies to the GTP binding protein $G_o$ antagonize noradrenaline-induced calcium current inhibition in NG 108-15 hybrid cells. Neuron 3:177–182
Milligan G, Streaty RA, Gierschik P, Spiegel AM, Klee WA (1987) Development of opiate receptors and GTP binding regulatory proteins in neonatal rat brain. J Biol Chem 262:8626–8630
Mullaney I, Magee AI, Unson CG, Milligan G (1988) Differential regulation of amounts of the guanine nucleotide binding proteins $G_i$ and $G_o$ in neuroblastoma hybrid cells in response to dibutyryl cyclic AMP. Biochem J 256:649–656
Munshi R, Pang IH, Sternweis PC, Linden J (1991) $A_1$-Adenosine receptors of bovine brain couples to guanine nucleotide proteins $G_{i1}$, $G_{i2}$ and $G_o$. J Biol Chem 266:22285–22289
Murtagh JJ Jr, Eddy R, Shows TB, Moss J, Vaughan M (1991) Different forms of $G_o\alpha$ mRNA arise by alternative splicing of transcripts from a single gene on human chromosome 16. Mol Cell Biol 11:1146–1155
Offermanns S, Gollasch M, Hescheler J, Spicher K, Schmidt A, Schultz G, Rosenthal W (1991a) Inhibition of voltage-dependent $Ca^{++}$ currents and activation of pertussis toxin sensitive G proteins via muscarinic receptors in GH3 cells. Mol Endocrinol 5:995–1002
Offermanns S, Schultz G, Rosenthal W (1991b) Evidence for opioid receptor-mediated activation of G proteins, $G_o$, $G_{i2}$ in membranes of neuroblastoma X glioma (NG 108-15) hybrid cells. J Biol Chem 266:3365–3368
Olate J, Martinez S, Purcell P, Jorquera H, Codina J, Birnbaumer L, Allende JE (1990) Molecular cloning and sequence of four different cDNA species coding for $\alpha$ subunits of G proteins from Xenopus laevis oocytes. FEBS Lett 268:27–31
Peraldi S, Nguyen Than Dao B, Brabet P, Homburger V, Rouot B, Toutant M, Bouillé C, Assenmacher I, Bockaert J, Gabrion J (1989) Apical localization of the $\alpha$ subunit of GTP binding protein $G_o$ in choroïdal and ciliated ependymocytes. J Neurosci 9:806–814
Pituello F, Homburger V, Audigier Y, Bockaert J, Duprat AM (1991) Expression of the guanine nucleotide binding protein $G_o$ correlates with the state of neural competence in the amphibian embryon. Devel Biol 145:311–322
Price SR, Murtagh JJ Jr, Tsuchiya M, Serventi IM, Van Meurs KP, Angus W, Moss J, Vaughan M (1990) Multiple forms of $G_o\alpha$ mRNA: analysis of the 3′-untranslated regions. Biochemistry 29:5069–5076
Richardson RM, Mayanil CSK, Hosey MM (1992) Subtype-specific antibodies for muscarinic cholinergic receptors. II. Studies with reconstituted chick heart receptors and the GTP binding protein, $G_o$. Mol Pharmacol 40:908–914
Quan F, Wolfgang WJ, Forte MA (1989) The Drosophila gene coding for $\alpha$ subunit of a stimulatory G protein is preferentially expressed in the nervous system. Proc Natl Acad Sci USA 86:4321–4325
Rouot B, Charpentier N, Chabbert C, Carrette J, Zumbihl R, Bockaert J, Homburger V (1992) Specific antibodies against $G_o$ isoforms reveal the early expression of the $G_{o2}\alpha$ subunit and appearance of $G_{o1}\alpha$ during neuronal differentiation. Mol Pharmacol 41:273–280
Scherer NM, Toro MJ, Entman ML, and Birnbaumer L (1987) G protein distribution in canine cardiac sarcoplasmic reticulum and sarcolemma:

comparison to rabbit skeletal mouse membranes and to brain and erythrocyte G proteins. Arch Biochem Biophys 259:431–440
Schmidt A, Hescheler J, Offeremanns S, Spicher K, Hinsch KD, Klinz FJ, Codina J, Birnbaumer L, Gausepohl H, Frank R, Schultz G, Rosenthal W (1991) Involvement of pertussis toxin-sensitive G proteins in the hormonal inhibition of dihydropyridine-sensitive $Ca^{++}$ currents in an insulin-secreting cell line (RINm5F). J Biol Chem 266:18025–18033
Schmidt CJ, Garen-Faziot S, Chow Y-K, Neer EJ (1989) Neuronal expression of a newly identified Drosophila melanogaster G protein $\alpha o$ subunit. Cell Regul 1:125–134
Spicher K, Klinz FJ, Rudolph U, Codina J, Birnbaumer L, Schultz G, Rosenthal W (1991) Identification of the G protein $\alpha$ subunit encoded by $\alpha o_2$ cDNA as a 39 kDa pertussis toxin substrate. Biochem Biophys Res Commun 175:473–479
Strathmann M, Wilkie TM, Simon MI (1990) Alternative splicing produces transcripts encoding two forms of the $\alpha$ subunit of GTP binding $G_o$. Proc Natl Acad Sci USA 87:6477–6481
Strittmatter SM, Valenzuela D, Kennedy TE, Neer EJ, Fishman MC (1990) $G_o$ is a major growth cone protein subject to regulation by GAP 43. Nature 344:836–841
Taussig R, Sanchez S, Rifo M, Gilman AG, Belardetti F (1992) Inhibition of the w-conotoxin-sensitive calcium current by distinct G proteins. Neuron 8:799–809
Terashima T, Katada T, Oinuma M, Inoue Y, Ui M (1987) Immunohistochemical localization of guanine nucleotide binding proteins in retina. Brain Res 410:97–100
Thambi NC, Quan F, Wolfgang WJ, Spiegel A, Forte M (1989) Immunological and molecular characterization of a $G_o$-like protein in the Drosophila central nervous system. J Biol Chem 264:18552–18560
Toselli M, Lang J, Costa T, Lux HD (1989) Direct production of voltage-dependent calcium channels by muscarinic activation of a pertussis toxin-sensitive G protein in hippocampal neurons. Pflügers Arch 415:255–261
Toutant M, Gabrion M, Vandaele S, Peraldi-Roux S, Barhanin J, Bockaert J, Bouot B (1990) Cellular distribution and biochemical characterization of G proteins in skeletal muscle: comparative location with voltage-dependent calcium channels. EMBO J 9:363–369
Tsukamato T, Toyama R, Itoh H, Kozasa T, Matsuoko M, Kaziro Y (1991) Structure of the human gene and two rat cDNAs encoding the $\alpha$ chain of GTP binding regulatory proteins $G_o$: two different mRNAs are generated by alternative splicing. Proc Natl Acad Sci USA 88:2974–2978
Ueda H, Harada H, Nozaki M, Katada T, Ui M, Satoh M, Takagi H (1988) Reconstitution of rat brain $\mu$-opioid receptors with purified guanine nucleotide binding regulatory proteins $G_i$ and $G_o$. Proc Natl Acad Sci USA 85:7013–7017
Van Dongen AMJ, Codina J, Olate J, Mattera R, Joho R, Birnbaumer L, Brown AM (1988) Newly identified brain potassium channels gated by the guanine nucleotide binding protein $G_o$. Science 242:1433–1436
Vincent SR, Hope BT, Drinnan SL, Reiner PB (1990) G protein mRNA expression in immunohistochemically identified dopaminergic and noradrenergic neurons in the rat brain. Synapse 6:23–32
Wolfgang WJ, Quan F, Goldsmith P, Unson C, Spiegel A, Forte M (1990) Immunolocalization of G protein $\alpha$ subunits in the Drosophila central nervous system. J Neurosci 10:1014–1024
Wolfgang WJ, Quan F, Thambi N, Forte M (1991) Restructured spatial and temporal expression of G protein $\alpha$ subunit during Drosophila embryogenesis. Development 113:527–538
Worley PF, Baraban JM, Van Dop C, Neer EJ, Snyder SH (1986) $G_o$, a guanine nucleotide binding protein: immunohistochemical localization in rat resembles

distribution of second messenger systems. Proc Natl Acad Sci USA 83:4561–4565
Yatani A, Codina J, Brown AM, Birnbaumer L (1987) Direct activation of mammalian atrial muscarinic potassium channels by GTP regulatory protein Gk. Science 235:207–211
Yoon J, Shortridge RD, Bloomquist BT, Schneuwly S, Perdew MH, Pak WL (1989) Molecular characterization of Drosophila gene encoding $G_o\alpha$ subunit homolog. J Biol Chem 264:18536–18543

CHAPTER 72

# Involvement of Pertussis-Toxin-Sensitive G-Proteins in the Modulation of $Ca^{2+}$ Channels by Hormones and Neurotransmitters

W. ROSENTHAL

## A. Introduction

In recent years, the modulation of ion channels via G-protein-coupled receptors has become a focus of interest in many laboratories. This review deals with the modulation of voltage-dependent $Ca^{2+}$ channels by the nonretinal G-proteins $G_i$ and $G_o$, which are substrates for the main exotoxin of *Bordetella pertussis*, pertussis toxin. Whereas members of the $G_i$ family are found in all mammalian cells, $G_o$-type G-proteins appear to be limited to neuronal and some nonneuronal secretory cells. Pertussis toxin catalyzes the ADP-ribosylation of a cysteine residue near the C terminus of the polypeptide chain forming G-protein $\alpha$ subunits. G-proteins modified by the toxin loose their ability to interact with activated receptors. For further details on G-proteins, ADP-ribosylating bacterial toxins and the involvement of the cholera-toxin-sensitive G-protein $G_s$ in the stimulation of voltage-dependent $Ca^{2+}$ channels see other chapters in this volume. This review is not comprehensive; it is meant to highlight the achievements and the remaining ambiguities in the field. Previous reviews covering the subject of this contribution have appeared (e.g., DOLPHIN 1990; MILLER 1990; ROSENTHAL et al. 1988b, 1989, 1990, 1992; SCHULTZ et al. 1990).

## B. Inhibitory Modulation of Voltage-Dependent $Ca^{2+}$ Channels

### I. Occurrence; Physiological Significance

The first report on receptor-mediated inhibition of voltage-dependent $Ca^{2+}$ currents appeared in 1981, when DUNLAP and FISCHBACH discovered that catecholamines, known to stimulate $Ca^{2+}$ currents in cardiac cells (REUTER 1983), inhibit $Ca^{2+}$ currents in neuronal cells. Subsequently, this observation was extended to various types of neuronal cells. An inhibitory, somatostatin-induced modulation of $Ca^{2+}$ currents in a nonneuronal cell, the mouse pituitary AtT-20 cell, was reported by LUINI et al. (1986). In the same year, DUNLAP's group showed that the inhibition of neuronal $Ca^{2+}$ currents by noradrenaline (acting on $\alpha_2$-adrenergic receptors) and GABA $\gamma$-aminobutyric acid (acting on $GABA_B$ receptors) was affected by guanine

**Table 1.** Pertussis toxin-sensitive inhibition of voltage-dependent $Ca^{2+}$ channels

| Cell | Receptor or agonist | Reference |
|---|---|---|
| Mouse dorsal root ganglion neurons | α2-adrenergic, $GABA_B$ | HOLZ et al. 1986 |
| Rat dorsal root ganglion neurons | $GABA_B$ | DOLPHIN and SCOTT 1987 |
| | Neuropeptide | EWALD et al. 1988 |
| Rat hippocampal neurons | Muscarinic | TOSELLI et al. 1989 |
| Human neuroblastoma (SH-SY5Y) cells | μ-Opioid | SEWARD et al. 1991 |
| Mouse/rat neuroblastoma x glioma hybrid cells (NG 108-15) | δ-Opioid | HESCHELER et al. 1987 |
| | $\alpha_2$-Adrenergic | MCFADZEAN et al. 1989 |
| | Somatostatin | TAUSSIG et al. 1992 |
| | Cannabinoids | MACKIE and HILLE 1992 |
| Rat sympathetic ganglion neurons | Muscarinic | WANKE et al. 1987 |
| | Somatostatin | IKEDA and SCHOfiELD 1989 |
| | $\alpha_2$-Adrenergic | SCHOFIED 1990 |
| Rat parasympathetic (nodose) ganglion neurons | κ-Opioid | GROSS et al. 1990 |
| Guinea pig parasympathetic (submucosal) ganglion neurons | $\alpha_2$-Adrenergic, δ-Opioid, Somatostatin | SUPRENANT et al. 1990 |
| Snail neurons | Dopamine | HARRIS-WARRICK et al. 1988 |
| Mouse pituitary (AtT-20) cells | Somatostatin | LEWIS et al. 1986 |
| Rat pituitary ($GH_3$) cells | Somatostatin | ROSENTHAL et al. 1988a |
| | $M_4$-muscarinic | OFFERMANNS et al. 1991b |
| Rat anterior pituitary cells | Dopamine | LLEDO et al. 1992 |
| SV 40-transformed hamster pancreatic B cells (HIT) | $\alpha_2$-Adrenergic | KEAHEY et al. 1989a |
| | Somatostatin | HSU et al. 1991 |
| Rat insulinoma (RINm5F) cells | $\alpha_2$-Adrenergic, somatostatin | SCHMIDT et al. 1991 |
| Bovine adrenal chromaffin cells | $\alpha_2$-Adrenergic, opioid | KLEPPISCH et al. 1992 |
| Rat pheochromocytoma (PC 12) cells | $\alpha_2$-Adrenergic, muscarinic, opioid | GOLLASCH et al. 1991b |
| Rat calcitonin-secreting (rMTC 44-2) cells | Somatostatin | SCHERÜBL et al. 1992 |

nucleotides and pertussis toxin, suggesting the involvement of a pertussis-toxin-sensitive G-protein (HOLZ et al. 1986). LEWIS et al. (1986) reported similar findings with regard to $Ca^{2+}$ currents in AtT-20 cells. More recently, a pertussis toxin-sensitive inhibition of $Ca^{2+}$ currents has been demonstrated in a number of neuronal cells and in various nonneuronal secretory cells (Table 1). The latter group includes rat anterior pituitary cells, a rat ($GH_3$) and a mouse (AtT-20) pituitary cell line, rat insulinoma (RINm5F) cells, hamster SV 40-transformed pancreatic B (HIT) cells, bovine adrenal chromaffin cells, rat pheochromocytoma (PC12) cells, and a rat calcitonin-secreting cell line (rMTC 44-2). Thus, this pathway appears to be a regulatory principle of all neuronal and neuroendocrine cells and of pituitary cells.

The $Ca^{2+}$ current inhibition is induced by secretion-inhibiting receptor agonists. In neuronal cells, the corresponding receptors are localized presynaptically. The inhibition of neuronal $Ca^{2+}$ currents and of $Ca^{2+}$ currents in chromaffin cells leads to a decreased influx of $Ca^{2+}$ during a depolarizing pulse and thereby attenuates the release of neurotransmitter or hormones, respectively (see, e.g., LIPSCOMBE et al. 1989; RANE et al. 1987 and references cited therein). In the other nonneuronal secretory cells, the decrease in $Ca^{2+}$ currents by inhibitory receptor agonists contributes to a decrease in cytosolic $Ca^{2+}$ (e.g., LUINI et al. 1986) and consequently to an attenuation of basal secretion or of a secretory response elicited *via* stimulatory receptors coupled to G-proteins (see Sect. C). For further considerations of functional aspects, see ROSENTHAL et al. (1988b) and references cited therein.

## II. Effects of Receptor Agonists, Pertussis Toxin, and Guanine Nucleotides

In most studies reviewed here, the whole-cell variation of the patch-clamp technique was applied (HAMILL et al. 1981); $Ca^{2+}$ currents were measured under voltage-clamp conditions, using $Ba^{2+}$ as a charge carrier. In this experimental setup, the reduction of $Ca^{2+}$ currents, induced by numerous receptor agonists (see Table 1), occurs without detectable delay and is maximal within about a minute. The $Ca^{2+}$ current increases immediately upon removal of receptor agonist from the bath solution and reaches control values after about a minute. Incubation of cells with pertussis toxin (holotoxin) prior to the experiment or intracellular application of the activated toxin greatly reduces or abolishes the receptor-mediated $Ca^{2+}$ current inhibition. A few reports have described a pertussis-toxin-insensitive inhibition of neuronal $Ca^{2+}$ currents or of current components by muscarinic agents and catecholamines (e.g., BEECH et al. 1992; SONG et al. 1989). Poorly or nonhydrolyzable GTP analogs, GTP$\gamma$S (guanosine-5′-0-[3-thiotriphosphate]) or GMP(NH)P (guanosine-5′-0-[2-thiodiphosphate]), present at concentrations above $50\,\mu M$ in the pipette solution, cause a strong inhibition of $Ca^{2+}$ currents, which is slow in the absence of receptor agonist ($t_{1/2} \gg 1\,\text{min}$). Dialysis of cells with the GDP analog GDP$\beta$S (guanosine-5′-[$\beta\gamma$-imino]triphosphate) generally renders cells unresponsive to receptor agonists; an exception is the somatostatin-induced inhibition of $Ca^{2+}$ currents in rat sympathetic neurons (IKEDA and SCHOFIELD 1989). A slow decrease in neuronal $Ca^{2+}$ currents is also observed upon a fast (stepwise) increase in the concentration of intracellular GTP analogs, as has been demonstrated by the use of caged compounds which are activated by light (DOLPHIN et al. 1988). Cells dialyzed with GTP analogs ($\geq 1\,\mu M$ in the pipette solution) respond to receptor agonists with a rapid irreversible inhibition of $Ca^{2+}$ currents. There are conflicting reports regarding the effect of pertussis on the inhibition of $Ca^{2+}$ currents by GTP analogs.

For example, the toxin abolishes the receptor-mediated inhibition of $Ca^{2+}$ currents in NG108-15 cells (identical with 108CC15 cells), but does not interfere with the GTPγS-induced inhibition of $Ca^{2+}$ currents (HESCHELER et al. 1988b); in rat dorsal root ganglion neurons, pertussis toxin abolishes the inhibition of $Ca^{2+}$ currents by a $GABA_B$ receptor agonist and GTPγS (DOLPHIN and SCOTT 1987). For further details on the effects of guanine nucleotides on $Ca^{2+}$ currents, see references in Table 1.

Although activation of a G-protein requires the exchange of GDP for GTP on the α subunit, addition of GTP to the pipette solution is not required in order to observe a G-protein-mediated inhibition of $Ca^{2+}$ channels. In fact, in most studies pipette solutions containing ATP but not GTP were used. Two explanations may account for this observation: a) ATP or its analog ATPγS (adenosine-5′-0-[3-thiotriphosphate]) can serve as a substrate in a reaction catalyzed by a GTP-forming nucleoside phosphokinase. A G-protein-associated form of the enzyme may transfer the terminal (thio)phosphate group of ATP (ATPγS) onto GDP bound to the G-protein α subunit, thereby yielding an active α subunit (GROSS et al. 1991; JAKOBS and WIELAND 1989; see also Chap. 74 by N. KIMURA); b) due to tight binding or compartmentalization, the concentration of GTP at the inner face of the plasma membrane is maintained sufficiently high to allow activation of G-proteins.

## III. Types of $Ca^{2+}$ Channels Affected by Inhibitory Receptor Agonists

Depending on their biophysical and biochemical properties, voltage-dependent $Ca^{2+}$ channels have been divided into three classes. These are low-voltage-activated or T-type channels, high voltage-activated, dihydropyridine-sensitive or L-type channels, and high voltage-activated, ω-conotoxin-sensitive or N-type channels (NOWYCKY et al. 1985; FOX et al. 1987; MCCLESKEY et al. 1987; PLUMMER et al. 1989). Whereas the L- and T-type channels are found in a wide variety of nonneuronal and neuronal cells, N-type channels are present in neuronal cells and also in PC12 and adrenal chromaffin cells (e.g., GOLLASCH et al. 1991b; KLEPPISCH et al. 1992; PLUMMER et al. 1989). The discovery of a new class of neuronal $Ca^{2+}$ channels (P-type channels) which are sensitive to a funnel web spider toxin but insensitive to dihydropyridines and to ω-conotoxin (LLINAS et al. 1989) and the emerging molecular diversity of $Ca^{2+}$ channel subunits (see Chap. 57 by F. HOFMANN et al.) may indicate that the above classification is far too simple.

In most neuronal cells, the $Ca^{2+}$ current component reduced by receptor agonists is blocked by ω-conotoxin, suggesting the involvement of N-type channels (e.g., GROSS and MACDONALD 1987; SEWARD et al. 1991; TAUSSIG et al. 1992; TOSELLI et al. 1989; WANKE et al. 1987; for review see TSIEN et al. 1988). Consistent with this assumption, caged GTP analogs abolish transient $Ca^{2+}$ currents (i.e., currents through N- or T-type channels) in chick dorsal root ganglion neurons, whereas sustained $Ca^{2+}$

currents (i.e., currents through L-type channels) are less affected (Dolphin et al. 1988). An inhibitory modulation of N-type channels is also evident from data obtained on the level of single channels (Lipscombe et al. 1989): in sympathetic neurons, noradrenaline decreases the mean open time of N-type channels, whereas L-type channels are not significantly affected by the catecholamine. The same study shows that N-type channels are the dominant entry pathway for $Ca^{2+}$ and thereby crucial for the release of neurotransmitter.

Other types of $Ca^{2+}$ channels have also been reported to be involved in the release of neurotransmitter and to be modulated via inhibitory receptors. Besides inhibiting $Ca^{2+}$ currents through N-type channels, opioids inhibit $Ca^{2+}$ currents through T-type channels in rat dorsal root ganglion neurons (Schroeder et al. 1991) and in NG108-15 cells (Kasai 1992). Similarly, the $GABA_B$-receptor agonist baclofen and the adenosine $A_1$ receptor agonist 2-chloroadenosine inhibit low-threshold $Ca^{2+}$ currents (i.e., currents through T-type channels) and high-threshold $Ca^{2+}$ currents (i.e., currents through N- or L-type channels) in rat dorsal root ganglion neurons (for review, see Dolphin and Scott 1990). Furthermore, high- and low-threshold $Ca^{2+}$ currents are inhibited by dopamine in dorsal root ganglion neurons of chick (Marchetti et al. 1986). In rat dorsal root ganglion neurons, the modulation of $Ca^{2+}$ currents through T-type channels *via* G-proteins appears to be complex: they are inhibited or stimulated by baclofen employed at a high ($100\,\mu M$) or a low concentration ($2\,\mu M$), respectively; likewise, GTP$\gamma$S released from a caged precursor inhibits $Ca^{2+}$ currents through T-type channels if employed at a high concentration ($20\,\mu M$) and stimulates them if employed at a low concentration ($6\,\mu M$) (Scott et al. 1990). Only the inhibitory modulations are sensitive to pertussis toxin.

An inhibitory modulation of neuronal $Ca^{2+}$ currents through L-type channels has mainly been proposed by Dunlap and coworkers. Working with dorsal root ganglion neurons of chick, they found an inhibition of $Ca^{2+}$ currents and $K^+$-evoked transmitter release by dihydropyridines and an inhibition of $Ca^{2+}$ currents through L-type channels by noradrenaline and GABA (the latter acting on $GABA_B$ receptors; Rane et al. 1987, and refs. cited therein). An interaction of G-proteins and neuronal L-type channels is also suggested by the finding that GTP$\gamma$S promotes the stimulatory effect of dihydropyridines and other classes of $Ca^{2+}$ channel ligands on $Ca^{2+}$ currents through L-type channels in rat dorsal root ganglion neurons; the effect of the nucleotide is abolished by pertussis toxin (Scott and Dolphin 1987).

All nonneuronal cells exhibiting a pertussis toxin-sensitive inhibition of $Ca^{2+}$ currents (see Table 1) possess dihydropyridine-sensitive $Ca^{2+}$ currents. $Ca^{2+}$ currents through N-type channels are not detectable in AtT20 cells (Luini et al. 1986) and RINm5F cells (Schmidt et al. 1991). According to one report (Suzuki and Yoshioka 1987), a $Ca^{2+}$ current component in $GH_3$ cells is sensitive to $\omega$-conotoxin; however, subsequent studies failed to detect an effect of $\omega$-conotoxin on $Ca^{2+}$ currents in this cell type

(OFFERMANNS et al. 1991a; GOLLASCH et al. 1991a). ω-Conotoxin-sensitive $Ca^{2+}$ currents are detectable in adrenal chromaffin cells (KLEPPISCH et al. 1992) and undifferentiated and differentiated PC12 cells (GOLLASCH et al. 1991b; PLUMMER et al. 1989). The portion of the $Ca^{2+}$ current reduced by inhibitory receptor agonists is sensitive to dihydropyridines in AtT20 cells (LUINI et al. 1986), $GH_3$ cells (OFFERMANNS et al. 1991a), adrenal chromaffin cells (KLEPPISCH et al. 1992), undifferentiated PC12 cells (GOLLASCH et al. 1991b), RINm5F cells (SCHMIDT et al. 1991), and HIT cells (KEAHEY et al. 1989a,b). Only in differentiated PC12 cells, internally applied GTPγS, and, by inference, inhibitory receptor agonists appear to inhibit $Ca^{2+}$ currents through N-type channels (PLUMMER et al. 1989). The functional significance of L-type channels for the secretory response is underlined by the fact that dihydropyridines inhibit $Ca^{2+}$ dependent secretion in all groups of nonneuronal cells mentioned above (see refs. cited above and in Table 1). Taken together the results obtained with neuronal and nonneuronal cells, it appears likely that one type of inhibitory receptor is involved in the modulation of N-type $Ca^{2+}$ channels in neuronal cells and in the modulation of L-type $Ca^{2+}$ channels in nonneuronal cells.

## IV. Mechanistic Aspects

### 1. Cyclic Nucleotides

$Ca^{2+}$ current-inhibiting receptor agonists also inhibit adenylyl cyclase. Like the inhibition of $Ca^{2+}$ currents, the inhibition of the enzyme is sensitive to pertussis toxin; the identities of the involved G-protein (possibly a member of the $G_i$ family) and the active subunits (α-subunits vs. βγ complexes) are still a matter of controversy (see Chap. 70 by J. HILDEBRANDT). Considering the fact that cardiac $Ca^{2+}$ currents are stimulated by cAMP-dependent protein kinase (see Chap. 57 by F. HOFMANN et al.), it is tempting to speculate that a decrease of intracellular cyclic AMP and a consequent decrease in the activity of cAMP-dependent protein kinase are causal for a receptor-mediated decrease in $Ca^{2+}$ currents. Although cAMP-dependent protein kinase increases neuronal $Ca^{2+}$ currents (GROSS et al. 1990; for review, see DOLPHIN 1990; DOLPHIN and SCOTT 1990), a cAMP-dependent step is apparently not involved in the receptor-mediated inhibition of $Ca^{2+}$ currents. As reviewed in detail (SCHULTZ et al. 1990; see also refs. in Table 1), the extent of $Ca^{2+}$ current inhibition by receptor agonists is not diminished in cells loaded with cAMP or superfused with forskolin, a direct stimulator of adenylyl cyclase.

The $Ca^{2+}$ current inhibition also appears to occur independently of cGMP: intracellularly applied cGMP (SCHULTZ et al. 1990) or extracellularly applied nitroprusside, a potent stimulator of soluble guanylyl cyclase (KASAI and AOSAKI 1989), do not affect $Ca^{2+}$ currents in NG108-15 cells or chick dorsal root ganglion neurons, respectively. In contrast, voltage-dependent

$Ca^{2+}$ channels in snail neurons are stimulated by intracellularly applied cGMP-dependent protein kinase (PAUPARDIN-TRITSCH et al. 1986).

### 2. Protein Kinase C and Fatty Acids

Modulators of protein kinase C, phorbol esters and diacylglycerols, affect the activity of voltage-dependent $Ca^{2+}$ channels in a wide variety of cells (for reviews, see KACZMAREK 1988; ROSENTHAL 1990). Since these compounds can reduce neuronal $Ca^{2+}$ currents independently of protein kinase C activation (HOCKBERGER et al. 1989), results are to be interpreted with caution. The inhibition of $Ca^{2+}$ currents in chick dorsal root ganglion neurons by a diacylglycerol (RANE and DUNLAP 1986) led to the hypothesis that protein kinase C is involved in the receptor-mediated inhibition of $Ca^{2+}$ currents. Based on long-term effects of phorbol esters on the $Ca^{2+}$ current inhibition by neuropeptide Y in rat dorsal root ganglion neurons, a role for protein kinase C in the inhibition of a $Ca^{2+}$ current component was also postulated by EWALD and coworkers (1988). However, based on the present knowledge, an involvement of protein kinase C in the inhibitory pathway is unlikely: (a) According to KASAI and AOSAKI (1989), $Ca^{2+}$ currents in chick dorsal root ganglion neurons are not affected by protein kinase C activating phorbol esters or diacylglycerols. Likewise, $Ca^{2+}$ currents in rat dorsal root ganglion neurons (DOLPHIN et al. 1989) and in rat sympathetic ganglion neurons (WANKE et al. 1987) are insensitive to modulators of protein kinase C; (b) hormones and neurotransmitters which inhibit $Ca^{2+}$ currents *via* pertussis toxin-sensitive G-proteins do not activate phosphatidylinositol 4,5-bis-phosphate-hydrolyzing phospholipase C, a generator of protein kinase C-activating diacylglycerol. An exception may be bradykinin, which has been reported to stimulate phospholipase C in NG108-15 cells in a pertussis toxin-sensitive manner (HIGASHIDA et al. 1986), although the pertussis toxin sensitivity of this pathway has been disputed (GRANDT et al. 1986; OSUGI et al. 1987). According to one report (EWALD et al. 1989), bradykinin inhibits $Ca^{2+}$ currents in rat dorsal root ganglion neurons; the effect is sensitive to pertussis toxin. According to another report (DOLPHIN et al. 1989), the peptide has no effect on $Ca^{2+}$ currents in this cell type. In NG108-15 cells, bradykinin inhibits $Ca^{2+}$ currents in a pertussis toxin-insensitive manner (OFFERMANNS et al. 1991b; TAUSSIG et al. 1992). Thus, the occurrence and the pertussis toxin sensitivity of the bradykinin-induced inhibition of voltage-dependent $Ca^{2+}$ channels requires further investigation; (c) protein kinase C-activating diacylglycerols or a precursor (phosphatidic acid) are also formed by receptor and G-protein-mediated activation of phosphatidylcholine-hydrolyzing phospholipase C or D, respectively (see Chap. 67 by J. EXTON). Typically, receptor agonists inducing hydrolysis of phosphatidylcholine also induce hydrolysis of phosphatidylinositol 4,5-bis-phosphate, thereby elevating cytosolic $Ca^{2+}$ and stimulating secretion. Thus, secretion-inhibiting receptor agonists mediating pertussis toxin-sensitive inhibition of $Ca^{2+}$ currents are unlikely to be involved in the formation of diacylglycerols by hydrolysis of phosphatidylcholine; (d) arachidonic acid and other fatty acids, formed by activation of phospholipase $A_2$ or subsequent to phosphoinositide hydrolysis by diacylglycerol lipase, also activate protein kinase C (NISHIZUKA 1988); in addition, direct effects of fatty acids on ion channels have been reported (for review see ORDWAY et al. 1991). Receptor-mediated stimulation of phospholipase $A_2$ is, at least in some instances, sensitive to pertussis toxin (AXELROD et al. 1988). In chick (KASAI and AOSAKI 1989) and rat dorsal root ganglion neurons (DOLPHIN et al. 1989), arachidonic acid and inhibitors of phospholipases do not affect $Ca^{2+}$ currents. These findings indicate that arachidonic acid or its metabolites are not involved in the pertussis toxin-sensitive $Ca^{2+}$ current inhibition. There is, however, a report on the inhibition of $Ca^{2+}$ currents by a cis fatty acid (oleate) in neuroblastoma cells (LINDEN and ROUTTENBERG 1989). As was mentioned above, elevation of intracellular cGMP does not modify neuronal $Ca^{2+}$ currents. Thus, activation of soluble guanylyl cyclase by arachidonic acid (GERZER et al. 1986) with subsequent elevation of cGMP and activation of cGMP-dependent protein kinase is not a likely mechanism underlying the receptor-mediated inhibition of $Ca^{2+}$ currents.

### 3. Evidence for a Membrane-Delimited Pathway

Since known intracellular signal molecules or protein kinases are apparently not involved in the inhibitory $Ca^{2+}$ current modulation, it is reasonable to assume that the signal transduction pathway is confined to the plasma membrane. Using the inside-out patch variation of the patch-clamp technique, the groups of Birnbaumer and Brown have shown that the hormonal stimulation of $K^+$ channels via pertussis toxin-sensitive G-proteins of the $G_i$ family or L-type $Ca^{2+}$ channels via the cholera toxin-sensitive G-protein $G_s$ do not require cytosolic components (BROWN and BIRNBAUMER 1990). Although the pertussis-toxin-sensitive inhibition of $Ca^{2+}$ currents has not yet been demonstrated in isolated membrane patches, a membrane-delimited pathway is supported by the finding that superfusion of neuronal cells with noradrenaline or baclofen inhibits whole-cell $Ca^{2+}$ currents but does not affect $Ca^{2+}$ channels occluded by a patch pipette, i.e., $Ca^{2+}$ channels in cell-attached patches (FORSCHER et al. 1986; GREEN and COTRELL 1988; LIPSCOMBE et al. 1989). Proof for a direct interaction of G-proteins with $Ca^{2+}$ channels awaits reconstitution experiments with purified or recombinant components (receptors, G-proteins, and $Ca^{2+}$ channels).

## V. Identification of the Involved G-Protein

### 1. Occurrence of $G_o$

As far as this has been tested, $G_o$ is present in all cells exhibiting the pertussis toxin-sensitive inhibition of $Ca^{2+}$ currents. Cells containing $G_o$ at easily detectable levels include neuronal cells (ASANO et al. 1988), NG108-15 cells (MILLIGAN et al. 1986; OFFERMANNS et al. 1991b), rat pituitary cells (ASANO et al. 1988), $GH_3$ cells (ROSENTHAL et al. 1988a), RINm5F cells (SCHMIDT et al. 1991), HIT cells (SCHMIDT et al. 1992), bovine adrenal chromaffin cells (KLEPPISCH et al. 1992), PC12 cells (GOLLASCH et al. 1991b), and finally parafollicular cells (calcitonin-secreting C cells) of the thyroid (ASANO et al. 1988).

### 2. Reconstitution Experiments with Native and Recombinant G-Proteins, Transfected Cells

In pertussis toxin-treated cells, various preparations of purified $G_o$, applied into the interior of the cell via the patch pipette, reconstitute the ability of receptor agonists to inhibit voltage-dependent $Ca^{2+}$ currents. In NG108-15 cells, the $\alpha$ subunit of $G_o$ is ten times more potent than heterotrimeric $G_i$ in reconstituting the opioid-induced $Ca^{2+}$ current inhibition; it fully restores the opioid effect if it is included in the pipette solution (with or without $\beta\gamma$ complex) at a concentration of 0.4 n*M*; the retinal G-protein transducin and $\beta\gamma$ complex without an $\alpha$ subunit are inactive (HESCHELER et al. 1987). Intracellular injection of mammalian $G_o$ $\alpha$ subunit restores the dopamine-induced inhibition of $Ca^{2+}$ currents in snail neurons injected with activated pertussis toxin (HARRIS-WARRICK et al. 1988). In pertussis toxin-treated rat dorsal root ganglion neurons, the $G_o$ $\alpha$ subunit (10 n*M* in the pipette solution) causes a complete reappearance of the $Ca^{2+}$ current inhibition by neuropeptide Y (EWALD et al. 1988). In subsequent studies, the same group compared the reconstituting activity of purified or recombinant $G_i$ or $G_o$ $\alpha$ subunits (EWALD et al. 1989; LINDER et al. 1990). Whereas the $G_o$ $\alpha$ subunit was superior in reconstituting the $Ca^{2+}$ current inhibition induced by neuropeptide Y, $G_i$ or $G_o$ $\alpha$ subunits were approximately equally effective in restoring the bradykinin-induced inhibition of $Ca^{2+}$ currents (see above). Likewise, $G_i$ (purified from guinea pig heart) and $G_o$ (purified from bovine brain) appeared to be equally effective in reconstituting the inhibitory $Ca^{2+}$ current modulation by noradrenaline, enkephalin, and somatostatin in guinea pig submucosal neurons (SUPRENANT et al. 1990). Similar to the findings made with NG108-15 cells, heterotrimeric $G_o$ or its $\alpha$ subunit are more potent than $G_i$ in restoring the muscarinic inhibition of $Ca^{2+}$ currents in rat hippocampal neurons; a complete recovery is observed at $G_o$ concentrations $\geq$1 n*M* in the pipette solution (TOSELLI et al. 1989).

On the protein level, the $G_o$ $\alpha$ subunit occurs in two forms, the $G_{o1}$ $\alpha$ subunit (identical with the $G_{oA}$ $\alpha$ subunit) and the $G_{o2}$ $\alpha$ subunit (identical with the $G_{oB}$ $\alpha$ subunit); the two forms are derived from alternative splicing of a single transcript (see Chap. 10 by H. ITOH and Y. KAZIRO). TAUSSIG and coworkers (1992) established a NG108-15 cell line expressing a mutant, pertussis-toxin-resistant $G_{o1}$ $\alpha$ subunit. In the wild-type and the mutant cell line, opioids, somatostatin, and adrenaline inhibit voltage-dependent $Ca^{2+}$

currents in a pertussis-toxin-sensitive manner (TAUSSIG et al. 1992), and in membranes from wild-type cells, these receptor agonists activate G-proteins of the $G_i$ and the $G_o$ type (OFFERMANNS et al. 1991b). In the mutant cell line, pertussis toxin abolishes the $Ca^{2+}$ current inhibition by somatostatin, but not by an opioid or noradrenaline, indicating that the latter two agonists act *via* $G_{o1}$ and that somatostatin acts via a distinct G-protein, i.e., *via* $G_{o2}$ or a G-protein of the $G_i$-family. Interestingly, $G_{o2}$ is involved in the somatostatin-induced inhibition of $Ca^{2+}$ current in $GH_3$ cells (KLEUSS et al. 1991; see below).

Functions of exogenous proteins added to or expressed in a biological system are not necessarily identical with their functions "in situ." However, observations made in different cell lines suggest that endogenous $G_o$ is the physiological mediator between inhibitory receptors and voltage-dependent $Ca^{2+}$ channels (see below).

### 3. Antibodies

Antibodies against $G_o$ $\alpha$ subunits suppress the inhibitory modulation of $Ca^{2+}$ currents if dialyzed into the cell via the patch pipette. An antibody raised against a peptide common to the two splice variants of the $G_o$ $\alpha$ subunit attenuates the adrenaline-induced inhibition of $Ca^{2+}$ currents in NG108-15 cells; cells injected with antipeptide antibodies against $G_i$ $\alpha$ subunits respond normally (MCFADZEAN et al. 1989). Similar findings were made in snail neurons in which an affinity-purified antibody generated against the purified $G_o$ $\alpha$ subunit interfered with the ability of dopamine to inhibit $Ca^{2+}$ currents; other antibodies were not tested (HARRIS-WARRICK et al. 1988). In rat pituitary cells, the same antibody reduces the inhibition of $Ca^{2+}$ currents but not the stimulation of $K^+$ currents *via* $D_2$ dopamine receptors; antibodies against the $\alpha$ subunits of $G_{i1}$, $G_{i2}$, and $G_{i3}$ are inactive with regard to $Ca^{2+}$ current inhibition; the antibody against the $G_{i3}$ $\alpha$ subunit, however, affects the dopamine-induced stimulation of $K^+$ currents (LLEDO et al. 1992).

### 4. $G_o$-Activating Receptors

Inhibitory receptor agonists interact with endogenous $G_o$ $\alpha$ subunits in membrane preparations. This was shown by the effect of receptor agonists on photolabeling of G-protein $\alpha$ subunits with the GTP analog [$\alpha$-$^{32}$P]GTP azidoanilide and identification of photolabeled $\alpha$ subunits by antipeptide antibodies (ROSENTHAL et al. 1990; OFFERMANNS et al. 1991c). In all cells tested including NG108-15 cells (OFFERMANNS et al. 1991b), RINm5F cells (SCHMIDT et al. 1991), HIT cells (SCHMIDT et al. 1992), and $GH_3$ cells (OFFERMANNS et al. 1991a), $Ca^{2+}$ current-inhibiting receptor agonists stimulated photolabeling of both $G_i$ and $G_o$ $\alpha$ subunits. In NG108-15 cell membranes, the ability of receptor agonists to stimulate photolabeling of G-protein $\alpha$ subunits correlated well with their ability to inhibit $Ca^{2+}$

currents (D-Ala$^2$-D-Leu$^5$-enkephalin ≫ somatostatin > adrenaline ≫ bradykinin); in contrast, the ability of receptor agonists to stimulate photolabeling of the $G_{i2}$ $\alpha$ subunit did not match their ability to inhibit $Ca^{2+}$ currents (D-Ala$^2$-D-Leu$^5$-enkephalin ≫ somatostatin ≅ adrenaline ≅ bradykinin; OFFERMANNS et al. 1991b).

## 5. Antisense Oligonucleotides

In pituitary $GH_3$ cells, the expression of G-protein subunits is transiently switched off by intranuclear injection of oligonucleotides complementary to a selective sequence found in a G-protein subunit encoding mRNA ("antisense" oligonucleotides). Only the transient suppression of $G_o$ $\alpha$ subunits (both splice variants; see below) but not that of $G_i$ or $G_s$ $\alpha$ subunits largely reduced or abolished the inhibition of $Ca^{2+}$ current by somatostatin and the muscarinic agonist carbachol (KLEUSS et al. 1991). These experiments establish the essential role of $G_o$ in the receptor-mediated inhibition of $Ca^{2+}$ currents. The data differ from those obtained in reconstitution experiments in which both $G_i$ and $G_o$ were able to reconstitute the inhibitory $Ca^{2+}$ current modulation, albeit with different efficiencies (see above). They are consistent with a study made with rat pituitary tumor cells, in which the expression of $G_o$ $\alpha$ subunits was correlated with the ability of dopamine to inhibit prolactin secretion (COLLU et al. 1988).

The use of antisense oligonucleotides also allows the selective suppression of the two splice variants of the $G_o$ $\alpha$ subunit, the $G_{o1}$ and the $G_{o2}$ $\alpha$ subunit. The results obtained with subtype-specific antisense oligonucleotides show that the $G_{o1}$ $\alpha$ subunit mediates the muscarinic

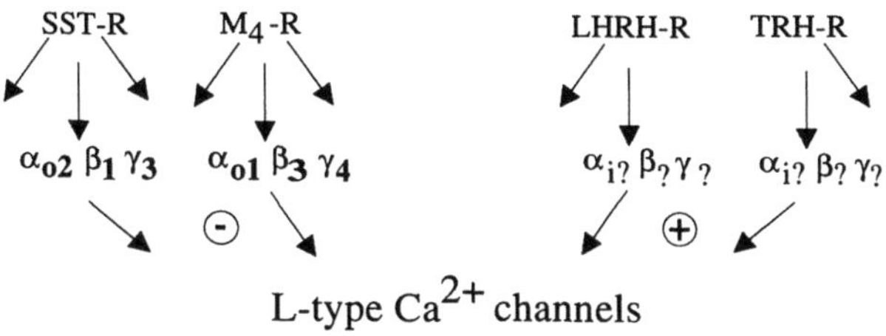


**Fig. 1.** Receptor-mediated modulations of voltage-dependent $Ca^{2+}$ channels in $GH_3$ cells. It is assumed that signal transduction from an activated receptor to $\alpha$ voltage-dependent $Ca^{2+}$ channel requires a heterotrimeric G-protein of defined composition of subtypes of subunits. Identified subtypes of subunits are shown in *bold* characters; *question marks* indicate that the involved subtypes remain to be determined. The model implies convergent signal transduction on the level of the effector, i.e., the voltage-dependent $Ca^{2+}$ channel of the L-type. Divergent signal transduction on the level of G-proteins (see Sect. D) is indicated by *nonvertical arrows* (*upper row of arrows*). *SST-R*, receptor for somatostatin; *$M_4$-R*, $M_4$ muscarinic receptors (see reference 4 in KLEUSS et al. 1992); *LHRH-R*, receptor for LHRH; *TRH-R*, receptor for TRH; "+", stimulation; "−", inhibition. For further explanations, see text

inhibition of $Ca^{2+}$ currents, whereas the $G_{o2}$ $\alpha$ subunit mediates the $Ca^{2+}$ current inhibition by somatostatin (Fig. 1; KLEUSS et al. 1991). In addition, the experimental approach has been used to assign functions to subtypes of G-protein $\beta$ and $\gamma$ subunits. In contrast to $G_o$ $\alpha$-subunits, the subtypes of these two subunits are encoded by different genes (see Chap. 10 by H. ITOH and Y. KAZIRO). Injected into nuclei of $GH_3$ cells, antisense oligonucleotides predicted to hybridize with all subtypes of $\beta$ subunits abolish the response to either receptor agonist, thereby confirming the assumption that $\beta$ subunits are required for the interaction of a G-protein with an activated receptor (KLEUSS et al. 1991, 1992). Suppression of the individual subtypes revealed that $\beta_1$ and $\beta_3$ subunits are required for signal transduction initiated by somatostatin and carbachol, respectively (see Fig. 1); suppression of $\beta_2$ or $\beta_4$ subunits had no effect on $Ca^{2+}$ current inhibition by either receptor agonist (KLEUSS et al. 1992). Recently obtained data show that antisense oligonucleotides specific for the $\gamma_3$ subunit selectively abolish the inhibition of $Ca^{2+}$ currents by somatostatin and that antisense oligonucleotides specific for the $\gamma_4$ subunit selectively suppress the effect of carbachol (see Fig. 1; KLEUSS et al. 1993). The discovery that not only subtypes of $\alpha$ subunits, but also subtypes of $\beta$ and $\gamma$ subunits, exhibit specificity with regard to receptors is of general significance for G-protein-mediated signal transduction. The data are consistent with the hypothesis that only a certain combination of subtypes of $\alpha$, $\beta$, and $\gamma$ subunits can successfully transduce a signal from an activated receptor to an effector. The assignment of a specific function to a heterotrimeric G-protein of a defined composition of subtypes of subunits may also help to explain tissue-specific responses and solve the discrepancy between the large number of G-protein-coupled receptors and the relatively few G-proteins as defined by their $\alpha$ subunits (see Chaps. 1, 11, 54 BIRNBAUMER; 10 ITOH and KAZIRO; 45 FINDLAY AND D. DONELLY).

## C. Stimulatory Modulation of Voltage-Dependent $Ca^{2+}$ Channels

### I. Occurrence; Physiological Significance

In neuronal and some nonneuronal secretory cells (e.g., adrenal chromaffin cells, insulin-secreting B cells of the pancreas) activation of voltage-dependent $Ca^{2+}$ channels is secondary to depolarization of the plasma membrane by G-protein-independent mechanisms; stimulation of voltage-dependent $Ca^{2+}$ channels via receptors coupled to G-proteins may augment or maintain the G-protein-independent influx of $Ca^{2+}$. In other nonneuronal secretory cells, the membrane potential and voltage-dependent $Ca^{2+}$ channels are mainly controlled by secretion-stimulating hormones acting on G-protein-coupled receptors (HESCHELER et al. 1988a; ROSENTHAL et al.

1988a,b). The involvement of pertussis toxin-sensitive G-proteins in the stimulation of voltage-dependent $Ca^{2+}$ currents by stimulatory hormones has been shown to occur in pituitary and adrenocortical cells. Like the receptor-mediated inhibition of $Ca^{2+}$ currents, the receptor-mediated stimulation of $Ca^{2+}$ currents, as measured in the whole-cell configuration of the patch-clamp technique (HAMILL et al. 1981), is rapid in onset and fully reversible.

In rat pituitary ($GH_3$) cells, luteinizing hormome-releasing hormone (LHRH) stimulates $Ca^{2+}$ currents under standard conditions, i.e., using $Ba^{2+}$ as a charge carrier (ROSENTHAL et al. 1988a). Under these conditions thyrotropin-releasing hormone (TRH), another secretion-stimulating hormone, inhibits $Ca^{2+}$ currents (GOLLASCH et al. 1991a). However, TRH increases currents through voltage-dependent $Ca^{2+}$ channels, if intracellular $Ca^{2+}$ stores are depleted and $Na^+$ is used as a charge carrier. The data indicate that TRH induces a massive release of $Ca^{2+}$ from intracellular stores and that the resulting increase in cytosolic $Ca^{2+}$ causes inactivation of voltage-dependent $Ca^{2+}$ channels, which are activated by TRH (or LHRH) at low to moderately elevated concentrations of cytosolic $Ca^{2+}$. Both the stimulatory and the inhibitory effect of TRH are abolished by loading cells with GDP$\beta$s, suggesting the involvement of G-proteins in either pathway (GOLLASCH et al. 1991a).

In murine adrenocortical (Y1) cells (HESCHELER et al. 1988a) as well as in bovine (COHEN et al. 1988) and porcine adrenal glomerulosa cells (HESCHELER et al. 1988a), angiotensin II, the major stimulator of aldosterone secretion, increases voltage-dependent $Ca^{2+}$ currents. Angiotensin II also stimulates $Ca^{2+}$ currents in $GH_3$ cells (ROSENTHAL et al. 1988a).

LHRH, TRH, and angiotensin II are typical examples of hormones that activate phosphatidylinositol 4,5-bis-phosphate-hydrolyzing phospholipase C with subsequent activation of protein kinase C by diacylglycerol and release of $Ca^{2+}$ from intracellular stores by inositol-1,4,5-trisphosphate. The resulting increase in cytosolic $Ca^{2+}$ appears to be crucial for the secretory response. However, numerous studies show that an increased influx of $Ca^{2+}$ through voltage-dependent $Ca^{2+}$ channels contributes to the stimulation of secretion by the three hormones (see references cited in HESCHELER et al. 1988a; ROSENTHAL et al. 1988a). In fact, the release of luteinizing hormone by LHRH in rat pituitary cells can be uncoupled from the production of inositol phosphates (HAWES et al. 1992).

## II. Effects of Pertussis Toxin and Guanine Nucleotides

In $GH_3$ and Y1 cells, the stimulatory effects of LHRH, TRH, and angiotensin II are abolished by treatment of cells with pertussis toxin (HESCHELER et al. 1988a; ROSENTHAL et al. 1988a; GOLLASCH et al. 1991a). In contrast, the TRH-induced $Ca^{2+}$ current inhibition is not affected by the toxin (Gollasch et al. 1991a). The pertussis toxin sensitivity of the angiotensin II-induced stimulation of $Ca^{2+}$ channels in bovine and porcine adrenal glomerulosa cells remains to be tested. The data indicate that pertussis toxin-sensitive G-proteins are involved not only in the inhibitory, but also in the stimulatory modulation of voltage-dependent $Ca^{2+}$ currents. The inhibitory and the stimulatory pertussis toxin-sensitive pathways may coexist within one cell, e.g., the $GH_3$ cell. This is also evident from the finding that both inhibitory and stimulatory hormones stimulate pertussis toxin-sensitive GTPases in membranes of $GH_3$ cells (Offermanns et al. 1989).

## III. Types of $Ca^{2+}$ Currents Affected by Stimulatory Receptor Agonists

In Y1 and $GH_3$ cells, stimulatory hormones increase a $Ca^{2+}$ current component with biophysical and pharmacological properties of L-type channels; another low-threshold, fast-inactivating current component, which may represent a current through T-type channels, is not sensitive to hormones (Hescheler et al. 1988a; Rosenthal et al. 1988a). In contrast, angiotensin II-sensitive $Ca^{2+}$ currents in bovine adrenal glomerulosa cells appear to represent currents through T-type channels (Cohen et al. 1988).

## IV. Mechanistic Aspects

Several lines of evidence suggest that the stimulatory modulation of $Ca^{2+}$ currents in adrenocortical and pituitary cells does not involve an increase in cAMP and subsequent stimulation of cAMP-dependent protein kinase as does the stimulation of L-type $Ca^{+}$ currents in cardiac and skeletal muscle (see Chap. 57 by F. Hofmann et al.): a) inclusion of cAMP in the pipette solution or extracellularly applied forskolin do not stimulate $Ca^{2+}$ currents in Y1 or $GH_3$ cells (Hescheler et al. 1988a; Rosenthal et al. 1988a). There is, however, evidence for a stimulatory effect of cAMP on $Ca^{2+}$ currents in AtT-20 cells (Lewis et al. 1986 and references cited therein), and in $GH_3$ cells, phosphorylation of L-type $Ca^{2+}$ channels by cAMP-dependent protein kinase may be a prerequisite for their ability to respond to changes in the membrane potential (Armstrong and Eckert 1987); b) LHRH, TRH, and angiotensin II do not elevate cAMP levels in pituitary cells. In particular, LHRH and TRH do not affect adenylyl cyclase activity in membranes of $GH_3$ cells (Rosenthal et al. 1989a), and angiotensin II inhibits adenylyl cyclase in membranes of adrenocortical cells, $GH_3$ cells, and other pituitary cells (Enjalbert et al. 1986; Rosenthal et al. 1988a; Woodcock and McLeod 1986). Regardless of these observations, stimulation of adenylyl cyclase is generally mediated by a cholera-toxin-sensitive G-protein and not by a pertussis-toxin-sensitive G-protein (see Chap. 55 Premont et al.).

It is unlikely that activation of phosphatidylinositol 4,5-bis-phosphate-hydrolyzing phospholipase C is involved in the $Ca^{2+}$ current stimulation by LHRH, TRH, and angiotensin II since the G-protein involved in this pathway is insensitive to pertussis toxin in adrenal glomerulosa cells (ENYEDI et al. 1986; KOJIMA et al. 1986), $GH_3$ cells (MARTIN et al. 1986), and other pituitary cells (ENJALBERT et al. 1986; NAOR et al. 1986). In contrast, the stimulation of $^{45}Ca^{2+}$ influx by angiotensin II in adrenal glomerulosa cells is sensitive to pertussis toxin (KOJIMA et al. 1986).

An involvement of products formed by other phospholipases (see Sect. B) in the pertussis toxin-sensitive stimulation of $Ca^{2+}$ currents remains to be examined. So far, it has been shown that a protein kinase C-activating phorbol ester does not affect $Ca^{2+}$ currents in Y1 cells, nor does cGMP affect the angiotensin II-induced stimulation of $Ca^{2+}$ currents in this cell type (HESCHELER et al. 1988a).

Two reports suggest the involvement of intracellular signal molecules in the hormone-activated $Ca^{2+}$ entry into pituitary gonadotrophs. Evidence for the involvement of a readily diffusable messenger comes from experiments by MASON and WARING (1986). Applying the cell-attached variation of the patch-clamp technique, they found that $Ca^{2+}$-permeable cation channels occluded by the patch pipette are stimulated by LHRH applied to the bath solution. It is, however, not clear whether these hormone-sensitive cation channels represent typical voltage-dependent $Ca^{2+}$ channels. According to SHANGOLD et al. (1988), protein kinase C is involved in the LHRH-induced $Ca^{2+}$ influx. They reported that a protein kinase C-activating phorbol ester promotes the portion of the LHRH-induced increase in cytosolic $Ca^{2+}$, which depends on dihydropyridine-sensitive $Ca^{2+}$ influx. Pertussis toxin was employed in neither of the two studies.

## V. Identity of the G-Protein Involved

In contrast to the direct evidence for the involvement of $G_o$ in the inhibitory $Ca^{2+}$ current modulation, there is only indirect evidence for the identity of the pertussis toxin-sensitive G-protein involved in the stimulatory $Ca^{2+}$ current modulation. In nonretinal cells, there are two groups of pertussis toxin-sensitive G-proteins: the ubiquitous members of the $G_i$ family ($G_{i1}$, $G_{i2}$, $G_{i3}$) with $\alpha$ subunits derived from different genes (see Chaps. 10 by H. ITOH and Y. KAZIRO) and the two forms of $G_o$ (see Sect. B), which are only present in neuronal cells and in some nonneuronal endocrine cells. In this context, the $GH_3$ cells is a particular interesting model, providing evidence for the participation of pertussis toxin-sensitive G-proteins in the stimulation and inhibition of $Ca^{2+}$ currents. Since the opposite modulation of $Ca^{2+}$ currents occur within one cell, they must be mediated by distinct G-proteins. Depending on the activated receptor, one or the other form of $G_o$ mediates the inhibitory modulation of $Ca^{2+}$ currents in the $GH_3$ cell (see Sect. B). Thus, the only remaining candidates for the stimulatory pathway are

G-proteins of the $G_i$ family (see Fig. 1). This also applies to Y1 cells which contain G-proteins of the $G_i$ family, but not $G_o$ (Hescheler et al. 1988a; Bloch et al. 1989). Proof for a role of $G_i$-type G-proteins in the stimulatory modulation of voltage-dependent $Ca^{2+}$ channels in certain endocrine cells may come from experiments with antisense oligonucleotides, analogous to those described above for the inhibitory pathway (see Sect. B).

# D. Conclusion

The modulation of voltage-dependent $Ca^{2+}$ channels is only one signal transduction pathway used by hormones and neurotransmitters controlling secretion. Inhibitory receptors of neuronal and nonneuronal cells modulate at least three effectors *via* pertussis toxin-sensitive G-proteins (Schmidt et al. 1991; Suprenant et al. 1992; see also references cited therein): (1) inhibition of adenylyl cyclase leads to a decrease in cytosolic cAMP, resulting in a decreased activity of cAMP-dependent protein kinase; (2) stimulation of various types of $K^+$ channels causes hyperpolarization of the plasma membranes and thereby closure of voltage-dependent $Ca^{2+}$ channels; and (3) inhibition of voltage-dependent $Ca^{2+}$ channels also takes place without the involvement of a hyperpolarizing current (see Sect. B). The stimulation of $K^+$ channels (for review see Brown and Birnbaumer 1990) and the inhibition of voltage-dependent $Ca^{2+}$ (see Sect. B), both of which contribute to a decrease in $Ca^{2+}$ influx through voltage-dependent $Ca^{2+}$ channels, do not involve a cAMP-dependent intermediate step. It follows that the three pathways are activated independently of each other. Recently obtained data show that a single type of receptor can activate all three pathways (Suprenant et al. 1992). The data suggest branching of signal transduction on the level of G-proteins. This assumption is supported by the finding that inhibitory receptors interact with multiple pertussis-toxin-sensitive G-proteins in membranes of NG 108-15 cells (Offermanns et al. 1991b), RINm5F cells (Schmidt et al. 1991), and $GH_3$ cells (Offermanns et al. 1991a).

Inhibitory receptor agonists appear to generally inhibit adenylyl cyclase and voltage-dependent $Ca^{2+}$ currents; a stimulation of $K^+$ channels is observed in most, but not all, cells. For example, somatostatin stimulates $K^+$ currents in pituitary $GH_3$ cells (Yatani et al. 1990), but not in calcitonin-secreting rMTC 44-2 cells (Scherübl et al. 1992). Thus, the contribution of the individual pathway to the ultimate cellular response may vary from cell type to cell type. Conceivably, the relative contribution of each pathway may be determined by the tissue-specific expression of subtypes of G-protein subunits ($\alpha$, $\beta$, and $\gamma$), which differ with regard to their receptor specificity (see Sect. B) and which may also differ with regard to their effector specificity. In addition, tissue-specific expression of subtypes of effector ($Ca^{2+}$ channel) subunits (see Chap. 57 by F. Hofmann et al.)

with different specificities for G-protein subunits may contribute to tissue-specific responses. This hypothesis is supported by the recent finding that isoforms of G-protein-regulated enzymatic effectors (adenylyl cyclase, phospholipase C) are differentially regulated by G-protein subunits (see Chaps. 67 EXTON; 59 GIERSCHICK and CAMPS; 70 HILDEBRANDT; 55 PREMONT et al.).

The secretion-stimulating hormones that activate voltage-dependent $Ca^{2+}$ channels through pertussis toxin-sensitive G-protein are known to elevate cytosolic $Ca^{2+}$ by the release of $Ca^{2+}$ from inositol-1,4,5-trisphosphate-sensitive stores (see Sect. C). However, an increased influx of $Ca^{2+}$ through voltage-dependent $Ca^{2+}$ channels contributes considerably to the secretory response (see refs. in HESCHELER et al. 1988a; ROSENTHAL et al. 1988a). The demonstration of a stimulatory effect of these hormones on voltage-dependent $Ca^{2+}$ channels strongly supports this assumption (see Sect. C). Similar to signal transduction by inhibitory receptors, signal transduction initiated by stimulatory receptors branches on the level of G-proteins: pertussis-toxin-insensitive and pertussis-toxin-sensitive G-proteins mediate hydrolysis of phosphoinositides and stimulation of voltage-dependent $Ca^{2+}$ channels, respectively. It is left to future work to determine in a given cell type the relative contribution to the secretory response of pathways activated by secretion-modulating hormones and neurotransmitters. Another task will be the identification of subtypes of G-protein and effector subunits involved in the receptor-mediated modulation of secretion.

*Acknowledgements.* I am indebted to my colleagues at the Institut für Pharmakologie and the Institut für Biochemie und Molekularbiologie, Freie Universität Berlin, for the years of collaboration on the subject of this review and for making available data on the receptor specificity of G-protein $\gamma$ subunits prior to publication. I thank Dr. L. Birnbaumer, Baylor College of Medicine, Houston, Texas, for support and helpful discussions during preparation of the manuscript and Anita Seibold, Baylor College of Medicine, Houston, Texas, for critical reading of the manuscript. Own work reported herein was supported by the Deutsche Forschungsgemeinschaft. The author is a recipient of a Heisenberg Fellowship from the Deutsche Forschungsgemeinschaft.

## References

Armstrong D, Eckert R (1987) Voltage-activated calcium channels that must be phosphorylated in order to respond to membrane depolarization. Proc Natl Acad Sci USA 84:2518–2522

Asano T, Semba R, Kemiya N, Ogasawara N, Kato K (1988) $G_o$, a GTP-binding protein: immunohistochemical localization in the rat. Neurochemistry 50:1164–1169

Axelrod J, Burch RM, Jelsema CL (1988) Receptor-mediated activation of phospholipase $A_2$ via GTP-binding proteins: arachidonic acid and its metabolites as second messengers. Trends Neurosci 11:117–123

Beech DJ, Bernheim L, Hille B (1992) Pertussis toxin and voltage-dependence distinguish multiple pathways modulating calcium channels of rat sympathetic neurons. Neuron 8:97–106

Bloch DB, Bonventre JV, Neer EJ, Seidman JG (1989) The G-protein $\alpha_o$ subunit alters morphology, growth kinetics and phospholipid metabolism of somatic cells. Mol Cell Biol 9:5434–5439
Brown AM, Birnbaumer L (1990) Ionic channels and their regulation by G-protein subunits. Annu Rev Physiol 52:197–213
Cohen C, McCarthy R, Barrett P, Rasmussen H (1988) Ca channels in adrenal glomerulosa cells: $K^+$ and angiotensin II increase T-type Ca channel current. Proc Natl Acad Sci USA 85:2412–2416
Collu R, Bouvier C, Lagace G, Unson CG, Milligan G, Goldsmith P, Spiegel A (1988) Selective deficiency of guanine nucleotide-binding protein $G_o$ in two dopamine-resistant pituitary tumors. Endocrinology 122:1176–1178
Dolphin AC (1990) G-protein modulation of calcium channels in neurons. Annu Rev Physiol 52:243–255
Dolphin AC, Scott RH (1987) Calcium channel currents and their inhibition by (−)-baclofen in rat sensory neurons: modulation by guanine nucleotides. J Physiol (Lond) 386:1–17
Dolphin AC, Scott RH (1990) Modulation of neuronal $Ca^{2+}$ currents by G-protein activation. In: Nahorski SR (ed) Transmembrane signalling. Intracellular messengers and implications for drug development. Wiley, London, pp 101–117
Dolphin AC, Wootton JF, Scott RH, Trentham DR (1988) Photoactivation of intracellular guanosine triphosphate analogues reduces the amplitude and slows the kinetics of voltage-activated calcium channel currents in sensory neurons. Pflugers Arch 411:628–636
Dolphin AC, McGuirk SM, Scott RH (1989) An investigation into the mechanisms of inhibition of calcium channel currents in cultured sensory neurones of the rat by guanine nucleotide analogues and (−)-baclofen. Br J Pharmacol 97:263–273
Dunlap K, Fischbach GD (1981) Neurotransmitters decrease the calcium conductance activated by depolarization of embryonic chick sensory neurons. J Physiol (Lond) 317:519–535
Enjalbert A, Sladeczek F, Guillon G, Bertrand P, Shu C, Epelbaum J, Garcia-Sainz A, Jard S, Lombard C, Kordon C, Bockaert J (1986) Angiotensin II and dopamine modulate both cAMP and inositol phosphates in adrenal glomerulosa cells. J Biol Chem 261:4071–4075
Enyedi P, Musci I, Hunyady L, Catt KJ, Spät A (1986) The role of guanyl nucleotide-binding proteins in the formation of inositol phosphates in adrenal glomerulosa cells. Biochem Biophys Res Commun 140:941–947
Ewald DA, Sternweis PC, Miller RJ (1988) Guanine nucleotide-binding protein $G_o$-induced coupling of neuropeptide Y receptors to $Ca^{2+}$ channels in sensory neurons. Proc Natl Acad Sci USA 85:3633–3637
Ewald DA, Pang I-H, Sternweis PC, Miller RJ (1989) Differential G-protein-mediated coupling of neurotransmitter receptors to $Ca^{2+}$ channels in rat dorsal root ganglion neurons. Neuron 2:1185–1193
Forscher P, Oxford GS, Schulz D (1986) Noradrenaline modulates calcium channels in avian dorsal root ganglion cells through tight receptor-channel coupling. J Physiol (Lond) 379:131–144
Fox AP, Nowycky M, Tsien RW (1987) Single-channel recording of three types of calcium channels in chick sensory neurons. J Physiol (Lond) 394:173–200
Gerzer R, Brash A, Hardman J (1986) Activation of soluble guanylate cyclase by arachidonic acid and 15-lipoxygenase products. Biochim Biophys Acta 886:383–389
Gollasch M, Haller H, Schultz G, Hescheler J (1991a) Thyrotropin-releasing hormone induces opposite effects on $Ca^{2+}$ channel currents in pituitary cells by two pathways. Proc Natl Acad Sci USA 88:10262–10266
Gollasch M, Hescheler J, Spicher K, Klinz F-J, Schultz G, Rosenthal W (1991b) Inhibition of $Ca^{2+}$ channels via $\alpha_2$-adrenergic and muscarinic receptors in pheochromocytoma (PC-12) cells. Am J Physiol 260:C1282–1289

Grandt R, Greiner C, Zubin P, Jakobs KH (1986) Bradykinin stimulates GTP hydrolysis in NG 108-15 membranes by a high-affinity, pertussis toxin-insensitive GTPase. FEBS Lett 196:279–283

Green K, Cottrell G (1988) Actions of baclofen on components of the Ca-current in rat and mouse DRG neurons in culture. Br J Pharmacol 94:235–245

Gross RA, Macdonald AL (1987) Dynorphin A selectively reduces a large transient (N-type) calcium current of mouse dorsal root ganglion neurons in cell culture. Proc Natl Acad Sci USA 84:5469–5473

Gross RA, Moises HC, Uhler MD, Macdonald AL (1990) Dynorphin A and cAMP-dependent protein kinase independently regulate calcium currents. Proc Natl Acad Sci USA 87:7025–7029

Gross RA, Uhler MD, Macdonald AL (1991) The reduction of neuronal calcium currents by ATP$\gamma$S is mediated by a G protein and occurs independently of cyclic AMP-dependent protein kinase. Brain Res 535:214–220

Hamill OP, Marty A, Neher E, Sakmann B, Sigworth FJ (1981) Improved patch clamp techniques for high resolution current recording from cells and cell-free membrane patches. Pflugers Arch 391:85–100

Harris-Warrick RM, Hammond C, Paupardin-Tritsch D, Homburger V, Rouot B, Bockaert J, Gerschenfeld HM (1988) An $\alpha_{40}$ subunit of a GTP-binding protein immunologically related to $G_o$ mediates a dopamine-induced decrease of $Ca^{2+}$ current in snail neurons. Neuron 1:27–32

Hawes BE, Waters SB, Janovick JA, Bleasdale JE, Conn PM (1992) Gonadotropin-releasing hormone-stimulated intracellular $Ca^{2+}$ fluctuations and LH release can be uncoupled from inositol phosphate production. Endocrinology 130:3475–3483

Hescheler J, Rosenthal W, Trautwein W, Schultz G (1987) The GTP-binding protein, $G_o$, regulates neuronal calcium channels. Nature 325:445–447

Hescheler J, Rosenthal W, Hinsch K-D, Wulfern M, Trautwein W, Schultz G (1988a) Angiotensin II-induced stimulation of voltage-dependent $Ca^{2+}$ currents in an adrenal cortical cell line. EMBO J 7:619–624

Hescheler J, Rosenthal W, Wulfern M, Tang M, Yajima M, Trautwein W, Schultz G (1988b) Involvement of the guanine nucleotide-binding protein, $N_o$, in the inhibitory modulation of neuronal calcium channels. Adv Second Messenger Phosphoprotein Res 21:165–174

Higashida H, Streaty RA, Klee W, Nirenberg M (1986) Bradykinin-activated transmembrane signals are coupled via $N_o$ or $N_i$ to production of inositol 1,4,5-trisphosphate, a second messenger in NG108-15 neuroblastoma-glioma hybrid cells. Proc Natl Acad Sci USA 83:942–946

Hockberger P, Toselli M, Swandulla D, Lux HD (1989) A diacylglycerol analogue reduces neuronal calcium currents independently of protein kinase C activation. Nature 338:340–342

Holz IV GG, Rane SG, Dunlap K (1986) GTP-binding proteins mediate transmitter inhibition of voltage-dependent calcium channels. Nature 319:670–672

Hsu WH, Xiang H, Rajan AS, Kunze DL, Boyd III AE (1991) Somatostatin inhibits insulin secretion by a G-protein-mediated decrease in $Ca^{2+}$ entry through voltage-dependent $Ca^{2+}$ channels in the beta cell. J Biol Chem 266:837–843

Ikeda SR, Schofield GG (1989) Somatostatin blocks a calcium current in rat sympathetic ganglion neurons. J Physiol (Lond) 409:221–240

Jakobs KH, Wieland T (1989) Evidence for receptor-regulated phosphotransfer reactions involved in activation of adenylate cyclase inhibitory G-protein in human platelet membranes. Eur J Biochem 183:115–121

Kaczmarek LK (1988) The regulation of neuronal calcium and potassium channels by protein phosphorylation. Adv Second Messenger Phosphoprotein Res 22:113–138

Kasai H (1992) Voltage-and time-dependent inhibition of neuronal calcium channels by a GTP-binding protein in a mammalian cell line. J Physiol (Lond) 448:189–209

Kasai H, Aosaki T (1989) Modulation of Ca-channel current by an adenosine analog mediated by a GTP-binding protein in chick sensory neurons. Pflugers Arch 414:145–149

Keahey HH, Boyd III AE, Kunze DL (1989a) Catecholamine modulation of calcium currents in clonal pancreatic $\beta$-cells. Am J Physiol 257:C1171–C1176

Keahey HH, Rajan AS, Boyd III AE, Kunze DL (1989b) Characterization of voltage-dependent calcium currents in a $\beta$-cell line. Diabetes 38:188–193

Kleppisch T, Ahnert-Hilger G, Gollasch M, Spicher K, Hescheler J, Schultz G, Rosenthal W (1992) Inhibition of voltage-dependent $Ca^{2+}$ channels via $\alpha_2$-adrenergic and opioid receptors in cultured bovine adrenal chromaffin cells. Pflugers Arch 421:131–137

Kleuss C, Hescheler J, Ewel C, Rosenthal W, Schultz G, Wittig B (1991) Assignment of G-protein-subtypes to specific receptors inducing inhibition of calcium currents. Nature 353:43–48

Kleuss C, Scherübl H, Hescheler J, Schultz G, Wittig B (1992) Different $\beta$-subunits determine G-protein ineraction with transmembrane receptors. Nature 358: 424–426

Kleuss C, Scherübl H, Hescheler J, Schultz G, Wittig B (1993) Selectivity in signal transduction determined by $\gamma$-subunits of heterotrimeric G-proteins. Science 259:832–834

Kojima I, Shibata H, Ogata E (1986) Pertussis toxin blocks angiotensin II-induced calcium influx but not inositol trisphosphate production in adrenal glomerulosa cells. FEBS Lett 204:347–351

Lewis DL, Weight FF, Luini A (1986) A guanine nucelotide-binding protein mediates the inhibition of voltage-dependent calcium current in a pituitary cell line. Proc Natl Acad Sci USA 83:9035–9039

Linden DJ, Routtenberg A (1989) Cis-fatty acids, which activate protein kinase C, attenuate $Na^+$ and $Ca^{2+}$ currents in mouse neuroblastoma cells. J Physiol (Lond) 419:95–119

Linder ME, Ewald DA, Miller RJ, Gilman AG (1990) Purification and characterization of $G_{o\alpha}$ and three types of $G_{i\alpha}$ in Escherichia coli. J Biol Chem 265:8243–8251

Lipscombe D, Kongsamut S, Richard WT (1989) $\alpha$-Adrenergic inhibition of sympathetic neurotransmitter release mediated by modulation of N-type calcium channel gating. Nature 340:639–642

Lledo PM, Homburger V, Bockaert J, Vincent J-D (1992) Differential G protein-mediated coupling of $D_2$ dopamine receptors to $K^+$ and $Ca^{2+}$ currents in rat anterior pituitary cells. Neuron 8:455–463

Llinas R, Sugimori M, Lin J-W, Cherksey B (1989) Blocking and isolation of a calcium channel from neurons in mammals and cephalopods using a toxin fraction (FTX) from funnel-web spider toxin. Proc Natl Acad Sci USA 86: 1689–1693

Luini A, Lewis D, Guild S, Schofield G, Weight F (1986) Somatostatin, an inhibitor of ACTH secretion, decreases cytosolic free calcium and voltage-dependent calcium currents in a pituitary cell line. Neuroscience 6:3128–3132

Mackie K, Hille B (1992) Cannabinoids inhibit N-type calcium channels in neuroblastoma glioma cells. Proc Natl Acad Sci USA 89:3825–3829

Marchetti C, Carbone E, Lux HD (1986) Effects of dopamine and noradrenaline on Ca channels of cultured sensory and sympathetic neurons of chick. Pflugers Arch 406:104–111

Mason W, Waring D (1986) Patch clamp recordings of single ion channel activation by gonadotropin releasing hormone in ovine pituitary gonadotrophs. Neuroendocrinology 43:205–219

Martin TFJ, Bajjalieh SM, Lucas DO, Kowalchyk JA (1986) Thyrotropin-releasing hormone stimulation of polyphosphoinositide hydrolysis in $GH_3$ cell membranes is GTP dependent but insensitive to cholera or pertussis toxin. J Biol Chem 261:10041–10049

McCleskey EW, Fox AP, Feldman DH, Cruz LJ, Oliver BM, Tsien RW, Yoshima D (1987) ω-Conotoxin: direct and persistent blockade of specific types of calcium channels in neurons but not in muscle. Proc Natl Acad Sci USA 84:4327–4331

McFadzean I, Mullaney I, Brown DA, Milligan G (1989) Antibodies to the GTP binding protein, $G_o$, antagonize noradrenaline-induced calcium current inhibition in NG108-15 hybrid cells. Neuron 3:177–182

Miller RJ (1990) Receptor-mediated regulation of calcium channels and neurotransmitter release. FASEB J 4:3291–3299

Milligan G, Gierschik P, Spiegel AM, Klee WA (1986) The GTP-binding regulatory proteins of neuroblastoma x glioma, NG 108-15, and glioma C6, cells. FEBS Lett 195:225–230

Naor Z, Azrad A, Limor R, Zakut H, Lotan M (1986) Gonadotropin-releasing hormone activates a rapid $Ca^{2+}$-independent phosphodiester hydrolysis of polyphosphoinositides in pituitary gonadotrophs. J Biol Chem 261:12506–12512

Nishizuka Y (1988) The molecular heterogeneity of protein kinase C and its implication for cellular regulation. Nature 334:661–665

Nowycky MC, Fox AP, Tsien RW (1985) Three types of neuronal calcium channels with different calcium agonist sensitivity. Nature 316:440–443

Offermanns S, Schultz G, Rosenthal W (1989) Secretion-stimulating and secretion-inhibiting hormones stimulate high-affinity pertussis-toxin-sensitive GTPases in membranes of a pituitary cell line. Eur J Biochem 180:283–287

Offermanns S, Gollasch M, Hescheler J, Spicher K, Schmidt A, Schultz G, Rosenthal W (1991a) Inhibition of voltage-dependent $Ca^{2+}$ currents and activation of pertussis toxin-sensitive G-proteins via muscarinic receptors in $GH_3$ cells. Mol Endocrinol 5:995–1002

Offermanns S, Schultz G, Rosenthal W (1991b) Evidence for opioid receptor-mediated activation of the G-proteins, $G_o$ and $G_{i2}$, in membranes of neuroblastoma x glioma (NG108-15) hybrid cells. J Biol Chem 266:3365–3368

Offermanns S, Schultz G, Rosenthal W (1991c) Identification of receptor-activated G-proteins by the use of the photoreactive GTP analog, [$\alpha$-$^{32}$P]GTP azidoanilide. Methods Enzymol 195:286–301

Ordway RW, Singer JJ, Walsh JV (1991) Direct regulation of ion channels by fatty acids. Trends Neurosci 14:96–100

Osugi T, Imaizum T, Mizushima A, Uchida S, Yoshida H (1987) Role of a protein regulating guanine nucleotide binding in phosophoinositide breakdown and calcium mobilization by bradykinin in neuroblastoma x glioma hybrid NG108-15 cells: effects of pertussis toxin and cholera toxin on receptor-mediated signal transduction. Eur J Pharmacol 137:207–218

Paupardin-Tritsch D, Hammond C, Gerschenfeld HM, Nairn AC, Greengard P (1986) cGMP-dependent protein kinase enhances $Ca^{2+}$ current and potentiates the serotonin-induced $Ca^{2+}$ current increase in snail neurons. Nature 323:812–814

Plummer MR, Logothetis DE, Hess P (1989) Elementary properties and pharmacological sensitivities of calcium channels in mammalian and peripheral neurons. Neuron 2:1453–1463

Rane SG, Dunlap K (1986) Kinase C activator 1,2-oleoylacetylglycerol attenuates voltage-dependent calcium current in sensory neurons. Proc Natl Acad Sci USA 83:184–188

Rane SG, Holz IV GG, Dunlap K (1987) Dihydropyridine inhibition of neuronal calcium current and substance P release. Pflugers Arch 409:361–366

Reuter H (1983) Calcium channel modulation by neurotransmitters, enzymes and drugs. Nature 301:569–574

Rosenthal W (1990) Signal transduction and ion channel activity. In: Habenicht A (ed) Growth factors, differentiation factors and cytokines. Springer, Berlin Heidelberg New York, pp 427–440

Rosenthal W, Hescheler J, Hinsch KD, Spicher K, Trautwein W, Schultz G (1988a) Cyclic AMP-independent dual regulation of voltage-dependent $Ca^{2+}$ currents by LHRH and somatostatin in a pituitary cell line. EMBO J 7:1627–1633
Rosenthal W, Hescheler J, Trautwein W, Schultz G (1988b) Receptor and G-protein-mediated control of voltage-dependent calcium channels. Cold Spring Harbor Symp Quant Biol 53:247–254
Rosenthal W, Offermanns S, Hescheler J, Spicher K, Hinsch K-D, Rudolph U, Schultz G (1989) Involvement of pertussis toxin-sensitive G-proteins in the modulation of voltage-dependent $Ca^{2+}$ channels by extracellular signals. In: Gehring U, Helmreich E, Schultz G (eds) Molecular mechanisms of hormone action. (Colloquium Mosbach, 40) Springer, Berlin Heidelberg New York, pp 139–146
Rosenthal W, Hescheler J, Eckert R, Offermanns S, Schmidt A, Hinsch K-D, Spicher K, Trautwein W, Schultz G (1990) Pertussis toxin-sensitive G-proteins: participation in the modulation of voltage-dependent $Ca^{2+}$ channels by hormones and neurotransmitters. Adv Second Messenger Phosphoprotein Res 24:89–94
Rosenthal W, Kleuss C, Wittig B, Hescheler J, Schultz G (1992) Approaches to study the interaction of G-proteins with $Ca^{2+}$-channels. In: Glossmann H, Striessnig J (eds) Methods in Pharmacology, vol 7. Plenum, New York (in press)
Scherübl H, Hescheler J, Schultz G, Kliemann D, Zink A, Ziegler R, Raue F (1992) Inhibition of $Ca^{2+}$-induced calcitonin secretion by somatostatin: roles of voltage-dependent $Ca^{2+}$ channels and G-proteins. Cell Signal 4:77–85
Schmidt A, Hescheler J, Offermanns S, Spicher K, Hinsch K-D, Codina J, Birnbaumer L, Gausepohl H, Frank R, Schultz G, Rosenthal W (1991) Involvement of pertussis-toxin-sensitive G-proteins in the hormonal inhibition of dihydropyridine-sensitive $Ca^{2+}$ currents in an insulin-secreting cell line (RINm5F). J Biol Chem 266:18025–18033
Schmidt A, Otto G, Spicher K, Rosenthal W (1992) G-proteins involved in the receptor-mediated inhibition of insulin secretion. Naunyn Schmiedebergs Arch Pharmacol 345 [Suppl]:R 49
Schofield GG (1990) Norepinephrine inhibits $Ca^{2+}$ current in rat sympathetic neurons via a G-protein. Eur J Pharmacol (Mol Pharmacol section) 207:195–207
Schroeder JE, Fischbach PS, Zheng D, McCleskey EW (1991) Activation of mu opioid receptors inhibits transient high- and low-threshold $Ca^{2+}$ currents but spares a sustained current. Neuron 6:13–20
Schultz G, Rosenthal W, Hescheler J, Trautwein G (1990) Role of G-proteins in calcium channel modulation. Annu Rev Physiol 52:275–292
Scott RH, Dolphin A (1987) Activation of G-proteins promotes agonist responses to calcium channel ligands. Nature 330:760–762
Scott RH, Wootton JF, Dolphin AC (1990) Modulation of T-type calcium channel currents by photoactivation of intracellular guanosine 5′-O(3-thio) triphosphate. Neuroscience 38:285–294
Seward E, Hammond C, Henderson G (1991) Mu-opioid receptor-mediated inhibition of N-type calcium channel current. Proc R Soc Lond [Biol] 244:129–135
Shangold G, Murphy S, Miller R (1988) Gonadotropin-releasing hormone-induced $Ca^{2+}$ transients in single identified gonadotrophs require both intracellular $Ca^{2+}$ mobilization and $Ca^{2+}$ influx. Proc Natl Acad Sci USA 85:6566–6570
Song S-Y, Saito K, Noguchi K, Konishi S (1989) Different GTP-binding proteins mediate regulation of calcium channels by acetylcholine and noradrenaline in sympathetic neurons. Brain Res 494:383–386
Suprenant A, Shen K-Z, North RA, Tatsumi H (1990) Inhibition of calcium currents by noradrenaline, somatostatin, and opioids in guinea-pig submucosal neurons. J Physiol (Lond) 431:585–608

Suprenant A, Horstman D, Akbarali H, Limbird LE (1992) A point mutation of the $\alpha_2$-adrenoceptor that blocks coupling to potassium but not to calcium channels. Science 257:977–980
Suzuki N, Yoshioka T (1987) Differential blocking action of synthetic $\omega$-conotoxin on components of $Ca^{2+}$ channel current in clonal $GH_3$ cells. Neurosci Lett 75:235–239
Taussig R, Sanchez S, Rifo M, Gilman AG, Belardetti F (1992) Inhibition of the $\omega$-conotoxin-sensitive calcium current by distinct G-proteins. Neuron 8:799–809
Toselli M, Lang J, Costa T, Lux HD (1989) Direct modulation of voltage-dependent calcium channels by muscarinic activation of pertussis toxin-sensitive G-protein in hippocampal neurons. Pflugers Arch 415:255–261
Tsien RW, Lipscombe D, Madison DV, Bley KR, Fox AP (1988) Multiple types of neuronal calcium channels and their selective modulation. Trends Neurosci 11:431–437
Wanke E, Ferroni A, Malgaroli A, Ambrosini A, Possan T, Meldolesi J (1987) Activation of muscarinic receptor selectively inhibits a rapidly inactivated $Ca^{2+}$ current in rat sympathetic neurons. Proc Natl Acad Sci USA 84:4313–4317
Woodcock E, McLeod J (1986) Adenylate cyclase inhibition is not involved in adrenal steroidogenic response to angiotensin II. Endocrinology 119:1697–1702
Yatani A, Birnbaumer L, Brown AM (1990) Direct coupling of the somatostatin receptor to potassium channels by a G protein. Metabolism 39:91–95

CHAPTER 73

# Regulation of Cell Growth and Proliferation by $G_o$

S.D. KROLL and R. IYENGAR

## A. Introduction

One function controlled by intercellular communication and cell surface signal transduction is the regulation of cellular growth and proliferation. Many proteins that regulate cellular growth are components of well-studied signal transduction pathways. Normal counterparts of most nonnuclear oncogene products function as ligands, receptors, or intracellular effectors of signaling pathways (BISHOP 1991). The heterotimeric G-protein coupled pathway is one such signal transduction pathway utilized by several cellular mitogens (POUYSSEGUR 1990). The sensitivity of this mitogenic signaling to pertussis toxin in some systems indicates that members of the $G_i/G_o$ family may be involved. Studies in our laboratory have focused on developing an understanding a role for $G_o$ in regulating the entry into the cell cycle as well as cell proliferation. The implications of these studies are discussed here.

## B. The $G_o$-Protein

$G_o$ is a heterotrimeric GTP binding protein that was purified as the major pertussis toxin substrate in neural tissue (STERNWEIS and ROBISHAW 1984). $G_o$ represents up to 0.5%–1% of the membrane-bound protein in brain (NEER et al. 1984), yet is poorly expressed in peripheral tissues. $G_o$ appears to be heterogeneously distributed in brain tissue and thus may account for even a higher percentage of membrane protein in some central nervous system structures. $G_o$ appears to be more abundant in cortex, thalamus, hypothalamus, and cerebellum and of lower abundance in the medullary and pons regions (GIERSCHIK et al. 1986b; ASANO et al. 1987). $G_o$ $\alpha$ subunit immunoreactivity is observed in brain tissue from many different species, including human, pig, rat, chicken, frog, snail, and locust (HOMBURGER et al. 1987). It appears to be the only pertussis toxin substrate in invertebrate neural tissue (HARRIS-WARRICK et al. 1988). $G_o$ immunoreactivity was also observed in peripheral neuronal tissue including spinal cord (GIERSCHIK et al. 1986b, sciatic nerve (HOMBURGER et al. 1987), and adrenal medulla and trachea (MUMBY et al. 1986; TOUTANT et al. 1987). Other nonneuronal sites of $G_o$ immunoreactivity have been observed in anterior pituitary (HOMBURGER et al. 1987), heart and kidney medulla (HUFF et al. 1985), olfactory neuroepithelium (ANHOLT et al. 1987), pancreas (HSU et al. 1990), and ovary (MUMBY et al. 1987). The peripheral distribution of $G_o$ immunoreactivity may in part be due to copurification of regional innervation. This is suggested by studies in which it was observed that $G_o$-$\alpha$ immunoreactivity increased during the cardiac parasympathetic innervation development (LUETJE et al. 1987; LIANG et al. 1986).

Cloning of a partial $G_o$-$\alpha$ was carried out from brain (ITOH et al. 1986). Sequence information was generated from tryptic degradation of the purified $G_o$-$\alpha$, and a full length clone was isolated from rat olfactory neuroepithelium (JONES and REED 1987) and bovine retina (VAN MEURS et al. 1987). The clones displayed a high homology with $G_i$-$\alpha$ (82%), 76% homology with transducin-$\alpha$, and 50% homology with $G_s$-$\alpha$. The highly conserved regions have been mapped as the regions of GTP binding and hydrolysis (DEVER et al. 1987). In addition, two immunologically distinct forms of the $G_o$ protein have been isolated (GOLDSMITH et al. 1989; PADRELL et al. 1991). These proteins display similar GDP release and GTP$\gamma$S binding rates, although the two species posess different GTP$\gamma$S binding kinetics in the absence of $Mg^{2+}$ (PADRELL et al. 1991).

The biological functions of $G_o$ have been difficult to define. Reconstitution experiments have demonstrated that $G_o$ can couple to muscarinic receptors (KUROSE et al. 1986) and $\alpha_2$-adrenergic receptors (CERIONE et al. 1986). $G_o$ was capable of reconstituting coupling of f-Met-Leu-Phe receptors and phospholipase C activation (KIKUCHI et al. 1986) in HL-60 cells, although $G_o$ immunoreactivity was not found in these cells. Such regulation of phospholipase C has also been reported for $G_i$ proteins. $G_o$ is not part of an adenylyl cyclase inhibitory pathway in *cyc*- membranes (ROOF et al. 1985) or purified $\alpha_s$-activated adenylyl cyclase (KATADA et al. 1986; Lindner et al. 1990) or in heterologously reconstituted systems (WONG et al. 1992). In neuroblastoma/glioma hybrid cells, $G_o$ inhibits $Ca^{2+}$ channels (HESCHELER et al. 1987; HARRIS-WARRICK et al. 1988; KLEUSS et al. 1991), although it is as yet uncertain whether this is a direct effect (HILLE 1992). Expression of the $\alpha$ subunit of $G_o$ in adrenal cells decreases the mitogenesis in these cells. Finally, it has been shown that $G_o$ (both forms) can specifically enhance muscarinic stimulation of phospholipase C in *Xenopus* oocytes. The use of antisense oligonucleotides to $G_o$-$\alpha$ subunit indicates that it carries a major component of the signal from expressed $\alpha_1$-adrenergic receptors to the phospholipase C pathway in *Xenopus* oocytes (BLITZER et al. 1992). In addition, activated $G_o$-$\alpha$ subunit stimulates the oocyte phospholipase C as assessed by the $Ca^{2+}$ dependent $Cl^-$ current in *Xenopus* oocytes (MORIARTY et al. 1990).

## C. The $G_o$-Protein And Cell Cycle Regulation in the *Xenopus* Oocyte

Many mitogenic hormones stimulate inositol phospholipid hydrolysis and specifically hydrolysis of phosphatydylinositol bisphosphate ($PIP_2$) (PARDEE 1989). This hydrolysis yields two intracellular second messengers, inositol trisphosphate ($IP_3$) and diacylglycerol (DAG). $IP_3$ binds to intracellular receptors on specialized components of the endoplasmic reticulum, promoting the release of intracellular calcium (BERRIDGE and IRVINE 1984). DAG activates a cytoplasmic protein kinase with serine/threonine kinase activity, called protein kinase C (PKC; NISHIZUKA 1984). Phospholipase C is the enzyme that catalyzes the hydrolysis of $PIP_2$ (REBECHI and ROSEN 1987; WILSON et al. 1985). This enzyme has been found to exist in several distinct forms, indicating various modes of regulation (RHEE et al. 1992).

The *Xenopus* oocyte has emerged as a model system for testing stimulation of cellular growth and proliferation (SMITH 1989). Fully grown amphibian oocytes are growth arrested in the late $G_2$ phase prior to meiosis I. These cells must progress to the second meiotic metaphase fertilization by sperm to proceed. For the maturation event to occur, in vivo, gonadotropic hormone stimulates the release of progesterone from ovarian follicular cells. This progesterone is thought to directly stimulate the maturation of the oocyte, causing the oocyte nuclear membrane to breakdown. It has also been suggested that acetylcholine released from the neural inputs (DASCAL 1987) can, through the activation of phospholipase C, potentiate the progesterone stimulation of maturation at submaximal concentrations of progesterone. Maturation is characterized by migration of the nuclear contents to the animal pole of the oocyte. This migration displaces pigment on the oocyte surface causing a white spot to form in the darkly pigmented animal pole. This white spot formation is a readily discernable feature of the mature oocyte and has been termed germinal vesicle breakdown (DUMONT 1972). Surgically removed oocytes can undergo the maturation process by extracellular application of progesterone (MALLER and KREBS 1977). Maturation is typically observed to occur about 6–18 h after progesterone addition. In addition to the ease with which growth control can be measured, the *Xenopus* oocyte is extremely useful because of its large size. *Xenopus* oocytes are routinely used for microinjection studies. Thus the capability of various cellular entities to trigger reenry into the cell cycle can be directly measured by microinjection into the oocyte by assaying for oocyte maturation.

Many different agents stimulate oocyte maturation (SMITH 1989). This has led to the hypothesis that oocytes may contain separate pathways be which growth resumption can be stimulated. Phorbol esters induce oocyte maturation (STITH and MALLER 1987) suggesting that activation of the PKC pathway may play a role in oocyte maturation induction. Microinjection of the catalytic fragment of the PKC enzyme induces germinal vesicle migration; however, dissolution of the vesicle does not occur (MURAMATSU et al. 1989). Phorbol ester induced maturation is inhibited by forskolin, exogenous cAMP, and 3-isobuty-1-methyl-xanthine (a cAMP phosphodiesterase inhibitor) in *Rana pipiens* oocytes. Several different oncogene products have been shown to stimulate oocyte maturation, including the $p21^{ras}$ protein (BIRCHMEIER et al. 1985), truncated forms of the epidermal growth factor receptor (OPRESKO and WILEY 1990) and $p39^{mos}$ (SAGATA et al. 1988). Other oncogene products fail to induce oocyte maturation, as in the case of *src* kinase (SPIVACK et al. 1984).

We have found that the activated form of the $G_o$ protein $\alpha$ subunit ($\alpha_o^*$) can induce *Xenopus* oocyte maturation (KROLL et al. 1991). Microinjection of $\alpha_o^*$ at a final intracellular concentration of 3 n*M* can induce maturation in defolliculated oocytes similar to that seen by incubation of oocytes in a bath of 10 $\mu M$ progesterone. This effect was specific for the $G_o$-$\alpha$ subunit as the GTP$\gamma$S activated forms of the $G_{i1}$, $G_{i2}$, and $G_{i3}$ $\alpha$ subunits did not induce oocyte maturation, when microinjected at similar concentrations. The effect was not seen with herterotrimeric $G_o$, nor was it attenuated by the addition of excess resolved $\beta\gamma$ subunits to the $\alpha_o^*$. The $\alpha^*$ response was attenuated, however, when the $\alpha_o^*$ subunit was specifically immunodepleted with an antipeptide antibody against a unique region of the $G_o$-$\alpha$ subunit, prior to microinjection.

The downstream effectors of $\alpha_o^*$ induced oocyte maturation were explored. By stimulating phospholipase C, $\alpha_o^*$ causes the release of $Ca^{2+}$ from $IP_3$-senstive store and also induces stimulation of protein kinase C activity. This stimulation of PKC activity is required for oocyte maturation induced by $\alpha_o^*$. Coinjection of a peptide inhibitor of PKC (19–36) blocks the $\alpha_o^*$ induced maturation. The sequence of the inhibitory peptide corresponds to the pseudosubstrate region within the primary sequence of the PKC enzyme. The PKC species cloned from *Xenopus* oocytes contain this sequence in their cDNAs. Progesterone induced oocyte maturation is not

**Table 1.** Protein kinase C involvement in regulation of *Xenopus* oocyte maturation

| Stimulation of maturation | Agent tested | Effect | Interpretation of result |
|---|---|---|---|
| $\alpha_o$* | PCK (19–36) peptide | Blockade of maturation | PKC activity required for $\alpha_o^*$-induced maturation |
| Progesterone | " | *No* blockade of maturation | PKC activity *not* required for progesterone-induced maturation |
| $p21^{ras}$ | " | *No* blockade of maturation | PKC activity *not* required for $p21^{ras}$ induced maturation |

inhibited by the microinjection of PKC (19–36). A somewhat surprising finding was that *ras* induced maturation was also not inhibited by the PKC (19–36) peptide (Table 1). These results suggesting that the progesterone-, *ras*-, and $\alpha_o^*$-induced maturation pathways are different up to at least the level of PKC.

Progesterone-induced oocyte maturation requires protein synthesis, as demonstrated by cycloheximide inhibition of progesterone-induced oocyte maturation (Wasserman and Masui 1975). Kinetic analysis of cycloheximide treatment has allowed the identification and characterization of a specifically translated and phosphorylated protein from maturing oocytes. This protein has been found to be the cellular homolog of the viral (v-*mos*) oncogene product (Sagata et al. 1989a). Microinjection of antisense oligonucleotides to the c-*mos* transcript inhibit progesterone induced maturation (Sagata et al. 1989b). Immature *Xenopus* oocytes contain a large amount of the mos mRNA species which is translated upon progesterone treatment (Sagata et al. 1988). Microinjection of excess c-*mos* mRNA into oocytes is sufficient to induce maturation (Sagata et al. 1989a). Thus, *mos* protein appears to be at least one of the specifically translated mRNA species that elicits oocyte maturation.

We find that $\alpha_o^*$-induced oocyte maturation can also be inhibited by preincubation of the oocytes with cycloheximide. Thus, the $\alpha_o^*$ pathway for maturation contains a necessary protein synthesis step(s). Prior injection of antisense oligonucleotides against the c-*mos* transcript also abolish $\alpha_o^*$-induced oocyte maturation. In addition, $\alpha_o^*$ induces the synthesis and phosphorylation of c-*mos* protein through protein kinase C since coinjection of the PKC (19–36) peptide with $a_o^*$, blocks the $\alpha_o^*$ induction of *mos*. From these experiments we have concluded that the synthesis and presumably the activity of *mos* protein is required for induced $\alpha_o^*$ maturation.

Early experiments had shown that downstream of the step(s) that require protein synthesis, oocytes undergoing maturation contain a cytoplasmic factor whose activity correlates with entry into M phase. The factor was called maturation-promoting factor (MPF). It has been purified by Maller and coworkers (1988). It contains a ubiquitous Ser-Thr kinase analogous to the yeast *cdc2* kinase (Lewin 1990). This protein kinase is essential for triggering the entry of both somatic and germ cells to enter the cell cycle. The regulatory component of MPF is called cyclin, a protein essential for the cell cycle (Murray and Kirschner 1989). Cyclin is phosphorylated during the cell cycle, and hence it appeared possible that the *mos* protein may regulate the activity of MPE through phosphorylation of cyclin. However, the ability of cyclin to trigger oocyte maturation does not appear to be the directly regulated by its phosphorylation state (Izumi and Maller 1991). Irrespective of the number of steps between *mos* protein and MPF it appears certain that expression of *mos* results in activation of MPF. Thus $\alpha_o^*$ through at least three Ser-Thr protein kinases triggers oocyte maturation. This is shown schematically in Fig. 1.

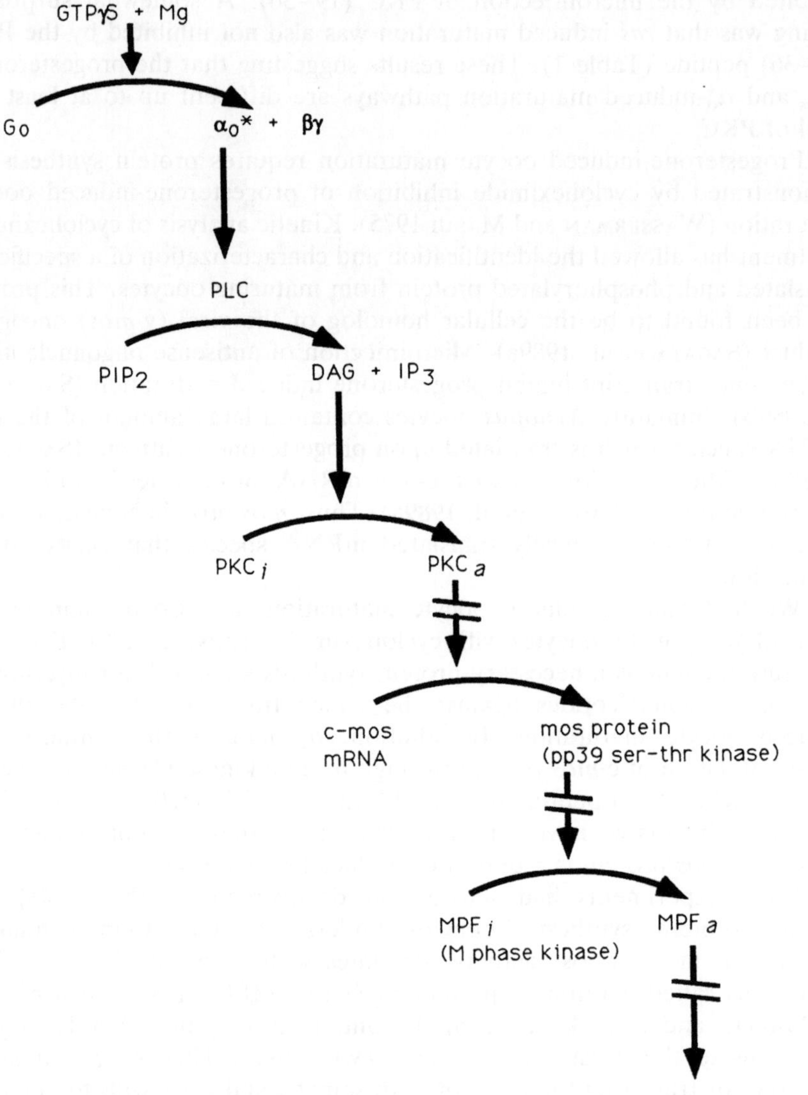


**Fig. 1.** The pathway utilized by $G_o$ to trigger oocyte maturation. This scheme is not comphrehensive; rather, it identifies currently known Ser-Thr protein kinases that are utilized in the $G_o$ signaling pathway. *Breaks in vertical arrows* indicate currently no direct evidence to indicate direct linkage between these steps. The position of each of these elements with respect to the others is supported by experimental evidence

## D. Regulation of Oocyte Maturation by Multiple Pathways

The observation that PKC (19–36) peptide blocks only the $\alpha_o^*$-induced maturation suggests that there are several distinct pathways through which the signals for triggering oocyte maturation can be communicated. Even though the intial steps in the various signaling systems may be different, the final effect is still the same, as assessed by germinal vesicle dissolution in the sliced sections of mature oocytes and the appearance of condensed chromosomes indicative of metaphase II. The convergence of the signals appears to occur at the level of *mos* protein synthesis. Experiments in our laboratory have shown that antisense oligonucleotides to the *mos* transcript block maturation induced by ras, progesterone, and $\alpha_o^*$. This is in agreement from previously published results from VANDEWOUDE and coworkers that placed *mos* protein downstream of c-*ras* in oocyte maturation (SAGATA et al. 1988). Our current knowledge of the three pathways that can induce oocyte maturation is summarized in Fig. 2.

Convergence of the $G_o$-$\alpha$, progesterone, and *ras* maturation pathways at the synthesis of *mos* protein presents a potentially crucial site at which growth signals may come together. The limited distribution may make the *mos* protein an entity unique to oocytes. However, the activation of an

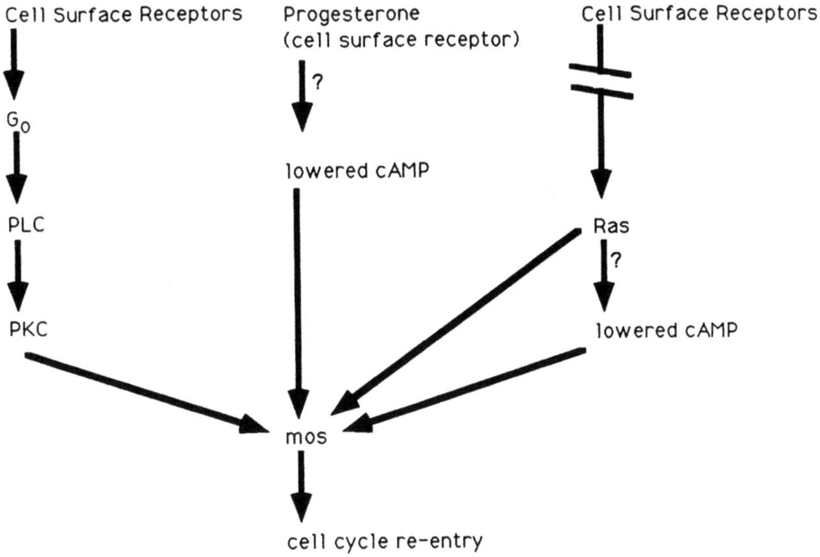


**Fig. 2.** Multiple pathways for oocyte maturation. The three distinct pathways that require synthesis of *mos* protein to induce oocyte maturation are shown. Since PKC (19–36) peptide blocks only $\alpha_o^*$-induced maturation, it is presumed that protein kinase C activity is not required for the other pathways. *Question marks* (*ras* and progestorone pathways) indicate our current lack of understanding of whether lowering of cellular cAMP levels is the sole signal in these pathways that triggers oocyte maturation through the synthesis of *mos* protein

analogous serine/threonine kinases downstream from PKC activation is possible in somatic cells. There have been suggestions of kinase cascades in cellular growth control (KAMATA and FUNG 1990; BURKHAND and TRAUGH 1983) as many growth-associated proteins are both kinases and phosphoproteins. It has been reported that raf, a cytoplasmic Ser-Thr kinase is downstream of PKC in NIH-3T3 cells and in T cells (KOLCH et al. 1991; SEIGEL et al. 1990). Ser-Thr kinases such as *raf*, *pim*, and *mos* are oncogenic in somatic cells, indicating that it is possible that similar such pathways may exist in these cells. In addition, it has been shown that *mos* is activated by phorbol ester stimulation of protein kinase C in NIH-3T3 cells (AL-BAGDADI et al. 1990). Thus, it appears possible that a cascade of protein kinases may be involved in signaling of cells to reenter the cell cycle. Although the identity of the members are different, such a kinase cascade is similar to what is used to regulate glycogenolysis in liver. The basic format for signal transmission utilized in both systems may be the same, even though there may be mechanistic differences at a given step. This format utilizes upstream protein kinases to activate downstream protein kinases that can modify specific physiological targets resulting in altered physiology in response to the external signal.

## E. Proliferation of Mammalian Cells by Activated $G_o$

The observations that activated $G_o$ can trigger the reentry of oocytes into the cell cycle raised the possibility that $G_o$ may be involved in the regulation of proliferation in mammalian cells as well. In the rat brain $G_o$ is a major component of the growth cone and could conceivably be involved in communicating growth and differentiation signals. The role $G_o$ plays in generating a proliferative signal can be explicitly tested by using standard fibroblasts assay systems in vitro. A convinient way to introduce activated $G_o$-$\alpha$ into mammalian cells is by transfection of the cDNA encoding a mutant $G_o$-$\alpha$, whose GTPase activity has been inactivated. This can be done by mutating a conserved glutamine within the GTP bindng/hydrolysis domain. The mutation of this residue (Q61) in *ras* results in a oncogenic form of the *ras* protein (BARBACID 1987). The Q205L mutant of $G_o$-$\alpha$ (Q205L$\alpha_o$), is analogous to the Q61 mutation of *ras*.

We constructed the Q205L mutant of $G_o$-$\alpha$ by oligonucleotide directed in vitro mutagenesis. The Q205L$\alpha_o$ mutant was inserted downstream of the MMTV-LTR in a commercially available mammalian transfection vector. This allows the expression of the mutant $G_o$-$\alpha$ to be induced by dexamethasone. The Q205L$\alpha_o$ mutant stimulated mitogenesis in transfected cells in a dexamethasone-dependent manner, while the wild-type $\alpha_o$ did not. Surprisingly, cells expressing Q205L-$\alpha_o$ did not show enhanced basal phospholipase C activity. It appears that the mitogenic effects of $\alpha_o$ in mammalian cells are largely independent of phospholipase C stimulation.

Expression of the Q205L-$\alpha_o$ mutant resulted in transformation of NIH 3T3 cells by several criteria. The cells appear to have lost contact inhibition as assessed by their capability to form foci in confluent cultures. When grown on soft agar NIH 3T3 cells expressing Q205L-$\alpha_o$ were able to form colonies. The colony formation rate was 1%–2%. We also tested whether the Q205L-$\alpha_o$ transfected NIH-3T3 cells could induce tumors in Nu/Nu mice. It was found that vector and WT$\alpha_o$ transfected cells induced tumors at very low frequency (0%–16% of mice), while cells when transfected with Q205L-$\alpha_o$ induced tumors in more than 80% of mice injected. The latency periods for both colony formation in soft agar and tumorigenesis in mice showed some interesting features. It was found that Q205L-$\alpha_o$ had to be expressed for nearly 2 weeks prior to plating the cells in soft agar to obtain colony formation. However, after the cells were plated a large number of colonies were formed in about 2 weeks even in the absence of dexamethasone, which is needed to induce the synthesis of the mutant $\alpha_o$ protein. Similarly, after the intial 2-week induction, the latency period with which Q205L-transformed cells induced tumors was comparable to that seen with *ras* transformed cells. These data suggest but do not prove that transformation of NIH-3T3 cells by Q205L-$\alpha_o$ may involve cellular processes that result in commitment to transformation. Once such commitment has been acheived, the continued presence of the intial signal from Q205L-$\alpha_o$ may not be necessary to maintain the transformed phenotype. One possibility that cannot be excluded at this point is that Q205L-$\alpha_o$ subunit acts in concert with other spontaneous mutations either by stimulating the clonal expansion of preexisting mutations in NIH-3T3 cells or inducing new mutations through enhanced growth rate and reduced fidelity of gene replication. Indeed, the mitogenic effect of Q205L-$\alpha_o$ in transfected cells suggests that the oncogenic potential of this gene may in part be due to expansion of other preexisting or newly acquired mutations.

## F. Specificity of Transformation by Signaling Through G-Protein Pathways

An interesting feature of Q205L-$\alpha_o$ induced transformation is that it is cell type specific. Expression of Q205L-$\alpha_o$ in RAT-1 fibroblasts does not result in transformation. The failure of Q205L-$\alpha_o$ to transform RAT-1 fibroblasts may be due to the lower number of endogenous mutations found in this cell type. Both NIH-3T3 and RAT-1 cells are mesenchyme-derived cells that undergo many changes in gene expression during the course of terminal differentiation. RAT-1 cells appear to be less on the verge of transformation, as indicated by the lack of spontaneous foci formation. However, this may not be the reason why Q205L-$\alpha_o$ does not transform RAT-1 cells. Another mutant G-protein, Q204L-$\alpha_{i2}$, can transform RAT-1 fibroblasts but is unable to transform NIH-3T3 cells (Gupta et al. 1992).

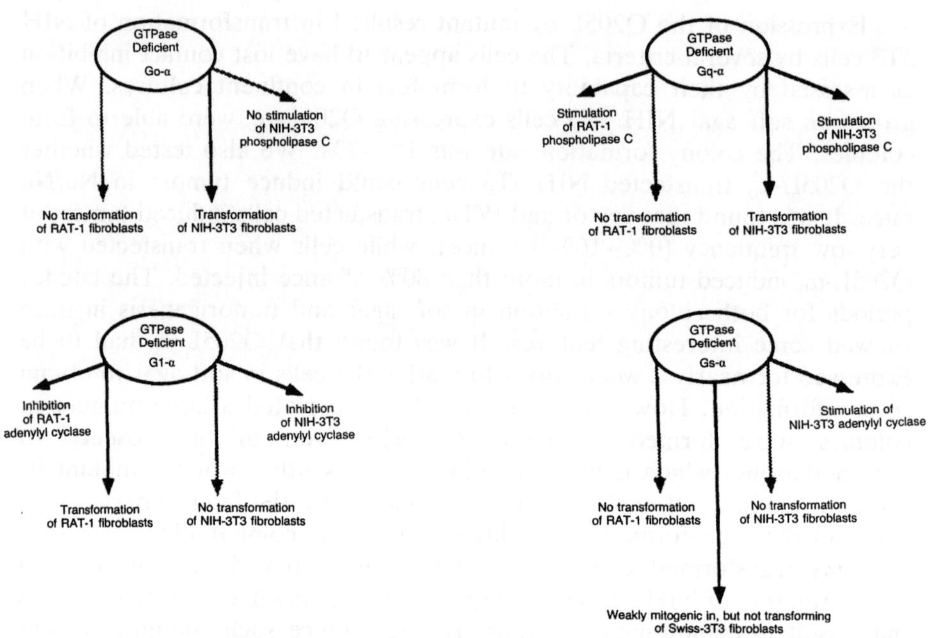


**Fig. 3.** Biochemical and transforming properties of the activated G-$\alpha$ subunits. All of the G-$\alpha$ subunits described here are mutated forms that are GTPase deficient. The inactivation of GTPase activity renders the protein continuously active. Please note the lack of correlation between the observed biochemical activities of these proteins and their transforming capabilities. These observations suggest that hithetto unrecognized pathways may be involved in growth signaling by G-proteins

The Q204L-$\alpha_{i2}$, however, inhibits adenylyl cyclase in both RAT-1 and NIH-3T3 cells, indicating that a reduction of adenylyl cyclase activity is not sufficient to induce transformation. These observations suggest that there may be hitherto unrecognized effectors through which these G-proteins transmit their growth signals. Such a hypothesis has gained further credence from studies in our laboratory which have shown that the Q209L-$\alpha_q$ stimulates phospholipase C activity in both NIH-3T3 cells and RAT-1 fibroblasts but only transforms NIH-3T3 cells (DE VLVO et al. 1992). Thus G-protein-induced transformation in general may be cell-type specfic. The capability of individual G-proteins to transform a certain cell type may depend on the preexisting growth control signaling pathways that exist in these cells. The mitogenic and transforming properties of the various activated G-$\alpha$ subunits are summarized in Fig. 3.

## G. Desensitization and Growth Signaling Through G-Protein Pathways

The potential of continuous signaling through G-protein pathways to consitute an oncogenic signal is noteworthy. Most G-protein-mediated signal transduction pathways in the cell contain substantial regulatory controls to prevent overamplification. Receptors are the primary targets of such negative feedback regulation. Several mechanisms, including uncoupling by phosphorylation, sequestration from the cell surface, and down-regulation, regulate both receptor number and function at the cell surface. In contrast, the G-proteins and other downstream components of signaling pathways do not appear to be as extensively regulated by negative or positive feedback mechanisms. Hence, a mutated G-protein that is persistently active can continuously deliver signals. Since it is now apparent that multiple pathways may be turned on by one G-protein (Birnbaumer 1990), it is possible that a mutation in a growth promoting G-protein will be far enough upstream to affect several pathways. Such prolonged and coordinated signals could result in commitment to unregulated growth and subsequently to neoplastic transformation.

*Acknowledgements.* The research in our laboratories is supported by NIH grants Ca44998 and DK38761 and a grant from the American Cancer Society CD-518.

### References

Al-Bagdadi F, Singh B, Arlinghaus RB (1990) Evidence for involvement of protein kinase C pathway in the activation of $p37^{v\text{-}mos}$ protein kinase. Oncogene 5:1251–1257

Burkhand SJ, Traugh JA (1983) Changes in ribosomal functions by cAMP dependent and independent phosphorylations of ribosomal protein S6. J Biol Chem 258:14003–14008

Anholt RRH, Mumby SM, Stoffers DA, Griard PR, Kuo JF, Snyder SH (1987) Transduction proteins of olfactory receptor cells: identification of guanine nucleotide binding proteins and protein kinase C. Biochemistry 25:788–795

Asano T, Semba R, Ogasawara N, Kato K (1987) Highly sensitive immunoassay for the $\alpha$ subunit of the GTP binding protein GO and its regional distribution in bovine brain. J Neurochem 48:1617–1623

Barbacid M (1987) ras Genes. Ann Rev Biochem 56:779–827

Berridge MJ, Irvine RF (1984) Inositol trisphosphate: a novel second messenger in cellular signal transduction. Nature 312:315–321

Birchmeier C, Broek D, Wigler M (1985) RAS proteins can induce meiosis in xenopus oocytes. Cell 43:615–621

Birnbaumer L, Abramowitz J, Brown AM (1990) Receptor-effector coupling by G proteins. Biochem biophys Acta 1031:163–184

Bishop JM (1991) Molecular themes in oncogenesis. Cell 64:235–249

Blitzer RD, Omri G, DeVivo M, Carty DJ, Premont RT, Codina J, Birnbaumer L, Cottechia S, Caron MG, Lefkowitz RJ, Landau EM, Iyengar R (1993) Coupling of the expressed $\alpha$1b-adrenergic receptor to the phospholipase C pathway in Xenopus oocytes: the role of $G_o$. J Biol Chem 268:7532–7537

Burkhand SJ, Traugh JA (1983) Changes in ribosome function by cAMP dependent and cAMP independent phosphorylation of ribosomal protein S6. J Biol Chem 258:14003–14008
Cerione RA, Regan JW, Nakata H, Codina J, Benovic JL, Gierschik P, Somers RL, Birnbaumer L, Lefkowitz RJ, Caron MG (1986) Functional reconstitution of the $\alpha_2$-adrenergic receptor with guanine nucleotide regulatory proteins in phospholipid vesicles. J Biol Chem 261:3901–3909
Dascal N (1987) The use of Xenopus oocytes for the study of ion channels. CRC Critical Rev Biochem 22:317–387
Dever TE, Glynias MJ, Merrick WC (1987) GTP binding domain: three consensus sequence elements with distinct spacing. Proc Natl Acad Sci USA 84:1814–1818
De Vivo M, Chen J, Codina J, Iyengar R. (1992) Enhanced phospholipase C stimulation and transformation in NIH-3T3 cells expressing Q209L-$G_q$ $\alpha$ subunit. J Biol Chem 267:18263–18266
Dumont JN (1972) Oogenesis in Xenopus laevis. J Morphol 136:153–180
Gierschik P, Falloon J, Milligan G, Pines M, Gallin JI, Spiegel A, (1986a) Immunochemical evidence for a novel pertussis toxin substrate in human Neutrophils. J Biol Chem 261:8058–8052
Gierschik P, Milligan G, Pines M, Goldsmith P, Codina J, Klee N, Spiegel A (1986b) Use of specific antibodies to quantitate the guanine nucleotide binding protein go in brain. Proc Natl Acad Sci Acad Sci USA 83:2258–2262
Goldsmith P, Backlund PS Jr, Rossiter K, Carter A, Milligan G, Unson C, Speigel A (1988) Purification of heterotrimeric G proteins from brain: identification of a novel form of $G_o$. Biochem 27:7085–7090
Gupta SK, Gallego C, Lowdnes JM, Pleiman CM, Sable C, Eisfelder BJ, Johnson GL (1992) Analysis of fibroblast transformation potential of GTPase deficient gip2 oncogenes. Mol Cell Biol 12:190–197
Harris-Warrick R, Hammond C, Paupardin-Tritsch D, Homburger V, Rouot B, Boekaert J, Gershengeld HM (1988) An $\alpha_{40}$ subunit of a GTP binding protein immunologically related to $G_o$ mediates a dopamine induced decrease in snail neurons. Neuron 1:27–32
Hescheler J, Rosenthal W, Trautwein W, Schultz G. (1987) The GTP-binding protein $G_o$ regulates neuronal calcium channels. Nature 325:445–447
Homburger V, Brabet P, Audigier Y, Pantoloni C, Bockaert J, Rouot B. (1987) Immulogical localization of the GTP binding protein $G_o$ in different tissues in vertebrates and invertebrates. Mol Pharmacol 31:313–319
Hsu WH, Rudolph U, Sanford J, Bertrand P, Olate J, Nelson C, Moss LG, Boyd AG, Codina J, Birbaumer L. (1990) Molecular cloning of a novel splice variant of the a-subunit of the mammalian $G_o$ protein. J Biol Chem 265:11220–11226
Huff RM, Aston JM, Neer E (1985) Physical and immunological characterization of a guanine nucleotide binding protein purifed from bovine cerebral cortex. J Biol Chem 260:10864–10871
Jones D, Reed RR (1987) Molecular cloning of five GTP biding portein cDNA species from rat olfactory neuroepithelium. J Biol Chem 262:14241–14249
Izumi T, Maller JL (1991) Phosphorylations of Xenopus cyclins B1 and B2 are not required for cell cycle transitions. Mol Cell Biol 8:3860–3867
Kamata T, Kung H-F (1990) Modulation of maturation and ribsomal portein S6 phosphorylation in Xenopus oocytes by microinjection of oncogenic ras porteins and protein kinase C. Mol Cell Biol 10:880–886
Katada T, Oinuma M, Ui M (1986) Mechanisms for inhibition of the catalytic activity of adenylate cyclase by the guanine nucleotide-binding proteins serving as the substrate of islet-activating protein, pertussis toxin. J Biol Chem 261:5215–5221
Kikuchi A, Kozawa O, Kaibuchi K, Ui M (1986) Direct evidence for involvement of a guanine nucleotide-binding protein in chemotactic peptide stimulated formation of inositol bisphosphate and trisphosphate in differentiated human leukemic (HL-60) cells. J Biol Chem 261:11558–11562

Kleus C, Hescheler J, Ewel C, Rosenthal W, Schultz G, Wittig B (1991) Assignment of G protein subtypes to specific receptors inducing inhibition of calcium currents. Nature 353:43–48

Kolch W, Heidecker G, Lloyd P, Rapp UR (1991) Raf-1 protein kinase is required for growth of induced NIH/3T3 cells. Nature 283:426–429

Kroll SD, Omri G, Landau EM, Iyengar R (1990) Activated $\alpha$ Subunit of the $G_o$ protein induces Xenopus oocyte. Proc Natl Acad Sci USA 88:5182–5186

Kroll SD, Chen J, DeVivo M, Carty DJ, Buku A, Premont RT, Iyengar R (1992) The Q205L mutant of $G_o$-$\alpha$ subunit expressed in NIH-3T3 cells induces transformation. J Biol Chem submitted

Kurose H, Katada T, Haga T, Haga K, Ichiyama A, Ui M (1986) Functional interaction of purified muscarinic receptors with purified inhibitory guanine regulatory proteins reconstituted in phospholipid vesicles. J Biol Chem 261:6423–6428

Liang BT, Helmreich MR, Neer EJ, Galper JB (1986) Development of muscarinic cholinergic inhibition of adenylyl cyclase in embryonic chick heart. J Biol Chem 261:9011–9021

Lewin B. (1990) Driving the cell cycle: M-phase kinase and its partners. Cell 64:235–249

Lindner ME, Ewald DA, Miller RJ, Gilman AG (1990) Purification and characterization of $G_o$-$\alpha$ and three types of $G_i$-$\alpha$ after expression in E coli. J Biol Chem 265:8243–8251

Lohka MJ, Hayes MK, Maller JL (1988) Purification of maturation promoting factor, an intracellular regulator of early mitotic events. Proc Natl Acad Sci USA 85:3009–3013

Luetje CW, Gierschik P, Milligan G, Unson C, Spiegel A, Nathanson NM (1987) Tissue specific regulation of GTP binding protein: relationship to muscarinic acetylcholine receptor mediated responses Biochem 26:4876–4884

Maller JL, Krebs EG (1977) Progesterone-stimulated meiotic cell division in Xenopus oocytes. J Biol Chem 252:1712–1717

Mathie A, Berheim L, Hille B (1992) Inhibition of N and L type calcium channels by muscarinic activation in rat sympathetic neurons. Neuron 8:907–914

Moriarty TM, Padrell E, Carty DJ, Omri G, Landau EM, Iyengar R (1990) Go protein as signal transducer in the pertussis toxin sensitive phosphatydylinositol pathway. Nature 343:79–82

Mumby SM, Kahn RA, Manning DR, Gilman AG (1986) Antisera of designed specificity for subunits of guanine nucleotide binding regulatory proteins. Proc Natl Acad Sci USA 83:265–269

Mumby SM, Pang IH, Gilman AG, Sternweis PC (1988) Chromatographic resolution and immunological identification of the $\alpha_{40}$ and $\alpha_{41}$ subunits of guanine nucleotide regulatory proteins from bovine brain. J Biol Chem 263:2020–2026

Muramatsu M, Kaibuchi K, Arai K (1989) A protein kinase C cDNA without the regulatory domain is active after transfection in vivo in the absence of phorbol ester. Mol Cell Biol 9:831–836

Murray AM, Kirschner MW (1989) Cyclin synthesis drives the early embryonic cell cycle. Nature 339:275–280

Neer EJ, Lok JM, Wolf LG (1984) Purification and properties of the inhibtiory guanine nucleotide regulatory subunit of the brain adenylate cyclase. J Biol Chem 259:14222–14229

Nishizuka Y (1984) The role of protein kinase C in cell surface signal transduction. Nature 308:693–698

Opresko LK, Wiley HS (1990) Functional reconstitution of the epidermal growth factor receptor system in Xenopus oocytes. J Cell Biol 111:1661–1671

Padrell E, Carty DJ, Moriarty TM, Hildebrandt JD, Landau EM Iyengar R (1991) Two forms of bovine brain $G_o$ that stimulate the inositol trisphosphate-mediated $Cl^-$ current in Xenopus oocytes. J Biol Chem 266:9771–9777

Pardee (1989) $G_1$ events and regulation of cell proliferation. Science 246:603–608
Pouyssegur R (1990) G proteins in growth factor action. In: Iyengar R, Birnbaumer L (eds) G proteins. Academic, New York, pp 555–566
Rebechi MJ, Rosen OM (1987) Purification of a phosphoinositide specific phospholipase C from Bovine brain. J Biol Chem 262:12526–12532
Rhee SG, Choi K-D (1992) Regulation of Inositol phosphate specific phospholipase C isozymes J Biol Chem 267:12393–12396
Roof D, Applebury ML, Sternweis PC (1985) Relationships within the family of GTP-binding proteins isolated from bovine central nervous system. J Biol Chem 260:16242–16249
Sagata N, Oskarsonn M, Copeland T, Brumbaugh J, VandeWoude GF (1988) Function of c-mos proto-oncogene product in meiotic maturation in Xenopus oocytes, Nature 335:519–525
Sagata N, Daar I, Oskarsonn M, Showalter SD, VandeWoude GF (1989a) The product of the mos proto-oncogene as a candidate 'initiator' for oocyte maturation. Science 245:643–646
Sagata N, Watanabe N, VandeWoude GF, Ikawa Y (1989b) The c-mos proto-oncogene product is a cytostatic factor responsible for meiotic arrest in vertebrate eggs. Nature 342:512–518
Siegel JN, Klausner RD, Rapp UR, Samelson LE (1990) T cell antigen receptor engagement stimulates c-raf phosphorylation and induces c-raf associated kinase activity via a protein kinase C dependent pathway. J Biol Chem 265:18472–18480
Smith LD (1989) The induction of oocyte maturation: transmembrane signalling events and regulation of the cell cycle. Development 107:685–699
Smith SB, White HD, Siegel JB, Krebs EG (1981) Cyclic AMP dependent protein kinase I: cyclic nucleotide binding, structural changes and release of the catalytic subunits. Proc Natl Acad Sci USA 78:1591–1595
Sternweis PC, Robishaw JD (1984) Isolation of two proteins with high affinity for guanine nucleotides from membranes of the bovine brain. J Biol Chem 259:13806–13813
Spivack JG, Erikson RL, Maller JL (1984) Microinjection of $pp60^{v\text{-}src}$ into Xenopus oocytes increases phosphorylation of ribosomal protein S6 and accelerates the rate of progesterone induced meiotic maturation. Mol Cell Biol 4:1631–1634
Stith BJ, Maller JM (1987) Induction of meiotic maturation by 12-0-tetradecanoylphorbol-13-acetate. Exp Cell Res 167:514–523
Toutant M, Aunis D, Bockaert J, Homburger V, Rouot B (1987) Presence of three pertussis toxin substrates and $G_o\alpha$ immunoreactivity in both plasma and granule membranes of chromaffin cells. FEBS Lett 222:51–55
Van Meurs KP, Angus W, Lavu S, Kung HF, Czarnecki SK, Moss J, Vaughan M (1987) Deduced amino acid sequence of bovine retinal: similarities to other guanine nucleotide binding proteins. Proc Nat Acad Sci USA 84:3107–3111
Wasserman WJ, Masui Y (1975) Effects of cycloheximide on a cytoplasmic factor initiating meiotic maturation in Xenopus oocytes. Exp Cell Res 91:381–388
Wilson DB, Bross TE, Hofmann SC, Majerus BW (1985) Hydrolysis of polyphosphoinositides by purified sheep seminal vesicle phospholipase C enzymes. Cell 43:615–621
Wong Y-H, Conklin BR, Bourne, HR (1992) Gz-mediated inhibition of cyclic AMP accumulation. Science 255:339–342

CHAPTER 74

# Role of Nucleoside Diphosphate Kinase in G-Protein Action

N. KIMURA

## A. Introduction

Nucleoside diphosphate kinase (NDP kinase, EC 2.7.4.6.) is a ubiquitous enzyme which catalyzes a transfer of the terminal phosphate of 5′-triphosphate nucleotides to 5′-diphosphate nucleotides (PARKS and AGARWAL 1973). Under physiological conditions, NDP kinase works to form triphosphate nucleotide pools at the expense of ATP. The products of the enzyme reaction are widely utilized for syntheses of essential cellular compounds such as nucleic acids, proteins, polysaccharides, and lipids.

The NDP kinase is generally considered to be one of the "housekeeping" enzymes with no regulatory role in cellular functions since the enzyme possesses high activity compared to other nucleotide metabolizing enzymes and displays low substrate specificity. However, recent reports demonstrate its possible regulatory role in (d)NTP-requiring reactions such as DNA, RNA, and protein syntheses. One of the remarkable features of this enzyme seems to be its association with proteins presenting strict specificity for guanine nucleotides. These proteins include microtubules (NICKERSON and WELLS 1984), initiation factor eIF2 (WALTON and GILL 1975), and GTP-binding (G) proteins (KIMURA and SHIMADA 1988b, 1990) involved in signal transduction. Possible functional interaction between NDP kinase and GTP-binding proteins has been collectively reviewed (OTERO 1990; LACOMBE and JAKOBS 1992). This chapter summarizes the role of NDP kinase in membrane signaling systems with special reference to the action of G-protein $G_s$ in hormone-sensitive adenylyl cyclase systems and describes some of the recent advances in the research of NDP kinase.

## B. General Model of Membrane Signaling Systems Involving G-Proteins

Many of the membrane-signaling systems consist of a receptor, guanine nucleotide binding protein(s), and effector (GILMAN 1987; BOURNE et al. 1990; BIRNBAUMER et al. 1990). G-proteins, which comprise $\alpha$, $\beta$, and $\gamma$ subunits, are located and mediate signals between receptor and effector. In these systems, upon agonist binding, the activated receptor induces GDP release from G-proteins and subsequent GTP binding, that is, GDP-GTP exchange reaction, leading to activation of G-proteins and in turn effectors. GTP hydrolysis by intrinsic GTPase activity of the G-protein turns off this activation. The GDP-GTP exchange reaction is generally accepted as a sole mechanism to activate heterotrimeric as well as low molecular weight G-proteins.

## C. Role of NDP Kinase in Membrane Signaling Systems

### I. Evaluation of the Effect of GDP in Comparison with GTP

It is well known that hormones can activate membrane adenylyl cyclase from various mammalian tissues in the presence of GDP, as well as GTP, with no ATP-regenerating system (RODBELL et al. 1971a; HANOUNE et al. 1975; KIMURA and NAGATA 1977; IYENGAR and BIRNBAUMER 1979). This observation seemed to contradict the GDP-GTP exchange model because the model requires GTP in the bulk phase. Thus the question of whether GDP has the same properties as GTP in hormone action was investigated.

The glucagon-sensitive as well as other hormone-sensitive adenylyl cyclase absolutely requires GTP or GDP for activation (RODBELL et al. 1971a; KIMURA and NAGATA 1977). A major problem in assessing the effect of GDP was the possible involvement of GTP formation catalyzed by membrane associated (m) NDP kinase during adenylyl cyclase assay. Attempts to circumvent this problem by the use of transphosphorylation-resistant analogs such as App(NH)p and GDP$\beta$S often provided misleading results because of the presence of unexpected impurities (KIMURA and NAGATA 1979; KIMURA et al. 1985; MURPHY and STANSFIELD 1983). We successfully introduced UDP, which suppressed NDP kinase activity by forming a UDP-enzyme abortive complex (KIMURA and SHIMADA 1983). Glucagon plus GDP-stimulated activity was preferentially blocked by the presence of UDP with no change in glucagon plus GTP, glucagon plus Gpp(NH)p, or NaF-stimulated activities (KIMURA and SHIMADA 1983). It was unambiguous that GDP itself was not capable of mediating the hormonal signal to adenylyl cyclase but rather acted as a competitive inhibitor of GTP with an approximate $K_i$ of $1\,\mu M$. Furthermore, the experimental data demonstrated that a small amount of GTP formed from the added GDP (less than 5%) was enough to produce a hormonal response even in the presence of a large amount of GDP, a competitive inhibitor of GTP. Figure 1 shows a schematic model of our working hypothesis of the role of mNDP kinase in the membrane signaling system (KIMURA 1987).

The NDP kinase activity in membranes was small but resistant to solubilization by high ionic strength, EDTA, and sonic oscillation followed by ultracentrifugation. However, it was readily released by detergents, suggesting close association with membranes (KIMURA and SHIMADA 1983).

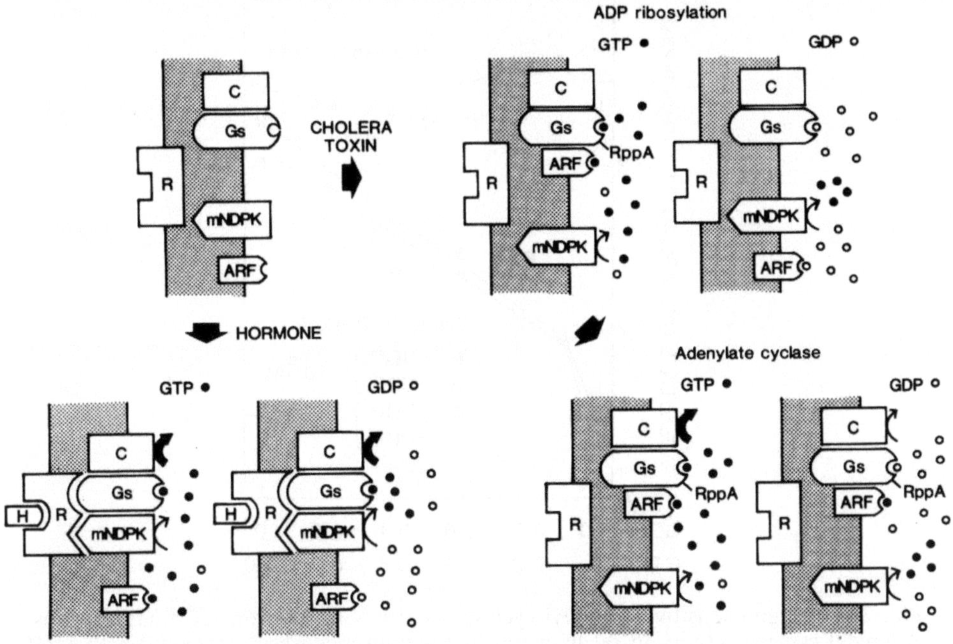


**Fig. 1.** A working hypothesis on the role of mNDP kinase in the hormone-sensitive adenylyl cyclase system. See text for detailed explanation. *R*, Receptor; *C*, catalyst (adenylyl cyclase); *Gs*, stimulatory G-protein; *ARF*, ADP ribosylation factor; *mNDPK*, membrane-associated NDP kinase; *H*, hormone; *closed circle*, GTP; *open circle*, GDP

## II. Role of mNDP Kinase in Signal Transduction

To investigate the mechanism by which small amounts of GTP formed from the added GDP produces such a substantial hormonal response, we conducted an experiment in which the hormonal response was examined at various GTP/GDP ratios but with constant total guanine nucleotide concentrations (10 and 100 $\mu M$) in the presence of UDP (Fig. 2). The results revealed that adenylyl cyclase activity was dependent on the GTP/GDP ratio rather than on the absolute GTP concentration, and further that glucagon stimulation was no longer detectable when the ratio decreased below 0.1. This observation seemed to be important because when mNDP kinase was active, the hormonal response was apparent with much lower GTP/GDP values; time studies demonstrated that the hormonal response with added GDP, if assessed as steady-state rate, was rapid in onset and seemed unlikely to depend on the GTP/GDP ratio in the bulk phase. This discrepancy postulated that there could be a mechanism by which the GTP formed via mNDP kinase is effectively supplied for hormonal activation of

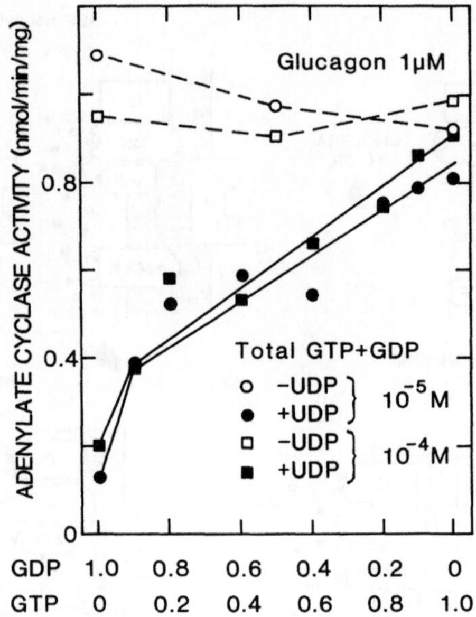


**Fig. 2.** Glucagon-sensitive adenylyl cyclase activity with different GTP/GDP ratios. Adenylyl cyclase activity in rat liver purified membranes was determined with 1 $\mu M$ glucagon and 10 or 100 $\mu M$ tatal (GTP + GDP) at different GTP/GDP ratio in the absence (*dashed lines*) and presence (*solid lines*) of UDP as an inhibitor of mNDP kinase

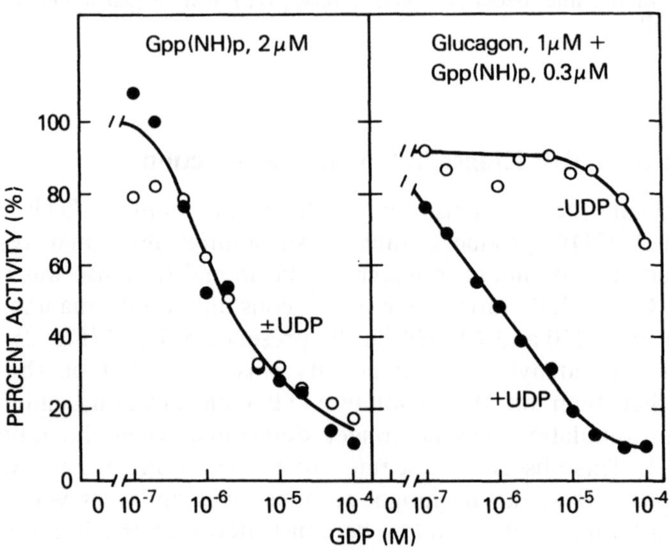


**Fig. 3.** Comparison of the inhibitory effect of GDP for Gpp(NH)p-stimulated (*left*) and glucagon plus Gpp(NH)p-stimulated (*right*) adenylyl cyclase activity. Adenylyl cyclase activity was determined with indicated additions in the absence and presence of 10 m$M$ UDP

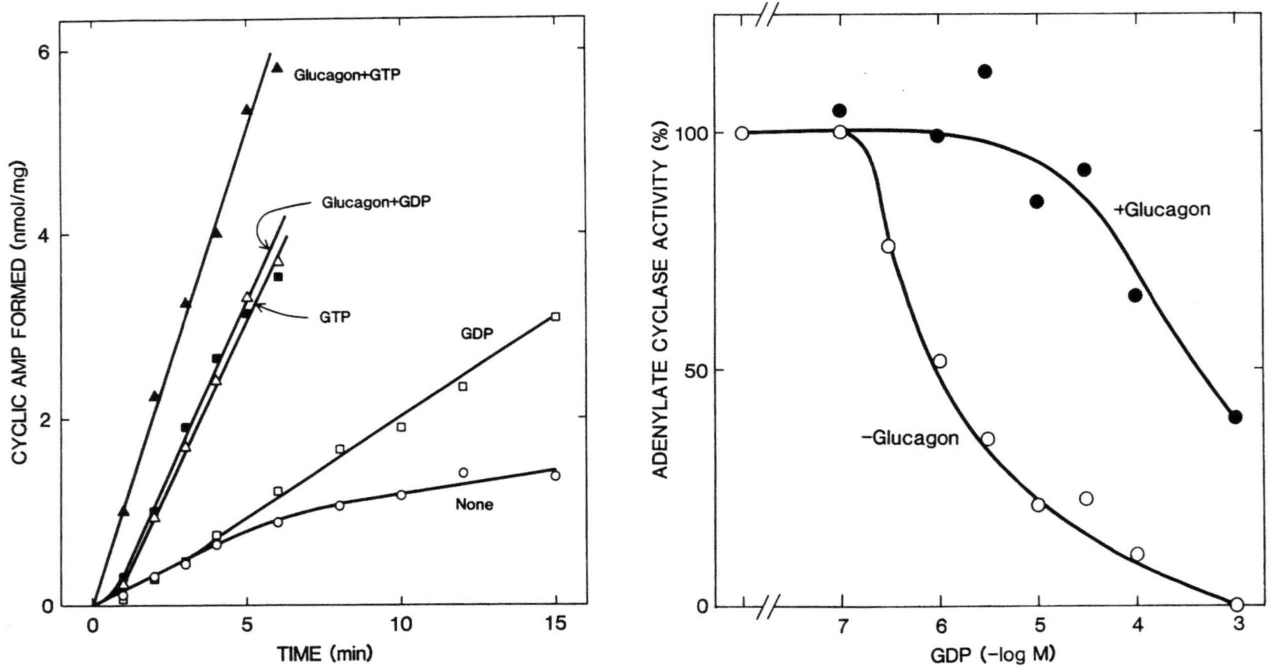


**Fig. 4.** Comparison of glucagon and cholera toxin actions on adenylyl cyclase activity in rat liver membranes. Top, reactions were started by adding cholera toxin-treated membranes into prewarmed incubation mixture and aliquots were withdrawn at indicated times for cyclic AMP determination. Glucagon, 1 $\mu M$; GTP and GDP, 10 $\mu M$. Bottom, adenylyl cyclase activity in the cholera toxin-treated membranes was determined with 0.3 $\mu M$ GTP along with various concentrations of GDP in the absence and presence of 1 $\mu M$ glucagon

adenylyl cyclase. The possibility was further examined in the following experiment.

While GDP was inhibitory for Gpp(NH)p-stimulated cyclase activity irrespective of whether NDP kinase activity was inhibited or not, the inhibitory effect of GDP for glucagon plus Gpp(NH)p-stimulated activity was apparent only under when NDP kinase activity was blocked (Fig. 3; KIMURA and SHIMADA 1983). It was clear that the hormone drastically (by about two orders of magnitude) reduced the inhibitory action of added GDP without altering binding affinity of the G-protein, by making use of GTP formed via mNDP kinase.

## III. Comparison Between Hormone and Cholera Toxin Actions

Cholera toxin can activate adenylyl cyclase in a GTP-dependent but receptor-independent manner through ADP ribosylation of $G_s$. Thus comparison of the actions of glucagon and cholera toxin was undertaken to solve the question of whether GTP supply via mNDP kinase operates specifically through receptor activation (KIMURA and SHIMADA 1985). The toxin-treated adenylyl cyclase showed a high steady-state rate by the addition of GTP alone (Fig. 4, left panel), which occurred with the same dose dependency as in the presence of glucagon. Under these conditions, added GDP caused no adenylyl cyclase activation and required further addition of glucagon for the activation (Fig. 4, left panel). The toxin-treated adenylyl cyclase was sensitive to the inhibitory effect of GDP, which was reduced by the coexistence of glucagon (Fig. 4, right panel). It should be noted that neither glucagon addition nor cholera toxin pretreatment altered mNDP kinase activity. These results demonstrated that activation of adenylyl cyclase ($G_s$) by the GTP produced via mNDP kinase from added GDP was a receptor-related phenomenon. The receptor-dependent introduction of the GTP formed via mNDP kinase seemed unlikely to occur in the case of another GTP-binding protein, ADP ribosylation factor (KIMURA and SHIMADA 1986).

## IV. Interaction Between mNDP Kinase and $G_s$, and Its Regulation

The purified rat liver membranes contained NDP kinase at a concentration of 1 pmol/mg protein (KIMURA and SHIMADA 1988a). The value was in good agreement with those of high-affinity glucagon receptor (1–2 pmol/mg; RODBELL et al. 1971b) and $G_s$ (0.2–1 pmol/mg; BOKOCH et al. 1984), suggesting a stoichiometric interaction of the mNDP kinase with a component(s) of adenylyl cyclase system. Thus the possible direct interaction between them was investigated by immunoprecipitation using anti-NDP kinase antibodies.

The $G_s$ in rat liver membranes was radiolabeled by ADP ribosylation using [$^{32}$P]NAD and cholera toxin. The membranes were then treated with or without glucagon and guanine nucleotides, followed by solubilization with detergents. A small portion (~5%) of the $G_s$ extracted with octylglucoside from membranes was immunoprecipitable by the antibodies. Its amount was decreased by receptor stimulation, and the decrease was prevented by the coexistense of an antagonist (KIMURA et al. 1988b). It follows that $G_s$ and membrane associated NDP kinase may form a complex which is sensitive to receptor regulation. The size of the complex was estimated to be 12 S by sucrose density gradient centrifugation (Fig. 5a; KIMURA et al. 1990). There has been no evidence to suggest direct interaction between mNDP kinase and hormone receptor. The ternary complex (hormone receptor–G-protein) extracted by digitonin seemed unlikely to be associated with NDP kinase (Fig. 5b).

## V. Regulatory Mechanism of G-Protein by NDP Kinase

Whether G-proteins are regulated by mNDP kinase by direct phosphate transfer to GDP bound to G-proteins or by altering the GTP pool in the immediate vicinity of the G-proteins is of primary importance (Fig. 6). While the latter mechanism requires the GDP-GTP exchange reaction to operate, the former does not. The following observations seem to prefer the latter mechanism. (a) G-proteins in membranes and in solution bear the GDP-GTP exchange reaction. (b) Membrane-associated $G_s$ (adenylyl cyclase) requires exogenous GTP or GDP to be activated in addition to ATP and $Mg^{2+}$. (c) Hormone-dependent activation of adenylyl cyclase occurs with GTP even when mNDP kinase activity is blocked, and it depends on GTP/GDP ratio. Further, considering the reaction mechanism of NDP kinase, whether the enzyme can recognize the substrate (GDP) bound to G-proteins remains uncertain.

On the other hand, if one presumes the direct phosphorylation of GDP bound to G-protein by NDP kinase, the reaction sequence would be as follows: NDP kinase forms a phosphorylated intermediate, followed by direct phosphate transfer into GDP bound to G-protein. The resulting GTP is hydrolyzed by intrinsic GTPase of the G-protein, leading to accumulation of inorganic phosphate (Pi) as an end product. This seemed to be the case in an experiment performed with purified G-proteins and NDP kinase; Pi accumulation proceeded with no appreciable GDP release from the G-protein when ATP and NDP kinase were present together with G-protein (KIKKAWA et al. 1990). In contrast, Pi accumulation along with simultaneous GDP release as a result of the GDP-GTP exchange reaction occurred by the addition of exogenous GTP. The possibility that released GDP was immediately converted to GTP by NDP kinase and the resulting GTP rapidly reassociated with G-protein remains to be rigorously examined.

Whichever mechanism is operating, at least in membrane preparations, interaction between the adenylyl cyclase system ($G_s$) and mNDP kinase seems to occur in a receptor-dependent manner (Fig. 6). Questions of how

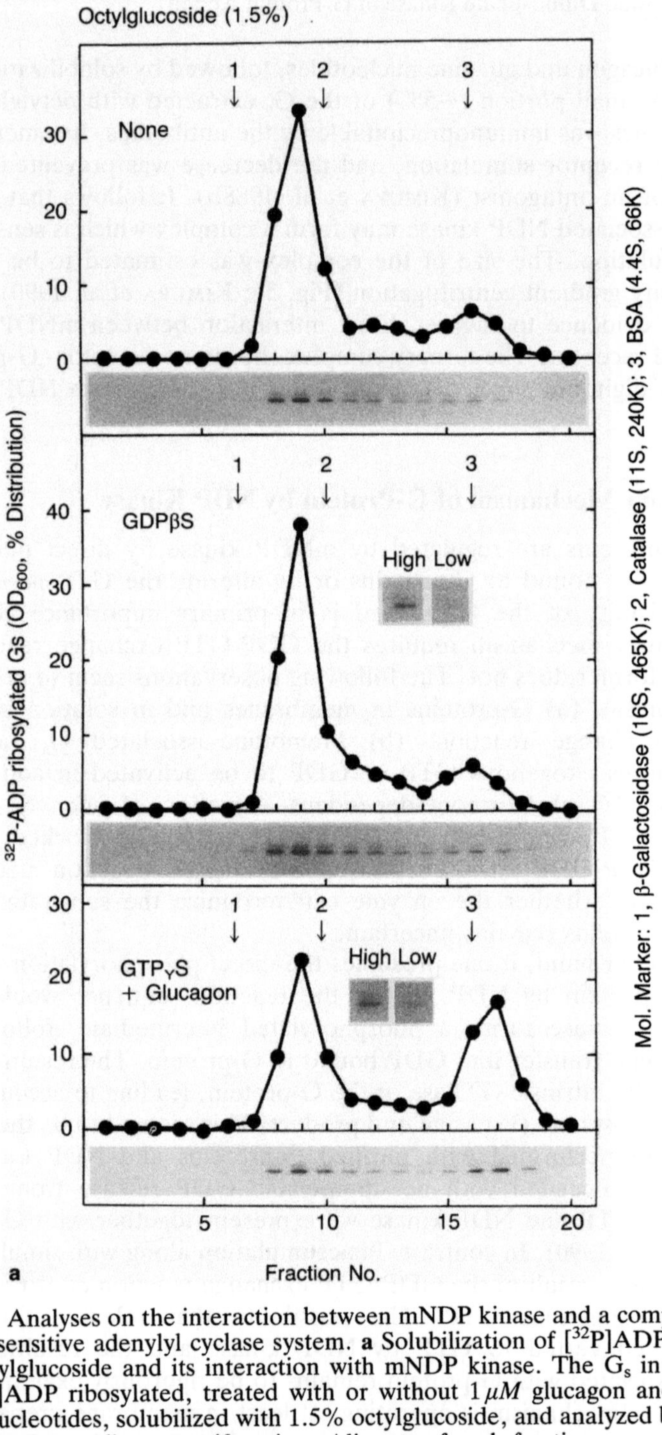


**Fig. 5a,b.** Analyses on the interaction between mNDP kinase and a component(s) of hormone-sensitive adenylyl cyclase system. **a** Solubilization of [$^{32}$P]ADP-ribosylated $G_s$ by octylglucoside and its interaction with mNDP kinase. The $G_s$ in membranes were [$^{32}$P]ADP ribosylated, treated with or without 1 $\mu M$ glucagon and/or 100 $\mu M$ guanine nucleotides, solubilized with 1.5% octylglucoside, and analyzed by 5%–20% sucrose density gradient centrifugation. Aliquots of each fraction were subjected to SDS-PAGE, followed by autoradiography. *Numbers with arrows*, the position of marker proteins. *Insets*, Fractions 7–11 (high molecular weight complexed form of

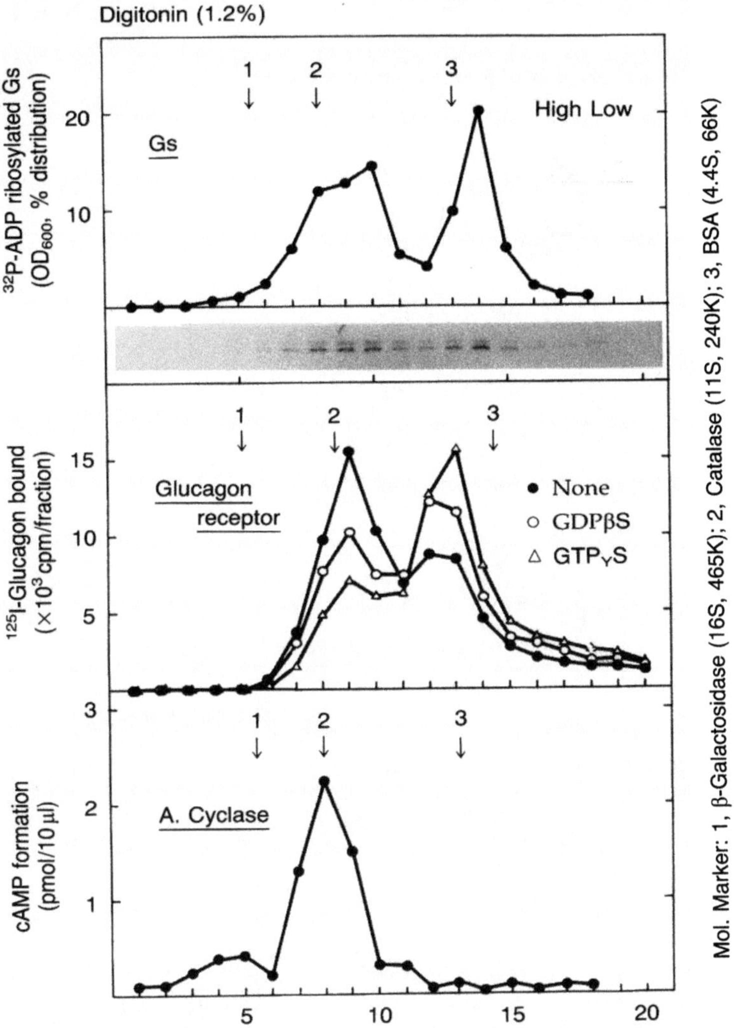


**b**

$G_s$) and 14–18 (free form of $G_s$ and/or $G_s\alpha$) were separately pooled and treated with anti-NDP kinase antibodies for 2 h at 0°C. Immune complexes were adsorbed to and eluted from Pansorbin and analyzed by SDS-PAGE, followed by autoradiography. *High*, *low*, complexed and free forms of $G_s$, respectively. It is evident that the $G_s$ in the high molecular weight fractions was immunoprecipitated by the antibody. **b** Solubilization of [$^{32}$P]ADP-ribosylated $G_s$, $^{125}$I-labeled glucagon-receptor-$G_s$ ternary complex and adenylyl cyclase activity by digitonin. *Top panel*, the [$^{32}$P]ADP ribosylated membranes were treated with 100 $\mu M$ GDP$\beta$S for 5 min at 30°C, solubilized with 1.2% digitonin, and centrifuged. *Inset*, the result of immunoprecipitation of $G_s$ by anti-NDP kinase antibodies. Detailed explanation is given in **a**. In this case, the $G_s$ in the high molecular weight fractions was not immunoprecipitated. *Middle panel*, membranes were incubated with $^{125}$I-labeled glucagon for 20 min at 30°C. After washing and solubilization, the glucagon-receptor complex was treated without or with 100 $\mu M$ GDP$\beta$S or GTP$\gamma$S for 5 min at 0°C, and applied on a sucrose density gradient. *Bottom panel*, membranes were treated with 100 $\mu M$ GDP$\beta$S for 5 min at 30°C and then solubilized. Adenylyl cyclase activity in the presence of 1 $\mu M$ forskolin was determined

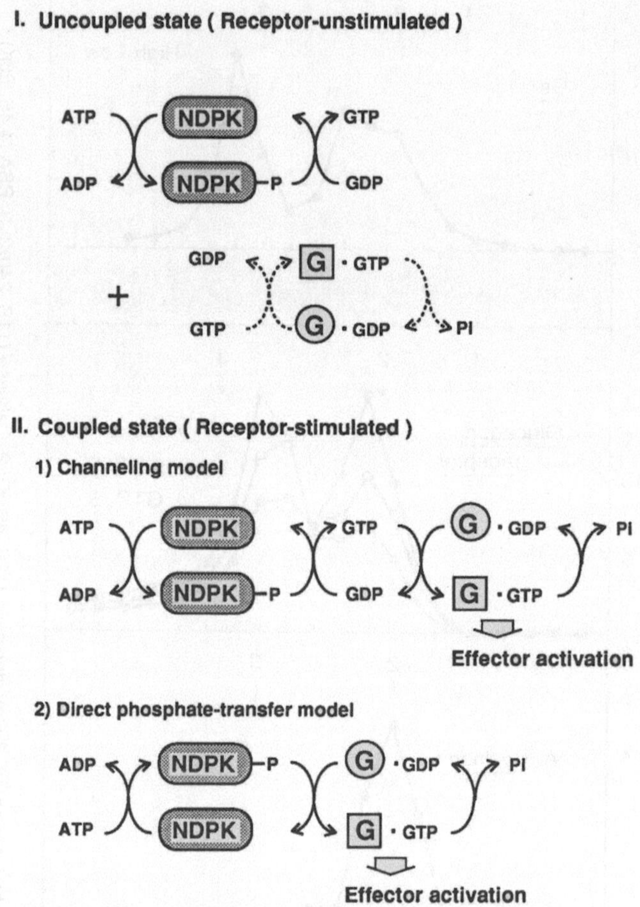


**Fig. 6.** A possible receptor-mediated regulation of GTP-binding proteins by mNDP kinase. See text for detailed explanation

NDP kinase recognizes specific target among various G-proteins including those with low molecular weight, and how hormone influences this interaction await further investigation.

## VI. Physiological Relevance of G-Protein Regulation by mNDP Kinase

Whether mNDP kinase regulates G-protein action under physiological conditions is of great importance. We previously investigated the molecular basis of an enhanced hormonal response of cultured cells induced by a drug treatment (KIMURA and JOHNSON 1983) and observed a close relationship between increased guanine nucleotide dependent adenylyl cyclase activity and increased NDP kinase activity in membranes. NDP kinase activity in the cell cytosol was unchanged by the drug treatment. In order to understand causal relationship between mNDP kinase and signaling systems, manipulation of the intracellular NDP kinase by, for example, injection of specific inhibitors, antibodies, or transfection and expression of NDP kinase cDNA in the cell would be promising.

## D. Properties of NDP Kinases and Their Structure

Membrane-associated and cytosolic NDP kinases from rat liver are quite similar, if at all different, in their physical, chemical, and enzymic properties (KIMURA and SHIMADA 1988a). The molecular weight of the polymeric form was approximately 100 kDa, consisting of five to six 18-kDa monomer subunits. Questions of number of subunits, subunit heterogeneity, and intracellular localization remain unsolved. Recently, two isoforms were demonstrated in human tissues by cDNA cloning techniques and confirmed by peptide analysis. We found two isoforms ($\alpha$ and $\beta$) in rat tissues.

Rat NDP kinase $\alpha$ isoform (KIMURA et al. 1990) contained a sequence like a glycine-rich loop generally observed in nucleotide binding proteins, the conserved sequence of ATP binding region (AMKXL), a leucine zipperlike sequence, and a hydrophobic claster. By comparison of NDP kinase homologs from lower to higher organisms, His-118 conserved throughout is probably a candidate phosphorylation site of the high-energy phosphorylated enzyme intermediate. It was suggested that the translated enzyme suffers proteolytic cleavage followed by modification at the $\alpha$-$NH_2$ group of the newly produced $NH_2$-terminal amino acid residue. Biochemical and histochemical analyses showed that NDP kinase was located mainly in the cell cytosol with small amounts in nuclei, mitochondria, and plasma membranes. The mechanism by which NDP kinase is associated with membranes remains uncertain.

## E. Novel Roles of NDP Kinases in Cellular Functions

Upon cDNA cloning and determination of amino acid sequence of NDP kinases from various origins, it turned out that the enzyme was identical to the nm23 protein (STEEG et al. 1988; LEONE et al. 1991), a candidate mammalian cancer metastasis suppressor gene product, and the awd protein, which is responsible for *Drosophila* morphogenesis (DEAROLF et al. 1988). Low nm23 mRNA expression in high metastatic tumor cells was established in experimental animal model systems as well as in human cases, including breast cancer (BEVILACQUA et al. 1989). We demonstrated a negetive correlation between immunostaining intensity of human cancer tissues with anti-NDP kinase antibodies and reccurence (HIRAYAWA et al. 1991) or metastasis (NAKAYAMA et al. 1992). Measurement of mRNA or protein content of nm23/NDP kinase is considered to be useful for diagnosis and prognosis. On the other hand, the awd protein is associated with the tubulin network and plays a role in chromosome condensation (BIGGS et al. 1990). Involvement of NDP kinase in cell growth and differentiation is suggested. A recent report (OKABE-KADO et al. 1992) describes identity of a differentiation-inhibiting factor of mouse myeloid leukemia cells to NDP kinase.

## F. Concluding Remarks

I have emphasized that receptor activation by ligand binding brings about more than the GDP-GTP exchange reaction of G-proteins in membrane-signaling systems, which would allow G-protein as well as effector systems to operate effectively in vivo. NDP kinase seems to work not only as a housekeeping enzyme but also as a regulatory one in relation to signal transduction, tumor metastasis, morphogenesis, cell growth and differentiation. It can therefore be envisioned that NDP kinase may play a pivotal role as a link between these various cellular phenomina.

### References

Bevilacqua G, Sobel ME, Liotta LA, Steeg PS (1989) Association of low nm23 RNA levels in human primary infiltrating ductal breast carcinomas with lymph node involvement and other histopathological indicators of high metastatic potential. Cancer Res 49:5185–5190

Biggs J, Hersperger E, Steeg PS, Liotta LA, Shearn A (1990) A Drosophila gene that is homologous to a mammalian gene associated with tumor metastasis codes for a nucleoside diphosphate kinase. Cell 63:933–940

Birnbaumer L, Abramowitz J, Brown AM (1990) Receptor-effector coupling by G proteins. Biochim Biophys Acta 1031:163–224

Bokoch GM, Katada T, Northup JK, Ui M, Gilman AG (1984) Purification and properties of the inhibitory guanine nucleotide-binding regulatory component of adenylate cyclase. J Biol Chem 259:3560–3567

Bourne HR, Sanders DA, McCormick F (1990) The GTPase superfamily: a conserved switch for diverse cell functions. Nature 348:125–132

Dearolf CR, Tripoulas N, Biggs J, Shearn A (1988) Molecular consequences of $awd^{b3}$, a cell-autonomous lethal mutation of Drosophila induced by hybrid dysgenesis. Dev Biol 129:169–178

Gilman AG (1987) G proteins: transducers of receptor-generated signals. Ann Rev Biochem 56:615–649

Hanoune J, Lacombe M-L, Pecker F (1975) The epinephrine-sensitive adenylate cyclase of rat liver plasma membranes. Role of guanyl nucleotides. J Biol Chem 250:4569–4574

Hirayama R, Sawai S, Takagi Y, Mishima Y, Kimura N, Shimada N, Esaki Y, Kurashima C, Utsuyama M, Hirokawa K (1991) Positive relationship between expression of anti-metastatic factor (nm23 gene product or nucleoside diphosphate kinase) and good prognosis in human breast cancer. J Natl Cancer Inst 83:1249–1250

Iyengar R, Birnbaumer L (1979) Coupling of the glucagon receptor to adenylyl cyclase by GDP: evidence for two levels of regulation of adenylyl cyclase. Proc Natl Acad Sci USA 76:3189–3193

Kikkawa S, Takahashi K, Takahashi K, Shimada N, Ui M, Kimura N, Katada T (1990) Conversion of GDP into GTP by nucleoside diphosphate kinase on the GTP-binding proteins. J Biol Chem 265:21536–21540

Kimura N (1987) Possible involvement of nucleoside diphosphate kinase in membrane signal-transduction (in Japanese). Seikagaku 59:1227–1233

Kimura N, Johnson GS (1983) Increased membrane-associated nucleoside diphosphate kinase activity as a possible basis for enhanced guanine nucleotide-dependent adenylate cyclase activity induced by picolinic acid treatment of simian virus 40-transformed normal rat kidney cells. J Biol Chem 258:12609–12617

Kimura N, Nagata N (1977) The requirement of guanine nucleotides for glucagon stimulation of adenylate cyclase in rat liver plasma membranes. J Biol Chem 252:3829–3835

Kimura N, Nagata N (1979) Mechanism of glucagon stimulation of adenylate cyclase in the presence of GDP in rat liver plasma membranes. J Biol Chem 254:3451–3457

Kimura N, Shimada N (1983) GDP does not mediate but rather inhibits hormonal signal to adenylate cyclase. J Biol Chem 258:2278–2283

Kimura N, Shimada N (1985) Differential susceptibility to GTP formed from added GDP via membrane-associated nucleoside diphosphate kinase of GTP-sensitive adenylate cyclases achieved by hormone and cholera toxin. Biochem Biophys Res Commun 131:199–206

Kimura N, Shimada N (1986) GTP-activated GTP binding protein (Gs) in membranes achieved by hormone plus GDP does not serve as a substrate for ADP-ribosylation by cholera toxin. Biochem Biophys Res Commun 134:928–936

Kimura N, Shimada N (1988a) Membrane-associated nucleoside diphosphate kinase from rat liver. Purification, characterization and comparison with cytosolic enzyme. J Biol Chem 263:4647–4653

Kimura N, Shimada N (1988b) Direct interaction between membrane-associated nucleoside diphosphate kinase and GTP-binding protein (Gs), and its regulation by hormones and guanine nucleotides. Biochem Biophys Res Commun 151:248–256

Kimura N, Shimada N (1990) Evidence for complex formation between GTP binding protein (Gs) and membrane-associated nucleoside diphosphate kinase. Biochem Biophys Res Commun 168:99–106

Kimura N, Shimada N, Nomura K, Watanabe K (1990) Isolation and characterization of a cDNA clone encoding rat nucleoside diphosphate kinase. J Biol Chem 265:15744–15749

Kimura N, Shimada N, Tsubokura M (1985) Adenosine 5′-($\beta$,$\gamma$-imino)triphosphate and guanosine 5′-o-(2-thiodiphosphate) do not necessarily provide non-phosphorylating conditions in adenylate cyclase studies. Biochem Biophys Res Commun 126:983–991

Lacombe ML, Jakobs KH (1992) Nucleoside diphosphate kinases as potential new targets for control of development and cancer. Trends Pharmacol Sci 13:46–48

Leone A, Flatow U, King CR, Sandeen MA, Margulies MK, Liotta LA, Steeg PS (1991) Reduced tumor incidence, metastatic potential, and cytokine responsiveness of nm23-transfected melanoma cells. Cell 65:25–35

Murphy GJ, Stansfield DA (1983) Problems associated with assessment of the effect of GDP upon hormone stimulation of adenylate cyclase. Biochem J 216:527–528

Nakayama T, Ohtsuru A, Nakao K, Shima M, Nakata K, Watanabe K, Ishii N, Kimura N, Nagataki S (1992) Immunohistochemical analysis of nucleside diphosphate (NDP) kinase expression in human hepatocellular carcinoma. J Natl Cancer Inst 84:1349–1354

Nickerson JA, Wells WW (1984) The microtubule-associated nucleoside diphosphate kinase. J Biol Chem 259:11297–11304
Okabe-Kado J, Kasukabe T, Honma Y, Hayashi M, Henzel WJ, Hozumi M (1992) Identity of a differentiation inhibiting factor for mouse myeloid leukemia cells with nm23/nucleoside diphosphate kinase. Biochem Biophys Res Commun 182:987–994
Otero ADS (1990) Transphosphorylation and G protein activation. Biochem Pharmacol 39:1399–1404
Parks RE Jr, Agarwal RP (1973) Nucleoside diphosphokinases. In: Boyer PD (ed) The enzymes, vol 8. Academic, New York pp 307–333
Rodbell M, Birnbaumer L, Pohl SL, Krans HMJ (1971a) The glucagon-sensitive adenyl cyclase system in plasma membranes of rat liver. V. An obligatory role of guanyl nucleotides in glucagon action. J Biol Chem 246:1877–1882
Rodbell M, Krans MJ, Pohl SL, Birnbaumer L (1971b) The glucagon-sensitive adenyl cyclase system in plasma membranes of rat liver. IV. Effects of guanyl nucleotides on binding of $^{125}$I-glucagon. J Biol Chem 246:1872–1876
Steeg PS, Bevilacqua G, Kopper L, Thorgeirsson UP, Talmadge JE, Liotta LA, Sobel ME (1988) Evedence for a novel gene associated with low tumor metastatic potential. J Natl Cancer Inst 80:200–204
Walton GM, Gill GN (1975) Nucleotide regulation of a eukaryotic protein synthesis initiation complex. Biochim Biophys Acta 390:231–245

CHAPTER 75
# G-Protein Regulation of Cardiac $K^+$ Channels

Y. KURACHI

## A. Introduction

Direct, or at least "membrane-delimited," G-protein-regulation of ion channel function is one of the most exciting recent topics in cellular electrophysiology (BROWN and BIRNBAUMER 1990; HILLE 1992). It has been shown that pertussis toxin (PTX, or islet-activating protein) sensitive G-proteins ($G_K$) are involved in the muscarinic acetylcholine (ACh) receptor dependent activation of a specific inward-rectifying $K^+$ ($K_{ACh}$) channel current in cardiac atrial whole cells (PFAFFINGER et al. 1985; BREITWIESER and SZABO 1985). The concept that $G_K$ directly activates the $K_{ACh}$ channel was further strengthened when the channel was activated in cell-free inside-out patches of the atrial cell membrane by intracellular GTP (with agonists) (KURACHI et al. 1986a,b,c), its non-hydrolyzable analogues (KURACHI et al. 1986b,c), and purified or recombinant G-protein subunits (in the absence of agonists; LOGOTHETIS et al. 1987; YATANI et al. 1987, 1988b; KURACHI et al. 1989a). Since then it has been disclosed that various receptors activate $K^+$ channels by similar mechanisms (NORTH et al. 1987), and that various ion channels can be directly regulated by G-proteins (BROWN and BIRNBAUMER 1990). Therefore, direct G-protein regulation of ion channel function is one of the most general and important cell signaling mechanisms. In this chapter, I focus on the molecular mechanisms underlying direct G-protein activation of the cardiac $K_{ACh}$ channel and modulations of the system by various substances in the cardiac atrial cell membrane (KURACHI 1989, 1990; KURACHI et al. 1992).

## B. Involvement of G-Protein in Muscarinic Activation of the $K_{ACh}$ Channel

The ACh activation of the $K_{ACh}$ channel was prevented by pretreating atrial cells with PTX, suggesting the involvement of PTX-sensitive G-proteins in the signaling (PFAFFINGER et al. 1985; see also BREITWIESER and SZABO 1985). In the inside-out patch condition with agonists (ACh or adenosine) in the pipette solution, the $K_{ACh}$ channels were activated by intracellular GTP, which was irreversibly inhibited by the A protomer of PTX with nicotinamide-adenine dinucleotide (NAD; Fig. 1A; KURACHI et al. 1986b).

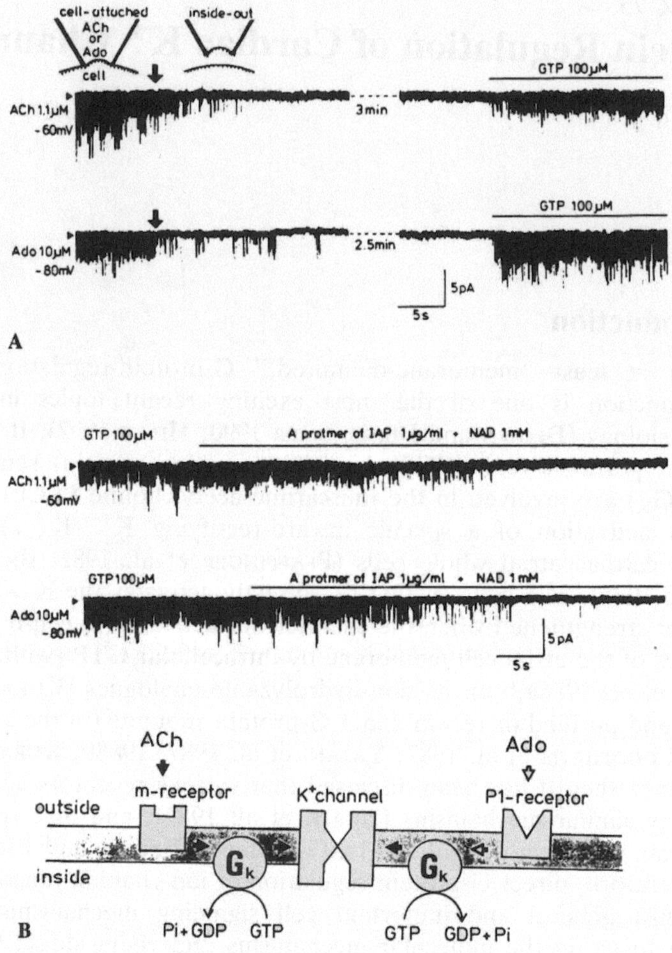


**Fig. 1. A** Activation of a $K^+$ channel by adenosine (*Ado*) and ACh requires intracellular GTP and is blocked by pertussis toxin. The guinea pig atrial cells were bathed in internal solution. The concentration of agonists, i.e., ACh and Ado, and the patch membrane potential are indicated at each current trace. *Arrows* (above each trace), the formation of an "inside-out" patch. Upon forming an "inside-out" patch, the $K_{ACh}$ activity disappeared quickly. When the internal solution containing GTP (100 $\mu M$) was perfused, the channel activity resumed abruptly. With 100 $\mu M$ GTP present in the intracellular side of the membrane, the channel activity persisted. When the active (*A*) protomer of pertussis toxin (or islet-activating protein, *IAP*) with NAD 1 m*M* was further added to the internal solution containing GTP, the channel activity was gradually blocked by IAP within 1–3 min. The blocking effect of IAP was irreversible. When the A protomer was perfused in the absence of NAD, the activation of the channel was not blocked. (From KURACHI et al. 1986b). **B** Simplified scheme of purinergic and muscarinic activation of a $K^+$ channel in the atrial cell membrane. In the cardiac atrial cell membrane, two different membrane receptors ($P_1$ purinergic and muscarinic ACh receptors) link with a $K^+$ channel via GTP-binding proteins, $G_i$ and/or $G_o$. This scheme does not represent any quantitative relations between each component. (From KURACHI et al. 1986b)

Since PTX specifically ADP-ribosylates and inhibits the functions of the $G_i$ family (KATADA and UI 1982; FLORIO and STERNWEIS 1985), these observations may indicate that a G-protein named $G_K$ couples the m-ACh and adenosine receptors to the $K_{ACh}$ channel in the atrial cell membrane, and that this system does not involve intracellular soluble second messengers (Fig. 1B). In the absence of agonists, although GTP failed to activate the

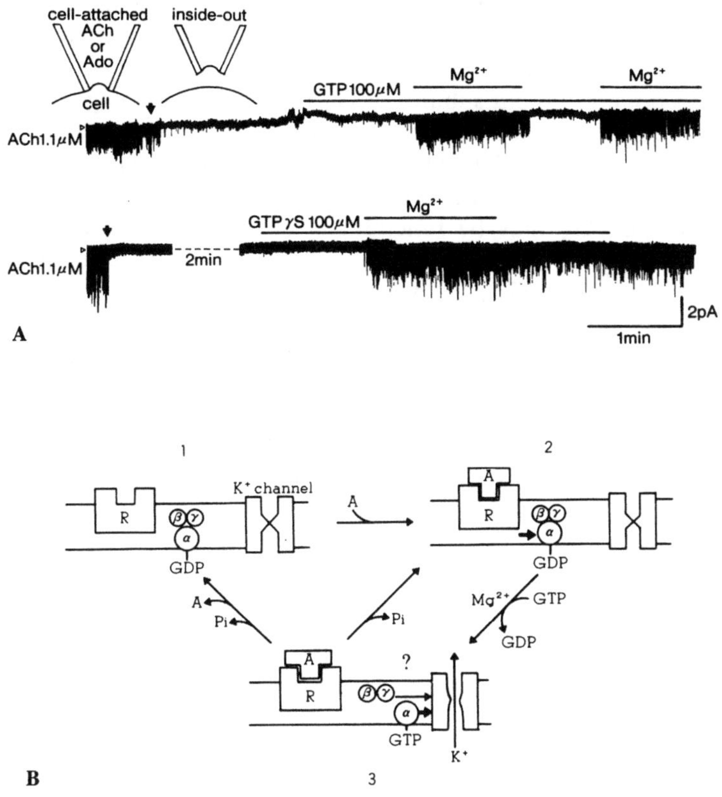


**Fig. 2. A** Effects of $Mg^{2+}$, GTP, and GTPγS on the K$^+$ channel activation. The atrial cells were bathed in internal solution. The concentration of ACh and adenosine (Ado) in the pipettes is indicated at each current trace, and the membrane potential of the patch was −60 mV in all cases. *Arrows* (above each trace), formation of an "inside-out" patch; *bars* (above each current trace), the protocols of perfusing $Mg^{2+}$, GTP, and GTPγS. $Mg^{2+}$ in the internal solution was raised by adding $MgCl_2$ (5 m*M*) to the solution; *open triangles* (at each current trace), zero current level. The results were essentially the same with ACh and with Ado. (From KURACHI et al. 1986c). **B** Simplified scheme of activation of a K$^+$ channel regulated by GTP-binding protein in the atrial cell membrane. *R*, a membrane receptor (m-ACh or Ado receptor); *A*, an agonist (ACh or Ado); $\alpha$, $\beta$, $\gamma$, the subunits of the GTP-binding protein. This scheme does not represent any quantitative relations and does not exclude the possibility of several other steps between each components. (From KURACHI et al. 1986c)

channel in the internal solution containing aspartate as anion, GTP-$\gamma$S and GppNHp, nonhydrolyzable GTP analogues, gradually increased the channel openings (KURACHI et al. 1986b). $AlF_4^-$ could also activate the $K_{ACh}$ channel in the absence of agonists, which was prevented by pretreating the patch with GDP-$\beta$S (KURACHI et al. 1987).

Intracellular $Mg^{2+}$ is essential for GTP activation of the $K_{ACh}$ channel (Fig. 2A; KURACHI et al. 1986c). Based on this observation, a simplified model has been proposed for the molecular mechanism underlying $G_K$ activation of the $K_{ACh}$ channel in the analogy of G-protein regulation of adenylyl cyclase (Fig. 2B). In the absence of agonists, $G_K$ remains in a trimeric complex, and the $K_{ACh}$ channel is closed (state 1). In this state, GDP may be bound to the $\alpha$ subunit of $G_K$ ($G_{K\alpha}$). When an agonist binds to the membrane receptor, some signal is transmitted to $G_K$ (state 2). If $Mg^{2+}$ is present in the intracellular side of the membrane, GTP binds to $G_{K\alpha}$, probably in exchange for GDP, and activates $G_K$ (state 3; turn-on reaction). During activation, $G_K$ may be functionally dissociated into its subunits of $G_{K\alpha\text{-}GTP}$ and $G_{K\beta\gamma}$. Either subunit may activate the $K_{ACh}$ channel. When GTP, which is bound to $G_{K\alpha}$, is hydrolyzed to GDP, $G_K$ is deactivated and goes back to state 1 or state 2 (turn-off reaction). Since GTP$\gamma$S and GppNHp are not hydrolyzed to GDP, it is expected that $G_K$, when activated by GTP$\gamma$S or GppNHp, may stay in state 3, and the openings of the $K_{ACh}$ channel remain persistently stimulated (BREITWIESER and SZABO 1985; KURACHI et al. 1986c).

The GTP-dependent channel activation, as well as $AlF_4^-$-dependent activation, requires intracellular $Mg^{2+}$ (KURACHI et al. 1986c, 1987), while G-protein subunit activation of the channel does not require it (LOGOTHETIS et al. 1987; KURACHI et al. 1989a). These observations are consistent with the notion that dissociation of the trimeric G into $G_\alpha$ and $G_{\beta\gamma}$ is $Mg^{2+}$ dependent (GILMAN 1987).

# C. Physiological Mode of G-Protein Activation of the $K_{ACh}$ Channel

Intracellular GTP activates the $K_{ACh}$ channel in a highly positive cooperative manner (KURACHI et al. 1990; ITO et al. 1991). Figure 3A shows the effects of GTP on the $K_{ACh}$ channel in the inside-out patch condition. Even in the absence of agonist, GTP caused some activation of the $K_{ACh}$ channel in the internal solution containing 130 m$M$ chloride ion, which was much less than that of the maximal activation induced by GTP$\gamma$S (~20%). This agonist-independent activation of the $K_{ACh}$ channel was due to the basal turn-on reaction of $G_K$, which required the coupling of the m-ACh and adenosine receptors and $G_K$ (ITO et al. 1991). In the presence of an agonist (ACh), GTP was much more efficient in activating the $K_{ACh}$ channel. Channel openings were evoked with 1 $\mu M$ GTP in the absence of agonist and 0.1 $\mu M$ GTP with 1 $\mu M$ ACh. Channel openings dramatically increased

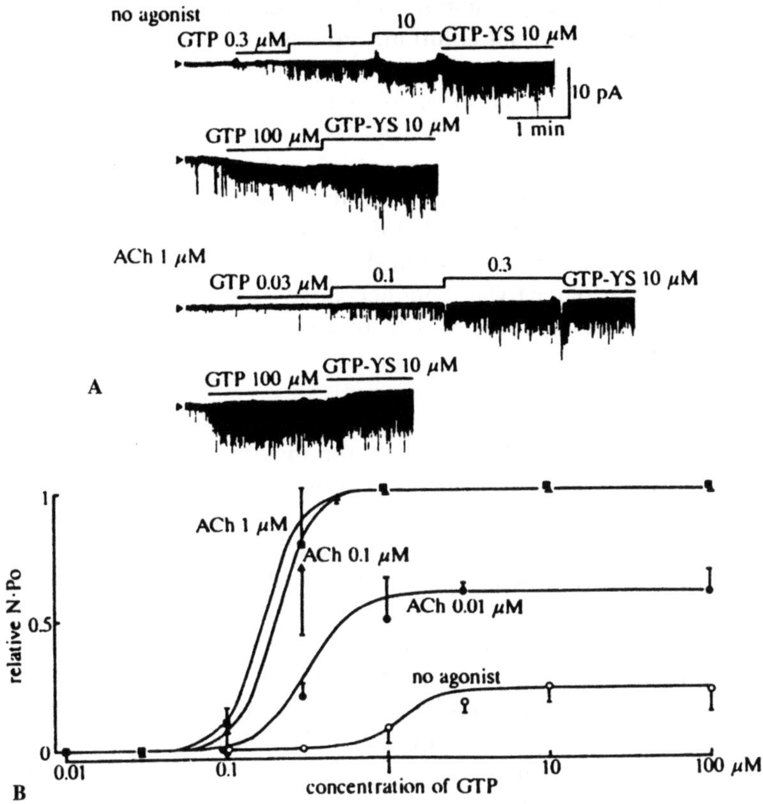


**Fig. 3A,B.** Concentration-dependent effect of GTP on the muscarinic $K^+$ channel in the absence and presence of ACh. **A** Examples of inside-out patch experiments. The concentration of ACh in the pipettes is 0 or $1\,\mu M$ as indicated. *Bars* (above each trace), the protocol of perfusing various concentrations of GTP and $10\,\mu M$ GTPγS. The membrane potential was $-80$ mV. Note that three to ten times increase in GTP concentration resulted in a dramatic increase of $N \cdot P_o$ of $K_{ACh}$, indicating the existence of a highly cooperative process. **B** The relationship between the concentration of GTP and the relative $N \cdot P_o$ of $K_{ACh}$ with reference to the $N \cdot P_o$ induced by GTPγS in each patch. *Symbols*, *bars*, mean ± SD. The continuous curves are fit by the Hill equation with the least-squares method. *Empty circles*, *filled circles*, *filled triangle*, *filled squares*, $0\,\mu M$ ($n = 7$), $0.001\,\mu M$ ($n = 6$), $0.1\,\mu M$ ($n = 6$), and $1\,\mu M$ ($n = 6$) ACh, respectively. (From Ito et al. 1991)

with increasing concentrations of GTP. They reached a maximal value at approximately $3-10\,\mu M$ with $0\,\mu M$ ACh in the pipette, and $0.3-1\,\mu M$ GTP with $1\,\mu M$ ACh. The maximal channel activity induced by GTP in the absence of agonist was much less than, and that in the presence of $1\,\mu M$ ACh was equivalent to, the GTPγS ($10\,\mu M$) induced channel activity. Note, again, that these experiments were performed using the internal solution containing 130 m$M$ chloride ion. We recently found that chloride ion disturbs the turn-off reaction of $G_K$, probably by inhibiting the intrinsic GTPase activity of $G_{K\alpha}$ (Nakajima et al. 1992; Gilman 1987). Thereby, weak activation of the $K_{ACh}$ channel could be induced by intracellular GTP alone in the chloride-internal solution, although the basal turn-on reaction of $G_K$ is slow. In the sulfate ion or aspartate-internal solution, the channel activation by GTP in the absence of agonists is not observed (see Sect. E.I.4).

Figure 3B shows the GTP concentration-dependent activation of the $K_{ACh}$ channel in the presence of various concentrations of ACh in the chloride-internal solution (ITO et al. 1991). The relative $N \cdot P_o$ (N represents the number of channels in each patch, $P_o$ represents the open probability of each channel in the patch) was obtained in reference to the $N \cdot P_o$ induced by $10\,\mu M$ GTP-$\gamma$S. Each curve was fit by the Hill equation using the least-squares method:

$$y = V_{MAX}/\{1 + (K_d/[GTP])^H\}$$

where $y$ = the relative $N \cdot P_o$, $V_{MAX}$ = the maximal $N \cdot P_o$, $K_d$ = the GTP concentration at the half-maximal activation of the channel, and $H$ = the Hill coefficient.

As the concentration of ACh in the pipette was raised from 0 to 0.01, 0.1, or $1\,\mu M$, the following points were observed: (a) the threshold concentration of GTP necessary to induce openings of the channel decreased, (b) the GTP concentration for half-maximal activation of the channel ($K_d$) decreased, and (c) the maximal relative $N \cdot P_o$ ($V_{max}$) increased; however, (d) the Hill coefficient of the fit curve was constant around 3 and was independent of the ACh concentration. These results indicate that ACh binding to the receptor increased both the maximal response and the apparent affinity of the $K_{ACh}$ channel for GTP, which may be due to the facilitation of functional dissociation of $G_K$ induced by agonist binding to the muscarinic ACh receptors (KUROSE et al. 1986), and that the positive cooperative effect of GTP on the channel open probability may be derived from the intrinsic properties of the functional steps between G-protein subunits and the $K_{ACh}$ channel (KURACHI et al. 1990; ITO et al. 1991). The positive cooperative activation of the $K_{ACh}$ channel by G-protein subunit was also suggested from the whole cell clamp experiment (KARSCHIN et al. 1991). Although we cannot completely exclude the possibility of some unknown intermediate steps between the G-protein subunit and the $K_{ACh}$ channel, I rather prefer to hypothesize that multiple G-protein subunits activate the $K_{ACh}$ channel directly.

## D. Effects of G-Protein Subunits on the Cardiac $K_{ACh}$ Channel

Upon receptor stimulation, the heterotrimeric G-proteins are functionally dissociated into GTP-bound $G_\alpha$ ($G_{\alpha\text{-GTP}}$) and $G_{\beta\gamma}$. Both subunits may play a role in the signal transduction pathways. However, because of the smaller molecular heterogeneity of $G_{\beta\gamma}$ among various G-proteins, it had been thought that $G_{\alpha\text{-GTP}}$ mediates specific signals to effectors in transduction systems, such as $G_s$-adenylyl cyclase or transducin-cGMP phosphodiesterase (GILMAN 1987). Since the report by LOGOTHETIS et al. (1987) that $G_{\beta\gamma}$ purified from bovine brain activates the $K_{ACh}$ channel in the atrial cell membrane, specific roles of $G_{\beta\gamma}$ other than binding to $G_{\alpha\text{-GTP}}$ have been demonstrated. (a) JELSEMA and AXELROD (1987) and JELSEMA et al. (1989) showed that transducin $G_{\beta\gamma}$ ($G_{Tr\beta\gamma}$) activates phospholipase $A_2$ ($PLA_2$) in rod outer segments. Since the $G_{Tr\beta\gamma}$ activation of $PLA_2$ occurred even in the presence of GTP$\gamma$S, which prevents reassociation of G protein subunits, they concluded that the $G_{Tr\beta\gamma}$ activation was a direct effect and was not caused by the binding of the $G_{Tr\beta\gamma}$ with an inhibitory $G_{Tr\alpha}$, thereby causing a removal of inhibition. (b) KATADA et al. (1987) showed that the $Ca^{2+}$-calmodulin activation of rat brain adenylyl cyclase can be inhibited by 2–10 n$M$ porcine brain $G_{i\beta\gamma}$ or $G_{o\beta\gamma}$. TANG and GILMAN (1991) showed that type II and type IV adenylyl cyclases are further activated by $G_{\beta\gamma}$ when they are preactivated by GTP-$\gamma$S-bound $G_{s\alpha}$ (see also TANG and GILMAN 1992). (c) WHITEWAY et al. (1989) showed that in yeast, genetic mutants lacking the STE4 or STE18 genes (which encode for putative yeast $G_\beta$ and $G_\gamma$, respectively), are unable to respond to pheromone. On the basis of these experiments they proposed that the putative yeast $G_{\beta\gamma}$ is directly involved in initiating the pheromone response. Also, in mutants in which the SCG1 gene (putative $G_\alpha$) has been modified or is lacking, the pheromone pathway is constitutively activated, suggesting that the excess free $G_{\beta\gamma}$ stimulates the pathway. (d) CAMPS et al. (1992) showed that phospholipase C is stimulated by $G_{\beta\gamma}$. (e) PITCHER et al. (1992) showed that $G_{\beta\gamma}$ interacts directly with $\beta$-adrenergic receptor kinase. However, the exact roles of $G_{\beta\gamma}$ in the $K_{ACh}$ channel regulation in the atrial cell membrane still remain controversial (CODINA et al. 1987; KIRSCH et al. 1988; YATANI et al. 1990; LOGOTHETIS et al. 1987, 1988; KURACHI et al. 1989a, 1992; NANAVATI et al. 1990, KURACHI 1989, 1990).

GTP$\gamma$S-bound $G_\alpha$ ($G^*_\alpha$) and $G_{\beta\gamma}$ of PTX-sensitive G-proteins purified from brain and erythrocytes, and recombinant $G^*_{i\alpha}$s have been used to identify their roles in the activation of the $K_{ACh}$ channel in atrial myocytes. It has been reported that both $G^*_{i\alpha}$ (CODINA et al. 1987; YATANI et al. 1987, 1988b; LOGOTHETIS et al. 1988) and $G_{\beta\gamma}$ (LOGOTHETIS et al. 1987; KURACHI et al. 1989a; NANAVATI et al. 1990) can activate the $K_{ACh}$ channel. There are two main points of the considerable controversy regarding $G_{\beta\gamma}$ activation of the $K_{ACh}$ channel. (a) $G^*_{i\alpha}$ activates the $K_{ACh}$ channel at picomolar concentrations while $G_{\beta\gamma}$ activates the channel at nanomolar concentrations (CODINA et al. 1987; YATANI et al. 1987; LOGOTHETIS et al. 1987, 1988; KURACHI et al. 1989a). Thus, $G_{\beta\gamma}$ activation of the $K_{ACh}$ channel may be due to the contaminating $G^*_{i\alpha}$ in the $G_{\beta\gamma}$ preparation (CODINA et al. 1987). (b) Detergent (CHAPS), which was used to suspend $G_{\beta\gamma}$, activated the $K_{ACh}$ channel (KIRSCH et al. 1988; YATANI et al. 1990). Thus, $G_{\beta\gamma}$ activation may be due to the effects of CHAPS but not $G_{\beta\gamma}$ itself. We recently reexamined the specificity and concentration dependence of G-protein subunit effects on the $K_{ACh}$ channel activity (KOBAYASHI et al. 1990; ITO et al. 1992).

## I. Comparison Between the Regulation of Adenylyl Cyclase Activity and the $K_{ACh}$ Channel Activity by Purified G-protein Subunits

Five subtypes of $G_\alpha^*$ and one type of $G_{\beta\gamma}$ of the PTX-sensitive G-proteins were purified from bovine brain (KOBAYASHI et al. 1990). Using the anti-$G_\alpha$ subunit antibodies, three were identified as $G_{i1\alpha}$, $G_{i2\alpha}$, and $G_{i3\alpha}$ while the remaining two were classified as $G_{o\alpha}$s. One of the $G_{o\alpha}$-type proteins was abundant in the membranes from the brain (termed $G_{o\alpha}$) while the other ($G_{o2\alpha}$) differed from $G_{o\alpha}$ in its proteolytic digestion data. All purified $G_\alpha$s interacted with $G_{\beta\gamma}$ and served as substrates for PTX-catalyzed ADP-ribosylation. The contamination of $G_\alpha$s in the $G_{\beta\gamma}$ preparation was estimated

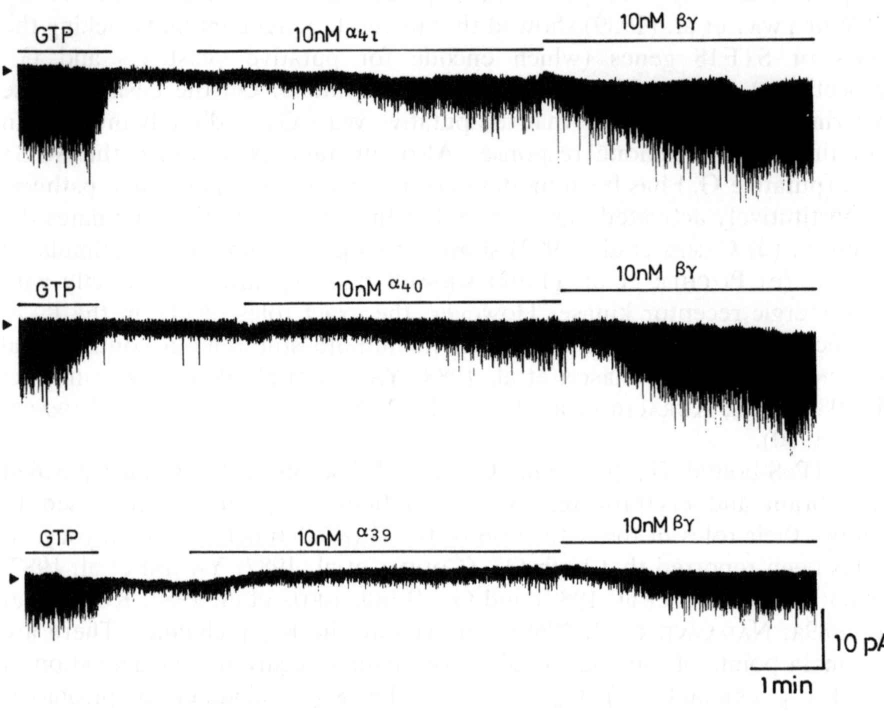


**Fig. 4.** Relative ability of activated $G_\alpha$ versus $G_{\beta\gamma}$ to activate the $K_{ACh}$ channel. Application of GTP to the bath caused maximal activation of the $K_{ACh}$ channel in inside-out patches (1 $\mu M$ ACh in the pipette). After removal of GTP, the addition of 10 n$M$ of the indicated GTP$\gamma$S-bound $G_\alpha$ subunit caused partial ($\alpha$40 and $\alpha$41) or no ($\alpha$39) activation of the channel. Subsequent addition of 10 n$M$ $G_{\beta\gamma}$ activated the channel maximally. The membrane potential was −80 mV. (From KURACHI et al. 1992)

to be less than 1%. We studied the effects of these purified $G^*_\alpha$s on adenylyl cyclase and the atrial $K_{ACh}$ channel. The nucleotide-bound forms of $G_{i1\alpha}$, $G_{i2\alpha}$, $G_{i3\alpha}$, and $G_{o2\alpha}$ inhibited the adenylyl cyclase activity of S40 *cyc*$^-$ membranes which had been reconstituted with GTP$\gamma$S-treated $G_s$. This inhibition appeared to be competitive with the activated $G_s$. The $G_{i1\alpha}$ had the most potent inhibitory activity. $G^*_{o\alpha}$ failed to inhibit the $G_s$-stimulated adenylyl cyclase activity.

The PTX-sensitive $G^*_\alpha$s (at 10 p*M*–10 n*M*) activated the atrial $K_{ACh}$ channel, with a lag time of 2–4 min only in 40 out of 124 inside-out patches (Fig. 4). The channel activity induced by various purified $G^*_\alpha$ of brain PTX-sensitive G-proteins at the concentration of 10 p*M*–10 n*M* were almost identical, i.e., their maximum activation of the channel was approximately 20% of the maximum channel activity induced by 10 $\mu$*M* GTP-$\gamma$S. $G^*_\alpha$s could never fully activate the channel in the patch even at a concentration of 10 n*M*. On the other hand, 10 n*M* $G_{\beta\gamma}$ activated the $K_{ACh}$ channel in 132 out of 134 patches without a lag time (<5 s). Activation of the channel by $G_{\beta\gamma}$ was equivalent to the activation by GTP$\gamma$S and much more potent than activation by any of the $G^*_\alpha$s (see the next section). In summary, $G^*_\alpha$ activation of the $K_{ACh}$ channel was inconsistent and weak, while $G_{\beta\gamma}$ activation was consistent and full. These results suggest that exogenously applied $G^*_\alpha$ and $G_{\beta\gamma}$ activate the $K_{ACh}$ channel by different mechanisms.

## II. Effects of $G_{\beta\gamma}$ on the $K_{ACh}$ Channel

### 1. Voltage-Dependent Properties of the $G_{\beta\gamma}$-Activated $K_{ACh}$ Channel

Figure 5 shows the properties of the $G_{\beta\gamma}$-activated $K_{ACh}$ channel in an "inside-out" patch of the guinea pig atrial cell membrane (Ito et al. 1992). Application of 10 n*M* $G_{\beta\gamma}$ to the patch induced persistent openings of a $K^+$ channel identical to the $K_{ACh}$ channel activated by GTP (Fig. 5A). The $G_{\beta\gamma}$-activated channel has a unit conductance of about 40–45 pS in symmetrical 150 m*M* $K^+$ and showed a strong inward-rectification with 2 m*M* $Mg^{2+}$ in the internal solution (Fig. 5B,C). The open-time histogram of the $G_{\beta\gamma}$-activated channel could be fit by a single exponential curve with a time constant of approximately 1 ms (Fig. 5D), similar to that induced by GTP (see also Logothetis et al. 1987).

As mentioned previously, activation of the $K_{ACh}$ channel by GTP required intracellular $Mg^{2+}$ (Kurachi et al. 1986c). In contrast, $G_{\beta\gamma}$ could activate the $K_{ACh}$ channel even in the 0 $Mg^{2+}$-EDTA internal solution (Ito et al. 1992; see also Kurachi et al. 1989a). The $G_{\beta\gamma}$-activated channel in the $Mg^{2+}$-free condition had the same unit conductance and the same mean open time. In the absence of $Mg^{2+}$ in the internal solution, the $G_{\beta\gamma}$-activated $K^+$ channel showed no inward rectification (Fig. 5B,C), and the channel activity decreased voltage-dependently when the holding potential was more positive than the equilibrium potential for $K^+$ ($E_k$ about 0 mV in the

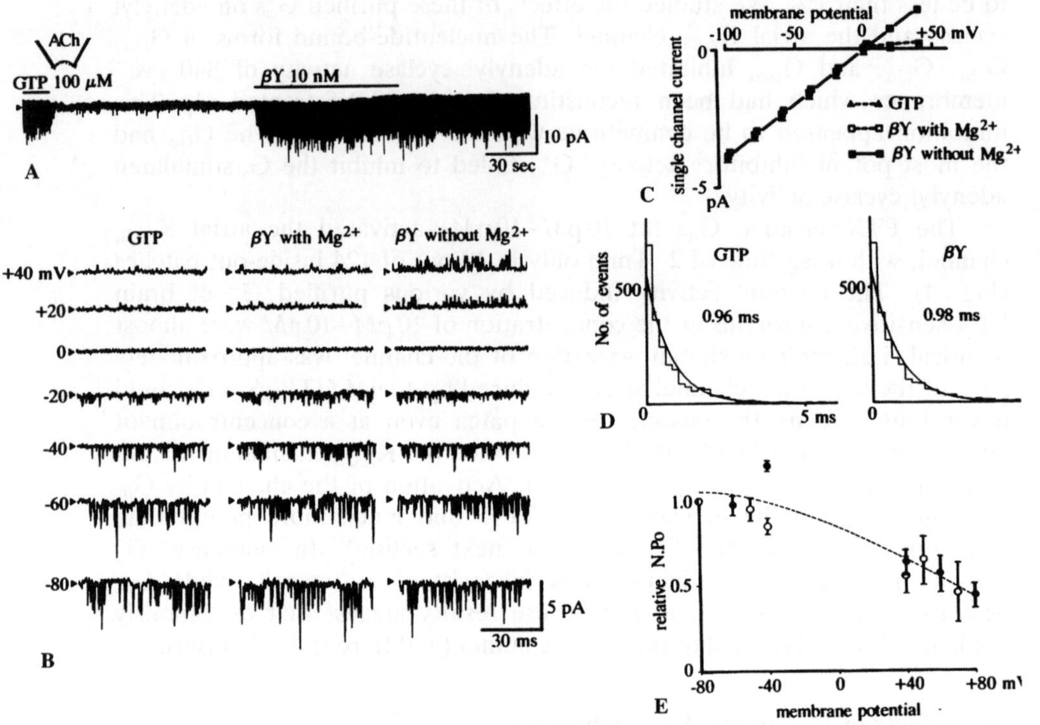


**Fig. 5A–E.** Activation of $K_{ACh}$ channel by GTP and $G_{\beta\gamma}$. **A** After forming an inside-out patch in internal solution containing 2 m*M* $MgCl_2$, 100 μ*M* GTP activated the $K_{ACh}$ channel with 0.3 μ*M* ACh in the pipette solution. After washing out GTP, channel activity disappeared. Subsequent application of $G_{\beta\gamma}$ activated the $K_{ACh}$ channel irreversibly. *Bars* (above the tracing), perfusing protocol. The patch was held at −80 mV. **B** Expanded recordings of the $K_{ACh}$ channel at various holding potentials induced by GTP, $G_{\beta\gamma}$ with 2 m*M* $Mg^{2+}$ or without $Mg^{2+}$ (indicated above each column). **C** Current-voltage (I/V) relationship of the $K_{ACh}$ channel. *Filled triangles*, *open squares*, *filled squares*, I/V relationship induced by GTP with $Mg^{2+}$ (2 m*M*), $G_{\beta\gamma}$ with $Mg^{2+}$ (2 m*M*), and $G_{\beta\gamma}$ without $Mg^{2+}$, respectively. Strong inward rectification was noted using GTP with $Mg^{2+}$ and $G_{\beta\gamma}$ with $Mg^{2+}$. **D** Open-time histograms of the $K_{ACh}$ channel currents induced by GTP and $G_{\beta\gamma}$ at −80 mV. **E** Voltage-dependent channel activity in the absence of $Mg^{2+}$ induced by $G_{\beta\gamma}$ (*filled circles*) and by GppNHp (*open circles*). The relative $N \cdot P_o$ was obtained in reference to the $N \cdot P_o$ induced by 10 n*M* $G_{\beta\gamma}$ or GppNHp (10 μ*M*) at −80 mV. The results were expressed as mean ± SD ($n = 3$ each). (From ITO et al. 1992)

symmetrical 150 m*M* $K^+$). The voltage dependence of the $G_{\beta\gamma}$-induced $K_{ACh}$ channel openings was identical with that of GppNHp- or GTPγS-induced channel openings (Fig. 5E). Thus, the $G_{\beta\gamma}$ activation of the $K_{ACh}$ channel showed the same conductance and kinetic properties as the $K_{ACh}$ channel activated by GTP analogues.

**2. Concentration Dependence of $G_{\beta\gamma}$ Activation of the $K_{ACh}$ Channel**

The quasi-steady-state relationship between the concentration of $G_{\beta\gamma}$ and the $K_{ACh}$ channel activity was obtained (Fig. 6; ITO et al. 1992). Various concentrations (10 p$M$–100 n$M$) of $G_{\beta\gamma}$ were sequentially applied to each patch. Steady-state activities of the channels were normalized to the $N \cdot P_o$ of the channel activity induced by 100 $\mu M$ GTP (in the presence of ACh in the pipette) or 10–100 $\mu M$ GTP-$\gamma$S (in the presence or absence of ACh in the pipette) in each patch. The minimal concentration of bovine brain $G_{\beta\gamma}$ to activate the channel was about 300 p$M$, which was ten times larger than that previously reported for $G_{\beta\gamma}$ obtained from rat brain preparation (KURACHI et al. 1989a). When the relationship was fit by the Hill equation, the Hill

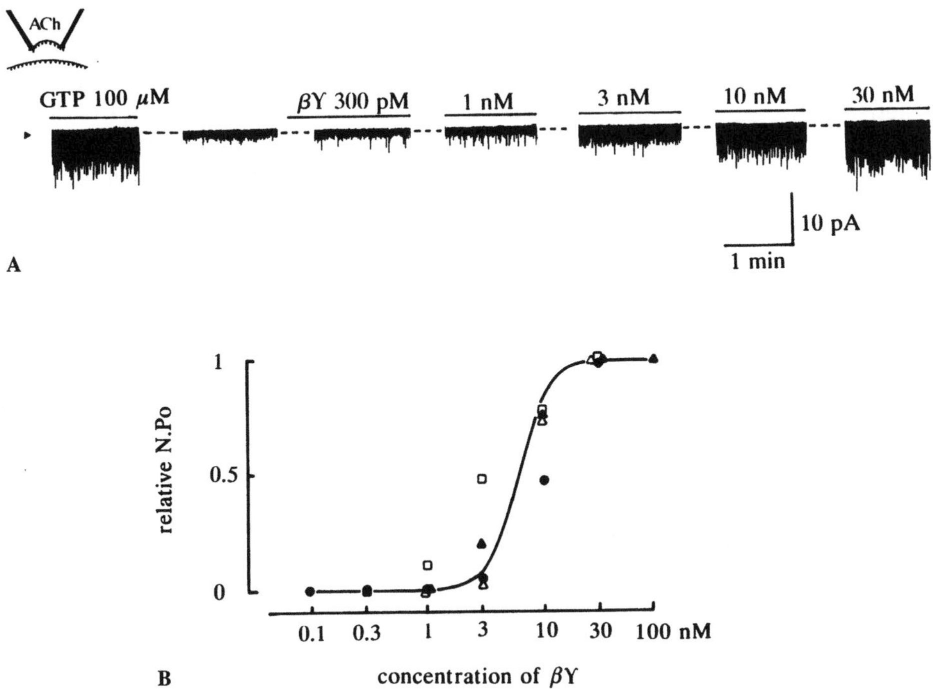


**Fig. 6. A** Concentration-dependent activation of the $K_{ACh}$ channel by $G_{\beta\gamma}$. After forming an inside-out patch, $G_{\beta\gamma}$ induced the $K_{ACh}$ channel activity in a concentration-dependent fashion. Either 100 $\mu M$ GTP (in the presence of 1 $\mu M$ ACh) in the beginning of the experiment or 10–100 $\mu M$ GTP$\gamma$S at the end of experiment were added to the internal side of the membrane to obtain maximal activation of the $K_{ACh}$ channel of each patch. The patch was held at −80 mV. **B** The relationship between the relative $N \cdot P_o$ of the $K_{ACh}$ channel and the concentration of $G_{\beta\gamma}$ obtained from four patches (represented by different symbols). The continuous line is a fit curve to the Hill equation using the nonlinear least-squares regression method. In the graph the Hill coefficient was 3.12, and the half-maximal channel activation occurred at 6 n$M$ $G_{\beta\gamma}$; $\gamma$ = 0.95. The $N \cdot P_o$ obtained from each concentration of $G_{\beta\gamma}$ was normalized with reference to the maximum $N \cdot P_o$ induced by 100 $\mu M$ GTP (with 1 $\mu M$ ACh) or GTP$\gamma$S in each patch. (From ITO et al. 1992)

coefficient was 3.12 and the half-maximal channel activation ($K_d$) occurred at 6 n$M$ $G_{\beta\gamma}$. This Hill coefficient value of 3.12 is almost the same as that for the relation in the GTP activation of the channel (value of 3; see above and Fig. 3B).

### 3. Specificity of $G_{\beta\gamma}$ Activation of the $K_{ACh}$ Channel

The hydrophobic $G_{\beta\gamma}$ was suspended in a detergent (CHAPS), which was reported to activate the $K_{ACh}$ channel (KIRSCH et al. 1988; YATANI et al. 1990). The concentration of CHAPS to suspend 10 n$M$ $G_{\beta\gamma}$ was 7–20 $\mu M$. Figure 7A shows that CHAPS (10–200 $\mu M$) alone did not activate the $K_{ACh}$ channel, but subsequent application of GTP or $G_{\beta\gamma}$ suspended in CHAPS activated the $K_{ACh}$ channel in the same patch, suggesting that (a) CHAPS itself does not activate the $K_{ACh}$ channel, and (b) the negative effect of CHAPS is not due to vesicle formation of the patch. The $G_{\beta\gamma}$ buffer solution also did not activate the $K_{ACh}$ channel. In Fig. 7B, the boiled $G_{\beta\gamma}$ did not activate the $K_{ACh}$ channel while native $G_{\beta\gamma}$ activated the channel in the same patch, indicating that a heat-labile substance in the preparation, but not a heat-stable substance such as CHAPS, is involved in activation of the

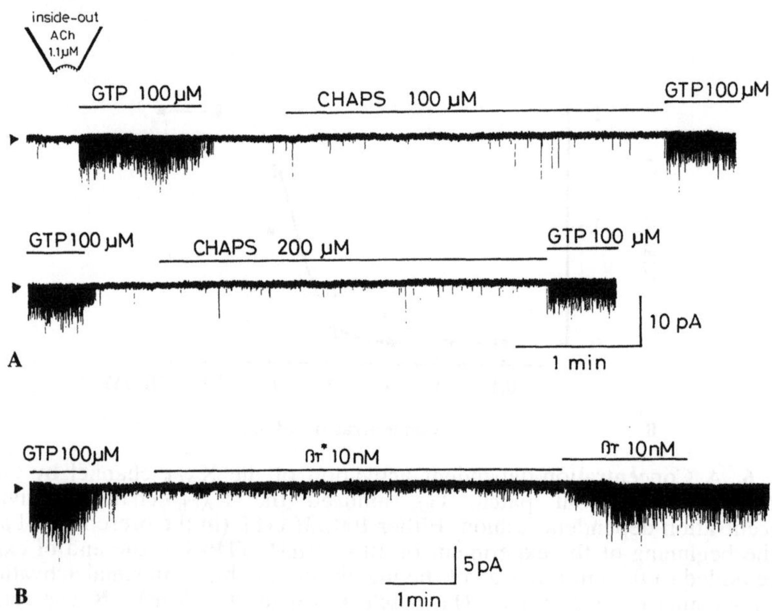


**Fig. 7.** Detergent CHAPS (**A**) and boiled $G_{\beta\gamma}$ (**B**) did not activate the $K_{ACh}$ channel. ACh (1.1 $\mu M$) in the pipette. The membrane potential was −80 mV. GTP (100 $\mu M$) induced brisk $K_{ACh}$ channel openings reversibly. CHAPS 100 or 200 $\mu M$ did not affect the $K_{ACh}$ channel activity (**A**). Boiled $G_{\beta\gamma}$ did not activate the channel, but subsequent application of nonboiled preparation induced the channel openings. (From KURACHI et al. 1992)

channel (KURACHI et al. 1989a; ITO et al. 1992; see also LOGOTHETIS et al. 1988; NANAVATI et al. 1990).

It has been reported that another detergent, Lubrol PX, does not activate the $K_{ACh}$ channel (KIRSCH et al. 1988) but rather blocked the channel activity at higher concentrations (LOGOTHETIS et al. 1988; KURACHI et al. 1989a). We tested the effects of $G_{\beta\gamma}$ suspended in Lubrol PX on the $K_{ACh}$ channel. The $K_{ACh}$ channel was activated dose-dependently by $G_{\beta\gamma}$ suspended in Lubrol PX. The concentration of Lubrol PX was $2.5 \times 10^{-4}\%$ in 10 n$M$ $G_{\beta\gamma}$ solution. High concentrations of Lubrol PX ($10^{-3}\%$) inhibited the $K_{ACh}$ channel activated by GTP with ACh in the pipette solution.

To eliminate the possible contamination of $G_{\alpha}^{*}$ in the $G_{\beta\gamma}$ preparation, $G_{\beta\gamma}$ were preincubated in the $Mg^{2+}$-free/EDTA-internal solution containing GDP or GDP$\beta$S for 24–48 h, as it is known that $G_{\alpha}^{*}$ is unstable under this condition (CODINA et al. 1984). The $G_{\beta\gamma}$ in the $Mg^{2+}$-free internal solution containing GDP or GDP$\beta$S activated the $K_{ACh}$ channel as potently as the control $G_{\beta\gamma}$ (ITO et al. 1992; KURACHI et al. 1989a). Under the same condition, neither $G_{i1\alpha}^{*}$, $G_{i2\alpha}^{*}$, nor $G_{o\alpha}^{*}$ activated the $K_{ACh}$ channel. These results may indicate that (a) the channel activation by the $G_{\beta\gamma}$ is not due to contamination of $G_{\alpha}^{*}$, and (b) the involvement of native G-protein subunit in the exogenous $G_{\beta\gamma}$ activation of the $K_{ACh}$ channel seems unlikely, since $Mg^{2+}$ and GTP are necessary for activation of G-protein (KURACHI et al. 1986c). Note also that activation of the $K_{ACh}$ channel by CHAPS required intracellular $Mg^{2+}$ by YATANI et al. (1990).

The specific effect of $G_{\beta\gamma}$ on the $K_{ACh}$ channel was further confirmed by the observation that $G_{\beta\gamma}$, when preincubated with excessive $G_{\alpha 39\text{-}GDP}$, could not activate the channel, while $G_{\beta\gamma}$ alone subsequently activated the channel (ITO et al. 1992; see also LOGOTHETIS et al. 1988). During preincubation, $G_{\beta\gamma}$ may have bound to $G_{\alpha 39\text{-}GDP}$ to form an inactive heterotrimer (GILMAN 1987). LOGOTHETIS et al. (1988) showed that $G_{\alpha 41\text{-}GDP}$ reversed the $K_{ACh}$ channel activation in a patch pretreated with $G_{\beta\gamma}$ or GTP$\gamma$S. This observation suggests that the functional activating arm of $G_K$ to the $K_{ACh}$ channel is $G_{\beta\gamma}$.

## 4. $G_{\beta\gamma}$ Activation of the $K_{ACh}$ Channel Is Not Mediated by Phospholipase $A_2$

KIM et al. (1989) proposed that $G_{\beta\gamma}$ activates the $K_{ACh}$ channel via stimulation of $PLA_2$. We tested this hypothesis by examining the effects of lipocortin I (a $PLA_2$ inhibitor), nordihydroguaiaretic acid (NDGA, a lipoxygenase inhibitor), and AA-861 (a specific 5-lipoxygenase inhibitor) on the $G_{\beta\gamma}$-activation of the $K_{ACh}$ channel (Fig. 8; ITO et al. 1992). None of these agents prevented the activation of the $K_{ACh}$ channel by $G_{\beta\gamma}$, suggesting that $G_{\beta\gamma}$ activation of the $K_{ACh}$ channel does not require direct activation of the $PLA_2$-eicosanoid cascade, and that $G_{\beta\gamma}$ may directly activate the $K_{ACh}$ channel. The molecular mechanism underlying arachidonic acid metabolite activation of the $K_{ACh}$ channel may differ from

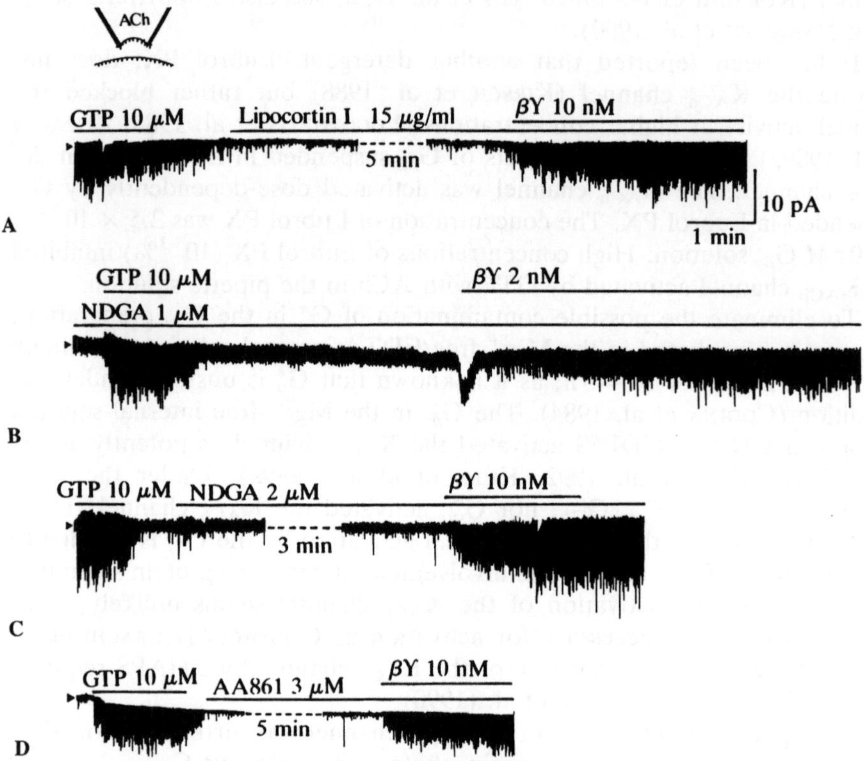


**Fig. 8A–D.** Effects of $PLA_2$ and lipoxygenase inhibitors on $G_{\beta\gamma}$ activation of the $K_{ACh}$ channel. Lipocortin I (a $PLA_2$ inhibitor, **A**), nordihydroguaiaretic acid (NDGA, a lipoxygenase inhibitor, **B**,**C**), and AA-861 (a 5-lipoxygenase inhibitor, **D**) were applied to the internal side of the patch membrane for 5–10 min before applying $G_{\beta\gamma}$, which did not prevent the $K_{ACh}$ channel activation by $G_{\beta\gamma}$. In **B** the cell was preincubated with NDGA (1 $\mu M$) for 15 min before forming an inside-out patch. Subsequent application of GTP also activated the $K_{ACh}$ channel. Pipette solution contained 0.3 $\mu M$ ACh. The holding potential was −80 mV. (From ITO et al. 1992)

that proposed by KIM et al. (1989; see Sect. F.I.1 for arachidonic acid metabolite modulation of $G_K$-$K_{ACh}$ channel system).

## 5. Antibody 4A Does Not Inhibit the Interaction Between $G_K$ and the $K_{ACh}$ Channel

YATANI et al. (1988a) reported that a monoclonal antibody (MAb4A) made against the $G_{Tr\alpha}$ blocks the GTP activation of the $K_{ACh}$ channel (agonists in the pipette) irreversibly in adult guinea pig atrial cells. In the experiments reported, the high $K_{ACh}$ channel activity induced by GTP (in the presence of agonist) was rapidly abolished when MAb4A was introduced to the bath. This inhibition of the endogenous G-protein–effector coupling required that

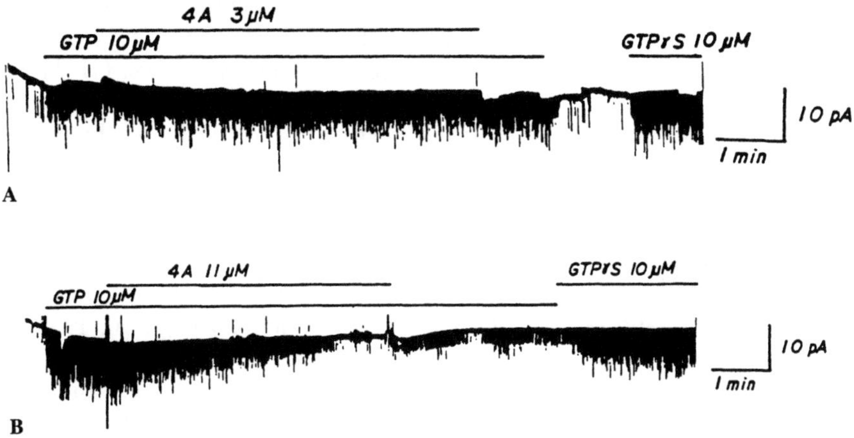


**Fig. 9A,B.** MAb4A failed to block the interaction between $G_K$ and the $K_{ACh}$ channel. In two different experiments, with 0.1 $\mu M$ ACh in the pipette, the $K_{ACh}$ channel in inside-out patches was activated by addition of 10 $\mu M$ GTP to the bath. **A** Addition of 3 $\mu M$ MAb4A (provided by Dr. H.E. Hamm) to the bath failed to inhibit the channel activity (up to 5 min). As Expected, GTP removal reduced activity, which could be increased upon addition of GTP-$\gamma$S. **B** The 11 $\mu M$ MAb4A gradually reduced GTP-induced activity. When the antibody was removed, thereby exposing the patch to GTP alone, activity slightly increased. Maximal activation could be achieved by exposing the patch to 10 $\mu M$ GTP$\gamma$S. (From Nanavati et al. 1990)

the G-protein is in its active GTP-bound state. Furthermore, the inhibition was irreversible since the channel could not be activated by GTP or GTP$\gamma$S even after the antibody had been removed from the bath. On the basis of these experiments, they concluded that the $G_\alpha$ subunit was the signal transducer in vivo (Yatani et al. 1988a). This report is the strongest evidence so far to support the concept that $G_{K\alpha}$ is the transducer of $G_K$ for the $K_{ACh}$ channel in vivo.

We attempted to reproduce these experiments in adult guinea pig atrial cells. Figure 9 shows the single channel currents recorded from two different inside-out patches. In both cases, 0.1 $\mu M$ ACh was included in the pipette, and 10 $\mu M$ GTP was added to the bath. The top trace (Fig. 9A) is from a patch in which exposure to MAb4A (3 $\mu M$) failed to block GTP activation. Activity gradually subsided as GTP was washed out but could be restored by adding 10 $\mu M$ GTP$\gamma$S ($n$ = 6). When the concentration of MAb4A was increased to 11 $\mu M$, MAb4A slowly reduced the $K_{ACh}$ channel activity, but upon addition of GTP$\gamma$S the channel was reactivated (Fig. 9B, $n$ = 6). Thus, the reduction in channel activity observed upon application of a high concentration of MAb4A was reversible. MAb4A did not affect the GTP-$\gamma$S-induced activation of the channel ($n$ = 4). These results suggest that MAb4A can uncouple the interaction between the receptor and $G_K$ but not that between $G_K$ and the $K_{ACh}$ channel. It was recently confirmed that

MAb4A uncouples only the interaction between membrane receptors and G-proteins but does not affect that between the G-proteins and the effectors, such as cGMP-phosphodiesterase and adenylyl cyclase (H. HAMM, personal communication). Therefore, the experiments using MAb4A cannot differentiate which subunit is the transducer of $G_K$ to the $K_{ACh}$ channel.

## III. Effects of G-Protein on the ATP-Sensitive K Channel

G protein was shown to activate the ATP-sensitive $K^+$ channel ($K_{ATP}$) in the cardiac cell (KIRSCH et al. 1990; TUNG and KURACHI 1990). We examined the effects of $G^*_{i\alpha}$ and $G_{\beta\gamma}$ on the $K_{ATP}$ channel activity (Fig. 10, ITO et al. 1992). We found that $G^*_{i1\ and\ 2\alpha}$ purified from bovine brain activated the $K_{ATP}$ channel in the guinea pig ventricular cell membrane, where the $K_{ACh}$ channel is not expressed. Since the $K_{ATP}$ channel in atrial myocytes is low in

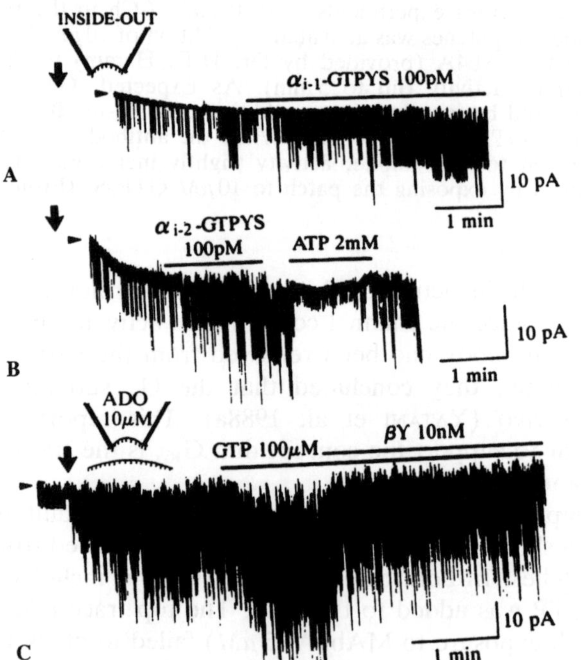


**Fig. 10.** Activation of the $K_{ATP}$ channel by $G^*_{i1\text{-}2\alpha}$ in ventricular myocytes. The inside-out patch from ventricular myocytes was formed in internal solution containing 100 $\mu M$ ATP and 0.5 m$M$ $MgCl_2$. When $G^*_{i\text{-}1,2\alpha}$ was added to the internal solution, burstlike openings of a $K^+$ channel with large conductance (~90 pS) appeared (**A**,**B**), which could be suppressed by 2 m$M$ ATP (**B**). *Arrows* (above each trace), where the inside-out patch was formed. **A**,**B** No agonist in the pipette solution. **C** 10 $\mu M$ Adenosine (*ADO*), *Arrowheads*, zero current level. The holding potential was −80 mV. *Bars* (above each current trace), the protocol for perfusing GTP, ATP, GTP-$\gamma$S-bound $G_{i\text{-}1,2\alpha}$ and $G_{\beta\gamma}$. (From ITO et al. 1992)

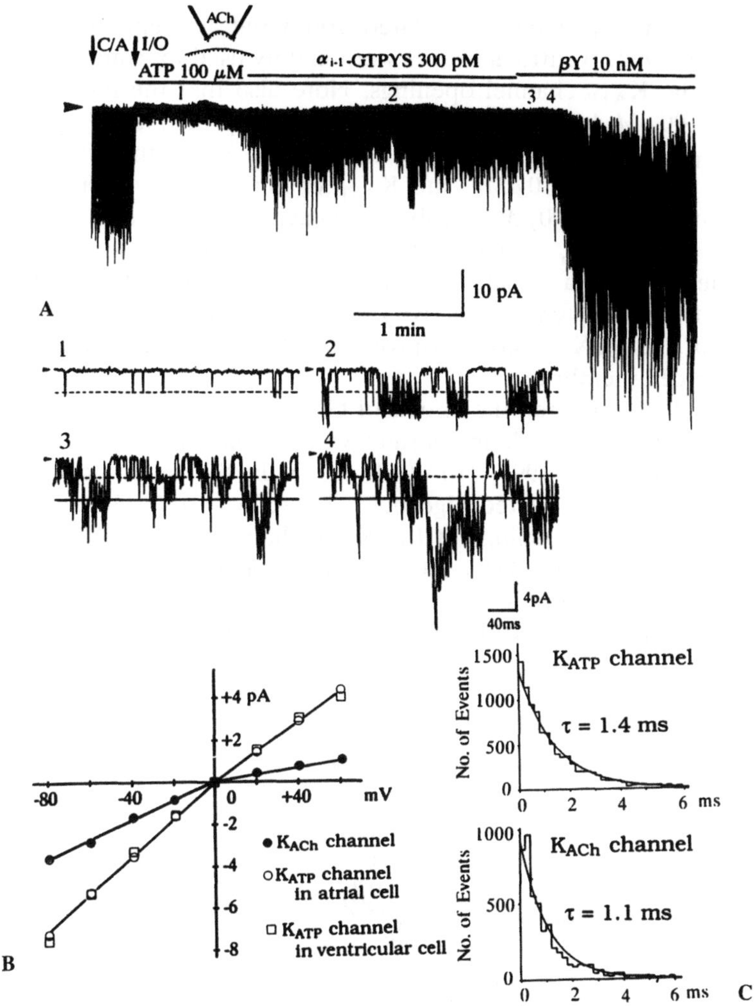


**Fig. 11. A** Effects of $G^*_{i\text{-}1\alpha}$ and $G_{\beta\gamma}$ on the $K_{ATP}$ and $K_{ACh}$ channels in the atrial cell membrane. The pipette solution contained $1\,\mu M$ ACh *Arrows* (above the current trace), the inside-out patch was formed in the internal solution containing $100\,\mu M$ ATP and $0.5\,mM$ $MgCl_2$. $G^*_{i\text{-}1\alpha}$ (300 p*M*) was first applied to the internal side of the patch, which clearly induced openings of the $K_{ATP}$ channel (~90 pS; *2*) without affecting background activity of the $K_{ACh}$ channel. Subsequently, $G_{\beta\gamma}$ (10 n*M*) was applied to the patch, which caused a dramatic increase of 45 pS $K_{ACh}$ channel openings in the same patch (*3*, *4*). *Numbers* (above the current trace), the location of each expanded current trace below. *Dotted line* (in the expanded current trace) the first level of the $K_{ACh}$ channel; *continuous line*, that of the $K_{ATP}$ channel. *Arrowhead* (at each trace), zero current level. **B** The current-voltage relation of $G^*_{i\text{-}1\alpha}$-induced $K_{ATP}$ channel in ventricular (*open squares*) and in atrial (*open circles*) cell membrane, and $G_{\beta\gamma}$-induced $K_{ACh}$ channel in the atrial cell membrane (*closed circles*). **C** The open time histograms of $G^*_{i\text{-}1\alpha}$-induced $K_{ATP}$ channel (in ventricle) and $G_{\beta\gamma}$-induced $K_{ACh}$ channel (in atrium) at −80 mV. (From Ito et al. 1992)

density and mostly dephosphorylated compared to ventricular myocytes (TUNG et al. 1992), ventricular cells were mainly used to examine the effects of $G^*_{i\alpha}$ on the $K_{ATP}$ channel openings. Note also that the internal solution contained 100 $\mu M$ ATP and 0.5 m$M$ $MgCl_2$ to keep the *phosphorylated* state of the $K_{ATP}$ channel (TUNG and KURACHI 1991). When the pipette solution contained adenosine (10 $\mu M$), the $K_{ATP}$ channel activity increased during application of GTP (100 $\mu M$), probably via activation of native G-proteins in the patch membrane. In contrast to the $K_{ACh}$ channel in the atrial cell membrane, further application of 10 n$M$ $G_{\beta\gamma}$ inhibited the GTP-induced increase of the $K_{ATP}$ channel activity to the baseline. These results indicate that $G_{\alpha}$s of the PTX-sensitive G-proteins activate the $K_{ATP}$ channel in the ventricular cell membrane.

We further compared the effects of $G^*_{i1\alpha}$ and $G_{\beta\gamma}$ on the $K_{ATP}$ and $K_{ACh}$ channels in the atrial cell membrane where both channels are expressed (Fig. 11; TUNG et al. 1991; ITO et al. 1992). In the cell-attached form the $K_{ACh}$ channel was activated vigorously by ACh (1 $\mu M$). In the inside-out patch condition the openings of the $K_{ACh}$ channel decreased to a minimal background level. When $G^*_{i\text{-}1\alpha}$ (300 p$M$) was applied to the interned side of the membrane, bursting $K^+$ channel openings with a conductance of about 90 pS were clearly induced. On the other hand, the background openings of

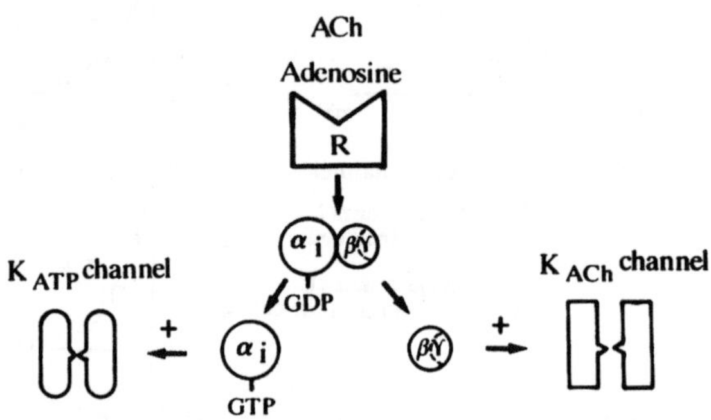


**Fig. 12.** Proposed mechanism of the PTX-sensitive G-protein subunit-activation of the $K_{ATP}$ and $K_{ACh}$ channels in cardiac cell membrane. Upon stimulation of the receptors by adenosine or ACh, PTX-sensitive G-proteins may be functionally dissociated into $G_{\alpha\text{-GTP}}$ and $G_{\beta\gamma}$. $G_{\alpha\text{-GTP}}$ may activate the $K_{ATP}$ channel, while $G_{\beta\gamma}$ activates the $K_{ACh}$ channel. This scheme does not represent any quantitative relationship between components and does not take into account possible intermediate steps between components. The former mechanism exists in both ventricular and atrial cells while the latter may exist in atrial but not in ventricular cells. Since cardiac myocytes contain millimolar concentrations of intracellular ATP, the G-protein-activation of the $K_{ATP}$ channel system may not be operative under physiological conditions. However, the system might play a significant role in the ischemia-induced shortening of the cardiac action potential. Although we cannot completely exclude the possibility that the $G_{i\alpha\text{-GTP}}$ pathway may partly contribute to the $G_K$-activation of the $K_{ACh}$ channel, the pathway cannot be the major regulatory mechanism for the $K_{ACh}$ channel. (From ITO et al. 1992)

the $K_{ACh}$ channel were not affected significantly. Openings of the 90 pS $K^+$ channel were blocked by 1 $\mu M$ glibenclamide (not shown), indicating that this is the $K_{ATP}$ channel. Subsequent application of $G_{\beta\gamma}$ (10 n$M$) to the patch dramatically increased openings of the 40- to 45-pS $K_{ACh}$ channel in the same patch membrane. The $G_{\beta\gamma}$-induced openings of the $K_{ACh}$ channel were not affected by glibenclamide (TUNG et al. 1991; ITO et al. 1992).

From these results we conclude that the G-protein subunit activates the $K_{ATP}$ channel and the $K_{ACh}$ channel differentially in the cardiac cell membrane: $G_{i\alpha\text{-}GTP}$ activates the $K_{ATP}$ channel, and $G_{\beta\gamma}$ activates the $K_{ACh}$ channel (Fig. 12). The former mechanism exists in both ventricular and atrial cells, while the latter may exist in atrial but not in ventricular cells. Since cardiac myocytes contain millimolar concentrations of intracellular ATP, the G-protein activation of the $K_{ATP}$ channel system may not be operative under physiological conditions. However, the system might play a significant role in the ischemia-induced shortening of the cardiac action potential. Although we cannot completely exclude the possibility that $G_{i\alpha\text{-}GTP}$ pathway may partly contribute to the $G_K$ activation of the $K_{ACh}$ channel, the pathway may not be the major regulatory mechanism for the $K_{ACh}$ channel.

## E. Stimulatory Modulation of the $G_K$-Gated Cardiac $K_{ACh}$ Channel

Besides the direct gating of the $K_{ACh}$ channel by G-protein, it has also recently been suggested that the G-protein-gated channel system can be modulated in a stimulatory manner by various substances, i.e., arachidonic acid (AA) and its metabolites and intracellular chloride ion (KURACHI et al. 1989b; NAKAJIMA et al. 1992).

### I. Arachidonic Acid and Its Metabolites

KURACHI et al. (1989b) reported that AA added to the bath can stimulate the $K_{ACh}$ channel activity in the cell-attached form in the absence of agonists in the pipette (Fig. 13). The channel stimulation persisted for over 10 min after washout of AA. AA can be converted to various biologically active substances, for example, to prostaglandins by cyclo-oxygenase and to leukotrienes (LTs) by 5-lipoxygenase (NEEDLEMAN et al. 1986). The effects of AA were prevented by NDGA (a lipoxygenase inhibitor) and AA-861 (a 5-lipoxygenase inhibitor, not shown), but were not affected by either indomethacin (a cyclo-oxygenase inhibitor) or baicalein (a 12-lipoxygenase inhibitor). Furthermore, this activation could be mimicked by metabolites of 5-lipoxygenase pathway, i.e., 5-hydroperoxyeicosatetraenoic acid (5-HPETE), LT $A_4$, and LT $C_4$, but not 12-HPETE, a metabolite of 12-lipoxygenase pathway (not shown). Therefore, 5-lipoxygenase metabolites of AA may stimulate the $K_{ACh}$ channel in a receptor-independent manner.

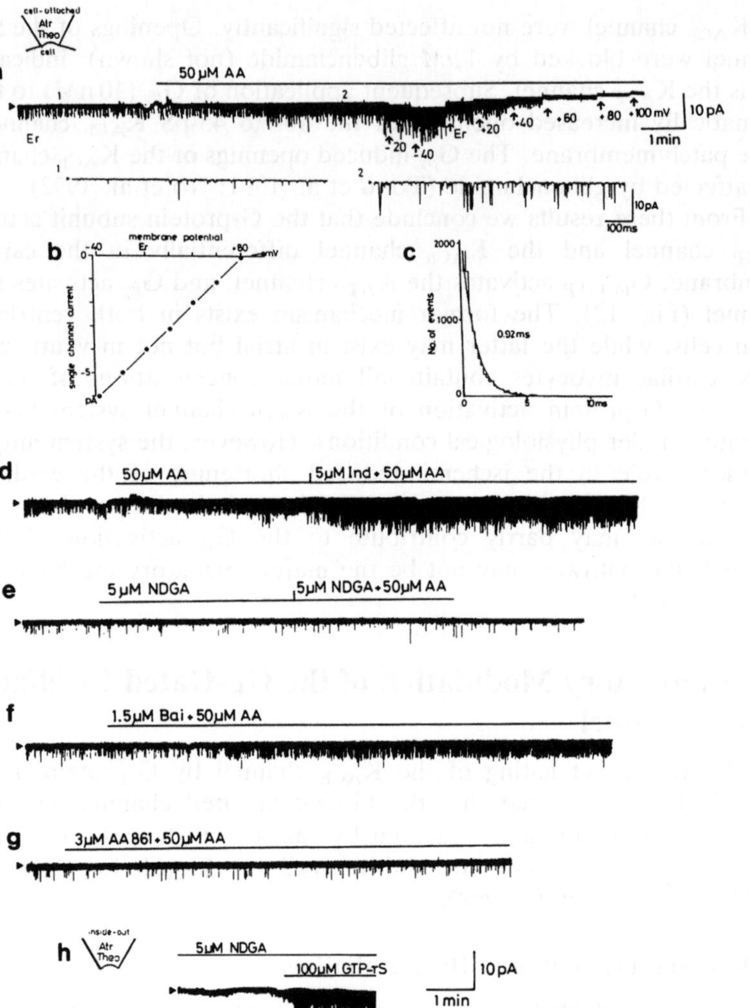


**Fig. 13a–h.** Arachidonic acid (*AA*) activation of the $K_{ACh}$ channel and the effects of inhibitors of the lipoxygenase and cyclo-oxygenase pathways. In the cell-attached patch, when AA (50 $\mu M$) was applied to the bath, it activated a $K^+$ channel (**a**). The pipette solution contained 10 $\mu M$ atropine and 100 $\mu M$ theophyline. The membrane potential of the patch is indicated below the current trace. The resting potential ($E_r$) of the cell in this case was −79 mV. (**b**) Current-voltage relationship. (**c**) Open-time histogram of the AA-induced $K^+$ channel at $E_r$. Effects of indomethacin (**d**), NDGA (**e**), baicalein (**f**), and AA861 (**g**) on the AA-induced activation of the $K_{ACh}$ channel in the cell-attached patches. (**h**) Effects of NDGA on the GTPγS-induced activation of the channel in the inside-out patch. (From KURACHI et al. 1989b)

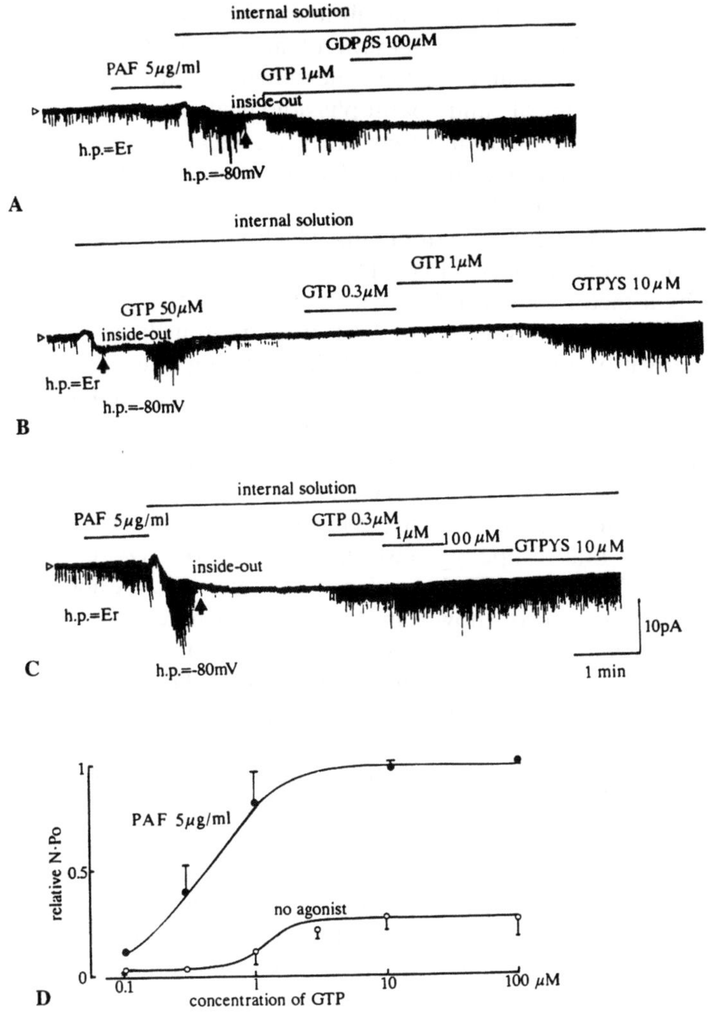


**Fig. 14A–D.** PAF-induced higher sensitivity of the $K_{ACh}$ channel to intracellular GTP. **A** PAF, added to the bath, induced openings of the $K_{ACh}$ channel in the cell-attached patch of atrial cell. Upon formation of the inside-out patch, the channel openings disappeared. GTP (1 $\mu M$) induced full openings of the channel in the absence of agonist in the pipette. The GTP-induced activation was blocked by GDP-$\beta$S. **B** In the inside-out patch without pretreatment of PAF (0.3–1 $\mu M$) GTP did not cause openings of the $K_{ACh}$ channel, but GTP-$\gamma$S (10 $\mu M$) activated the channel fully. **C** In the PAF-treated inside-out patch, GTP (>0.3 $\mu M$) caused activation of the channel in a concentration-dependent manner. **D** Concentration-dependent effect of GTP on the $K_{ACh}$ channel in the PAF-treated (*filled circles*) and untreated (*open circles*) cell membrane. *Symbols*, *bars*, the mean ± SD obtained from three (PAF-treated) and seven (untreated) different patches. The relative $N \cdot P_o$ of the $K_{ACh}$ channel at each concentration of GTP was obtained with reference to the maximum channel activity achieved by 10 $\mu M$ GTP$\gamma$S in each patch. (From NAKAJIMA et al. 1991)

AA is released from the cell membrane in response to various chemical and physical stimulations (NEEDLEMAN et al. 1986); AA metabolites are possibly common intracellular second messengers to the $K_{ACh}$ channel in a variety of physiological and pathophysiological regulations of cardiac excitation. In fact, it was found that the $K_{ACh}$ channel activity can also be stimulated by $\alpha$-adrenergic agonists (KURACHI et al. 1989c) and platelet-activating factor (PAF; NAKAJIMA et al. 1991) via the AA metabolite pathway. Among the various stimulants which we have studied so far, PAF is the most potent agonist. PAF ($5\,\mu$g/ml) increased the $K_{ACh}$ channel openings by a factor of $132 \pm 80$ over $N \cdot P_{o.back}$ in the cell-attached patch. Note that this stimulating effect of PAF was one-fourth to one-third of the maximal activation of the channel by ACh in the pipette.

The AA- and $LTC_4$-induced activation of the $K_{ACh}$ channel disappeared upon formation of the inside-out patch (KURACHI et al. 1989b). In these inside-out patches GTP alone caused full activation of the channel, which was inhibited by GDP$\beta$S. Both AA and $LTC_4$ failed to activate the $K_{ACh}$ channel when applied to the internal side of inside-out patches. These results suggest that 5-lipoxygenase metabolites of AA cause persistent stimulation of $G_K$. The GTP dependence of effects of AA and $LTC_4$ on the $K_{ACh}$ channel was also shown in frog atrial whole cells (SCHERER and BREITWIESER 1990). We recently examined effects of PAF treatment on the sensitivity of the $K_{ACh}$ channel to intracellular GTP (Fig. 14). The PAF-induced $K_{ACh}$ channel activation observed in cell-attached patches disappeared upon formation of inside-out patches. Neither ACh nor adenosine were present in the pipette solution. Yet upon application of low concentrations ($0.1-1\,\mu M$) of GTP, channel activity reappeared. Maximal channel activity was induced by $1-10\,\mu M$ GTP in these PAF-treated patches, which was inhibited by GDP$\beta$S (Fig. 14A,C). In the untreated patch low concentrations ($0.3-1\,\mu M$) of GTP did not cause significant openings of the $K_{ACh}$ channel (Fig. 14B). Figure 14D shows the relative activity of the $K_{ACh}$ channel at various concentrations of GTP in both PAF-pretreated and -untreated inside-out patches. The GTP-induced activation of the $K_{ACh}$ channel was clearly enhanced by pretreatment of the atrial cells with PAF. Thus, AA metabolites released upon stimulation of various receptors, such as PAF, probably stimulate $G_K$ but not the $K_{ACh}$ channel itself (Fig. 15).

## II. Phosphorylation

KIM (1990) reported that $\beta$-adrenergic agonists and dibutylic cAMP, acting through A-kinase, enhance the $K_{ACh}$ current in the cell-attached patches of neonatal rat atrial myocytes. The increase in the current resulted from a longer channel open time upon phosphorylation. The mean open time of the channel became approximately 5 ms from 1 ms during perfusion of isoproterenol. The closed time was also prolonged, which was less

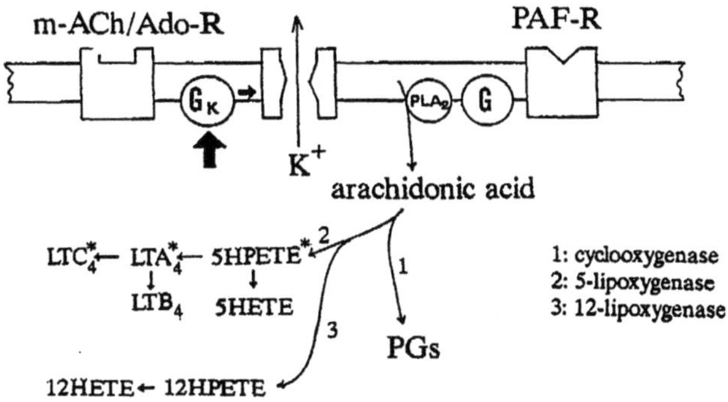


**Fig. 15.** The proposed signaling pathway underlying the receptor-dependent, arachidonic acid metabolite mediated stimulatory modulation of the $G_K$-gated $K_{ACh}$ channel. *Asterisks*, the 5-lipoxygenase metabolites which activated the $K_{ACh}$ channel. (From KURACHI et al. 1992)

prominent than the effects on the open time. In the inside-out patch, the catalytic subunit of A-kinase with ATP mimicked the effects of isoproterenol, which was reversed by alkaline phosphatase. KIM (1991) also showed that intracellular ATP itself could prolong the open time of the $K_{ACh}$ channel, which was not affected by H-8 (an A-kinase and C-kinase inhibitor). The ATP effect was postulated to be mediated by a membrane-bound unidentified protein kinase. However, in guinea pig atrial cells, isoproterenol and dibutylic and 8-bromo cAMP applied to the bath did not affect the ACh activation of the $K_{ACh}$ channel in the cell-attached patches, and intracellular ATP did not affect the open time of the $K_{ACh}$ channel in the inside-out patches (TUNG and KURACHI, unpublished observations; see also HEIDBÜCKEL et al. 1990 in guinea pig atrial myocytes; KAIBARA et al. 1991 in rabbit atrial myocytes). These results suggest that the protein kinase dependent phosphorylation of the $K_{ACh}$ channel may be age or species dependent. These possibilities have not yet been examined.

## III. NDP-Kinase

HEIDBÜCHEL et al. (1990) and KAIBARA et al. (1991) reported that millimolar concentrations of intracellular ATP activate the $K_{ACh}$ channel in the absence of an agonist in a $Mg^{2+}$-dependent manner. They interpreted this as due to NDP-kinase mediated phosphotransfer to $G_K$ from ATP. Since low concentrations of GDP inhibited this ATP-induced activation of the $K_{ACh}$ channel, this system may not play significant roles in the physiological regulation of the $K_{ACh}$ channel (KAIBARA et al. 1991). In these experiments, the internal solutions containing 20 or 130 m$M$ chloride ion were used. We recently found that chloride ion inhibited the intrinsic GTPase activity of $G_{K\alpha}$, thereby resulting in higher sensitivity of the $K_{ACh}$ channel to intracellular GTP in the chloride-rich internal solution than that in the sulfate or aspartate-internal solution (NAKAJIMA et al. 1992). When we used the sulfate or aspartate-internal solution, neither GTP (100 $\mu M$) nor ATP (1–2 m$M$) caused any significant openings of the $K_{ACh}$ channel in the absence of agonists in the pipette. These observations may indicate that activation of the $K_{ACh}$ channel either by the basal turn-on reaction of $G_K$ (ITO et al. 1991) or phosphotransfer to $G_K$ by NDP-kinase does not play a significant physiological role, but may be significant in the conditions where GTPase activity of $G_K$ is disturbed (NAKAJIMA et al. 1992; GILMAN 1987).

## IV. Intracellular Chloride

We recently examined effects of intracellular anions on activation of the $K_{ACh}$ channel in inside-out patches of guinea pig atrial cells (NAKAJIMA et al. 1992). With ACh (0.5 $\mu M$) in the pipette, 1 $\mu M$ GTP caused different

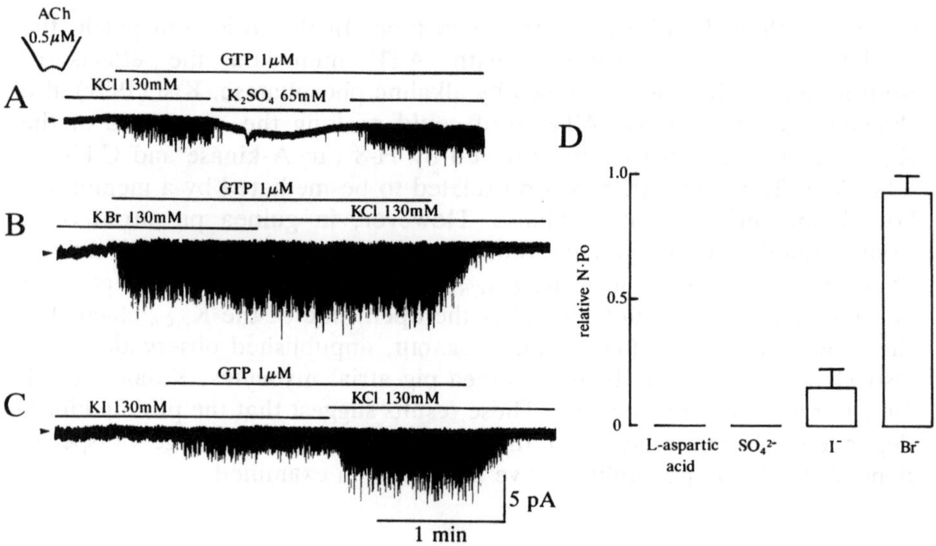


**Fig. 16A–D.** Effects of intracellular anions on the muscarinic $K^+$ channel in inside-out patches. **A**, **B**, **C** Inside-out patch experiments. The holding potential was −80 mV. The pipette solution contained 0.5 $\mu M$ ACh. *Bars* (above each current trace), the protocol for perfusing various internal solutions and GTP; *arrowheads*, zero current level. Effects of the $Cl^-$ and $SO_4^{2-}$-internal solutions on the 1 $\mu M$ GTP-induced activation of the $K_{ACh}$ channel were compared in **A**, $Br^-$ and $Cl^-$ in **B**, and $I^-$ and $Cl^-$ in **C**. Note that no anion by itself activated the channel. **D** Comparison of the effects of anions on the open probability of the $K_{ACh}$ channels activated by GTP (1 $\mu M$). The relative $N \cdot P_o$ of the $K_{ACh}$ channel activated by GTP (1 $\mu M$) in various internal anions ($SO_4^{2-}$, $Br^-$, $I^-$) and L-aspartatic acid was calculated with the reference to the $N \cdot P_o$ in the 130 m$M$ $Cl^-$-internal solution. The data for each symbol were obtained from four patches in each case. (From NAKAJIMA et al. 1992)

magnitudes of the $K_{ACh}$ channel activation in internal solutions containing different anions. The order of potency of anions to induce the $K_{ACh}$ channel activity by 1 $\mu M$ GTP was $Cl^- \geqslant Br^- > I^-$. In the sulfate or aspartic acid-internal solution, no channel opening was induced by 1 $\mu M$ GTP (Fig. 16). These differences in the GTP activation of the channel were due to the effects of various anions on the turn-off reaction of $G_K$. The turn-off reaction of $G_K$ estimated from the $K_{ACh}$ channel activity on washout of GTP in the chloride solution was much slower than that in the sulfate internal solution. The turn-on reaction of $G_K$ was not affected significantly by the anions in the internal solution containing $Mg^{2+}$.

Since in the heart a chloride channel is activated by various neurotransmitters, such as isoproterenol and histamine (HARVEY and HUME 1989; BAHINSKI et al. 1989), these agonists may affect the activity of the $K_{ACh}$ channel induced by ACh through an indirect effect on $[Cl^-]_i$. In addition, it is likely that a number of channels regulated by G-proteins other than the $K_{ACh}$ channel can also be modulated by intracellular anions. Thus, intracellular anions may act as a regulator of G-protein-gated channels in various kinds of cells. Further studies are, however, needed to elucidate the physiological and pathophysiological roles of anions on the regulation of the $K_{ACh}$ channel.

# F. Conclusion

In this chapter, I summarize mainly our recent results on the molecular regulation of the G-protein-gated $K_{ACh}$ channel and modulations of the system by various substances in the cardiac atrial cell membrane. In summary, $G_{\beta\gamma}$ activates the $K_{ACh}$ channel effectively and consistently, and may play a major role in the muscarinic activation of the $K_{ACh}$ channel in vivo. However, further studies are necessary to elucidate the molecular mechanisms underlying the interaction between $G_{\beta\gamma}$ and the $K_{ACh}$ channel.

*Acknowledgements.* This research was supported by NIH RO1 HL47360, AHA Grant in Aid 91013540, and AHA Established Investigatorship to Y. K.

## References

Bahinski A, Nairn AC, Greengard P, Gadsby DC (1989) Chloride conductance regulated by cyclic AMP-dependent protein kinase in cardiac myocytes. Nature 340:718–721

Breitwieser GE, Szabo G (1985) Uncoupling of cardiac muscarinic and β-adrenergic receptors from ion channels by a guanine nucleotide analogue. Nature 317:538–540

Brown AM, Birnbaumer L (1990) Ionic channels and their regulation by G protein subunits. Annu Rev Physiol 52:197–213

Camps M, Hou C, Sidiropoulous D, Stock JB, Jakob KH, Gierschik P (1992) Stimulation of phospholipase C by guanine-nucleotide-binding protein βγ subunits. Eur J Biochem 206:821–831

Codina J, Hildebrandt JD, Birnbaumer L (1984) Effects of guanine nucleotides and Mg on human erythrocyte $N_i$ and $N_s$, the regulatory components of adenylyl cyclase. J Biol Chem 259:11408–11418

Codina J, Yatani A, Grenet D, Brown A, Birnbaumer L (1987) The α subunit of the GTP binding protein $G_k$ opens atrial potassium channels. Science 236:442–445

Dunlap K, Holz GG, Rane SG (1987) G proteins as regulators of ion channel function. TINS 10(6):241–244

Fesenko EE, Kolesnikov SS, Lyubarsky AL (1985) Induction by cyclic GNP of cationic conductance in plasma membrane of retinal rod outer segment. Nature 313:310–313

Florio VA, Sternweis PC (1985) Reconstitution of resolved muscarinic cholinergic receptors with purified GTP-binding proteins. J Biol Chem 260:3477–3483

Gilman AG (1987) G proteins: transducers of receptor-generated signals. Annu Rev Biochem 56:615–649

Harvey RD, Hume JR (1989) Autonomic regulation of a chloride current in heart. Science 244:983–985

Heidbüchel H, Callewaert G, Vereecke J, Carmeliet E (1990) ATP-dependent activation of atrial muscarinic $K^+$ channels in the absence of agonist and G-nucleotides. Pflugers Arch 416:213–215

Hille B (1992) G Protein-coupled mechanisms and nerve signaling. Neuron 9:187–195

Ito H, Sugimoto T, Kobayashi I, Takahashi K, Katada T, Ui M, Kurachi Y (1991) On the mechanism of basal and agonist-induced activation of the G protein-gated muscarinic $K^+$ channel in atrial myocytes of guinea-pig heart. J Gen Physiol 98:517–533

Ito H, Tung RT, Sugimoto T, Kobayashi I, Takahashi K, Katada T, Ui M, Kurachi Y (1992) On the mechanism of G protein βγ subunit-activation of the muscarinic $K^+$ channel in guinea pig atrial cell membrane: comparison with the ATP-sensitive $K^+$ channel. J Gen Physiol 99:961–983

Jelsema CL, Axelrod J (1987) Stimulation of phospholipase $A_2$ activity in bovine rod outer segments by the βγ subunits of transducin and its inhibition by the α subunit. Proc Natl Acad Sci USA 84:3623–3627

Jelsema CL, Burch RM, Jaken S, Ma AD, Axelrod J (1989) Modulation of phospholipase $A_2$ activity in rod outer segments of bovine retina by G protein subunits, guanine nucleotides, protein kinases and calpactin. Neurol Neurobiol 49:25–41

Kaibara M, Nakajima T, Irisawa H, Giles W (1991) Regulation of spontaneous opening of muscarinic $K^+$ channels in rabbit atrium. J Physiol (Lond) 433:589–613

Karschin A, Ho BY, Labarca C, Elroy-Stein O, Moss B, Davidson N, Lester HA (1991) Heterologously expressed serotonin 1A receptors couple to muscarinic $K^+$ channels in heart. Proc Natl Acad Sci USA 88:5694–5698

Katada T, Ui M (1982) Direct modification of the membrane adenylate cyclase system by islet-activating protein due to ADP-ribosylation of a membrane protein. Proc Natl Acad Sci USA 79:3129–3133

Katada T, Kusakabe K, Oinuma M, Ui M (1987) A novel mechanism for the inhibition of adenylate cyclase via inhibitory GTP-binding proteins. Calmodulin-dependent inhibition of the cyclase catalyst by the βγ-subunits of GTP-binding proteins. J Biol Chem 262:11897–11900

Kim D (1990) β-Adrenergic regulation of the muscarinic-gated $K^+$ channel via cyclic AMP-dependent protein kinase in atrial cells. Circ Res 67:1292–1298

Kim D (1991) Modulation of acetylcholine-activated $K^+$ channel function in rat atrial cells by phosphorylation. J Physiol (Lond) 437:133–155

Kim D, Lewis DL, Graziadei L, Neer EJ, Bar-Sagi D, Clapham DE (1989) G-protein βγ-subunits activate the cardiac muscarinic $K^+$-channel via phospholipase $A_2$. Nature 337:557–560

Kirsch GE, Yatani A, Codina J, Birnbaumer L, Brown AM (1988) $\alpha$-Subunits of $G_k$ activates atrial $K^+$ channels of chick, rat, and guinea pig. Am J Physiol 254:H1200–H1205

Kirsch GE, Codina J, Birnbaumer L, Brown AM (1990) Coupling of ATP-sensitive $K^+$ channels to $A_1$ receptors by G proteins in rat ventricular myocytes. Am J Physiol 259:H820–H826

Kobayashi I, Shibasaki H, Takahashi K, Tohyama K, Kurachi Y, Ito H, Ui M, Katada T (1990) Purification and characterization of five different $\alpha$ subunits of guanine-nucleotide-binding proteins in bovine brain membranes. Eur J Biochem 191:499–506

Kurachi Y (1989) Regulation of G protein-gated $K^+$ channels. News Physiol Sci 4:158–161

Kurachi Y (1990) Muscarinic acetylcholine-gated $K^+$ channels in mammalian heart. In: Reuss L, Russell JM, Szabo G (eds) Regulation of potassium transport across biological membrane. University of Texas Press, Austin, pp 404–428

Kurachi Y, Nakajima T, Sugimoto T (1986a) Acetylcholine activation of $K^+$ channels in cell-free membrane of atrial cells. Am J Physiol 251:H681–H684

Kurachi Y, Nakajima T, Sugimoto T (1986b) On the mechanism of activation of muscarinic $K^+$ channels by adenosine in isolated atrial cells: involvement of GTP-binding proteins. Pflugers Arch 407:264–274

Kurachi Y, Nakajima T, Sugimoto T (1986c) Role of intracellular $Mg^{2+}$ in the activation of muscarinic $K^+$ channel in cardiac atrial cell membrane. Pflugers Arch 407:572–574

Kurachi Y, Nakajima T, Ito H (1987) Intracellular fluoride activation of muscarinic K channel in atrial cell membrane. Circulation 76:IV-105 (abstract)

Kurachi Y, Ito H, Sugimoto T, Katada T, Ui M (1989a) Activation of atrial muscarinic $K^+$ channels by low concentrations of $\beta\gamma$ subunits of rat brain G protein. Pflugers Arch 413:325–327

Kurachi Y, Ito H, Sugimoto T, Shimizu T, Miki I, Ui M (1989b) Arachidonic acid metabolites as intracellular modulators of the G protein-gated cardiac $K^+$ channel. Nature 337:555–557

Kurachi Y, Ito H, Sugimoto T, Shimizu T, Miki I, Ui M (1989c) $\alpha$-Adrenergic activation of the muscarinic $K^+$ channel is mediated by arachidonic acid metabolites. Pflugers Arch 414:102–104

Kurachi Y, Ito H, Sugimoto T (1990) Positive cooperativity in activation of the cardiac muscarinic $K^+$ channel by intracellular GTP. Pflugers Arch 416:216–218

Kurachi Y, Tung RT, Ito H, Nakajima T (1992) G protein activation of cardiac muscarinic $K^+$ channels. Prog Neurobiol 39:229–246

Kurose H, Katada T, Haga K, Haga A, Ichiyama A, Ui M (1986) Functional interaction of purified muscarinic receptors with purified inhibitory guanine nucleotide regulatory proteins reconstituted in phospholipid vesicles. J Biol Chem 261:6423–6428

Logothetis DE, Kurachi Y, Galper J, Neer EJ, Clapham DE (1987) The $\beta\gamma$ subunits of GTP-binding proteins activate the muscarinic $K^+$ channel in heart. Nature 325:321–326

Logothetis DE, Kim D, Northup JK, Neer EJ, Clapham DE (1988) Specificity of action of guanine nucleotide-binding regulatory protein subunits on the cardiac muscarinic $K^+$ channel. Proc Natl Acad Sci USA 85:5814–5818

Nakajima T, Sugimoto T, Kurachi Y (1991) Platelet-activating factor activates cardiac $G_K$ via arachidonic acid metabolites. FEBS Lett 289:239–243

Nakajima T, Sugimoto T, Kurachi Y (1992) Effects of anions on the G protein-mediated activation of the muscarinic $K^+$ channel in the cardiac atrial cell membrane: intracellular chloride-inhibition of the GTPase activity of $G_K$. J Gen Physiol 99:665–682

Nakamura T, Gold GH (1987) A cyclic nucleotide-gated conductance in olfactory receptor cilia. Nature 325:442–444

Nanavati C, Clapham DE, Ito H, Kurachi Y (1990) A comparison of the roles of purified G protein subunits in the activation of the cardiac muscarinic $K^+$ channel. In: Nathanson NM, Harden TK (eds) G proteins and signal transduction. Rockefeller University Press, New York, pp 29–41

Needleman P, Turk J, Jakschik BA, Morrison AR, Lefkowith JB (1986) Arachidonic acid metabolism. Annu Rev Biochem 55:69–102

North AR, Williams JT, Surprenant A, Christie MJ (1987) m and d receptors belong to a family of receptors that are coupled to potassium channels. Proc Natl Acad Sci USA 84:5487–5491

Pfaffinger PJ, Martin JM, Hunter DD, Nathanson NM, Hille B (1985) GTP-binding proteins couple cardiac muscarinic receptors to a K channel. Nature 317:536–538

Pitcher JA, Inglese J, Higgins JB, Arriza JL, Casey PJ, Kim C, Benovic JL, Kwatra MM, Caron MG, Lefkowitz RJ (1992) Role of $\beta\gamma$ subunits of G proteins in targeting the $\beta$-adrenergic receptor kinase to membrane-bound receptors. Science 257:1264–1267

Scherer RW, Breitwieser GE (1990) Arachidonic acid metabolites alter G protein-mediated signal transduction in heart. Effects on muscarinic $K^+$ channel. J Gen Physiol 96:735–755

Tang W-J, Gilman AG (1991) Type-specific regulation of adenylyl cylase by G protein $\beta\gamma$ subunits. Science 254:1500–1503

Tang W-J, Gilman AG (1992) Adenylyl cyclases. Cell 70:869–872

Tung RT, Kurachi Y (1990) G protein activation of cardiac ATP-sensitive $K^+$ channel (abstract). Circulation 82:III-462

Tung RT, Kurachi Y (1991) On the mechanism of nucleotide diphosphate activation of the ATP-sensitive $K^+$ channel in ventricular cell of guinea-pig. J Physiol (Lond) 437:239–256

Tung RT, Ito H, Takikawa R, Sugimoto T, Kobayashi I, Takahashi K, Katada T, Ui M, Kurachi Y (1991) Respective activation of ATP-sensitive and muscarinic K channels by G protein $\alpha$ and $\beta\gamma$ subunits (abstract). Biophys J 59:5a

Tung RT, Shen WK, Kurachi Y (1992) Comparison of ATP-sensitive K channel in atrial vs ventricular myocytes of guinea-pig. JACC 19(3):245A

White RE, Schonbrunn A, Armstrong DL (1991) Somatostatin stimulates $Ca^{2+}$-activated $K^+$ channels through protein dephosphorylation. Nature 351:570–573

Whiteway M, Hougan L, Dignard D, Thomas DY, Bell L, Saari GC, Grant FJ, O'Hara P, MacKay VL (1989) The STE4 and STE18 genes of yeast encode potential $\beta$ and $\gamma$ subunits of the mating factor receptor-coupled G protein. Cell 56:467–477

Yatani A, Codina J, Brown AM, Birnbaumer L (1987) Direct activation of mammalian atrial muscarinic potassium channels by GTP regulatory protein $G_k$. Science 235:207–211

Yatani A, Hamm H, Codina J, Mazzoni MR, Birnbaumer L, Brown AM (1988a) A monoclonal antibody to the $\alpha$ subunits of $G_k$ blocks muscarinic activation of atrial $K^+$ channels. Science 241:828–831

Yatani A, Mattera R, Codina J, Graf R, Okabe K, Padrell E, Iyengar R, Brown AM, Birnbaumer L (1988b) The G protein-gated atrial $K^+$ channel is stimulated by three distinct $G_{i\alpha}$-subunits. Nature 336:680–682

Yatani A, Okabe K, Birnbaumer L, Brown AM (1990) Detergents, dimeric $G_{\beta\gamma}$, and eicosanoid pathways to muscarinic atrial $K^+$ channels. Am J Physiol 258:H1507–H1514

CHAPTER 76

# Modulation of $K^+$ Channels by G-Proteins

L. BIRNBAUMER

## A. Direct Regulation of Ionic Channels by G-Proteins

### I. The Inwardly Rectifying "Muscarinic" $K^+$ Channel

#### 1. Experiments Leading to the Discovery of G-Protein Gating

The possibility of direct regulation of an ion channel by a G-protein, i.e., as shown in cell-free systems not involving soluble second messengers, and hence of G-protein-gated ion channels, emerged from studies on the mechanism by which muscarinic acetylcholine receptors (mAChR) activate the atrial muscarinic $K^+$ channel that mediates vagal regulation of chronotropy. Ion channels have since then been found that are affected either by $G_i$ ($G_i$-gated channels), $G_o$ ($G_o$-gated channels), or $G_s$ ($G_s$-gated channels). G-protein-gated $K^+$ channels are physiologically very relevant as modulators of cellular function. Activation of these $K^+$ channels causes cells to hyperpolarize and become less excitable. As a consequence secretion is attenuated in endocrine and nerve cells such as found in sympathetic ganglia (ECCLES and LIBET 1961; HARTZEL et al. 1977), parasympathetic ganglia (GRIFFITH et al. 1981; HILL-SMITH and PURVES 1978), and the central nervous system (NAKAJIMA et al. 1986; TRAUTWEIN and DUDEL 1958). In heart they cause a decrease in chronotropy (TRAUTWEIN and DUDEL 1958; GILES and NOBLE 1976; reviewed by HARTZELL 1981).

Acetylcholine inhibits adenylyl cyclase in heart membranes (MURAD et al. 1962; KUROSE and UI 1983; MATTERA et al. 1985) and clearly does so acting through a G-protein, as indicated among others by the fact that agonist binding is GTP-sensitive (BERRIE et al. 1979; ROSENBERGER et al. 1980). In one way or another it stood to reason that regulation of other functions by muscarinic receptors would also be mediated by a G-protein, and it seemed likely that these actions would involve changes in intracellular second messenger levels. Yet a lack of involvement of soluble second messengers (i.e., decrease in cAMP levels) in the action of muscarinic receptors on atrial cells had been proposed on grounds of the rapidity with which acetylcholine modifies $K^+$ channel activity (TRAUTWEIN et al. 1982; NARGEOT et al. 1983). Such a lack of involvement of a soluble second messenger was indeed proven first in electrophysiological experiments which

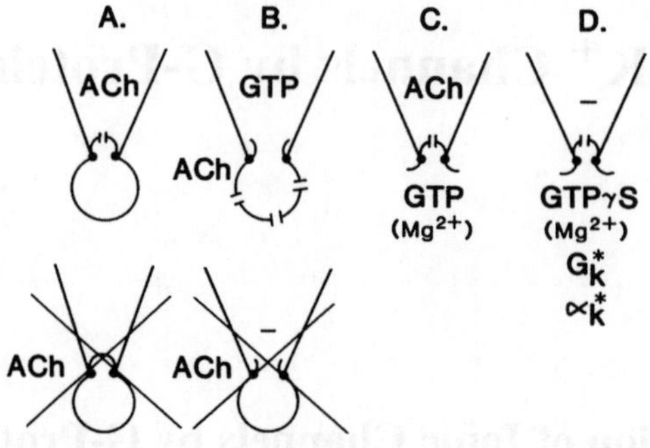


**Fig. 1.** Summary of experiments leading to the demonstration of direct (membrane-delimited and independent of soluble second messengers and protein phosphorylation) G-protein gating of $K^+$ channels. **A** Results from Soejima and Noma 1984; **B** experiments of Pfaffinger et al. 1985; **C,D** experiments shown in Fig. 2

used the giga-seal patch clamp approach of Neher, Sakman, and collaborators (Hamill et al. 1981) in the cell-attached configuration (Soejima and Noma 1984) and then in experiments in which the cell-attached broken-patch configuration was applied (Pfaffinger et al. 1985; Breitwieser and Szabo 1985). Direct regulation of the channel by a $G_i$-protein was then confirmed in experiments with inside-out membrane patches (Yatani et al. 1987a,b; Codina et al. 1987a,b; Kirsch et al. 1988; Mattera et al. 1988).

Figure 1 summarizes the type of experiments that demonstrated direct (or membrane-delimited) regulation of muscarinic $K^+$ channels by a G protein ($G_k$). The cell-attached patch clamp experiments showed that atrial muscarinic $K^+$ channels in the membrane patch are insensitive to muscarinic stimulation when acetylcholine is added to the bath but become activated when the agonist is superfused onto the outer surface of the isolated membrane patch (Fig. 1A; Soejima and Noma 1984). This indicated that the muscarinic receptors had to be in close proximity to the channel regulated by them and eliminated a soluble cytoplasmic second messenger as mediator of receptor activation. The actual coupling mechanism was not addressed by this experiment. Specifically, it left open the possibilities of direct receptor regulation of the channel (receptor-channel coupling), of activation of a G-protein by the receptor followed by direct G-protein-channel coupling, or of receptor-mediated formation, possibly also through activation of a G-protein, of a diffusible yet membrane-delimited second messenger such as could be diacylglycerol, and regulation of channel activity as a result of protein kinase C mediated phosphorylation.

The involvement of a G-protein in the coupling of heart muscarinic receptors to $K^+$ channels was established in whole-cell broken patch voltage clamp experiments in which atrial cells were "perfused" with pipette solutions containing GTP or the GTP analogue GMP-P(NH)P (Fig. 1B; PFAFFINGER et al. 1985; BREITWIESER and SZABO 1985). One set of experiments established that acetylcholine is able to increase the $K^+$ currents only when the pipette solution contains GTP, and that PTX treatment of cells abolishes the muscarinic response (PFAFFINGER et al. 1985). This extended the involvement of a PTX substrate in the action of the cardiac muscarinic receptor from its action to inhibit adenylyl cyclase to that of stimulating the $K^+$ channel. The other set of experiments used amphibian atrial cells and measured not only $K^+$ current but also $Ca^{2+}$ currents (BREITWIESER and SZABO 1985). In these cells acetylcholine induces a $K^+$ current (inwardly rectifying $I_{ACh}$) and attenuates $\beta$-adrenergic receptor induced increases in a specific $Ca^{2+}$ current (the so-called slow inward $I_{Ca}$). The pipette/cytoplasm exchange (cell dialysis) was not extensive, allowing for hormonal regulation of ion currents even without addition of GTP to the pipette solution. Addition of GMP-P(NH)P to the pipette had no effect unless hormones were added to the extracellular bathing fluid, presumably because of the inherent hysteresis (slowness) in the action of the nucleotide analogue in the absence of hormonal stimulation, slowness that was accentuated by the prevailing low (submillimolar) intracellular $Mg^{2+}$ concentrations. In the presence of hormones, the hysteresis in GMP-P(NH)P activation of G-proteins was reduced, as was the requirement for free $Mg^{2+}$ so that agonist-induced, antagonist-resistant, persistent activations of ion currents were obtained: ACh induced a persistent $I_{ACh}$ current and isoproterenol induced a persistent slow $I_{Ca}$. These results indicated involvement of a guanine nucleotide binding protein in the action of both types of receptors. Further, the persistent nature of the acetylcholine response indicated that the G-protein intervening between the muscarinic receptor and the $K^+$ channel, termed generically $G_k$ (BREITWIESER and SZABO 1985), resembles the stimulatory G-protein of adenylyl cyclase in its activation kinetics in that activation by GTP analogs is slow in the absence of hormonal stimulation and fast in its presence. In agreement with the postulate that muscarinic regulation of $K^+$ currents is independent of cAMP levels, stimulation of cells held with a GMP-P(NH)P-containing pipette in the presence of isoproterenol had no effect on the induction of $K^+$ currents by acetylcholine even though persistent effects of isoproterenol, and therefore activation of $G_s$, were obtained (BREITWIESER and SZABO 1985).

## 2. Direct Stimulation by hRBC $G_i$ and Its $\alpha$ Subunit

Direct action of a G-protein on the heart muscarinic receptor-sensitive $K^+$ channel was demonstrated in a cell-free system using inside-out membrane patches (Fig. 1C,D; YATANI et al. 1987a; CERBAI et al. 1988). In these experiments it was found that mere addition to the bath, i.e., to the cytoplasmic face of the $K^+$ channel containing membrane, of GTP$\gamma$S or an

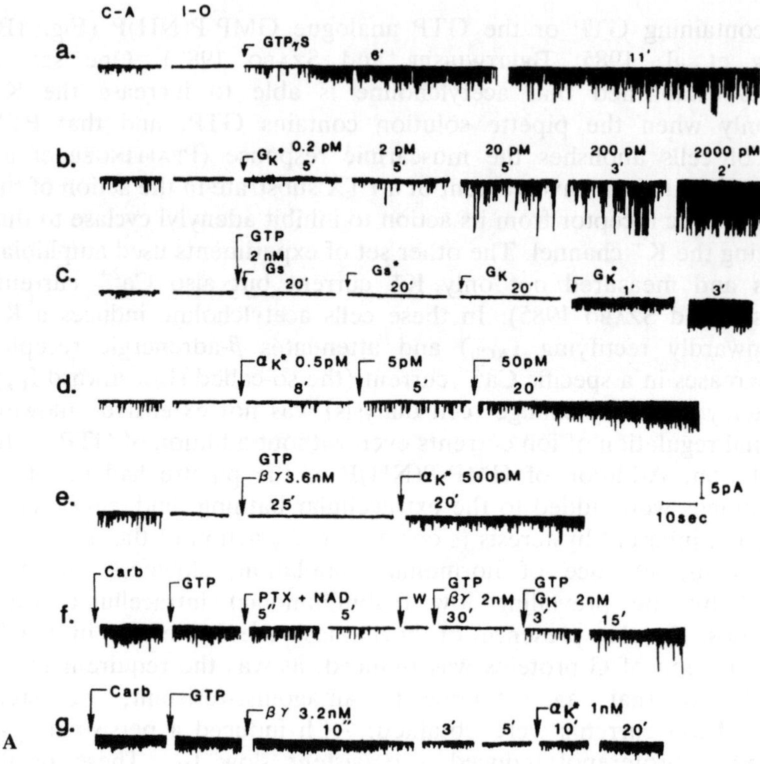


**Fig. 2A,B.** Properties of $G_k$-mediated regulation of G-protein-sensitive $K^+$ channels as seen in inside-out membrane patches of guinea pig atrial cells and $GH_3$ rat pituitary tumor cells. Each line represents a separate experiment in which single-channel $K^+$ currents were recorded before (cell-attached, *C-A*) and after membrane patch excision to the inside-out configuration (*I-O*). Records were obtained at holding potentials varied between −80 and −100 mV and using symmetrical 130–140 m*M* KCl or K-methanesulfonate solutions in 5 m*M* HEPES, pH 7.5 containing in addition either 1.8 m*M* $CaCl_2$ in the pipette or 5 m*M* EGTA and 2 m*M* $MgCl_2$ in the 100-μl bathing chamber. Other additions are shown or described for each experiments. *Numbers above records*, time elapsed between the indicated additions and the beginning of the segment of record shown. Routinely, the first addition was made between 5 and 10 min after patch excision, and subsequent additions were at 5- to 25-min intervals, depending on the purpose of the experiments; 5-min intervals were used when dose response relationships were studied, and 25-min intervals were used when substances added had no apparent effect. In some instances the bathing solutions were exchanged by perfusion at 1 to 2 ml/min. **A** Experiments with atrial membrane patches from adult guinea pigs. Cell were obtained by collagenase digestion and used without further culturing. *a*, Stimulation of single-channel $K^+$ currents by activation of the atrial $G_k$ protein with GTPγS (100 μ*M* in the bath). *b*, Stimulation of the atrial $K^+$ channels in a dose-dependent manner by exogenously added human erythrocyte PTX substrate referred to as $G_k$ (formerly referred to as $N_i$ or $G_i$), preactivated by incubation with GTPγS and $Mg^{2+}$, and dialyzed extensively to remove free GTPγS to ineffective levels ($G_k$*). Threshold effects of $G_k$* were obtained with 0.2–1 p*M* in separate membrane patches. *c*, The bathing solution contained 100 μ*M* GTP throughout; proteins were added at 2 n*M* each. Lack of stimulatory effects of 1 n*M* of either nonactivated or GTPγS-activated $G_s$ or of

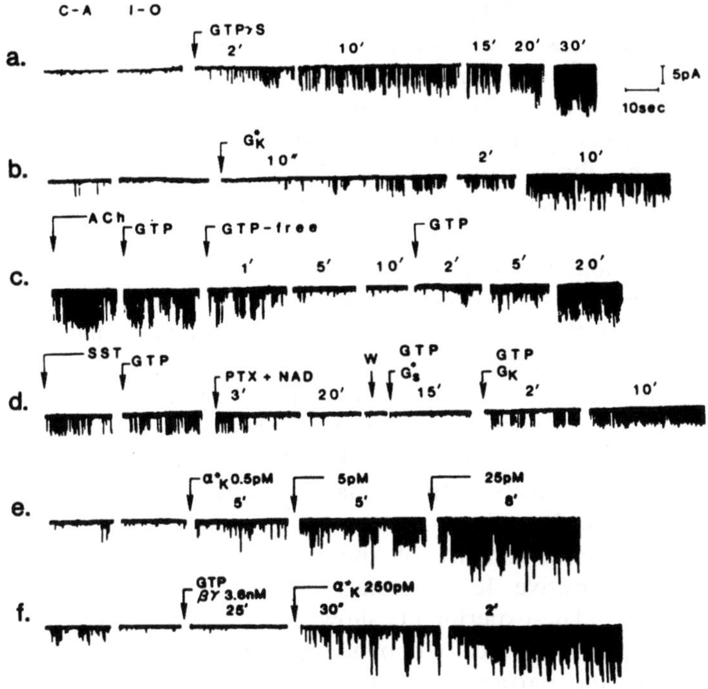


B

nonactivated $G_k$. *d*, Mimicry of the effect of GTP$\gamma$S-activated holo-$G_k$ by equally low concentrations of the resolved GTP$\gamma$S-activated $\alpha$ subunit of the protein ($\alpha_k$*). *e*, Lack of effect of resolved human erythrocyte $\beta\gamma$ exposed to a $G_k$*-sensitive membrane patch. *f*, Stimulation of $G_k$-sensitive $K^+$ channels by the muscarinic receptor agonist carbachol (*Carb*) present in the pipette throughout, maintenance of receptor–G-protein effector coupling after patch excision into bathing solution with 100 $\mu M$ GTP, and uncoupling of atrial $G_k$ by treatment with PTX and NAD added to the bathing solution. Lack of recoupling effect of resolved human erythrocyte $\beta\gamma$ and reconstitution of acetylcholine receptor-$K^+$ channel stimulation by addition of native unactivated human erythrocyte $G_k$ in the presence of GTP. Note that this contrasts the result in *c* and demonstrates that the exogenously added $G_k$ requires receptor participation for activation by GTP. *g*, Inhibition of GTP- and carabchol-stimulated $K^+$ channel activity by exogenously added $\beta\gamma$ and reversal of this inhibition by addition of $\alpha_k$ preactivated with GTP$\gamma$S ($\alpha_k$*). **B** Experiments with inside-out membrane patches from rat $GH_3$ pituitary tumor cells. Cells were grown as monolayers on cover slips and membrane patches excised from their upper membrane surface. *a*, effect of GTP$\gamma$S added to the bathing medium. *b*, Representative time course of activation of $GH_3$ cell $K^+$ channel by a saturating concentration (2 n*M*) of GTP$\gamma$S-activated human erythrocyte $G_k$ ($G_k$*). *c*, Dependence on GTP and reversibility of receptor-mediated stimulation of $G_k$-sensitive $K^+$ channel. Acetylcholine (*ACh*) was present in the pipette solution throughout; 100 $\mu M$ GTP was present in the bathing medium, removed, and readded as shown. *d*, Stimulation of $GH_3$ cell $G_k$-sensitive $K^+$ channels by somatostatin (*SST*). Demonstration of PTX sensitivity of the $GH_3$ $G_k$ protein and reconstitution of the signal transduction pathway by addition of unactivated native human erythrocyte $G_k$ in the presence of GTP. *e*,*f*, Mimicry of the effects of GTP$\gamma$S-activated $G_k$ by resolved GTP$\gamma$S-activated $\alpha$ subunit of human erythrocyte $G_k$ and lack of effect of resolved, $\alpha$ subunit-free $\beta\gamma$ dimers. (Adapted from YATANI et al. 1987a,b; CODINA et al. 1987a,b)

activated pertussis toxin (PTX) sensitive G-protein purified from human erythrocyte membranes results in $K^+$ channel activation. The effect of GTP$\gamma$S required $Mg^{2+}$ and proceeded with a lag, as expected from the activation kinetics of G-proteins. The effect of the PTX-sensitive G-protein preactivated with GTP$\gamma$S developed much faster. Since these studies were carried out in the absence of ATP (YATANI et al. 1987a), they ruled out involvement of a phosphorylation reaction in the action of muscarinic receptors and defined functionally a $G_k$, i.e., a stimulatory regulatory G-protein of a $K^+$ channel. Similar studies with pituitary $GH_3$ cells showed the existence in these cells of a similar $K^+$ channel stimulated by a PTX-sensitive G-protein that is under regulatory control of somatostatin and muscarinic receptors (YATANI et al. 1987b).

G-protein-regulated $K^+$ channels in atrial and pituitary membranes are highly selective with respect to the G-protein with which they interact. Highly purified $G_s$ even at 100-fold higher concentrations than needed to obtain half-maximal effects with $G_i$ had no effect on the $K^+$ channel (YATANI et al. 1987a,b).

Figure 2 summarizes many of the properties of receptor-mediated regulation of $G_i$-sensitive $K^+$ channels as seen in atrial and pituitary membrane patches. The experiments show that G-protein regulation of $K^+$ channels by receptors is critically dependent on GTP, that activation of $G_k$ by GTP requires the participation of agonist-occupied receptor, that the endogenous $G_k$ is uncoupled from regulation by receptor by PTX, and that a PTX-uncoupled system is readily reconstituted by addition of exogenous unactivated $G_k$, provided GTP is present in the bath. Further, resolved GTP$\gamma$S-activated $\alpha$ subunits of hRBC $G_i$ ($\alpha_i$*) mimic the actions of GTP$\gamma$S-activated $G_i$ ($G_i$*) with comparable potency, which resolved $\beta\gamma$ dimers do not (CODINA et al. 1987a,b; KIRSCH et al. 1988). Thus, the experiments also in-dicate that the most plausible mechanism used by muscarinic receptors to stimulate G-protein-gated $K^+$ channels is the catalysis of the activation of membrane $G_k$ by GTP and formation of activated free $\alpha_k$* which in turn acts as mediator to stimulate the $K^+$ channel.

## 3. Properties of the $G_i$-Stimulated $K^+$ Channel

The microscopic kinetic properties of the $G_i$-sensitive $K^+$ channels have been measured and are typical of the muscarinic" type of $K^+$ channels (YATANI et al. 1987a,b; CODINA et al. 1987a,b; KIRSCH et al. 1988; CERBAI et al. 1988). The effect of activated $G_i$ is clearly one of increasing the opening probability of the channel. Once stimulated by $G_i$, the channel opens in bursts and clusters of bursts, as illustrated by the NP (number of channels per patch × open probability of each channel in the patch) diaries of activity shown in Fig. 3. Frequency histograms of open times are well fit by first-order decay functions with time constants of about 1 ms. Open amplitude histograms are fit well by Gaussian frequency distributions. Slope

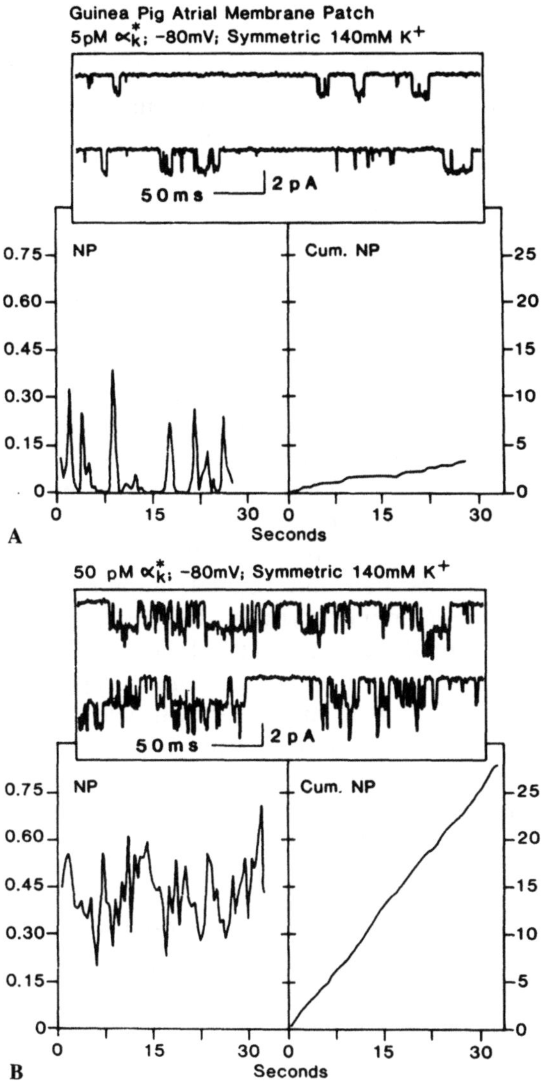


**Fig. 3A,B.** Single-channel $K^+$ currents in guinea pig atrial membranes after patch excision and addition of either 5 p*M* (**A**) or 500 p*M* (**B**) of GTP$\gamma$S-activated resolved $\alpha$ subunit of $G_i$ ($\alpha_k{}^*$). Quantification of activity: NP and cumulative NP values. Records of single-channel currents (*top insets*) were subdivided into consecutive 200-ms segments to quantify the proportion of time during which single-channel currents are observed in each segment. The variation of these values per 200 ms, representing the product of the opening probability (*Po*) of each of the channels present in the membrane patch times the number of channels in the patch (*N*) as a function of time are the NP diaries (*lower left*). Accumulation of NP values as a function of time gives cumulative NP values over the same time period (*Cum NP*). These are equivalent to a time course of channel activity (*lower right*, **A**, **B**). The slopes of the Cum NP curves (Cum NP/min) referred to maximum cumulative NP/min values that can be elicited from the same membrane patch provide individual points that can serve to construct dose-response curves for the action of a G-protein. Concentrations of $G_k{}^*$ and $\alpha_k{}^*$ giving half-maximal stimulations of guinea pig atrial $K^+$ channels vary between 5 and 60 p*M*, with a mean around 20 p*M* for either $G_k{}^*$ or $\alpha_k{}^*$

conductances calculated from data obtained with 130–140 m$M$ $K^+$ on both sides of the membrane range from ca. 40 pS in guinea pig atria to ca. 50 pS in $GH_3$ cells. It is not clear whether this means that there are multiple genes for these channels, or whether the slightly different single channel conductances observed in atrial and endocrine cells are due to cell-specific posttranslational modifications of the same channel molecules. In both systems the channels conduct very poorly in the outward direction. In both systems the microscopic properties of the channels stimulated by receptor plus GTP are indistinguishable from those seen on stimulation by GTP$\gamma$S or exogenous addition of preactivated $G_i$ or its activate $\alpha$ subunit. Figure 8 also expresses channel activation data as cumulative NP values and illustrates that when averaged over long periods of time this allows for comparison of activities on a quantitative basis, as may be required when comparing effects of increasing concentrations of holo-G or resolved $\alpha$ subunits.

Among the receptors that catalyze activation of $G_k$-type G-protein by GTP are the heart atrial muscarinic, adenosine and neuropeptide Y receptors (YATANI, BIRNBAUMER and BROWN, unpublished), hippocampal serotonin 1A and gamma-aminobutyric acid B receptors (NEWBERRY and NICOLL 1984; ANDRADE et al. 1986; SASAKI and SATO 1987; THALMANN 1988), and endocrine cell somatostatin and muscarinic receptors (YATANI et al. 1987b). The list is bound to grow and, as shown below, may involve more than one type of $K^+$ channel. Moreover, while it is clear at this time that in atrial and pituitary cells receptors cause activation of $G_i$-gated $K^+$ channels, studies in other cell types may reveal that $G_o$ rather than $G_i$ is involved in channel gating.

## 4. Identity of the $G_k$ that Gates the Muscarinic-Type $K^+$ Channels

Figure 4 illustrates that the GTP$\gamma$S-activated forms of the $\alpha$ subunits of all three known $G_i$-proteins are about equally effective in activating the atrial $K^+_{[Ach]}$ channel. These effects have also been obtained with recombinant $\alpha_i$ subunits, indicating that actions of these purified proteins are not due to contaminants (YATANI et al. 1988).

The following tests proved that the effects observed with each of the recombinant $\alpha_i$ subunits are intrinsic properties of the expressed molecules. (a) Boiling prior to addition to the bath obliterated the activity of each recombinant $\alpha_i$. (b) Threshold concentrations of GTP$\gamma$S for activation of atrial muscarinic $K^+$ channels in inside-out membrane patches when 100 n$M$ carbachol was in the pipette were between 10 and 100 n$M$, which is at least ten times the maximum GTP$\gamma$S added with saturating concentrations (1000 p$M$) of any of the GTP$\gamma$S-activated partially purified and ultrafiltrated recombinant $\alpha_i$ preparations. (c) Prior addition of 100 $\mu M$ GDP$\beta$S, which blocks carbachol-mediated effects of 10 $\mu M$ GTP, did not interfere with the effects of GTP$\gamma$S-activated recombinant $\alpha_{i1}$, recombinant $\alpha_{i2}$, or recombinant $\alpha_{i3}$.

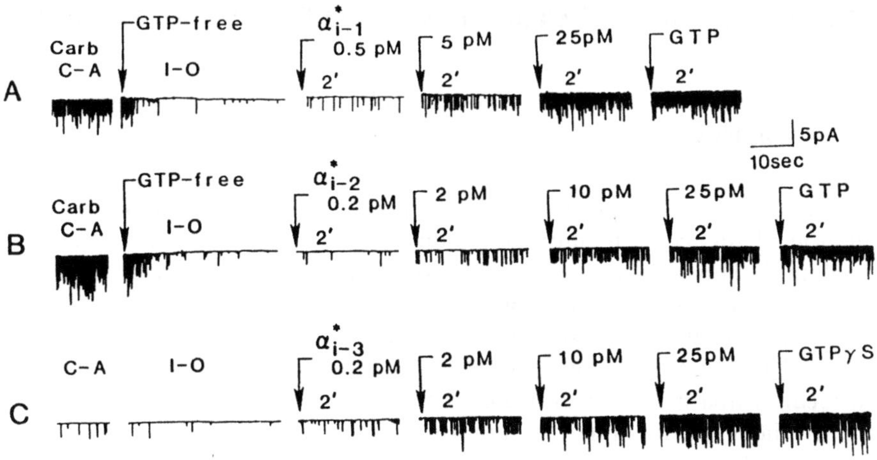


**Fig. 4.** Stimulation of single $K^+$ channel currents in guinea pig atrial membrane patches by increasing concentrations of bovine brain $\alpha_i$-1 (**A**) and human erythrocyte $\alpha_i$-2 (**B**) and $\alpha_i$-3 (**C**). $G_i\alpha$-2 and $G_i\alpha$-3 were purified from human erythrocyte membranes as described in BIRNBAUMER et al. (1988) and GRAF et al. (1993). $G_i\alpha$-1 was purified by E. PADRELL and R. IYENGAR as described in YATANI et al. (1988)

These data need to be contrasted to those obtained initially by Clapham, Neer, and collaborators (LOGOTHETIS et al. 1987). These authors reported first that neither bovine brain GTPγS-activated $\alpha_i$ ($\alpha_{41}$*) nor bovine brain GTPγS-activated $\alpha_o$ ($\alpha_{39}$*) stimulated atrial muscarinic $K^+$ channels. Subsequently, these same authors noted stimulation with the $\alpha_o$-type protein, while the $\alpha_i$-type molecule continued to fail in their hends to stimulate the muscarinic channel (LOGOTHETIS et al. 1988). In our hands, the purified bovine brain $G_o$ has only about 1% the $G_k$ activity of purified hRBC $G_i$, now known to be $G_{i3}$, and it has been impossible to determine whether this activity is due to a weak activity intrinsic to $G_o$ or to presence in the $G_o$ preparation of contaminating $G_i$-type protein(s) (YATANI et al. 1987a). Since we find that all three types of $G_i$ have $G_k$ activity, contamination with any $G_i$ (or $\alpha_i$) would suffice. In a more recent publication by one of the coauthors of the reports that failed to observe effects with the $\alpha_i$-type proteins, it is reported that all $\alpha_i$ proteins stimulate the muscarinic $K^+$ channel (KOBAYASHI et al. 1990), indicating that something must have been amiss with the G-proteins used in the original studies by Logothetis and collaborators.

## II. The ATP-Sensitive $K^+$ Channel: A Second $G_i$-Gated $K^+$ Channel

### 1. General Properties of the ATP-Sensitive $K^+$ Channel/Sulfonyliurea Receptor Complex

It is now commonly accepted that an ATP-sensitive $K^+$ channel is one of the components of the "glucose sensor" of pancreatic $\beta$ cells responsible for regulation of this cell's membrane potential (for details see PANTEN et al., this volume). This results in modulation of the activity of voltage-gated $Ca^{2+}$ channels, which upon activation promote first-phase secretory events. Insulin secretion is antagonized by hormones such as somatostatin of local origin and by adrenaline of adrenal origin, as well as by norepinephrine from local sympathetic innervations. The effects of somatostatin as well as those of the catecholamine, acting through $\alpha_2$-adrenergic receptors, are blocked by PTX treatment and involve a complex set of transmembrane effects that include reduction of adenylyl cyclase activity (KATADA and UI 1981), inhibition of L-type $Ca^{2+}$ channels (HSU et al. 1991), and, last but not least, stimulation of the ATP-sensitive $K^+$ channel (RIBALET and CIANI 1987; RIBALET et al. 1989). It is not clear at this time whether the different effects of these receptors are mediated by one and the same G-protein, or whether several G-proteins with differing effector specificity are involved.

It has been shown that the ATP-inhibited $K^+$ channel is also inhibited by sulfonylureas, and it is likely, although not yet proven, that the sulfonylurea receptor may be the ATP-sensitive $K^+$ channel proper. As reviewed by PANTEN and his coworkers (this volume), the regulation of this channel is complex, involving ATP, ADP-Mg, channel phosphorylation, and G-proteins. Inhibition by ATP can be mimicked by AMP-P(NH)P and does not appear to involve a phosphorylation event. Inhibition by sulfonylurea requires ADP-Mg, which appears to interact with the channel at a site allosteric to that of ATP. The binding affinity for sulfonylureas is decreased by Mg-ATP under phosphorylating conditions. Regulation by G-protein requires that the channel be first inhibited by ATP [or an analog such as AMP-P(NH)P] and may require prior phosphorylation of the channel, although this last point has been difficult to establish with certainty.

ATP-sensitive $K^+$ conductances are found not only in pancreatic cells but also in a variety of other tissues, including skeletal and smooth muscle and heart. As reported by NOMA in 1983, an ATP-sensitive $K^+$ channel having a 65-pS single-channel conductance can be readily identified in cell-attached as well as in inside-out membrane patches from guinea pig and rabbit myocytes. This channel exhibits inward rectification, albeit less marked than the that of the muscarinic channel, from which it can be readily distinguished because of its larger conductance. The $IC_{50}$ for inhibition by ATP is ca. $0.1\,mM$.

As seen in whole cell broken-patch recordings from neonatal rat ventricular myocytes, the $K^+$ current arises as a function of time presumably due to gradual depletion of ATP and is inhibited, like that of other tissues, by sulfonylurea drugs (e.g., tolbutamide). Addition of adenosine to the bath prior to ATP depletion causes a rapid rise in the $K^+$ current that, as the current obtained after ATP-depletion, can be inhibited by sulfonylurea (Kirsch et al. 1990). As seen in inside-out membrane patches, the rat myocyte ATP-sensitive channel has a single-channel conductance of 85 pS and mean open times of 1–2 ms and also exhibits inward rectification. These results are consistent with the existence in intact myocytes of an ATP-sensitive, sulfonylurea-sensitive $K^+$ channel, akin to that of pancreatic $\beta$ cells. This channel can be activated by adenosine, which is known to act through receptors that are coupled to effectors by G-proteins.

## 2. Identity of G-Proteins that Regulate the ATP-Sensitive $K^+$ Channel

Figure 5 illustrates the GTP-dependent activation of the AMP-P(NH)P-inhibited $K^+$ channel, both as seen in outside-out membrane patch with ATP and GTP in the pipette (no effect of adenosine was obtained in patches where GTP had been omitted from the pipette solution) and in-inside out membrane patches with adenosine or the adenosine analogue cyclohexyladenosine in the pipette.

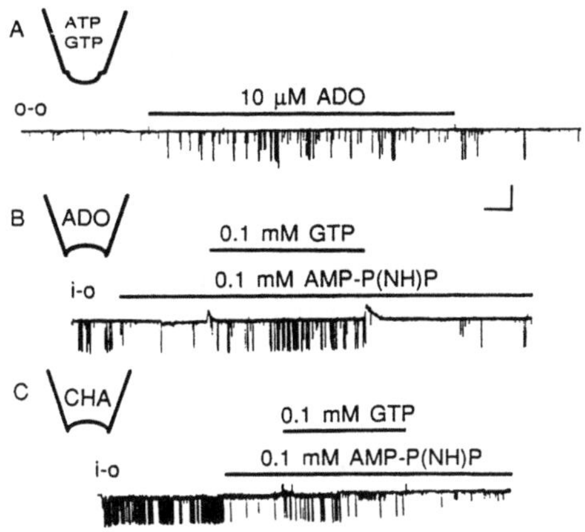


**Fig. 5A–C.** Dependence on GTP of the effects of adenosine (*ADO*) and cyclohexyladenosine (*CHA*) on the $K^+$[ATP] channel complex in excised membrane patches from neonatal rat ventricular myocytes. Recordings from outside-out patches were obtained with a pipette (intracellular) solution that included 140 m*M* KCl, 5 m*M* EGTA, 0.1 m*M* ATP, 0.1 m*M* GTP, 2 m*M* $Mg^{2+}$, and a bath (extracellular) modified Tyrode solution (including 137 m*M* NaCl, 5.4 m*M* KCl, 2.0 m*M* $CaCl_2$, 1.0 m*M* $MgCl_2$, 10 m*M* glucose, and 10 $\mu M$ tetrodoxtoxin). Recordings from inside-out membrane patches were obtained with the pipette (extracellular) modified Tyrode and bath (intracellular) solutions described above. Membrane patches were perfused with bath solutions having the additives as indicated. Holding potential was −80 mV; *vertical*, *horizontal bars*, 5 pA and 20 s, respectively. (Adapted from Kirsch et al. 1990)

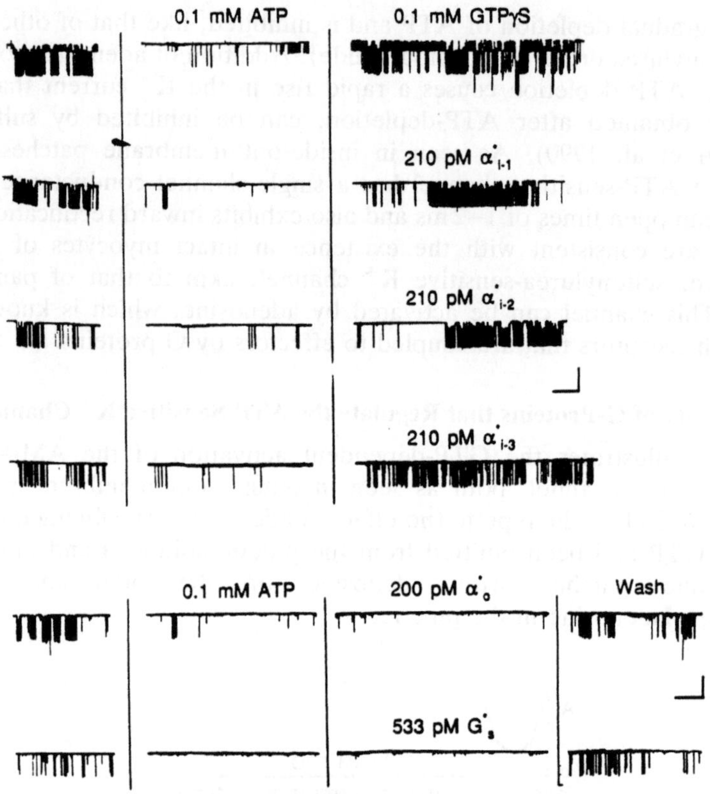


**Fig. 6.** Activation of the ATP-inhibited $K^+$[ATP] channel complex by GTP$\gamma$S and activated $G_i$ but by not activated $G_s$ or $G_o$, in inside-out membrane patches of neonatal rat ventricular myocytes. *Top four panels*, sensitivity to GTP$\gamma$S, and to GTP$\gamma$S-activated $\alpha_{i1}$, $\alpha_{i2}$, and $\alpha_{i3}$: *left*, $K^+$ channel currents after excision into ATP-free bath solution; *middle*, $K^+$ channel activity after addition of 0.1 m*M* ATP; *right*, $K^+$ channel activity after the indicated additives in the presence of ATP. All solutions contained 2.0 m*M* $MgCl_2$. *Bottom two panels*, insensitivity to the GTP$\gamma$S-activated form of $\alpha_{o1}$ ($\alpha_o^*$ in the figure) and $\alpha_s$. The design was the same as above, except that after failure to obtain a response with $\alpha_o$ and $\alpha_s$ the viability of their channel in the patch was tested by removal of both the G-protein $\alpha$ subunit and ATP (*wash*). Solutions were as in Fig. 5; holding potential was −80 mV; *vertical, horizontal bars*, 5 pA and 20 s, respectively. (Adapted from Kirsch et al. 1990)

As expected if this channel were to be regulated by a G-protein, addition of GTP$\gamma$S to inside-out membrane patches with ATP- or AMP-P(NH)P-inhibited $K^+$ channels, results in their activation. Upon testing for the molecular identity of the G-protein(s) that could mediate the effect of the adenosine receptors, it was established that any one of the $G_i$ $\alpha$ subunits but neither that of $G_s$ nor that of a $G_o$ protein mimicked the effects of adenosine receptors and GTP$\gamma$S (Fig. 6).

It follows from these studies that $G_i$-proteins are multifunctional, in that they can regulate more than one cellular function, two types of $K^+$ channels in this case. It also follows that inasmuch as these two types of $K^+$ channels is concerned, $G_{i1}$, $G_{i2}$, and $G_{i3}$ are isoforms, for all three act indistinguishably on both of channels.

## III. G-Protein Gating as a Tool To Discover Novel Ionic Channels: Neuronal $G_o$-Gated $K^+$ Channels

One of the propertics of the "muscarinic" $K^+$ channels is that they are essentially silent in the absence of stimulation by an activated G-protein ($G_k$). That is, in the absence of activated G-protein their $P_o$ is close to zero. The possibility existed that not only $G_i$-proteins regulate $K^+$ channels but also the structurally related $G_o$. Since nervous tissue is rich in $G_o$, central nervous system neurons, specifically hippocampal pyramidal cells, were placed into culture and studied for potential presence of both $G_i$- and $G_o$-gated $K^+$ channels. Findings with a highly purified preparation of one of the splice variants of bovine brain $G_o$, the GTP$\gamma$S-activated $G_{o1}$ ($G_{o1}$*) were of interest (VANDONGEN et al. 1988). They identified the existence of several novel G-protein-gated, more precisely $G_o$-gated, $K^+$ channels, which are distinct from $G_i$-gated $K^+$ channels. Thus, application of purified bovine brain $G_{o1}$* to the cytoplasmic aspect of inside-out membrane patches of cultured hippocampal pyramidal cells resulted in appearance of three new types of single-channel $K^+$ currents consistent with the existence of the three nonrectifying of $K^+$ channels having sizes of 13, 40, and 55 pS, respectively, plus an inwardly rectifying $K^+$ channel with a slope conductance of 40 pS. No such channel activities were observed with hRBC $G_{i3}$* or hRBC $\alpha_{i3}$*; $G_{o2}$ or $\alpha_{o2}$* have not been tested in this system as yet. The effect of increasing concentrations of bovine brain $G_o$* on the 55-pS type of $K^+$ channel is shown in Fig. 7A. Note that in contrast to earlier observations with the same preparation of $G_o$* added to guinea pig atrial membrane patches (see Fig. 2A) the hippocampal $K^+$ channel is highly sensitive to $G_o$*. Significant activation was obtained at 1 p*M*, and half-maximal effects were obtained at about 10 p*M*.

To confirm that the channels observed on addition of $G_o$* are indeed gated by $G_o$ and not by a contaminant, the G-protein specificity of the channel was studied by addition of partially purified recombinant $\alpha$* molecules, including not only those of $\alpha_o$ (type 1) but also the three $\alpha_i$. All of the above-mentioned types of $K^+$ channel were stimulated by recombinant $\alpha_o$*, under conditions where prior addition to one of the recombinant $\alpha_i$* preparations, active on guinea pig atrial muscarinic $K^+$ channels, had no effect. One such experiment, with a 40-pS inwardly rectifying $K^+$ channel, is shown in Fig. 7B. The $G_o$-gated channels were stimulated in the absence of $Ca^+$ or ATP. The presence of AMP-P(NH)P, added routinely to inhibit ATP-sensitive 70-pS $K^+$ channels, and of EGTA did not interfere with the actions of $G_o$ or recombinant $\alpha_o$. Thus, in hippocampal pyramidal cells of the rat, $G_o$ is a $G_k$, and the $K^+$ channels gated by it are several and differ from those present in atrial cells in various aspects, including G-protein specificity.

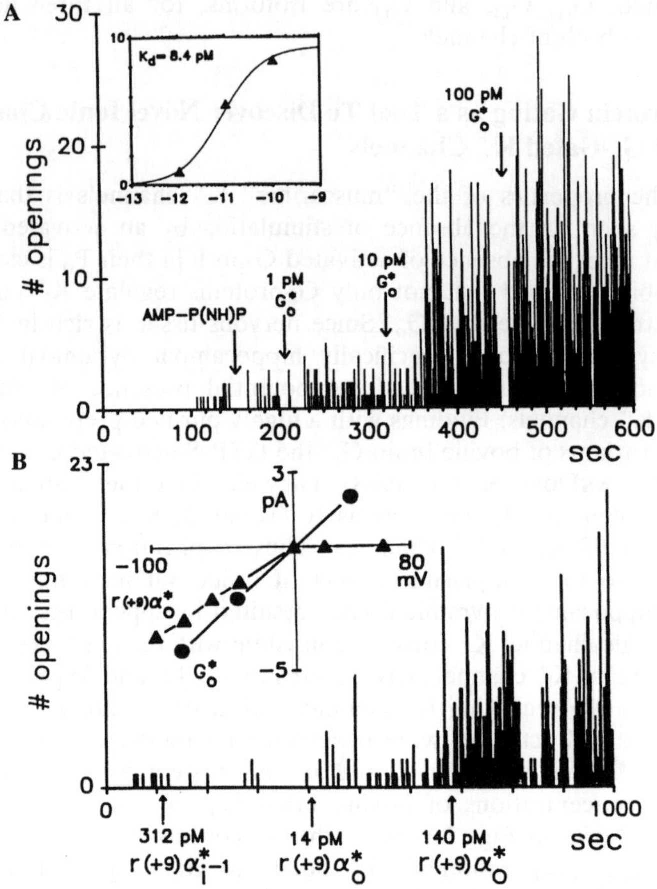


**Fig. 7.** Bovine brain $G_o$ (**A**) and recombinant $\alpha_o$, but not recombinant $\alpha_i$-1 (**B**), gate single $K^+$ channel currents in hippocampal pyramidal cells of neonatal rats. **A** The effect of increasing concentrations of $G_o{}^*$ on a 55-pS $K^+$ channel is shown by plotting the number of openings per 0.8 s as a function of time of continuous recording. *Inset*, openings per 0.8 s averaged for 1 min as a function of $G_o{}^*$ concentration. **B** Lack of effects of 312 p*M* recombinant $\alpha_i{}^*$-1 and stimulatory effect of increasing concentrations of recombinant $\alpha_o{}^*$ on a 40-pS $K^+$ channel, plotted as number of openings per 0.2 s as a function of the time of continuous recording shown. Single $K^+$ channel currents were recordeded from excised inside-out membrane patches as in Fig. 5 in the presence of 0.2 m*M* AMP-P(NH)P. Holding potential was −80 mV, and both pipette and bath solutions contained 140 m*M* K-methanesulfonate, 1 m*M* EGTA, 1 m*M* $MgCl_2$, and 10 m*M* HEPES adjusted to pH 7.4 with Tris base. *Inset*, Current-voltage relationships of two distinct $G_o$-gated $K^+$ channels recorded from hippocampal pyramidal cells. One, an inward rectifying channel was uncovered upon addition of GTPγS-activated $G_o$ ($G_o^*$); the other, a non-rectifying channel, was uncovered upon addition of a GTPγS-activated and partially purified recombinant $\alpha_o$ with a 9-amino acid N-terminal extension (r(+9)$\alpha_o^*$). (Adapted from VANDONGEN et al. 1988)

## B. Effect of $\beta\gamma$ Dimers: Inhibition versus Stimulation of the Muscarinic K+ Channel – A Persisting Controversy

Results obtained by Clapham, Neer, and collaborators on regulation of the atrial muscarinic K+ channel by $\beta\gamma$ dimers differed from results described above (and below), and follow-ups continue to do so. Thus, LOGOTHETIS et al. (1987, 1988) and KURACHI et al. (1989a) observed that bovine brain as well as human placental $\beta\gamma$ dimers stimulate atrial K+ channel activity (see also KURACHI, this volume). In contrast, we did not observe stimulation upon addition of hRBC $\beta\gamma$ dimers under assay conditions that transduce $\alpha_i$ effects into K+ channel stimulation. Instead, we noted that addition of $\beta\gamma$ dimers at high (2–4 n$M$) concentrations to atrial membrane patches stimulated by carbachol in the pipette and GTP in the bath caused inhibition

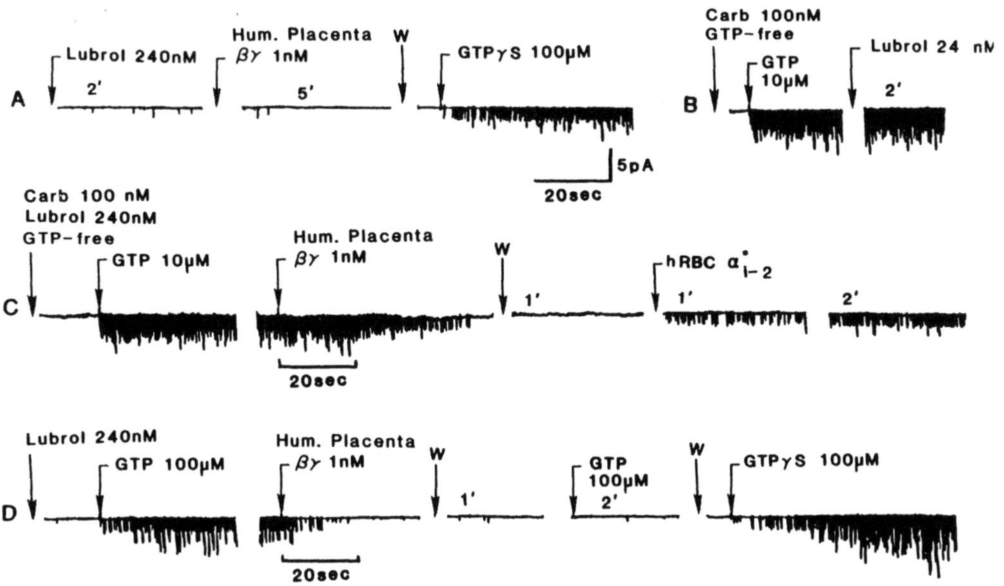


**Fig. 8.** Inhibition of G-protein-gated K+ channel activity by $\beta\gamma$ dimers. Inside-out membrane patches from adult guinea pig atrial cells were exposed to the bathing solution (140 m$M$ KCl, 2 m$M$ MgCL$_2$, 5 m$M$ EGTA, 10 m$M$ HEPES-K, pH 7.4) containing the additives shown in the figure. The pipette solution was identical to the bathing solution and contained 100 n$M$ carbachol when indicated. The composition of bathing solution was changed by a concentration-clamp method. Note that Lubrol PX and/or bovine serum albumin, used to maintain $\beta\gamma$ dimers in suspension, do not interfere with stimulation of activity by GTP (**D**) or GTP plus agonist (**B**,**C**) and that $\beta\gamma$ dimers have no effect on their own (**A**) but inhibit atrial K+ channel stimulation by the membrane $G_k$ (**C**,**D**). Inhibition is faster and is elicited with lower concentrations of $\beta\gamma$ dimers when K+ channels are operating under baseline conditions (GTP only, **D**; cessation of activity after 16 s) than when they are stimulated by agonist (carbachol plus GTP, **C**; cessation of activity after 50 s). *Numbers above records*, time elapsed in minutes between a solution change and the beginning of the record shown. (Adapted from OKABE et al. 1990)

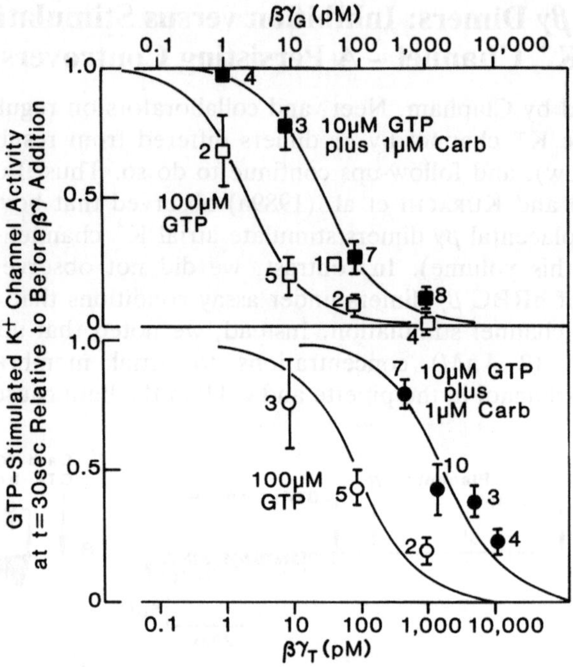


**Fig. 9.** Effect of agonist (carbachol) on the dose-dependent inhibiton by $\beta\gamma$ dimers of GTP-dependent $K^+$ channel activities in inside-out guinea pig atrial membrane patches by $\beta\gamma$ dimers. When $1\,\mu M$ carbachol was present in the pipette, GTP was $10\,\mu M$; in the absence of carbachol GTP was $100\,\mu M$. $\beta\gamma_G$, Data obtained when $\beta\gamma$ dimers derived from either human erythrocyte, human placenta, or bovine brain were pooled; $\beta\gamma_T$, data obtained with $\beta\gamma$ derived from bovine rod outer segments. (Adapted from OKABE et al. 1990)

of signal transduction. While this result was obtained in only two out of five initial experiments (CODINA et al. 1987b), it was later obtained consistently, especially when low concentrations of both GTP and carbachol were used.

A typical set of results obtained by us (OKABE et al. 1990) with $\beta\gamma$ dimers is illustrated in Figs. 8 and 9. In these experiments addition of $\beta\gamma$ dimers to inside-out membrane patches in which $K^+$ channels had been stimulated either by GTP alone (baseline activity) or by carbachol plus GTP (agonist-stimulated activity) resulted consistently in inhibition of $K^+$ channel activity. On their own, i.e., when added to silent patches in the absence of GTP, $\beta\gamma$ dimers had no effect (Fig. 8; CODINA et al. 1987b; KIRSCH et al. 1988; OKABE et al. 1990). Concentration-effect studies showed clearly that $\beta\gamma$ dimers are more potent in inhibiting agonist-indepenent than agonist-stimulated activity, and that this phenomenon applies not only to $\beta\gamma$ dimers suspended in Lubrol PX, such as those from human placenta, human erythroyctes, and bovine brain, but also to $\beta\gamma$ dimers presented to the patches in aqueous media, such as transducin $\beta\gamma$ (Fig. 9).

The fact that $\beta\gamma$ dimers inhibited GTP-dependent activity in the absence of agonist at lower concentrations as compared with inhibition in the presence of agonist was observed previously in a homologous signal transduction system in which purified $\beta$-adrenergic receptor, purified $G_s$ and a partially purified preparation of adenylyl cyclase free of receptor and $G_s$ were reconstituted into phospholipid vesicles (CERIONE et al. 1985; see also BIRNBAUMER 1987) in the absence and the presence of light-activated rhodopsin and $G_i$ (a source of $\beta\gamma$ dimers). This result serves to support the thesis that $\beta\gamma$ dimers act in intact membranes as suppressors of "noise" generated by agonist-unoccupied receptors.

The inhibitory effects of $\beta\gamma$ dimers reported here need to be contrasted to stimulatory effects obtained by Clapham, Neer, and their collaborators (LOGOTHETIS et al. 1987, 1988; KIM et al. 1989; ITO et al. 1991) expanded on by KURACHI (Chap. 75, this volume). The reasons for the discrepancy are not understood. It is not due to the detergent used, for, as mentioned above, inhibition was also obtained by us in the absence of Lubrol PX using $\beta\gamma$ dimers from transducin which are water soluble (Fig. 9). The idea that $\beta\gamma$ dimers may act by stimulation of arachidonic acid formation (KIM et al. 1989) is suspect, for in an adjacent report KURACHI et al. (1989b) show that to obtain effects with arachidonic acid and its metabolites it is necessary to add GTP to the bath, and that these effects are abolished upon addition of GDP$\beta$S. Yet the stimulatory effects of $\beta\gamma$ dimers occur in the absence of GTP (LOGOTHETIS et al. 1987, 1988; KIM et al. 1989). This then dissociates the stimulatory effects of $\beta\gamma$ dimers as reported by KIM et al. (1989) and LOGOTHETIS et al. (1987, 1988) from the effects of arachidonic acid (or its metabolites), as reported by KURACHI et al. (1989b).

## C. Conclusions

In the present chapter I have summarized most of the evidence that relates to G-protein-mediated modulation of two types of $G_i$-gated $K^+$ channels and of a class of $G_o$-gated $K^+$ channels. In all cases the effects appear to be independent of phosphorylation events, or classical formation of second messengers such as cAMP, or inositol trisphosphate. Our working hypothesis is that the effect of the G-proteins described here is due to direct protein-protein interaction, as opposed to formation of intermediary diffusible membrane-delimited second messengers. However, it needs to be pointed out that involvement of such diffusible membrane-delimited second messengers has not been fully ruled out. Moreover, even if diffusible second messengers were not involved in the action of the G-proteins, this would of course not impede that these channels could not be additionally regulated by second messengers and/or phosphorylations, such as seems to be the case for the ATP-sensitive $K^+$ channel. Full resolution of the biochemical mechanisms by which $K^+$ channels are regulated will probably have to await full purification of the channel protein(s), so that they may then be tested upon incorporation into lipid bilayers together with the purified G-protein(s). While this is currently not possible, it is to be hoped that molecular cloning and subsequent overexpression of the channel proteins, which normally constitute only 1/10000th–1/100000th of cell protein, may aid in this aim.

## References


Andrade R, Malenka RC, Nicoll RA (1986) A G protein couples serotonin and $GABA_B$ receptors to the same channels in hippocampus. Science 234:1261–1265

Berrie CP, Birdsall NJM, Burgen ASV, Hulme EC (1979) Guanine nucleotides modulate muscarinuc receptor binding in the heart. Biochem Biophys Res Commum 87:1000–1005

Birnbaumer L (1987) Which G protein subunits are the active mediators in signal transduction. Trends Pharmacol Sci 8:209–211

Birnbaumer L, Codina J, Mattera R, Yatani A, Graf R, Olate J, Sanford J, Brown AM (1988) Receptor-effector coupling by G proteins. Purification of human erythrocyte $G_i$-2 and $G_i$-3 and analysis of effector regulation using recombinant $\alpha$ subunits synthesized in E. coli. Cold Spring Harbor Symp Quant Biol 53:229–239

Breitwieser GE, Szabo G (1985) Uncoupling of cardiac muscarinic and beta-adrenergic receptors from ion channels by a guanine nucleotide analogue. Nature 317:538–540

Cerbai E, Kloeckner U, Isenberg G (1988) The $\alpha$ subunit of the GTP binding protein activates muscarinic potassium channels of the atrium. Science 240:1782–1784

Cerione RA, Staniszewski C, Caron MG, Lefkowitz RJ, Codina J, Birnbaumer L (1985) A role for $N_i$ in the hormonal stimulation of adenylate cyclase. Nature 318:293–295

Codina J, Yatani A, Grenet D, Brown AM, Birnbuamer L (1987a) The $\alpha$ subunit of the GTP-binding protein $G_k$ opens atrial potassium channels. Science 236:442–445

Codina J, Grenet D, Yatani A, Birnbaumer L, Brown AM (1987b) Hormonal regulation of pituitary $GH_3$ cell $K^+$ is mediated by its alpha subunit. FEBS Lett 216:104–106

Eccles RM, Libet B (1961) Origin and blockade of the synaptic responses of curarized sympathetic ganglia. J Physiol (Lond) 157:484–503

Giles W, Noble SJ (1976) Changes in membrane currents in bullfrog atrium produced by acetylcholine. J Physiol (Lond) 261:103–123

Graf R, Mattera R, Codina J, Evans T, Ho Y-K, Estes MK, Birnbaumer L (1989) Studies on the interaction of $\alpha$ subunits of G proteins with $\beta\gamma$ dimers. Eur J Biochem (in press)

Griffith WH III, Gallagher JP, Shinnick-Gallagher P (1981) Sucrose-gap recordings of nerve-evoked potentials in mammalian parasympathetic ganglia. Brain Res. 208:446–451

Hamill OP, Marty A, Neher E, Sakmann B, Sigworth FJ (1981) Improved patch-clamp techniques for high resolution current recording from cells and cell-free membrane patches. Pflugers Arch 391:85–100

Hartzell HC (1981) Mechanisms of slow postsynaptic potentials. Nature 291:539–544
Hartzell HC, Kuffler SW, Stickgold R, Yoshikami D (1977) Synaptic excitation and inhibition resulting from direct action of acetylcholine on two types of chemoreceptors on individual amphibian parasympathetic neurones. J Physiol (Lond) 271:817–846
Hill-Smith I, Purves RD (1978) Synaptic delay in the heart: an ionophoretic study. J Physiol (Lond) 270:31–54
Hsu WH, Xiang H, Rajan AS, Kunze DL, Boyd AE (1991) Somatostatin inhibits insulin secretion by a G-protein-mediated decrease in $Ca^{2+}$ entry through voltage-dependent $Ca^{2+}$ channels in the beta cell. J Biol Chem 266:837–843
Ito H, Sugimoto T, Kobayashi I, Takahashi K, Katada T, Ui M, Kurachi Y (1991) On the mechanism of basal and agonist-induced activation of the G-protein gated musarinic $K^+$ channel in atrial myocytes of guinea pig heart. J Gen Physiol 98:517–533
Katada T, Ui M (1981) Islet-activating protein; a modifier of receptor-mediated regulation of G rat islet adenylate cyclase. J Biol Chem 256:8310–8317
Kim D, Lewis DL, Graziadei L, Neer EJ, Bar-Sagi D, Clapham DE (1989) G protein $\beta\gamma$-subunits activate the cardiac muscarinic $K^+$ channel via phospholipase $A_2$. Nature 337:557–560
Kirsch G, Yatani A, Codina J, Birnbaumer L, Brown AM (1988) The alpha subunit of $G_k$ activates atrial $K^+$ channels of chick, rat and guinea pig. Am J Physiol 254 (Heart Circ Physiol 23): H1200-H1205
Kirsch G, Codina J, Birnbaumer L, Brown AM (1990) Coupling of ATP-sensitive $K^+$ channels to purinergic receptors by G-proteins in rat ventricular myocytes. Am J Physiol 259:H820–H826
Kobayashi I, Shibasaki H, Tohyama K, Kurachi Y, Itoh H, Ui M, Katada T (1990) Purification and characterization of five different $\alpha$ subunits of guanine nucleotide binding proteins in bovine brain membranes. Their physiological properties concerning the activities of adenylate cyclase and atrial muscarinic $K^+$ channels. Eur J Biochem 191:499–506
Kurachi Y, Ito H, Sugimoto T, Katada T, Ui M (1989a) Activation of atrial muscarinic $K^+$ channels by low concentrations of $\beta\gamma$ subunits of rat brain protein. Pflugers Arch 413:325–327
Kurachi Y, Ito H, Sugimoto T, Shimizu T, Miki I, Ui M (1989b) Arachidonic acid metabolites as intracellular modulators of the G protein-gated cardiac $K^+$ channel. Nature 337:555–557
Kurose H, Ui M (1883) Functional uncoupling of muscarinic receptors from adenylate Cyclase in rat cardiac membranes by the active component of islet-activating protein, pertussis toxin. J Cycl Nucl Protein Phosphoryl Res 9:305–318
Logothetis DE, Kurachi Y, Galper J, Neer EJ, Clapham DE (1987) The $\beta\gamma$ subunits of GTP-binding proteins activate the muscarinic $K^+$ channel in heart. Nature 325:321–326
Logothetis DE, Kim D, Northup JK, Neer EJ, Clapham DE (1988) Specificity of action of guanine nucleotide-binding regulatory protein subunits on the cardiac muscarinic $K^+$ channel. Proc Natl Acad Sci USA 85:5814–5818
Mattera R, Pitts BJR, Entman MS, Birnbaumer L (1985) Guanine nucleotide regulation of a mammalian myocardial receptor system. Evidence for homo- and heterotrophic cooperativity in ligand binding analyzed by computer assisted curve fitting. J Biol Chem 260:7410–7421
Mattera R, Yatani A, Kirsch GE, Graf R, Olate J, Codina J, Brown AM, Birnbaumer L (1988) Recombinant $\alpha_i$-3 subunit of G protein activates $G_k$-gated $K^+$ channels. J Biol Chem 264:465–471
Murad F, Chi Y-M, Rall TW, Sutherland EW (1962) Adenyl cyclase. III. The effect of catecholamine and choline esters on the formation of adenosine 3′, 5′-phosphate by preparations from cardiac muscle and liver. J Biol Chem 237:1233–1238

Nakajima Y, Nakajima S, Leonard RJ, Yamagucchi K (1986) Acetylcholine raises excitability by inhibiting the fast transient potassium current in cultured hippocampal neurons. Proc Natl Acad Sci USA 83:3022–3026

Nargeot J, Nerbonne JM, Engels J, Lester HA (1983) Time course of the increase in the myocardial slow inward current after a photochemically generated concentration jump of intracellular cAMP. Proc Natl Acad Sci USA 80:2395–2399

Newberry NR, Nicoll RA (1984) Direct hyperpolarizing action of baclofen on hippocampal pyramidal cells. Nature 308:450–452

Noma A (1983) ATP-regulated $K^+$ channles in cardiac muscle. Nature 305:147–148

Okabe K, Yatani A, Evans T, Ho Y-K, Codina J, Birnbaumer L, Brown AM (1990) $\beta\gamma$ Dimers of G proteins inhibit muscarinic $K^+$ channels in heart. J Biol Chem 265:12854–12858

Pfaffinger PJ, Martin JM, Hunter DD, Nathanason NM, Hille B (1985) GTP-binding proteins couple cardiac muscarinic receptors to a K channel. Nature 317:536–538

Ribalet B, Ciani S (1987) Regulation by cell metabolism and adenine nucleotides of a K channel in insulin-secreting B cells (RIN m5F). Proc Natl Acad Sci USA 84:1721–1725

Ribalet B, Ciani S, Eddlestone GT (1989) Modulation of ATP-sensitive K channels in RINm5F cells by phosphorylation and G proteins. Biophys J 55:587A.

Rosenberger LB, Yamamura HL, Roeske WR (1980) Cardiac muscarinic cholinergic receptor binding is regulated by $Na^+$ and guanyl nucleotides. J Biol Chem 255:820–823

Sasaki K, Sato M (1987) A single GTP-binding protein regulates $K^+$-channels coupled with dopamine, histamine and acetylcholine receptors. Nature 325:259–262

Soejima M, Noma A (1984) Mode of regulation of the ACh-sensitive K-channel by the muscarinic receptor in rabbit atrial cells. Pflugers Arch 400:424–431

Thalmann RH (1988) Evidence that guanosine triphosphate (GTP)-binding proteins control a synaptic response in brain: effect of pertussis toxin and GTP$\gamma$S on the late inhibitory postsynaptic potential of hippocampal $CA_3$ neurons. J Neurosci 8:4589–4602

Trautwein W, Dudel J (1958) Zum Mechanismus der Membranwirkung des Acetylcholin an der Herzmuskelfaser. Pflugers Arch 266:324–334

Trautwein W, Taniguchi J, Noma A (1982) The effect of intracellular cyclic nucleotides and calcium on the action potential and acetylcholine response of isolated cardiac cells. Pflugers Arch 392:307–314

VanDongen A, Codina J, Olate J, Mattera R, Joho R, Birnbaumer L, Brown AM (1988) Newly identified brain potassium channels gated by the guanine nucleotide binding (G) protein $G_o$. Science 242:1433–1437

Yatani A, Codina J, Brown AM, Birnbaumer L (1987a) Direct activation of mammalian atrial muscarinic K channels by a human erythrocyte pertussis toxin-sensitive G protein, $G_k$. Science 235:207–211

Yatani A, Codina J, Sekura RD, Birnbaumer L, Brown AM (1987b) Reconstitution of somatostatin and muscarinic receptor mediated stimulation of $K^+$ channels by isolated $G_k$ protein in clonal rat anterior pituitary cell membranes. Mol Endocrinol 1:283–289

Yatani A, Codina J, Imoto Y, Reeves JP, Birnbaumer L, Brown AM (1987c) A G protein directly regulates mammalian cardiac calcium channels. Science 238:1288–1292

Yatani A, Mattera R, Codina J, Graf R, Okabe K, Padrell E, Iyengar R, Brown AM, Birnbaumer L (1988) The G protein-gated atrial $K^+$ channel is stimulated by three distinct $G_i\alpha$-subunits. Nature 336:680–682

Yatani A, Okabe K, Codina J, Birnbaumer L, Brown AM (1990) Heart rate regulation by G proteins acting on the cardiac pacemaker channel. Science 249:1163–1166

CHAPTER 77

# ATP-Sensitive $K^+$ Channel: Properties, Occurrence, Role in Regulation of Insulin Secretion

U. PANTEN, C. SCHWANSTECHER, and M. SCHWANSTECHER

## A. Introduction

A $K^+$ channel inhibited by cytosolic ATP ($K_{ATP}$ channel) was first demonstrated in cardiac cells by NOMA (1983). This channel is also inhibited by sulfonylureas, is highly $K^+$ selective, and has a single channel conductance of 80 pS when measured in symmetrical high $K^+$ solutions (TRUBE and HESCHELER 1984; KAKEI and NOMA 1984; BELLES et al. 1987). Noma's discovery prompted the identification of a $K^+$ channel with similar properties in pancreatic $\beta$-cells (COOK and HALES 1984; ASHCROFT et al. 1984; STURGESS et al. 1985), skeletal muscle (SPRUCE et al. 1985, 1987; WOLL et al. 1989), smooth muscle (NELSON et al. 1990; KAJIOKA et al. 1991) and nerve cells (JONAS et al. 1991; POLITI and ROGAWSKI 1991; OHNO-SHOSAKU and YAMAMOTO 1992). This chapter confines itself to this $K^+$ channel, and we use the term $K_{ATP}$ channel with this restriction. Other types of ATP-sensitive $K^+$ channels have been found in various cells (ASHCROFT and ASHCROFT 1990). Several reviews concerning the $K_{ATP}$ channel have been published during the recent years (ASHCROFT 1988; ASHCROFT and RORSMAN 1991; ASHFORD 1990; ASHCROFT and ASHCROFT 1990; RORSMAN and TRUBE 1990; DUNNE and PETERSEN 1991; NICHOLS and LEDERER 1991; DAVIES et al. 1991; NOMA and TAKANO 1991; PANTEN et al. 1992; ASHCROFT and ASHCROFT 1992). Where appropriate, these reviews are cited to limit the primary references.

## B. Biophysical Properties

The $K_{ATP}$ channel is highly selective for $K^+$, the permeability of external $K^+$/external $Na^+$ ranging from 50:1 to 65:1 in skeletal muscle (SPRUCE et al. 1987; PARENT and CORONADO 1989) and 100:1 in hippocampal neurones (POLITI and ROGAWSKI 1991) to 145:1 in the pancreatic $\beta$-cell (ASHCROFT and RORSMAN 1991). External $Rb^+$ also permeates the channel but less well than $K^+$ (ASHCROFT and RORSMAN 1991; SPRUCE et al. 1987). In symmetrical high $K^+$ solutions (100–160 m$M$) the single-channel slope conductances for inward current flow (14°C–24°C) range from 44 to 80 pS (ASHCROFT and ASHCROFT 1990; WOLL et al. 1989; KAJIOKA et al. 1991; JONAS et al. 1991; OHNO-SHOSAKU and YAMAMOTO 1992). With normal ionic gradients ($[K]_o$ =

4–6 m$M$; $[K]_i$ = 120–155 m$M$) current flow is outward at potentials positive to the $K^+$ equilibrium potential and single-channel conductances of 15–35 pS (room temperature) have been obtained in inside-out patches in the presence of 0–1 m$M$ $Mg^{2+}$ (ASHCROFT and RORSMAN 1991; NICHOLS and LEDERER 1991; DAVIES et al. 1991; NELSON et al. 1990; KAJIOKA et al. 1991; OHNO-SHOSAKU and YAMAMOTO 1992). The characteristic kinetics of the $K_{ATP}$ channel currents and the voltage dependence of the open and closed time distributions have been discussed in detail in a number of recent reviews (ASHCROFT 1988; ASHCROFT and RORSMAN 1991; ASHCROFT and ASHCROFT 1990; DAVIES et al. 1991). In summary, the open state probability of the $K_{ATP}$ channel is essentially voltage independent and this $K^+$ channel cannot be classified as yet.

## C. Regulation of the $K_{ATP}$ Channel

### I. Inhibition by Intracellular Nucleotides

ATP induces closure of the $K_{ATP}$ channel when applied to the cytoplasmic side of the plasma membrane (Fig. 1). This effect does not require hydrolysis of the nucleotide since it occurs not only in the presence but also

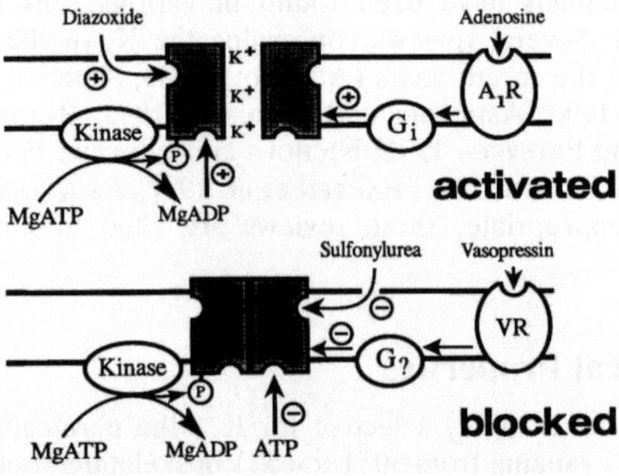


**Fig. 1.** Simplified model for regulation of the $K_{ATP}$ channel activity by ligands and G-proteins. Normal regulation requires phosphorylation (*P*) of the $K_{ATP}$ channel and/or separate receptor proteins by a tightly associated protein kinase. Channel-activation (+) is induced by interaction of adenosine with $A_1$ receptors ($A_1R$) or by MgADP. The response to MgADP is enhanced by diazoxide, which is ineffective in the absence of MgADP. Channel inhibition (−) is induced by interaction of vasopressin with its receptor (*VR*) or by free ATP. The inhibitory effect of free ATP on MgADP-induced channel activity is strongly enhanced by sulfonylureas. Sulfonylureas are only weakly inhibitory unless both free ATP and MgADP are present (SCHWANSTECHER et al. 1992a)

in the absence of internal $Mg^{2+}$ ions and is also observed for some nonhydrolyzable ATP analogues (ASHCROFT and ASHCROFT 1990). In inside-out membrane patches excised from cardiac cells, 25–75 $\mu M$ of ATP caused half-maximal inhibition in the absence of $Mg^{2+}$ (FINDLAY 1988; NICHOLS and LEDERER 1990). The corresponding half-maximally inhibitory concentrations ($IC_{50}$) were 54–135 $\mu M$ for skeletal muscle (SPRUCE et al. 1987; VIVAUDOU et al. 1991), 29 $\mu M$ for vascular smooth muscle (KAJIOKA et al. 1991), 35 $\mu M$ for vertebrate axon (JONAS et al. 1991) and 4–15 $\mu M$ for pancreatic $\beta$-cell (ASHCROFT and KAKEI 1989; SCHWANSTECHER et al., to be published). Apart from ATP a number of nucleotides are able to inhibit the $K_{ATP}$ channel. In inside-out patches of $\beta$-cells exposed to $Mg^{2+}$-free cytosol-like solution, $IC_{50}$ values of 17, 49, 60, 62, 93, 370, 830, 3000, over 3000 and over 5000 $\mu M$ have been determined for ATP$\gamma$S, ADP, AMP-PNP, ADP$\beta$S, dATP, UTP, dADP, GTP, GDP and UDP, respectively (PANTEN et al. 1990; SCHWANSTECHER et al. 1992a; SCHWANSTECHER et al., to be published). For cardiac muscle the order of inhibitory effectiveness in $Mg^{2+}$-free solution is ATP > AMP-PNP > ADP > CTP > GDP = AMP = ITP (LEDERER and NICHOLS 1989), and a similar order has also been observed for skeletal muscle (SPRUCE et al. 1987).

When a nucleotide is tested using solutions supplemented with $Mg^{2+}$, the inter-pretation is complicated by the simultaneous presence of free and Mg-bound nucleotide. Thus in the case of the ATP analogues ATP$\gamma$S, AMP-PCP, and AMP-PNP, the Mg complexes are more inhibitory than the free nucleotides (FINDLAY 1988; ASHCROFT and KAKEI 1989; SCHWANSTECHER et al. 1992a). On the other hand, in $\beta$-cells MgATP seems to be less effective than free ATP (ASHCROFT and ASHCROFT 1990; DUNNE and PETERSEN 1991). However, the presence of $Mg^{2+}$ causes enzymatic degradation of ATP and thereby leads to underestimation of the potency of ATP if significant amounts of cytoplasm are located between the ATP-containing solution and the intracellular face of the plasma membrane (NOMA and SHIBASAKI 1985; BELLES et al. 1987; NICHOLS and LEDERER 1990; SCHWANSTECHER et al. 1992b). Underestimation of potency is enhanced by the simultaneous formation of MgADP that activates the $K_{ATP}$ channel via a specific stimulatory receptor (see below).

## II. Activation by Intracellular Nucleoside Diphosphates

In insulin-secreting cells ADP, ADP$\beta$S, GDP and GDP$\beta$S reduce inhibition of $K_{ATP}$ channel activity induced by ATP in the presence of $Mg^{2+}$ (ASHCROFT and RORSMAN 1991). It has been suggested that these nucleoside diphosphates occupy a $Mg^{2+}$-dependent stimulatory receptor not identical with the aforementioned receptor for inhibitory nucleotides (Fig. 1; FINDLAY 1987). This interpretation is supported by recent observations (BOKVIST et al. 1991; SCHWANSTECHER et al. 1992a; HOPKINS et al. 1992). There is also evidence that a similar stimulatory receptor for Mg complexes of nucleoside diphosphates exists in cardiac and skeletal muscle cells (FINDLAY 1988; LEDERER and NICHOLS 1989; TUNG and KURACHI 1991; ALLARD and LAZDUNSKI 1992). It has been proposed that this type of activation is specific to the dephosphorylated state of the cardiac $K_{ATP}$ channel (TUNG and KURACHI 1991). However, protein phosphorylation has not been ruled out in this study. Furthermore, evidence for the stimulatory receptor in insulin-secreting cells has been obtained by experiments in which MgATP was not limiting as substrate for protein kinases (SCHWANSTECHER et al. 1992a).

## III. Activation by Intracellular MgATP

After excision of inside-out patches from membranes of insulin-secreting cells the activity of the $K_{ATP}$ channels rapidly declines (channel run-down) if MgATP is not present at the cytoplasmic face of the membrane (ASHCROFT and RORSMAN 1991). The run-down can be prevented or considerably slowed by application of MgATP at the cytoplasmic face, but not by free ATP or the Mg complexes of nonhydrolyzable ATP analogues. Therefore, protein dephosphorylation by endogenous phosphatases might be involved in run-down and MgATP might act by serving as substrate for protein phosphorylation required to maintain channel function (Fig. 1). In an insulin-secreting tumor cell line (RINm5F) the catalytic subunit of protein kinase A activates the $K_{ATP}$ channels (RIBALET et al. 1989). However, for normal $\beta$-cells the identity of the protein kinases mediating channel activation by MgATP is unknown. In inside-out patches from cardiac cells channel run-down is also rapid and can be prevented or reversed by MgATP (ASHCROFT and ASHCROFT 1990; NICHOLS and LEDERER 1991). Run-down is much less pronounced in skeletal muscle and smooth muscle.

## IV. Activation by G-Proteins

Clear evidence for G-protein activation of the $K_{ATP}$ channel has been presented by KIRSCH et al. (1990) in experiments with neonatal ventricular myocytes. Application of GTP$\gamma$S to the cytoplasmic side of excised membrane patches stimulated the $K_{ATP}$ channel in a $Mg^{2+}$-dependent manner. A similar effect was induced by the combined application of cytosolic GTP and extracellular adenosine or $N^6$-cyclohexyladenosine (a stable $A_1$ receptor agonist). The latter agents were ineffective in the absence of GTP. Channel activity was also enhanced by exposure of the inner side of the membrane to G-protein $\alpha_i$ subunits, preactivated with GTP$\gamma$S. The $\alpha_o$ subunit and $G_s$ were ineffective. These findings are strong evidence that $G_i$ induces activation of the $K_{ATP}$ channel via a membrane-delimited pathway and that the responses to $A_1$ receptor agonists are mediated by $G_i$ (Fig. 1).

$Mg^{2+}$-dependent activation of the $K_{ATP}$ channel by intracellular GTP$\gamma$S has also been found in T-tubule membranes from skeletal muscle (PARENT and CORONADO 1989) and in insulin-secreting RINm5F-cells (FINDLAY 1987).

In excised membrane patches from RINm5F-cells the combined application of intracellular GTP and extracellular galanin enhanced $K_{ATP}$ channel activity (DUNNE et al. 1989). Hence, a membrane-delimited pathway appears to couple galanin receptor stimulation to $K_{ATP}$ channel activation observed in intact RINm5F cells and rat pancreatic $\beta$-cells (DE WEILLE et al. 1988; DUNNE et al. 1989; LINDSKOG and AHREN 1991). However, unequivocal evidence is lacking that $K_{ATP}$ channels of mouse $\beta$-cells are influenced by galanin though galanin receptors are present in the cell membrane (ASHCROFT and RORSMAN 1991; LINDSKOG and AHREN 1991). In RINm5F-cells $K_{ATP}$ channels were also activated by somatostatin (ASHCROFT and ASHCROFT 1990).

## V. Inhibition by G-Proteins

Application of arginine-vasopressin to the extracellular side of outside-out patches excised from RINm5F cells caused closure of $K_{ATP}$ channels (MARTIN et al. 1989). In these experiments a high MgATP concentration was present at the cytoplasmic side of the patch and probably kept the GTP/GDP ratio high via transphosphorylation reactions. A soluble cytosolic messenger did not appear to be involved since arginine-vasopressin was ineffective when added to the bath solution in cell-attached patch experiments (Fig. 1). However, arginine-vasopressin did not close $K_{ATP}$ channels in mouse $\beta$-cells (GAO et al. 1990). Extracellular ATP inhibited the $K_{ATP}$ channel of insulin-secreting cells, an effect that may have been mediated by G-protein coupled to purinergic receptors existing in these cells (DUNNE and PETERSEN 1991).

## VI. Inhibition by Drugs

Sulfonylureas and their analogues stimulate insulin secretion by specific inhibition of the $K_{ATP}$ channel activity (STURGESS et al. 1985; PANTEN et al. 1992; ASHCROFT and ASHCROFT 1992). The receptor for these drugs is located in the $\beta$-cell plasma membrane and can be reached from both sides of the membrane and by lateral diffusion in the lipid phase of the membrane (TRUBE et al. 1986). There is compelling evidence that the antidiabetic effect of sulfonylureas results from inhibition of the $K_{ATP}$ channel in the $\beta$-cell (PANTEN et al. 1992). Whereas channel inhibition in cardiac muscle and adenohypophysis cells was induced by free sulfonylurea concentrations occurring in the plasma from treated patients (FINDLAY 1992; DE WEILLE et al. 1992), much higher concentrations were required for channel inhibition in skeletal muscle, smooth muscle and vertebrate axon (ASHCROFT and ASHCROFT 1990; DAVIES et al. 1991; JONAS et al. 1991). It is unclear whether the low potency of sulfonylureas reported for some non-$\beta$-cells reflects a particular feature of the sulfonylurea receptor in these cells or is due to experimental design.

A number of other drugs inhibit the $K_{ATP}$ channel; some of them have also been shown to stimulate insulin secretion from isolated pancreatic islets. Except for quinine, these drugs are only effective at concentrations toxic in humans: quinine (HENQUIN 1982; BOKVIST et al. 1990); sparteine (PAOLISSO et al. 1985; ASHCROFT et al. 1991); amantadine (GARRINO and HENQUIN 1987; ASHCROFT et al. 1991), chlorpromazine and some related phenothiazines (MÜLLER et al. 1991); thiopental and some other barbiturates (KOZLOWSKI and ASHFORD 1991); phentolamine and some related imidazoline derivatives (HENQUIN et al. 1982; CHAN et al. 1991); linogliride (RONNER et al. 1991); ligustrazine (PEERS et al. 1990); and M-56324 (a pyridine-carboxylic acid derivative; HOPKINS et al. 1990). All drugs except linogliride (only whole-cell currents were measured) have been shown to reduce the activity of the $K_{ATP}$ channel. Quinine, chlorpromazine, and thiopental do not inhibit the $K_{ATP}$ channel of insulin-secreting cells by interaction with the binding site for sulfonylureas (SCHWANSTECHER et al. 1992d).

## VII. Activation by Drugs

Diazoxide inhibits insulin secretion by increasing the activity of the $K_{ATP}$ channel (TRUBE et al. 1986; ASHCROFT and RORSMAN 1991; DUNNE and PETERSEN 1991). The $K_{ATP}$ channels in $\beta$-cells and cardiac myocytes can also be activated by high concentrations of pinacidil, cromakalim, RP49356 (thioformamide derivative) and nicroandil (ASHCROFT and ASHCROFT 1990; DUNNE and PETERSEN 1991; SHEN et al. 1991). In vascular smooth muscle, these novel drugs produce $K^+$ channel openings at lower concentrations than in $\beta$-cells and cardiac cells (EDWARDS and WESTON 1990). However, there is no unequivocal evidence that the $K_{ATP}$ channel is involved in the potent effects of $K^+$ channel openers in smooth muscle.

## VIII. Characteristics of the Sulfonylurea Receptor

High-affinity binding sites for sulfonylureas have been found in membranes from insulin-secreting cells, cardiac muscle, smooth muscle, and brain (KAUBISCH et al. 1982; GEISEN et al. 1985; ASHCROFT and ASHCROFT 1992). Correlations between the $K^+$ channel-blocking potencies of sulfonylureas and their affinities for binding to membranes from insulin-secreting cells suggest that the high-affinity binding site of a 140-kDa membrane protein might represent the sulfonylurea receptor (PANTEN et al. 1992; ASHCROFT and ASHCROFT 1992). The Mg complex of ATP or some other phosphate and thiophosphate group donating nucleotides decreases sulfonylurea-binding to membranes from pancreatic islets (SCHWANSTECHER et al. 1991; 1992e). A similar effect has also been observed in HIT T15-cells (an insulin-secreting cell line) and rat cerebral cortex (GOPALAKRISHNAN et al. 1991; NIKI and ASHCROFT 1991; SCHWANSTECHER et al. 1992c, 1992d, 1992f; BERNARDI et al.

1992). MgATP-induced inhibition of sulfonylurea binding is accompanied by enhanced displacement of sulfonylurea from its receptor by diazoxide or pinacidil (SCHWANSTECHER et al. 1991, 1992d; NIKI and ASHCROFT 1991). As the effects of MgATP are maintained after solubilization of microsomal membranes with nonionic or zwitterionic detergents, protein phosphorylation by a tightly associated protein kinase might modulate the binding properties of the sulfonylurea receptor (Fig. 1) (SCHWANSTECHER et al. 1992c).

The sulfonylurea receptor seems to be a member of the protein assembly organizing the $K_{ATP}$ channel (AGUILAR-BRYAN et al. 1992). Binding studies and patch-clamp experiments suggest that the $K_{ATP}$ channel is controlled via four separate binding sites for ATP (and related inhibitory nucleotides), MgADP (and related activating nucleotides), diazoxide (and pinacidil) and sulfonylureas (Fig. 1) (SCHWANSTECHER et al. 1992a,b,d,e). Phosphorylation of the $K_{ATP}$ channel and/or regulatory proteins appears to be necessary for diazoxide to be effective (KOZLOWSKI et al. 1989; DUNNE 1989). When tested as individual ligands under experimental conditions favoring protein phosphorylation, diazoxide is ineffective and tolbutamide is only weakly inhibitory (SCHWANSTECHER et al. 1992a,b). However, diazoxide sensitizes MgADP-induced channel activation whereas sulfonylureas sensitize the inhibitory effect of ATP on MgADP-induced channel activation (Fig. 1).

## D. Role of the $K_{ATP}$-Channel in Regulation of Insulin Secretion

In muscle and nerve cells, but not in $\beta$-cells, the levels of creatine kinase and phospho-creatine are high (PANTEN et al. 1986; GOSH et al. 1991; WALLIMANN et al. 1992). Thus, buffering of high ATP and low ADP concentrations in muscle and nerve cells appears to keep the $K_{ATP}$ channels silent, unless channel opening by transmitters, hormones, drugs or exhaustion of energy metabolism takes place (DAVIES et al. 1991; NICHOLS and LEDERER 1991). In the resting $\beta$-cell (presence of 3 m*M* glucose, 37°C) 6%–8% of all $K_{ATP}$ channels are still open (PANTEN et al. 1990; SCHWANSTECHER et al. 1992b) although the cytosolic ATP concentration lies in the millimolar range (GOSH et al. 1991). This probably reflects lower ATP concentration in the immediate vicinity of the $K_{ATP}$ channels (PANTEN and LENZEN 1988) and channel activation by MgADP and MgGDP (see Sect. CII).

Initiation of insulin release requires increase in the cytosolic $Ca^{2+}$ concentration of the $\beta$-cell (Fig. 2). Enhancement of $Ca^{2+}$ influx across the plasma membrane is induced by depolarization of the $\beta$-cell to the threshold potential at which voltage-dependent $Ca^{2+}$ channels are activated (ASHCROFT and RORSMAN 1991). The threshold potential is reached when

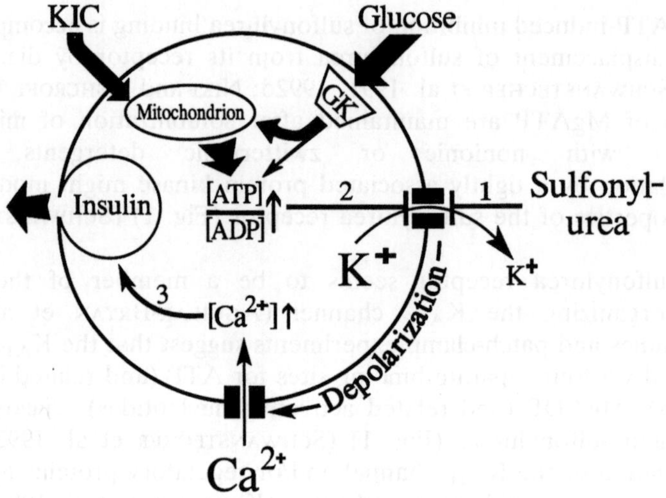


**Fig. 2.** Initiation of insulin release from the $\beta$-cell. Closure of $K_{ATP}$ channels by sulfonylureas (*1*) or by an increase in the ratio of the cytosolic concentrations of ATP and ADP (*[ATP]/[ADP]*) (*2*) leads to membrane depolarization, opening of voltage-dependent $Ca^{2+}$ channel and enhanced $Ca^{2+}$ influx. The resultant elevation of the cytosolic $Ca^{2+}$ concentration initiates insulin release (*3*). The [ATP]/[ADP] ratio rises when the mitochondrial energy metabolism is sufficiently activated by nutrient secretagogues (e.g., glucose or $\alpha$-ketoisocaproic acid, *KIC*; Panten and Lenzen 1988). KIC increases ATP production only by its intramitochondrial catabolism whereas glycolysis contributes to glucose-induced ATP production. Raising the extracellular glucose concentration enhances phosphorylation of glucose by glucokinase (*GK*), the rate-limiting enzyme for glucose utilization in the $\beta$-cell

less than 3% of the $K_{ATP}$ channels are open (COOK et al. 1988; PANTEN et al. 1990; SCHWANSTECHER et al. 1992b). Glucose and other insulin-releasing fuels produce channel closure and subsequent insulin release by enhancing the substrate pressure at the respiratory chain in the $\beta$-cell mitochondria (PANTEN and LENZEN 1988). The link between mitochondrial energy metabolism and $K_{ATP}$ channel activity is most probably brought about by the submembrane concentrations of free ATP, MgADP, and MgGDP. In glucose-stimulated $\beta$-cells the expected association between total cellular content of ATP and ADP and rate of insulin secretion has repeatedly been missing (GOSH et al. 1991). These observations suggest that the total cellular content of ATP and ADP is not representative of the submembrane concentrations (PANTEN and LENZEN 1988; ASHCROFT and ASHCROFT 1990). In $\beta$-cells exposed to maximally effective glucose concentrations (30–40 m*M*) sulfonylureas are still effective (PANTEN et al. 1986; HENQUIN 1988). This finding suggests that not all $K_{ATP}$ channels are closed by high glucose concentrations due to channel-activation induced by the rise in cytosolic $Ca^{2+}$ (HENQUIN 1990).

Insulin secretion is impaired in type II diabetic patients. The elucidation of the complex regulation of the $K_{ATP}$ channel may reveal some of the disturbances of the diseased $\beta$-cells.

## References


Aguilar-Bryan L, Nichols CG, Rajan AS, Parker C, Bryan J (1992) Co-expression of sulfonylurea receptors and $K_{ATP}$ channels in hamster insulinoma tumor (HIT) cells. J Biol Chem 267:14934–14940

Allard B, Lazdunski M (1992) Nucleotide diphosphates activate the ATP-sensitive potassium channel in mouse skeletal muscle. Pflugers Arch 422:185–192

Ashcroft FM (1988) Adenosine 5′-triphosphate-sensitive potassium channels. Ann Rev Neurosci 11:97–118

Ashcroft FM, Kakei M (1989) ATP-sensitive $K^+$ channels in rat pancreatic $\beta$-cells: modulation by ATP and $Mg^{2+}$ions. J Physiol (Lond) 416:349–367

Ashcroft FM, Rorsman P (1991) Electrophysiology of the pancreatic $\beta$-cell. Prog Biophys molec Biol 54:87–143

Ashcroft FM, Harrison DE, Ashcroft SJH (1984) Glucose induces closure of single potassium channels in isolated rat pancreatic $\beta$-cells. Nature 312:446–448

Ashcroft FM, Kerr AJ, Gibson JS, Williams BA (1991) Amantadine and sparteine inhibit ATP-regulated K-currents in the insulin-secreting $\beta$-cell line HIT-T15. Br J Pharmacol 104:597–584

Ashcroft SJH, Ashcroft FM (1990) Properties and functions of ATP-sensitive K-channels. Cellular Signalling 2:197–214

Ashcroft SJH, Ashcroft FM (1992) The sulfonylurea receptor. Biochim Biophys Acta 1175:45–59

Ashford MLJ (1990) Potassium channel and modulation of insulin secretion. In: Cook NS (ed) Potassium channels: structure, classification, function and therapeutic potential, Ellis Horwood, Chichester, pp 300–325

Belles B, Hescheler J, Trube G (1987) Changes of membrane currents in cardiac cells induced by long whole cell recordings and tolbutamide. Pflugers Arch 409:582–588

Bernardi H, Fosset M, Lazdunski M (1992) ATP/ADP binding sites are present in the sulfonylurea binding protein associated with brain ATP-sensitive $K^+$ channels. Biochemistry 31:6328–6332

Bokvist K, Rorsman P, Smith PA (1990) Block of ATP-regulated and $Ca^{2+}$-activated $K^+$-channels in mouse pancreatic $\beta$-cells by external tetraethylammonium and quinine. J Physiol (Lond) 423:327–343

Bokvist K, Ämmälä C, Ashcroft FM, Berggren P-O, Larsson O, Rorsman P (1991) Separate processes mediate nucleotide-induced inhibition and stimulation of the ATP-regulated $K^+$-channels in mouse pancreatic $\beta$-cells. Proc R Soc Lond [B] 243:139–144

Chan SLF, Dunne MJ, Stillings M, Morgan NG (1991) The $\alpha_2$-adrenoceptor antagonist efaroxan modulates $K^+{}_{ATP}$ channels in insulin-secreting cells. Eur J Pharmacol 204:41–48

Cook DL, Hales CN (1984) Intracellular ATP directly blocks $K^+$-channels in pancreatic B-cells. Nature 311:271–273

Cook DL, Satin LB, Ashford MLJ, Hales CN (1988) ATP-sensitive $K^+$ channels in pancreatic $\beta$-cells. Spare channel hypothesis. Diabetes 37:495–498

Davies NW, Standen NB, Stanfield PR (1991) ATP-dependent potassium channels of muscle cells: their properties, regulation, and possible functions. J Bioenerg Biomembr 23:509–535

De Weille JR, Schmid-Antomarchi H, Fosset M, Lazdunski M (1988) ATP-sensitive $K^+$ channels that are blocked by hypoglycemia-inducing sulfonylureas in insulin-secreting cells are activated by galanin, a hyperglycemia-inducing hormone. Proc Natl Acad Sci USA 85:1312–1316

De Weille JR, Fosset M, Epelbaum J, Lazdunski M (1992) Effectors of ATP-sensitive $K^+$ channels inhibit the regulatory effects of somatostatin and GH-releasing factor on growth hormone secretion. Biochem Biophys Res Commun 187:1007–1014

Dunne MJ (1989) Protein phosphorylation is required for diazoxide to open ATP-sensitive potassium channels in insulin (RINm5F) secreting cells. FEBS Lett 250:262–266

Dunne MJ, Petersen OH (1991) Potassium selective ion channels in insulin-secreting cells: physiology, pharmacology and their role in stimulus-secretion coupling. Biochim Biophys Acta 1071:67–82

Dunne MJ, Bullet MJ, Goudong L, Wollheim CB, Petersen OH (1989) Galanin activates nucleotide-dependent $K^+$ channels in insulin-secreting cells via a pertussis toxin-sensitive G-protein. EMBO J 8:413–420

Edwards G, Weston AH (1990) Potassium channel openers and vascular smooth muscle relaxation. Pharmc Ther 48:237–258

Findlay I (1987) The effects of magnesium upon adenosine triphosphate-sensitive potassium channels in a rat insulin-secreting cell line. J Physiol (Lond) 391:611–629

Findlay I (1988) $ATP^{4-}$ and ATP Mg inhibit the ATP-sensitive $K^+$ channel of rat ventricular myocytes. Pflugers Arch 412:37–41

Findlay I (1992) Inhibition of ATP-sensitive $K^+$ channels in cardiac muscle by the sulphonylurea drug glibenclamide. J Pharmacol Exp Ther 261:540–545

Gao Z-X, Drews G, Nenquin M, Plant TD, Henquin JC (1990) Mechanisms of the stimulation of insulin release by arginine-vasopressin in normal mouse islets. J Biol Chem 265:15724–15730

Garrino MG, Henquin JC (1987) Adamantance derivatives: a new class of insulin secretagogues. Br J Pharmacol 90:583–591

Geisen K, Hitzel V, Ökomonopoulos R, Pünter J, Weyer R, Summ HD (1985) Inhibition of $^3$H-glibenclamide binding to sulfonylurea receptors by oral antidiabetics. Arzneimittelforsch 35:707–712

Gopalakrishnan M, Johnson DE, Janis RA, Triggle DJ (1991) Characterization of binding of the ATP-sensitive potassium channel ligand, [$^3$H]glyburide, to neuronal and muscle preparations. J Pharmacol Exp Ther 257:1162–1171

Gosh A, Ronner P, Cheong E, Khalid P, Matschinsky FM (1991) The role of ATP and free ADP in metabolic coupling during fuel-stimulated insulin release from islet $\beta$-cells in the isolated perfused rat pancreas. J Biol Chem 266:22887–22892

Henquin JC (1982) Quinine and the stimulus secretion coupling in pancreatic B-cells: glucose-like effects on potassium permeability and insulin release. Endocrinology 110:1325–1332

Henquin JC (1988) ATP-sensitive $K^+$ channels may control glucose-induced electrical activity in pancreatic B-cells. Biochem Biophys Res Comm 156, 2:769–775

Henquin JC (1990) Glucose-induced electrical activity in $\beta$-cells. Feedback control of ATP-sensitive $K^+$ channels by $Ca^{2+}$. Diabetes 39:1457–1460

Henquin JC, Charles S, Nenquin M, Mathot F, Tamagawa T (1982) Diazoxide and D600 inhibition of insulin release: distinct mechanisms explain the specifity for different stimuli. Diabetes 31:776–783

Hopkins WF, Fatherazi S, Cook DL (1990) The oral hypoglycemic agent, U-56324, inhibits the activity of ATP-sensitive potassium channels in cell-free membrane patches from cultured mouse pancreatic B-cells. FEBS Lett 277:101–104

Hopkins WF, Fatherazi S, Peter-Riesch B, Corkey BE, Cook DL (1992) Two sites for adenine-nucleotide regulation of ATP-sensitive potassium channels in mouse pancreatic $\beta$-cells and HIT cells. J Membrane Biol 129:287–295

Jonas P, Koh D-S, Kampe K, Hermsteiner M, Vogel W (1991) ATP-sensitive and Ca-activated K channels in vertebrate axons: novel links between metabolism and excitability. Pflugers Arch 418:68–73

Kajioka S, Kitamura K, Kuriyama H (1991) Guanosine diphosphate activates an adenosine 5′-triphosphate-sensitive $K^+$ channel in the rabbit portal vein. J Physiol (Lond) 444:397–418

Kakei M, Noma A (1984) Adenosine 5′-triphosphate-sensitive single potassium channel in atrioventricular node cell of the rabbit heart. J Physiol (Lond) 352:265–284

Kaubisch N, Hammer R, Wollheim C, Renold AE, Offord RE (1982) Specific receptors for sulfonylureas in brain and in a $\beta$-cell tumor of the rat. Biochem Pharmacol 31:1171–1174

Kirsch GE, Codina J, Birnbaumer L, Brown AM (1990) Coupling of ATP-sensitive $K^+$ channels to $A_1$ receptors by G proteins in rat ventricular myocytes. Am J Physiol 259:H820–H826

Kozlowski RZ, Ashford MLJ (1991) Barbiturates inhibit ATP-$K^+$ channels and voltage-activated currents in CRI-G1 insulin-secreting cells. Br J Pharmacol 103:2021–2029

Kozlowski RZ, Hales CN, Ashford MLJ (1989) Dual effects of diazoxide on ATP-$K^+$ currents recorded from an insulin-secreting cell line. Br J Pharmacol 97:1039–1050

Lederer WJ, Nichols CG (1989) Nucleotide modulation of the activity of rat heart ATP-sensitive $K^+$ channels in isolated membrane patches. J Physiol (Lond) 419:193–213

Lindskog S, Ahren B (1991) Studies on the mechanism by which galanin inhibits insulin secretion in islets. Eur J Pharmacol 205:21–27

Martin SC, Yule DI, Dunne MJ, Gallacher DV, Petersen OH (1989) Vasopressin directly closes ATP-sensitive potassium channels evoking membrane depolarization and an increase in the free intracellular $Ca^{2+}$ concentration in insulin-secreting cells. EMBO J 8:3595–3599

Müller M, de Weille JR, Lazdunski M (1991) Chlorpromazine and related phenothiazines inhibit the ATP-sensitive $K^+$ channel. Eur J Pharmacol 198:101–104

Nelson MT, Huang Y, Brayden JE, Hescheler J, Standen NB (1990) Arterial dilations in response to calcitonin gene-related peptide involve activation of $K^+$ channels. Nature 344:770–773

Nichols CG, Lederer WJ (1990) The regulation of ATP-sensitive $K^+$ channel activity in intact and permeabilized rat ventricular myocytes. J Physiol (Lond) 423:91–110

Nichols CG, Lederer WJ (1991) Adenosine triphosphate-sensitive potassium channels in the cardiovascular system. Am J Physiol (Lond) 261:H1675–H1686

Niki I, Ashcroft SJH (1991) Possible involvement of protein phosphorylation in the regulation of the sulfonylurea receptor of a pancreatic $\beta$-cell line, HIT T15. Biochim Biophys Acta 1133:95–101

Noma A (1983) ATP-regulated $K^+$ channels in cardiac muscle. Nature 305:147–148

Noma A, Shibasaki T (1985) Membrane current through adenosine-triphosphate-regulated potassium channels in guinea pig ventricular cells. J Physiol (Lond) 363:463–480

Noma A, Takano M (1991) The ATP-sensitive $K^+$ channel. Jap J Physiol 41:177–187

Ohno-Shosaku T, Yamamoto C (1992) Identification of an ATP-sensitive $K^+$ channel in rat cultured cortical neurons. Pflugers Arch 422:260–266

Panten U, Lenzen S (1988) Alterations in energy metabolism of secretory cells In: Akkerman JWN (ed) Energetics of secretion responses, vol 2. CRC Press, Boca Raton, pp 109–123

Panten U, Zünkler BJ, Scheit S, Kirchhoff K, Lenzen S (1986) Regulation of energy metabolism in pancreatic islets by glucose and tolbutamide. Diabetologia 29:648–654
Panten U, Heipel C, Rosenberger F, Scheffer K, Zünkler BJ, Schwanstecher C (1990) Tolbutamide-sensitivity of the adenosine 5′-triphosphate-dependent $K^+$ channel in mouse pancreatic B-cells. Naunyn-Schmiedeberg Arch Pharmacol 342:566–574
Panten U, Schwanstecher M, Schwanstecher C (1992) Pancreatic and extrapancreatic sulfonylurea receptors. Horm Metab Res 24:549–554
Paolisso G, Nenquin M, Schmeer W, Mathot F, Meissner HP, Henquin JC (1985) Sparteine increases insulin release by decreasing the $K^+$ permeability of the B-cell membrane. Biochem Pharmacol 34:2355–2361
Parent L, Coronado R (1989) Reconstitution of the ATP-sensitive potassium channel of skeletal muscle. J Gen Physiol 94:445–465
Peers C, Smith PA, Nye PCG (1990) Ligustrazine selectively blocks ATP-sensitive $K^+$ channels in mouse pancreatic $\beta$-cells. FEBS Lett 261:5–7
Politi DMT, Rogawski MA (1991) Glyburide-sensitive $K^+$ channels in cultured rat hippocampal neurons: Activation by cromakalim and energy-depleting conditions. Mol Pharmacol 40:308–315
Ribalet B, Ciani S, Ribalet GT (1989) ATP mediates both activation and inhibition of K(ATP) channel activity via cAMP-dependent protein kinase in insulin-secreting cell-lines. J Gen Physiol 94:693–717
Ronner P, Higgins T, Kimmich GA (1991) Inhibition of ATP-sensitive $K^+$ channels in pancreatic $\beta$-cells by nonsulfonylurea drug linogliride. Diabetes 40:885–892
Rorsman P, Trube G (1990) Biophysics and physiology of ATP-regulated $K^+$-channels ($K_{ATP}$). In: Cook NS (ed) Potassium channels: structure, classification and therapeutic potential, Ellis Horwood, Chichester, pp 96–116
Schwanstecher C, Dickel C, Panten U (1992a) Cytosolic nucleotides enhance the tolbutamide sensitivity of the ATP-dependent $K^+$ channel in mouse pancreatic B cells by their combined actions at inhibitory and stimulatory receptors. Mol Pharmacol 41:480–486
Schwanstecher C, Dickel C, Ebers I, Lins S, Zünkler BJ, Panten U (1992b) Diazoxide-sensitivity of the adenosine 5′-triphosphate dependent $K^+$ channel in mouse pancreatic $\beta$-cells. Br J Pharmacol 107:87–94
Schwanstecher M, Löser S, Rietze I, Panten U (1991) Phosphate and thiophosphate group donating adenine and guanine nucleotides inhibit glibenclamide binding to membranes from pancreatic islets. Naunyn-Schmiedeberg Arch Pharmacol 343:83–89
Schwanstecher M, Behrends S, Brandt C, Panten U (1992c) The binding properties of the solubilized sulfonylurea receptor from a pancreatic B-cell line are modulated by the $Mg^{++}$-complex of ATP. J Pharmacol Exper Ther 262:495–502
Schwanstecher M, Brandt C, Behrends S, Schaupp U, Panten U (1992d) Effect of MgATP on pinacidil-induced displacement of glibenclamide from the sulphonylurea receptor in a pancreatic $\beta$-cell line and rat cerebral cortex. Br J Pharmacol 106:295–301
Schwanstecher M, Löser S, Brandt C, Scheffer K, Rosenberger F, Panten U (1992e) Adenine nucleotide-induced inhibition of binding of sulphonylureas to their receptor in pancreatic islets. Br J Pharmacol 105:531–534
Schwanstecher M, Schaupp U, Löser S, Panten U (1992f) The binding properties of the particulate and solubilized sulfonylurea receptor from cerebral cortex are modulated by the $Mg^{2+}$ complex of ATP. J Neurochem 59:1325–1335
Shen WK, Tung RT, Machulda MM, Kurachi Y (1991) Essential role of nucleotide diphosphates in nicorandil-mediated activation of cardiac ATP-sensitive $K^+$ channel. Circ Res 69:1152–1158
Spruce AE, Standen NB, Stanfield PR (1985) Voltage-dependent ATP-sensitive potassium channels of skeletal muscle membrane. Nature 316:736–738

Spruce AE, Standen NB, Stanfield PR (1987) Studies on the unitary properties of adenosine-5′-triphosphate-regulated potassium channels of frog skeletal muscle. J Physiol (Lond) 382:213–237
Sturgess NC, Ashford MLJ, Cook DL, Hales CN (1985) The sulphonylurea receptor may be an ATP-sensitive potassium channel. Lancet 8453:474–475
Trube G, Hescheler J (1984) Inward-rectifying channels in isolated patches of the heart cell membrane: ATP-dependence and comparison with cell-attached patches. Pflugers Arch 401:178–184
Trube G, Rorsman P, Ohno-Shosaku T (1986) Opposite effects of tolbutamide and diazoxide on the ATP-dependent $K^+$ channel in mouse pancreatic $\beta$-cells. Pflugers Arch 407:493–499
Tung RT, Kurachi Y (1991) On the mechanism of nucleotide diphosphate activation of the ATP-sensitive $K^+$ channel in ventricular cell of guinea-pig. J Physiol (Lond) 437:239–256
Vivaudou MB, Arnoult C, Villaz M (1991) Skeletal muscle ATP-sensitive $K^+$ channels recorded from sarcolemmal blebs of split fibers: ATP inhibition is reduced by magnesium and ADP. J Membr Biol 122:165–175
Wallimann T, Wyss M, Brdiczka D, Nicolay K, Eppenberger HM (1992) Intracellular compartmentation, structure and function of creatine kinase isoenzymes in tissues with high and fluctuating energy demands: the "phosphocreatine circuit" for cellular energy homeostasis. Biochem J 281:21–40
Woll KH, Lönnendonker U, Neumke B (1989) ATP-sensitive potassium channels in adult mouse skeletal muscle: Different modes of blockage by internal cations, ATP and tolbutamide. Pflugers Arch 414:622–628

CHAPTER 78

# Modulation of Maxi-Calcium-Activated K Channels: Role of Ligands, Phosphorylation, and G-Proteins

L. TORO and E. STEFANI

## A. Introduction

Calcium-activated K channels are present in a vast number of tissues, including: brain, skeletal and smooth muscles, epithelia, and secretory glands. They have been classified in terms of their conductance as "maxi" (BK; ≈150–300 pS), "intermediate" (IK; ≈100 pS), and "small" (SK; <100 pS) Ca-activated K ($K_{Ca}$) channels (for review see LATORRE et al. 1989; COOK and YOUNG 1990; MCMANUS 1991).

Maxi-$K_{Ca}$ channels have been well characterized in different types of cells. One technical advantage is their large conductance and their ease of reconstitution into lipid bilayers where long-lived experiments can be performed (>1 h). Maxi-$K_{Ca}$ channels are voltage dependent, and their open probability increases as a function of the internal $Ca^{2+}$ concentration. Depending on the tissue and experimental conditions, they require submicromolar or micromolar calcium concentrations to open. Some pharmacological characteristics useful to distinguish this type of channels are: (a) "fast" blockade by micromolar concentrations of external TEA resulting in a reduction of channel conductance (VERGARA et al. 1984; YELLEN 1984), (b) "slow" blockade by external charybdotoxin (nanomolar range) resulting in long-lived closed states (MILLER et al. 1985), (c)"slow" blockade by external iberiotoxin (nanomolar range; CANDIA et al. 1992), (d) blockade by internal barium (VERGARA and LATORRE 1983), (e) insensitivity to apamin (PÉREZ et al. in press), and (e) insensitivity to 4-aminopyridine (CASTLE et al. 1989; PÉREZ et al. in press).

It is now clear that BK channels are modulated by a variety of metabolites such as hormones, neurotransmitters, nucleotides, and lipids (for review see TORO and STEFANI 1991; TORO and SCORNIK 1991). However, the mechanisms involved in their regulation are not yet defined. Nevertheless, we and others have found that changes in intracellular $Ca^{2+}$ are not the sole cause of modifications in channel gating, but other mechanisms are likely to be involved, which may in turn change the channel affinity for $Ca^{2+}$. Among these mechanisms are: (a) ligand modulation, (b) phosphorylation/dephosphorylation cycles, and (c) G-protein gating (Fig. 1). It is important to stress that these mechanisms are not exclusive, and that in the cell system more than one may be responsible for channel regulation.

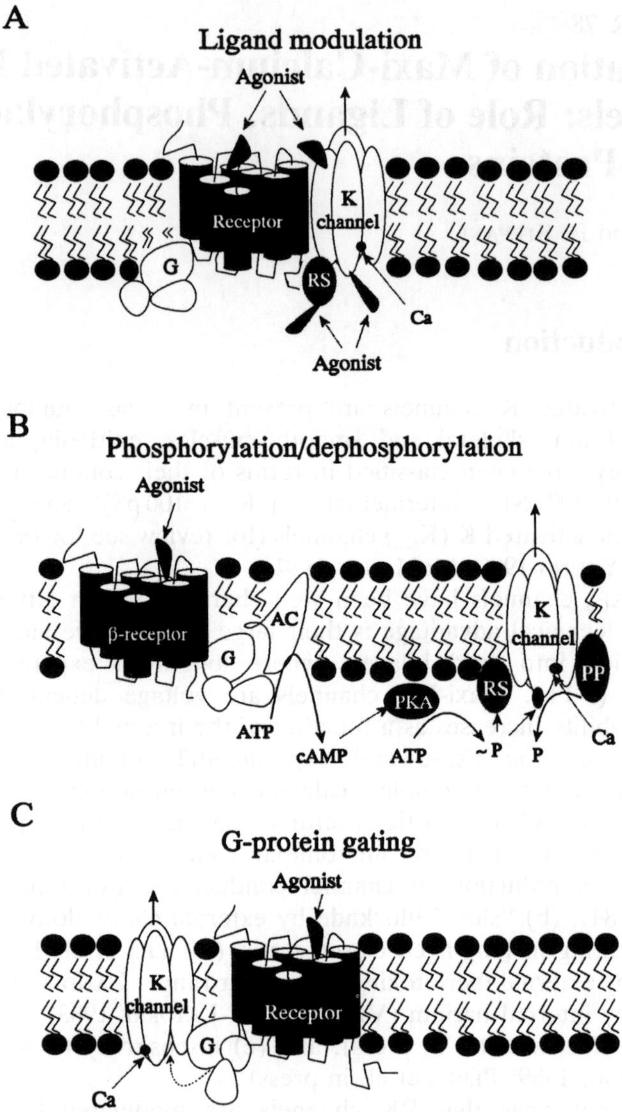


**Fig. 1A–C.** Putative mechanisms of $K_{Ca}$ channel regulation. **A** Ligand modulation. Agonists can bind to extracellular or intracellular binding sites producing changes in $K_{Ca}$ channel gating. The agonist binding site may be located, for example, on receptor proteins, the channel itself, or on a channel regulatory subunit (*RS*). **B** Phosphorylation/dephosphorylation. $K_{Ca}$ channels can be activated or inhibited by phosphorylation and dephosphorylation. A summary of putative events after stimulation of a β-receptor. $K_{Ca}$ channels or a regulatory subunit may be targets of protein kinase A (*PKA*). The phosphorylation state (and activity?) of the channel protein is defined by a balance between the activity of PKs and protein-phosphatases (*PP*). *AC*, Adenylyl cyclase; *G*, G-protein. **C** G-protein gating. $K_{Ca}$ channels may be activated by a G-protein coupled to a receptor (β-receptor?)

**Table 1.** Regulation properties of Ca-activated K channels

| Modulatory agent | Tissue or cell type | Effect | Mechanism proposed | Reference |
|---|---|---|---|---|
| Maxi-K(Ca) channels | | | | |
| Acetylcholine | Chromaffin cells | (+) | Ligand modulation | GLAVINOVIC (1990) |
| | Smooth muscle | (−) | G-protein gating? | COLE et al. (1989); KUME and KOTLIKOFF (1991) |
| Adenosine and atrial natriuretic peptide | Smooth muscle | (+) | GMP-mediated | WILLIAMS et al. (1988) |
| Adrenergic agents | Smooth muscle | (+) | Phosphorylation | SADOSHIMA et al. (1988); KUME et al. (1989) |
| | | (+) | G-protein gating | TORO et al. (1990b) |
| Angiotensin II | Smooth muscle | (−) | Ligand modulaton | TORO et al. (1990a) |
| Antidiuretic hormone | Kidney | (−) | Phosphorylation | GUGGINO et al. (1985) |
| Arachidonic acid and fatty acids | Smooth muscle | (+) | Ligand modulation | KIRBER et al. (1992) |
| ATP | Brain | (+) | Phosphorylation | CHUNG et al. (1991) |
| cGMP, GMP GDP, GTP | Smooth muscle | (+) | Ligand modulation? | WILLIAMS et al. (1988) |
| Endothelin | Smooth muscle | (+, −) | $[Ca^{2+}]_i$ | HU et al. (1991) KLÖCKNER and ISENBERG (1991) |
| $[H^+]$ | Epithelia | (−) | Ligand modulation | CHRISTENSEN and ZEUTHEN (1987) COPELLO et al. (1991) |
| | Pancreatic $\beta$-cells | (−) | Ligand modulation? | COOK et al. (1984) |
| | Skeletal muscle | (−) | Ligand modulation? | LAURIDO et al. (1991) |
| | Smooth muscle | (−) | Ligand modulation | KUME et al. (1990) |
| Secretin | Pancreatic duct cells | (+) | Phosphorylation | GRAY et al. (1990) |
| Somatostatin | Pituitary tumor cells | (+) | Dephosphorylation | WHITE et al. (1991) |
| Substance P | Smooth muscle | (+) | $[Ca^{2+}]_i$ | MAYER et al. (1989, 1990) |
| | | (−) | ? | MAYER et al. (1989) |
| Thromboxane $A_2$ | Smooth muscle | (−) | Ligand modulation | SCORNIK and TORO (1992) |
| Small-KCa channels | | | | |
| ADP | Oocytes | (+) | Ligand modulation? | YOSHIDA et al. (1990) |
| Histamine | HeLa Cells | (+) | $[Ca^{2+}]_i$ | SAUVÉ et al. (1987) |
| Gonadotrophin releasing hormone | Gonadotrophs | (+) | $[Ca^{2+}]_i$ | MASON and WARING (1986) SIKDAR et al. (1989) |
| Bradykinin | Endothelial cells | (+) | $[Ca^{2+}]_i$? | SAUVÉ et al. (1990) |
| | | (+) | G-protein gating? | VACA et al. (1992) |

(+), activation; (−) inhibition;

## B. Mechanisms of Metabolic Regulation of Maxi-$K_{Ca}$ Channels

Table 1 gives an overview of various metabolites capable to produce a change in $K_{Ca}$ channel (maxi and small types) activity; it also states the possible mechanisms involved in the modulatory process: (a) changes in intracellular calcium, $[Ca]_i$; (b) ligand modulation (direct action on the channel or a closely associated molecule); (c) phosphorylation/dephosphorylation cycles, and/or (d) G-protein gating (Fig. 1). It is important to stress that the physiological response of $K_{Ca}$ channels after a stimulus may involve more than one mechanism that can be either synergistic or antagonic. In this review we focus on the regulation of maxi-$K_{Ca}$ channels.

### I. Ligand Modulation

The mechanism of ligand modulation implies that the protein constituting the maxi-$K_{Ca}$ channel possesses a receptor site for its modulator or else that the binding of the agonist to a closely associated receptor molecule (e.g., a regulatory subunit, associated receptor) can cause a modification in channel gating (Fig. 1A). The existence of this type of mechanism has been experimentally supported not only using excised patches but with a more stringent control of the solutes in the microenvironment of the channel by using reconstituted channels in lipid bilayers. Metabolites found to modulate $K_{Ca}$ channels in this fashion are: (a) in lipid bilayers, angiotensin II (AgII; Toro et al. 1990), thromboxane $A_2$ ($TXA_2$; Scornik and Toro 1992), and protons (Laurido et al. 1991), and (b) in excised patches, nucleotides (Williams et al. 1988), arachidonic acid (Kirber et al. 1992), and protons (Cook et al. 1984; Christensen and Zeuthen 1987; Kume et al. 1990; Copello et al. 1991).

#### 1. Arachidonic Acid

Arachidonic acid (AA) has been shown to "directly" activate maxi-$K_{Ca}$ channels. Kirber et al. (1992) and Katz et al. (1990) demonstrated that maxi-$K_{Ca}$ channels from vascular smooth muscle can be activated by AA at concentrations ranging from 50 n*M* to 20 $\mu M$. Because this effect could be reproduced by fatty acids that are not substrates of the cyclo-oxygenase nor of the lipoxygenase metabolic pathways, it was suggested that AA may directly modulate maxi-$K_{Ca}$ channel activity (Kirber et al. 1992). This type of mechanism has also been postulated for the activation of a small conductance $K^+$ channel (≈30 pS) from smooth muscle of the toad *Bufo marinus* (Ordway et al. 1989).

The activation of maxi-$K_{Ca}$ channels by AA occurred in inside-out patches in the "absence" of nucleotides and calcium (5 m*M* EGTA; Kirber et al. 1992). Moreover, we have observed in lipid bilayers, in the absence of nucleotides, that AA stimulates $K_{Ca}$ channel activity from myometrial (unpublished observations) and coronary smooth muscle (TORO and SCORNIK 1991) in a dose-dependent fashion. Both lines of evidence are consistent with the idea that AA may modulate the channel or a closely associated molecule, and that this effect cannot be explained by changes in calcium, transduction pathways leading to phosphorylation, or G-proteins. However, it is possible that in the complex cellular machinery AA also serves as a second messenger stimulating a G-protein (ABRAMSON et al. 1991) which in turn triggers a metabolic cascade or activates the channel in a direct "membrane-delimited" way.

### 2. Angiotensin II and Thromboxane $A_2$

Maxi-$K_{Ca}$ channels from coronary smooth muscle, incorporated into lipid bilayers, are inhibited by the vasoconstrictors AgII (TORO et al. 1990) and $TXA_2$ (SCORNIK and TORO 1992).

AgII added to the external side of the channels produces a diminution of $P_O$. The $K_{1/2}$ of inhibition is 58 n$M$. Because the experiments were performed in the absence of added nucleotides (GTP or ATP) or $Mg^{2+}$, it is possible that AgII may have a direct action on the channel molecule or on a closely related molecule. It would be interesting to determine the site of modulatory-binding of AgII. Some possibilities are: (a) a receptor that belongs to the family of AgII receptors functionally coupled to the $K_{Ca}$ channel, (b) the channel protein itself, or (c) an unknown closely associated protein (e.g., regulatory subunit).

Recently, we have shown that $TXA_2$ receptors are involved in the regulation of $K_{Ca}$ channels from coronary smooth muscle (SCORNIK and TORO 1992). $K_{Ca}$ channels incorporated into lipid bilayers are inhibited by the $TXA_2$ mimetic U46619 in the nanomolar range (50–150 n$M$; Fig. 2). This inhibitory effect could be prevented or reversed by the $TXA_2$ receptor antagonist SQ29548, strongly indicating that a $TXA_2$ receptor is involved in the modulation of $K_{Ca}$ channel activity. Because these experiments were performed in the absence of added nucleotides (GTP, ATP, cAMP) it is unlikely that endogenous G-proteins or protein kinases are responsible for the inhibitory effect of $TXA_2$. Thus, the possibility exists (although remote) that the $K_{Ca}$ channel itself has a binding site for $TXA_2$ or most likely that a $TXA_2$ receptor protein coupled to the channel modulates its activity. Besides this mechanism, in the intact cell, it is possible that $TXA_2$ activates PLC in a parallel manner modulating $K_{Ca}$ channel gating through $G_{PLC}$ protein and/or PKC-dependent phosphorylation (Fig. 3).

### 3. Guanine Nucleotides

Maxi-$K_{Ca}$ channels from aortic smooth muscle are activated by guanine nucleotides GMP, cGMP, GDP, and GTP (WILLIAMS et al. 1988) in inside-

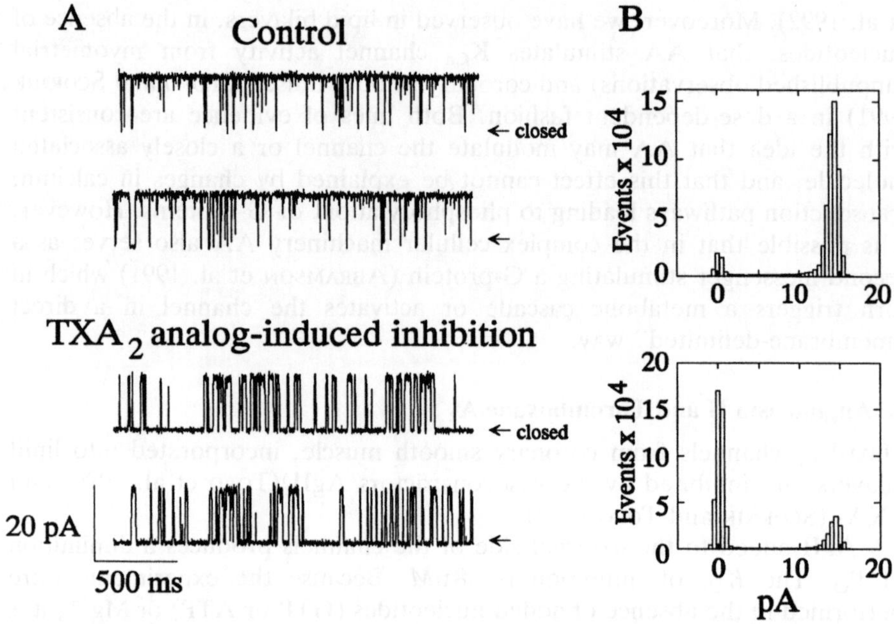


**Fig. 2A,B.** $K_{Ca}$ channels from coronary smooth muscle are inhibited by thromboxane $A_2$ mimetic U46619 in lipid bilayers. A receptor ($TXA_2$ receptor) may be in close association with the channel protein, explaining the coupling between the agonist binding and the channel response. **A** Channel recordings before and after inhibition with $TXA_2$ analog U46619 (100 n*M*). **B** Corresponding total point histograms. The peak at 0 pA corresponds to the channel in the closed state, while the other peak is the channel amplitude and corresponds to the open state

out patches. GMP and not cGMP was proposed to be the second messenger responsible for $K_{Ca}$ channel activation after stimulation with adenosine, atrial natriuretic factor (ANF), or nitroprusside. This conclusion was based on the fact that cGMP (100 $\mu M$–1 m*M*) could not fully mimic the activation of $K_{Ca}$ channels induced with adenosine, ANF, or nitroprusside. On the other hand, GMP (1–100 $\mu M$) could mimic this response. GMP activated $K_{Ca}$ channels by increasing their affinity for $Ca^{2+}$ and shifting their voltage activation curves towards more negative potentials. GDP and GTP (500 $\mu M$) could also enhance $K_{Ca}$ channel activity. Because the activation of $K_{Ca}$ channels was observed in inside-out patches, it is plausible that these nucleotides bind to the channel or to a closely associated protein causing its change in activity.

#### 4. Intracellular pH

Maxi-$K_{Ca}$ channels from several tissues can be modulated by intracellular pH through the binding of $H^+$ to the channel protein (Cook et al. 1984;

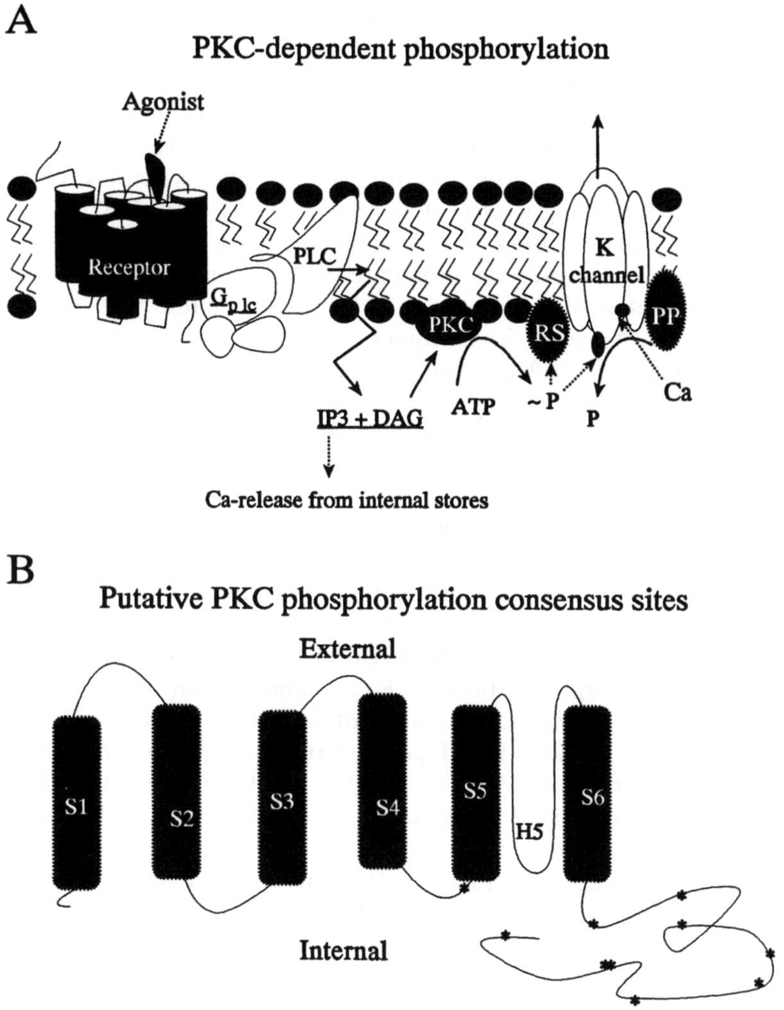


**Fig. 3A,B.** Putative modulation of $K_{Ca}$ channels via protein kinase C (*PKC*) dependent phosphorylation. **A** Metabolic pathway after an agonist binds to a receptor coupled to phospholipase C (*PLC*) via a G-protein ($G_{plc}$). Activation of PLC leads to breakdown of membrane phospholipids to inositol-1,4,5-triphosphate ($IP_3$) and diacylglycerol (*DAG*). Activation of PKC by DAG may induce phosphorylation of $K_{Ca}$ channels or a closely related molecule (RS?). Closely bound PP counteracts the modulatory effect of phosphorylation. **B** Putative array of the six transmembrane domains (S1–S6) and the "pore" region (*H5*) of one of the four (?) $K_{Ca}$ channel subunits. The amino acid sequence of *slowpoke* (component of $K_{Ca}$ channels from *Drosophila*) possesses ten consensus sites for PKC-dependent phosphorylation (*) (Atkinson et al. 1991)

CHRISTENSEN and ZEUTHEN 1987; KUME et al. 1990; COPELLO et al. 1991; LAURIDO et al. 1991). Acidification of the internal milieu decreases $K_{Ca}$ channel open probability without affecting its conductance. The reduction in channel activity results from both a reduction in the channel affinity for $Ca^{2+}$ and a parallel shift in the voltage activation curves to more positive potentials. It has been proposed that $H^+$ competes with $Ca^{2+}$ for the regulatory $Ca^{2+}$ binding sites (CHRISTENSEN and ZEUTHE 1987; KUME et al. 1990; COPELLO et al. 1991), but a detailed kinetic analysis has shown in skeletal muscle that protons modulate $K_{Ca}$ channel activity through an "alosteric" modification of the $Ca^{2+}$ binding sites and not through competition for the same site (LAURIDO et al. 1991). Thus, it is possible that changes in pH may induce conformational changes of the $K_{Ca}$ channel protein and modify its binding to $Ca^{2+}$ and its coupling with other regulatory molecules (e.g., neurotransmitters, hormones, and second messengers).

## II. Phosphorylation/Dephosphorylation Cycles

Many neurotransmitters and hormones may modulate the response of their effectors (ionic channels) via a common end that involves phosphorylation (by a kinase) of the effector or a closely related molecule. The kinase catalyzing the phosphorylation of $K_{Ca}$ channels that has received more attention is the cAMP-dependent protein kinase (PKA). Examples of agonists that may induce PKA-dependent phosphorylation are: $\beta$-adrenergic agents (SADOSHIMA et al. 1988; KUME et al. 1989) and secretin (GRAY et al. 1990). The general transduction scheme for PKA-dependent phosphorylation would be: (a) binding of the agonist to its receptor, (b) activation of a coupled G-protein, (c) activation of adenylyl cyclase and increase in cAMP levels, (d) activation of cAMP-dependent PKA, and (e) ATP-dependent phosphorylation of the effector (e.g., $K_{Ca}$ channel) or a closely related molecule (e.g., regulatory subunit; Fig. 1B).

The first direct evidence that a calcium-activated $K^+$ channel may be activated by PKA-mediated phosphorylation was put forward by EWALD et al. (1985), who demonstrated that an intermediate-$K_{Ca}$ channel (40–60 pS) from neurons of *Helix aspersa* could be activated by the kinase in isolated membrane patches and in bilayers. The modulation of maxi-$K_{Ca}$ channels via PKA-dependent phosphorylation has also been demonstrated. PKA may induce both an increase in channel open probability or an inhibition of channel activity. For example, maxi-$K_{Ca}$ channels from gonadotrophs of ovine pituitary (SIKDAR et al. 1989), a "type 2" maxi-$K_{Ca}$ channel from brain vesicles (REINHART et al. 1991), and maxi-$K_{Ca}$ channels from human nonpregnant myometrium (PÉREZ et al. 1992) are inhibited by PKA-dependent phosphorylation. On the other hand, upregulation of maxi-$K_{Ca}$ channels has been shown in a variety of tissues including aortic, tracheal, and colonic myocytes (SADOSHIMA et al. 1988; KUME et al. 1989; CARL et al.

1991), pancreatic duct cells (GRAY et al. 1990), brain (REINHART et al. 1991; CHUNG et al. 1991), and, our own observations, in coronary and pregnant uterine smooth muscle (PÉREZ G., SCORNIK F.S., STEFANI E., TORO L, unpublished). The activation of $K_{Ca}$ channels by phosphorylation increases the affinity of the channels for $Ca^{2+}$ without a modification of the Hill coefficient of the process (REINHART et al. 1991). Because the change in channel gating produced by phosphorylation is comparable with the one observed after $Ca^{2+}$ activation, it is possible that phosphorylation modulates the $Ca^{2+}$ sensor of $K_{Ca}$ channels. Therefore, phosphorylation or dephosphorylation of different sites in $K_{Ca}$ channels may contribute to the variety in $Ca^{2+}$ affinities found in $K_{Ca}$ channels of the same class.

The dual regulation by phosphorylation of the same class of channels may be explained by the existence of several isoforms of maxi-$K_{Ca}$ channels with different relevant phosphorylation sites, or else by the assumption that the regulatory proteins or subunits associated with the channel forming protein are different according to the tissue. In fact, several isoforms of maxi-$K_{Ca}$ channels have been observed in coronary smooth muscle (TORO et al. 1991) and at least two types in brain (FARLEY and RUDY 1988; REINHART et al. 1991). Moreover, molecular studies of *slowpoke* (cloned $K_{Ca}$ channel from *Drosophila*) transcripts indicate that a large number of $K_{Ca}$ channels isoforms may exist (LAGRUTTA et al. 1992).

Phosphorylation of maxi-$K_{Ca}$ channels by other protein kinases such as PKC has not been demonstrated directly. Modulation of $K_{Ca}$ channels by PKC-dependent phosphorylation is likely to occur since several agonists such as AgII, $TXA_2$, bradykinin, and histamine, known to increase lipid turnover via phospholipase C (PLC), control $K_{Ca}$ channel activity (BIRNBAUMER et al. 1989; TORO and STEFANI 1991). Moreover, the putative $K_{Ca}$ channel (*slowpoke*) cloned from *Drosophila* has more consensus sites for PKC- than for PKA-dependent phosphorylation (ATKINSON et al. 1991). The modulation of $K_{Ca}$ channel activity by these agonists could occur through the following scheme: agonist $\rightarrow$ receptor $\rightarrow$ $G_{plc}$-protein $\rightarrow$ PLC $\rightarrow$ $IP_3$ + diacylglycerol; $\{IP_3 \rightarrow$ release $Ca^{2+}\}$; $\{$diacylglycerol $\rightarrow$ activation of PKC $\rightarrow$ PKC-mediated phosphorylation$\}$. In this sequence of reactions modulation of $K_{Ca}$ channels by PKC-mediated phosphorylation would be the final step; however, it is obvious that other ways of regulation are possible as well. We have highlighted modulatory candidates other than $Ca^{2+}$ and phosphorylation needed to be explored: $G_{plc}$-protein, $IP_3$, and diacylglycerol (Fig. 3).

Similarly to other proteins regulated by phosphorylation, $K_{Ca}$ channels can also be dephosphorylated. Until now we have discussed the role of phosphorylation in the regulation of the activity of $K_{Ca}$ channels. In the following section we discuss how $K_{Ca}$ channels from different tissues are also modulated by dephosphorylation. A balance between these two chemical modifications on the channel protein or a closely associated molecule may be relevant to control cellular excitability and function.

### 1. Pituitary Maxi-$K_{Ca}$ Channels

Recently WHITE et al. (1991) proposed that maxi-$K_{Ca}$ channels from pituitary tumor cells ($GH_4C_1$ cells) are activated through protein dephosphorylation and inhibited by phosphorylation. Dephosphorylation and activation of $K_{Ca}$ channels were triggered after the cells were stimulated with somatostatin. The authors discarded a G-protein induced activation of $K_{Ca}$ channels through a direct "membrane-delimited" mechanism (BROWN and BIRNBAUMER 1990) because the somatostatin effects was observed in cell attached patches (somatostatin in the bath solution and not in the pipette), and because the somatostatin effect was not mimicked by addition of GTP in the cytoplasmic side of cell free patches. However, it is conceivable that a G-protein is involved in the process of somatostatin activation of $K_{Ca}$ channels because its effect was blocked by preincubation of the cells with pertussis toxin. Two hypothetic mechanisms have been proposed: (a) somatostatin activates a phosphatase responsible for the dephosphorylation of $K_{Ca}$ channels, or (b) somatostatin inhibits PKA activity responsible of maintaining the channel protein in a less active phosphorylated state. In any case, the role of the pertussis-sensitive G-protein remains to be determined. Furthermore, despite the good experimental data the interpretation of how regulatory systems work from isolated patches must be carefully evaluated since the microenvironment near the channel cannot be adequately controlled.

### 2. Brain Maxi-$K_{Ca}$ Channels

The modulation of brain $K_{Ca}$ channels (type 1 and type 2) by phosphorylation/dephosphorylation has been demonstrated by REINHART et al. (1991) using the lipid bilayer technique.

Type 1 maxi-$K_{Ca}$ channels (charybdotoxin sensitive) were subclassified according to their response (inhibition or activation) toward PKA-mediated phosphorylation. The predominant subclass was activated by phosphorylation (400 n*M* catalytic subunit of PKA) while the other was inhibited. Both the stimulatory and inhibitory actions of PKA catalytic subunit could be reversed by the action of the catalytic subunit of protein phosphatase $2_A$ but not by protein phosphatase 1, indicating that phosphorylation and dephosphorylation events were taking place in a specifc manner. On the other hand, type 2 maxi-$K_{Ca}$ channels (charybdotoxin insensitive) were inhibited by PKA and activated by subsequent dephosphorylation with phosphatase $2_A$. What the role of each of these channels is, and how phosphorylation or dephosphorylation determines their function, are questions that remain unanswered.

Regulation of $K_{Ca}$ channels by phosphorylation/dephosphorylation cycles was manifested when type-1 $K_{Ca}$ channels were phosphorylated with the catalytic subunit of PKA. It was observed that large variations of channel activity were produced, which on average reflected an increase in channel open probability. These variations were explained by the activity of an endogenous phosphatase in close association with the channel protein. Moreover, type 2 channels (inhibited by exogenous PKA) could be activated by ATP or ATP$\gamma$S indicating that another protein kinase different to PKA could be present in the bilayer system. Because the activation by ATP was also oscillatory, the presence of an endogenous phosphatase in close association with this type of channels was also suggested (CHUNG et al. 1991). Taken together, these findings suggest that modulation of $K_{Ca}$ channel activity may occur via cycles of phosphorylation and dephosphorylation, and support the idea that $K_{Ca}$ channels form stable complexes with other proteins that are maintained as such after incorporation into the lipid bilayer (TORO et al. 1990; CHUNG et al. 1991).

### 3. Colonic Maxi-$K_{Ca}$ Channels

The role of phosphorylation/dephosphorylation in $K_{Ca}$ channel activity has also been addressed in smooth muscle. CARL et al. (1991) have shown that $K_{Ca}$ channels from canine colon are activated by PKA-mediated phosphorylation, and that this activation can be enhanced by inhibition of phosphatases. Phosphatase inhibitors such as calyculin A and okadaic acid were tested. Because the experiments were performed in inside-out patches, the authors suggested that phosphorylation may take place on the channel protein itself or on a membrane-bound and closely associated protein. The fact that phosphatase inhibitors induced a further activation of previously phosphorylated $K_{Ca}$ channels gives evidence for the presence of active phosphatases in the vicinity of the channel protein. What determines the precise tuning of phosphorylation versus dephosphorylation remains to be elucidated.

### 4. Myometrial Maxi-$K_{Ca}$ Channels

We have recently reported that $K_{Ca}$ channels from nonpregnant human myometrium reconstituted in lipid bilayers are inhibited by PKA (PÉREZ et al. 1992) and activated by dephosphorylation (PÉREZ G., STEFANI E., TORO L., unpublished observations). Contrary to what has been shown in other smooth muscles (aorta, trachea, and colon; SADOSHIMA et al. 1988; KUME et al. 1989; CARL et al. 1991) addition of nanomolar concentrations of the catalytic subunit of PKA to the internal side of the channel, decreased channel open probability in an ATP-dependent fashion. This inhibition can be reversed by a specific PKA inhibitor IP20. The opposite finding can be explained as an overexpression of a certain isoform of $K_{Ca}$ channel or regulatory subunit in nonpregnant human myometrium that has an inhibitory phosphorylation site(s). In fact, new studies have shown that only a small proportion of channels (3 out of 15) is stimulated by PKA-dependent phosphorylation (PÉREZ G., STEFANI E., TORO L., unpublished observations). What determines the expression of a certain isoform(s) of $K_{Ca}$ channel or its regulatory protein(s)? How is the modulation via phosphorylation being regulated? Because the uterine muscle is under intense hormonal control we can speculate that inhibition or stimulation of $K_{Ca}$ channels by phosphorylation is under strict hormonal control as well.

In summary, maxi-$K_{Ca}$ channels may constitute a large family of K channels that can be inhibited and/or activated by phosphorylation or dephosphorylation. Furthermore, $K_{Ca}$ channels seem to be functionally coupled or in close association with protein kinases (PKC?) and protein phosphatases.

## III. G-Protein Gating

G-protein gating type of modulation was first proposed for the muscarinic $K^+$ channel from atrial cells (for review see Birnbaumer et al. 1987; Brown and Birnbaumer 1990). This mechanism requires that after the agonist is bound to its receptor, a coupled G-protein is activated, followed by the

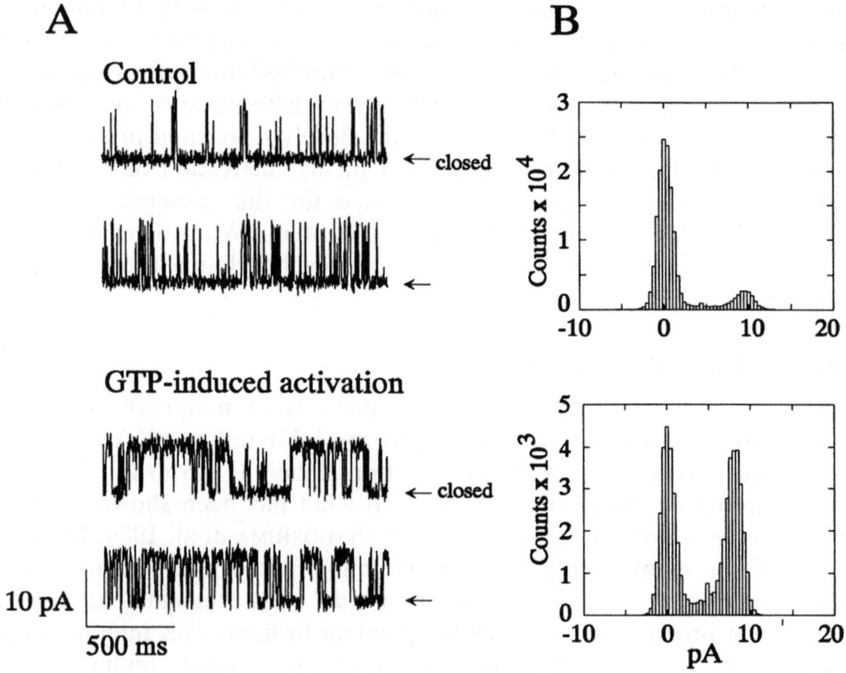


**Fig. 4A,B.** $K_{Ca}$ channels are modulated by G-proteins. Maxi-$K_{Ca}$ channels from myometrium reconstituted into lipid bilayers are activated after addition of GTP + Mg. Under these experimental conditions GTP + Mg are most likely activating an endogenous G-protein which promotes an increase in $K_{Ca}$ channel open probability. **A** Channel records before and after G-protein stimulation. **B** Corresponding total point histograms. The peak at 0 pA corresponds to the channel in the closed state, while the other peak is the channel amplitude and corresponds to the open state

direct interaction of the activated G-protein with the channel protein modifying its gating properties (Fig. 1C). Recent evidence has put forward the hypothesis that $K_{Ca}$ channels might be coupled and be substrates of G-protein modulation. The more direct evidence to this hypothesis resides in studies in lipid bilayers where we have demonstrated that $K_{Ca}$ channels from myometrial and coronary smooth muscles may be modulated by both endogenous G-proteins (GTP or GTP$\gamma$S mediated stimulation; Fig. 4) or exogenous purified G-proteins (TORO et al. 1990b; SCORNIK et al. 1992). To date, two types of agonists (acetylcholine and $\beta$-adrenergic agents) seem to be involved in the activation of G-proteins leading to a direct G-protein modulation of $K_{Ca}$ channels.

### 1. Muscarinic Regulation

$K_{Ca}$ channels from colonic (COLE et al. 1989; COLE and SANDERS 1989) and tracheal smooth muscle (KUME and KOTLIKOFF 1991) are inhibited by muscarinic agents.

COLE et al. (1989) observed that acetylcholine (ACh; 10$\mu M$) diminished $K_{Ca}$ channel activity when present in the bath and pipette solutions. This inhibition was absent in inside-out patches with ACh in the pipette (COLE et al. 1989) or when ACh was only applied to the bath but not to the pipette in the cell-attached mode. Furthermore, the inhibition of whole cell currents required GTP or nonhydrolyzable GTP analogs (GTP$\gamma$S or 5′-guanylylimidodiphosphate) in the pipette (COLE and SANDERS 1989). These results indicated that: (a) in inside-out patches essential internal components were lost (GTP?), preventing muscarinic inhibition from taking place, (b) in cell-attached patches that the coupling molecule (G-protein?) between the muscarinic receptor and the channel was functional only if they were in "close" proximity, and (c) that a GTP-dependent protein coupled the receptor occupancy with channel inhibition.

SIMILARLY, KUME and KOTLIKOFF (1991) demonstrated that $K_{Ca}$ channels are inhibited by extracellular metacholine (50$\mu M$) in outside-out patches. An interesting observation was that the inhibition could take place in the absence of added GTP in the pipette. However, this "basal" inhibition was larger and more consistent when GTP or GTP$\gamma$S was included in the pipette and was abolished in the presence of GDP$\beta$S. These results favored the idea that metacholine induced inhibition of $K_{Ca}$ channels through a GTP-dependent mechanism (G-protein). This idea implies that in outside-out patches, the intracellular milieu is not completely exchanged with the pipette content as was traditionally thought, but that a sufficient amount of nucleotides remains in the microenvironment of the G-protein and channel permitting the activation of the former and the inhibition of the latter.

The role of a G-protein in the muscarinic-induced inhibition of $K_{Ca}$ channels was confirmed by its inhibition with pertussis toxin. Intracellular dialysis or incubation of the cells with pertussis toxin suppressed the muscarinic-mediated inhibition of $K_{Ca}$ whole-cell (Cole and Sanders 1989) and single channel currents (Kume and Kotlikoff 1991). These experiments taken together with those previously described are consistent with a pertussis toxin sensitive G-protein having a role in the muscarinic inhibition of $K_{Ca}$ channels. In this context, the fact that ACh could not inhibit $K_{Ca}$ channels in inside-out patches when it was only present in the pipette, may be explained by the absence of GTP in the external solution necessary to activate the proximal G-protein. On the other hand, the requirement of ACh in the pipette in cell attached experiments supports the idea that: (a) after ACh binds to its receptor, a closely related G-protein is activated gating the channel in a direct mode, and (b) that a diffusible second messenger produced after the G-protein activation was not involved in the inhibition of $K_{Ca}$ channels.

## 2. Adrenergic Stimulation

$\beta$-Adrenergic stimulation causes an increase in $K_{Ca}$ channel activity in smooth muscle from aorta (Sadoshima et al. 1988), trachea (Kume et al. 1989), and uterus (Toro et al. 1990; Anwer et al. 1992). Because these channels are abundant in smooth muscle and their activation would lead to hyperpolarization of the cell membrane, it is very plausible that they play a major role during $\beta$-adrenergic induced relaxation. In tracheal smooth muscle isoprenalin causes ten times more relaxation than forskolin (which activates adenylyl cyclase; Honda et al. 1986) indicating that mechanism(s) other than PKA-dependent phosphorylation (of $K_{Ca}$ channels?) may be involved after $\beta$-adrenergic stimulation. In fact in cell-attached patches from the same tissue isoprenaline in the bath increased the Po of $K_{Ca}$ channels about seven times, while PKA-dependent phosphorylation in inside-out patches caused only a fourfold increase in channel activity (Kume et al. 1989). In this type of experiment it has been assumed that if the agonist is not required in the pipette to activate the channel, the most likely mechanism to occur is one which involves a second messenger such as cAMP and the subsequent activation of $K_{Ca}$ channels via PKA-dependent phosphorylation. Even though phosphorylation may partially explain these results, how can one explain that $\beta$-adrenergic stimulation is more potent in stimulating $K_{Ca}$ channel activity than the sole PKA-mediated phosphorylation? One can speculate that although the agonist was present in the bath and not in the pipette (close to the channel), it is possible that the tip of the pipette does not make an absolute barrier and that a receptor (near the pipette tip) is able to activate a G-protein coupled to the channel inducing its activation. Thus a G-protein gated mechanism may be parallel to PKA-dependent phosphorylation of $K_{Ca}$ channels after $\beta$-adrenergic stimulation.

We have tested the hypothesis that $\beta$-adrenergic stimulation of $K_{Ca}$ channels may involve a G-protein gating mechanism. We found that $K_{Ca}$

channels from myometrium or coronary smooth muscle incorporated into lipid bilayers were activated by GTP or GTP$\gamma$S solely in the presence of $Mg^{2+}$ from the "intracellular" side (Toro et al. 1990; Scornik et al. 1992; Fig. 4). In addition, experiments in the presence of AMP-PNP (adenylyl-imidodi-phosphate) a nonphosphorylating analog of ATP, or IP20 (inhibitor of PKA, unpublished observations) gave similar results indicating that activation of $K_{Ca}$ channels in our system was not due to "contaminant" or "endogenous" ATP, cAMP or protein kinase A. These results strongly suggested to us that $K_{Ca}$ channels were activated through a coupled GTP-dependent protein. Furthermore, in myometrium the activation of $K_{Ca}$ channels was potentiated by "extracellular" application of norepinephrine or the $\beta$-agonist isoproterenol, and partially inhibited by "internal" GDP$\beta$S (nonhydrolyzable analog of GDP that inhibits G proteins; Eckstein et al. 1979) or by the $\beta$-antagonist, propanolol ("external"). These results were consistent with our hypothesis and stressed the fact that the $\beta$-receptor, the stimulatory G-protein, and the $K_{Ca}$ channel were coupled and formed a stable association that was not disrupted after incorporation into the bilayer.

What is the nature of the G protein that gates $K_{Ca}$ channels, and how are $K_{Ca}$ channels gated? Experiments performed on $K_{Ca}$ channels from coronary smooth muscle, in lipid bilayers, indicate that the stimulatory G-protein may be $G_s$ (Scornik et al. 1992). Stimulation of $K_{Ca}$ channels by the activated $G\alpha_s$ subunit ($\alpha_s^*$) resembles the increase in channel activity observed with GTP$\gamma$S, strongly suggesting that the endogenous G protein is $G_s$. Both GTP$\gamma$S and $\alpha_s^*$ caused a shift in the voltage-activation curve toward more negative potentials making the channel to behave as if its affinity for $Ca^{2+}$ were higher (Toro et al. 1990; Scornik et al. 1992), and induced a change in kinetics that fundamentally consisted in a decrease in the time the channel remained in the long closed states.

In summary, two mechanisms for $\beta$-adrenergic stimulation of $K_{Ca}$ channels may be proposed: (a) direct G-protein gating and (b) phosphorylation via PKA. Certainly, it is possible that other parallel mechanisms may occur.

## C. Conclusions

Maxi $K_{Ca}$ channels constitute a class of ionic channels that are regulated by a great variety of inputs. At least three major mechanisms of action can be proposed to explain how $K_{Ca}$ channels modify their activity after a stimulus: (a) ligand modulation, (b) phosphorylation/dephosphorylation, and (c) G-protein gating.

A new concept of stable molecular interactions between channels and other membrane proteins has emerged from bilayer studies. It is now clear that after incorporation into the bilayer the molecular complex of proteins (e.g., receptors, G proteins, $K_{Ca}$ channels) does not diffuse away in the lipid environment, but proteins remain functionally coupled to each other.

Many questions wait to be explored at the molecular level, and structure-function studies of cloned $K_{Ca}$ channels promise to give crucial information about the mechanisms of regulation. For example, which are the phosphorylation sites of the channel involved in up-regulation and which in inhibition? Is the channel differentially modulated by various protein kinases? Where is the binding site for G-proteins located on the channel protein, if there is one? How does $H^+$ modify the channel $Ca^{2+}$ affinity or the binding of other ligand?

In spite of many unanswered questions, it is possible to conclude that $K_{Ca}$ channels may be up- or down-regulated by intracellular and extracellular metabolites (hormones, lipids, nucleotides) without an obligatory change in intracellular $Ca^{2+}$.

*Acknowledgments*. This work was supported by Grant-in-Aid 900963 from the American Heart Association, National Center (L.T.) and by grants HD-25616 and HL-37044 from NIH (E.S.)

## References

Abramson SB, Leszczynska-Piziak J, Weissmann G (1991) Arachidonic acid as a second messenger. Interactions with a GTP-binding protein of human neutrophils. J Immunol 147:231–236

Anwer A, Toro L, Oberti C, Stefani E, Sanborn BM (1992) Calcium-activated potassium channel in pregnant rat myometrium: modulation by a $\beta$-adrenergic agent. Am J Physiol 263:C1049–C1056

Atkinson NS, Robertson GA, Ganetzky B (1991) A component of calcium-activated potassium channels encoded by the Drosophila slo locus. Science 253:551–555

Birnbaumer L, vanDongen AMJ, Codina J et al. (1989) Identification of G protein gated and G protein modulated ionic channels. In: Armstrong CM, Oxford GS (eds) Secretion and its control. Rockefeller University Press, New York, pp 14–54

Brown AM, Birnbaumer L (1990) Ionic channels and their regulation by G protein subunits. Annu Rev Physiol 52:197–213

Candia S, Garcia ML, Latorre R (1992) Mode of action of iberiotoxin, a potent blocker of the large conductance $Ca^{2+}$-activated $K^+$ channel. Biophys J 63: 583–590

Carl A, Kenyon JL, Uemura D, Fusetani N, Sanders KM (1991) Regulation of $Ca^{2+}$-activated $K^+$ channels by protein kinase A and phosphatase inhibitors. Am J Physiol (Cell Physiol) 261:C387–C392

Castle NA, Haylett, DG, Jenkinson, DH (1989) Toxins in the characterization of potassium channels. Trends in Neurosc 12:59–65

Christensen O, Zeuthen T (1987) Maxi $K^+$ channels in leaky epithelia are regulated by intracellular $Ca^{2+}$, pH and membrane potential. Pflugers Arch 408:249–259

Chung S, Reinhart PH, Martin BL, Brautigan D, Levitan IB (1991) Protein kinase activity closely associated with a reconstituted calcium-activated potassium channel. Science 253:560–562

Cole WC, Carl A, Sanders KM (1989) Muscarinic suppression of $Ca^{2+}$-dependent K current in colonic smooth muscle. Am J Physiol 257:C481–C487

Cole WC, Sanders KM (1989) G proteins mediate suppression of $Ca^{2+}$-activated K current by acetylcholine in smooth muscle cells. Am J Physiol 257:C596–C600

Cook DI, Young JA (1990) Cation channels and secretion. In: Young JA, Wong TYD (eds) Epithelial secretion of water and electrolytes. Springer-Verlag, Berlin, Heidelberg, pp 15–38

Cook DL, Ikeuchi M, Fujimoto WY (1984) Lowering of pHi inhibits $Ca^{2+}$-activated $K^+$ channels in pancreatic B-cells. Nature 311:269–271

Copello J, Segal Y, Reuss L (1991) Cytosolic pH regulates maxi $K^+$ channels in Necturus gall-bladder epithelial cells. J Physiol (Lond) 434:577–590

Ewald DA, Williams A, Levitan IB (1985) Modulation of single $Ca^{2+}$-dependent $K^+$-channel activity by protein phosphorylation. Nature 315:503–506

Farley J, Rudy B (1988) Multiple types of voltage-dependent $Ca^{2+}$-activated $K^+$ channels of large conductance in rat brain synaptosomal membranes. Biophys J 53:919–934

Glavinovic MI (1990) Effect of acetylcholine on single $Ca^{2+}$-activated $K^+$ channels in bovine chromaffin cells. Neuroscience 39:815–822

Gray MA, Greenwell JR, Garton AJ, Argent BE (1990) Regulation of maxi-$K^+$ channels on pancreatic duct cells by cyclic AMP-dependent phosphorylation. J Membr Biol 115:203–215

Guggino SE, Suarez-Isla BA, Guggino WB, Sacktor B (1985) Forskolin and antidiuretic hormone stimulate a $Ca^{2+}$-activated $K^+$ channel in cultured kidney cells. Am J Physiol 249:F448–F455

Honda K, Satake T, Takagi K, Tomita T (1986) Effects of relaxants on electrical and mechanical activities in the guinea-pig tracheal muscle. Br J Pharmac 87:665–671

Hu S, Kim HS, Jeng AY (1991) Dual action of endothelin-1 on the $Ca^{2+}$-activated $K^+$ channel in smooth muscle cells of porcine coronary artery. Eur J Pharmacol 194:31–36

Kirber MT, Ordway RW, Clapp LH, Walsh JV Jr, Singer JJ (1992) Both membrane stretch and fatty acids directly activate large conductance $Ca^{2+}$-activated $K^+$ channels in vascular smooth muscle cells. FEBS Lett 297:24–28

Klöckner U, Isenberg G (1991) Endothelin depolarizes myocytes from porcine coronary and human mesenteric arteries through Ca-activated chloride current. Pflugers Arch 418:168–175

Kume H, Takai A, Tokuno H, Tomita T (1989) Regulation of $Ca^{2+}$-dependent $K^+$-channel activity in tracheal myocytes by phosphorylation. Nature 341:152–154

Kume H, Takagi K, Satake T, Tokuno H, Tomita T (1990) Effects of intracellular pH on calcium-activated potassium channels in rabbit tracheal smooth muscle. J Physiol (Lond) 424:445–457

Kume H, Kotlikoff MI (1991) Muscarinic inhibition of single $K_{Ca}$ channels in smooth muscle cells by a pertussis-sensitive G protein. Am J Physiol (Cell Physiol) 261:C1204–C1209

Lagrutta A, Bond CT, Warren RA, North RA, Adelman JP (1992) Diversity of slowpoke potassium channel transcripts. Biophys J 61:A377

Latorre R, Oberhauser A, Labarca P, Alvarez O (1989) Varieties of calcium-activated potassium channels. Annual Review of Physiology 51:385–399

Laurido C, Candia S, Wolff D, Latorre R (1991) Proton modulation of a $Ca^{2+}$-activated $K^+$ channel from rat skeletal muscle incorporated into planar bilayers. J Gen Physiol 98:1025–1043

Mason WT, Waring DW (1986) Patch clamp recordings of single ion channel activation by gonadotrophin-releasing hormone in ovine pituitary gonadotrophs. Neuroendocrinology 43:205–219

Mayer EA, Loo DDF, Kodner A, Reddy SN (1989) Differential modulation of $Ca^{2+}$-activated $K^+$ channels by substance P. Am J Physiol 257:G887–G897

Mayer EA, Loo DDF, Snape WJ, Jr, Sachs G (1990) The activation of calcium and calcium-activated potassium channels in mammalian colonic smooth muscle by substance P. J Physiol (Lond) 420:47–71

McManus OB (1991) Calcium-activated potassium channels: regulation by calcium. J Bioenerg Biomembr 23:537–560

Miller C, Moczydlowski E, Latorre R, Phillips M (1985) Charybdotoxin, a protein inhibitor of single $Ca^{2+}$-activated $K^+$ channels from mammalian skeletal muscle. Nature 313:316–318
Ordway RW, Walsh JV Jr, Singer JJ (1989) Arachidonic acid and other fatty acids directly activate potassium channels in smooth muscle cells. Science 244:1176–1179
Pérez G, Toro L, Erulkar SD, Stefani E (1993) Characterization of large conductance calcium-activated potassium channels from human myometrium. Am J Obstet Gynecol 168:652–660
Perez G, Toro L, Stefani E (1992) Protein kinase A (PKA) modulates calcium activated potassium ($K^{Ca}$) channel from human myometrium. Biophys J 61:A255
Reinhart PH, Chung S, Martin BL, Brautigan DL, Levitan IB (1991) Modulation of calcium-activated potassium channels from rat brain by protein kinase A and phosphatase 2A. J Neurosci 11:1627–1635
Sadoshima J-I, Akaike N, Kanaide H, Nakamura M (1988) Cyclic AMP modulates Ca-activated K channel in cultured smooth muscle cells of rat aortas. Am J Physiol 255:H754–H759
Sauve R, Chahine M, Tremblay J, Hamet P (1990) Single channel analysis of the electrical response of bovine aortic endothelial cells to bradykinin stimulation: contribution of a $Ca^{2+}$-dependent $K^+$ channel. J Hypertension 8:S193–S201
Sauve R, Simoneua C, Parent L, Monette R, Roy G (1987) Oscillatory activation of calcium-dependent potassium channels in HeLa cells induced by histamine $H_1$ receptor stimulation: a single channel study. J Membr Biol 96:199–208
Scornik FS, Codina J, Birnbaumer L, Stefani E, Toro L (1992) Activation of calcium activated potassium K(Ca) channels from pig coronary artery by GTP$\gamma$S may involve a G protein. Biophys J 61:A254
Scornik F, Toro L (1991) Modulation of Ca-activated K channels from coronary smooth muscle. In: Sperelakis N, Kuriyama H (eds) Ion channels of vascular smooth muscle cells and endothelial cells. Elsevier Science Publishing Company, Inc, New York, pp 111–124
Scornik FS, Toro L (1992) U46619, a thromboxane A2 agonist, inhibits $K_{Ca}$ channel activity from pig coronary artery. Am J Physiol 262:C708–C713
Sikdar SK, McIntosh RP, Mason WT (1989) Differential modulation of $Ca^{2+}$-activated $K^+$ channels in ovine pituitary gonadotrophs by GnRH, $Ca^{2+}$ and cyclic AMP. Brain Res 496:113–123
Toro L, Amador M, Stefani E (1990a) Ang II inhibits calcium-activated potassium channels from coronary smooth muscle in lipid bilayers. Am J Physiol 258:H912–H915
Toro L, Ramos-Franco J, Stefani E (1990b) GTP-dependent regulation of myometrial $K_{Ca}$ channels incorporated into lipid bilayers. J Gen Physiol 96:373–394
Toro L, Vaca L, Stefani E (1991) Calcium activated potassium channels from coronary smooth muscle reconstituted in lipid bilayers. Am J Physiol Heart Circ Physiol 260:H1779–1789
Toro L, Stefani E (1991) Calcium-activated $K^+$ channels: metabolic regulation. J Bioenerg Biomembr 23:561–576
Vaca LA, Schilling WP, Kunze DL (1992) G-protein-mediated regulation of a $Ca^{2+}$-dependent $K^+$ channel in cultured vascular endothelial cells. Pflugers Arch 422:66–74
Vergara C, Moczydlowski E, Latorre R (1984) Conduction, blockade and gating in a $Ca^{2+}$-activated $K^+$ channel incorporated into planar lipid bilayers. Biophys J 45:73–76
Vergara C, Latorre R (1983) Kinetics of $Ca^{2+}$-activated $K^+$ channels from rabbit muscle incorporated into planar bilayers. Evidence for a $Ca^{2+}$ and $Ba^{2+}$ blockade. J Gen Physiol 82:543–568

White RE, Schonbrunn A, Armstrong DL (1991) Somatostatin stimulates $Ca^{2+}$-activated $K^+$ channels through protein dephosphorylation. Nature 351:570–573

Williams DL Jr, Katz GM, Roy-Contancin L, Reuben JP (1988) Guanosine 5′-monophosphate modulates gating of high-conductance $Ca^{2+}$-activated $K^+$ channels in vascular smooth muscle cells. Proc Natl Acad Sci USA 85:9360–9364

Yellen G (1984) Ionic permeation and blockade in Ca-activated K-channels of bovine chromaffin cells. J Gen Physiol 84:157–186

Yoshida S, Plant S, McNiven AI, House CR (1990) Single $Ca^{2+}$-activated $K^+$ channels in excised membrane patches of hamster oocytes. Pflugers Arch 415:516–518

CHAPTER 79

# Regulation of the Endosomal Proton Translocating ATPase ($H^+$-ATPase) and Endosomal Acidification by G-Proteins

R.W. GURICH and T.D. DUBOSE, JR.

## A. Introduction

This chapter focuses on the role of G-proteins in the regulation of endosomal acidification, a process mediated by a proton translocating ATPase ($H^+$-ATPase). Initially we review general aspects of endocytosis with emphasis on the role of acidification in this process. Since it has been appreciated only recently that G-proteins play a regulatory role in the acidification of endosomes derived from the renal cortex, it is necessary to review endocytosis in the kidney. Finally, we discuss in detail the transport pathways involved in endosomal acidification that may be regulated by G-proteins, as elucidated by recent studies in our laboratory.

## B. Endocytosis

### I. General

Endocytosis, as defined classically, is a process through which a variety of extracellular compounds can be taken up across cell membranes and stored transiently in a class of intracellular vesicles, referred to as "endosomes" (SILVERSTEIN et al. 1977; STEINMAN et al. 1983). Endocytosis can be described as one of two general types: (a) fluid phase or (b) adsorptive (SILVERSTEIN et al. 1977; STEINMAN et al. 1983). In fluid-phase endocytosis compounds that do not interact specifically with plasma membrane glycoproteins are internalized into the cell along with bulk fluid (COHN and MELLMAN 1982). Fluid-phase endocytosis is a constitutive process that cannot be modified by extracellular processes (COHN and MELLMAN 1982). Adsorptive endocytosis occurs when ligands, such as lectins and toxins, bind specifically to a large number of cell surface receptors allowing the ligand/receptor complex to be internalized into noncoated membrane invaginations (GONATAS et al. 1977; JOSEPH et al. 1978; MONTESANO et al. 1982). Receptor-mediated endocytosis is a variant of adsorptive endocytosis in which physiologically active ligands bind to a relatively small number of cell surface receptors and the ligand/receptor complex is internalized into clathrin coated pits (STEINMAN et al. 1983; GOLDSTEIN et al. 1985; HELENIUS et al. 1983; PASTAM and WILLINGHAM 1983). Although the initial internalization events of adsorptive and receptor-mediated endocytosis differ, it appears that the coated pit pathway and the noncoated vesicle pathways converge in the "early endosome" compartment (TRAN et al. 1987; for review of the nomemclature of endocytosis, see GRUENBERG and HOWELL 1989; KORNFELD and MELLMAN 1989). Initial sorting of internalized membrane components occurs in the early endosome (PARTON and SIMONS 1991).

Receptor-mediated endocytosis is a process common to many, if not all, cell types (SILVERSTEIN et al. 1977; PASTAM and WILLINGHAM 1983; GOLDSTEIN et al. 1979; KAHN 1976). However, depending on the specific ligand and cell type, the means by which substances are endocytosed can vary considerably. For instance, after internalization, low-density lipoprotein (LDL) is dissociated from its receptor in the endosomal compartment (GOLDSTEIN et al. 1985). The receptor is then recycled back to the cell surface while LDL is targeted to lysosomes for degradation (GOLDSTEIN et al. 1985). Conversely, transferrin remains associated with its receptor in early endosomes and the entire complex is recycled back to the cell surface (HOPKINS 1983; KARIN and MINTZ 1981; VAN RENSWOUDE et al. 1982). A third variant of endocytosis is represented by epidermal growth factor (EGF). EGF and its receptor are internalized, but its receptor is not recycled. Rather, both compounds are targeted to lysosomes for degradation (FINE et al. 1981; GORDEN et al. 1978). Lastly, immunoglobulins such as IgA are endocytosed, "transcytosed" across the cell, and secreted intact into the bile (LIMET et al. 1985).

## II. The Kidney

The kidney is a heterogeneous organ that contains a variety of polar epithelial cell types with several different functions. Although endocytosis has not been as well characterized in the kidney as in nonpolar cells, most of the endocytic functions mentioned above have also been described in the kidney (for a review see PARTON and SIMONS 1991). However, due to the highly varied and unique physiologic demands imposed on the kidney, certain endocytic functions are highly developed and deserve special mention. The first process to be discussed is the endocytosis of filtered proteins by the cells of the proximal convoluted tubule. The glomerular capillary membrane serves as an effective barrier against the loss of large molecular weight proteins and other macromolecules into the urine (MAACK et al. 1985). However, a significant but relatively small amount of albumin and other low molecular weight proteins are filtered through the glomerulus into Bowman's space. These filtered proteins are reclaimed by the proximal tubule by endocytosis (WALL and MAACK 1985). Specifically, the filtered proteins adhere to the apical membrane of proximal tubule cells and are internalized into endosomes where initial degradation occurs. Subsequently, partially degraded proteins are directed to lysosomes where final degradation is accomplished. The degradation products are then released across the basolateral membrane into the systemic circulation. The unique ability of the proximal tubule cell to endocytose, nonspecifically, a variety of substances, provides a convenient means by which to label endocytic compartments in vivo. Compounds infused intravenously such as horseradish peroxidase (STRAUSS 1964), the pH sensitive compound fluorescein isothiocyanate dextran (FITC-dextran; LENCER et al. 1990; SABOLIC et al. 1985; GURICH et al. 1991), and the chloride sensitive dye 6-methoxy-*N*-3-sulfopropyl quinolinium (SPQ; BAE and VERKMAN et al. 1990) are taken up by the kidney and stored transiently in endosomes. These endosomes can then be separated from other intracellular organelles or plasma membranes allowing transport experiments to be performed in vitro.

Another function of the kidney is to participate in the regulation of acid-base balance by modulation of acid excretion and bicarbonate absorption. Well-documented responses of the kidney to acid-base perturbations include changes in the activity of acid-base transporters such as the $Na^+/H^+$ antiporter, the $Na^+/HCO_3$ cotransporter, the $Cl/HCO_3$ exchanger, and the proton translocating ATPase ($H^+$-ATPase; ALPERN et al. 1991). While the exact mechanism by which the activity of these transporters is modulated is not known, there is evidence that endocytic removal and/or exocytic insertion of some of these transporters into the apical or basolateral membrane, as appropriate, could be involved. For example, it has been demonstrated in turtle bladder (GLUCK et al. 1982; STETSON and STEINMETZ 1983) and in the mammalian kidney collecting tubule (MADSEN and TISHER 1986; SCHWARTZ and AL-AWQATI 1985) that changes in ambient $PCO_2$ or systemic pH can modulate exocytosis of $H^+$-ATPase-containing vesicles into the plasma membrane. This exocytic process produces an increase in pump density and an increase in acid secretion (BROWN 1989). Endocytic removal of the pumps likely reverses the process. In addition, an electroneutral $Na^+/H^+$ exchanger has been detected in endosomal vesicles from rabbit and rat proximal tubules (GURICH and WARNOCK 1986; SABOLIC and BROWN 1990; HILDEN et al. 1990). This exchanger could be related to the $Na^+/H^+$ antiporter on the apical membrane, and thus it is possible that an endocytic/exocytic process could regulate activity of the $Na^+/H^+$ antiporter. However, the endosomal $Na^+/H^+$ has not been observed by other investigators (YE et al. 1989) in proximal tubule endosomes, nor have studies been performed to determine whether this transporter per se is actually inserted into or removed from the luminal membrane.

The kidney also responds to changes in volume status by altering water reabsorption. Under influence of antidiuretic hormone, water channels are inserted into or removed from the apical membrane of principal cells of the collecting duct (BROWN 1989). Vesicles that contain water channels have been isolated from toad bladder (WANG et al. 1991) and renal papilla (LENCER et al. 1990a). Unlike other endosomes, these vesicles do not contain an $H^+$-ATPase and thus do not likely acidify the interior. However, vesicles that contain water channels and an $H^+$-ATPase have been recovered from the renal proximal tubule (YE et al. 1989). Thus, it is not certain that acidification, which is important for other endosomal functions (see below), is necessary for water channel trafficking.

## C. Endosomal Acidification

From the preceding discussion it is apparent that endosomal function varies considerably not only between different cell types, but within a particular cell type as well. It is generally accepted that different endocytic pathways merge in the early endosomal compartment and that this compartment is responsible for initiation of the complex task of sorting endocytosed material (PARTON and SIMONS 1991). The process of endosomal sorting, as well as other endosomal functions, i.e., initial degradation of endocytosed material, is dependent on effective acidification of the endosomal compartment (MELLMAN et al. 1986). For instance, it has been demonstrated that acidification is necessary for the dissociation of LDL from its receptor (GOLDSTEIN et al. 1985). Conversely, acidification of early endosomes may actually increase the affinity of the transferrin receptor for transferrin and thereby prevent the degradation of transferrin (DAUTRY-VARSAT 1983). Endosomal acidification is also necessary for the pathologic internalization of viruses and toxins into the cell cytosol (GONATAS et al. 1977; JOSEPH et al. 1978; MONTESANO et al. 1982; MARSH et al. 1983; YOSHIMURA et al. 1982). Studies of the endocytosis of viruses and toxins have led to the development of mutant cell lines in which it has been demonstrated clearly that impairment of acidification disrupts normal endosomal function and intracellular vesicular trafficking (KLAUSNER et al. 1987; ROBBINS et al. 1983; ROBBINS et al. 1989; ROFF et al. 1986). In the kidney it has been demonstrated that alkalinization of endosomal and lysosomal compartments halts the degradation of endocytosed proteins (CAMARGO et al. 1989). In addition to alkalinization it has been shown that activation of G-proteins by GTP-$\gamma$-S can disrupt endocytosis and trans-Golgi recycling of endosomal vesicles (MAYORGA et al. 1989). Thus it appears reasonable to assume that there may be an important relationship between G-proteins and endosomal acidification with respect to endosomal function.

It is generally appreciated that endosomal acidification, as well as acidification of other intracellular organelles such as lysosomes and the golgi apparatus, is mediated primarily by a proton-translocating ATPase ($H^+$-ATPase; MELLMAN et al. 1986). This enzyme is a member of the vacuolar class of ATPases and uses the energy of hydrolysis of ATP to translocate protons across plasma and/or cell membranes (for review see STONE 1988).

This pump lowers pH to 6.0–6.5 in early endosomes, approximately 5.5 in late endosomes, and approximately 4.5 in lysosomes (Parton and Simons 1991). Although the $H^+$-ATPase has not been purified from endosomes, the enzyme has been purified from clathrin-coated vesicles (Xie and Stone 1986; these are likely early endosome precursors) as well as other intracellular organelles (Moriyama and Nelson 1987) and renal plasma membranes (Gluck and Caldwell 1987). This enzyme is composed of at least five subunits and interorgan heterogeneity of pump subunit composition has been demonstrated. In addition, it has been postulated that subunit heterogeneity could produce differences in proton transport properties (Gluck and Caldwell 1987; Wang and Gluck 1990). This in turn could produce different functional properties in organelles containing the $H^+$-ATPase. With regard to endosomes, heterogeneity of the regulation of acidification by cAMP has been demonstrated (Gurich and DuBose 1989). Moreover, direct measurements of individual endosomal pH have also demonstrated significant heterogeneity between endosomes (Shi et al. 1991). Thus it appears that heterogeneity of endosomal function is mirrored by heterogeneity of endosomal acidification.

In addition to the $H^+$-ATPase, endosomes and other acidic vesicles contain a chloride entry pathway that provides charge compensation for the electrogenic $H^+$-ATPase. While not demonstrated conclusively, it is assumed that the chloride entry pathway is probably a chloride conductive channel. In this regard, it has been demonstrated, using the patch-clamp technique in purified endosomal vesicles from rabbit renal cortex, that endosomes contain a 100 pS $Cl^-$ channel (Reeves and Gurich 1991). This channel is inhibited by the $Cl^-$ channel blocker IAA-94. Furthermore, other investigators have labelled rat renal endosomes in vivo with the $Cl^-$ sensitive dye SPQ and have demonstrated that these endosomes contain a chloride conductance (Bae and Verkman 1990) which was activated by protein kinase A. Since GTP-$\gamma$-S did not activate this conductance, it is unlikely that G-proteins regulate this pathway. This finding suggests that the endosomal chloride channel differs from other renal chloride channels that have been demonstrated to be regulated by G-proteins (Schwiebert et al. 1990).

Although the $H^+$-ATPase and the chloride channel provide a general mechanism for the acidification of endosomal vesicles, they do not explain the marked differences in the intravesicular pH of early endosomes, late endosomes, and lysosomes (Parton and Simons 1991). As mentioned, heterogeneity of proton pump subunits or different isoforms of a particular subunit could explain variation in proton pumping capacity (Gluck and Caldwell 1987; Wang and Gluck 1990). Alternatively, the presence of transport proteins other than the $H^+$-ATPase and the chloride channel could influence intravesicular pH. In this regard studies have suggested that endosomes derived from several cell types contain $Na^+/K^+$ ATPase activity. Based on the known stoichiometry of the $Na^+/K^+$ ATPase, an interior-positive membrane potential would be anticipated, which would inhibit proton pumping by an electrogenic $H^+$-ATPase (CORLEY et al. 1989). However, studies of endosomes derived from renal proximal tubule cells have failed to verify the existence of a $Na^+/K^+$ ATPase in these membranes (GURICH and WARNOCK 1986; SABOLIC and BROWN 1990; HILDEN et al. 1990). Thus, it appears doubtful that the $Na^+/K^+$ ATPase plays any role in the regulation of endosomal acidification. An additional transporter, the $Na^+/H^+$ exchanger, has also been detected in endosomes from renal proximal tubule cells (GURICH and WARNOCK 1986; SABOLIC and BROWN 1990; HILDEN et al. 1990), but the function of this transporter has not yet been defined conclusively. However, the $K_m$ for $Na^+$ in endosomal membranes, equal to 10 m*M*, is approximately the same as the cytsolic $Na^+$ concentration. Thus, it is possible that a favorable inward $Na^+$ gradient across the endosomal membrane coupled with an outward proton gradient could drive $Na^+/H^+$ exchange and thereby alkalinize the interior of the endosome. Lastly, it is possible that second messengers such as cAMP could modulate endosomal acidification in vivo. As mentioned, cAMP has been demonstrated to regulate the rat renal endosomal chloride conductance in vitro (BAE and VERKMAN 1990). cAMP has also been demonstrated to inhibit the rabbit renal endosomal [$H^+$]ATPase in vitro (GURICH and DUBOSE 1989). It is interesting to speculate that regulation of endosomal acidification by cAMP in vivo could explain the heterogeneity of endosomal pH. Moreover, cAMP, through its effects on endosomal acidification, could serve as a regulator of endocytosis and intracellular vesicular trafficking.

## I. Potential Role for G-Proteins in Endosomal Acidification

The observation that cAMP can modulate acidification of endosomes from rat and rabbit kidney not only identifies one potential in vivo regulator of acidification but also raises the possibility that G-proteins, through their known ability to regulate cAMP production, could be involved in the regulation of endosomal acidification. A number of studies have demonstrated the presence of several heterogeneously distributed G-proteins in the kidney. Briefly, the $\alpha$ subunits of $G_{i\text{-}3}$, $G_{i\text{-}2}$, and $G_o$ have been demonstrated in the glomerulus and related structures (GUPTA et al. 1991). In another study the subunits of $G_s$ were localized to the basolateral membranes of thick ascending limb and cortical collecting duct cells (STOW et al. 1991). In addition, $\alpha_{i\text{-}2}$ activity was observed in principal cells of the collecting duct. Interestingly, $\alpha_{i\text{-}3}$ in the proximal tubule was concentrated in subapical invaginations, the same region where early endosomes are observed (PARTON and SIMONS 1991) and corresponds to the region of the proximal tubule cell that contains the highest density of $H^+$-ATPase (BROWN et al. 1988). However, when endosomal vesicles were prepared and the membranes probed by western blotting, $\alpha_{i\text{-}3}$ was not detected (STOW et al. 1991). A third study found large quantities of $G_s$ and $G_{i\text{-}3}$ in brush-border membranes from the proximal tubule (BRUNSKILL et al. 1991). Since this is the membrane across which the bulk of endocytosis in the proximal tubule is accomplished, it is reasonable to speculate that the same G-proteins, which reside in the apical membrane, could also be localized transiently in endosomal membranes. Taken together, these studies suggest a spatial relationship between G-proteins, $H^+$-ATPase, and endocytosis.

We have recently investigated the role of G-proteins in the regulation of endosomal acidification. In our studies, endosomes were prepared from rabbit renal cortex by differential and sucrose gradient centrifugation (GURICH et al. 1991). Endosomes in this preparation are most enriched in fraction 1. Enzyme markers for Golgi, endoplasmic reticulum, mitochondria, lysozyme, or plasma membranes are not detectable in this fraction. Moreover, recent flow cytometry studies indicate that virtually all the vesicles in this fraction are endosomes (T.E. HAMMOND and R.W. GURICH, unpublished observations). Basolateral membranes are most enriched in fractions 5 and 6 whereas the brush border, i.e., apical membranes, are most enriched in fractions 8 and 9. ADP-ribosylation studies were then performed in fraction 1 endosomes using pertussis and cholera toxin.

The top panel of Fig. 1 depicts the results from experiments to determine if cholera toxin substrates were present in endosomes derived from kidney cortex. Two cholera toxin substrates with a molecular weight between 41 and 43 kDa are observed in endosomes. These substrates

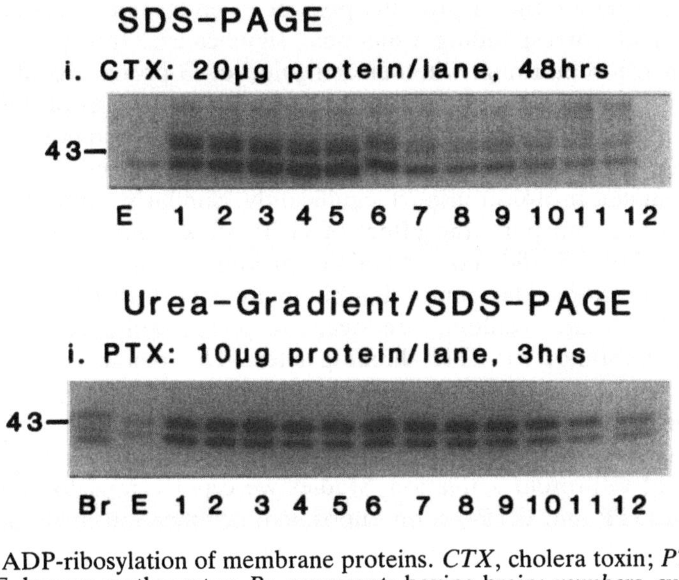


**Fig. 1.** ADP-ribosylation of membrane proteins. *CTX*, cholera toxin; *PTX*, pertussis toxin; *E*, human erythrocytes; *Br*, represents bovine brain; *numbers*, sucrose density gradient fractions

comigrate with $G_{\alpha s}$ short and $G_{\alpha s}$ long, respectively, which have been documented to reside in erythrocyte membrane controls (E). These two substrates were also present in the basolateral membrane region of the gradient. Since endosomes are also present in the basolateral membrane region, it is not known with certainty whether or not these substrates are contained within endosomes, basolateral membranes, or both.

The bottom panel displays the results from ADP-ribosylation experiments using pertussis toxin (PTX). "Br" represents membranes derived from bovine brain and "E", human erythrocytes. The autoradiograph reveals two PTX substrates in the endosomal fractions, which have a molecular weight of about 43 kDa. The top band in this lane comigrates with either $\alpha_{i\text{-}1}$ or $\alpha_{i\text{-}3}$ while the middle band could represent $\alpha_{o\text{-}1}$ or $\alpha_{i\text{-}2}$. When compared to bovine brain and human erythrocyte controls, the top band likely represents $\alpha_{i\text{-}3}$. The bottom band could represent $\alpha_{o\text{-}1}$ or $\alpha_{i\text{-}2}$. In additional studies, however, using anti-$\alpha_o$ anti-sera, it was not possible to detect significant amounts of any $\alpha_o$ species in endosomal membranes (data not shown). Thus, the PTX substrate represented by the bottom band in Fig. 1 most likely represents $\alpha_{i\text{-}2}$. Similar substrates were also observed in both the basolateral region as well as the brush border region of the gradient. Taken together, these experiments demonstrate the presence of at least four G-proteins in endosomal membranes ($G_{\alpha s}$ long and short, $\alpha_{i\text{-}3}$ and $\alpha_{i\text{-}2}$).

## II. Effects of G-Proteins on Endosomal Acidification

To determine whether the G-proteins present in endosomes derived from renal cortex had corresponding functional significance, transport studies were performed to determine whether activation of G-proteins could affect endosomal proton transport (acidification). Since multiple G-proteins are present in these membranes, initial studies employed activation of G-proteins nonspecifically by GTP and GTP-$\gamma$-S. As displayed in Fig. 2A, GTP-$\gamma$-S stimulated proton transport significantly. Similarly, while GTP also stimulated proton transport, the effect of GTP-$\gamma$-S was more pronounced. Conversely, GDP-$\beta$-S did not stimulate proton transport but, rather, inhibited acidification. This minor inhibition may be due to competition between GDP-$\beta$-S and residual, "endogenous" GTP, which resides in the endosomal membrane, for GTP binding sites. The residual GTP could provide tonic activation of endosomal acidification. GDP-$\beta$-S, by competing with GTP for binding sites, would be expected to block this tonic activation. Since it is possible that GTP or GTP-$\gamma$-s could stimulate the $H^+$-ATPase independent of G-protein activation, studies were performed to determine the effects of GTP and GTP-$\gamma$-S on endosomal acidification in the absence of ATP. As displayed in Fig. 3, GTP and GTP-$\gamma$-S were not capable of supporting proton transport in the absence of ATP. Taken together with

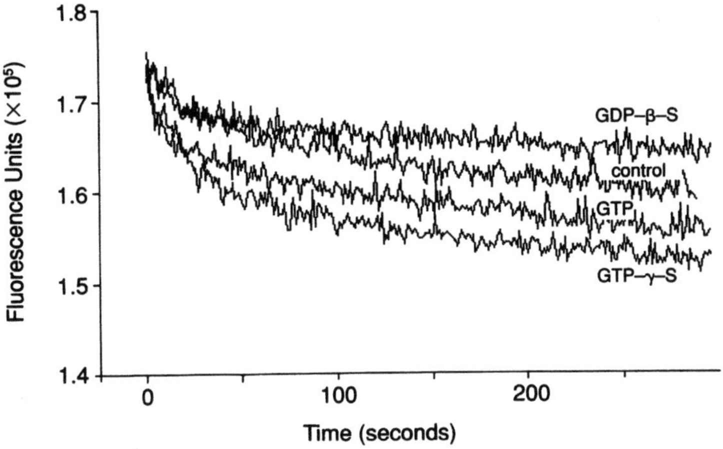


**Fig. 2.** Effects of GTP and its analogues on endosomal acidification. *Abscissa*, time, in seconds; *ordinate*, fluorescence units. A downward deflection of the FITC-dextran tracing represents acidification of the endosomal interior

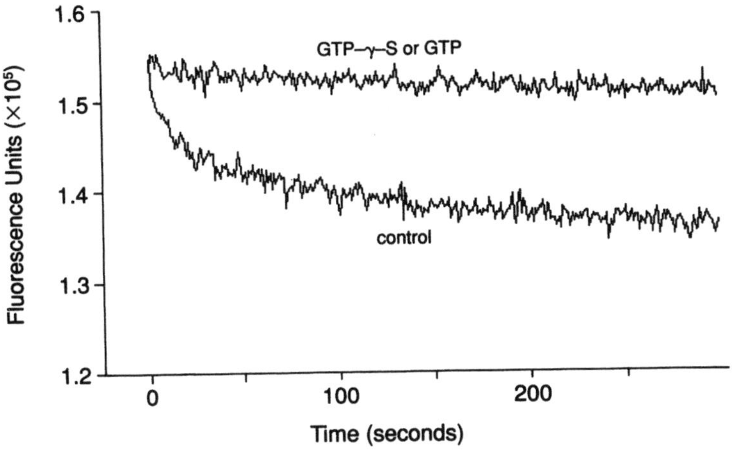


**Fig. 3.** Effects of GTP-$\gamma$-S and GTP on endosomal acidification in the absence of ATP. *Axes*, as in Fig. 2

Fig. 2, these studies indicate that nonspecific activation of G-proteins can regulate endosomal acidification in vitro.

There are several potential transport pathways that could contribute to endosomal acidification. Thus, the findings noted above could be the result of an effect on the $H^+$-ATPase, $Cl^-$ channel, proton leak pathway, or any combination thereof. Therefore, we next attempted to define the specific

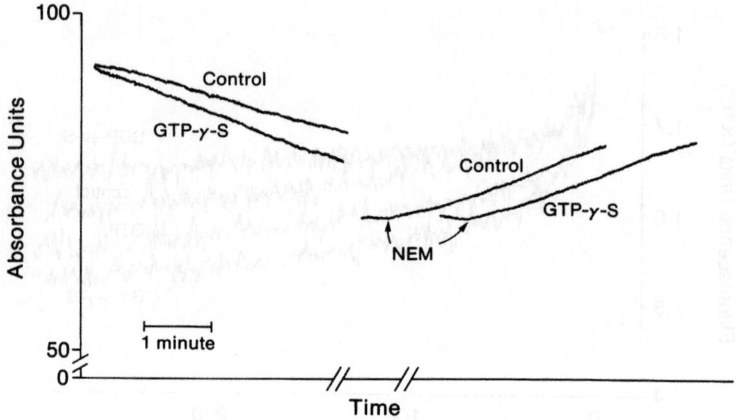


**Fig. 4.** Effects of GTP-$\gamma$-S on endosomal leak rates. These results were obtained using acridine orange rather than FITC-dextran. *Abscissa*, time, in minutes; *ordinate*, arbitrary absorbance units

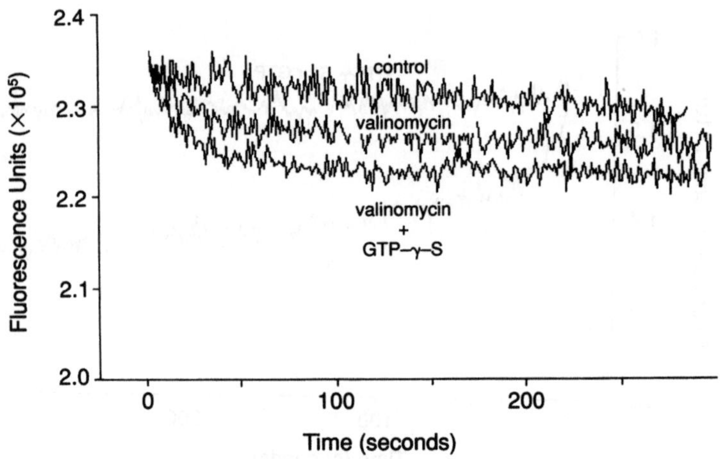


**Fig. 5.** Effects of GTP-$\gamma$-S on endosomal acidification under voltage-clamped conditions. FITC-dextran tracing. *Axes*, as in Fig. 2

pathway for G-protein interaction. First to determine whether G-proteins affected proton leak rates, equivalent pH gradients were generated by the $H^+$-ATPase in the presence or absence of GTP-$\gamma$-S (Fig. 4). *N*-Ethyl maleimide (NEM), a nonspecific inhibitor of the $H^+$-ATPase, was added to inhibit the $H^+$-ATPase. The subsequent proton leak rates, defined as the upward deflection of the tracing induced by NEM, were similar in the presence or absence of GTP-$\gamma$-S. Thus, these studies suggest that G-proteins do not affect endosomal proton leak pathways.

Studies to determine the effects of G-proteins on endosomal chloride channels were based on the following hypothesis. If stimulation of acidification of GTP-$\gamma$-S were observed under conditions in which chloride was removed and thus could not contribute to acidification, it could be inferred that the effects of G-proteins are on the $H^+$-ATPase directly, rather than through effects on the chloride conductance. The results of these studies are depicted in Fig. 5. Charge compensation in these experiments was provided by potassium/valinomycin voltage clamping (GURICH and WARNOCK 1986). GTP-$\gamma$-S plus valinomycin stimulated proton transport significantly when compared to valinomycin alone. Thus, stimulation of acidification was observed under conditions in which the $Cl^-$ contribution to acidification was eliminated. Therefore, the effects of GTP-$\gamma$-S appear to be

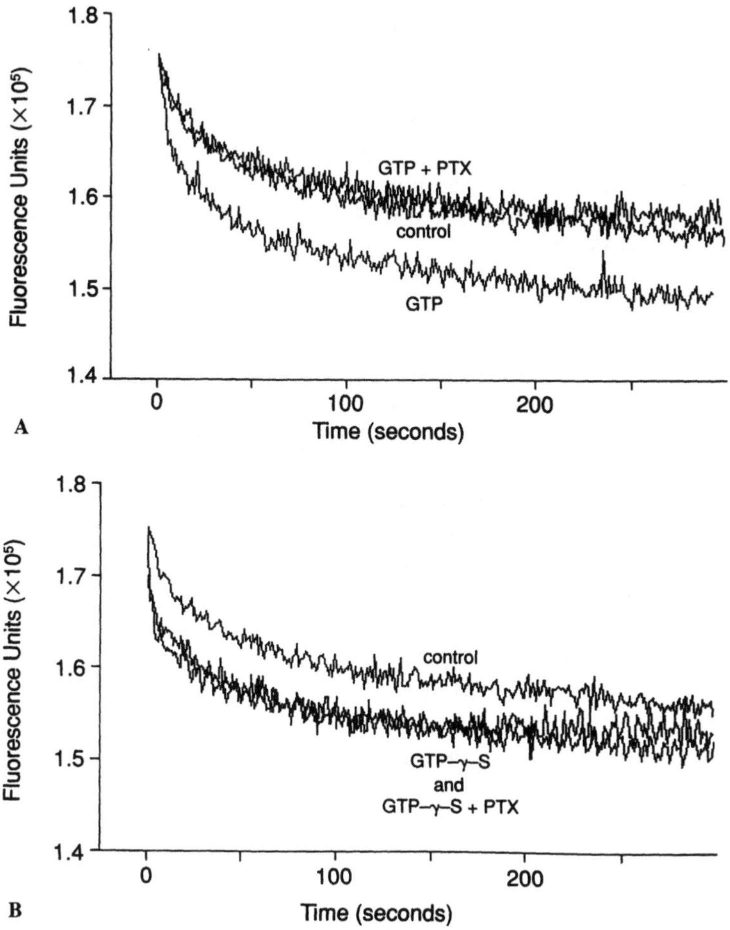


**Fig. 6A,B.** Effects of pertussis toxin (*PTX*) on endosomal acidification. **A** Effects of PTX in the presence of GTP. **B** Effects of PTX in the presence of GTP-$\gamma$-S. FITC-dextran tracing. *Axes*, as in Fig. 2

on the $H^+$-ATPase per se. These results are consistent with those reported previously by other investigators who used the $Cl^-$-sensitive dye SPQ to label rat endosomes in vivo (BAE and VERKMAN 1990). In these studies it was found that GTP-$\gamma$-S had no effect on conductive $Cl^-$ entry.

Although the previous studies indicate a role for G-proteins in the regulation of endosomal acidification, they do not indicate the specific G-protein involved. Since the pattern of stimulation of acidification by GTP and its analogues, i.e., GTP-$\gamma$-S > GTP > control > GDP-$\beta$-S was similar to the pattern of activation of ion channels by $G_i$ species, the effects of PTX were investigated. PTX, per se, had no effect on endosomal acidification (not shown). However, PTX blocked the stimulation of acidification by GTP (Fig. 6A). Conversely, PTX had no effect on the irreversible stimulation of acidification in the presence of GTP-$\gamma$-S (Fig. 6B). These results indicate that the G-protein that is involved in the regulation of endosomal acidification is a pertussis toxin substrate. The reversible activation of this protein by GTP can be blocked by PTX, whereas the irreversible activation by GTP-$\gamma$-S cannot be blocked by PTX. Since $\alpha_o$ species were not detected in these membranes, the G-protein(s) involved in the stimulation of proton transport is either $G_{i\text{-}2}$ or $G_{i\text{-}3}$. Preliminary results from our laboratory using purified subunits, indicate that $\alpha_{i\text{-}3}$ stimulates proton transport in a manner similar to GTP and GTP-$\gamma$-S.

## D. Conclusions

Taken together, these studies indicate that larger molecular weight G-proteins can regulate acidification in endosomes derived from the proximal tubule in vitro. The physiologic significance of this observation has not been established, however. It is conceivable that larger molecular weight G-proteins, through their effects on acidification, could influence intracellular vesicular trafficking, protein degradation and/or receptor recycling. Alternatively, in the renal proximal tubule, where an $H^+$-ATPase has been shown to contribute to reclamation of filtered bicarbonate (PREISIG et al. 1987), G-protein regulation of the $H^3$-ATPase could serve as an important determinant of luminal acidification. A net increase in acidification could be achieved through direct stimulation by $G_{i\text{-}3}$ of existing proton pumps or by exocytic insertion of "new" pumps into the membrane in a manner analagous to that observed in the collecting duct (BROWN 1989). It is possible that endosomal G-proteins other than $G_{i\text{-}3}$ i.e., $G_{i\text{-}2}$, $G_{short}$, or $G_{long}$, whose function have not been determined, could mediate the exocytic process.

In addition to the question of physiologic significance, there are other issues raised by our studies. For instance, it is not known whether G-proteins activate the $H^+$-ATPase directly, as has been described for certain ion channels, or through traditional second messenger pathways. Although cAMP has been demonstrated to regulate endosomal acidification, it has not been demonstrated that G-proteins mediate the cAMP effect. Similarly, it is not known whether endosomal G-proteins activate other second messenger pathways such as phospholipases, and whether such activation has an effect on endosomal acidification.

The presence of G-proteins in endosomal membranes, the stimulation of endosomal acidification by G-proteins and the regulation of acidification by cAMP all suggest a ligand/receptor interaction that is coupled to G-proteins. However, a candidate ligand/receptor complex that G-proteins might couple to the $H^+$-ATPase has not been identified. Given the variety of filtered substances that are endocytosed by the proximal tubule, it is difficult to identify any one receptor or class of receptors that would serve as likely candidates. Transferrin, which is endocytosed into many, if not all mammalian cells, could be a possibility since transferrin receptors are present in proximal tubule. Alternatively, angiotensin II, which has potent effects on sodium transport (DOUGLAS et al. 1990) and bicarbonate reabsorption (LIU and COGAN 1989) by the proximal tubule, could be involved in the endocytic process. Both sodium transport and bicarbonate reabsorption are known to be mediated by both G-proteins and cAMP. Until these types of relationships can be explored by more direct studies, the mechanism of the regulation of endosomal acidification by G-proteins in the kidney and the relevance of this process to endocytosis in other cell types remains to be elucidated, however.

## References

Alpern RA, Stone DK, Rector FC (1991) Renal acidification mechanisms. In: Brenner BM, Rector FC (eds) The kidney. Saunders, Philadelphia, p 337

Bae HR, Verkman AS (1990) Protein kinase A regulates chloride conductance in endocytic vesicles from proximal tubule. Nature (London) 348:637–639

Brown D (1989) Vesicle recycling and cell-specific function in kidney epithelia cells. Ann Rev Physiol 51:771–787

Brown D, Hirsch S, Gluck S (1988) Localization of a proton-pumping ATPase in rat kidney. J Clin Invest 82:2114–2126

Brunskill N, Bastani B, Hayes C, Morrisson J, Klahr S (1991) Localization and polar distribution of several G-proteins subunits along nephron segments. Kidney Int 40:997–1006

Camargo MSF, Sumpio BE, Maack T (1989) Renal hydrolysis of absorbed protein: influence of load and lysosome pH. Am J Phys 247 F656–664

Cohn ZA, Steinman RA (1982) Phagocytosis and fluid-phase pinocytosis. In: Evered D, Collins G (eds) CIBA Foundation Symposium on Membrane Recycling. Pitman Books Ltd, London, pp 15–34

Corley C, Sipe DM, Murphy RF (1989) Regulation of endocytic pH by the $Na^+/K^+$ ATPase in living cells. Proc Natl Acad Sci USA 86:544–548

Dautry-Varsat A, Ciechanover A, Lodish H (1983) pH and the recycling of the transferrin during receptor mediated endocytosis. Proc Natl Acad Sci USA 80:2258–2262

Douglas JG, Romero M, Hopfer U (1990) Signaling mechanisms coupled to the angiotensin receptor of proximal tubular epithelium. Kidney Int 38 [Suppl 30]:43–47

Fine R, Goldenberg R, Sorrentino J, Herschman H (1981) Subcellular structures involved in internalization and degradation of epidermal growth factor. J Supramol Struct 15:235–251

Gluck S, Cannon C, Al-Awqati Q (1982) Exocytosis regulates urinary acidification in turtle bladder by rapid insertion of $H^+$ ATPase into the luminal membrane. Proc Natl Acad Sci USA 79:4327–4331

Gluck S, Caldwell J (1987) Immunoaffinity purification and characterization of vacuolar $H^+$ ATPase from bovine kidney. J Biol Chem 262:15780–15789

Goldstein JL, Anderson RG, Brown MS (1979) Coated pits, coated vesicles and receptor mediated endocytosis. Nature 279:679–685

Goldstein JL, Brown MS, Anderson REW, Russell DW, Schneider WJ (1985) Receptor-mediated endocytosis: concepts emerging from the LDL receptor system. Annu Rev Cell Biol 1:1–39

Gonatas NK, Kim SU, Stieber A, Avrameas S (1977) Internalization of lectins in neuronal GERL. J Cell Biol 73:1–13

Gorden P, Carpentier JL, Cohen S, Orci L (1978) Epidermal growth factor: morphological demonstration of binding, internalization and lysosome association in human fibroblasts. Proc Natl Acad Sci USA 75:5025–5029

Gruenberg J, Howell KE (1989) Membrane traffic in endocytosis: insights from cell-free assays. Ann Rev Cell Biol 5:453–481

Gupta A, Bastani B, Purcell H, Hruska KA (1991) Identification and localization of pertussis toxin sensitive GTP-binding proteins in bovine kidney glomeruli. J Am Soc Neph 2(2):172–178

Gurich RW, Warnock DG (1986) Electrically neutral $Na^+/H^+$ exchange in endosomes obtained from rabbit renal cortex. Am J Physiol 251:F702–F709

Gurich RW, DuBose Jr TD (1989) Heterogenety of cAMP effect on endosomal protein transport. Am J Physiol 251:F777–F784

Gurich RW, Codina J, DuBose Jr TD (1991) A potential role for guanine nucleotide-binding protein in the regulation of endosomal acidification. J Clin Invest 87:1547–1552

Helenius A, Mellman I, Wall D, Hubbard A (1983) Endosomes. Trends Biochem Sci 8:245–249

Hilden SA, Ghoshroy KB, Madias NE (1990) $Na^+/H^+$ exchange, but not $Na^+/K^+$ ATPase is present in endosome-enriched microsomes from rabbit renal cortex. Am J Physiol 258:F1311–F1319

Hopkins CR (1983) Intracellular routing of transferrin and transferrin receptors in epidermoid carcinoma A431 cells. Cell 35:321–330

Joseph KC, Kim SU, Stieber A, Gonatas NK (1978) Endocytosis of cholera toxin in neuronal GERL. Proc Natl Acad Sci USA 75:2815–2819

Kahn CR (1976) Membrane receptors for hormones and neurotransmitters. J Cell Biol 70:261–286

Karin M, Mintz B (1981) Receptor mediated endocytosis of transferrin in developmentally totipotent mouse teratocarcinoma stem cells. J Biol Chem 256:3245–3252

Klausner RD, van Renswoude J, Kempf C, Rao K, Bateman JL, Robbins AR (1987) Failure to release iron from transferrin in a chinese hamster ovary cell mutant pleiotropically defective in endocytosis. J Cell Biol 98:1078–1101

Kornfeld S, Mellman I (1989) The biogenesis of lysosomes. Ann Rev Cell Biol 5:483–526

Lencer WI, Weyer P, Verkman AS, Ausiello DA, Brown D (1990) FITC-dextran as a probe for endosome function and localization in kidney. Am J Physiol 258:C309–C317

Lencer WI, Verkman AS, Amin Arnaout, Ausiello DA, Brown D (1990a) Endocytic vesicles from renal papilla which retrieve the vasopressin sensitive water channel do not contain a functional $H^+$ ATPase. J Cell Biol 111: 379–389

Limet JW, Quintart J, Schneider YJ, Courtoy PJ (1985) Receptor-mediated endocytosis of polymeric IgA and galactosylated serum albumin in rat liver. Eur J Biochem 176:539–548
Liu FY, Logan MG (1989) Angiotensin II stimulates early proximal bicarbonate absorption in the rat by decreasing cyclic adenosine monophosphate. J Clin Invest 84:83–91
Maack T, Park CH, Camargo MJF (1985) Renal filtration, transport and metabolism of proteins. In: Seldin DW, Giebisch G (eds) The kidney: physiology and pathophysiology. Roman, New York, p 1773
Madsen K, Tisher CC (1980) Structural-functional relationships along the distal nephron. Am J Physiol 250:F1–F15
Marsh M, Bolzau E, Helenius A (1983) Penetration of semliki forest virus from acid prelysosomal vacuoles. Cell 32:931–940
Mayorga LS, Diaz R, Stahl PD (1989) Regulatory role for GTP-binding proteins in endocytosis. Science 244:1475–1477
Mellman I, Fuchs R, Helenius A (1986) Acidification of the endocytic and exocytic pathways. Annual Rev Biochem 55:633–700
Montesano R, Roth J, Robert A, Orci L (1982) Non-coated membrane invaginations are involved in binding and internalization of cholera and pertussis toxins. Nature 296:651–653
Moriyama Y, Nelson N (1987) The purified ATPase from chromaffin granule membrane is an anion dependent proton pump. J Biol Chem 262:9175–9180
Parton RG, Simons K (1991) Endocytosis in the kidney: insights from the MDCK cell systems. In: Tisher CL, Madsen KM (eds) Seminars in nephrology, vol 11. Saunders, Philadelphia, p 440
Pastam I, Willingham MC (1983) Receptor mediated endocytosis: coated pits, receptosomes and the Golgi. Trends Biochem Sci 8:250–254
Preisig PA, Ives HE, Cragoe EJ, Alpern RJ, Rector FC (1987) Role of the $Na^+/H^+$ antiporter in rat proximal tubule bicarbonate absorption. J Clin Invest 80:970–978
Reeves B, Gurich RW (1991) Characterization of renal endosomal chloride channels. J Am Soc Nephrol 2(3):749
Robbins AR, Peng SS, Marshall JL (1983) Mutant Chinese hamster ovary cells pleiotropically defective in receptor mediated endocytosis. J Cell Biol 96:1064–1071
Robbins AR, Oliver C, Bateman JL, Krag SS, Galloway CJ, Mellman I (1989) A single mutation in Chinese hamsters ovary cells impairs both Golgi and endosomal function. J Cell Biol 99:1296–1308
Roff LF, Fuchs R, Mellman I, Robbins AR (1986) Chinese hamster ovary cell mutants with temperature-sensitive defects in endocytosis. J Cell Biol 103:2283–2797
Sabolic I, Hause W, Burckhardt G (1985) ATP-dependent $H^+$ pump in membrane vesicles from rat kidney cortex. Am J Phys 248:F835–F844
Sabolic J, Brown D (1990) $Na^+(Li^+)$-$H^+$ exchange in rat renal cortical vesicles with endosomal characteristics. Am J Physiol 258:F1245–F1253
Schwartz GJ, Al-Awqati Q (1985) Carbon dioxide causes exocytosis of vesicles containing $H^+$ pumps in isolated perfused proximal and collecting tubules. J Clin Invest 75:1638–1644
Schwiebert E, Light D, Fejes-Toth G, Naray-Fejes-Toth A, Stanton WB (1990) A GTP-binding protein activates chloride channels in a renal epithelium. J Biol Chem 265:7725–7728
Shi LB, Fushimi K, Bae HR, Verkman AS (1991) Heterogeneity in ATP-dependent acidification in endocytic vesicles from kidney proximal tubule. Biophys J 59:1208–1217
Silverstein SC, Steinman RM, Cohn ZA (1977) Endocytosis. Ann Rev Biochem 46:669–722

Steinman RM, Mellman I, Muller W, Cohn Z (1983) Endocytosis and the recycling of the plasma membrane. J Cell Biol 96:1–27
Stetson DL, Steinmetz PR (1983) Role of membrane fusion in $CO_2$ stimulation of proton secretion by turtle bladder. Am J Physiol 245:C113–C120
Stone DK (1988) Proton translocating ATPases: issues in structure and function. Kidney Int 33:767–779
Stow JL, Sabolic I, Brown D (1991) Heterogeneous localization of G-protein $\alpha$-subunits in rat kidney. Am J Physiol 261:F831–F840
Strauss W (1969) Occurrence of phagosomes and phagolysosomes in different segments of the nephron in relation to the reabsorption, transport, digestion, and extrusion of intravenously injected horseradish peroxidase. J Cell Biol 21:295–308
Tran D, Carpentier JL, Sawano F (1987) Ligands internalized through coated or non-coated invaginations follow a common intracellular pathway. Proc Natl Acad Sci USA 89:7957–7961
Van Renswoude J, Bridges K, Harford J, Klausner R (1982) Receptor mediated endocytosis of transferrin and the uptake of $F_e$ in K562 cells. Proc Natl Acad Sci USA 79:6186–6190
Wall DA, Maack T (1985) Endocytic uptake, transport and catabolism of proteins by epithelial cells. Am J Phys 248:C12–C20
Wang, Shi LB, Verkman AS (1991) Functional water channels and proton pumps are in separate populations of endocytic vesicles in toad bladder granular cells. Biochem 30:2888–2894
Wang ZQ, Gluck S (1990) Isolation and properties of bovine kidney brush border vacuolar $H^+$-ATPase. J Biol Chem 265:21957–21965
Xie XS, Stone DK (1986) Isolation and reconstitution of the clathrin-coated vesicle proton translocating complex. J Biol Chem 261:2492–2495
Ye R, Shi LB, Lencer WI, Verkman AS (1989) Functional colocalization of water channel and proton pumps in endosomes from kidney proximal tubule. J Gen Phys 93:885–902
Yoshimura A, Kuroda K, Kawasaki K, Yamashima S, Maeda T, Ohnishi S (1982) Infectious cell entry mechanism of influenza virus. J Virol 43:289–293

CHAPTER 80

# cAMP-Independent Regulation of Adipocyte Glucose Transport Activity and Other Metabolic Processes by a Complex of Receptors and Their Associated G-Proteins

C. LONDOS and I.A. SIMPSON

## A. Introduction

Many contributions to this volume describe the molecular progress on GTP-binding proteins and on the interactions of the G-proteins with specific receptors and their various targets. By contrast, the information described in this chapter is at a level less well-defined biochemically since it is based on cellular studies, and any conclusions regarding G-protein involvement are based on inferences drawn from the "black box" level of the intact adipose cell. On the other hand, the wealth of information on the modulation of a wide variety of metabolic events in adipocytes by a complex of G-protein-linked receptors, such as the $\beta$-adrenergic and adenosine $A_1$ receptors, provides a basis for probing for targets of these G-proteins which mediate these metabolic effects. Since, as is discussed below, both the $\beta$-adrenergic and adenosine receptors produce opposing effects on adenylyl cylcase activity in adipose cells, it is important first to discriminate between events secondary to the changes in cAMP and events independent of cAMP. Insulin stimulates the translocation of glucose transporters to the plasma membrane, and a primary focus of our efforts, reviewed in this report, has been to establish the cAMP-independent modulation of plasma membrane-bound glucose transporters by the same receptor/G-protein complexes that regulate adenylyl cyclase. We shall also touch on the evidence that the same receptor complex intervenes in a number of biological events at the plasma membrane, such as insulin-like growth factor II (IGF-II)/mannose-6-phosphate receptor binding and insulin receptor signaling. These interactions suggest a comprehensive role for complexes of receptor/G-proteins in the coordinating regulation of energy metabolism in adipose cells.

Glucose transport into cells is mediated by a family of transporter proteins designated GLUT1–GLUT5, based on the order of cloning of their cDNAs (for review see BELL et al. 1990). In cells in which insulin significantly stimulates glucose transport activity, i.e., white and brown adipose cells, heart and skeletal muscle, two transporter isoforms are found, GLUT1 and GLUT4, of which the latter is the more abundant (HOLMAN et al. 1990; ZORZANO et al. 1989). In all of the above cells, insulin has been shown to stimulate transport by initiating a translocation of transporter molecules from an intracellular pool to the plasma membrane (SUZUKI and KONO 1980; CUSHMAN and WARDZALA 1980; WARDZALA and JEANRENAUD 1981; SLOT et al. 1991a). However, most of the biochemical characterization of the translocation phenomenon has been deduced from studies on primary rat adipocytes due to their ease of isolation and subfractionation.

Adipose cells have long been a model for investigating receptor/G-protein regulation of adenylyl cyclase. The cyclase is regulated by two opposing circuits, one containing an ensemble of stimulatory receptors ($R_s$), for example, $\beta$-adrenergic, adrenocorticotropic hormone (ACTH), glucagon, and thyroid-stimulating hormone, and the other of inhibitory receptors ($R_i$), such as adenosine $A_1$, prostaglandin, nicotinic acid, $\alpha_2$-adrenergic, and their associated G-proteins, known generically as $G_s$ and $G_i$. It is of historical interest that the idea of a "transducing" system interposed between the hormone receptors and the cAMP-generating component first arose from the finding that the various stimulatory receptors converged on a common target, the cyclase (BIRNBAUMER and RODBELL 1969).

Clinical studies had suggested that catecholamines inhibit insulin-stimulated glucose disposal in peripheral tissues. However, a direct demonstration of catecholamine-mediated inhibition of insulin-stimulated glucose transport in the isolated adipocyte proved elusive. A breakthrough in this area was provided by TAYLOR et al. (1976) who demonstrated that the catecholamine effect is markedly antagonized by adenosine. This nucleoside, which is found in the incubation medium of isolated adipocytes, has profound effects on cellular metabolism at extremely low concentrations. Subsequently, the efforts of a number of laboratories have confirmed that activation of the adenosine receptor negates the inhibitory responses of $\beta$-adrenergic agents on glucose transport (see SIMPSON and CUSHMAN 1986, for review), and that careful management of ambient adenosine was essential for examining control of glucose transport activity by a variety of receptors. It should be noted also that in addition to adenosine, agents for all other $R_i$ receptors both augment glucose transport activity and counteract the inhibitory actions of all $R_s$ receptor agonists on transport activity. That is, either nicotinic acid or prostaglandin $E_2$ as well as adenosine antagonize the inhibitory actions of either ACTH or glucagon as well as $\beta$-adrenergic effectors, i.e., isoproterenol.

## B. Lack of a Relationship Between cAMP and Glucose Transporter Activity

Although TAYLOR and HALPERIN (1979) noted inconsistencies between cAMP levels in adipocytes and corresponding glucose transport activity, they and most other investigators assumed that the opposing actions of $\beta$-adrenergic agents and adenosine on transport result from their ability to

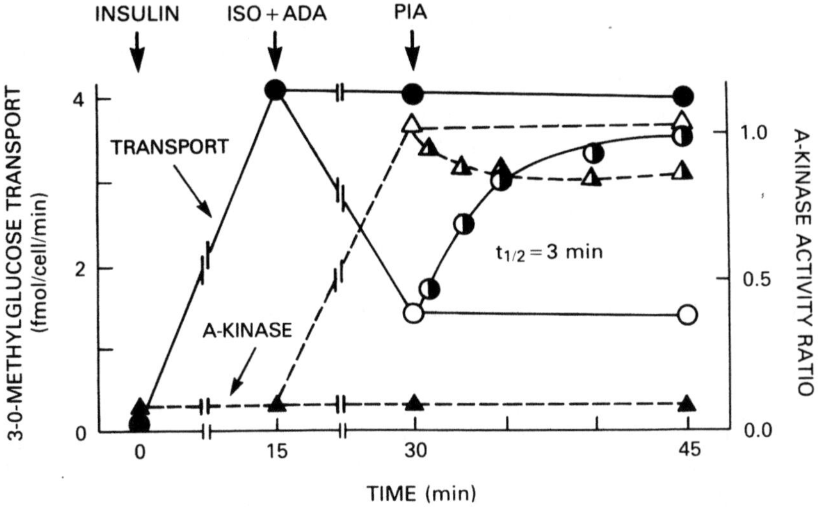


**Fig. 1.** PIA reversal of isoproterenol inhibition of glucose transport activity is independent of changes in A-kinase activity. The data compare 3-O-methylglucose transport activity (*circles*) and A-kinase activity (*triangles*) in isolated rat adipose cells. Cells were incubated for 15 min to initiate the glucose transport response. Subsequently, isoproterenol (*ISO*) and adenosine deaminase (*ADA*) were added to provide simultaneous ß-adrenergic stimulation and deamination of adenosine in the medium, respectively. Note that the profound inhibition of transport activity by isoproterenol (○) was prevented by PIA (●), the deaminase-resistant adenosine $A_1$ receptor agonist. The critical observation is that addition of PIA at 30 min reversed the isoproterenol-inhibited transport (◑) but had only a minimal effect on A-kinase activity (▲). From such data is was concluded that adenosine receptor effects on glucose transport activity are independent of changes in cAMP. Data from Kuroda et al. (1987)

modulate cAMP and thus cAMP-dependent protein kinase (A-kinase) activity. Key observations indicating that alterations in cAMP might not be responsible for the transport changes were reported by Kuroda et al. (1987), who monitored A-kinase activity and glucose transport activity under carefully defined conditions for maintaining the kinase activity at steady-state levels (Honnor et al. 1985). It was found that adenosine or phenylisopropyladenosine (PIA) augments (30%) insulin-stimulated glucose transport activity without any change in the A-kinase activity. Moreover, a marked dissociation between transport and A-kinase was observed upon examination of PIA reversal of isoproterenol-induced inhibition of transport activity; this phenomenon was revealed upon sequential addition of reagents as depicted in Fig. 1. In these experiments insulin was added to initiate the glucose transport response in the presence of adenosine. Subsequently, isoproterenol and adenosine deaminase were added to promote the $\beta$-adrenergic inhibition of transport activity. Adenosine deaminase converts adenosine to inosine which is inactive at adenosine receptors, and the presence of isoproterenol without an adenosine receptor agonist produces maximal activation of adenylyl cylcase leading to cAMP concentrations in the cell far in excess of those required to activate fully the A-kinase. Finally, PIA, a potent deaminase-resistant adenosine $A_1$ receptor agonist, was added, which completely reversed the isoproterenol-induced transport inhibition. However, A-kinase remained nearly fully active because of the initial overshoot of cAMP. This sequential addition of reagents established that adenosine receptor mediated reversal of the $\beta$-adrenergic receptor effect is cAMP independent and strongly suggested that $\beta$-adrenergic inhibition of glucose transport activity is not mediated by an elevation of cAMP.

Ultimately, more direct evidence against a role for cAMP in $\beta$-adrenergic inhibition of glucose transport was provided by a comparison in $^{32}$P-loaded rat adipocytes among phosphorylation of GLUT4, transport activity, and the activity state of A-kinase (Nishimura et al. 1991). Although isoproterenol inhibits glucose transport activity in the absence but not in the presence of adenosine (Fig. 2a), under both conditions the $\beta$-adrenergic agonist stimulates A-kinase activity, as assessed by determining the phosphorylation state of the major cellular A-kinase substrate perilipin (Nishimura et al. 1991). In addition to the failure of $\beta$-adrenergic receptor-mediated transport inhibition to correlate with activation of A-kinase, measurement of phosphate incorporation into the transporter molecule argues against a phosphorylation event in transport inhibition (Fig. 2b–d). In the absence of isoproterenol stimulation GLUT4 was only partially phosphorylated (0.1–0.2 mol/mol), as seen also by James et al. (1989), and addition of isoproterenol produced only a modest (40%) increase in phosphate incorporation. However, this increase was evident both in the presence and absence of adenosine, whereas transport inhibition was seen only in the absence of the nucleoside. Finally, the bulk of A-kinase phosphorylation of GLUT4 occurs in intracellular membranes, whereas it is the plasma membrane-bound transporters that are inhibited by isoproterenol. Such data cast serious doubt on a role for A-kinase-mediated phosphorylation in $\beta$-adrenergic inhibition of glucose transport activity.

## C. G-Proteins in Glucose Transporter Regulation

Since all of the $R_s$ and $R_i$ receptors noted above exert their regulatory actions on adenylyl cyclase through G-proteins, it was important to determine whether they modulate their cAMP-independent effects on the glucose transport process by an analogous mechanism. One method for determining whether certain G-proteins are involved in a given process is to monitor the effects of the bacterial toxins, cholera and pertussis, which ADP-ribosylate $G_s$ and $G_i$-1, -2, -3/$G_o$, respectively. Since intoxication of adipocytes presents special problems, it was necessary to devise procedures

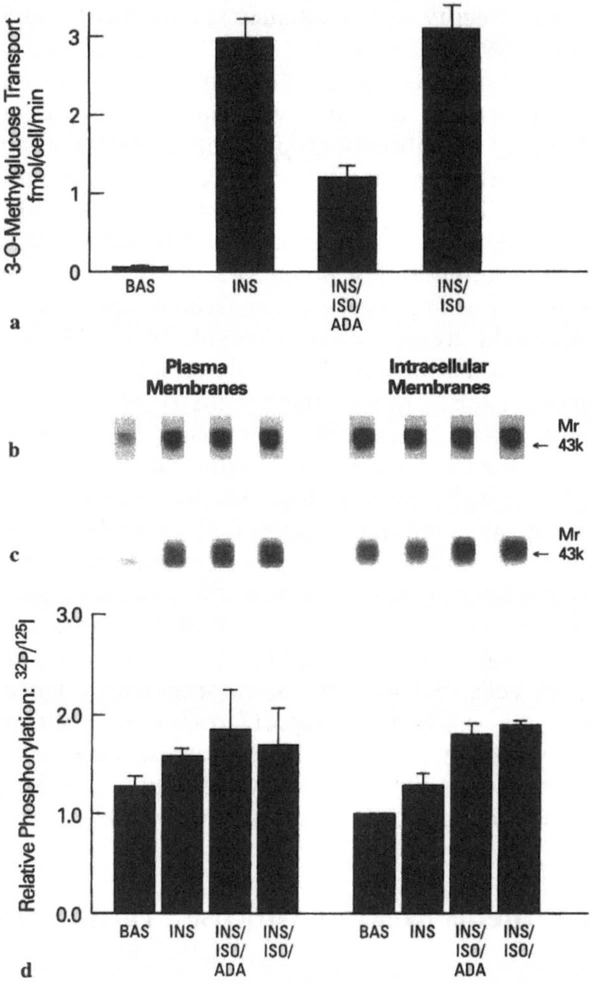


**Fig. 2a–d.** Lack of correlation between isoproterenol inhibition of glucose transport activity and phosphorylation of GLUT4. **a** 3-O-Methylglucose transport activity. **b** GLUT4 distribution, Western blotting. **C** Phosphorylation state of GLUT4. **d** Relative phosphorylation of GLUT $^{32}P/^{125}I$. Adipose cells were equilibrated with $[^{32}P_i]$ and subsequently exposed to the indicated conditions. Glucose transport activity was measured in parallel incubations. Cells were fractionated and GLUT4 glucose transporters immunoprecipitated with specific antibodies. The immunoprecipitate was divided, one fraction was separated by SDS-PAGE and $^{32}P$ incorporation detected by autoradiography. The remaining portion was Western blotted to determine the subcellular distribution of GLUT4. Isoproterenol inhibited insulin-stimulated glucose transport activity by 64% in the absence but not in the presence of adenosine receptor agonists (PIA). GLUT4 was partially phosphorylated (0.1–0.2 mol/mol) in basal cells when A-kinase activities were suppressed, and the level of phosphorylation was unchanged by insulin. Isoproterenol increased the extent of phosphorylation in the intracellular GLUT4 transporters (0.3–0.4 mol/mol) but not plasma membranes, and was equally effective in the presence or absence of adenosine receptor agonists. Thus changes in phosphorylation state of GLUT4 do no appear to account for the modulation of transport activity. (Data from Nishimura et al. 1991)

that also preserved cell integrity and maintained responsiveness to $R_s$ and $R_i$ ligands (HONNOR et al. 1992).

To assess the effects of cholera and pertussis intoxication on insulin-stimulated glucose transport activity, we chose three criteria: (a) augmentation by PIA, (b) inhibition by isoproterenol, and most importantly (c) prevention of the isoproterenol inhibition by PIA. These phenomena are evident in control cells (Fig. 3). Following treatment with cholera toxin there was a modest decrease in transport activity, but surprisingly a loss of responsiveness to isoproterenol. As noted previously, the isoproterenol effect was significantly less in cholera toxin treated cells than in either control or pertussis toxin treated cells (HONNOR et al. 1992). Such an observation is contrary to that observed typically on adenylyl cyclase where cholera toxin promotes isoproterenol stimulation (see HONNOR et al. 1992; Fig. 2). On the other hand, as expected, there was full retention of the sensitivity of PIA to augment transport activity following cholera toxin treatment of cells. Conversely, pertussis toxin treatment resulted in a loss of PIA augmentation but full retention of isoproterenol inhibition, which was no longer blocked by the adenosine receptor agonist. These data clearly established the involvement of toxin sensitive G-proteins in both the $R_s$-mediated inhibition and the $R_i$-mediated augmentation of glucose transport activity. Currently, several forms of $G_s$ and $G_i$-1, -2, and -3 are known to be present in rat adipose cells, but we have no information of which of these isoforms mediates the various transport responses (MILLIGAN and SAGGERSON 1991; WATKINS et al. 1989). In contrast to the toxin effects on other receptors, neither toxin interferes with the ability of a maximal insulin cocnentration to translocate glucose transporters to the plasma membrane.

## D. How Do G-Proteins Mediate Glucose Transporter Activity?

To summarize briefly, the complex of receptors that both stimulate and inhibit adenylyl cyclase activity in adipocytes by a G-protein-mediated process inversely regulate insulin-stimulated glucose transport activity. These $R_s$- and $R_i$-induced changes in transport activity are both cAMP independent and mediated by toxin-sensitive G-proteins. Furthermore, the receptors appear to modify transport activity at the level of the plasma membrane since they do not alter the subcellular location of the transporters (KURODA et al. 1987; NISHIMURA et al. 1991; VANNUCCI et al. 1992).

The key question that arises is whether or not transporter molecules interact directly with the G-proteins. Certain structural and chemical features might suggest a direct interaction. Both the cyclase and the various glucose transporter isoforms share a similar structural motif, i.e., 12 membrane-spanning regions. Also, both contain high-affinity binding sites for forskolin (LAURENZA et al. 1989). Finally, estimates for the number of

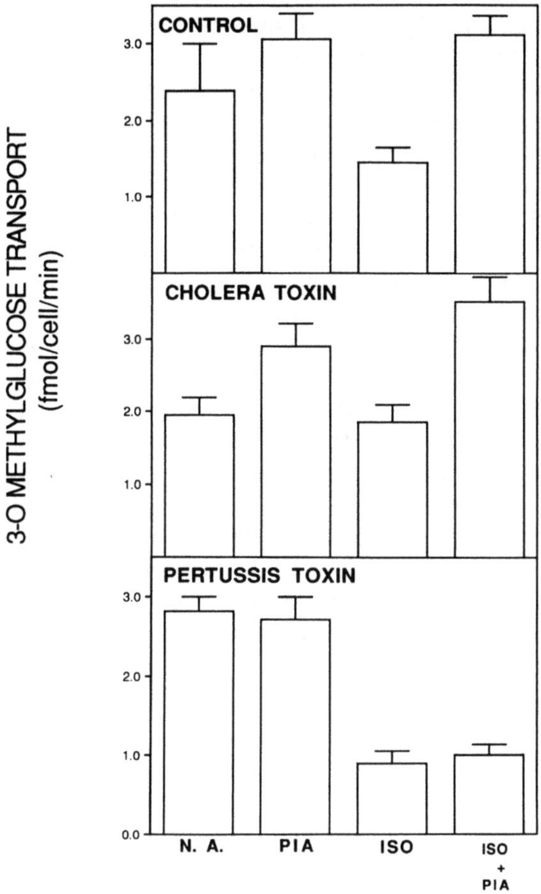


**Fig. 3.** Effects of cholera and pertussis toxins on glucose transport activity in isolated rat adipose cells. Isolated rat adipocytes were preincubated without toxin (control) or with cholera or pertussis toxin. After washing the cells were exposed to insulin for 15 min to initiate the transport response. Subsequently, the effects of various on 3-O-methylglucose transport activity were examined. *N.A.*, No additions; *PIA*, the adenosine receptor agonist; *ISO*, isoproterenol, a ß-adrenergic receptor agonist; *PIA + ISO*, the combination of receptor agonists. Note that PIA prevented or reversed the isoproterenol-mediated inhibition in control and cholera toxin treated cells but not that in pertussis toxin treated cells. Similarly, pertussis toxin prevented the PIA augmentation of transport activity evident under the other two conditions. On the other hand, the PIA effect was not altered by cholera toxin, which inhibited transport activity and obviated any further effect of isoproterenol. Analysis by paired *t* test revealed that the loss of the isoproterenol effect by cholera toxin is statistically significant. Thus, the data show that both the ß-adrenergic and adenosine receptor mediated modulations of insulin-stimulated glucose transport activity are mediated by toxin-sensitive G-proteins. (Data from HONNOR et al. 1992)

G-protein molecules in adipocyte plasma membranes are comparable to those for glucose transporter molecules, i.e., approximately 30 pmol/mg protein. On the other hand, there is a growing body of data that argues to the contrary. First, whereas adenylyl cyclase, ion channels, and phospholipases respond to G-protein-mediated regulation in seconds, the modulation of transport occurs with a half-time of the order of 1–2 min (KURODA et al. 1987). Second, we have failed in attempts to coprecipitate transporters/G-protein complexes with antibodies against either transporters or a variety of G-protein $\alpha$ subunits (S. VANNUCCI and I.A. SIMPSON, unpublished). Finally, in $K^+$-depleted cells (H. NISHIMURA and I.A. SIMPSON, unpublished) or at temperatures below 25°C (JOOST et al. 1987), isoproterenol effects on transport activity are lost, but adenylyl cyclase regulation remains intact. Such data suggest an intervening step between G-protein activation and glucose transporter responses.

More recently, insights into the mechanism of transporter modulation have been gained with the use of a novel exofacial photolabel that labels only those transporters accessible to the extracellular medium (VANNUCCI et al. 1992). Isoproterenol is found to occlude access to the transporter without causing any significant loss of transporters from the plasma membrane. Conversely, adenosine enhances access to the transporters and blocks the

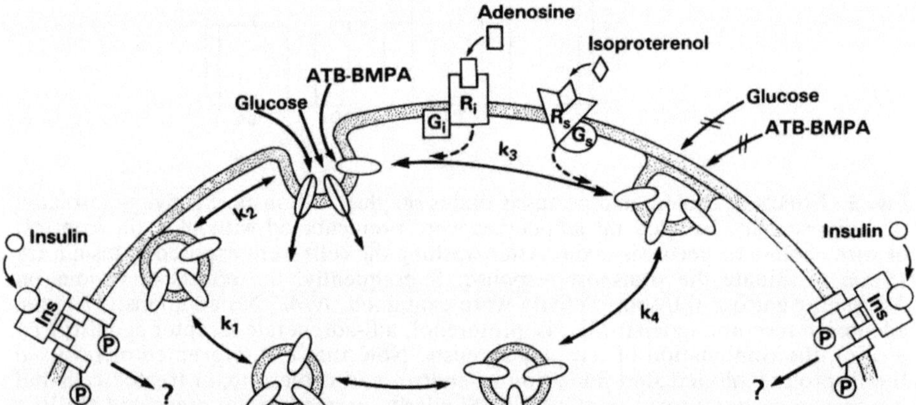


**Fig. 4.** Model depicting regulation of glucose transport activity in the adipose cell. This schematic model depicts our current thoughts on the mechanism by which adenosine and isoproterenol modulate insulin-stimulated glucose transport activity. Insulin is envisaged to promote translocation by altering $k_1$ and $k_2$. In the plasma membrane, the transporter can exist in two states; one is accessible to glucose and the photolabel 2–4-(1-azi-2,2,2-trifluoethyl)benzoyl-1,3-bis(D-mannos-4-yloxy)-2-propylamine (*ATB-BMPA*), and the other is inaccessible to extracellular substrates and consequently nonfunctional. The interconversion between the two states ($k_3$) is modulated by $R_iG_i$ ligands promoting the open, active state and the $R_sG_s$ ligands the occluded, nonactive state. (From VANNUCCI et al. 1992).

actions of isoproterenol. As depicted in Fig. 4, we envisage the changes in transporter accessibility to reflect structural changes in the plasma membrane, but conformational changes of transporters obviously cannot be ruled out.

## E. Other $R_sG_s$- and $R_iG_i$-Mediated Processes in Adipocytes

Remarkably similar observations to those on transporter activity have been reported for isoproterenol and adenosine effects on IGF-II/mannose-6-phosphate binding to intact adipocytes (LÖNNROTH et al. 1988). Again, without a change in receptor number in the plasma membrane, isoproterenol decreases and adenosine augments insulin-stimulated IGF-II binding (H. NISHIMURA and I.A. SIMPSON, unpublished). The absence of any structural similarity between glucose transporters and IGF-II receptors, which possesses only one membrane-spanning region, further argues for G-protein regulation of a more global membrane event rather than direct interaction with the respective proteins. One possibility might be modulation of coated pit formation. IGF-II receptors have long been known to recycle through coated pits. More recently, GLUT4 has been detected in coated pits (SLOT et al. 1991b; ROBINSON et al. 1992). Interestingly, $K^+$ depletion, which blocks clathrin assembly (DELUCA-FLAHERTY et al. 1990), prevents the actions of isoproterenol on both glucose transport activity and IGF-II receptor binding. Thus, the $R_sG_s$ and $R_iG_i$ modulation of glucose transport activity and IGF-II binding above might reflect a cAMP-independent alteration in membrane structure.

In addition to insulin, glucose transport in adipocyte has been shown to be stimulated modestly by agents that stimulate protein kinase C, for example, vasopressin, oxytocin, tumor-promoting phorbol esters (for review, see EGAN et al. 1990; SALTIS et al. 1991). In contrast to insulin, which does not require G-proteins for glucose transporter translocation, the protein kinase C stimulators exhibit an absolute requirement for $R_i/G_i$ activation in order to translocate the transporters. We have found that either the absence of an $R_i$ ligand or intoxication with pertussis toxin blocks transporter translocation induced by the protein kinase C activators, data which again suggest a membrane-stabilizing role for $G_i$ (J. SALTIS and I.A. SIMPSON, unpublished).

As recently reviewed by KLEIN et al. (1991), another important target for $R_s$ and $R_i$ receptor modulation is the adipocyte insulin-signaling system. While this phenomenon has been studied primarily with the glucose transport response, it is evident also with other insulin responses, such as antilipolysis (LONDOS et al. 1985). $R_i$ receptors enhance sensitivity to insulin, $R_s$ receptors produce the opposite effect, and an $R_i$ agonist counteracts an $R_s$ agonist. Previously, we found that the $R_s/R_i$ effects on insulin inhibition of lipolysis are independent of cAMP, and further studies indicate that the leftward dose-response shift by adenosine in the glucose transport response is also cAMP independent (J.J. EGAN, I.A. SIMPSON, and C. LONDOS, unpublished). Moreover, the adenosine effect on the insulin dose-response shift is abolished by pertussis toxin, indicating that this too is mediated by a G-protein. Recently (KLEIN et al. 1991; WEBER et al. 1991) have demonstrated that adenosine and isoproterenol effects are readily observed in insulin receptor phosphorylation and tyrosine kinase activity, indicating that the insulin receptor itself is a downstream target of the $R_sG_s$ and $R_iG_i$ circuits.

## F. Conclusions and Speculations

We have reviewed briefly the evidence that at least three plasma membrane proteins (glucose transporters, IGF-II, and insulin receptors) are subject to cAMP-independent regulation by the same receptor/G-protein complexes that also regulate adenylyl cylcase in adipocytes. Other likely targets for such regulation are protein kinase C substrates through which agents such as vasopressin stimulate glucose transporter translocation. Given the relatively few processes studied in detail, it is remarkable that four different potential G-protein targets have been identified. It appears likely that all $R_s$ and $R_i$ receptors as well as their respective G-proteins participate in these cAMP-independent effects, and in all cases the two circuits, $R_sG_s$ and $R_iG_i$, produce counteractive effects. One might envisage a situation in which the two circuits impinge on a series of different targets, one for glucose transporter modulation, another for IGF-II receptor regulation, and so on. However, it would seem more reasonable to postulate that the receptor-linked G-proteins modify a single target which in turn interacts with the various plasma membrane proteins. For example, as noted above, clathrin assembly might underlie some of the effects we have described, or other cytoskeletal elements may be involved. It is also possible that polymeric G-protein assemblies (RODBELL et al., this volume) interacting with cytoskeletal components underlie global membrane changes. These may be responsible for the cAMP-independent, G-protein-mediated changes in a number of plasma membrane proteins.

Finally, the physiological consequences of the regulatory processes discussed in this paper may not be trivial. Take, for example, the actions of adenosine on modulation of insulin-stimulated glucose transport activity. Acting through $G_i$ the adenosine receptor both increases sensitivity to insulin initiation of glucose transporter translocation to the plasma membrane and increases the intrinsic activity of the plasma membrane-bound transporter by approximately 30%. In the face of insulin concentrations in the low physiological range, the combined effect of these two actions is to increase glucose uptake many-fold over what is seen in the absence of adenosine. The case for the importance of such regulation becomes compelling when one takes into consideration the fact that interstitial adenosine concentrations in adipose tissue are sufficient to activate adipocyte adenosine receptors (LÖNNROTH et al. 1989).

## References


Bell GI, Kayano TJB, Buse JB, Burant CF, Takeda J, Lin D, Fukumoto H (1990) Molecular biology of mammalian glucose transporters. Diabetes Care 13:198–208

Birnbaumer L, Rodbell M (1969) Adenyl cyclase in fat cells. J Biol Chem 244:3477–3482

Cushman SW, Wardzala LJ (1980) Potential mechanism of insulin action on glucose transport in the isolated rat adipose cell. J Biol Chem 255:4758–4762

DeLuca-Flaherty C, McKaY DB, Parham P, Hill BL (1990) Uncoating protein (hsc70) binds a conformationally labile domain of clathrin light chain $LC_a$ to stimulate ATP hydrolysis. Cell 62:875–887

Egan JJ, Saltis J, Wek SA, Simpson IA, Londos C (1990) Insulin, vasopressin and oxytocin stimulate protein kinase C activity in adipocyte plasma membranes. Proc Natl Acad Sci USA 87:1052–1056

Holman GD, Kozka IJ, Clark AE, Flower CJ, Saltis J, Habberfield, AD, Simpson, IA, Cushman SW (1990) Cell surface labeling of glucose transporter isoform GLUT-4 by bis-mannose photolabel. J Biol Chem 265:18172–18179

Honnor RC, Dhillon GS, Londos C (1985) cAMP-dependent protein kinase and lipolysis in rat adipocytes. I. Cell preparation, manipulation, and predictability in behavior. J Biol Chem 260:15122–15129

Honnor RC, Naghshineh S, Cushman SW, Wolff J, Simpson IA, Londos C (1992) Cholera and pertussis toxins modify regulation of glucose transport activity in rat adipose cell: evidence for mediation of a cAMP-independent process by G-proteins. Cell Signal 4:87–98

James DE, Hiken J, Lawrence JC Jr (1989) Isoproterenol stimulates phosphorylation of the insulin-regulatable glucose transporter in rat adipocytes. Proc Natl Acad Sci USA 86:8368–8372

Joost HG, Weber TM, Cushman SW, Simpson IA (1987) Activity and phosphorylation state of glucose transporters in plasma membranes from insulin-, isoproterenol-, and phorbol ester-treated rat adipose cells J Biol Chem 262: 11261–11267

Klein HH, Matthaei S, Drenkhan M, Ries W, Scriba PC (1991) The relationship between insulin binding, insulin activation of insulin-receptor tyrosine kinase, and insulin stimulation of glucose uptake in isolated rat adipocytes. Biochem J 274:787–792

Kuroda M, Honnor RC, Cushman SW, Londos C, Simpson IA (1987) Regulation of insulin-stimulated glucose transport in the isolated rat adipocyte: cAMP-independent effects of lipolytic and antilipolytic agents. J Biol Chem 262:245–253

Laurenza A, McHugh E, Sutkowski M, Seamon KB (1989) Forskolin: a specific stimulator of adenylyl cyclase or a diterpene with multiple sites of action? TIPS 10:442–447

Londos C, Cooper DMF, Rodbell M (1981) Receptor-mediated stimulation and inhibition of adenylate cyclase: the fat cell as a model system. Adv Cyclic Nucleotide Res 14:163–171

Londos C, Honnor RC, Dhillon GS (1985) cAMP-dependent protein kinase and lipolysis in rat adipocytes: 3) Multiple modes of insulin regulation of lipolysis and regulation of insulin responses by adenylate cyclase regulators. J Biol Chem 260:15139–15145

Lönnroth P, Appell KC, Wesslau C, Cushman SW, Simpson IA, Smith U (1988) Insulin-induced subcellular redistribution of insulin-like growth factor II receptor in the rat adipose cell: counterregulatory effects of isoproterenol, adenosine, and cAMP analogs. J Biol Chem 263:15386–15391

Lönnroth, P, Jansson P-A, Fredholm BB, Smith U (1989) Microdialysis of intercellular adenosine concentration in subcutaneous tissues in humans. Am J Physiol 256:E250–E255

Ludvigsen K, Jarrett L, McDonald JM (1980) The characterization of catecholamine stimulation of glucose transport by rat adipocytes and isolated plasma membranes. Endocrinology 106:786–790

Milligan G, Saggerson ED (1990) Concurrent up-regulation of guanine-nucleotide-binding proteins Gi1 alpha, Gi2 alpha and Gi3 alpha in adipocytes of hypothyroid rats. Biochem J 270:765–769

Nishimura H, Saltis J, Habberfield AD, Garty NB, Greenberg AS, Cushman SW, Londos C, Simpson IA (1991) Phosphorylation state of the GLUT4 isoform of the glucose transporter in subfractions of the rat adipose cell: Effects of insulin, adenosine, and isoproterenol. Proc Natl Acad Sci USA 88:11500–11504

Robinson LJ, Pang S, Harris DS, Heuser J, James DE (1992) Translocation of the glucose transporter (GLUT4) to the cell surface in permeabilized 3T3-L1 adipocytes: effects of ATP, insulin, and GTPγS and localization of GLUT4 to clathrin lattices. J Cell Biol 117:1181–1196

Ros M, Northup JK, Malbon CC (1988) Steady-state levels of G-proteins and (beta)-adrenergic receptors in rat fat cells. Permissive effects of thryoid hormones. J Biol Chem 263:4362–4368

Saltis J, Habberfield AD, Egan JJ, Londos C, Simpson IA, Cushman SW (1991) Role of protein kinase C in the regulation of glucose transport in the rat adipose cell. Translocation of glucose transporters without stimulation of glucose transport activity. J Biol Chem 266, 261–267

Simpson IA, Cushman SW (1986) Hormonal regulation of mammalian glucose transport. Annu Rev Biochem 55:1059–1089

Simpson IA, Cushman SW, Egan JJ, Habberfield AD, Londos C, Nishimura H, Saltis J (1990) Hormonal regulation of glucose transport in the rat adipose cell. Biochem Soc Trans 18:1133–1135

Slot JW, Geuze HJ, Gigengack S, James, DE, Lienhard GE (1991a) Translocation of the glucose transporter GLUT4 in cardiac myocytes of the rat. Proc Natl Acad Sci USA 88:7815–7819

Slot JW, Geuze HJ, Gigengack S, Lienhard GE, James DE (1991b) Immunolocalization of the insulin regulatable glucose transporter in brown adipose tissue of the rat. J Cell Biol 113:123–135

Suzuki K, Kono T (1980) Evidence that insulin causes translocation of glucose transport activity to the plasma membrane from an intracellular storage site. Proc Natl Acad Sci USA 77:2542–2545

Taylor WM, Halperin MC (1979) Stimulation of glucose transport in rat adipocytes by insulin, adenosine, nicotinic acid, and hydrogen peroxide. Role of adenosine 3′:5′-cyclic monophosphate. Biochem J 178:381–389

Taylor WM, Mak MM, Halperin ML (1976) Effect of 3′:5′-cyclic AMP on glucose transport in rat adipocytes. Proc Natl Acad Sci USA 73:4359–4363

Vannucci SJ, Nishimura H, Satoh S, Cushman SW, Holman GD, Simpson IA (1992) Cell surface accessibility of GLUT4 glucose transporters in insulin-stimulated rat adipose cells. Modulation by isoproterenol and adenosine. Biochem J 288:325–330

Wardzala LJ, Jeanrenaud B (1981) Potential mechanism of insulin action on glucose transport in the isolated rat diaphragm. Apparent translocation of intracellular transport units to the plasma membrane. J Biol Chem 256:7090–7093

Watkins DC, Northup JK, Malbon CC (1989) Pertussis toxin treatment in vivo is associated with a decline in G-protein beta-subunits. J Biol Chem 264:4186–4194

Weber TM, Joost HG, Kuroda M, Cushman SW, Simpson IA (1991) Subcellular distribution and phosphorylation state of insulin receptors from insulin- and isoproterenol-treated rat adipose cells. Cell Signal 3:51–58

Zorzano A, Wilkinson, W, Kothliar N, Thiodis G, Wardzinkski BE, Rouho AE, Pilch PF (1989) Insulin-regulated glucose uptake in rat adipocytes is mediated by two transporter isoforms present in at least two vesicle populations. J Biol Chem 264:12358–12363

# Subject Index

# Springer-Verlag and the Environment

We at Springer-Verlag firmly believe that an international science publisher has a special obligation to the environment, and our corporate policies consistently reflect this conviction.

We also expect our business partners – paper mills, printers, packaging manufacturers, etc. – to commit themselves to using environmentally friendly materials and production processes.

The paper in this book is made from low- or no-chlorine pulp and is acid free, in conformance with international standards for paper permanency.

MIX
Papier aus verantwortungsvollen Quellen
Paper from responsible sources
FSC® C105338

If you have any concerns about our products,
you can contact us on
**ProductSafety@springernature.com**

In case Publisher is established outside the EU,
the EU authorized representative is:
**Springer Nature Customer Service Center GmbH**
**Europaplatz 3, 69115 Heidelberg, Germany**

Printed by Libri Plureos GmbH
in Hamburg, Germany